ESTATES GAZETTE LAW REPORTS

Estates Gazette Law Reports

1986

Volume 1

Edited by
J Muir Watt O.B.E., M.A.,
of the Inner Temple, barrister

THE ESTATES GAZETTE LIMITED
151 WARDOUR STREET, LONDON W1V 4BN

First Published 1986
Reprinted 1986

ISBN for complete set of 2 volumes: 0 7282 0100 3
ISBN for this volume: 0 7282 0101 1

Typesetting by Digital Graphics
Printed and bound at The Bath Press, Avon

CONTENTS

SUPPLEMENT

Cases not reported in "Estates Gazette"

PREFACE

This first volume of Estates Gazette Law Reports for 1986 follows the pattern set by the two 1985 volumes which began this new series of case reports. The publishers are pleased to record that the series itself and the style and presentation of the volumes have been very well received by both branches of the legal profession and by the surveying profession and it is clear that a need for speedy publication of property reports in hardbound and easy-to-handle form has been met.

Published in the present volume are 77 complete case reports which have appeared in the weekly issues of **Estates Gazette** during the six months January to June 1986. They include nine decisions of the Lands Tribunal and two decisions of leasehold valuation tribunals. In addition, six cases which have not appeared in **Estates Gazette** are fully reported in the 32-page supplement which is printed on green paper.

The total number of reports in this volume is therefore 83.

There has been no slackening in the number of decided cases on points concerning rent review clauses and 17 such cases are reported herein, the same number as in the corresponding volume of 1985.

Citation of the reports in this volume will be:
[1986] 1 EGLR (page no)

TABLE OF CASES

Titles of cases shown in bold type
Names of parties reversed shown in ordinary type

INDEX OF SUBJECT-MATTER

AGRICULTURE

Court of Appeal

December 19 1985

(Before Sir John ARNOLD P, Lord Justice PURCHAS and Lord Justice RALPH GIBSON)

TROOP AND ANOTHER v GIBSON AND ANOTHER

Estates Gazette March 15 1986

277 EG 1134-1144

Agricultural holding — Assignment of tenancy in breach of a restriction on assignment — Plea of estoppel by convention — Appeal from an order made by county court judge in favour of landlords — A tenancy granted in 1923 contained a provision that the tenant should not part with the possession of any part of the holding without the consent in writing of the landlords — An assignment of the tenancy took place with consent, as a result of which the tenancy came to be vested in a father and two of his sons jointly — Subsequently the 1923 tenancy agreement containing the restriction on assignment was lost sight of, so that arbitrations on rent took place in 1970 and 1974 on the basis, accepted by both parties, that there was no written tenancy agreement and no restriction on assignment — In 1980 the survivor of the three joint tenants assigned the tenancy to his nephew without seeking any consent from the landlords — The effect of such an assignment, being from an older to a much younger man, would be to reduce the value of the landlords' reversion — When this assignment was notified to the landlords they claimed that it was in breach of the restriction in the 1923 agreement, that the breach was not remediable, and that it had materially prejudiced their interest — A notice to quit was given based on Case E in section 2(3) of the Agricultural Holdings (Notices to Quit) Act 1977 and the matter proceeded to arbitration — The arbitrator stated a special case for the opinion of the county court — The judge decided the questions raised in favour of the landlords and the tenant appealed — The Court of Appeal held that the restriction on assignment in the 1923 agreement continued to apply; that the assignment to the nephew in 1980 was in breach of the restriction (rejecting an argument that an assignment of the whole was not in breach of a covenant against parting with possession of a part); and that the breach was not capable of being remedied — But they held that the landlords were estopped from relying on the breach of the covenant against assignment — This was because the principle of estoppel by convention applied — During the negotiations and dealings between the parties, particularly during the rent arbitration proceedings in 1974, a convention was established, namely, a common underlying assumption by both parties that no written agreement containing a restriction on assignment was in existence — The landlords were therefore estopped from enforcing against the tenant a restriction in a tenancy agreement which both parties had assumed during the material times not to exist — The parties had gone to arbitration on the rent in the belief that there was no written agreement, an assumption which was to the detriment of the tenant as the absence of restrictions would tend to increase the rent — Among cases considered were *Amalgamated Investment & Property Co Ltd* v *Texas Commerce International Bank Ltd, Field* v *Barkworth* and *Scala House & District Property Co Ltd* v *Forbes* — Tenant's appeal allowed

This was an appeal against a decision of Judge Kellock at Worksop County Court on a case stated by the arbitrator under the Agricultural Holdings Act 1948. The case raised questions relating to the tenancy of a farm known as Manor Farm, Harwell Everton, Nottinghamshire. The appellant Arthur Gibson was the present tenant of the holding. The appellant Herbert Gibson was the uncle of Arthur Gibson and the assignor of the tenancy to the latter. T Troop and G Troop were the landlords, respondents to this appeal.

H W Burnett QC and A de Freitas (instructed by Hayes Son & Richmond, of Gainsborough) appeared on behalf of the appellants; Derek Wood QC and Wayne Clark (instructed by Burges Salmon & Co, of Bristol) represented the respondents.

Giving judgment, SIR JOHN ARNOLD P said: This is an appeal from an order made in the Worksop County Court by His Honour Judge Kellock on November 5 1984.

The learned judge had before him a special case stated by Mr G M Cooper for the opinion of the county court. The case was stated under the Agricultural Holdings Act 1948 and raised a number of questions relating to the tenancy of a holding known as Manor Farm, Harwell Everton, Nottinghamshire. The relevant history of this holding is the following. By an agreement dated February 28 1923 and made between Albert Smith Denton of the one part and Joseph Gibson of the other part the holding was let on a tenancy which commenced on April 6 1923 and was made terminable by either party giving to the other not less than six calendar months' notice to expire on April 6 in any year. The rent was £100 for the first and second years of the tenancy and £130 for the third and following years and it was provided that the tenant (Joseph Gibson) should reside upon the holding and not part with the possession of any part thereof without the written consent of the landlord (Albert Smith Denton), the subletting of the cottages on the holding for short periods being excepted. In the year 1938 the gates and the roof of a shed on the holding were in disrepair and the matter was mentioned by Joseph Gibson to Henry Spencer & Sons, the landlord's agents, and by a letter dated November 8 1938 the agents wrote to Joseph Gibson reporting on the result of their approach to the landlord's solicitors. This letter refers to the agreement of February 28 1923 and quotes therefrom an extract of the repair obligations. Joseph Gibson had three sons and some time in or before the year 1939 two of them, Walter Gibson and Herbert Gibson, were taken into partnership by their father in the farming business carried on by him on the holding. By an agreement endorsed on the tenancy agreement of February 28 1923, it was agreed between Harry Arbuthnott Spencer, described as the landlord, and Joseph Gibson, described as the tenant, that as and from April 6 1939 Messrs Walter Gibson and Herbert Gibson were to be joint tenants with their father. This agreement was drafted in such a way as to provide for the signatures of Walter Gibson and Herbert Gibson to be added, but it was not signed by them but only by Harry Arbuthnott Spencer and Joseph Gibson. It is, however, in my view entirely clear that the endorsed agreement constituted a written consent of the landlord to an assignment by Joseph Gibson to himself and Walter Gibson and Herbert Gibson for the purposes of the tenancy agreement.

By a memorandum of agreement dated January 27 1940 between Joseph Gibson, Walter Gibson and Herbert Gibson it was recited that the holding was held by Joseph Gibson as tenant and that the parties had for some time past been carrying on the farming business on the holding in partnership. This appears to me to provide convincing evidence that there had by January 27 1940 been no devolution of the tenancy from Joseph Gibson to the three partners.

On November 26 1946 there was an appointment of Tom Shearman as arbitrator under the provisions of the Agricultural Holdings Act 1923 and the Arbitration Acts 1889-1934 to determine and award the amount of rent to be paid as and from April 6 1947 in respect of the holding and this was signed by Willie Ward, evidently as landlord, and Joseph, Walter and Herbert Gibson, evidently as tenants, and an award was made by Tom Shearman on February 10 1948 determining the annual rental as from April 6 1947 at £108.

It is evident from the transactions in 1946 to 1948 that the tenants had by this time come to be Joseph, Walter and Herbert Gibson. The matter was considered by the learned judge in his judgment and he concluded that there was in the course of these transactions a representation to the landlord that Joseph Gibson and his two sons were joint tenants on the original terms contained in the tenancy agreement of February 28 1923 and that all parties proceeded on this basis. There has been some argument in this court as to the proper inference to be drawn from the events of 1946 to 1948 as to the devolution of the tenancy down to that time. In view of the endorsed agreement of May 6 1939 my view is that there must have been an assignment of the tenancy from Joseph Gibson to himself and Walter Gibson and Herbert Gibson at some time after May 6 1939 and before November 26 1946, and not earlier than January 27 1940. I am therefore in agreement with the learned judge in concluding that Joseph, Walter and Herbert Gibson held the farm in 1946 on the terms set out in the tenancy agreement of February 28 1923.

Provision was made by section 8 of the Agricultual Holdings Act 1948 for arbitration of the question what rent should be payable in respect of an agricultural holding as from the next ensuing day on which the tenancy of the holding could have been determined by notice to quit, and on reference the arbitrator was to determine what rent should be payable. This section was amended by the Agriculture Act 1958 by section 2 of which there was inserted in section 8 of the Agricultural Holdings Act 1948 a provision that the rent properly payable in respect of an agricultural holding should be the rent at which having regard to the terms of the tenancy (other than those relating to rent) the holding might reasonably be expected to be let in the open market by a willing landlord to a willing tenant. This amendment was effective from August 1 1958. There was a rent arbitration as to this holding in 1970 between the executors of Willie Ward deceased as landlord and Messrs Gibson and Sons, ie the partners Walter and Herbert Gibson, as tenants in which Messrs Shearmans acted as agents for the landlord. The arbitration hearing took place on April 29 1970 and Shearmans in their statement of case on behalf of the landlord included under the heading "Facts" the statement "Tenancy Agreement believed none". This belief appears to have been shared by the tenants and constituted the basis on which the arbitration proceeded. The amendment effected in 1958 to the Agricultural Holdings Act 1948 being in operation, it was of course a matter of importance in the rent arbitration that any terms of the tenancy, other than in relation to rent, should have been known to the arbitrator, and the events of 1970 suggest that by this time both landlord and tenant proceeded on the basis that there were no special terms applicable to the tenancy of the holding.

The letter dated November 8 1938 from Henry Spencer & Sons to Joseph Gibson had evidently been retained by the latter and on January 14 1971 Clay Allison & Clark, solicitors acting for the partnership, wrote to Henry Spencer & Sons that their clients were anxious to get hold of a copy of the original tenancy agreement, if one were available, and referred to the fact that Mr Gibson had brought to them the letter dated November 8 1938 "which refers to the tenancy agreement being in the hands of the trustees' solicitors of Albert Smith Denton, deceased". Henry Spencer & Sons referred the Gibsons' solicitors to solicitors Broomhead Pye-Smith & Reed, who had acted for the Smith Denton trustees, and the Gibsons' solicitors made inquiry of them, stating in a letter dated January 18 1971: "We are anxious to get hold of a copy of the tenancy agreement as our clients have lost theirs and there is a possible sale of the farm to be made shortly and they wish to refer to this agreement." There was no result from this correspondence such as to produce the missing tenancy agreement. The present landlords, the respondents to this appeal, purchased the holding in 1971 and on March 16 1972 their solicitors wrote to the Gibsons' solicitors: "We are instructed by our clients that they have agreed with Mr Gibson that there should be a written tenancy agreement to commence on April 6 next" enclosing a draft for approval. This draft consisted of a printed form which included a provision:

> the tenant shall not without the written consent of the landlord during his tenancy assign underlet or part with the possession of the said messuage, farm buildings, cottages, land and premises or any part thereof and shall reside in the farmhouse.

On March 22 1972 the landlords' agents wrote to Herbert Gibson, by then the sole survivor of the partners, demanding a reference to arbitration of the rent payable from the next ensuing day on which the tenancy could be terminated by notice to quit. On March 29 1972 Herbert Gibson's solicitors wrote to the landlords' solicitors that they had taken their client's instructions on the draft tenancy agreement which had been submitted and returned it amended and, after referring to the notice of reference to arbitration received by Herbert Gibson, inquiry was made whether it was worthwhile postponing the making of a fresh tenancy agreement until the following year; and by a letter dated April 11 1972 the landlords' solicitors agreed to this course. In his judgment, the learned judge said that, as Herbert Gibson had recently been willing apparently to sign the new agreement with a non-assignment clause in it, it seemed likely that at the time of the 1974 arbitration (to be mentioned shortly) he was assuming he was bound by a similar term in the tenancy written or oral. The respondents to this appeal have not attempted to justify this conclusion and in my view it is plainly unsupportable. On February 13 1974 the respondent landlords by their agents gave notice of a demand for a further arbitration under section 8 of the Agricultural Holdings Act 1948 as to the rent to be paid for the holding as from the next date on which the tenancy could be determined by notice to quit and on this occasion Shearmans acted for the tenant and the landlords were represented by Mr C W Thompson [FRICS] of Brackett Moon & Lee. The arbitrator was Mr Colin Thomson [FRICS] and the hearing took place on December 6 1974. Mr Thomson's note of the hearing is in evidence and it records that in opening the landlords' case Mr C W Thompson said that "there was no written agreement". This was accepted by the tenant and the arbitration proceeded on that basis and the arbitrator made his award "in the absence of a written tenancy agreement".

The appellant Arthur (usually known in the family as Peter) Gibson is the son of the third son of Joseph Gibson, that is the son who was not a partner with his father and his brothers. Arthur Gibson worked for the partnership for some time and in 1980 it was desired by Herbert Gibson and himself that he, Arthur Gibson, should take over the farming business. Pursuant to this intention, on May 30 1980 Herbert Gibson assigned the holding to Arthur Gibson by an assignment which recited that by an oral agreement the holding stood demised to Herbert Gibson and Walter Gibson (since deceased) on an agricultural tenancy from year to year and that the benefit of such tenancy was now vested in Herbert Gibson solely. Notice of this assignment was given to the respondent landlords by letter dated June 2 1980. On October 10 1980 the landlords' solicitors wrote to the tenant's solicitors that they had been consulted by the landlords with regard to the notice of assignment and continued: "There is a written tenancy agreement and it would appear that your client's assignment is a breach of covenant contained therein." They intimated an intention that the landlords would commence forfeiture proceedings and they enclosed a copy of the tenancy agreement dated February 28 1923. On November 17 1980 in reply to that letter Herbert Gibson's solicitors wrote that he had absolutely no knowledge of the tenancy agreement of 1923 and that he denied that he was in any way subject to its terms and that he had always understood that he held the farm under an unwritten tenancy agreement. Reference was made to the arbitrations of 1970 and 1974 and to the statement made therein on behalf of the landlords. By a letter dated August 19 1981 the landlords' solicitors wrote to the tenant's solicitors contending that the breach constituted by the assignment to Arthur Gibson was an irremediable breach and stating that they had served a notice to quit on both the assignee and assignor. Counternotices were served and the matter referred to arbitration. Mr G M Cooper [FRICS] was appointed as arbitrator and on January 6 1984 he stated a special case for the opinion of the county court. In that special case the arbitrator found (*inter alia*) as follows:

> Mr Tom Shearman junior in his oral evidence stated that in the 1970 rental arbitration he acted for the landlord and in the 1974 arbitration for the tenant. I accept his evidence that in the course of these arbitrations he sought to determine as far as reasonably possible whether or not there was a written

> agreement and was satisfied there was not. He further stated that in his opinion the landlord also accepted this position. I admit as hearsay evidence Mr Colin Thomson's notes when he acted as arbitrator in 1974 and that he was satisfied there was no written agreement. Mr Thomson was not called. I found as a fact that whatever may have been the prior knowledge of the parties at the arbitrations in 1970 and 1974 the position was at both dates that there was no known or ascertainable agreement. I found that no one knew or could have known whether the written agreement would be found if it still existed. It is clear from Mr Shearman's evidence (and I so hold) that he (and therefore his client Herbert Gibson) relied upon the absence of any special terms in considering whether to assign the tenancy and that the assignment of the tenancy in breach (as it now appears) of the covenant against assignment is an action by Herbert to Herbert's detriment.

In the special case the arbitrator submitted for the opinion of the court the following questions of law:

(1) Whether Herbert Gibson ought to be regarded as at any stage bound by the terms of the 1923 agreement (a) by reason of the estoppel, (b) by reason of surrender and regrant.

(2) If the answer to either limb of question (1) is in the affirmative, then whether the landlords were at the date of assignment of the tenancy by Herbert Gibson to Arthur Gibson estopped from relying upon the restriction upon assignment contained in the 1923 agreement.

(3) If the answer to question (2) is in the negative, then whether the said assignment was made in breach of the 1923 agreement.

(4) If the answer to question (3) is in the affirmative, then whether such a breach was or was not capable of remedy within the meaning of Case E of section 2(3) of the Agricultural Holdings (Notices to Quit) Act 1977.

Mr Cooper referred to this in his findings, of which I have quoted a part, and stated that in answer to question (2) "the court may require to consider whether it is a fair and proper inference from the" paragraphs in question "that the parties regulated their arrangements on the footing at least that there were no special terms unless and until established otherwise" and he referred to *Amalgamated Investment & Property Co Ltd* v *Texas Commerce International Bank Ltd* [1982] QB 84. The arbitrator also found that the value of the landlords' freehold interest in the holding subject to the tenancy of Herbert Gibson would have been considerably more than the value subject to the tenancy of the assignee Arthur Gibson. The learned judge in his judgment, after stating the relevant facts and the questions which had been raised by the arbitrator in the special case, concluded, first, that Herbert Gibson was at the time of the assignment bound by the terms of the 1923 agreement and held, as is not in dispute in this court, that the assignment was a breach of the provision against parting with possession in the 1923 agreement. He also held that this breach was not capable of being remedied and that the breach by means of the assignment materially prejudiced the landlord, Arthur Gibson being a much younger man than Herbert Gibson. He then turned to the question of estoppel and proceeded as follows:

> The first question is whether the statements made in 1970 and on which the arbitrator acted amounted to a clear and unequivocal representation that there were no special terms. In my view the representation by the landlord that there was no written agreement is not unequivocal. It may mean, as was the case, that the written agreement cannot be found. In any event the tenant in 1971 had enquired through his solicitor about the whereabouts of the written agreement and therefore cannot at that time have been in any doubt as to the position. Herbert Gibson in his written evidence states, in spite of this correspondence, that he did not know of the written agreement and therefore seeks to rely on the representations made in the 1974 arbitration. It is apparent that the present landlord who by then had purchased the farm did not know of the written agreement and the arbitrator's hand-written note, admitted in evidence in the present proceedings, shows that their representative put their case on the basis of "no written agreement" or "in the absence of a written tenancy agreement". Again in my view this is equivocal, especially when taken with the recent correspondence. It may mean "in the absence of a written agreement being available" or "the letting was oral". Even if it means that there was an oral agreement it cannot be said to clearly negative special terms, such as a limitation on assignment, when the specific question was not before the arbitrator.

The learned judge then held that the landlord was not estopped from relying on the 1923 agreement and answered the questions put by the arbitrator, (1) Yes. (2) No. (3) Yes. (4) Not capable of remedy.

The tenant appeals to this court pursuant to leave given by the learned judge. His first ground is that there was no justification for the conclusion that the submission to arbitration in 1946 estopped Herbert Gibson from contending that he was not at that time a tenant jointly with his father and his brother Walter. In my view this question does not arise, since the conclusion that Joseph Gibson was G
such a tenant and was bound by the terms of the 1923 tenancy agreement does not properly depend upon an estoppel but upon the inference that the benefit of that tenancy had by 1946 been assigned to the father and the two sons. The second ground, which has not been sustained by argument, was to the effect that the assignment was not a breach, since the covenant contained only so far as relevant an embargo against parting with possession. It has, however, been argued that there was no breach in this case because the assignment was an assignment of the whole holding and the covenant was against parting with possession only of part of the holding. As regards this ground I am quite unconvinced by it. The opposite was decided, in my judgment rightly, at first instance in *Field* v *Barkworth* on
October 23 1985.* The third ground was that the breach was not H
incapable of being remedied and that accordingly the requirements of Case E in section 2(3) of the Agricultural Holdings (Notices to Quit) Act 1977 were not satisfied. This court is bound by the decision in *Scala House & District Property Co Ltd* v *Forbes* [1974] QB 575 which, in my view, is indistinguishable. The remaining ground concerns estoppel. The learned judge dealt with the case exclusively as one of estoppel by representation, but it was argued in this court, as it had been in his court, on the basis alternatively of estoppel by representation and estoppel by convention. So far as estoppel by representation is concerned, there has been extensive argument as to the necessity for any representation upon which an estoppel is
founded being clear and unequivocal so that it will be reasonably J
understood as amounting to the assertion relied upon by the claimant to the benefit of the estoppel. In this connection reference has been made particularly to *Marquis of Bute* v *Barclays Bank Ltd* [1955] 1 QB 202, *Woodhouse AC Israel Cocoa Ltd SA* v *Nigerian Produce Marketing Co Ltd* [1972] AC 741, *Bremer Handelsgesellschaft mbH* v *Vanden Avenne-Izegem PVBA* [1978] 2 Lloyd's Rep 109 and *Bremer Handelsgesellschaft mbH* v *C Mackprang Jr* [1979] 1 Lloyd's Rep 221, in relation to the issue whether the requirement is satisfied if there be more than one reasonable interpretation of the words used to make the representation, one of them being that relied upon by the party claiming the benefit of estoppel, or whether this must be the only reasonable interpretation. For my part I do not think that it is
necessary to resolve this question in the present appeal. In the first K
place I do not think that the representation relied upon in this case is susceptible of more than one interpretation. Where in an arbitration in which it is proper to take into account, as it was here, by reason of the 1958 Act amendment to the Agricultural Holdings Act 1948, the terms of a tenancy agreement between the parties to the arbitration, a representation that there is no written tenancy agreement can, in my view, mean only that there is no written agreement operating between those parties in relation to the subject-matter of the arbitration. In the second place the question does not arise, in my view, where the nature of the estoppel is that of estoppel by convention, since this is of a consensual character and the terms of the convention, just as those of a contract once the language is established by the evidence, must be L
interpreted by the court and the only true meaning is that decided upon by the court. In this case, for the reasons which I have already indicated, the only true interpretation can be that they embraced a convention that there was no written agreement operative to effect the relations between them as landlord and tenant respectively of the holding. It seems to me that the estoppel which arises in this case is truly of the nature of an estoppel by convention, the convention being that there was no written tenancy agreement in the sense that I have just mentioned. It is true that that convention was initiated by a representation made on behalf of the landlords and perfected when it was acquiesced in and accepted on behalf of the tenant, but this does not affect, in my view, its fundamental character as an estoppel by
convention. The case does not seem to me to be distinguishable in M
principle from *Amalgamated Investment & Property Co Ltd* v *Texas Commerce International Bank Ltd* [1982] QB 84. In that case the doctrine of estoppel which was applied by the Court of Appeal was variously indicated. It is described by Lord Denning MR at p 121 in somewhat universal terms, but the other two members of the court defined it more narrowly by adopting in effect the passages from *Spencer Bower and Turner on Estoppel by Representation*, cited by Eveleigh LJ at p 126 and by Brandon LJ at pp 130 to 131. It is argued by the respondents that this principle can have no reference to the present appeal both because there was no course of dealing between these parties in which the convention was adopted and because the

*Editor's note: Reported at p 46 *post* and (1985) 277 EG 193.

transaction to which it is sought to apply the convention is not the transaction in which it was adopted. Neither of these distinctions is in my view valid.

The ambit and context of the 1974 arbitration constituted in my view a perfect course of dealing between the parties to establish the convention for relevant purposes. As regards the identity of the transaction the application of the doctrine in the *Amalgamated Property* case illustrates in my view that the doctrine extends to the kind of application which is relied upon in the present case. The issue in the *Amalgamated Property* case was whether some money which was in the possession of the respondent bank could properly be retained by it as being subject to the charge which had been recognised as existing by convention between the parties. The convention had been established, as explained by Brandon LJ at p 129, by the conduct of transactions between the parties from 1974 to 1976 in a common (but mistaken) belief as to the extent of the relevant hypothecation, but it was in 1977, well after the adoption by the parties of their conventional stance, that the chargee purported to appropriate the money in question to the satisfaction of his relevant charge. Nevertheless the chargor was held estopped from reversing that transaction. This was because the inevitable consequence of the convention adopted in the earlier years was that the bank would be entitled to act as it did. So in the present case the parties proceeded in the 1974 arbitration by the convention that there was no written tenancy agreement, the existence of which alone could have constituted the 1980 assignment a cause of forfeiture, and the nexus between the two transactions is in my judgment as close as that between the relevant transactions in the *Amalgamated Property* case. Moreover, in my view the finding by the arbitrator in para (xv) of the special case constituted a finding, relevantly binding on the county court and on this court of a reliance upon the convention sufficient to sustain the estoppel, since the absence of any written tenancy agreement, testified by the convention, is altogether sufficient in the circumstances of this case to exclude the existence of any embargo upon assignment or parting with possession. I would therefore uphold the estoppel as barring the respondent's claim and allow this appeal.

I should add that it is evident from the course of this litigation that the effect of the convention is now altogether spent, with the consequence that no such estoppel as I have upheld can continue to operate.

Agreeing that the appeal should be allowed, PURCHAS LJ said: I agree on one only of the five grounds upon which the appellants relied. The general background history of this matter has already been described in the judgment just delivered and need not be repeated. Since 1923 Harwell Manor Farm has been farmed by the Gibson family as tenants. The relationship started by an agreement in writing of February 28 1923 ("the 1923 agreement") which was made between one Denton and Joseph Gibson ("Joseph"). In January 1940 Joseph's two sons, to whom I shall refer as "Walter" and "Herbert", and Joseph himself signed a partnership agreement which recited that the land on which the partnership was to farm was held by Joseph as tenant on a yearly tenancy.

On November 26 1946 Denton's successor in title, one Ward, agreed with Joseph, Walter and Herbert as joint tenants to the appointment of an arbitrator for the purposes of fixing a rent. One of the issues raised in this appeal is the manner in which the three Gibsons had become joint tenants in place of Joseph as sole tenant. Further arbitration proceedings took place in 1970 and between December 1974 and January 1975 in which the joint tenants were named as a partnership entitled "J Gibson and Sons" (Joseph Gibson in fact died in 1955). In May 1980 Herbert, being the sole survivor of the original partnership, assigned his "tenancy" to his nephew Arthur Gibson ("Arthur") and on June 2 1980 notified the landlords, Messrs Troop, who had become the reversioners. The landlords served a notice under section 146 of the Law of Property Act 1925 ("the 1925 Act") on October 10 1980. This was served on Herbert. The landlords followed this notice on August 19 1981 by a notice to quit pursuant to Case E of section 2(3) of the Agricultural Holdings (Notices to Quit) Act 1977 ("the 1977 Act"). This was dated to expire on April 6 1983 and was served on Herbert and Arthur. As a result of counternotices served by Herbert and Arthur the arbitration from which this appeal arises took place. The hearing was during November 1983. On November 7 1983 the arbitrator produced a draft award as a result of which the parties asked for a formal statement of case for the opinion of the county court. This referred four questions to the county court and it is from his answers to these questions that this appeal is brought.

Before coming to the four questions I should remark that among the papers before us there was included the draft award to which I have already referred. Although there are no direct inconsistencies between the case stated and the draft award there are differences in emphasis on certain aspects which promoted some discussion during argument. It does not appear that the draft award was before the learned judge, who correctly restricted his consideration of the questions to the findings contained in the case stated. I have formed the firm conclusion that this is the correct approach and that the draft award is not relevant for consideration on this appeal. The term of the 1923 agreement upon which this appeal turns is contained in clause 6, relevant parts of which for convenience of reference I set out here:

> THE Tenant shall reside upon the holding and not part with the possession of any part thereof without the written consent of the Landlord but nothing herein contained shall preclude the tenant from subletting the said cottages on a monthly or shorter tenancy.

Besides this clause the agreement contained numerous other covenants and reservations to which it is not necessary to refer in detail. It was, however, by no means a simple letting of an agricultural holding.

The four questions which were referred to the learned judge were:

(1) Whether Herbert ought to be regarded as at any stage bound by the terms of the 1923 agreement (a) by reason of estoppel; or (b) by reason of surrender and regrant.

(2) If the answer to either limb of question (1) is in the affirmative, whether the landlords were at the date of the assignment of the tenancy by Herbert to Arthur estopped from relying upon the restriction upon assignment contained in the 1923 agreement.

(3) If the answer to question (2) is in the negative, then whether the assignment was in breach of the 1923 agreement.

(4) If the answer to question (3) is in the affirmative, then whether such a breach was or was not capable of remedy within the meaning of Case E of section 2(3) of the 1977 Act.

The learned judge answered these questions as follows: (1) Yes. (2) No. (3) Yes. (4) Not capable of remedy. Mr Burnett, who appeared for the appellants, submitted that the learned judge was in error in his answer to each of the questions. Although the main discussion on this appeal has centred on the second answer, namely estoppel, it is convenient to deal shortly with the other three points made by Mr Burnett. On the first question Mr Burnett submitted that the learned judge was incorrect in holding that it did not matter whether the tenancy enjoyed by the partnership was created by assignment or the 1923 agreement or whether by a subsequent informal oral agreement between the landlord and the partners resulting from a surrender and regrant. Mr Burnett submitted that in the case of a new oral tenancy no special clause removing the tenants' basic right to assign should be imported. In support of this contention it was submitted that when Herbert came to assign he mentioned his "oral tenancy". With respect to the submissions made by Mr Burnett I cannot agree with them. I agree, for the reasons given by the learned President, that at some time before 1946 but subsequent to 1939 the brothers and the father had become joint tenants in some form and that that tenancy must be assumed to contain a clause, if by assignment the same clause or if by regrant a similar clause, restricting the right to part with possession of the land. Mr Burnett sought to distinguish the authority of *Rodenhurst Estates Ltd* v *W H Barnes Ltd* [1936] 2 All ER 3 on the basis that the partnership agreement into which the brothers and the father entered in 1940 "wiped the slate clean" and, in considering the terms of the joint tenancy which had been established by 1946, it was not relevant to look back at the 1939 addendum as this was not signed by Herbert and Walter. In my judgment, one effect of the addendum is that it is evidence of the landlord's, one Spencer, consenting to an assignment by Joseph to create a joint tenancy by assignment including himself, Herbert and Walter. It must also be inferred that any assignment to the partnership, as it would have been with the consent of the landlord, would have included a term restricting the right to part with possession. The same would apply to a new joint tenancy created by surrender and regrant. The judge was, in my view, correct in holding that the precise mechanics were not important.

On the fourth question Mr Burnett courageously submitted that the court could distinguish the case under consideration from the authority of *Scala House & District Property Co Ltd* v *Forbes* [1974]

1 QB 575, which otherwise would be binding upon us. His submission was based on a construction of section 2(3), Case D and Case E of the 1977 Act. Case E, under which the notice was served by the landlords in this case, provides:

Case E — at the date of the giving of the notice to quit the interest of the landlord in the agricultural holding to which the notice relates had been materially prejudiced by the commission by the tenant of a breach, which was not capable of being remedied, of any term or condition of the tenancy that was not inconsistent with the fulfilment by the tenant of his responsibilities to farm in accordance with the rules of good husbandry and it is stated in the notice that it is given by reason of the matter aforesaid.

The submission was that the words "not capable of being remedied" in the 1977 Act could be construed in a manner different from the construction placed upon the words of section 146 of the 1925 Act "is capable of remedy". If this were so, then the court freed from the direct authority of the *Scala House* case to the effect that a breach of a covenant not to assign was incapable of remedy could construe the words of Case E in a far more restricted manner. Mr Burnett submitted that for the purpose of Case E a breach would be capable of remedy if as a matter of theoretical possibility the tenant could by some means, including the co-operation of third parties obtained if necessary by financial or other reward, restore the position. The test must, however, depend on the nature of the covenant and its breach rather than the particular circumstances of the case. Thus if an assignor could possibly induce the assignee to consent to the revocation of the assignment then the breach could be remedied and a notice in that event ought to have been served not under Case E but under Case D of section 2(3) of the 1977 Act. In this case an undertaking to reassign the tenancy was forthcoming from Arthur. But on a matter of pure construction Mr Burnett was constrained to concede that his submission in fact involved the proposition that in order for a breach to be irremediable, even if a breach of a negative covenant, it would have to be shown that the breach was incapable of remedy under any possible combination of circumstances even if this involved obtaining the consent of an independent and free agent who might well be acting contrary to his interests. I regret that, in my judgment, this submission fails. I am unable to see any ground for distinguishing as a matter of construction the effect of Case E of the 1977 Act and the words of section 146 of the 1925 Act. For the purposes of construing two statutes which are clearly *in pari materia* I consider that we are bound by the authority of the *Scala House* case and that, therefore, the breach of the covenant not to part with possession is irremediable. Even if this were not so, I could not agree with Mr Burnett's submission as a matter of construction free from authority. The construction for which he contended was in my view far too restricted.

The remaining submission, apart from the issue of estoppel, involved an assertion that the decision of Nicholls J (as he then was) in the case of *Field* v *Barkworth* was wrong. There is no authority in point, probably because the matter is so clear. Mr Burnett's argument was based on the mirror-image proposition from which stemmed the line of authorities which indicated that a covenant against parting with possession of the whole of the property did not prevent an assignment in part. In my judgment, this argument does not have any force. I agree that the judgment of Nicholls J in *Field* v *Barkworth* is correct.

I now turn to the question of estoppel. Although the findings have already been described by my lord it is convenient shortly to set out the crucial matters. The important findings are (xiv) and (xv) in the case stated, which result in the finding in (xvi):

(xiv) I find as a fact that whatever may have been the prior knowledge of the parties at the arbitration in 1970 and 1974 the position was at both dates that there was no known or ascertainable agreement. I find that no one knew or could have known whether the written agreement would be found if it still existed.

(xv) It is clear from Mr Shearman's evidence (and I so held) that he (and therefore his client Herbert Gibson) relied upon the absence of any special terms in considering whether to assign the tenancy and that the assignment of the tenancy in breach (as it now appears) of the covenant against assignment is an action by Herbert to Herbert's detriment, I say this especially in view of the finding I made as to the consequences of breach.

(xvi) Paragraphs (ix) to (xv) above inclusive raise the second question upon which the opinion of the court is sought. In order to answer this question the court may require to consider whether it is a fair and proper inference from these paragraphs that the parties regulated their arrangements on the footing at least that there were no special terms unless and until established otherwise (see *Amalgamated Investment & Property Co Ltd* v *Texas Commerce International Bank Ltd* [1982] QB 84).

Notwithstanding the specific finding in the case stated and other G references to the *Amalgamated Investment & Property Co Ltd* case the learned judge seems to have concentrated solely upon the requirement for a "clear and unequivocal representation that there were no special terms" such as would be required under the authorities to establish estoppel by representation. To be fair to the learned judge, he may have been induced into this attitude by the emphasis placed by learned counsel on this particular aspect; but it is common ground that the appellants did rely also upon estoppel by convention. There may not be so much importance to be attached to the distinction between the two so-called types of estoppel. The law as it has developed now tends against a series of different estoppels in their own compartments but, as Mr Burnett submitted, it is to the H principle that the court should direct its attention. In certain circumstances of estoppel by representation a clear and unequivocal statement upon which the person relying upon the estoppel acted to his detriment may be an important feature; but where both parties engage upon a course of negotiations or transactions representing mutually the one to the other that a certain state of affairs is accepted as regulating their conduct then the necessity for clear and unequivocal statements becomes of less importance. The crucial requirement for convention estoppel is that at the material time both parties should be of a like mind. Thus the difficulties over ambiguity with which their lordships were concerned in the case of *Woodhouse AC Israel Cocoa Ltd SA* v *Nigerian Produce Marketing Co Ltd* [1972] AC 741 do not arise where both parties enter into a course of J negotiations under a common and agreed assumption of fact.

The effect of the two important cases *Bremer Handelsgesellschaft mbH* v *Vanden Avenne-Izegem PVBA* [1978] 2 Lloyd's Rep 109 and *Bremer Handelsgesellschaft mbH* v *C Mackprang Jr* [1979] 1 Lloyd's Rep 221 is that where parties have conducted themselves so as to lead one reasonably to believe that reliance will not be placed upon a defect in goods, whether known at the time or not by the person making the representation or conducting himself so as to give that impression, then that party cannot, thereafter, go back upon the conduct or representation (see per Shaw LJ in *Bremer* v *Mackprang* at p 230):

Accordingly, so it seems to me, whether the conduct of a contracting party may amount to a waiver must be determined by reference to all the prevailing K circumstances. It need not be such as to amount virtually to an express declaration that this or that right is waived or surrendered. If in the prevailing conditions affecting the position of the parties to a contract the conduct of one of them affords a reasonable foundation for the inference that he is prepared to forgo any right or rights he may have in a certain regard and the other contracting party does draw that inference and persists in the residual contractual relationship upon that basis, then whether it be regarded as waiver or estoppel the forgoing of those rights cannot thereafter be gainsaid . . . Their counsel sought to reinforce his argument against the implication of waiver by laying emphasis on the buyers' ignorance as to whether, on the facts if and when they became known, the sellers could claim to be exonerated from liability for failing to deliver as required by their contract. I see no reason, however, to limit the effect of waiver to rights known to exist. It may be embracing enough, and so intended, as to forgo rights which *might* exist in L regard to a particular contract or in a particular context.

The last passage, cited from the judgment of Shaw LJ, seems to me to be particularly apposite in the circumstances prevailing in the 1970 and 1974 arbitrations the conduct of which took place against the common belief that there was no written agreement and no special terms affecting the tenancy. It remains only to refer to the speech of Lord Denning MR in the *Amalgamated Property Co* case at p 122:

When the parties to a transaction proceed on the basis of an underlying assumption — either of fact or of law — whether due to misrepresentation or mistake makes no difference — on which they have conducted the dealings between them — neither of them will be allowed to go back on that assumption when it would be unfair or unjust to allow him to do so. If one of them does seek to go back on it, the courts will give the other such remedy as M the equity of the case demands.

and per Eveleigh LJ who cites a passage from *Spencer Bower and Turner on Estoppel by Representation* at page 126B:

When the parties have acted in their transaction upon the agreed assumption that a given state of facts is to be accepted between them as true, then as regards that transaction each will be estopped against the other from questioning the truth of the statement of facts so assumed.

The limitation on the broadness of the concept of estoppel mentioned by Eveleigh LJ at p126 F is, in my judgment, important:

Estoppel operates so as to prevent a party from denying a representation or an assumed state of facts in relation to the transaction supported by that representation or assumed state of facts. The estoppel does not go beyond the

A transaction in which it arose. The representation or assumed state of facts are not to be held irrefutable beyond the purpose for which the representation or assumption was made. In the present context the representation is not made for the purpose of establishing its own truth but as a part of the whole transaction. An assumption is not to be treated as having the effect of an assumpsit.

The finding inferentially recorded by the arbitrator in the case stated at para (xvi) in inviting the court to consider whether this inference is fairly drawn places the basic finding firmly and clearly within the concept of estoppel by convention as described in the authorities to which I have just referred. I am firmly of the view that the primary findings of fact set out in the case stated, to which my lord has already referred, justify such an inference. Over this period
B of time the parties were content to and did negotiate the new rent on the basis that there were no special terms controlling the tenancy. This would increase the value of the holding and, therefore, the rent awarded and so the tenant acted to his detriment in this respect. It is not clear at what point in time the landlords became aware of the 1923 agreement. Whenever it was, in my judgment they were not entitled to rely upon it for the purposes of forfeiture for breach of the covenant not to part with possession while the tenant was still conducting his affairs in the context of the assumption, jointly made for the purposes of negotiations in the past, that there were no such special terms and that the written tenancy did not exist. As Eveleigh LJ emphasised, however, in his judgment, estoppel is of limited
C application. The equitable relief afforded thereby will persist only so long as it is necessary to protect a tenant who has altered his position on the basis of the common mistake or common misrepresentation — however it may be put. Now that the written tenancy has been declared and rediscovered, the effect of the estoppel will no longer run and Arthur in his turn, now knowing the true state of affairs, will not be able to avail himself of the equitable relief which has protected the assignment from his uncle to himself. In my judgment, therefore, this appeal succeeds on the ground of estoppel alone and I would concur that the appeal should be allowed.

Agreeing, RALPH GIBSON LJ said: I wish to add one comment upon the question as to the degree of clarity and unequivocalness required of an estoppel if it is to be the effective foundation of an
D estoppel by representation of fact, or of promissory estoppel, or the origin of an estoppel by convention. I agree that it is not necessary to resolve this question in this appeal because the representation relied upon was, in my judgment, capable of only one meaning in the context, namely that there was no written agreement operating between the parties. That representation was the origin of the estoppel by convention in this case. I agree further that where reliance is placed upon estoppel by convention, where both parties have engaged upon a course of negotiation or transactions representing mutually the one to the other that a certain state of affairs is accepted as regulating their conduct, then the necessity for proof of some clear and unequivocal statement becomes of less importance. The court
E must determine what the state of affairs is which the parties have accepted and decide whether there is sufficient certainty and clarity in the terms of the convention to give rise to any enforceable equity. For my part I think that the extent to which the importance of clear and unequivocal statements is reduced in cases of estoppel by convention is probably small. In all cases the representation or statement must be sufficiently clear; and, since the doctrine of estoppel, when applied, deprives a party of the ability to enforce a legal right for the period of time and to the extent required by the equity which the estoppel has raised, the clarity required will seldom fall below what is unequivocal for the relevant purpose. In this case I have no doubt that in justice the landlord could not be permitted to enforce against the tenant the terms of the written agreement in respect of an act done while the
F tenant, in reliance upon the mutually accepted state of affairs, believed that no written tenancy agreement regulated their relationship.

The appeal was allowed and the following answers given to the questions raised before the county court judge:
Question (1): Yes, at every stage since at least November 26 1946.
Question (2): Yes.
Questions (3) and (4) do not arise.

The appellants were awarded costs, as between party and party, in the Court of Appeal and below. Common fund taxation was ordered for legal aid purposes. Leave to appeal to the House of Lords was refused.

Court of Appeal

March 12 1986

(Before Lord Justice SLADE, Lord Justice STOCKER and Sir Roualeyn CUMMING-BRUCE)

FEATHERSTONE AND OTHERS v STAPLES AND OTHERS

Estates Gazette May 24 1986
278 EG 867-878

Agricultural Holdings (Notices to Quit) Act 1977, section 2(1) — Validity of counternotice — Whether a counternotice under section 2(1) was valid when given by two only out of three joint tenants, the third being a company wholly owned by the landlords — Partnership agreement between the two individual joint tenants and the company joint tenant provided expressly that no counternotice could be served without the consent of the company — Under tenancy agreements land was let to the individuals and the company to farm in partnership, although the company played no actual part in the farming business and had no interest in it save for a small entitlement to profits — Landlords served notices to quit on the joint tenants and at the same time the company served on the individual partners notice of dissolution of partnership — The two individual joint tenants, but not the company, served counternotices on the landlords in purported compliance with section 2(1) of the 1977 Act — Landlords brought an action claiming that the counternotices were invalid, relying on the general rule of statutory construction that in the case of a joint tenancy "the tenant" means all the joint tenants — Nourse J held that the counternotices were valid and effective and the plaintiff landlords brought the present appeal — Court of Appeal first considered earlier case law, such as *Leek and Moorlands Building Society* v *Clark*, from which the principle to be derived seemed to be that all joint tenants must concur in the renewal of a tenancy; on this principle, as a counternotice was designed to secure (contingently) a renewal, all joint tenants should join in it — Later decisions were divided into two groups, those taking a strict and those taking a more flexible view of joint proprietorship — The former included *Jacobs* v *Chaudhuri, T M Fairclough & Sons Ltd* v *Berliner* and *Newman* v *Keedwell* — The latter included *Howson* v *Buxton, Lloyd* v *Sadler*, and the decision of Nourse J in the present case — Nourse J, distinguishing *Newman* v *Keedwell* and deriving guidance from *Lloyd* v *Sadler*, decided that the true purpose of section 2(1) of the 1977 Act was to give security to those who needed it and that the individual joint tenants in the present case, like the occupying tenant in *Lloyd* v *Sadler*, needed security — Court of Appeal did not feel able to accept Nourse J's flexible construction of "the tenant", but in the end arrived at the same conclusion, that the counternotices were valid, by a different route — The court accepted that "the tenant" in the case of a joint tenancy must mean all the joint tenants — It might be possible, in cases where one of the joint tenants was the landlord in person, to regard "the tenant" as meaning all the joint tenants except the landlord, but it would be difficult to extend the definition to cover a company merely controlled by the landlord — The problem in the present case could, however, be resolved if the restrictive condition in the partnership agreement, prohibiting a counternotice without the company's consent, were found to be void as against public policy — In that case the two individual joint tenants would have been able to serve the counternotices on behalf of the partnership — On the principle of *Johnson* v *Moreton*, a condition in a tenancy agreement by which three joint tenants, one of whom was also the landlord, undertook not to serve a counternotice would be void as against public policy —

Although issues of public policy were difficult, it would seem immaterial that the condition was in an associated partnership agreement rather than in the tenancy agreement and was in favour of a landlord-controlled company — Hence the counternotices were valid — Appeal dismissed

The appellants, plaintiffs in the action before Nourse J, were the trustees of the will of the late Col Hugo Meynell and were the freehold owners of three areas of agricultural land at East Ferry and Wildsworth in Lincolnshire, which had been let on annual tenancies to the respondents (defendants below). The respondents were two brothers, R H and J D Staples, and a company called Laughton Contracting Co Ltd. This company was wholly owned by the appellants, all the appellants being shareholders in it and two of them being directors. Although the company was a party it did not, for obvious reasons, take any part in the proceedings.

The judgment of Nourse J was reported at [1985] 1 EGLR 1; (1985) 273 EG 193. Derek Wood QC and Miss Joanne Moss (instructed by Pinsent & Co, of Birmingham) appeared on behalf of the appellants; Harold Burnett QC and Anthony De Freitas (instructed by Hayes, Son & Richmond, of Gainsborough, Lincs) represented the respondents, R H and J D Staples.

Giving judgment, SLADE LJ said: This is an appeal by the four plaintiffs in landlord and tenant proceedings from a judgment of Nourse J given on October 31 1984. The principal question in the case is whether a counternotice, purporting to be served on behalf of a partnership of three persons and requiring that section 2(1) of the Agricultural Holdings (Notices to Quit) Act 1977 ("the 1977 Act") shall apply to a notice to quit, is valid, where it is given by only two of the three partner-tenants, but the third is a company wholly owned by the landlord, whose consent to the service of any such notice is required by the express terms of the partnership agreement.

The plaintiffs are the trustees of the will of the late Col Hugo Meynell. They are the freehold owners of all the land in question. The three defendants are Mr R H Staples and his brother, Mr J D Staples, to whom I shall refer as "the Staples", and a company wholly owned by the plaintiffs called Laughton Contracting Co Ltd ("Laughton"). All the plaintiffs are shareholders in Laughton and they include two of its four directors.

On September 12 1977 the Staples and Laughton entered into a written partnership agreement by which they agreed that they should form a partnership to farm Bramley Farm, Laughton, Lincolnshire (which the Staples were already farming) as from September 29 1977 under the name "Staples Brothers", provided that a tenancy of the farm was granted to the partners and any such tenancy should form partnership property. Clause 3 provided:

The partnership shall commence on September 29 1977 and shall continue for five years and thereafter unless and until dissolved by any Partner giving to the other Partners not less than 12 months written notice of dissolution such notice to take effect on the first anniversary of September 29 next following the date of the notice Provided always that Laughton may not serve notice of dissolution except following a breach of this agreement by another Partner, or if a notice to quit has been given in respect of the new tenancy or any extension thereof Provided further that Laughton may serve notice of dissolution during the initial term of 5 years if such a notice to quit has been given and either any report shall have been commissioned or published on the law or relating to security of tenure of agricultural tenants or legislation or other governmental act shall have been announced or introduced in Parliament with a view to altering such law.

I take the following summary of the later provisions of the partnership agreement verbatim from the learned judge's judgment:

Clause 6 provided that Laughton should not be liable to contribute capital to the partnership. Clause 7 provided that profits should be computed on the same basis as that employed in the computation of the profits of the Staples for previous accounting periods, save that no item of expenditure on capital equipment in excess of a certain amount was to be debited to profit and loss account without Laughton's consent in writing. The profits were to belong to the partners, as to an amount equal to the rent for the time being payable, to the Staples in such proportions as they might from time to time agree, and subject thereto to Laughton and the Staples in the proportions $7\frac{1}{2}\%$ and 92% respectively. Losses were to be borne by the partners in the same proportions as they were entitled to share profits. By clause 10 each of the Staples agreed diligently to attend to the partnership business, to devote a reasonable amount of time and attention thereto subject only to annual and other holidays on a basis to be agreed between them, and to farm the land in accordance with the rules of good husbandry and to observe the terms of the relevant tenancy or tenancies. Clause 11 was in the following terms: "The Staples shall together have complete conduct of all negotiations leading to the grant of the new tenancy and thereafter in connection therewith, including **G** any rent review (and the service of notice and settlement of compensation under Section 34 Agricultural Holdings Act 1948 ('the Act') but after 29th September 1982 no partner may serve a counternotice under Section 24 of the Act without the consent of Laughton".

Clause 12 provided that the Staples should have complete conduct of the business of the partnership. Clause 13 provided that on the death, bankruptcy or retirement of either of the Staples his share in the partnership should accrue to the other of them on payment as therein mentioned. Clause 14 gave the Staples certain rights to require Laughton to retire from the partnership in the event of its winding up, whereupon Laughton's share would accrue to them without payment save for its share of undrawn profits. Clause 15 provided in effect for the provisions of the Partnership Act 1890 as to winding up to apply on a dissolution of the partnership, except that any surplus was to be divided between the Staples in the same proportions as they were entitled to share **H** profits. These provisions can be summarised by saying that Laughton was to play no part in the partnership business and to have no interest in it save for its small share of profits.

On November 23 1977 the plaintiffs and the three partners signed a tenancy agreement whereby the partners were granted a yearly, non-assignable tenancy of Bramley Farm from September 29 1977 at an initial yearly rent of £5,328.

On October 6 1978 the Staples and Laughton entered into a second written partnership agreement whereby they agreed that they should form a partnership to farm further land at Wildsworth, Lincolnshire, as from November 25 1977, provided that a tenancy of the land was granted to the partners and any such tenancy should form **J** partnership property. Though there were certain small differences between the provisions of the two partnership agreements, I think that in all respects material for present purposes their provisions were the same; like the learned judge, I will therefore hereafter refer to the provisions of the two agreements by reference to those contained in that dated September 12 1977.

The terms of the partnership agreements thus clearly contemplated two things. First, at all times relevant for present purposes, Laughton was to have the right to serve 12 months' notice of dissolution to take effect on the first anniversary of September 29 next following the date of the notice if notice to quit had been given in respect of the new tenancy or any extension thereof. Second, no partner was to be free to serve a counternotice under section 24 of the 1948 Act without **K** Laughton's consent.

Also, on October 6 1978, the plaintiffs and the three partners signed two tenancy agreements whereby the partners were granted yearly, non-assignable tenancies of certain lands at Wildsworth from November 25 1977 at initial yearly rents of £1,587 and £1,856.

In due course the tenancy of some parts of the land at Wildsworth was surrendered. In the result, the three areas of land still subject to the three tenancies as at September 1983 comprised about 210 acres.

The partnership agreements show that it was originally intended that there should be two separate partnerships, one to farm two of the areas under the name "Staples Brothers" and the other to farm the third under the name "Staples Brothers (Wildsworth) Partnership". **L** However, the accounts suggest that the two partnerships were treated as one, and, as the judge said, the point is of no importance.

The accounts of the partnership for the six years to September 29 1983 and the annual reports and accounts of Laughton for the five years to March 31 1982 are in evidence. The former accounts show that up to September 29 1983 the partnership either made a loss or did not make sufficient profits to exceed the amount of the rents for the time being payable in respect of the tenancies, with the result that no share of profits ever became payable to Laughton. It appears that the only activity of Laughton was trading in partnership with the Staples.

By three notices to quit dated September 26 1983 and served on the same day, the plaintiffs gave to the Staples and Laughton notice to quit the three areas of land on September 29 1984. By two notices **M** dated September 28 1983 and served on the same day, Laughton gave the Staples notice of dissolution of the two partnerships, also to take effect on September 29 1984.

On or about October 14 1983, without Laughton's consent, the Staples served on the plaintiffs three counternotices. Each was headed "Agricultural Holdings (Notices to Quit) Act 1977" and specified the area of land in respect of which it was given. The material part of each notice was in these terms:

Pursuant to the powers and provisions of the above Act WE HEREBY GIVE YOU NOTICE that we require that sub-s (1) of s 2 of the above Act shall apply to your Notice to Quit in respect of the above land served on us and dated 26th September, 1983. This Notice is not to be taken as any admission on our part

A as to the validity of your Notice to Quit and it is given without prejudice to any points arising thereon.

Each notice was dated October 14 1983 and signed by the two Staples "On behalf of" the relevant partnership, which was described as "Tenant".

On November 11 1983 the plaintiffs issued proceedings against the Staples and Laughton by which they sought declarations that the three counternotices were of no effect and that the three tenancy agreements would accordingly determine on September 29 1984. The Staples counterclaimed for a declaration that the three counternotices were valid.

At the trial of the action, which took place in the Vacation Court B on September 5 1984, Laughton took no part in the proceedings and no oral or affidavit evidence was adduced on either side. As the learned judge recorded, there was no agreed statement of facts and the only materials placed before him were the pleadings and an agreed bundle consisting of the documents already mentioned and two letters of September 6 1977 (which have not been referred to in argument before us). As regards any matters of fact, we are therefore obliged, as he did, to proceed simply on the basis of the documentary evidence before us. Except in so far as this appears from the documents, we have no evidence of the background to the execution of the partnership agreements and tenancy agreements.

Relevant statutory provisions

C Section 2 of the 1977 Act, of which the relevant provisions were formerly contined in section 24 of the 1948 Act (referred to in clause 11 of the first partnership agreement) imposes a number of restrictions on the operation of a notice to quit an agricultural holding in a case where the tenant is unwilling to accept the notice. Thus section 2(1) provides:

Where — (a) notice to quit an agricultural holding or part of an agricultural holding is given to the tenant thereof; and (b) not later than one month from the giving of the notice to quit the tenant serves on the landlord a counter-notice in writing requiring that this subsection shall apply to the notice to quit, then, subject to subsection (2) below, the notice to quit shall not have effect unless the Tribunal consent to its operation.

Section 2(2) provides that "Subsection (1) shall not apply in any of D the Cases set out in subsection (3) below."

Section 2(3) lists a number of particular cases, none of which have any relevance on the facts of the present case, save that a reference to a "sole surviving tenant" in Case G makes it clear that Parliament implicitly recognised the obvious possibility that a tenancy agreement might name two or more persons as joint tenants.

The general pattern of these three subsections is thus as follows. In any of the cases set out in subsection (3) the restrictions imposed by section 2 on the operation of notices to quit do not apply at all. In other cases, it is a condition precedent to their application that, not later than one month from the giving of the notice, "*the tenant*" (the italics are mine) serves on the landlord the requisite written E counternotice. If in such other cases "the tenant" has duly done this, the notice to quit will be inoperative unless the agricultural land tribunal consents to its operation. If the tribunal does not consent to its operation, the tenancy will continue, together with the mutual rights and obligations of the respective parties under the tenancy agreement.

The 1977 Act itself contains no definition of the word "tenant" and therefore gives no explicit guidance as to the manner in which the word is to be construed when the provisions of section 2(1)(b) fall to be applied to a case where there is more than one tenant. However, section 12(2) (a) of the 1977 Act provides, except where the context otherwise requires, that that Act and the Agricultural Holdings Act 1948 shall be construed as one. Section 94(1) of the 1948 Act, so far as F material, provides that, unless the context otherwise requires, "tenant" means: "the holder of land under a contract of tenancy . . .". By virtue of section 1(1)(b) of the Interpretation Act 1889 the singular phrase "the holder of land" would include the plural "unless the contrary appears".

It is not in dispute that a counternotice in proper form served by one or more of several tenants of land on behalf of all the tenants *with the actual authority of the other tenants* will suffice to satisfy the requirements of section 2(1)(b) of the 1977 Act. That much is common ground. The question of statutory construction that arises in the present case concerns the effect of the service of such a counternotice by some only of the tenants *without* such actual authority, and indeed contrary to the wishes of the other tenant(s).

Some earlier decisions

At this point it is convenient to refer to some earlier decisions which illustrate the position at common law of joint lessees and have recently been helpfully reviewed by Mr Donald Rattee QC sitting as a deputy judge in *Parsons* v *Parsons* [1983] 1 WLR 1390. In *Doe d Aslin* v *Summersett* (1830) 1 B & Ad 135 it was held that a notice to quit signed by one of several joint tenants on behalf of the others is sufficient to determine a tenancy from year to year, as to all. Lord Tenterden CJ, in the course of his judgment (at p 140), said:

Upon a joint demise by the joint-tenants upon a tenancy from year to year, the true character of the tenancy is this, not that the tenant holds of each the share of each so long as he and each shall please, but that he holds the whole of all so long as he and all shall please; and as soon as any one of the joint-tenants gives a notice to quit, he effectually puts an end to that tenancy; the tenant has a right upon such a notice to give up the whole, and unless he comes to a new arrangement with the other joint-tenants as to their shares, he is compellable so to do.

That case was decided before the reform of the law relating to the joint ownership of land by the Law of Property Act 1925. However, it was considered in a subsequent case, *Leek and Moorlands Building Society* v *Clark* [1952] 2 QB 788, where it was relied on to support an argument that one of two joint owners of a tenancy could bind the other by a unilateral surrender of the tenancy. The Court of Appeal rejected this argument, holding that a tenancy held jointly could be surrendered only by the act of all the joint tenants. It distinguished the decision in *Doe d Aslin* v *Summersett*, but cast no doubt on its correctness. Somervell LJ in the course of delivering the judgment of the court said (at p 793):

A periodic tenancy renews itself unless either side brings it to an end. But if one of two or more joint lessees does not desire it to continue, we would have thought that it was in accordance with Lord Tenterden's principle, and with common sense that he should be able to make that effective.

The Court of Appeal thus clearly accepted the position at common law that a periodic tenancy is to be continued into the next period only if all the holders of the joint tenancy or the reversion so wish and that, if one does not, he is at liberty, on his own, to determine the tenancy.

More recently in *Greenwich London Borough Council* v *McGrady* (1982) 81 LGR 288* this court applied the same principle in holding that one of two joint weekly tenants could give an effective notice to the landlords to terminate the tenancy. Sir John Donaldson MR, after citing from the judgments in *Leek and Moorlands Building Society* v *Clark*, said (at p 290):

In my judgment it is clear law that if there is to be a surrender of a joint tenancy, that is, a surrender before its natural termination, then all must agree to the surrender. If there is to be a renewal, which is the position at the end of each period of a periodic tenancy, then again all must concur.

The substantial effect of the service of a valid counternotice by or on behalf of joint lessees pursuant to section 2(1)(b) of the 1977 Act is to achieve a renewal of the tenancy, subject only to the tribunal's withholding its consent to the operation of the notice to quit. If, therefore, the legislature, in enacting section 2(1)(b), had in mind the well-established principles at common law as set out in the *Leek and Moorlands Building Society* case, one might have expected it *prima facie* to contemplate that, if there was to be an effective counternotice in the case of a joint tenancy, then all tenants must concur. However, where the phrase "the tenant" is used in a statute, the proper meaning to be attributed to it must no doubt depend on its context. In two cases which were referred to by the learned judge in the court below and decided in the context of particular statutes other than the 1977 Act, the court construed it as meaning in effect "the joint tenants or any one or more of them". The first was *Howson* v *Buxton* (1928) 97 LJKB 749. That case concerned section 12 of the Agricultural Holdings Act 1923, of which subsection (7) provided:

Compensation shall not be payable under this section — . . . (b) unless the tenant has, not less than one month before the termination of the tenancy, given notice in writing to the landlord of his intention to make a claim for compensation under this section . . .

The tenancy in that case was held by two persons, but only one of them had any interest in the chattels for which compensation was sought and it was he alone who gave notice to the landlord under section 12(7). The Court of Appeal held that the notice was valid. Scrutton LJ said at p 753:

It is clear that one mischief that it was intended to avoid was the loss caused to people who had to remove their household goods or sell their agricultural implements at the termination of the tenancy, and it would be an extremely

*Editor's note: Also reported at (1982) 267 EG 515.

common thing that one tenant might own the household goods and the agricultural implements, and that he might suffer loss by having to remove them, and that the other joint tenant might suffer no loss at all. That is the mischief which Parliament intended to deal with and to provide a remedy for. I think I am justified in adopting the construction that where there is a tenant, although he is only one of the tenants who suffers damage by having to move his household furniture and his agricultural implements, that tenant, although he is only one of the joint tenants, can recover by giving a notice.

Sankey LJ, though giving additional reasons, agreed (at p 757) with the reasons given by Scrutton LJ. Greer LJ reached the same conclusion, but on the grounds that the giver of the notice was agent for both tenants.

However, the context of section 12(7) of the 1923 Act was a special one, as appears from the following observations in the judgment of Winn LJ in *Jacobs* v *Chaudhuri* [1968] 2 QB 470 at p 496, in which the Court of Appeal, by a majority, held that one of two joint tenants was not "the tenant" within the meaning of section 24(1) of the Landlord and Tenant Act 1954 and therefore could not apply for the grant of a new tenancy:

First, the statute which was construed, the Agricultural Holdings Act, 1923, imposed a general obligation upon landlords of agricultural holdings to pay compensation of the kind to which it referred, and the provision that compensation should not be payable "unless the tenant has . . . given notice" represents an exception from this general provision, and, therefore, requires to be strictly construed; second, the subject-matter of the right granted to the "tenant" was his personal belongings in the form of household goods, farm produce, etc, and not the tenancy; it follows that he was entitled to compensation *qua* owner of those goods, not *qua* owner of the tenancy; third, no continuing or bilateral set of obligations were involved; only a single unilateral obligation to pay money; fourth, as Scrutton LJ recognised, one of several tenants could effectively claim only in respect of his own separate property and any notice given by any tenant would refer only to *his* property.

The second of the two cases particularly relied on by Nourse J was *Lloyd* v *Sadler* [1978] 1 QB 774. In that case the issue was whether one of two joint holders of a protected tenancy, who alone remained in occupation of the premises, was entitled to claim the benefit of section 3(1)(a) of the Rent Act 1968, which provided: ". . . after the termination of a protected tenancy of a dwelling-house the person who, immediately before that termination, was the protected tenant of the dwelling-house shall, if and so long as he occupies the dwelling-house as his residence, be the statutory tenant of it; . . .".

The Court of Appeal answered this question in the affirmative. Megaw LJ (at pp 786-787) said:

In my opinion, the judgment of Scrutton LJ in *Howson* v *Buxton* shows that, where the strict application of the doctrine of joint tenancy would lead to unreasonable results, or results which the legislature is unlikely to have intended, it is permissible for the court to conclude that the legislature did not so intend but that, instead, in such a case, the phrase "the tenant" where there is a joint tenancy, is to be read as meaning "the joint tenants or any one or more of them".

There is thus authority that the doctrine of joint participation by joint tenants is not a sacrosanct or immutable doctrine of statutory interpretation, where such phrases as "the tenant" and "the tenancy" are used. Where, then, is the line to be drawn? When does a suggested exception become a heresy? In the present case I believe that, for the purposes of section 3(1)(a) of the Act of 1968, the freedom from strict doctrinal restraint should, as in *Howson* v *Buxton*, be held to apply.

Lawton LJ (at p 789) said:

As the Act gives protection to persons, not to legal concepts such as joint tenants, I am entitled, in my judgment, to construe section 3(1)(a) so as to allow one of two joint tenants to become "the statutory tenant".

A little later (at p 790) Lawton LJ said:

The object of the Act of 1968 was to give security of tenure to persons, and one of two joint tenants might have wanted it and the other not. The mischief for which Parliament provided a remedy was eviction for reasons other than those which the Act deemed good. If Miss Sadler has to leave the flat because Miss Lunt did not want to stay there any longer, that is not a reason which was specified in the Act. I am satisified that section 3(1)(a) can be construed so that one of two joint tenants can become "the statutory tenant". Any other construction would defeat one of the objects of the Act.

Shaw LJ (at p 791) said:

As has been emphasised in the preceding judgments, the primary object of section 3(1)(a) of the Act of 1968 is to provide security of tenure after the term granted by a protected tenancy has expired. Such security is a matter of personal and individual interest. Even though that interest is not shared by or common to all the erstwhile joint tenants, one or some of them may be in need of the security which the Act affords.

The decisions in *Lloyd* v *Sadler* and *Howson* v *Buxton* illustrate that there is no immutable doctrine that the phrase "the tenant", when appearing in a statute, must always be construed as referring to G
all the joint tenants in any case where a joint tenancy is involved. In two further cases, however, one finds the court construing the phrase in accordance with what Megaw LJ in *Lloyd* v *Sadler* (at p 783) described as "the ordinary law as to joint tenancy, as it affects rights of property".

In *T M Fairclough & Sons Ltd* v *Berliner* [1931] 1 Ch 60 Maugham J held that where there are joint lessees, relief against forfeiture cannot be granted on the application of only one of them under section 146(2) of the Law of Property Act 1925. He said (at p 66):

In the present case in my judgment it is not accurate to say that the second defendant is the lessee. The lessees here are joint lessees — the first defendant and the second defendant. Sub-s 5 of s 146 of the Law of Property Act, 1925, H
states that "'Lessee' includes an original or derivative under-lessee, and the persons deriving title under a lessee"; but it does not in any way lead to the conclusion that if there are two or more joint lessees one of them can apply to the Court for relief; . . . if I were to accede to the application on behalf of the second defendant, the result would be that the first defendant would continue to be liable under the onerous covenants contained in these leases, including the onerous obligation to pay rent up to the termination of the leases without, it may be, any prospect of being able to recoup himself by the use of the premises.

The earlier decision which is closest of all to that in the present case is that of Fox J (as he then was) in *Newman* v *Keedwell* (1978) 35 P & CR 393*. The facts of that case appear from the headnote to the report:

On April 25, 1973 the plaintiff landlords, as owners of a freehold agricultural J
farm, entered into a written agreement to grant to the defendants, as joint tenants, a tenancy of the farm from year to year "until determined by either the landlords or the tenants on December 25 in any year by 12 months notice in writing". On January 6, 1976, the plaintiffs gave to the defendants written notice to quit. On January 30 the second defendant served on each of the plaintiffs a counternotice purporting to require that section 24(1) of the Agricultural Holdings Act 1948 should apply to the notice to quit. At the time the first defendant wished the tenancy to be terminated and he gave no authority to the second defendant to serve the counternotices.

Fox J granted the plaintiff's application for declarations that the notice to quit was valid and that the counternotices were of no effect. The second defendant, in the course of argument, relied on *Howson* v *Buxton* (*supra*), but the learned judge thought that decision K
distinguishable from the case before him (just as I have already concluded it is distinguishable from the present case). He continued (at pp 397-398):

Prima facie it seems to me unlikely that, if a statute or a lease permits a tenant to take some step which may directly alter the duration of the tenancy and there are in fact beneficial joint tenants, it could have been intended that it should be competent to one of the joint tenants to take that step without the authority of the other . . .

In the present case the effect of the service of the notice to quit by the landlord is that the tenancy will determine unless a valid counternotice is given by the tenant. If a valid counternotice is given then the tenancy will continue (unless the tribunal consents to the operation of the notice to quit). Thus the subsisting position after the service of the notice to quit is that the tenancy was L
determined unless the counternotice is given. It seems to me that if the subsisting position is to be displaced the authority of both joint tenants is necessary because of the serious effect which it has upon their obligations. If the tenancy continues both joint tenants have a continuing liability in respect of the rent and the covenants in the agreement. It is difficult to see why one joint tenant should have these burdens forced upon him against his will when the position in law is that the tenancy cannot continue unless "the tenant" serves a counternotice. It seems to me that the fair reading of "tenant" in section 24(1) is, in the circumstances of this case, "the joint tenants"; what it may mean in other sections of the Act I need not consider.

A little later, having cited a passage from the judgment of Maugham J in the *Fairclough* case, Fox J observed (at p 399):

In my view that applies here. It was not at the time of the service of the M
counternotices, and is not at present, the wish of the first defendant that the tenancy should continue. If the counternotices are valid then, subject to the power of the tribunal to give consent to the operation of the notice to quit, the tenancy will continue, but what continues is the old tenancy. It is not suggested that there is any power in the court to direct the grant of a tenancy to one only of the joint tenants or to release a joint tenant from liability under the covenants in the tenancy agreement. Indeed in many cases it would be quite unfair on the landlord to force him to accept the covenants of one only of the joint tenants.

In the result, Fox J concluded that the second defendant was not competent to give a valid notice, except with the authority of the first defendant.

*Editor's note: Also reported at (1977) 244 EG 469.

The judgment of Nourse J

In argument before Nourse J, counsel appearing for the Staples put their case in two alternative ways. First, he submitted that, on the true construction of the 1977 Act, the giving of an effective counternotice under section 2(1) does not require the concurrence of a corporate joint tenant which is wholly owned by the landlord. Second, he submitted that, notwithstanding the provisions of clause 11 earlier referred to, the Staples did have authority to give the counternotices on behalf of all three partners.

Nourse J did not find it necessary to deal with the second of these submissions because he accepted the first. Having quoted the provisions of section 2(1) of the 1977 Act, he continued ([1985] 1 EGLR 1 at p 2):

> The general rule of statutory construction is well settled. In the case of a joint tenancy "the tenant" means all the joint tenants and not just one or some of them. On this principle a counternotice served by only one of two individual joint tenants of an agricultural holding has been held to be of no effect; see *Newman* v *Keedwell* (1977) 244 EG 469. On the other hand, the Court of Appeal has twice departed from the general rule in cases where they thought that it would lead to unreasonable results which the legislature is unlikely to have intended.

After making a brief reference to *Howson* v *Buxton,* the learned judge cited in full the passages which I have cited from the judgments of this court in *Lloyd* v *Sadler.* He observed (at [1985] 1 EGLR p 3):

> In *Lloyd* v *Sadler* the Court of Appeal thought it was fairer to deprive the landlord of one source of rent than to deprive the occupying joint tenant of her protection. In *Newman* v *Keedwell* the decision went in favour of the landlord because the continuation of the joint tenancy would have been unfair, not to him but to the joint tenant who wished that it should come to an end. I must now apply the principles which are to be deduced from these two authorities to the circumstances of the present case.

The learned judge thus did not question the correctness of the decision in *Newman* v *Keedwell* on its facts. By necessary implication, he accepted that when section 24(1) of the 1948 Act (the predecessor of section 2(1) of the 1977 Act) fell to be applied to the particular facts of *Newman* v *Keedwell,* it was correct for Fox J to treat the word "tenant" as bearing the narrower sense "the joint tenants" instead of the wider sense "the joint tenants or any one or more of them". However, the effect of his decision was that, in cases where a joint tenancy is involved, the word "tenant", as appearing in the relevant subsections, should be given a flexible meaning, so that, while in some cases it would fall to be read in the narrower sense, in others it should be read in the wider sense, bearing in mind the purpose of the subsection. He considered that the fair reading of "tenant" in the subsection, in the circumstances of the present case, was in a wider sense.

The reasoning which led the learned judge to his conclusion (see [1985] 1 EGLR at p 3) may, I think, be summarised as follows. By serving notices of dissolution which had the effect of dissolving the partnerships on September 29 1984, Laughton had elected not to continue in occupation of the land. By virtue of their rights to any surplus, the Staples could procure that the full benefit of the tenancies, together with all rights attaching to them, could vest in them to the exclusion of Laughton. The true purpose of section 2(1), he considered, is "to give security to those who need it". In his view there was no distinction in principle between the purpose of that subsection and section 3(1)(a) of the Rent Act 1968. The Staples, like the occupying tenant in *Lloyd* v *Sadler,* needed security; Laughton, on the other hand, like the non-occupying tenant in *Lloyd* v *Sadler,* did not. To hold that the Staples are not protected by section 2(1) would, in the learned judge's view, defeat its purpose. To hold that the counternotices were valid and that the tenancies continued would involve no deprivation of the plaintiffs, since Laughton would remain liable to payment and to perform the tenants' other obligations. *Newman* v *Keedwell* was distinguishable because, on the facts of that case, Fox J had considered that the continuation of its obligations under the tenancies was not a burden which ought to be forced on the dissenting tenant against his will. The present case, however, was a different one, because the continuation of its obligations under the lease would not prejudice or be unfair to Laughton in any real sense, in as much as the persons to whom these obligations are owed own all the shares in Laughton and include two of its four directors and would have nothing to gain by enforcing them. The learned judge summarised his conclusions thus (at [1985] 1 EGLR p 3):

> In the circumstances, I am of the opinion that *Newman* v *Keedwell* is distinguishable on the facts. Like Mr Langdon-Davies I think that it may be significant that Fox J restricted himself to the fair reading of "tenant" in section 24(1) in the circumstances of that case; see (1977) 244 EG at p 472. I do not think that it is the fair reading in the circumstances of this case. Being unable to seen any distinction in principle between the purposes of the two enactments, and being bound to apply the decision in *Lloyd* v *Sadler,* I find myself at a point at which I might not have been able to arrive without its guidance. Mr Langdon-Davies' first submission succeeds. I hold that the three counternotices dated October 14 1984 are valid.

Issue of statutory construction

The plaintiffs' submission on the issue of statutory construction is in essence that the learned judge misconstrued section 2(1) of the 1977 Act in holding that "the tenant" in the case of a joint tenancy can, in some cases, mean some (as distinct from all) of the joint tenants. They say that he ought to have held that "the tenant" in the case of a joint tenancy must mean all the joint tenants. They accept that a counternotice duly served by one or more joint tenants with the authority of the other joint tenants will be effective to bind both those other joint tenants and the landlord. However, they submit that a counternotice served by one or more joint tenants without the authority of the others can never be an effective counternotice for the purpose of section 2(1)(b) so as to bind either the landlord or the other joint tenants.

In considering these submissions, I for my part find the decisions in *Howson* v *Buxton* and *Lloyd* v *Sadler* of only very limited assistance. They certainly show that it is permissible to construe the phrase "the tenant" as meaning "the joint tenants or any one or more of them" when that word appears in a statutory provision and the particular provision indicates the legislature's intention that it shall bear that meaning in any case where joint tenants are involved. However, both those decisions concerned statutes different from the 1977 Act (in the case of *Howson* v *Buxton* very different). And in neither of them did the court attribute the flexible meaning to the word "tenant" which the learned judge favoured in the present case — flexible in the sense that in some circumstances, such as those of *Newman* v *Keedwell,* it would bear the narrower sense to which I have referred above, while in other circumstances it would bear the wider sense.

With great respect to the learned judge, I find difficulty in accepting his flexible construction of the phrase "the tenant" as meaning (in effect), in the case of joint tenants, simply those of the joint tenants who "need security of tenure". There seems to me at least two formidable objections to it. First, its impact would be vague and uncertain in its operation. Let it be supposed that there are three tenants of an agricultural holding and that, having served a notice to quit, the landlord then receives a counternotice purporting to be served on behalf of all three tenants, but signed by only two of them, while the third informs him that they had no authority to sign it on his behalf. It could well be said that at least in one sense the first two tenants "need security", to use the learned judge's phrase. Nevertheless, the legislature, in my view, cannot have contemplated that in such circumstances it should be incumbent on the landlord (who has no association with any of the tenants) to investigate the equities as between the respective tenants with a view to ascertaining whether or not the continuation of the tenancy would be unfair to the joint tenant who wishes that it should come to an end, and thus whether or not he must regard the counternotice as a valid one.

Second, with respect, I think the learned judge's construction does not take sufficient account of the fact that in cases where there are several tenants, only some may "need security" in the sense that for their own good reasons they wish the tenancy to continue, while others, for their own, no less good, reasons, may wish the tenancy to come to an end. The dispute or difference of opinion between the tenants may well be a genuine one based on substantial grounds. I can see no sufficient reason why the legislature, in enacting section 2(1)(b), should have contemplated that in such circumstances those tenants who wish the tenancy to continue should have the statutory right, by the service of a counternotice, to bind the other tenants to the continuation of the tenancy against their will; the legislature would, I think more probably, have envisaged that in any such case unanimous action by the tenants was essential if a counternotice was to be served. It should be added that if unanimous action cannot be achieved in any other way, it will still be open to the tenants who wish a counternotice to be served to seek a court order to compel the co-operation of their cotenants: see *Harris* v *Black* (1983) 46 P&CR 366 and *Sykes* v *Land* (1984) 271 EG 1264. The tenants who desire the continuation of the tenancy will therefore not inevitably be left without redress, though it is true that the time available to them to make an effective application of this nature may be very short.

Subject to a possible qualification referred to later in this judgment, I have come to the clear conclusion that, in the context of section 2(1)(b) of the 1977 Act, the phrase "the tenant", in the case of any joint tenancy, must mean "the joint tenants" and not "the joint tenants or any one or more of them". I do not read Fox J's similar construction of the phrase in *Newman* v *Keedwell* as having been limited to the particular circumstances of that case, apart from the particular circumstance that the land in question was let to joint tenants. I find his reasoning compelling and respectfully adopt it, merely adding three points. First and foremost, a clear line of authority, exemplified by the judgments of this court in the *Leek and Moorland Building Society* case and more recently in the *Greenwich London Borough Council* case, shows that at common law, if there is to be a renewal of a periodic tenancy held by joint tenants at the end of one of its periods, then all tenants must concur. The substantial effect of a valid counternotice under the 1977 Act (unless the tribunal consents to the operation of the notice to quit) is to renew the tenancy. In these circumstances, it is, in my view, *prima facie* most unlikely that the legislature would have intended that in the case of a joint tenancy a valid counternotice could be served without the concurrence of all the joint tenants. Second, as I think Mr Burnett on behalf of the Staples accepted, the phrase "the tenant" in section 2(1)(a) of the 1977 Act must in the case of a joint tenancy mean "the joint tenants". *Prima facie* the same phrase, immediately following in section 2(1)(b), must bear the same meaning. Third, Megaw LJ in *Lloyd* v *Sadler* [1978] 1 QB 774 at p 786 referred to the decision of Fox J in *Newman* v *Keedwell* with apparent approval, but said that by reason of the nature of the issues in that case it did not assist on the issue before him.

However, the facts of *Newman* v *Keedwell* did not contain a special feature of the case now before us; the joint tenants did not include the landlord himself or a company controlled by him. This leads me to consider the possible qualification referred to above. Accepting that the phrase "the tenant" in section 2(1)(b) ordinarily means all the joint tenants in any case where a joint tenancy is involved, does the subsection, by necessary implication, exclude the landlord himself (or a company controlled by him) if he or it happens to be one of the joint tenants?

There are, in my opinion, quite strong grounds for saying that in the particular context where the landlord himself is one of the tenants (not the present case) the word "tenant" in section 2(1)(b) would be capable of bearing more than one meaning, namely, (1) all the joint tenants or (2) all the joint tenants other than the landlord himself; and that the court should prefer the second construction as leading to a "sensible and just result complying with the statutory objective" (see *Johnson* v *Moreton* [1980] AC 37 at p 50 per Lord Salmon) instead of adopting a construction which would invalidate a counternotice given in such circumstances, unless the landlord had joined in or consented to its service on himself. On the other hand, I think that the engrafting of further exceptions by a process of statutory interpretation on the ordinary meaning of the phrase "the tenant", to cover the case (such as the present) where one of the joint tenants was merely a company controlled by the landlord or was otherwise associated with the landlord, would present rather greater difficulties, if only because of problems of definition and deciding where to draw the line.

However, I find it unnecessary to express any concluded view on the question whether, as a matter of construction, any exceptions should be engrafted on to what I regard as the ordinary meaning of the phrase "the tenant", as used in section 2(1)(b), in the case where the tenants are more than one in number. Even assuming for the present (as I will assume) that as a matter of statutory interpretation the phrase must mean *all* the joint tenants, in any case where joint tenants are involved, I am still of the opinion that, for the reasons to be given later in this judgment, the counternotices in the present case must be treated as having been served with the authority of Laughton, even though that authority was not in fact given.

The respondents' notice

I now turn to the reamended respondents' notice in which the Staples' submissions on this question of authority are to the following effect:

1. That the first and second Defendants as partners of the third Defendants were authorised by section 5 of the Partnership Act 1890 and by the terms of the partnership agreements to serve counternotices under section 2(1) of the Agricultural Holdings (Notice to Quit) Act 1977 on behalf of the partnership including the third Defendants unless that authority was lawfully limited.
2. That the clauses in the partnership agreements which purported to limit the G Defendants' power to serve such notices were void as being contrary to the terms of section 2(1) of the said Act and against public policy.
3. That at all material times to the knowledge of the Plaintiff the third Defendants as trustees of the assets of the partnership including the tenancy were under a duty to preserve the tenancy alternatively to give effect to the wishes of the first and second Defendants as the ultimate beneficiaries by joining in the service of or ratifying any counternotices served by the first and second Defendants.

As has already appeared, under the partnership agreements it had been specifically agreed that after September 29 "no partner may serve a counternotice ... without the consent of Laughton". I will for convenience refer to this as "the restrictive condition". If the restrictive condition was a valid contractual term, the Staples were H acting in clear breach of it when they served the counternotices. Correspondingly, on the assumption that the restrictive condition was valid, there could have been no possible question of Laughton acting in breach of a fiduciary duty owed to the Staples (in the manner asserted in para 3 of the respondents' notice) in refusing to join in the service of the counternotices or to ratify them; by the terms of the condition the parties had given or purported to give Laughton the unrestricted right to take this course; Laughton, on this assumption, would have been in no breach of duty owed to the Staples, legal or equitable.

The contentions advanced in para 3 of the respondents' notice thus entirely depend on the Staples first establishing the invalidity of the restrictive condition. J

The same comment applies to the contentions advanced in para 1. I am inclined to think that the combined effect of section 5 of the Partnership Act 1890 and the provisions of the partnership agreements would have sufficed to give the Staples express authority to serve the counternotice on behalf of the partnership if the partnership agreements had not included the restrictive condition. However, they did include it. If any of the arguments advanced by the Staples in their respondents' notice are to prevail, they must, in my judgment, first establish the invalidity of that condition. Mr Burnett's argument to this effect gives rise to issues which are, to my mind, by far the most difficult in this case. Its sheet anchor has been the decision in *Johnson* v *Moreton* [1980] AC 37 where the House of K Lords unanimously held unenforceable a clause in a lease in the following terms:

> The tenant agrees to give possession of the whole of the farm to the landlords immediately upon the determination of the term hereby granted and not in any event to serve a counternotice under section 24(1) of the Agricultural Holdings Act 1948 or to take any steps to claim the benefit of any statutory provision granting security of tenure which may be in force at the time of the determination thereof.

On the face of it this clause was, as Lord Hailsham observed (at p 54):

> an open, not to say brazen, attempt to get round the provisions of the agricultural holdings legislation both as it had existed up to that point and as Parliament might enact it in the future at least so far as it provided security of L tenure to the tenant.

The routes by which their lordships concluded that the clause was unenforceable were not quite identical, but, for present purposes, were adequately reflected in the following passages from the speech of Lord Salmon. At p 51 D-E he said:

> I do not consider that any question of implication arises in relation to section 24(1). Its meaning is plain and unambiguous. It gives the tenant a statutory option to be exercised within one month of receiving a notice to quit. The option is to go voluntarily or to serve a counternotice and remain in possession unless the Agricultural Land Tribunal exercises its very restricted powers of allowing the notice to quit to become effective. The option cannot be exercised any sooner or any later than the subsection prescribes. Nor can it, in my view, be renounced by the tenant in advance. The language of the section M makes this plain. Moreover the statutory option was conferred on tenant farmers, not for their personal protection alone, but for the public good.

At pp 52G-53A Lord Salmon said:

> The security of tenure which tenant farmers were accorded by the Act of 1947 was not only for their own protection as an important section of the public, nor only for the protection of the weak against the strong; it was for the protection of the nation itself. This is why section 31(1) of the Act of 1947, reproduced by section 24(1) of the Act of 1948, gave tenant farmers the option to which I have referred and made any agreement to the contrary void. If any clause such as clause 27 was valid landlords might well insist upon a similar clause being introduced into every lease; and prospective tenants, having no money with which to buy the land they wanted to farm, would, in reality, have had little choice but to agree. Accordingly if clause 27 is enforceable the

A security of tenure which Parliament clearly intended to confer, and did confer upon tenant farmers for the public good would have become a dead letter.

Mr Burnett has submitted that essentially the same reasoning as was applied by the House of Lords in *Johnson* v *Moreton* renders void and unenforceable the restrictive condition in each of the partnership agreements. He also reminded us of the recent decision in *Street* v *Mountford* [1985] 1 AC 809 in which Lord Templeman, in the course of delivering the leading speech holding that the appellant was not a licensee but a tenant, said (at p 825G-H):

> Although the Rent Acts must not be allowed to alter or influence the construction of an agreement, the court should, in my opinion, be astute to detect and frustrate sham devices and artificial transactions whose only object
> B is to disguise the grant of a tenancy and to evade the Rent Acts.

Mr Burnett specifically disclaimed any suggestion that the agreements in the present case were sham devices, but, further or alternatively to his submission as to the invalidity of the restrictive condition itself, he submitted more broadly that these were artificial transactions whose only object was to evade section 2(1) of the 1977 Act by depriving the Staples of the protection which that Act would otherwise have given them, and that the court should accordingly be astute to frustrate such object.

Mr Wood, for his part, realistically accepted that, for the purpose of his argument in answer to the respondents' notice, he would not seek to rely on any distinction between the plaintiffs and their wholly owned company Laughton. For this purpose he was content that they
C should be regarded as one and the same.

I would like, first, to deal briefly with Mr Burnett's submissions based on the alleged "artificiality" of the transactions. Whatever be the proper meaning to be attached to that word, I am not prepared to attach that label to the transactions now under consideration on the basis of the evidence before the court. As Mr Wood pointed out, and as I think is common knowledge, arrangements are not infrequently made by virtue of which landowners enter into a partnership with one or more other persons on the basis that the partnership will be granted a tenancy of an agricultural holding and that the landowner himself will be what is colloquially known as a sleeping partner. Quite apart from any considerations relating to security of tenure, there may well
D be good and sufficient reasons (whether of a commercial, family, fiscal or practical nature) why all interested parties should regard such an arrangement as sensible and beneficial. There is nothing contrary to public policy in any arrangements of this nature as such.

In the present case the onus of showing that these arrangements, to all of which they were party, were artificial transactions would fall on the Staples. As I have already said, no oral or affidavit evidence was adduced before the learned judge and there was no agreed statement of facts. It has been suggested that the documentation in the present case itself shows that there were no good reasons for the introduction of Laughton as a cotenant and copartner beyond the desire to circumvent the provisions of section 2(1)(b) of the 1977 Act. Conceivably, if this suggestion had been explored in oral evidence, it
E might have been borne out. As things are, however, I think that in the absence of any finding or evidence to the contrary, the plaintiffs are entitled to ask the court to assume in their favour that there were good and sufficient reasons for the introduction of Laughton as a cotenant beyond the desire to circumvent the provisions of section 2(1)(b) of the 1977 Act, even though the last five lines of clause 3 of the first partnership agreement (quoted above) clearly showed that the parties had in mind the current law relating to the security of agricultural tenants. For the rest of this judgment I will so assume.

However, the plaintiffs are still faced with a formidable hurdle. The House of Lords decision in *Johnson* v *Moreton* establishes beyond argument that if "A" grants a tenancy of an agricultural holding to "B" and the tenancy agreement contains a condition by
F which "B" agrees never to serve a counternotice under the relevant legislation, that condition is void and unenforceable. It must, I think, inevitably follow from this decision that if "A" were to grant a tenancy of a holding to "A", "B" and "C" and the tenancy agreement were to contain a condition by which "A", "B" and "C" as tenants agreed with "A" as landlord never to serve a counternotice without "A"'s consent, that condition would be no less void and enforceable.

The present case is not quite the same as that. In essence it is a case where "A" has granted a tenancy of the holding to "A", "B" and "C" and, by a separate partnership agreement, "B" and "C" have agreed with "A" that they will not serve a counternotice without "A"'s consent. (In this present context, as I have indicated, Mr Wood does not rely on the point that the tenancy was in fact granted by "A" not to "A" but to "A"'s wholly owned company, along with "B" and "C".)

Does the introduction of a partnership give rise to quite different legal consequences? Mr Wood has strongly submitted that it does. As I have already indicated, he has stressed the various good reasons, quite apart from any desire to circumvent the provisions of section 2(1) of the 1977 Act, why the parties might have wished to enter into the transactions in the present case. He further submitted that the tenancy agreements were entered into simply to facilitate the partnerships, that each tenancy was protected during its currency and that the documentation clearly demonstrated the partners' intentions that on the dissolution of the partnerships the tenancies should come to an end. He further contended that there is nothing contrary to section 2(1) of the 1977 Act or to public policy in one partner or tenant being at liberty to refuse to participate in the giving of a notice which will perpetuate against his will a legal estate vested in him.

In the course of the very able arguments submitted by counsel on both sides, my mind has fluctuated greatly on all the principal issues in the present case, and I regard this issue of public policy as particularly difficult. In this context I feel some sympathy with the contention that a landlord who chooses to grant a tenancy to a farming partnership of which he is a member should not be forced into the position of having to submit to a continuation of the tenancy for an indefinite period even after the dissolution of the partnership, and that public policy does not require the discharge of his cotenants from their contractual obligations under a restrictive condition such as this.

Nevertheless, as I have already indicated, if there had been no partnership agreements, and the tenancy agreements themselves in the present case had contained a restrictive condition that the three tenants would not serve a counternotice without the plaintiffs' consent, I think it clear that the decision in *Johnson* v *Moreton* would have compelled the court to hold such a condition unenforceable as being "an open, not to say brazen, attempt to get round the provisions of the agricultural holdings legislation". I am not persuaded that any different result ensues in the present case because the three tenants had formed themselves into a partnership and the restrictive condition happened to be contained in the partnership agreement. I am driven to the conclusion that, if a landowner chooses to grant other persons a tenancy of agricultural land (whether or not including himself as a tenant), public policy (affirmatively) requires that those other tenants should have authority, or be treated as having authority, to serve an effective counternotice under section 2(1) of the 1977 Act on behalf of all the tenants without his concurrence, and thus (negatively) requires the avoidance of any contractual condition, whether express or implied and whether contained in the tenancy agreement itself or in a partnership agreement or elsewhere, which purports to deny those other tenants such authority. I might add that any contrary decision of this court would be likely to open the door to widespread evasion of the 1977 Act to the detriment of the security of tenure which Parliament clearly intended to confer on agricultural tenants: (see and compare *Johnson* v *Moreton* [1980] AC at pp 52G-53A per Lord Salmon).

I therefore conclude that the particular restrictive condition agreed between the parties and any other provision in the documentation before this court which might otherwise be said to have limited the authority of the Staples to serve valid counternotices on behalf of the partnerships was void and unenforceable. It follows that, in my judgment, these counternotices were valid. I agree in the result with the decision reached by the learned judge. I would dismiss this appeal.

STOCKER LJ and SIR ROUALEYN CUMMING-BRUCE agreed and did not add anything.

The appeal was dismissed with costs. Leave to appeal to the House of Lords was granted. Legal aid taxation of the respondents' costs was ordered.

AUCTIONS

Court of Appeal

February 27 1986

(Before Lord Justice FOX, Lord Justice DILLON and Lord Justice NEILL)

SPRIGGS v SOTHEBY PARKE BERNET & CO LTD

Estates Gazette May 31 1986

278 EG 969-974

Auction — Theft of valuable diamond while on view on day before auction sale — Appeal by owner from dismissal of his action against auctioneers — Deputy judge, Sir Douglas Frank QC, had held that auctioneers were protected by exclusion clause in contract disclaiming responsibility, but he also held that in any case they had not been negligent — Diamond had been stolen from a viewing ring while the attention of a senior porter, who was on his own in the centre of the ring, was momentarily distracted by a request from a viewer on the other side of the ring — Four security guards were stationed in the room during viewing time, their duty being not only to watch persons who had items of jewellery in their possession for viewing but also to prevent persons from leaving the room with items which had not been handed back — Despite these precautions the thief escaped with the diamond — The instructions for sale signed by the appellant owner contained an exclusion clause agreeing that the auctioneers were "not responsible for loss or damage of any kind, whether caused by negligence or otherwise, the property being accepted at owner's risk" — The instructions mentioned that goods could be insured against risks while on the auction premises, if vendors so desired, at a quoted premium rate, but the appellant had given specific instructions that he did not require insurance — The Court of Appeal considered first the issue of negligence and on this issue differed from the trial judge, who decided that in all the circumstances the respondent auctioneers took reasonable care — The court were unable to accept that the system employed by the respondents in July 1977 was a reasonably safe system in the circumstances — It was inevitable that the single porter inside the viewing ring would have to turn his back on customers with valuable lots in their possession; and the fields of vision of the security guards were not likely to remain unobstructed when the room was full or even partially full — The respondents had not, therefore, shown that they had taken all reasonable care, in their capacity as bailees for reward, to keep the diamond safe — The court then considered whether the respondents were protected by the exclusion clause — They rejected various criticisms against its validity and effectiveness put forward by the appellant, such as a suggestion that it was no more than a warning and so insufficient to exclude liability and that the clause could not exempt the respondents from liability where the fundamental purpose of the sale by auction had been frustrated by a prior theft — Held that the respondents were entitled to rely on the exclusion of liability in the contract with the appellant — Appeal dismissed

This was an appeal by the plaintiff, Mr John Gordon Spriggs, a F
businessman, from a decision of Sir Douglas Frank QC, sitting as a deputy judge of the Queen's Bench Division, dismissing the plaintiff's claim against the defendants, the present respondents, Sotheby's, for damages as a result of a theft on the respondents' premises of a 9.46-carat diamond on a viewing day before the auction sale in which the diamond was to have been included. The judgment of Sir Douglas Frank was reported at (1984) 272 EG 1171.

Igor Judge QC and Nigel Baker (instructed by Horwood & Co, of Leicester) appeared on behalf of the appellant; Cyril Newman QC and B Stancombe (instructed by Hewitt, Woolacott & Chown) represented the respondents.

Giving the first judgment at the invitation of Fox LJ, NEILL LJ
said: This is an appeal from the judgment of Sir Douglas Frank sitting G
as a judge of the High Court whereby he dismissed a claim by Mr John Gordon Spriggs against Sotheby Parke Bernet & Co Ltd (Sotheby's), the well-known auctioneers.

The learned judge delivered his reserved judgment on Friday July 13 1984 following a hearing which lasted for three days between June 11 and 13 1984.

Mr Spriggs is a businessman. He lives in Leicester. In 1962 he bought some precious stones from a diamond merchant. One of the stones was a large diamond weighing 9.46 carats. He paid £7,800 for this diamond which he bought as an investment. For the next 15 years Mr Spriggs kept the diamond at the bank having insured it for the
sum of £10,000. Early in 1977 he decided to have the diamond valued H
with a view to selling it. He took it with another stone to a company in Market Street, Leicester, called Tarratt Ltd.

There the stone was valued by Mr Christopher Tarratt. He valued it on a retail replacement basis in the sum of £22,500. Mr Tarratt put his valuation in writing, describing the stone as a single brilliant cut diamond weighing 9.46 carats and stating that the valuation was for insurance. The written valuation was dated January 31 1977. Following his visit to Tarratt Ltd Mr Spriggs put the diamond back in the bank.

Three months later, at the beginning of May 1977, Mr Spriggs decided to sell the diamond. He told his insurance brokers, took the diamond out of the bank again and on May 5 drove to London. On
reaching London he went to Sotheby's in New Bond Street. There he J
saw Mr Hinks, the director in charge of jewellery. A discussion took place which lasted for about half an hour. At the end of the discussion, Mr Spriggs left his diamond with Mr Hinks on the understanding that it would be put up for sale by auction at the next important sale of jewels, which was due to take place on Thursday July 7.

Before he left Sotheby's, Mr Spriggs signed a document which Mr Hinks produced during the course of the discussion. At the trial there was an issue as to whether an oral agreement was reached before the document was produced and signed, but it is now accepted on behalf of Mr Spriggs that the contract between the parties was on the terms
contained in the document which Mr Hinks produced. I shall have to K
refer to the terms of this document later. At this stage it is sufficient to note that it was agreed that a reserve of £9,000 would be placed on the diamond and that Mr Spriggs did not require Sotheby's to insure the diamond for him.

Following the discussion, Mr Spriggs left Sotheby's premises and went home. In due course he received the catalogue for the sale which included his diamond as lot 290, describing the stone in capital letters as follows: "AN IMPORTANT UNMOUNTED CIRCULAR-CUT DIAMOND WEIGHING 9.5 cts APPROXIMATELY". The catalogue which gave notice of the sale stated that the items would be on view at least three days previously to the sale.

Unfortunately, however, on the day before the diamond was to be put up for sale by auction it was stolen. The circumstances of the theft were described in a memorandum dated July 12 1977, which was prepared by Mr Cunningham, the chief security guard at Sotheby's. The relevant part of the memorandum was in these terms:

> About 2.10 pm on Wednesday 6th July 1977 Lot 290 of Sale 7.7.77, an unmounted diamond of 9.5 carats, value approximately £10,000 was stolen from view in the New Gallery under the following circumstances:
>
> About 2.10 pm a man (described) came to the viewing ring at the point nearest to the door and was served by Dan O'Connor singly with five or six lots, including Lots 157 and 290 (both unmounted diamonds) which he asked to view. Lot 290 was the second lot which he viewed. At the end of the group he again asked to see Lot 290 and was given it on a pad by Dan, who was then called by a viewer on the other side of the ring. He cannot say if this man, also unknown, was an accomplice or not but I think it unlikely. The time was then 2.10 pm. Within seconds a Frenchman who was viewing, drew the attention of Dan O'Connor to the fact that the diamond was missing. Due to the language barrier, it was some time before Dan got the message.
>
> Dan told the guards and they made a search for a man whom they had seen leave via the Entrance Gallery. It was later established that they saw the wrong man and that the real thief left by the front entrance.
>
> I was called and saw Dan O'Connor and the four guards and obtained details.
>
> On the following day I traced the Frenchman, a M Castro. I spoke to him through an interpreter and got an outline of his version. He had to leave for Madrid but will call and see me again on July 18. It seems that the thief had an accomplice. There were at least 20 people in the gallery at the time.
>
> We now have a clear message that the viewing conditions must be drastically overhauled.

When Mr Spriggs heard of the theft he was naturally very upset. In due course he consulted his solicitors, who wrote to Sotheby's claiming compensation. The solicitors' letter dated April 6 1978 included these paragraphs:

> We have had the opportunity of investigating the circumstances in which the diamond stone was stolen and indeed are grateful to you for the co-operation you gave our agent in reference thereto. It is clear however from these inquiries that all reasonable steps were not taken by you to ensure the safekeeping of the stone. In these circumstances our client holds you liable in respect of his loss.
>
> The diamond stone in question was, as we are instructed you are aware, valued by adjusters . . . in January 1977 at a figure of £22,500. Our client feels this to be the minimum figure realisable on a sale of the stone. Our client has of course the benefit of insurance up to a figure of £10,000 but undoubtedly his insurers equally will seek reimbursement of this sum. We believe Mr Spriggs's insurers may have already been in touch with you direct already with regard to this.

Sotheby's, however, took the view that they had taken all reasonable care of the diamond and that in addition they were entitled to rely on clause 2 of the "Instructions for Sale", which were printed on the back of the document which Mr Spriggs signed on his visit to New Bond Street on May 5 1977.

The parties were unable to reach a compromise and in due course, on October 3 1978, a writ was issued. Nearly six years later the action came on for trial.

It is common ground that in order to escape liability for the loss of the diamond it was necessary for Sotheby's to establish how the loss occurred and also either (a) that the loss occurred despite the exercise by them of reasonable care in looking after the stone or (b) that despite a lack of reasonable care any liability for negligence had been excluded by the terms of the contract between themselves and Mr Spriggs.

Before I examine these issues, however, I should give some further account of the circumstances in which the theft occurred.

The viewing took place in a room called the New Gallery on the first floor of Sotheby's premises in New Bond Street. The New Gallery was built for the purpose of viewing valuable objects and is well lit. It is about 20 ft square. A sketch plan of the New Gallery is included in the bundle of documents, and the position of the room in relation to the rest of the premises is indicated by the plans.

It will be seen that the entrance to the room is by a pair of doors forming a single opening at the top of and to the right of a flight of stairs leading from the ground floor. It will also be seen that on the far side of the room from the entrance a door leads into a small storage room which has no other exit.

On July 6 1977 the jewellery for viewing was placed in showcases arranged in a ring in the centre of the room. Two of these showcases were high cases though their position was not precisely as shown on the plan.

The exterior diameter of the ring was about 10 ft; the interior diameter was about 6 ft. There was an alarm button on the inside of the ring and two further alarm buttons adjacent to and on either side of the entrance doors.

Around the walls of the room there were other showcases containing porcelain which was to be included in an auction sale due to take place on some later occasion.

Mr O'Connor was on duty in the centre of the ring. His main task as a porter was to assist persons who wished to view particular items of jewellery by handing such items to them for inspection. In the case of persons who were known to Sotheby's either as dealers or as regular customers, Mr O'Connor was authorised to let them have more than one item or a tray of items at a time. In the case of anyone who was not known to Sotheby's, however, Mr O'Connor was only allowed to make one item available at a time.

In addition to assisting customers who wished to view the jewellery, Mr O'Connor had the further task of ensuring that the items which had been handed over for viewing were duly returned after inspection.

The main responsibility for security, however, was in the hands of four security guards who were stationed in the room during the viewing period, two of these guards being near the entrance to the room. The task of the security guards was not only to watch persons who had items of jewellery in their possession for viewing but also to prevent anyone leaving the room with items which had not been handed back.

There was no restriction on the number of persons who could be in the room and it will at once be apparent that at any one time a number of items might have been out of the cases for viewing by different customers at various points on the outside of the ring. It will also be apparent that having handed an item of jewellery to one customer, Mr O'Connor might be asked by another customer at the other side of the ring to take some other item from a showcase to enable the second customer to view it. In that event Mr O'Connor would have had to turn his back on the first customer.

At the trial, however, Mr Cunningham said in evidence that the guards had instructions to watch a person viewing an item if the porter's attention had been distracted by going to serve a second customer.

It is clear that on the occasion of the theft, Mr O'Connor's attention was diverted from the man who was viewing lot 290 by another customer who asked to be shown some other item which he wished to view. In order to assist this other customer, Mr O'Connor had to go to a showcase, abstract the item in question and then hand it to the customer at some other point on the ring.

It seems that if the system had been working properly one of the security guards would then have paid particular attention to watching the man with lot 290. But, as Mr Cunningham accepted in cross-examination, the system (if such it was) did not work on this occasion. Furthermore there was no evidence that there was any recognised method whereby Mr O'Connor could communicate with the guards to alert them to the fact that his attention had been diverted from a particular customer nor was any evidence adduced at the trial by any of the guards as to any observations they might have made on the day in question.

The issue of reasonable care

The learned judge held that Sotheby's were excused from liability for any loss or damage whether caused by negligence or otherwise by reason of the terms of the contract between them and Mr Spriggs. I shall have to return to this aspect of the case a little later. But he also held that Sotheby's took reasonable care "having regard to all the circumstances".

He explained the basis for this conclusion in these words:*

> One of their directors had overall responsibility for security. They employed a retired detective chief inspector as chief security officer and for supervising the jewellery viewing room five security guards. It is true that since this theft the defendants have made changes to increase security, but, of course, it is always possible to make changes to increase security; but whatever improvements are effected, there can never be a guarantee of absolute security. It is a matter of looking at the arrangements as they were immediately before the theft and not of being wise after the event. I accept that the duty to take reasonable care, if there is one, applies to the defendants' servants. It can, of course, be said that Mr O'Connor should not have left the stone unwatched, notwithstanding that he was called to another viewer. However, it seems to me that that lapse, if such it be, has to be looked at in the context of his job. It has to be borne in mind that if jewels are to be sold, then

*Editor's note: See (1984) 272 EG 1171 at p 1174.

reasonable facilities have to be provided for potential bidders to view them. It seems to me impossible to expect any man to maintain such a vigilance as to prevent a theft which occurred in a matter of seconds. Furthermore, it was reasonable to expect that if such a theft occurred, the security guards would be able to catch the culprit. All in all, I have come to the conclusion that there was such a combination of circumstances here as could not reasonably have been foreseen by the defendants and their servants to have occurred in such a way as to breach their security arrangements.

Counsel for Mr Spriggs criticised this finding by the judge. He submitted that the system in operation at the time of viewing was inadequate or alternatively that on July 6 1977 the system broke down. He pointed out that the records of earlier thefts which had occurred between 1961 and 1969 showed that there was a plain risk that jewellery might be stolen during viewing and that there was an obligation on Sotheby's to establish and maintain an adequate system of security. The system was defective in that there was no limit on the number of persons who could view items at once, so that if there was only one porter in the ring he was bound to have his attention diverted from time to time from any particular customer and on occasion would have to turn his back on an unknown customer who had in his possession (as in the present case) a tiny object of substantial value.

There was no established procedure for alerting the security guards to watch a particular viewer from whom the porter had to turn away nor was there any sufficient evidence that in a crowded room the security guards had a clear enough view of what was taking place at the ring.

Furthermore, none of the guards had been called to give evidence, so that Sotheby's could not discharge the burden of proving that the system was operating satisfactorily at the relevant time: indeed Mr Cunningham admitted in cross-examination that it had broken down.

Moreover, it was to be remembered that this stone was one of the most valuable items in the forthcoming sale.

Counsel for Sotheby's on the other hand placed great weight on the fact that some risk to security had to be accepted if potential bidders for the jewellery were to be given a suitable opportunity to examine the items which they might wish to purchase. Sotheby's had to strike a difficult balance. If the conditions of security which were imposed were too stringent, potential bidders would be discouraged and persons such as Mr Spriggs who wished to sell items of jewellery would suffer as a result.

He drew attention to the arrangements which were made for security in the New Gallery, to the fact that after the thefts between 1961 and 1969 a number of changes had been made and to the fact that in the period since 1969 no items of jewellery had been stolen despite the fact that auctions had been held at monthly intervals during the eight years or so since the last theft took place.

Furthermore, said counsel, it was a mistake to regard the contract between Sotheby's and Mr Spriggs as an ordinary contract of bailment; Sotheby's had taken possession of the diamond as agents for sale and the bailment prior to the auction was merely an incidental feature of the contract.

It had also to be remembered that in the room at the time there were likely to be, and were on the relevant occasion, a number of potential bidders in addition to the security guards and the porter. The attention of all these people was likely to be attracted by any attempted theft. Indeed, it was a Frenchman who had been looking at some other item who gave the alarm in the instant case.

Finally, it should not be overlooked, it was said, that the judge had the opportunity of observing the witnesses called on behalf of Sotheby's and of making his assessment of them.

For my part I am unable to accept the submission that the obligations of Sotheby's in regard to the security of the diamond were other than those of a bailee for reward, though plainly the question whether reasonable care was taken has to be judged in the context of the relevant circumstances. The duty of a bailee for reward is to take reasonable care in regard to the goods entrusted to him; the steps to be taken to fulfil that duty may vary from time to time and in the light of the function which the bailee has to perform in relation to the goods; the nature of the duty was put in these words by Lord Denning MR in *Morris* v *CW Martin & Sons Ltd* [1966] 1 QB 716 at p 726:

> Once a man has taken charge of goods as a bailee for reward, it is his duty to take reasonable care to keep them safe: and he cannot escape that duty by delegating it to his servant. If the goods are lost or damaged, whilst they are in his possession, he is liable unless he can show — and the burden is on him to show — that the loss or damage occurred without any neglect or default or misconduct of himself or of any of the servants to whom he delegated his duty G . . . The bailee, to excuse himself, must show that the loss was without any fault on his part or on the part of his servants. If he shows that he took due care to employ trustworthy servants, and that he and his servants exercised all diligence, and yet the goods were stolen, he will be excused: but not otherwise.

In some cases the duty to take reasonable care will involve the establishment of a safe system of security, but the establishment of a safe system is not a sufficient discharge of the duty if the system is not followed with a proper degree of diligence. Thus in *Global Dress Co* v *Boase & Co Ltd* [1966] 2 Lloyd's Rep 72, the defendants' system of security was as good as any other at Liverpool docks, but they failed to establish that at the time of the theft the system was being operated satisfactorily. Salmon LJ said this at p 77:

> I accept the fact that these defendants are experienced and careful master H porters. No criticism is made of their system: it is said to be as good as that of any other master porters in the dock. It is not, however, sufficient for them to prove that their system, as a system, was impeccable. Part of the system was to have a watchman on duty in the shed. The defendants have to show that that watchman, for whom they were responsible, carefully performed his duty.

I return to the instant case. This court is reluctant to interfere with the finding of a judge on the question whether a system of protection was adequate in all the circumstances. But the facts in this case were not really in dispute. Moreover we are concerned not with a question of primary fact but with the inferences to be drawn from the evidence as admitted and found. With all due deference to the learned judge, I am unable to accept that the system employed by Sotheby's in July J 1977 for the viewing of jewellery was a reasonably safe system in the circumstances.

Some of the items to be viewed were valuable but extremely small. Sotheby's placed a value of £10,000 to £12,000 on this unmounted diamond. Some of the persons viewing the lots, including the thief, were strangers to Sotheby's. The only person with a clear and uninterrupted view of persons at the side of the ring was the single porter, Mr O'Connor, but it was inevitable that from time to time he would have to divert his attention from one customer in order to deal with a request from one or more other customers, and that, in doing so, he would have to turn his back on one or more customers who still had valuable lots in their possession.

The security guards were, it is true, an important line of defence, K but I cannot see that their fields of vision would have been unobstructed if the room were full or even partially full. Moreover, though no doubt they could have played a useful part in apprehending a thief if he had been detected, I do not find it in any way surprising that the guards apparently failed to observe the initial taking in this case. Nor do I think that Sotheby's were entitled to place more than slight reliance on the vigilance of other potential bidders who might be in the room.

Accordingly I have reached the clear conclusion that Sotheby's have failed to discharge the burden of proving that they had a safe system in operation in July 1977. Furthermore, in the absence of any evidence from any of the guards who formed the second line of defence, I am not satisfied that Sotheby's sufficiently demonstrated L that the system was functioning properly on July 6.

For these reasons I, for my part, would hold that Sotheby's had not shown that they exercised all reasonable care to keep the diamond safe. It therefore becomes necessary to consider the second issue.

Clause 2 of the conditions

It is now no longer in dispute that the terms of the contract between Mr Spriggs and Sotheby's included those set out in the document which Mr Spriggs and Mr Hinks signed on May 5 1977. In that document the reserve price was stated to be £9,000 and Mr Hinks recorded the fact that Mr Spriggs did not wish Sotheby's to arrange insurance. Immediately above the space where Mr Spriggs signed his M name on the front of the document were the printed words: "I have read and agree to the Instructions for Sale as detailed on the reverse of this form." On the reverse of the document under the heading "INSTRUCTIONS FOR SALE" there were eight printed paragraphs following some preliminary words. Clause 2 was in these terms:

> I accept that, whilst you take all reasonable care in handling the property on behalf of clients, you are not responsible for loss or damage of any kind whether caused by negligence or otherwise, the property being accepted at owner's risk.
>
> Nevertheless, property accepted for sale and taken in at your premises will be insured with underwriters at my cost against all risks of physical loss or damage and subject to the insurance terms and conditions usual in your business, for the amount for the time being estimated by you to be the sale

room value of the property until sold. The rate is 50p per £100 on the hammer price (minimum premium 50p).

Transit insurance can be arranged on receipt of instructions before despatch.

It will be appreciated that the second para of this clause, which related to insurance at Sotheby's premises, was in effect excluded in relation to the contract between Mr Spriggs and Sotheby's because he had given specific instructions that he did not want insurance to be arranged. Nevertheless I am quite satisfied that for the purpose of construing the clause and in particular the first para it is necessary to read and construe the clause as a whole.

The case for Sotheby's both before the trial judge and in this court was that even if the loss had occurred as a result of lack of reasonable care by them they were nevertheless entitled to avoid liability for the loss by reason of a specific term (clause 2) in their contract with Mr Spriggs. The judge accepted this argument. In a short passage in his judgment he said this:

> In my opinion the words in the exclusion clause here are clear and as a matter of construction the defendants were not to be liable for any loss or damage whether caused by negligence or otherwise.

Counsel for Mr Spriggs on the other hand submitted that Sotheby's could not rely on clause 2 so as to escape liability.

I shall attempt to summarise as shortly as possible the detailed arguments which were advanced in support of this submission. The arguments were on the following lines:

(1) That it was necessary to give due weight to the opening words of clause 2 " . . . whilst you take all reasonable care in handling the property on behalf of clients . . . "; these words either set out the standard of care which Sotheby's were to observe as a matter of contractual obligation or they amounted, or could be construed as amounting, to no more than an assertion of fact or a statement as to Sotheby's general practice.

(2) That if the opening words imposed some contractual obligation, then the following words " . . . you are not responsible for loss or damage of any kind whether caused by negligence or otherwise . . . " should be ignored on the ground of repugnancy. On this point our attention was drawn to the principle of law recognised by the Privy Council in *Forbes* v *G I T* [1922] 1 AC 256 at p 259:

> If in a deed an earlier clause is followed by a later clause which destroys altogether the obligation created by the earlier clause, the later clause is to be rejected as repugnant and the earlier clause prevails. In this case the two clauses cannot be reconciled and the earlier provision in the deed prevails over the later. Thus if A covenants to pay £100 and the deed subsequently provides that he shall not be liable under his covenant, that later provision is to be rejected as repugnant and void, for it altogether destroys the covenant.

On this aspect of the case we were also referred to *Swan Hunter and Wigham Richardson Ltd* v *France Fenwick Tyne & Wear Co Ltd* [1953] 1 WLR 1026 and to a passage in the judgment of Devlin J in *Firestone Tyre & Rubber Co Ltd* v *Vokins & Co Ltd* [1951] 1 Lloyd's Rep 32 at p 39.

(3) That if in the alternative the opening words amounted to no more than an assertion of fact or a statement of general practice, then the whole clause was merely a notice or warning, or at any rate was capable of being so construed, and accordingly was not sufficient to exclude liability. This argument was based on the contention that sufficient content could be given to clause 2 by construing it as a warning or notice, and support for the argument was sought to be derived from the decisions of this court in *Olley* v *Marlborough Court Ltd* [1949] 1 KB 532 and *Hollier* v *Rambler Motors (AMC) Ltd* [1972] 2 QB 71.

In the former case the defendants relied on a notice in the bedroom of a private residential hotel: "Proprietors will not hold themselves responsible for articles lost or stolen, unless handed to manageress for safe custody"; in the latter case the defendants relied on the following words on the forms signed by the customer when from time to time he had brought his car in for repair or servicing: "The company is not responsible for damage caused by fire to customers cars on the premises. Customers cars are driven by staff at owners' risk".

In both cases the Court of Appeal held that the words relied on were no more than a warning and that they were not sufficiently clear to exclude liability.

In *Olley* v *Marlborough Court Ltd* Denning LJ put the matter as follows at p 550:

> Ample content can be given to the notice by construing it as a warning that the hotel company is not liable, in the absence of negligence. As such it serves a useful purpose. It is a warning to the guest that he must do his part to take care of his things himself, and, if needs be, insure them. It is unnecessary to go further and to construe the notice as a contractual exemption of the hotel company from their common law liability for negligence.

In addition counsel submitted that it was sufficient for his purpose that the clause was capable of being construed as merely a notice or warning. He drew our attention to the judgment of Stamp LJ in *Hollier's* case at p 83 where he said:

> . . . I would hold that, where the words relied upon by the defendants are susceptible either to a construction under which they become a statement of fact in the nature of warning or to a construction which will exempt the defendant from liability or negligence, the former construction is to be preferred.

(4) That whether or not the opening words had any contractual effect the following words were not sufficient to exclude liability because they were ambiguous and because they *could* be construed as excluding the negligence only of persons other than Sotheby's.

The basis for this argument was that Sotheby's could have excluded liability for their own negligence by an express reference to their "own negligence", but that in the absence of such express words sufficient content could be given to the exclusory words by restricting their meaning to the negligence of third parties.

It was further argued that as the exclusory words were capable of bearing either meaning they should be construed *contra proferentes* in the more restrictive sense.

(5) That furthermore whether or not the opening words had any contractual effect, Sotheby's could not rely on the following words to exclude liability because the reader of the clause as a whole would be lulled into a sense of security by the assertion of reasonable care contained in the opening words.

(6) That in any event the exclusory words in the clause could not apply in the circumstances of the instant case because the breach was so serious as to be fundamental.

The contract between the parties, it was argued, was to secure the sale of the diamond by auction and accordingly, as no sale of the diamond had taken place and Sotheby's themselves had delivered the diamond into the hands of the thief, an exemption clause could not exempt Sotheby's from so fundamental a failure to perform the contract.

In the course of the argument on this final point our attention was drawn to the relevant authorities: *Levison* v *Patent Steam Carpet Cleaning Co Ltd* [1978] QB 69; *Suisse Atlantique Société d'Armement Maritime SA* v *NV Rotterdamsche Kolen Centrale* [1967] 1 AC 361 and *Photo Production Ltd* v *Securicor Transport Ltd* [1980] AC 827.

To my mind it is possible to deal with these several arguments, attractively deployed as they were, quite shortly. As I see it, there are two questions to be answered:

(a) What is the proper construction of clause 2?

(b) Does the clause, when properly construed, apply to the facts of the instant case?

I turn to the question of construction.

Looking at the clause as a whole I am quite unable to attach to the opening words: "Whilst you take all reasonable care in handling the property on behalf of clients . . . " any obligation in the nature of a contractual warranty or promise. Counsel for Mr Spriggs specifically disclaimed that he was putting forward any case based on misrepresentation and accordingly I think these words can only be construed as a statement as to general practice.

What then is the effect of the following words?

It seems to me that the words ". . . whether caused by negligence or otherwise . . . " are quite clear and are adequate to exclude any liability for any act of negligence by Sotheby's or by any person for whom Sotheby's would be responsible. I cannot construe these words as a mere notice or warning nor in the context do I consider that the word "negligence" can be restricted to the negligence of third parties.

Furthermore, in the absence of any plea of misrepresentation or estoppel, I do not consider that Sotheby's are debarred from relying on the exclusory words by the mere presence of the opening words.

The ordinary reader of the clause as a whole would not be lulled into a sense of security but would recognise that Sotheby's were seeking to exclude a liability which in the absence of the exclusory words they would incur as bailees.

I turn to the second question.

I recognise that there is some attraction in the argument that as the

A contract was to effect the sale of the diamond by auction an exemption clause should not exempt Sotheby's from liability if the fundamental purpose of the contract has been wholly frustrated by a prior theft. It seems to me, however, that this argument ignores the fact that any "loss" of the object to be auctioned will frustrate the sale. Accordingly I am satisfied that the clause when properly construed is apt to cover a loss both before as well as after the sale has taken place.

Furthermore, in handing the diamond to the thief, Sotheby's were in my view acting within the scope of the contract with Mr Spriggs which was to effect the sale of the diamond by auction. It was a necessary preliminary to this sale that the stone should be available for inspection and be inspected by potential bidders. I cannot regard any action which they took as being so wholly outside the scope of the contract as to disentitle them from relying on the exclusion clause. Indeed at the relevant time they were seeking, albeit negligently as I see it, to make sure that a satisfactory sale of the diamond took place on the following day. C

Furthermore, it is important to bear in mind that the decision of the House of Lords in the *Photo Production* case "gave the final quietus to the doctrine that a 'fundamental breach' of contract deprived the party in breach of the benefit of clauses in the contract excluding or limiting his liability": see *George Mitchell (Chesterhall) Ltd* v *Finney Lock Seeds Ltd* [1983] 2 AC 803 at p 813, *per* Lord Bridge.

It is necessary to construe an exclusion clause in its context and to avoid giving any strained interpretation to the words used. The clause when so construed must then be applied to the facts of the particular case.

Accordingly I consider that Sotheby's are entitled to rely on the terms of the contract between them and Mr Spriggs so as to exclude liability for the loss of the diamond. D

I would dismiss the appeal.

FOX and DILLON LJJ agreed and did not add anything.

The appeal was dismissed with costs.

Court of Appeal

November 29 1985

(Before Sir John DONALDSON MR, Lord Justice MAY and Lord Justice GLIDEWELL)

PALATINE GRAPHIC ARTS CO LTD v LIVERPOOL CITY COUNCIL

Estates Gazette February 1 1986

277 EG 412-416

Compensation for acquisition of premises by agreement but on compulsory purchase compensation basis — In present case claimants' lease had only a nominal value but a substantial sum was due by way of compensation for disturbance — Part of the sum payable for disturbance was in respect of works which attracted a regional development grant under the Industry Act 1972 amounting to 22% of the capital expenditure involved — The issue was as to whether the amount of the grant should be deducted from the compensation payable for disturbance — The Lands Tribunal decided that the amount of the grant should not be deducted and the acquiring authority appealed — The authority, relying on *Horn* v *Sunderland Corporation*, argued that if the claimants received both the cost of carrying out the works as disturbance compensation and the regional development grant, the result would "transgress the principle of equivalence" by paying the claimants more than their loss — The authority also cited in their favour by analogy the decisions in *British Transport Commission* v *Gourley* and *West Suffolk County Council* v *W Rought Ltd* about the deduction of tax in the assessment of damages for loss of profits — The Court of Appeal, however, rejected these submissions and held that the case was governed by the principles applied by the House of Lords in *Parry* v *Cleaver*, when they refused to deduct from the damages awarded to a policeman injured in a car accident the amount of a police disability pension to which he had contributed — Applying these principles, it would be wrong in the present case to deduct the regional development grant from the disturbance compensation (1) because it would be contrary to public policy and (2) because such a grant is wholly different in kind from compensation for disturbance — Indeed, it is not compensation at all but an incentive for industrial expansion — Authority's appeal dismissed

This was an appeal by case stated by Liverpool City Council, the acquiring authority, from a decision of the Lands Tribunal (Mr C R Mallett FRICS) holding that the amount of a regional development grant should not be deducted by the authority from the compensation for disturbance payable under certain heads, following the acquisition of the factory held under lease by the respondents (claimants) at 5 Hood Street, Liverpool.

Michael Kershaw QC and Miss Jane Tracy Forster (instructed by W I Murray, City Solicitor, Liverpool) appeared on behalf of the appellants; Andrew Gilbart (instructed by Lacies & Co, of Manchester) represented the respondents.

Giving the first judgment at the invitation of the Master of the F
Rolls, GLIDEWELL LJ said: This is an appeal by way of case stated from a decision of the Lands Tribunal (Mr C R Mallett FRICS) dated June 11 1984 of a disputed issue as to the compensation payable by Liverpool City Council for the acquisition of premises formerly occupied by the respondent company, Palatine Graphic Arts. The question at issue as set out in the case stated is:

> whether an amount equivalent to a regional development grant paid by the government, or some lesser amount, should be deducted from the compensation payable by the local authority in respect of disturbance following the acquisition of the claimants' property.

Palatine are long-established printers, specialising in the high- G
quality printing of labels and packaging. Until 1980 they occupied, under a lease expiring on December 31 1983, a factory at 5 Hood Street, near the city centre of Liverpool. They used machines of considerable value, and the process necessitated special equipment for dust extraction, treatment of effluent, silver recovery and lighting.

The city council desired to acquire the Hood Street premises in order to carry out a scheme of redevelopment. The council could have used powers of compulsory purchase. However, the parties reached agreement that Palatine should sell to the council their interest in the premises at a price equivalent to the compensation which would have been payable if the acquisition had been by way of compulsory purchase order. H

In order to carry on their business, Palatine needed to acquire alternative premises. It was agreed that such premises need not have been in Liverpool. Palatine used two printing processes, letter press and offset-litho. By 1980, the offset-litho process dominated. In relation to this, the case stated contains the following extract from an agreed statement of facts:

> This has led to difficulties in the Liverpool area in training and maintaining sufficient qualified staff in the offset litho process, said to be due in part to the attitude of the local trade unions which sought to retain the letter press process. Although the business has been established in Liverpool for many years local trade has declined considerably so that now about 80% of the custom is outside the Liverpool area and 60-70% outside Merseyside and the surrounding area of Lancashire. J
> At the material time it would have eased customer and staff problems to relocate the claimants' business in Manchester or elsewhere. Against this was the advantage of staying in Liverpool which was a special development area for the purpose of regional development grants paid by Government as part of their incentive to industry scheme introduced by the Industry Act 1972. In special development areas grants of 22% were made towards investment in new buildings, or works to new or existing buildings, and new plant or machinery. The cost thereof, before deducting the grant, could be offset against tax. At the material date Manchester was in an intermediate area where 20% grants were available, but only in respect of building works.
>
> In the event, the claimant company decided to retain their business in Liverpool at a single storey purpose built factory unit in Fox Street and spent some £250,000 on the basic shell, a further £200,000 on fitting out and about £500,000 on plant and machinery, half of which was on hire purchase or lease. K
> All or most of these expenditures qualified for a regional development grant of 22%.

Palatine entered into an agreement for the erection and acquisition of the Fox Street factory on March 6 1978 and completed the move to those premises on July 1 1980.

It has long been accepted law (and it is agreed by both parties to the present dispute) that the owner of an interest in land which is compulsorily acquired is entitled to be compensated both for the value of his interest as such and for any loss he may suffer as a result of being dispossessed of his land. This latter category is compendiously called "compensation for disturbance". It embraces, among other matters, the cost of removal to new premises and any loss of the profits of a business proved to be attributable to the dispossession.

Although the principle of disturbance compensation was generally accepted before the passage of the Lands Clauses Consolidation Act 1845 — see eg *Jubb* v *Hull Dock Co* (1846) 9 QB 443, based upon a Private Act of earlier date — the right to compensation for disturbance is not expressly granted by statute. Neither the 1845 Act nor its successor, the Compulsory Purchase Act 1965, gives such a right in terms. The reason is always said to be that the dispossessed owner is entitled to receive compensation based upon the value to him of his land and premises. Since he is likely to incur expense and loss if he leaves the land and moves elsewhere, or indeed if he ceases to carry on his business as a result of the compulsory acquisition, the value to him may comprehend not merely the value of his interest in the land as such but also the right to be compensated for that consequential loss. As Scott LJ said in the leading case of *Horn* v *Sunderland Corporation* [1941] 2 KB 26 at p 45:

> None the less, the owner in a proper case — that is, in a case where he really does incur a loss of money by disturbance due to the taking over and beyond the loss for which he is to be reimbursed in respect of the land taken — is entitled, because it has to do with the land, to have that element of personal loss taken into the reckoning of the fair price of the land, as has been held by the courts from a very early stage.

This right to compensation for disturbance, though not deriving from statute, has been preserved by statute. Section 5 of the Land Compensation Act 1961, which repeats a similar provision in the Acquisition of Land (Assessment of Compensation) Act 1919, provides:

> Compensation in respect of any compulsory acquisition shall be assessed in accordance with the following rules . . .
> (2) The value of land shall, subject as hereinafter provided, be taken to be the amount which the land if sold in the open market by a willing seller might be expected to realise . . .

The proviso which preserves disturbance compensation is in rule (6):

> The provisions of rule (2) shall not affect the assessment of compensation for disturbance or any other matter not directly based on the value of land.

It is, of course, necessary for the open market value of the land acquired and the disturbance compensation to be assessed separately, but once agreed or awarded, the sums are aggregated and the total is the price at which the land is conveyed to the acquiring authority: see *Horn* v *Sunderland Corporation*.

In the present case, almost the whole of the price to be paid consists of the compensation for disturbance. The value of Palatine's interest in their leasehold is agreed at £5. Taking into account the second matter on which Palatine succeeded before the Lands Tribunal and which is not in dispute before us, various items of disturbance are also agreed in the total sum of £83,324.82.

The dispute centres on the remaining items. In order to equip their new premises for their machinery and processes, Palatine carried out various works at a total cost of £102,748 which, it is agreed, properly fall within "disturbance". However, the city council argued that they are not obliged to pay the whole of this sum by way of compensation since in respect of this expenditure Palatine applied for and received regional development grant at the rate of 22%, namely £22,605. This last sum the city council seek to deduct from the compensation they are otherwise liable to pay.

The member of the Lands Tribunal found in favour of Palatine on this issue, hence this appeal.

At the relevant date, regional development grant was payable under section 1 of the Industry Act 1972, which provided:

> *Grants for buildings, machinery, plant and certain works;*
> (1) The Secretary of State may make a grant to a person towards approved capital expenditure incurred by that person, being expenditure of any description in column 1 of the Table below.
> (2) The amount of a grant under this section shall be the prescribed percentage of the expenditure in respect of which it is made.

Subsection (3) and the table which followed prescribed the percentage appropriate for Palatine's works as 22%.

Both counsel are agreed that there is no direct authority on the question whether regional development grant should be deducted from the compensation. Mr Kershaw for the city council relies upon another principle conveniently expressed in *Horn* v *Sunderland Corporation*. In his judgment at [1941] 2 KB 26 at pp 48 and 49 Scott LJ summarised the principles upon which compensation for disturbance should be assessed. At p 49 he said:

> . . . on a compulsory sale the principle of compensation will include in the price of the land, not only its market value, but also personal loss imposed on the owner by the forced sale, whether it be the cost of preparing the land for the best market then available, or incidental loss in connection with the business he has been carrying on, or the cost of reinstatement, because otherwise he will not be fully compensated. (6) But here we come to the other side of the picture. The statutory compensation cannot, and must not, exceed the owner's total loss, for, if it does, it will put an unfair burden on the public authority or other promoters who on public grounds have been given the power of compulsory acquisition, and it will transgress the principle of equivalence which is at the root of statutory compensation, the principle that the owner shall be paid neither less nor more than his loss.

Mr Kershaw submits that if Palatine receive both the whole of the cost of carrying out the necessary works as disturbance compensation and the sum of £22,605 as RDG, they will have been paid more than their loss. To ensure that they receive no more, the amount of the RDG must be deducted.

Mr Gilbart for Palatine starts his submission by arguing that compensation for compulsory purchase, including compensation for disturbance, is to be assessed on the same basis as would damages at common law. In other words, if the acquiring authority had no power of compulsory acquisition, but had simply evicted the landowner from his property, he would have a right of action in trespass, and the basis upon which damages would be assessed for a trespass which deprived him of his property permanently provides the test for the assessment of compensation.

In my judgment, this submission is correct. In *Ricket* v *Metropolitan Railway Co* (1865) 34 LJQB 257, in the Court of Exchequer Chamber, Erle CJ, giving the judgment of the majority, said at p 261:

> . . . the company, claiming to take land by compulsory powers, expel the owner from his property and are bound to compensate him for all the loss incurred by the expulsion; and the principle of compensation then is the same as in trespass for expulsion, and so it has been determined in *Jubb* v *Hull Dock Co*.

Although this passage in the judgment was strictly obiter dictum, it has been generally followed and applied.

Mr Kershaw submits that, even if the principles are the same as those for common law damages, compensation ought to be assessed after deduction of RDG on the same basis as assessment of damages or loss of future earnings in a personal injury case, where the relevant figure is that of the earnings after deduction of tax. The principle in *British Transport Commission* v *Gourley* [1956] AC 185 was followed by the House of Lords in *West Suffolk County Council* v *W Rought Ltd* [1957] AC 403, which decided that, if compensation for loss of future profits as part of disturbance on compulsory purchase was not itself subject to tax, then the assessment ought to be based on the net figure of the profits after deduction of tax.

Mr Gilbart replies that the taxation cases are concerned with deciding what is the true loss suffered by a plaintiff at common law or a claimant to compensation on compulsory purchase. This is distinct from the present situation, in which the claimants have received following the compulsory purchase an additional payment from another source. The principles to be applied in this case, he submits, are those derived from the decision of the House of Lords in *Parry* v *Cleaver* [1970] AC 1. In that case the plaintiff police constable was severely injured in a car accident for which the defendant was liable. As a result of his injuries he was discharged from the police force. During his service he had made compulsory contributions to a police pension fund, and on his discharge he was entitled to and did commence to draw a disability pension from the fund. It was argued on behalf of the defendant that the amount of the pension should be deducted from the plaintiff's loss of earnings in assessing damages. By a majority the House of Lords rejected this argument. At the beginning of his speech Lord Reid said at p 13:

> Two questions can arise. First, what did the plaintiff lose as a result of the accident? . . . And secondly, what are the sums which he did in fact receive as a result of the accident but which he would not have received if there had been no accident? And then the question arises whether the latter sums must be deducted from the former in assessing the damages.

His Lordship then said that *Gourley*'s case dealt with the first question but not at all with the second. He said in relation to the second question:

> The common law has treated this matter as one depending on justice, reasonableness and public policy.
> So I must inquire what are the real reasons, disregarding technicalities, why these two classes of receipts are not brought into account. (pp 13-14).

Lord Reid considered whether a sum received following the accident as a result of benevolence would have to be taken into account in assessing damages. He said:

> It would be revolting to the ordinary man's sense of justice, and therefore contrary to public policy, that the sufferer should have his damages reduced so that he would gain nothing from the benevolence of his friends or relations or of the public at large, and that the only gainer would be the wrongdoer.

> We do not have to decide in this case whether these considerations also apply to public benevolence in the shape of various uncovenanted benefits from the welfare state, but it may be thought that Parliament did not intend them to be for the benefit of the wrongdoer.

> As regards moneys coming to the plaintiff under a contract of insurance, I think that the real and substantial reason for disregarding them is that the plaintiff has bought them and that it would be unjust and unreasonable to hold that the money which he prudently spent on premiums and the benefit from it should enure to the benefit of the tortfeasor.

Later at p 15D he said:

> If remoteness has any relevance here it is quite a different kind of remoteness — the connection or absence of connection between the source of the benefit and the source of the wages. But what has that got to do with the defendant? It is rational to make the extent of the defendant's liability depend on remoteness from his point of view — on what he knew or could or should have foreseen. But it is, to my mind, an irrational technicality to make that depend on the remoteness or closeness of relationship between the plaintiff's source of loss and source of gain. Surely the distinction between receipts which must be brought into account and those which must not must depend not on their source but on their intrinsic nature.

It will be seen that Lord Reid based his decision that the pension moneys were not deductible in assessing damages on two reasons: (i) because deduction would be contrary to public policy and (ii) because the pension moneys and the earnings loss were intrinsically different.

Mr Gilbart submits and I agree that both reasons apply in the present case. On the facts found in this case the prospect of obtaining regional development grant was an important inducement to Palatine to relocate themselves in Liverpool rather than moving elsewhere in Lancashire. To the extent that the amount of RDG was deducted from their compensation, that inducement would be lessened. That seems to me to be contrary to the policy of assisting development areas of which the payment of RDG was a part. This was one of the reasons which led to the decision of the member of the Lands Tribunal that the RDG should not be deducted from disturbance compensation, and in this respect he was in my view correct.

Second, the loss caused by disturbance on compulsory purchase and the payment of RDG are different in kind. The loss on disturbance flowed from the fact that the landowner had been forced to give up occupation of his land and premises as a result of the acquisition of his interest. The RDG was paid in respect of part of the expenditure incurred when moving into new premises.

For both these reasons I therefore conclude that the decision of the Lands Tribunal in this case was correct and that the compensation for disturbance should be assessed without any deduction in respect of the payment of regional development grant. I should add for completeness that Mr Gilbart before the Lands Tribunal advanced another argument, why RDG should be disregarded, namely that its payment was not as of right but a matter of discretion for the Secretary of State. The member of the Lands Tribunal based his decision in part on this reason. Before us Mr Gilbart did not advance this argument. In this respect only I disagree with the decision of the Lands Tribunal. Discretionary or not, RDG was in fact paid to Palatine, and if for other reasons it should have been taken into account the discretionary nature of the Secretary of State's power to pay RDG would in my view be irrelevant.

However, for the other reasons set out above I am of the opinion that the decision of the Lands Tribunal was correct.

I would dismiss the appeal.

MAY LJ agreed with the judgment of Glidewell LJ and also with the judgment of the Master of the Rolls below and did not add anything.

Agreeing that the appeal should be dismissed, SIR JOHN DONALDSON MR said: The leading authority on compensation for disturbance by compulsory purchase is *Horn* v *Sunderland Corporation* [1941] 2 KB 26 and, in particular, the judgment of Scott LJ at p 40. This establishes that such compensation has no existence separate from the price of the land, but is an element in assessing that price. It also establishes, at p 49, that:

> the statutory compensation cannot, and must not, exceed the owner's total loss, for, if it does, it will put an unfair burden on the public authority . . . and it will transgress the principle of equivalence which is at the root of statutory compensation, the principle that the owner shall be paid neither less nor more than his loss.

Scott LJ went on to say that:

> the enunciation of this principle, the most fundamental of all, is easy enough. Its justice is self-evident, but its application to varying facts is apt to be difficult. It is not easy to spell out of it a general criterion which will afford a practical test in all cases.

He then illustrated this by reference to the facts of that appeal, where the compulsory sale had enabled the owner to realise the full value of the land as building land, which he could not, or would not, have done but for the compulsory sale. He said that it was a false interpretation of the compulsory purchase legislation to hold that in all circumstances "disturbance" loss must be added to "the actual price of the land", by which I understand him to mean its full market value and, on the facts of that case, he held that compensation based on market value as building land would adequately make good the owner's loss of the land valued on the basis of its actual user as agricultural land together with disturbance loss.

Although the appellants submit that this points conclusively to a decision that the owner must give credit for regional development grant to the extent that it is related to work done or expenditure incurred which would also form the basis of a claim for disturbance loss, I do not so understand it. *Horn*'s case decides no more than that there shall not be over-compensation — "the owner shall be paid neither less nor more than his loss". So be it, but it is necessary to determine what exactly was the respondents' loss by way of disturbance.

A similar problem can arise in the assessment of damages and I have derived considerable assistance from the decision of the House of Lords in *Parry* v *Cleaver* [1970] AC 1 and, in particular, from the speech of Lord Reid at p 13. He said that three questions arose: (i) "What did the plaintiff lose as a result of the accident? What are the sums which he would have received but for the accident but which by reason of the accident he can no longer get?" (ii) "What are the sums which he did in fact receive as a result of the accident but which he would not have received if there had been no accident?" (iii) "To what extent must the latter sums be deducted from the former in the assessment of damages?"

On the facts of this case, there is no dispute but that the respondents incurred disturbance expenditure in the amount, I think, of £147,478, although the precise figures do not matter, and that this expenditure would not have been incurred but for the compulsory purchase. There is also no dispute that some of this expenditure attracted regional development grant. The sole question in dispute is whether the disturbance loss is properly to be regarded as being the disturbance expenditure as abated *pro tanto* by the regional development grant. If it is, then the amount of the compulsory purchase price can only take account of the disturbance expenditure so abated, as otherwise it would amount to over-compensation and offend against the principles enunciated in *Horn*'s case. If it is not, the compulsory purchase price can take account of the full disturbance expenditure, since that, and not the abated sum, represents the disturbance loss.

It is at this point that it is necessary to take a closer look at the nature of a regional development grant, just as the House of Lords took a closer look at the nature of a police pension in *Parry* v *Cleaver* (*supra*). A regional development grant is *not* paid in compensation for dispossession or disturbance. It is payable whether or not there is a change of ownership or location, so long as it relates to expenditure of a relevant kind incurred in a relevant geographical area. It is therefore different in kind from compensation for disturbance. Indeed it is not compensation at all. Regional development grants are, as one of the Department of Industry booklets rightly describes them, "Incentives for Industry in the Areas for Expansion".

This analysis leads me to conclude that in the instant and similar cases, (i) the person whose land is compulsorily acquired incurs disturbance expenditure in re-establishing his business, (ii) other things being equal, this expenditure is the same wherever he incurs it, (iii) this expenditure is *prima facie* the measure of his disturbance loss, (iv) his disturbance loss is not reduced if he chooses to incur the disturbance expenditure in a particular area and is rewarded — not compensated — for so doing by the receipt of an incentive payment in the form of a regional development grant. In principle this is no different from the mail order customer who buys goods for £100 and has his name entered in a draw for a prize of £100. The goods will cost him £100, whether he wins or loses. If he wins, he will have paid £100 for the goods and received a prize of £100. He will not have received the goods for nothing.

One can reach the same conclusion by a slightly different route. It is the dispossessed owner who chooses where to re-establish himself. Although regional development grants were not available when the relevant land compensation statute was enacted in 1961, it can hardly have been the intention of Parliament that incentives to establish or re-establish a business in a particular area should be taken into account in assessing the compulsory purchase price, since to do so would involve giving an incentive to the compensating authority which cannot choose where the business is to be established and taking away from the person who can so choose. This would be pure "Alice in Wonderland".

The appeal was dismissed with costs. An application for leave to appeal to the House of Lords was refused.

Court of Appeal

November 28 1985

(Before Lord Justice OLIVER, Lord Justice RALPH GIBSON and Mr Justice ANTHONY LINCOLN)

GREATER LONDON COUNCIL v HOLMES

Estates Gazette February 15 1986

277 EG 641-644

Land Compensation Act 1973 — Entitlement to home loss payment — Appeal by applicant against judge's decision in favour of acquiring authority — Appeal raised two questions: (1) Whether, at date of appellant's displacement, the land, which had been compulsorily acquired, was for the time being held by the respondent authority for the purpose for which it was acquired, and (2) whether the appellant's displacement was "in consequence of the carrying out of redevelopment on the land" — The facts were that at the time of acquisition of the site it had been the intention of the respondent authority to clear the site and build local authority housing on it — The plan was, however, subsequently changed and it was decided to dispose of the land in the open market for housing by a private developer — Held, on the first question, disagreeing in this respect with the judge's decision, that although the original concept was the provision of local authority housing, the broad "purpose" for which the land was acquired was that it should be cleared for the erection of houses and at the material time it was still held for that purpose — Held further, on the second question, again disagreeing with the judge's decision, that the appellant was displaced in consequence of redevelopment carried out by the respondent authority — It had been argued on behalf of the authority that demolition carried out to clear the land for sale to a purchaser was not redevelopment, which meant putting up new buildings — The Court of Appeal, rejecting this argument, considered that the demolition which resulted in the appellant's displacement was an essential step in the process of redevelopment which was begun by the respondents but was intended to be completed by the purchaser — Appeal allowed and the questions raised as to the appellant's entitlement to a home loss payment answered in his favour

This was an appeal from a decision of Judge Sir William Stabb QC, sitting as a High Court judge, by Leonard Thomas Holmes, who had, until displaced, been the occupant of a mobile home on a site at 37 Clark Street, London E1, in the borough of Tower Hamlets. The judge had answered in favour of the respondents, the Greater London Council, certain questions raised by an originating summons as to the entitlement of the appellant to a home loss payment under Part III of the Land Compensation Act 1973.

D R Watkinson (instructed by May Maughan, Tower Hamlets Law Centre) appeared on behalf of the appellant; M G Anthony (instructed by R A Lanham, Greater London Council) represented the respondents.

Giving judgment, OLIVER LJ said: This is an appeal from an order of His Honour Judge Sir William Stabb QC, made on February 17 1984, by which he answered in favour of the respondent, the Greater London Council, certain questions raised by an originating summons as to the entitlement of the appellant to a "home loss" payment under the provisions of the Land Compensation Act 1973.

The summons, the evidence in support of which consisted simply of an agreed statement of facts with some supporting documents, raised two questions, the first of which was expressed in very general terms, that is to say, whether the appellant was entitled to a payment as a person displaced from his home in consequence of the carrying out of redevelopment on the site on which his home was situate. The second question was whether displacement consequential on a decision of the respondent to clear the land and sell with vacant possession constituted displacement from a dwelling on land "in consequence of the carrying out of redevelopment on the land".

As regards the first question it is agreed that the real issue raised is a much narrower one, namely that of the purpose for which the land was held by the respondent at the time when the appellant was displaced.

The section of the Land Compensation Act 1973 with which this appeal is concerned is section 29(1)(c). It appears in Part III of the Act, which is headed "Provisions for Benefit of Persons displaced from land: Home loss payments", and is in the following terms:

29. (1) Where a person is displaced from a dwelling on any land in consequence of (a) the compulsory acquisition of an interest in the dwelling; (b) the making, passing or acceptance of a housing order, resolution or undertaking in respect of the dwelling; (c) where the land has been previously acquired by an authority possessing compulsory purchase powers or appropriated by a local authority and is for the time being held by the authority for the purposes for which it was acquired or appropriated, the carrying out of redevelopment on the land, he shall, subject to the provisions of this section and section 32 below, be entitled to receive a payment (hereafter referred to as a "home loss payment") from the acquiring authority, the authority who made the order, passed the resolution or accepted the undertaking or the authority carrying out the redevelopment, as the case may be.

The remaining subsections of the section deal with the qualifications for entitlement and it is unnecessary to refer to them because it is common ground that, subject to the two questions raised on the appeal, the appellant was otherwise qualified to receive a payment.

The instant case is not one where the land concerned had been "appropriated" by the local authority. It was land which was compulsorily acquired and it is agreed that, subject to the question of whether the displacement was in consequence of redevelopment by the respondent, the only question which requires to be answered is whether at the time of the displacement the land so acquired was, on the true construction of the section, being held by the respondent for the purposes for which it was acquired.

The relevant facts as they appear from the agreed statement are as follows: The appellant became the tenant of a mobile home at 37 Clark Street, London E1 (in the Borough of Tower Hamlets) in April 1967 and until displaced in April 1980 occupied that site as his only home. Clark Street runs parallel to Ashfield Street and the block of land between the two (which included no 37) was known as "the Ashfield Street site". It had been compulsorily acquired in 1963 and we have been given to understand (though this does not appear in the agreed statement) that it was acquired under powers conferred by the Housing Act 1957.

It may be convenient to refer to the relevant provisions of that Act at this stage, since they form the basis of one of Mr Watkinson's submissions on behalf of the appellant.

Section 96 of the Act provides that "a local authority shall have power under this Part of this Act (a) to acquire any land, including any houses or buildings thereon, as a site for the erection of houses...".

Section 97 authorises the acquisition either by agreement or by compulsory purchase and provides that, with the consent of the minister, land may be acquired for the statutory purposes even if not immediately required for these purposes if it appears to the minister that it is likely to be required for those purposes within 10 years from the date on which he confirms the compulsory purchase order.

Section 96(e) authorises acquisition for the purposes of the sale or lease of the land under powers conferred by para (a) of subsection (1) or by subsection (2) of section 105 of the Act — a paragraph which provides that where land has been acquired for the purposes of the Act it may, with ministerial consent, be sold or leased for the purpose and subject to conditions as to the erection of numbers and types of houses specified by the local authority. Section 105(1)(b) also contains a general power, with the consent of the minister, to sell the land.

Returning to the agreed facts, it appears that the immediate provision of housing was satisfied, at any rate so far as the appellant was concerned, by the provision of a mobile home, but para 5 of the agreed statement indicates what, so it is said, the respondent's ultimate intention was.

Para 5 of the statement reads as follows:

> At the time of acquisition of the site, it was the intention of the plaintiffs to clear the site and to build housing thereon as is shown by the outline planning brief dated April 4 1979, a copy of which is appended hereto at Appendix 2.

How the intention of the respondent in 1963 is manifested by a document brought into being some 16 years later is not entirely clear, but we have to accept the agreed statement at its face value and it is not in dispute that the intention with which the land was acquired was the clearance of the site and the provision of local authority housing. It is unnecessary to refer in any detail to the housing brief, but the following paragraphs may be material.

Para 1.4 is headed "Site Availability" and reads:

> The site can be made available when required subject to the usual arrangements for programming the rehousing and clearance.

Para 2.1 describes the type of housing proposed, that is, a mixture of flats and houses. Section 3.2 is headed "Statutory Position" and reads:

> These sites are vested in this Council and are zoned in the Initial Development Plan for public open space. As the IDP is still in the statutory document for the area, the proposed housing development will conflict with the provisions of the plan and will have to be advertised.

Part 4 of the brief indicates a programme providing for the scheme and estimate to be produced by April 1980. By July 1979, however, the picture had changed and on July 16 1979 the respondent's housing development committee ordered all work on the project to be stopped. A report on August 31 1979 listed a number of options for the development of the land, one of which was sale on the open market for housing use. On February 11 1980 the housing policy committee resolved that the Ashfield Street Scheme be not proceeded with and that the central area planning committee be informed that the sites "are surplus to, or are not required for the purposes of the Council's own housing programme". It recommended that the site be disposed of by "open market sale for housing for sale".

That was confirmed by the central area planning committee on February 13 1980. Thereafter the appellant was visited by a housing welfare officer and sent a leaflet explaining his possible entitlement to home loss and disturbance payments but prefaced by a caution that the issue of the leaflet did not necessarily imply any liability on the part of the respondent for such payment. The appellant then vacated the site on April 12 1980 and was rehoused. Thereafter, as I understand it, the respondent proceeded with the demolition of buildings on the site and its clearance in order to enable it to be sold with vacant possession for housing development by private developers.

The first question raises a short but not altogether easy point of construction against the background facts of this case, and that is simply what is meant by the words "the purposes for which it was acquired". The respondent's contention, which succeeded before the judge, is that "purpose" is not to be distinguished from "intention". The "intention" of the local authority here was that the land should be cleared and redeveloped by the erection of houses by or under the control of the local authority in accordance with a scheme which they subsequently evolved but which had the important characteristic that the title to the land at that time was intended to remain with the local authority and the housing provided on it was to be part of the local authority's housing stock. Thus, Mr Anthony argues, when the respondent resolved in 1980 to clear the land and dispose of it in the market, this was a complete change of intention which resulted in the land now being held by it for the purpose of sale and disposal. The learned judge accepted that argument.

He said: G

> It seems to me that the Greater London Council by abandoning their original intention to redevelop the land, and by resolving that the land should be sold to a purchaser for the purpose of that purchaser building houses, were divorcing themselves from the whole original concept. They no longer could be said to be holding the land for the purpose for which it was acquired, which I hold to be for the purpose of redevelopment by themselves.

Mr Watkinson submits that this approach puts altogether too narrow a construction upon the critical phrase "purpose for which it was acquired". "Purpose" may sometimes equate precisely with intention, but in the context of this Act, which is to compensate *(inter alia)* people displaced as a result of the exercise of statutory powers for the disturbance resulting from their being deprived of their homes following acquisition, one has to look not at the precise scheme which H the acquiring authority may have had in mind at the particular time or any subsequent time for the carrying out of the statutory purpose, but at the general purpose upon which reliance was placed in making the original acquisition.

Here the acquisition was for the purpose specified in section 96(a) of the Act of 1957 to which reference has already been made. That purpose still exists and the fact that the acquiring authority has proposed to carry it out by clearing the site and selling to a private developer for him to erect the houses does not alter the fact that the land was at the material time held as a site for the erection of houses.

Speaking for myself, I do not derive a great deal of assistance from a consideration of the Housing Act 1957 or the precise statutory J purpose which justified the original acquisition of the site by the respondent in 1963. If "purpose" in the Land Compensation Act 1973 was intended to refer to the statutory justification for the original acquisition, then I am persuaded by Mr Anthony's argument that, on that analysis, the "purpose" of selling the land in 1980 was not the same purpose as the purpose of acquisition in 1963. If one assumes that the statutory "purpose" of acquisition was that described in section 96(a) of the 1957 Act, that must, as it seems to me, be construed, as a matter of necessary implication, as the purpose of erecting houses by the local authority for it contrasts with the purpose specified in sections 96(e) and 105(a).

A decision to sell for private development outside the control of the local authority would involve invoking the provisions of section K 105(1)(b) and that would, in my judgment, be a different purpose.

On the other hand, I am not persuaded that the expression "the purposes for which it was acquired" in section 29 (1)(c) of the Act of 1973 falls to be construed in the narrow way suggested by Mr Anthony as synonymous with the intention of the acquiring authority to carry out a particular scheme of development. One has, as it seems to me, to look at the Act as a whole and to give a purposive construction to its provisions. The legislature clearly intended to produce a scheme for the payment to a person who was compelled to vacate his home as a result of development carried out under compulsory powers of compensation for the loss of his home and of assistance with the expenses necessarily entailed in moving. It is to be L noted in this connection that the conditions for payment of disturbance allowances in section 37 of the Act are the same as those for home loss payments under this section.

The matter is to a large extent one of impression, as so many points of construction are, but although the arguments are finely balanced, I have not, for my part, felt able to take the same restrictive view of the word "purpose" as that adopted by the learned judge. While it is true that the original concept was, no doubt, the provision of local authority housing, the broad "purpose" for which the land was acquired was that it should be cleared for the erection of houses.

The fact that the respondent abandoned the idea of erecting houses itself and proceeded to move the tenants with a view to clearing the site so that someone else could erect houses there does not, in my M judgment, involve the conclusion that the land was not at the material time held for the purpose for which it was acquired. So far as the tenant is concerned, he has been removed and disturbed in order to enable the original purpose of housing development to be carried out and it would, in my judgment, defeat the purpose of the Act to hold him deprived of compensation for his loss because the acquiring authority has chosen to stop its own operations at the point of completion of clearance and to arrange for the process of housing development to be completed by someone else. I find it difficult to accept that the legislature, in framing a statute designed to provide compensation for disturbance on redevelopment pursuant to compulsory powers, could have envisaged that payment could so

A easily be avoided by taking all the necessary steps to permit the proposed redevelopment to take place and then handing the actual process of erecting houses to someone else by way of compensation-free realisation. While, therefore, I have not found the question an easy one, I would answer the first question in the sense opposite to that urged by Mr Anthony and found by the learned judge.

That, however, does not dispose of the matter, because there remains the further question whether the displacement of the appellant was in fact the consequence — and Mr Anthony stresses these words — of the carrying out of the development on the land *by* the respondent. As he points out, the machinery of the Act is to enable the displaced person to obtain his home loss payment from the authority which carries out the redevelopment and from no one else. B If therefore one finds that the appellant was displaced by something legitimately called "redevelopment" one still has to find, before he establishes his claim, that that redevelopment was carried out by the respondent.

The argument here is that the mere act of demolishing so as to provide vacant possession for a purchaser who is going to carry out housing development is not itself redevelopment. The point is encapsulated in the following extract from the judgment of the learned judge:

> Furthermore, in my judgment, a step in connection with or for the purpose of the sale cannot properly be said to be part of a process of redevelopment; it is part of the process of the sale and not of the redevelopment, which, as I have C said, must be a redevelopment carried out by the authority who made the payment, as the concluding words of section 29(1) make clear.
>
> If the words "the carrying out of redevelopment" are to be given their ordinary meaning, as it was said that they should by Lord Widgery CJ in *R v Corby District Council* [1975] 1 WLR 735, then I would consider those words to mean, or at least to include, the putting up of new buildings. Here the Greater London Council are not to put up any new buildings, and therefore cannot be said to be carrying out any redevelopment. Any other construction, with all respect to Mr Watkinson's able and closely reasoned argument, would in my view be putting an unjustifiable strain on the words of the section.

Mr Watkinson relies heavily upon the case of *R* v *Corby District Council* and the subsequent case of *Follows* v *Peabody Trust* [1983] D 10 Housing LR 62 as establishing that demolition, so long as it forms merely a step in a larger process of redevelopment, can itself constitute redevelopment for the purpose of the section.

In the former case the tenant of an aluminium bungalow had been displaced and rehoused following the demolition of his bungalow as part of a scheme of clearance prompted by the fact that a great many of the similar bungalows on the estate had become unfit.

Lord Widgery CJ at p 738 said:

> Mr Knight contends, and I think this is perhaps his best point, that when one looks at the language of section 29 it clearly contemplates a causal connection between the displacement of the tenant and the factor causing displacement, which is referred to later in the subsection. He says that these tenants were not displaced because of the carrying out of the redevelopment on their land. He E says that Mr McLean was displaced because his bungalow proved unfit. It is an interesting argument, but for my part I do not think it stands up to consideration for very long. I do not see why in principle, or, as Bridge J put it in argument, as a result of any social principle, a distinction should be made in these cases between the private tenant who is evicted because his house is unsuitable and the local authority tenant who has to give up possession of his house for the same reason. It seems to me that the same principle should apply to each, and I have no difficulty, speaking for myself, in saying that the carrying out of redevelopment in this section should be given its ordinary common sense meaning and can go back to and include the act of demolition which precedes the substitution of new buildings which are to come under a redevelopment scheme.

The same approach is reflected in *Follows* v *Peabody Trust* where Cumming-Bruce LJ at pp 73-4 said:

> F The next question is the question did the plaintiff prove that there was a carrying out of redevelopment? Demolition is not synonymous with redevelopment. There may be demolition and no redevelopment; there may be demolition entirely unconnected with any scheme of redevelopment. At the other end of the spectrum there may be demolition which is only undertaken by the owner as a step in a process of redevelopment. It is a question of fact in any case where demolition is in contemplation whether that demolition at that date may sensibly and reasonably be held to be one step in a process of redevelopment, or whether it is demolition without sufficient nexus with redevelopment so that it is not, for the purposes of these two subsections, to be regarded as the carrying out of redevelopment. In every case it must be a question of fact to decide on which side of the line a demolition is proved to fall at the relevant date.

Similarly, Sir Denys Buckley at p 80 observed:

> The judge held that the words "carrying out of redevelopment" refer to a continuing process and one wide enough to cover any material stage in the achievement of contemplated redevelopment. The use of the word "redevelopment" infers that the relevant land is not a virgin site but has already been developed in the past, where, as in this case, an authority acquired property which is encumbered with unwanted buildings, with a view to its redevelopment and the continued existence of those buildings would inhibit the proposed redevelopment, the demolition of those buildings may easily be found to constitute an integral part of the development, notwithstanding that at the date of demolition it may be uncertain when the redevelopment is likely to be completed and even if it may then be uncertain how it may be completed or whether it will ever be completed. Whether there has been a step which constituted an act of redevelopment is, I would agree with my lord, one of fact to be determined in the light of the surrounding circumstances and this, in my view, would be nonetheless true because the early taking of such a step as demolition might have financial advantages of a short-term nature unrelated to the achievement of the projected development such, for example, as matters of good estate management.

As Mr Anthony points out, however, both those cases were cases where the authority concerned was itself proceeding with the subsequent stages of the redevelopment, albeit possibly much later on. Here, by contrast, the demolition was an act designed to clear the land for sale and subsequent redevelopment by a purchaser and he argues that the contrast points to the conclusion for which he contends.

That argument has its attraction, but again I have not myself felt able to accept it. Whether demolition per se constituted a "redevelopment" for the purposes of the Act must be a question to be decided on the facts of the particular case.

There is nothing in the agreed facts here to lead to the conclusion that demolition and clearance was an isolated measure prerequisite to the disposal of the land and unrelated to the subsequent erection of houses upon it. It was, as I see it, something done to enable the respondent to dispose of the land more advantageously for the purposes of enabling it to be developed as a housing estate by the purchaser.

I do not think that the question of the status of demolition can be divorced from the first question which I have already sought to answer. If one asks: "Was the land still held at the date of the respondent's decision for the purposes of housing redevelopment?" I, for the reasons already given, think that that question must be answered affirmatively. If that be right, then it seems to me that the demolition which resulted in the appellant's being displaced was an essential step in the process of redevelopment which at that stage was being carried out by the respondent but, following a sale, was intended to be completed by the purchaser. It was done to enable or encourage the purchaser to carry out the original purpose of acquisition and I would hold that it had itself the character of redevelopment and redevelopment carried out by the respondent.

I would accordingly allow this appeal and substitute affirmative answers to both questions raised by the summons.

RALPH GIBSON LJ and ANTHONY LINCOLN J agreed and did not add anything.

The appeal was allowed with costs. Legal aid taxation was ordered.

Court of Appeal

March 19 1986

(Before Lord Justice LAWTON, Lord Justice PARKER and Lord Justice NOURSE)

ABBEY HOMESTEADS (DEVELOPMENTS) LTD v NORTHAMPTONSHIRE COUNTY COUNCIL AND ANOTHER

Estates Gazette June 14 1986

278 EG 1249-1254

Compensation for compulsory purchase — Basis of valuation of land — Appeal from decision of Lands Tribunal in favour of developers on preliminary issues — Essential question, although approached "through a tangle of statutory and contractual provisions", was whether an area of 1.23 hectares, part of land which under an agreement pursuant to section 52 of the Town and Country Planning Act 1971 was to

be "reserved for school purposes", should be valued as freely available for residential development or as subject to a permanent covenant restricting its use to such purposes only — Lands Tribunal decided that the reservation did not create a restrictive covenant running with the land; that the restriction was of limited duration and did not preclude a valuation under rule (2) of section 5 of the Land Compensation Act 1961; and alternatively that if it did create a restrictive covenant it was one which should be discharged under section 84 of the Law of Property Act 1925 (an application for which had been made by the developers) — Held on appeal to the Court of Appeal that the reservation did create a permanent restrictive covenant running with the land; and that Lands Tribunal was in error in holding that the restriction ought to be discharged — The court rejected submissions that the reservation was not a negative restriction and that it was of temporary effect only — Appeal by district council allowed, the land to be valued as subject to a restrictive covenant — An appeal by the county council from a decision of the tribunal refusing it leave to be joined as an objector to the section 84 application was not challenged in the Court of Appeal, and this appeal was dismissed

This was an appeal by case stated by the Northamptonshire County Council and the South Northamptonshire District Council from a decision of the Lands Tribunal (V G Wellings QC) determining certain preliminary issues in favour of developers, Abbey Homesteads (Developments) Ltd, respondents to this appeal.

Gavin Lightman QC, J P Whittaker and Miss Elizabeth Jones (instructed by Shoosmiths & Harrison, of Northampton) appeared on behalf of the appellants; Martin Graham QC, M R West and Miss B Schwehr (instructed by Dickins & Batty, of Potters Bar) represented the respondents.

Giving judgment, LAWTON LJ said: This is an appeal, by way of a case stated, by the Northamptonshire County Council (the county council) and the South Northamptonshire District Council (the district council) from a decision of the Lands Tribunal made on November 9 1984 whereby certain preliminary issues were decided in favour of Abbey Homesteads (Developments) Ltd (the developers). Stripped of its legal niceties, the appeal raises three questions. On what basis is 1.23 hectares of land at Towcester, Northamptonshire, which the county council have compulsorily purchased from the developers, to be valued? Is the compensation to be assessed on the basis that the land is available to the developers for residential development or, as both the county and district councils submit, as land which was "reserved for school purposes". The answers to these questions have to be sought through a tangle of statutory and contractual provisions.

The development plan of the county council, drawn up in the 1960s, envisaged that the population of Towcester would increase with the inevitable result that there would be a need for more houses, schools, open spaces and services of various kinds. The developers saw opportunities for them in the Towcester area. Some time before 1976 they bought 13 hectares (34.3 acres) of agricultural land to the south-east of the Brackley Road (the A43). Between 1969 and 1975 the developers negotiated with the county council, the then planning authority, about the development of this land. After the reorganisation of local government in 1976 the planning authority became the district council. The developers negotiated with it for planning permission. The negotiations ended on June 4 1976 with the making of an agreement pursuant to section 52 of the Town and Country Planning Act 1971 and the grant on the same day of outline planning permission for "residential development (343 units), incorporating associated open space, a site for a primary school and local shopping provision".

Section 52 of the 1971 Act provides:

(1) A local planning authority may enter into an agreement with any person interested in land in their area for the purpose of restricting or regulating the development or use of the land, either permanently or during such period as may be prescribed by the agreement; and any such agreement may contain such incidental and consequential provisions (including provisions of a financial character) as appear to the local planning authority to be necessary or expedient for the purposes of the agreement.

(2) An agreement under this section with any person interested in land may be enforced by the local planning authority against persons deriving title under that person in respect of that land, as if the local planning authority were possessed of adjacent land and as if the agreement had been expressed to be made for the benefit of such land.

It follows that if the agreement was one restricting "the development or use of the land" the district council could enforce it as a restrictive covenant running with the land. The district council contended before this court that the agreement did restrict the use of part of the land to "school purposes" and, in consequence, created a restrictive covenant running with the land which they could enforce pursuant to section 52(2) of the 1971 Act. The developers contended that it was not, and, even if it was, it expired with the compulsory purchase on July 7 1983 by the county council, as the local education authority, for the building of a primary school. Further, it was void for uncertainty.

The agreement recited that the developers had applied for planning permission for residential development of the land referred to in specified planning applications and that the district council wished to secure that the development of the land should be in the interests of the proper planning of the area and be restricted and regulated in accordance with the provisions of the agreement. In my judgment the proper planning of a site as large as 34.3 acres and the building of 343 dwelling-houses on it would require the making of provision for a school to be built and maintained and for open spaces.

Clause 1 of the agreement was in these terms:

Pursuant to the provisions of section 52 of the Town and Country Planning Act 1971 the company agrees declares and covenants with the District Council that from the date on which planning permission is granted to either of the applications the land shall be subject to the conditions restricting and regulating the development specified in the First Schedule hereto in addition to any conditions or limitations subject to which planning permission may be granted as set out in the Second Schedule together with any further conditions or limitations imposed on any appeal by the Secretary of State for the Environment.

Schedule 1 dealt with various planning matters such as the construction of a road and a bridge. Para 5 provided as follows: "An area of 1.3 hectares adjacent to the playing field and amenity open space areas shall be reserved for school purposes."

The plans annexed to the agreement did not specify where the 1.3 hectares were to be reserved by the developers; but the plan annexed to the outline planning permission given the same day did so in these terms: "An area of 1.30 hectares in the approximate position indicated on the deposited plan 710/2 adjacent to the playing field open space areas shall be reserved for school purposes . . . ". The developers can never have been in doubt as to where they were expected to reserve land for school purposes. Mr Graham took the point on behalf of the developers that the plan 710/2 could not be used to construe para 5 of the First Schedule. That may be; but it makes no difference, because the developers were entitled to build 343 houses on the site and it was for them, subject to the terms of the outline planning permission, to decide where on the land to reserve 1.30 hectares for school purposes.

On July 7 1983 the county council made a compulsory purchase order of 1.23 hectares stating it was for a primary school for 245 children. On June 6 1984 it gave development permission for the erection of a primary school and unit for 315 children. This number was probably greater than was necessary to accommodate the children on the developers' estate. It was intended to accommodate children from an adjacent housing estate as well as from the developers' estate. In my judgment, this fact does not affect the planning purpose which para 5 was intended to achieve.

I should add that on May 25 1983 the developers applied to the district council for a certificate of alternative development. The district council granted that certificate. The county council have appealed against the grant. That appeal has not yet been heard and awaits the outcome of this appeal.

On September 30 1983 the developers applied to the Lands Tribunal pursuant to section 84(1) of the Law of Property Act 1925 (as amended) for an order that the restriction imposed by para 5 of the First Schedule to the agreement should be discharged wholly. Two grounds were put forward. First, that there had been a change in the character of the property because the making of the compulsory purchase order had rendered the purpose of the restriction obsolete; and second, no one entitled to the benefit of the restriction would be prejudiced by its discharge.

The county council applied to be joined as an objector pursuant to section 84(3A) of the 1925 Act as amended. The Lands Tribunal refused it leave to do so. The county council has appealed from that

A decision, but in this court Mr Gavin Lightman on its behalf has not sought to challenge the Lands Tribunal's decision on that issue. Accordingly the county council's appeal is dismissed.

The Lands Tribunal decided that para 5 of the First Schedule did not create a restrictive covenant running with the land, that such restriction as it did contain was of limited duration and in consequence as a matter of law it did not preclude a valuation under rule (2) of section 5 of the Land Compensation Act 1961. This meant that the 1.23 hectares should be valued without reference to the restriction for school purposes. Alternatively the Lands Tribunal decided that if para 5 did create a restrictive covenant running with the land it had become obsolete because of the compulsory purchase
B order, the entry into possession by the county council and the grant of authorisation of development for school purposes. The district council have challenged all these decisions.

The fundamental finding of the Lands Tribunal was that para 5 did not create a restrictive covenant running with the land. The learned member, who signed the case stated, attached importance to the absence in para 5 of any words such as "and shall not be used for any other purpose" and to what he considered to be uncertainty in the identification of the 1.30 hectares. Mr Graham adopted this reasoning. In my judgment it is faulty. On a natural reading of para 5 and of clause 1 of the agreement the developers covenanted with the district council for the benefit of the land that 1.30 hectares should *not* be used other than for school purposes. This covenant was just as
C restrictive as the seminal one in *Tulk* v *Moxhay* (1848) 2 Ph 774 which was to keep the plot and land forming Leicester Square Gardens as it was at the time of the conveyance. Under para 5 all the developers had to do was to refrain from building on the 1.30 hectares. Of course, if the identity of the land could not be ascertained, as Mr Graham submitted it could not be, then the covenant would not bite on any part of the 34 acres. But in this case the identity of the 1.30 hectares could be ascertained, and was intended to be ascertained, by the developers themselves. Their outline planning permission gave them guidance as to the area in which they were not to build houses: they were left by the agreement and the outline planning permission to decide what land should be reserved and this they did. Once they had done so the land was identified.

D Para 5 was unlimited in time: no period was prescribed for its duration as section 52 of the 1971 Act allowed. Mr Graham, adopting the reason of the learned member in the case stated, submitted that para 5, on its proper construction, was intended to come to an end when the 1.23 acres were acquired by the county council for school purposes, since its planning objective was then accomplished. I do not agree. As I have already stated, para 5 was put into the First Schedule with the object of securing the proper planning of the area. As long as there were 343 houses on the estate there would be need for a primary school somewhere on it. The need would be a continuing one. Proper planning for the area would not have been achieved had the covenant allowed the county council, or a private person, to start
E a school on the 1.30 acres, and then, after a short time, close it down and use the buildings for some other purpose.

The construction as to duration which I adjudge should be given to para 5 provides the reason why the Lands Tribunal misdirected itself in deciding that if, contrary to its decision on construction, para 5 did create a restrictive covenant running with the land, it was obsolete because its planning purpose had been achieved and in consequence its discharge would not injure the district council. As there was a continuing need for a primary school on the estate, any restrictive covenant running with the land which was intended to ensure that the need could be met was one for a proper planning purpose. As, in my opinion, para 5 did achieve a proper planning purpose, Mr Graham's submission that any attempt after the county council's entry on to the
F 1.23 hectares to enforce it would be *ultra vires* is misconceived. Enforcement would achieve a proper planning purpose.

For these reasons I would allow the district council's appeal and declare that para 5 of the First Schedule to the agreement did create a restrictive covenant running with the land and that it should not be discharged.

Agreeing, PARKER LJ said: Section 52 of the Town and Country Planning Act 1971 empowers a local planning authority to enter into an agreement with any person interested in land in their area for the purpose of *restricting* or regulating *the development or use* of the land either permanently or during such period as may be prescribed by the agreement. (Emphasis above and hereafter is mine.)

On June 4 1976 the respondents and the district council entered into an agreement which was not only in fact made pursuant to that section but specifically recited: (a) that the district council was the local planning authority for the purposes of section 52, (b) that the respondents had applied for planning permission for residential development on their land, (c) that the district council wished to secure that the development of the land should in the interests of proper planning "be *restricted* and regulated in accordance with this agreement", (d) that for that purpose the respondents were willing to enter the agreement "pursuant to the provisions of section 52 of the Town and Country Planning Act 1971" *restricting* and regulating the development of the land in the manner hereafter appearing.

Furthermore, the respondents by Clause 1 agreed, covenanted and declared with the district council pursuant to section 52 that "*the land* shall be subject to the conditions *restricting* and regulating the development specified in the First Schedule". Para 5 of that Schedule provides simply:

> An area of one point three (1.30) hectares adjacent to the playing field and amenity open space areas shall be reserved for school purposes.

In the light of those provisions it appears to me plain beyond all argument that it was the common intention of the parties to restrict the development of 1.30 hectares of the respondents' land in the position specified on a permanent basis. It is also clear from clause 2 of the agreement that the respondents' covenants were intended to bind their successors and from section 52 itself that such covenants were intended to be enforceable by the district council also on a permanent basis.

It is therefore somewhat surprising to find that the respondents should have contended that para 5 is too uncertain to be enforceable, that it is positive and not restrictive and cannot therefore run with the land and that it is in any event of limited duration and expires when the planning objective, namely, the acquisition of the land for a school, has been achieved.

All three contentions are in my judgment without foundation. The general position of the land to which para 5 applied was readily ascertained by the map specifically referred to in the agreement itself (see para 12 of the Second Schedule). Its precise boundaries were, it is true, not defined, and it is said that, this being so and there being no formula to resolve any dispute, para 5 fails for uncertainty. I see no need for any formula, for I see no room for any dispute. It was for the respondents to fix the precise boundaries and so long as the area was in the position defined the district council had no right to object. This objection fails.

Next it is said that despite the repeated expressions of intention to restrict development and the express provision in clause 1 that the land was to be subject to restrictions, para 5 was positive and not intended to run with the land. It is said to be positive because it involved a positive obligation to define the area and reserve it. I have had some difficulty in following this argument and I reject it without hesitation. One has only to ask the question "were the respondents free to build residential houses on the land?" to get the answer, "No they were not!" If that is not restrictive I do not know what is. As to the suggestion that the covenant was not intended to run with the land, I regard it as being disposed of conclusively by the provisions of clauses 1 and 2, the recitals and the provisions of section 52.

There remains on this part of the case only the contention that the covenant was of limited duration. The alternatives provided for by section 52 are either permanence or for a period prescribed by the agreement. No period was prescribed in the agreement. Permanence therefore prevails, at least unless there can be found elsewhere in the agreement or from admissible surrounding circumstances some cogent basis for importing a limitation. I can find none.

The respondents' contention that the covenant if it existed was of limited duration is, as I understood it, based on the notion that its purpose was merely to enable the local education authority to acquire the land for a school and accordingly that once that authority had acquired it the covenant was spent. If this was indeed the contention it is in my judgment misconceived. It was for the district council to determine, as a matter of planning, where they wished a school to be. It was and is the duty of the local education authority to secure that there are available for its area sufficient primary and secondary schools. It is, however, not obliged itself to provide all the schools and it is specifically empowered by section 6(1) of the Education Act 1953 to make arrangements for the provision of primary and secondary schools not maintained by it. There is therefore no reason why the local education authority should not have made arrangements with, for example, some charitable body that it should purchase the land in question and provide a school. If such a body did purchase the land but the covenant then lapsed that body would be possessed of land upon which there was no fetter. Furthermore, even if the purchaser is the local education authority it by no means follows that the covenant has no further use, for such authority might, after a time, decide that in accordance with its duties it could close down one school and wished to close down this particular school. In such circumstances the covenant could remain of much importance to the district council.

If then, as I think, the covenant binds the land, what the respondents had, as soon as the land was identified, was land subject to a permanent covenant which would clearly affect its value in the open market. They contend, however, that it must be valued on the basis that it could be sold on the open market free of the covenant.

The only bases upon which, in my view, such a contention could succeed are: (i) if the covenant had been discharged before the acquisition of the land under the compulsory purchase order, which it had not, or (ii) if the circumstances were such as to justify discharge at that time.

The learned member of the Lands Tribunal considered that it would be or would have been appropriate to discharge the covenant and this court is asked whether he erred in law in the grounds upon which he took that view. In my view he did. The fact that at a particular time the owner or prospective owner of the land does not wish to breach the covenant or may even be under a statutory duty to comply with it does not render the covenant obsolete. In the same way the fact, if it be a fact, that the discharge of the covenant would not at the present time injure the district council is not to the point. It could clearly in certain circumstances be of value to them and this being so it does not seem to me possible to say that its discharge will not injure them.

For the foregoing reasons as well as those appearing in the judgments of Lawton and Nourse LJJ, which I have read in draft, I also would allow the appeal of the district council. I agree that the appeal of the county council should be dismissed and I agree with the order proposed.

Also agreeing, NOURSE LJ said: The first group of questions which arise on this appeal is concerned with the construction and effect of the section 52 agreement which was made between the district council and the developers on June 4 1976. Section 52 and the principal provisions of the agreement have already been referred to in the judgments of Lawton and Parker LJJ and I need not repeat them.

The first step must be to construe the material condition in para 5 of the First Schedule to the agreement. In clause 1 of the agreement, that and the other conditions specified in the First Schedule are described as conditions "restricting and regulating the development". Accordingly, the effect of a condition requiring a specified area to be "reserved for school purposes" is that the development must be so restricted and regulated that that area is left available for those purposes. Where then is the specified area to be? The condition prescribes that it is to be adjacent to the playing field and amenity open-space areas. It assumes that the situation of those areas is already known. In order to acquire that knowledge you must look, not at the small-scale plan attached to the agreement, but at the one of larger scale which was attached to the developers' application for planning permission. That application is referred to in recital (3) to the agreement and the plan is referred to in para 12 of the Second Schedule (see below). Both the application and the plan are thus admissible for the purposes of construing the agreement. On looking at the plan, you see that the site proposed for a school is indeed adjacent to, which means close to, both areas. It actually adjoins the open space area. It does not adjoin the playing field area.

Condition 5 thus requires that the school site shall be close to both areas. It does not require that it shall adjoin either of them. There is therefore some latitude in the matter and the only real question, so far as certainty is concerned, is how is the precise location to be determined? In my view that is not a question which admits of any doubt. The agreement was one whereby the developers agreed that their development of the land would be subject to certain conditions and limitations which were required by the district council. Provided that those conditions and limitations were observed, the developers remained masters of the situation and could develop the land as they chose. When, therefore, you find a condition requiring that a specified area shall be "reserved" for certain purposes, it can only mean that its location is to be determined by the developers, subject to any requirement imposed by the district council. The only such requirement is that it shall be close to two other designated areas. G That requirement having been imposed, the parties cannot be taken to have intended that some different or additional requirement should be imposed, namely that the district council is to be party to the determination of the precise location. In my judgment the effect of condition 5 is exactly that which it would have had if the words "in such position as the company shall determine" had been added to it.

So construed, condition 5 cannot be void for uncertainty and the decision of this court in *Bushwall Properties Ltd* v *Vortex Properties Ltd* [1975] 1 WLR 1649, on which Mr Graham so strongly relied, is readily distinguishable. I should add that, although a dispute could well have arisen as to whether the site chosen by the developers was adjacent to the two other areas, the mere possibility of such a dispute did not render the condition uncertain, since it was of a nature which H the court could, and would if necessary, have resolved.

The next question is whether the obligation imposed by condition 5, which admittedly is in form positive, is in substance positive or negative. If, as I have earlier identified it, the effect of the condition is that the development must be so restricted and regulated that the specified area is left available for school purposes, the substance of it is that the development must not be carried out so as to interfere with that requirement. That is in substance a negative and not a positive obligation.

Then, it must be asked, did the parties intend that condition 5 should create a restrictive covenant whose burden was to run with the land, or did they intend that the obligation of it should exist only in J contract? The agreement was expressed to be made pursuant to the provisions of section 52 which, by subsection (1), empowers a local planning authority to enter into agreements "for the purpose of restricting or regulating the development or use of the land" either permanently or for some limited period. As to that, it is worth observing that a developer is by definition someone who will often soon part with his interest in the land, and it is difficult to see how its use can be permanently restricted or regulated by obligations existing solely in contract. Moreover, as was pointed out by Lord Greene MR in *Ransom & Luck Ltd* v *Surbiton Borough Council* [1949] Ch 180 at p 194, the provision which is now embodied in section 52(2) is clearly designed to overcome the normal rule that a restrictive covenant can only be enforced against successors in title of the covenantor if it was K made for the benefit of land which is retained by the covenantee. There would be little point in that provision if section 52(1) was not concerned, at least in part, with covenants enforceable against successors in title of the covenantor.

In the circumstances, the agreement having been expressed to be made pursuant to the provisions of section 52, it seems that it must be approached in the expectation that it may have been intended to impose one or more restrictive covenants whose burden runs with the land. And in clause 1 it is seen that "the land" is to be subject to the conditions restricting and regulating the development which are specified in the First Schedule. How then can it be said that such of those conditions are as negative in substance were not intended to run L with the land, but to exist only in contract?

Mr Graham sought to answer this question with a series of submissions, many of which appeared necessarily to involve the proposition that an agreement made pursuant to section 52 ought not to be construed in accordance with those principles which govern the construction of all other contracts, but in some special way appropriate to its planning subject-matter. Any such proposition which goes beyond a recognition that general principles of construction will always allow for the subject-matter of a particular contract must be firmly rejected. An application of those general principles makes it necessary for me to mention only one of Mr Graham's submissions on this part of the case. He said that since the material words in clause 1 of the agreement refer only to conditions M restricting and regulating the development of the land, and not its use, the First Schedule conditions are shown to be concerned only with planning matters and must therefore have been intended to exist only in contract. It is possible that that submission might have had some force had it not been for the immediately following words in clause 1, sc: "in addition to any conditions or limitations subject to which planning permission may be granted as set out in the Second Schedule". Para 12 of the Second Schedule is in these terms: "An area of 1.30 hectares in the approximate position indicated on the deposited plan 710/2 adjacent to the playing field and amenity open space areas shall be reserved for school purposes such area to be additional to that referred to in Condition 9(b)". The wording of that condition is no different from that of condition 5. I would certainly be prepared to accept that condition 12, like the other conditions in the Second Schedule, is concerned only with planning matters. But that reinforces, rather than detracts from, the view that condition 5 is concerned with something else. The learned member of the Lands Tribunal thought that the purpose of condition 5 was the same as that of condition 12 but, with respect, that does not explain the presence of both of them, nor of the two distinct sets of words in clause 1. Reading the agreement as a whole, I am in no doubt, on general principles of construction, that Mr Graham's submission is incorrect and that the burden of condition 5 was intended to run with the land.

The final question in the first group is whether the effect of condition 5 is permanent or temporary and, if the latter, for what period. The learned member of the Lands Tribunal thought that it was to endure only until the planning objective was achieved by the provision of a school site of the specified area. He did not, I think, give any reason of his own for that view. No doubt he accepted the argument of Mr Graham, which was to the effect that once the compulsory purchase order had been made and confirmed the land could not be used for anything but school purposes and the promise contained in condition 5 had therefore been fulfilled. I believe that that argument again involves a failure to give proper effect to condition 5 as distinct from condition 12. Section 52 expressly provides that a restriction may be either permanent or "during such period as may be prescribed by the agreement". On general principles of construction a condition whose effect is not expressly limited must be taken to be unlimited in point of time. Again there would be little point in having a restrictive covenant whose burden was intended to run with the land if it was not intended to run permanently.

For these reasons I am of the opinion that the submissions of the district council as to the construction and effect of condition 5 are correct. On that footing Mr Graham raised an entirely fresh point on this appeal, with which it is convenient to deal at this stage. His submission, as I understood it, was that if condition 5 had the effect for which the district council contends then it conferred on the district council a planning gain, ie a gain which could not lawfully have been secured by conditions attached to the planning permission, and that it was accordingly *ultra vires* and unenforceable. There seem to me to be a number of conclusive objections to that submission, but I do not think that I should take time to state them, because Mr Graham was unable to satisfy me that the submission was one in respect of which the district council might not have wished to adduce evidence if it had been advanced in the right place at the right time. Mr Lightman's consent to this point being taken at this stage was naturally subject to that important proviso.

Having held that condition 5 created a valid and enforceable restrictive covenant of permanent duration, I now turn to the developers' application to discharge it under section 84(1)(a) or (c) of the Law of Property Act 1925. That subsection, so far as material, provides that the Lands Tribunal shall have power to discharge or modify a restrictive covenant on being satisfied:

(a) that by reason of changes in the character of the property or the neighbourhood or other circumstances of the case which the Lands Tribunal may deem material, the restriction ought to be deemed obsolete . . . ; or (c) that the proposed discharge or modification will not injure the persons entitled to the benefit of the restriction.

Although on the view there taken as to the construction and effect of condition 5 it was unnecessary for the Lands Tribunal to express a view on the application under section 84(1), the learned member did give an alternative decision on it in the following terms:

it would be my decision in the alternative, that the making and confirmation of the compulsory purchase order, the service of the notice to treat, the entry into possession of the land acquired by the County Council and the grant of planning permission to it for school purposes, have rendered the restriction obsolete . . . I would also be of the opinion that the proposed discharge would not injure the District Council because, notwithstanding what is said in its notice of objection, its purpose, as enshrined in the agreement, had been achieved.

With the greatest of respect to the learned member, it seems to me that these views are another product of the basic failure to treat condition 5 as being a restrictive covenant of full effect and independent of condition 12.

With regard to para (a) the obsoleteness, if it exists, must have come about by reason of changes in the character of the property or the neighbourhood "or other circumstances of the case which the Lands Tribunal may deem material". The general words must be construed *ejusdem generis* with the particular words, and it seems to me that the matters relied on by the Lands Tribunal come nowhere near the genus. Moreover, the test for obsoleteness is whether the original object of the restriction can still be achieved; see *Re Truman, Hanbury Buxton & Co Ltd's Application* [1956] 1 QB 261 and *Driscoll* v *Church Commissioners for England* [1957] 1 QB 330. In this case we are presented with the compelling fact that not only is the original object capable of being achieved; it is in course of actual and flourishing achievement.

With regard to para (c), Mr Graham submitted that the persons entitled to the benefit of the restriction are the district council who, since they are not the local education authority, cannot be injured by the proposed discharge of condition 5. I do not think that that can be correct. It would involve too narrow a view of the district council's function. Moreover, I think that in a case like this, where there is an obvious difficulty in fitting section 52(2) and para (c) together, the persons entitled to the benefit of the restriction must be taken to be those who benefit from the permitted use of the land which is subject to it. They are the children who go to school there and the generations to come, all of whom would be injured by the proposed discharge. In any event, there is authority for the view that para (c) is only a long stop against vexatious objections to extended user; see *Ridley* v *Taylor* [1965] 1 WLR 611 at p 622F, *per* Russell LJ. I certainly do not think that para (c) is a ground on which it would be proper to discharge the restriction in the present case.

For these reasons, which are substantially the same as those which were advanced in the clear and concise argument of Mr Lightman in this court, I, too, would allow the appeal of the district council and make the declarations proposed by Lawton LJ.

I agree that the appeal of the county council must be dismissed.

The appeal of the district council was allowed with costs and the appeal of the county council dismissed with costs; directions given as to reserved costs below. Appropriate declarations made. Leave to appeal to the House of Lords refused.

A petition for leave to appeal was subsequently refused by the House of Lords.

For further cases on this subject see p 205

ESTATE AGENTS

Queen's Bench Division

October 18 1985

(Before Mr Charles WHITBY QC, sitting as a deputy judge of the division)

ERIC V STANSFIELD (A FIRM) v SOUTH EAST NURSING HOME SERVICES LTD

Estates Gazette January 25 1986

277 EG 311-316

Claim by plaintiff estate agents against defendant purchasers for an introductory and retaining fee of 2% for acting on behalf of the defendants in the purchase of a large property to be used for the provision of shelter and nursing services — Defendants, in resisting the claim, alleged that the plaintiffs were disentitled to commission by reason of breaches of professional society and common law rules — It was alleged that the plaintiffs, without the defendants' consent, had been acting at the same time for other clients interested in the property, that this had forced up the price and had consequently enhanced the sum claimed as commission — It was submitted that there was thus a conflict of interest situation contrary to the rules of the plaintiffs' professional society (Incorporated Society of Valuers and Auctioneers) and to a term implied at common law — After reviewing the evidence the judge found that no actual conflict of law situation had arisen — Plaintiffs had been in touch with two other persons possibly interested in the property but neither had evinced any firm intention of proceeding — As soon as the defendants had shown a definite interest in purchasing the property the plaintiffs regarded themselves as acting for the defendants and no one else — The judge found as a fact that the defendants had been fully aware that other persons might be interested in the property and had indeed arranged to view it on the same day as the defendants — He held further, however, that the duty to disclose the existence and position of other parties arose only if they had shown an "active interest", ie had evinced a real determination to instruct the agent to act for them in the matter, which was not the case here — Plaintiffs had acted in good faith and were not in breach of professional society or common law rules — Professional society rules considered and conflict of interest discussed — Authorities on disentitlement to commission mentioned — Plaintiffs entitled to commission — Defendants' counterclaim for damages abandoned

In this action Eric V Stansfield FSVA, a firm of estate agents, sued South East Nursing Home Services Ltd under an agreement in writing for an introductory and retaining fee of 2% of the purchase price, in respect of the purchase by the defendants of Edith Edwards House, Park Road, Banstead, Surrey, for £284,284.

L Marsh (instructed by John Smythe & Co, of Sutton) appeared on behalf of the plaintiffs; J W Haines (instructed by Brand & Co, of Cheam) represented the defendants.

Giving judgment, MR CHARLES WHITBY QC said: In this interesting case the plaintiffs are a firm of estate agents and valuers carrying on practice in Carshalton, Surrey, under the title of Eric V
Stansfield (a firm) and as appears from a document on p 10 of the F
agreed bundle, and as is demonstrated by their headed notepaper, they practise under the style of Eric V Stansfield FSVA (Fellow of the Incorporated Society of Valuers and Auctioneers). By their statement of claim, which was endorsed on the writ of summons, they claim the sum of £6,538-odd for an introductory and retaining fee of 2% of the purchase price plus VAT arising out of an agreement in writing dated November 3 1982 in respect of the purchase by the defendants of premises known as Edith Edwards House, Park Road, Banstead, Surrey, for the sum of £284,284 freehold. The purchase took place in or about the spring of 1983.

The defendants by their defence in the action admit the written G
agreement but say it was subject to implied terms: first, that the plaintiffs' conduct would at all times be governed by the relevant rules of the ISVA. After a slight hiccup as to which were the relevant rules, it appears that they were those set out in the society's year book for the year 1977-78. It is conceded that such an implied term existed. They also maintained in their defence that there was a further term to be implied at common law to give business efficacy to the contract, that the plaintiffs would not act for any party in the transaction in addition to the defendants or their associated companies and that in breach of those terms, or one of them, the plaintiffs had acted for a Mr Pitman and a Dr Madina without the consent of the defendants.

The defendants maintained by their defence and counterclaim as
originally served that by reason of those breaches they took over the H
burden of the negotiations with the vendors of the property as from about November 5 1982, that the plaintiffs' conduct was unprofessional and unethical in so far as it was likely to result and did result in an increased interest in the premises with a higher purchase price having to be paid than would otherwise have been the case and hence a corresponding increase in the 2% fee claimed by the plaintiffs, and they stated in that defence and counterclaim that in the premises no fee was payable and, by the counterclaim, now abandoned, that they had sustained damage as a result of the said breaches to the tune of £74,284, being the difference between a bid of £210,000 which they had originally made early in November, which
they said would have fructified, and the ultimate purchase price paid J
by them of £284,284.

That counterclaim has been abandoned, but it is instructive to see how the defence was first put. The defendants now restrict their defence to the proposition that the plaintiffs' breach of duty in acting for Mr Pitman and Dr Madina without the full knowledge and consent of the defendants was so serious as to disentitle the plaintiffs to their fees. It is to be noted from the agreed bundle D1, and from copied extracts from the plaintiffs' day book P1, that the plaintiffs were as active on behalf of the defendants in and about the negotiations for the purchase of the property as, in effect, the defendants allowed them to be. The defendants took over direct negotiations themselves with Cluttons, agents for the vendors of the
property, in or about the middle of November, rather later than was K
alleged in the pleadings, but the day book P1 demonstrates that the plaintiffs were continuing to act on the defendants' behalf. Nevertheless, the defendants say that the plaintiffs were not entitled to their fees.

The matter came about, as I find, in this way. The defendants and their associated company AKQ Enterprises Ltd were interested in purchasing large premises in the Banstead-Cheam area which might prove useful to them in their business of providing shelter and nursing services for the sick and advising others thereon. Dr Madina and his partner Dr Kurwa were also interested in buying property which might be suitable as a nursing home provided that the size and type of the building suited their purposes and pockets. Mr Pitman, who was the managing proprietor of homes for the elderly in the Sutton area, was a possible interested party also. The plaintiffs by their employee, Mr Vickers, knew about this and were aware of these matters. All these gentlemen gave evidence before me in the hearing.

The property ultimately purchased by the defendants was advertised in the national press on October 27 1982. This was Edith Edwards House School, Park Road, Banstead, Surrey. Messrs Cluttons, chartered surveyors of Grosvenor Street in London, were the agents for the vendors and they disclosed details of the property to the plaintiffs on the understanding that they, the plaintiffs, would not be entitled to any commission from the vendors or from Cluttons if they introduced a purchaser. The plaintiffs had to look for their remuneration in this matter elsewhere and when Cluttons sent details of the property Cluttons had written on the documents "For retained clients only".

Mr Vickers proceeded to approach Dr Madina, Mr Pitman and Mr Cole of the defendant company, who had previously been employed by the plaintiffs for about two years, from 1975 to 1977 or thereabouts, and was a highly experienced man in the estate agency world. What Mr Vickers said to these parties came in effect to this, in my judgment, although no doubt he expressed himself in different language, particularly to Mr Cole, who literally spoke his own language: "I will disclose the details of this property which might interest you but I am getting no commission or sub-commission from the vendors or their agents so I shall have to ask you to sign a document which will entitle us to act for you if you decide to buy the property and sets out the commission which will be payable to us if you succeed in purchasing the property and if you wish to go ahead with purchasing the property we will of course act for you."

Each signed the relevant document. Mr Pitman signed it on November 2, Dr Madina on November 3 and Mr Walker, a director of the defendant company, also on November 3. There is a dispute as to whether Mr Vickers informed each signatory of the existence and possible interest, or potential interest, of the other signatories, or potential signatories, as the case may be, for he did not approach them all at once. I will come to that later. Suffice it to say that in my judgment the plaintiffs were saying: "We will act for you as your agents for the purchase of this property if you wish. You must not ask anybody else to act for you. If you purchase the property these are our terms of commission: 2% etc to be payable if you succeed in a desire to purchase". I am quite satisfied that the plaintiffs did not regard the signatories to that document as being under any obligation whatsoever to follow up the matter but that if they did proceed to negotiate for purchase they must act through the agency of the plaintiffs and pay the agreed fees, if successful in purchasing. This in my view was a purely introductory and preliminary step which envisaged a situation when the defendants, for example, might take active steps to purchase the property, or that some other signatory might do so. In fact Dr Madina never did instruct the plaintiffs to act for him in that sense. When he gave evidence he said: "After seeing the property on November 3" — I quote him exactly — "I had no further dealings with the estate agents", for the simple reason that the property was too big and too ambitious a project for him and his partner. It is in my view significant that there is no evidence that he ever took any steps to retrieve his so-called retainer, nor is there any evidence that the plaintiffs felt in any way aggrieved at his unheralded and unannounced withdrawal from the scene. I do not believe that the plaintiffs ever acted for Dr Madina within the meaning of either of the terms which the defendants say ought to be implied into this agreement.

Mr Pitman, whose recollections of this and other matters was vague, perhaps understandably so, and hazy, thought that he might have mentioned to Mr Vickers what figure he might go to but could not remember giving him any specific instructions to bid for the property, and indeed no bid was ever put forward by him or on his behalf. I am entirely satisfied that no situation arose whereby, by acting for the defendants in the purchase of the property, the plaintiffs created an actual conflict of interest situation *vis-à-vis* the other two signatories to the document dated November 2 1982. I accept Mr Vickers' evidence that about a week later Mr Pitman was on the telephone testing out the temperature of the water and trying to discover the amount of any other bids made and that he was told by Mr Vickers that he could not possibly divulge to him any bid made by the defendants. The reason for that was that in Mr Vickers' mind the plaintiffs now regarded themselves as acting in and about the purchase of the property solely for the defendants, whom they knew meant business, because, as is shown by a letter sent by Mr Cole to Mr Vickers, whom he addressed as "Dear Doug," the defendants had made a firm bid of £210,000, subject to contract, for that property. This is what Mr Vickers said in evidence about that conversation: "I did not regard Mr Pitman as our client once we were dealing with the offer from the defendants". Again, I do not accept that the plaintiffs acted for Mr Pitman in this transaction within the meaning of the terms to be implied into the agreement, or said to be implied into it, at any time. Mr Vickers, Mr Graham Stansfield, and Mr K W Forbes, a chartered surveyor who gave expert evidence on behalf of the plaintiffs in his capacity as professional services officer of the ISVA, accepted, as is indeed obvious when one thinks about it, that an estate agent cannot act for more than one party when a conflict of interest situation arises, or may arise if the two clients start bidding against each other, except in what appears to me to be the unlikely event that each party agrees to him so acting, or, to put it in more homely language, once you are definitely acting for a client on instructions to negotiate for the purchase of a property you cannot in practice act for anybody else without the consent of all concerned.

The defendants, who, like I suspect most people in this type of business, are highly sensitive to the leakage of information, suspected that information was being leaked in respect of their bids for the property; they had bid £210,000, as I have stated, and their bid was improved to one of £258,000 on December 3, and ultimately by means of an informal tender in a sealed envelope, to £284,284. Mr Walker told me that Cheam and its environs is something of a village. The buzz gets around; there is a lot of gossip. The defendants' own letters betrayed that they were very keen indeed to buy this property. Each letter to Cluttons making an offer contained an entreaty that they should be told if the offer proved to be unacceptable so that they could consider bidding more. So anxious were they to clinch the deal that on December 3 Mr Cole and a colleague, Mr Trotman, actually visited Cluttons' offices and saw Mr De Boinville — who gave evidence before me, which I accept, that this was an unusual thing — armed with several letters metaphorically up Mr Cole's sleeve, whereby he might hoist the offer if necessary or desirable, if it meant finalising the transaction. There may well have been leakage about the defendants' bids, it would not surprise me at all if there was, and it may have come from more than one source, but one thing is clear on the evidence, that there is absolutely no evidence that the plaintiffs either deliberately or inadvertently, by a partner or by any employee, leaked any information confidential to their clients, the defendants.

The defendants decided that they would conduct the negotiations themselves direct with Cluttons because they thought this would be more efficient, there would be less delay, there would be less chance of leakage of information and, according to Mr Cole and Mr Walker, because they suspected that the plaintiffs were acting for others. In fact by then they were not doing so in any meaningful sense of the term. The defendants never taxed them with the matter because, if I may use the expression, "they did not want to rock the boat"; they did not want anything to happen which might interfere with their strong desire to get these premises. Because they undertook the burden of the negotiations (although, as I have said, the plaintiffs' documents show that they did all that was required of them and were ready to do more) the defendants began to harbour and nourish a grievance that the amount of commission claimed was unreasonably high, and, because I assume in the defendants' favour that the counterclaim as originally framed was a *bona fide* one, they thought that by acting for other parties the plaintiffs had caused the price to be forced up and hence the amount of their commission to get to the substantial amount that it did. So when the sealed offer of £284,284 carried the day and in due course the plaintiffs presented their bill for their fees in or about March 1983, contracts having been exchanged and completion imminent, the defendants were in no mood to pay it.

By this time Mr Cole had already been sent, on February 25, to talk to Mr Graham Stansfield, a partner in the plaintiff firm, about this matter. A note made by Mr Stansfield of that conversation is as follows:

> Phoned re Edith Edwards School House

that is to say Mr Cole phoned re Edith Edwards School House

> and advised that contracts exchanged. Wished to see me to discuss a few matters. Appointment arranged for 4.15 today. Later called: We can erect boards. Completion March 31. Sale price £284,284. Liaise on newspaper features.

That note did not represent a precis of all the conversation that took place, as both Mr Cole and Mr Stansfield agree. Mr Cole certainly either suggested or agreed that the plaintiffs could erect appropriate boards on the premises, there was a conversation too about possible publicity campaigns in the press, there was some conversation about the wisdom of the tactical approach of Cluttons to the whole matter in originally pitching the guide price at the comparatively low sum of £180,000, and there was some discussion about the plaintiffs' fees. Mr Cole says that he criticised the plaintiffs for having more than one retaining client and that as a result the defendants felt bound to take over the negotiations, doing a lot of extra work, and that in the circumstances a proportion of the fee ought to be waived. Mr Stansfield, whom I regard as a most reliable, careful and completely honest witness, said all that happened was that his former employee or colleague attempted to beat him down on the fees and tried to put pressure on him by suggesting that the plaintiffs might have difficulty in enforcing their claim to fees because the document dated November 2 was signed on behalf of one company and the purchase of the property would be in the name of another. This was a thoroughly bad point as well as being an unmeritorious one, because a glance at the document shows that the fee was payable if an associated company completed. When Mr Cole was shown the document in question he abandoned that argument. When asked "Were you put up to this by Mr Walker?" he smiled. However, I am not prepared to find that Mr Walker was a party to that unsavoury manoeuvre. I believe the reason why it does not find its way into the book is that Mr Stansfield did not bother to put it down; he regarded it as a hopeless attempt to beat him down on his fees, and I do not believe that Mr Cole at the meeting made any criticism of the conduct of the plaintiffs in this matter.

On March 21 Mr Walker spoke to Mr Stansfield on the telephone and took a different tack, suggesting on behalf of the defendants, as I find, for the first time that the plaintiffs had acted unethically in acting for more than one party as they had signed forms from two other applicants. Mr Stansfield's response at that time, and his response till as recently as yesterday when he was giving evidence, was that in his view his firm had not acted unprofessionally in any way as no other party evinced any firm intention in the matter and no conflict situation arose even on the horizon. Mr Stansfield's note of the conversation I believe to be an accurate one and that states that as soon as the plaintiffs realised that the defendants were interested in purchasing the property and negotiating for it they had acted for the defendants and no one else and had not divulged to any other party any of the defendants' business. He then consulted Mr Forbes to see if his reading of the situation could be sustained and after speaking to Mr Forbes he refused to reduce his fees.

Mr Vickers gave evidence for the plaintiffs. He was not a good witness in the lawyers' sense of the term in so far as evincing any natural aptitude for that exacting role, but he impressed me as being a most honest one. I was impressed by the fact that at one time he thought that Mr Cole attended a meeting at the premises to view on November 3. Particulars of the reply and defence to counterclaim were delivered nailing the colours of the plaintiffs to that mast and a certain amount was made of this during the trial. As soon as Mr Vickers realised he had been in error he withdrew the suggestion that Mr Cole had been present and apologised. I am entirely satisfied that this was the reaction of an honest man and not a ploy to try to get favour with the court. He told me that he told Mr Cole on November 2 that he was contacting two of their clients, Mr Pitman and Dr Madina. This is what he said:

> Mr Cole said if the premises were right he would discuss fees with Mr Walker. He phoned back and said they wanted to look at the premises. I told him that a meeting had been set up, or a view had been set up with Cluttons for the next day. I told him that I had introduced the property to two others who would be wanting to view that day. Mr Cole said "Who are they?" I said "Mr Pitman and Dr Madina". I said I had introduced the property to them as clients. He said that the defendants were keen and they could come at the same time the next day.

He said:

> I told Mr Cole about Dr Madina and Mr Pitman. He said he had no objection to them being present. He said he knew of them.

Mr Cole says that he was never told any such thing and that he was unaware of the interest of Dr Madina and Mr Pitman, that he was unaware that they were also clients.

Having seen and heard the rival contestants on this issue, I prefer the evidence of Mr Vickers who, as I say, was an honest witness and a reliable one. I do not believe that Mr Cole was at all surprised to hear that these two other clients were interested parties and were coming to the meeting, and indeed I think he might have thought that it was such **G** a commonplace matter that he never even bothered to tell Mr Walker about it. Mr Vickers also said that he believes that he told Dr Madina and Mr Pitman about it too. I am not so sure that he did. After all, Doug Cole was an old friend of his and a fellow estate agent, or worked in an estate agent's office, and he may well have told him more about the matter, I do not know. He may well have told Mr Pitman and Dr Madina as well. Dr Madina was very vague about it. He knew there were other interested parties there but I do not think it really bothered him. Mr Pitman's attitude is coloured by some hostility to the plaintiffs, whom he regarded as having exhibited some dilatoriness in their dealings with him, unreasonably so in my view because I am quite satisfied that he never gave them any firm instructions. Indeed, there is some irony in a situation where the **H** defendants say they need not pay because the plaintiffs were acting, contrary to their interests, for Mr Pitman, and Mr Pitman criticises the plaintiffs because they were not acting for him. However that may be, I have no doubt that Mr Cole knew that when Mr Walker visited the premises on the 3rd (Mr Cole could not do so because of a previous engagement) there were going to be, or likely to be, two other clients of the plaintiffs viewing as well. I am not satisfied that Mr Walker knew that. But whether Mr Cole was a director of the defendant company, as Mr Walker says, or whether he became so later, he had sufficient standing in the company for his knowledge to be that of the company.

These, then, are the basic facts of the matter either agreed or found **J** by me on the written and oral evidence. In the circumstances the defendants resist the plaintiffs' claim on the basis that from the moment when the form was signed they were fully active as agents in and about the purchase of the property and that a duty arose there and then, under the rules and under the implied term at common law, to disclose to each signatory the full position, interest and identity of the others, and that they should not have continued to act for any of them except with full consent. They say that if Mr Walker had known of the existence and interest of Dr Madina and Mr Pitman he would not have signed the document. I do not accept that, although I do accept that in hindsight Mr Walker may have convinced himself of it. This is all to some extent speculative, but it seems to me the probabilities are that if Mr Walker had said "I am not signing this if **K** you have got two other clients", and had convinced the plaintiffs, as was indeed the fact, that he really meant business, they would have acted for him and acted for him alone, as indeed transpired. The defendants say that there was a duty to disclose and that the breach of it was so serious a matter that it went to the root of the contract so as to disentitle the plaintiffs to their fees.

Since I have found that Mr Cole knew perfectly well that Mr Vickers had two other clients viewing the property with the defendants or likely to view the property with the defendants on November 3 and knew of the existence of Mr Pitman and Dr Madina, this really disposes of the defence in this case and the plaintiffs are entitled to judgment. However, in deference to the cogent arguments **L** ably addressed to me by both counsel, I will approach the case on the hypothesis that the disclosure was not made to the defendants and consider the position which would have arisen in that contingency, in particular: 1. Were the plaintiffs in breach of an implied term in failing to make full disclosure of the existence and position of Dr Madina and Mr Pitman to the defendants? 2. If so, was that breach so grave as to disentitle them to their commission? As to (1), it is necessary to look at the relevant rules of the society, which are conveniently set out in the amended defence and counterclaim under the general heading 1(i):

> It is at all times the first duty of a member to protect and promote the legitimate and ethical interests of his client to the utmost of his ability consistent with a professional standard of behaviour towards the other parties **M** in any transaction. (ii) No member shall act for two parties in the same transaction save with the full knowledge and consent of both parties.

It seems to me that if anybody acts as is forbidden by (ii) he is also in breach of (i). No doubt (ii) is set out separately because to put oneself in the position of acting, for example, for vendor and purchaser is a temptation to which many have fallen and is an obvious example of getting into a conflict of interest situation. There has indeed been some discussion as to whether (ii) does deal with the position whereby if an agent acts for two competing parties, eg two purchasers as opposed to vendor and purchaser, he is in breach. I think that he is, and even if he was not he would be in breach of (i) if he acts for both parties in the same transaction in circumstances which might lead to a

A conflict of interests. As I say, the plaintiffs accept that a term was to be implied that the plaintiffs' actions would be governed by the rules prevailing. I have no doubt that it would be a breach of those rules for an agent to act for more than one party who had evinced a definite determination to bid for the purchase of property, and indeed the evidence for that is really not in dispute, unless with full disclosure to both parties, if there were two, they nevertheless consented to his so acting, which I have said seems to me to be normally a most unlikely event to occur. When did the duty arise to make that disclosure: before asking for Mr Walker's signature; immediately after it; and, if so, when? On this topic, which involves the construction of the rules, I have been helped by the evidence of Mr Stansfield and Mr Vickers
B and by the expert evidence of Mr Forbes. I have come to the conclusion that that duty arises as soon as a client demonstrates a real intent to buy; for example, in this case, takes up the option afforded to him in the written agreement and, in effect, instructs the agents to go ahead with the purchase. At that moment the agents must not act for any other party unless they make full disclosure of it and get permission to do so. I do not think that that stage had been reached in this case on November 3, and I do not believe that the plaintiffs were in breach of the professional rules. On a proper construction of the written document of November 2, signed by Mr Walker on November 3, the defendants, in consideration of being given details of property in which they might be interested, were bound to instruct the plaintiffs to act if indeed they decided to purchase the property
C and not in any other circumstances. The further and better particulars of the reply to the defence and counterclaim put the point aptly where it is said:

> Mr Vickers, a manager employed by the plaintiffs, telephoned a Mr Cole and informed him there was a property for sale.

I paraphrase:

> Mr Vickers informed Mr Cole: (1) that the plaintiffs would require the defendants to retain the plaintiffs in the event of the defendants deciding to purchase the property and that the plaintiffs' fees would be 2% of the amount of the purchase price plus VAT; and (2) that Mr Vickers had already contacted two other prospective clients, namely Mr Pitman and Dr Madina.

It may well be that in hindsight it would have been prudent as well
D as courteous for the plaintiffs to have disclosed to Mr Cole or to Mr Walker what the position was with regard to Mr Pitman and Dr Madina (I think that Mr Stansfield probably in his heart of hearts will accept that now), just as it would have been I think prudent and sensible to have told Mr Pitman formally that they were no longer concerned with any possible interest he might have in the premises when he spoke to them a few days after November 3, but I do not believe they were in breach of their professional duty in the circumstances. I do not think that they did act for Mr Pitman or Dr Madina within the meaning of the rules or the implied terms said to be in the contract. Mr Forbes said this:

> It would be prudent to let everybody know what is happening even if, as in this
> E situation, a conflict did not arise. A man who enters into a retaining agreement such as that set out in the agreed bundle may be or may become actively interested. There is no duty to disclose at that stage unless the client has shown active interest. Then the agent must not act for the others without consent.

By "active interest", as I say, he is evincing a real determination to instruct the agent to act for him in and about the matter. It would be strange if the plaintiffs in a situation like this in effect had to play a game of a sort of musical chairs, pirouetting, asking one possibly interested client, waiting to see how interested he was, then, finding that he was not, pass to a second, possibly to a third and fourth before finding anybody who was really interested and by then, contrary to the plaintiffs' interest and the interest of that client, the property
F might have gone elsewhere. I think that the construction I have put on the rule is a sensible one and helps to give business efficacy rather than remove it.

As for the common law position, it was pleaded as being an implied term of the agreement to give business efficacy thereto that the plaintiffs would not act for any party in the transaction in addition to the defendants or their associated companies. Counsel for the plaintiffs, Mr Marsh, said that that was much too wide, and I agree with him. For example, it does not even accept that they would be entitled to act provided there was disclosure and consent. However that may be, I do not think there is any important difference here between the term to be applied at common law and under the rules, and in any event I do not accept that the plaintiffs did act within the meaning of any term to be implied for anybody in addition to the defendants or their associated companies.

Supposing I am wrong, however, and not only, contrary to my findings, did the plaintiffs fail to give full disclosure to the defendants of the possible interest of Mr Pitman and Dr Madina but also they were in breach of the implied terms or one of them. If so, was that breach sufficiently serious to disentitle them to their fees? If they were in breach it would be because the principle of disclosure is so potentially important that it must be imported into this written agreement even at that early stage. That demonstrates the importance of the principle but not necessarily the seriousness of the breach. I entirely acquit the plaintiffs in this case of any *mala fides* whatsoever in their whole conduct of this affair and from the authorities which have been cited to me this is an important, possibly crucial, point. My attention has been drawn in particular to the case of *Andrews* v *Ramsay & Co* [1903] 2 KB 635, to *Hippisley* v *Knee Brothers* [1905] 1 KB 1, to *Keppel* v *Wheeler* [1927] 1 KB 577, to *Bowstead on Agency*, and to a recent case of *Robinson Scammell & Co* v *Ansell* reported in [1985] 7 CL 466.* The general principle which seems to me to emerge from these authorities is that an agent who has acted in good faith is normally entitled to his commission for work done. There is clearly a world of difference from the situation that obtained in *Andrews* v *Ramsay*, where an agent took a secret profit, and this case, or from a situation where an agent otherwise sets out to act against his principal's interest, although circumstances I think may obtain where a breach made even in good faith may be so grave as to go to the root of the contract and disentitle the agent to his fees. The facts in this case do not, in my judgment, come into that category and in my view the plaintiffs are entitled to their commission.

An alternative claim that the defendants were entitled to damages for such breaches really was abandoned in this case when the counterclaim was abandoned. Mr Haines submitted at one stage that he was entitled to a diminution, to pay an introductory fee only of 1% perhaps, by reason of the breaches of the implied term, but he was constrained finally to concede that he could not succeed in that context.

Accordingly there must be judgment for the plaintiffs for the amount claimed.

Judgment was given for the plaintiffs for £8,536.62, including interest, and for costs. The defendants' counterclaim for damages had been abandoned.

---

*Editor's note: This case is fully reported at [1985] 2 EGLR 41; (1985) 275 EG 369.

Court of Appeal

November 18 1985

(Before Sir John ARNOLD P and Lord Justice PURCHAS)

WITHEY ROBINSON (A FIRM) v EDWARDS

Estates Gazette February 22 1986

277 EG 748-749

Estate agency — Professional charges — Instructions by client intending to purchase a night-club to firm of chartered surveyors "to provide valuations orally of the premises and business, to prepare an inventory of fixtures and fittings, take details of trading figures and generally to assist" the client in negotiations with the vendor — The purchase was effected and the plaintiff agents sent their bill to the defendant client based on former RICS Scale 17(b) (a basis which had previously been described to the defendant) for the sum of £5,980 — Defendant denied that the employment was subject to any terms of remuneration save for an implied term that the plaintiffs should receive a reasonable remuneration for work done — After some earlier abortive proceedings the dispute was referred to the county court — County court judge found that there was no express or implied term that the remuneration should be on the RICS scale, but held that the plaintiffs were entitled to be paid on a *quantum meruit* basis — On this basis he calculated that the plaintiffs were entitled to £80 per hour for 45 hours, which, after adding VAT, came

to £4,140 — On appeal to the Court of Appeal the defendant argued that the agents, having consistently claimed on the basis of the RICS scale, could not rely on a *quantum meruit* ground (although he himself considered it to be the correct method) — Held that the county court judge was perfectly entitled to take the course he did — Both contentions were before him and were argued, although the plaintiffs had not themselves put forward the *quantum meruit* basis as an alternative — Appeal dismissed

This was an appeal by Christopher Edwards from a decision of Judge Gosnay, at Bradford County Court, in an action by Withey Robinson, a firm of chartered surveyors, for professional fees alleged to be due from Mr Edwards for services in connection with his purchase of a night-club.

S Spencer (instructed by Edwards & Lockwood, of Guisley, Leeds) appeared on behalf of the appellant; Miss A Woolley (instructed by Sugden & Spencer, of Bradford) represented the respondents.

Giving judgment, SIR JOHN ARNOLD P said: In the early summer of 1983 Mr Edwards, the defendant, and the appellant in this case, was minded to purchase a night-club which he had seen advertised for sale. He had, at the stage at which the plaintiffs came on the scene, already entered into some discussions with the vendor of this night-club, a Mr Abrahams. The defendant dealt, so far as the plaintiffs are concerned, with a man, Paul Withey, who was one of the partners. It was eventually agreed between the defendant and Mr Withey that the plaintiff firm — a firm of chartered surveyors — should be employed by the defendant to do certain work in connection with the proposed purchase. The work is described in the defendant's defence as being:

to provide valuations orally of the premises and business, to prepare an inventory of fixtures and fittings, take details of trading figures and generally to assist the defendant at negotiations conducted by the defendant with the vendors.

No one has suggested that that is other than an accurate description of what the defendant employed the plaintiffs to do.

In the event the purchase went through. On December 23 1983 the plaintiffs delivered to the defendant a bill in the sum of £5,980 for professional charges for the various operations set out in that bill. The plaintiffs had earlier on — on October 17 1983 — written a letter to the defendant, arising out of a telephone conversation which they had had together, in which the plaintiffs described the basis of their charges. That basis was revealed as being a scale numbered 17(b), which is among the Royal Institution of Chartered Surveyors' scales of charges, a scale which is said in that letter to be applicable for negotiating the purchase of a particular property on clients' instructions, freehold or leasehold.*

It is apparent from the coincidence of the figures in the letter of October 17 and the invoice of December 23 that the two are the same and, therefore, so far as relevant to this appeal, that the invoice dated December 23 1983 is precisely based on scale 17(b) of the scales of charges of the Royal Institution of Chartered Surveyors. So when the bill was not paid and the time came at which the plaintiffs were minded to sue the defendant, a specially endorsed statement of claim revealed exactly the same figure of £5,980 and referred back referentially to the invoice of December 23 1983, so that the figure in the specially endorsed statement of claim is a figure which, by the means I have described, can be seen to be one based on the figures in the scale referred to.

The actual wording of the statement of claim is:

The plaintiff's claim is for the sum of £5,980 being the cost of work done on behalf of the defendant and to his order,

and then for the make-up of the sum of £5,980 the particulars refer the reader back to the invoice of December 23 1983. The matter being then pleaded in the High Court, on March 23 1984 the defendant filed a defence in which he denied that he was indebted to the plaintiffs in the sum claimed of £5,980. He then described the circumstances in which the work was done and the employment formed and the pleading continues thus:

It is denied that the employment was subject to any terms as to remuneration save for an implied term that the plaintiff firm should receive a reasonable payment for work done. The sum claimed is unreasonable and exorbitant. The defendant says that no more than 20 hours of the plaintiff firm's time can have been spent in the employment alleged herein; the hourly rate of pay thereby produced of £250 is manifestly excessive. G

"Thereby produced" means by dividing the number of hours into the sum claimed by the plaintiffs in the action.

The pleading continues:

The defendant denies the allegation if it be made, that the plaintiff firm was employed by him upon terms that it would be entitled to a commission on the purchase price or a remuneration being a percentage of the purchase price calculated in accordance with the RICS (or any other) scale.

No reply was filed, and therefore the matters alleged in the defence remained at issue. After an abortive attempt to obtain summary judgment under Order 14 by the plaintiffs, the case was remitted for trial to the Bradford County Court, and in due course the Bradford H
County Court heard the case and gave judgment.

The learned judge from whose decision this appeal comes, after lightly sketching in the history of the case, concluded first of all that there was no express or implied term that the remuneration of the plaintiff firm should be on the RICS scale. He then proceeded thus:

I turn to the question whether the plaintiff was entitled to a *quantum meruit*. Counsel for the defendant says that on the pleadings the plaintiff is not entitled to rely on a *quantum meruit* although the defence does say that it is the appropriate method. The defendant says that either the plaintiff succeeds in showing that the RICS scale is the appropriate one or he fails completely. I think it is open to the plaintiff on the statement of claim to claim a *quantum meruit*. The statement of claim says that it is for the cost of work done for the defendant and to his order. That is really accepted. The method of calculation J
of the amount in the statement of claim may be wrong, but in view of the wording in the defence and other matters I find that it is open to the plaintiff to claim a *quantum meruit*. It has not been suggested by the defendant that the work done in this case was of no value to him or useless . . . I find that the plaintiff was entitled to *quantum meruit* for his services.

He then considers the evidence which had been given on the one side and the other and he came to the conclusion that, on a *quantum meruit* basis, the proper charge was at a rate of £80 an hour, and he then held that there was effectively 45 hours' work carried out by the plaintiffs in the course of the employment. By multiplying the two figures of 45 hours by £80 an hour he comes out, having added value added tax, at an amount of £4,140, which was the amount of his award against which this appeal is brought. K

The case which is put forward on the appeal by the defendant is very much the same case which was put forward to the learned judge in accordance with the account of the matter which I have read from the learned judge's judgment. The substance of the case is that the plaintiffs, having tied themselves — as they plainly did as a matter of pleading — to a justification of the figure which they put forward as the appropriate amount that they ought to be paid for what they described as "the cost of work done on behalf of the defendant and to his order", to the RICS scale, they did not put forward as an alternative basis of quantification any *quantum meruit* basis. That was introduced into the case by the defendant's defence and so, says the defendant, the learned judge was not entitled to make an award based on that alternative quantification of the defendant. L

It is plain that at the stage at which the pleadings closed, and at the stage at which the matter came on before the learned judge, the plaintiffs were saying that they were entitled to be remunerated under the contract and the defendant was not disputing that the plaintiffs were entitled to some remuneration under the contract, nor did that position change throughout the trial; but the plaintiffs said throughout that they were entitled to be remunerated on the basis of an express or implied term in the contract; that they should be entitled to recover the appropriate amount designated by the Royal Institution of Chartered Surveyors' scale. The defendant was saying that the plaintiffs' remuneration was limited to that which was appropriate to be computed on a *quantum meruit* basis. The learned judge on that point of issue decided in favour of the defendant, and M
then quantified the amount which derived as a sum of money from the acceptance of that approach, and gave judgment in that sum.

There seems to be no doubt, to my way of thinking, that both contentions were available for argument before the learned judge, and both were argued, the defendant's contention successfully and the plaintiffs' unsuccessfully. So there emerged at the end of the day what, in the learned judge's judgment, was the appropriate basis upon which the amount claimable by the plaintiffs under the contract ought to be computed. I cannot, for my part, see why that was not a perfectly legitimate course of dealing, notwithstanding that it involved deciding the quantification in a sense and on a basis opposed by the plaintiffs and advocated by the defendant, and without the

* Editor's note: Most of the scales of charges of the Royal Institution of Chartered Surveyors concerned with valuation, sales, lettings and management of property, including scale 17, were abolished by the institution in 1982.

A plaintiffs ever having adopted by way of amendment of their pleadings as an alternative view the basis that was put forward by the defendant. Both points were, as a matter of pleading, before the learned judge and he took his pick in favour of the defendant's quantification basis.

I cannot see that there was anything wrong in that course and I would accordingly dismiss this appeal.

PURCHAS LJ agreed and did not add anything.

The appeal was dismissed with costs.

B Queen's Bench Division

March 21 1986

(Before His Honour Judge TIBBER, sitting as a judge of the High Court)

BRIAN COOPER & CO v FAIRVIEW ESTATES (INVESTMENTS) LTD

Estates Gazette June 7 1986

278 EG 1094-1104

Estate agents' commission — Entitlement to commission for C letting a property — Terms accepted were "We . . . offer a full-scale letting fee to your company should you introduce a tenant by whom you are unable to be retained, and with whom we have not been in previous communication and who subsequently completes a lease" — The facts were that the plaintiff agents were the first to introduce the company which eventually completed the lease to the defendant lessors, but there was a period during which the company for various reasons ceased to be looking for a property — During this period contact between the plaintiffs and the company was interrupted — When the company's interest in finding a D property revived, an introduction to the property was made by a different firm of agents and through a different officer of the company — The judge found as a fact that the effective cause of the letting was the introduction by the second firm of agents — This finding, however, did not settle the question of the entitlement to commission — The effective cause in this case was irrelevant owing to the terms of the plaintiffs' employment — The words "should you introduce a tenant who subsequently completes a lease" were the very negation of causation — The terms contemplated that commission would be earned by the first agent who introduced somebody who subsequently completed the lease — The formula was E evidently framed precisely to obviate arguments as to who was the effective cause of the transaction — Plaintiffs satisfied all the terms — They introduced a tenant by whom they were not retained, with whom the defendants had not been in previous communication, and who subsequently completed the lease — Judgment for plaintiffs

In this case the plaintiffs, Brian Cooper & Co, estate agents, sued the defendants, Fairview Estates (Investments) Ltd, for commission claimed to have been earned by the introduction of a company, Metier Management Systems Ltd, as the lessees of an office building, Fairview House, at Station Road, Hayes, Middlesex.

F David Eady QC and Stephen Suttle (instructed by H Davis & Co) appeared on behalf of the plaintiffs; Miss Elizabeth Gloster (instructed by Lovell, White & King) represented the defendants.

Giving judgment, JUDGE TIBBER said: Once again, the court is being asked to construe a contract by which somebody retains an agent to let a property for them. One would have thought that the number of variations in this type of contract were limited and had come to an end, but it is plainly not so and, although the courts have since the early parts of this century busily spent time construing such contracts, they continue so to do.

In this case there are three agents involved. There is Mr Victor Spencer of Brian Cooper & Co, who is the plaintiff. Without doubt, he was the first agent on the scene. There was subsequently an agent called Phillips Roth & Co, and somewhere, somewhat mysteriously, Druce & Co appear and disappear. I can say from the outset that although I am a little mystified by the role which Druce & Co played in the transaction which I have to consider, it is not a matter which I think has any great relevance to the points on which I have to make a decision.

It is Mr Spencer of Brian Cooper & Co who is making a claim for commission in this case, and the first thing I must do is to review the course of the negotiations and the facts.

It all started in April 1982. I say in parentheses, as in all these cases where one is investigating a course of conduct between parties, people are asked to remember the contents of telephone conversations three and four years ago. Life is very difficult for them when they are asked to do that. They do their best to remember, but the sort of people who are involved in this case are constantly, as I understand it, on the telephone and it is difficult for them to remember accurately what took place. It is so easy to transpose back into 1982 ideas and notions which have in fact only arisen since then. They look back and say what they must have done and what they did do. It becomes very difficult to fix exactly what was said as long ago as that. There are documents, of course, and they help.

In March 1982 Metier wished to acquire new premises. In the person of Mr Clark, they approached Mr Spencer of Brian Cooper to ask for his assistance in finding accommodation. Mr Clark — and I do not mean any disrespect to him — was a fairly lowly figure in the Metier organisation. He was the publications manager. He was not a director. He formed no part of what I have learned in this company to call the "executive", because it is an American company; it is what I think we would call here the "board". He had apparently been deputed to do something about finding some accommodation and he approached Mr Spencer whom he had known for a number of years.

On March 18 1982 Mr Spencer wrote to Mr Clark and sent him some information about available buildings and there the matter rested for a very short time. On March 22 1982 the defendants came on the scene. They are, as I understand it, a company which builds and then lets or sells property. I think they let residential as well as industrial property, but we are only concerned here with industrial or commercial property. The defendants were in the process of building, at Station Road, Hayes, an office building, and they sent to some 300 agents in the West and West End of London the circular letter which appears on p 4 of the agreed bundle. They enclosed brochures and they set out in the letter the terms on which they would pay commission to an agent who found a customer for them:

> We . . . offer a full-scale letting fee to your company should you introduce a tenant by whom you are unable to be retained, and with whom we have not been in previous communication and who subsequently completes a lease.

I shall have to come back to that letter at a later stage.

Shortly after receipt of that letter, Mr Spencer discussed the building at Hayes with Mr McCulloch, who for the purposes of this case is Fairview; he is the senior development surveyor there. On April 15 Mr Spencer wrote a letter to Mr McCulloch. He referred to their "several telephone conversations" and used the letter to confirm "my introduction to you of Metier". He said, as he had arranged apparently with Mr Ross or Mr Clark of Metier:

> The Company hope to arrange for the building to be inspected . . . and I will . . . arrange a suitably convenient time. In the meantime I look forward to receiving floor plans that have been requested by the Company.

Then, again, comes one of the contract terms:

> . . . I am not retained by Metier and in the event that the Company do complete a lease on the building I would wish to take advantage of your offer of full-scale commission.

On the same day he wrote to Mr Ross of Metier, pushing, so to speak, the building and setting out certain relevant facts about it, among others the rent asked, which he puts at £10 per sq ft. In fact, I think £10.50 was being asked. He says that the developers "are certainly prepared to negotiate" and describes the basis on which such negotiations could be done. Mr McCulloch says that at that stage the developers certainly were not prepared to negotiate. They only became prepared to negotiate rather later when it was apparent that the market was, as he put it, "thin". But that is what Mr Spencer put in his letter, and I dare say he thought that once he could get the parties talking something could be done.

The floor plans were duly obtained and there was a meeting on April 23 at the site, the building being at this stage incomplete. Present at that meeting were Mr McCulloch, Mr Spencer, Mr Clark and Mr Ross. Be it noted again that neither Clark nor Ross had any executive role in Metier; they were fairly low down in the scale. I think it is right to say at this stage that at no stage did they report either the existence of Fairview House or its suitability to any of the directors of the company. It is put in this way, that there was never any high-level interest shown by Metier in Fairview House until early in 1983. By "high-level interest", what I mean is that no director became involved or even knew about it until that stage.

Contact with Mr Clark and Mr Ross about Fairview House was maintained by Mr Spencer. He obtained, for example, the rating information that they asked for.

By August 1982 or thereabouts — and I use a phrase which was used by the witness — hatches were battened down at Metier. Metier is apparently a company which goes through periodic cash-flow crises because of the nature of its product, which is something to do with computers, the market for which is apparently volatile. They were going through a period when it began to seem as though it was not right for them to be interested in relocating themselves.

From about June 1982 there were between two and four conversations between Mr Spencer and Mr Ross concerning Fairview House. I am not sure who initiated them. I think they were mainly initiated by Mr Spencer, who wanted Mr Ross to come to some sort of view about Fairview House and pass it on to the directors.

On October 11 Mr Spencer wrote for the first time to Mr Hood. Mr Hood was in a rather different position from Mr Clark and Mr Ross. He was the financial director of Metier. Apparently, what had happened was that at some stage Mr Spencer had had a conversation with Mr Ross, who had told him that Metier were still not suited in their search for office accommodation and, realising — as of course he would, being an experienced and astute estate agent — that Mr Ross and Mr Clark were really too low down in the hierarchy to be dealt with satisfactorily, Mr Spencer finally managed to get in touch with Mr Hood and wrote the letter which is at p 40 of the bundle. That is not a letter which relates to Fairview House; indeed, it is pushing an entirely different property. The letter shows his contact with Metier and it shows that Mr Spencer realised, or thought he realised, that Metier were still looking for properties and if he could not satisfy them with Fairview House he would try to satisfy them with another property.

There matters rested until December 21 1982. What happened was this: on December 2 1982, or just before, Mr McCulloch had telephoned Mr Spencer. He not only telephoned Mr Spencer. He says he telephoned a number of other agents who had also shown an interest in Fairview House. He telephoned those agents, as I understand it, because they had shown an interest in Fairview House and because they had not attended some sort of cocktail party that Fairview had given at the premises in order to try to whip up some interest in them, because they were sticking: they were not being let. I think that the building was very much nearer completion at that stage, although it had not been completed. Mr Spencer had not been able to get to that party, so Mr McCulloch telephoned him. The effect of the conversation was "You might like to get back to Metier and see whether they are still interested in this property." Further, he was offering now not scale commission, but double scale commission.

And so on December 2 Mr Spencer wrote to Mr McCulloch confirming that conversation. That conversation is relied on by the plaintiff as showing that Fairview recognised that he was the contact man so far as Metier were concerned. On the same day, December 2, Mr Spencer wrote to Mr Hood enclosing the new brochure and pointing out to Mr Hood that Fairview House had been inspected by Mr Ross and Mr Clark "earlier this summer during its construction". In fact, it was April; it was not really early summer. He pointed out that the building had now been completed:

> I have been requested by the developers to enquire if it would be of interest as the terms . . . have changed more favourably . . .

The new terms are set out, and on the second page he says:

> I know that you and your co-directors have not seen the building and would genuinely suggest that you carry out an inspection . . . I look forward to hearing from you and in any event would be pleased to know what is actually happening with your office moving, as it is many months since we have been in contact.

He says that that "many months" really refers to contact with Hood and not with the company, because he had had some telephone conversations with either Mr Ross or Mr Clark during the summer. I accept that to be the position.

Then Mr Spencer received from Mr Hood the letter which seemed to shut the matter down completely:

> For a variety of operational reasons, we are not now contemplating a relocation in the near term, so in consequence we have no interest in any properties. As and when the situation changes, you are assured we will contact you and in the meantime we thank you for your interest in us. G

After December 21 little or nothing was done so far as contact between Mr Spencer and Fairview was concerned until in March 1983 Mr Spencer received another letter from Fairview. Again, I think I am right in saying that it was a circular letter, but this particular one was "personalised" in the sense that it was sent to Mr V Spencer and I think a number of similar letters were sent to other agents. Mr Spencer had shown direct interest in Fairview House, and Fairview sent a new brochure, pointed out that the offices were now ready for occupation and put in this sentence:

> I confirm that, should your Company introduce a tenant by whom you are unable to be retained and with whom we have not been in previous communication, then, should your applicants enter into a lease, we are pleased to pay a double scale letting commission. H

There is nothing really very new about that, although I shall have to consider that at a later stage as well.

On receipt of that letter, Mr Spencer rang Metier, and after what I think were two more abortive attempts he finally succeeded in speaking to Mr Hood, he said, and his account of the conversation was this:

> I said that Mr Clark suggested I should telephone.

When I spoke of "abortive attempts", I meant that Mr Spencer had telephoned Metier and had been able to speak only to Mr Clark or Mr Ross; that was not satisfactory. J

> Mr Hood — or Mr Clark, I am not sure which — said they were looking at a property in Uxbridge, Harefield Place. We discussed it and I

Mr Spencer

> said it was not suitable for you. And he agreed. I then mentioned Fairview House, and he said he had the building in mind but it was a question of location because their present building was in Stonebridge Park and they did not want to move far because of the problems of recruiting staff.

Mr Hood cannot recall that conversation. He is not prepared to say it did not take place. What he said, somewhat acidly, was that he cannot recall conversations which took place on the telephone three years ago. What he does say is: K

> I would not have said at that stage that I had the property in mind. In May 1983 I was not conscious of Station Road as a property the company was interested in. It may have been in my mind, as were other properties. I do not recall a conversation with Mr Spencer. If he says he spoke, I cannot argue.

That is when he made the remark about not recalling phone conversations three years ago.

> But I cannot think that in May 1983 I would have said we had any specific property in mind.

It is put that if Mr Hood had said that, Mr Spencer, whom I have already categorised as an experienced and astute estate agent, would have almost certainly followed up that telephone conversation with a letter to Mr Hood. No such letter exists. Nevertheless, I think the conversation did take place. I think Fairview House was probably mentioned. I do not accept that Mr Hood said he had Fairview House in mind, because I do not think from his account of the matter that he could possibly have done so. There had been no report by Mr Ross and Mr Clark to Mr Hood. There apparently was some system of filing: references were made to a property file. But then some things went into a proper file, some things went into a different file, which was apparently the waste-paper basket. He may have heard the name, but he certainly did not have it in mind, in my view, as a property that Metier might well end up taking. L

That conversation seems to be the end of Mr Spencer's contact with Metier. He says that he had several conversations with Mr McCulloch of Fairview during that summer, in the course of which he said that Metier were still not suited and he was convinced they would end up in Fairview House, as he always had been. He says that he had recognised from the beginning that that was the place that was ideally suited to them and, though they might not know it, he did, as an experienced estate agent who knew what they were looking for and knew what Fairview House was. That, after all, is his occupation — matching up people to what they need. M

Mr McCulloch does not accept that these conversations took place and what Mr Spencer says is:

> Well, I did not actually ring him about Fairview House. What happened was that we were discussing rent reviews of some of Fairview's properties which were coming up in that area, and he asked me my view about rents in the area.

A In the course of those conversations I would raise this point about Fairview House and Metier not being suited, what was happening, or something of that sort.

I think again those conversations probably did take place: I am not sure how many. I am equally sure that the mention of Fairview House was almost casual. It had to be really. What could Mr McCulloch do? He was not the person to push, as Mr Spencer well knew. The conversations were primarily directed to something else; that is to say, the question of rent reviews or matters of that sort. Mr Spencer says Fairview House was mentioned, that Metier were not suited and so on. I think such conversations probably took place. I do not suppose they made much impact on Mr McCulloch because, after all, Mr Spencer was not reporting anything that was of any interest to Mr
B McCulloch. He was not reporting interest being shown by Metier; he was merely saying that they were not yet suited.

Metier's financial position improved and their interest in finding some new property revived in the spring of 1983; this time, however, in the person of Mr Gunn. Mr Gunn was the finance and administration manager of Metier. He said that at that stage he had never heard of Fairview House and did not in fact hear of Fairview House until he was introduced to it by Mr Roth, whom he retained in May 1983 to help the company find a suitable property.

That amazes Mr Spencer, but I accept that it is so. This company, Metier, seems to be very compartmentalised. The matter had never got beyond Mr Ross and Mr Clark in the earlier stages of the
C negotiation except in that conversation with Mr Hood to which I have already referred. As I have said, I do not think that Mr Hood had really heard of Fairview in the sense of considering it as a property to which the company might eventually move. I accept that Mr Gunn had not heard of Fairview House until he was introduced to it by Mr Roth.

In August 1983 a Mr Lodge comes on to the scene. I am not sure what his position is technically, but he is known as "the Boss". He is the head, I think, of the American side of Metier. On August 8 he sent to Mr Gunn, with copies to other people, a memorandum which was received on August 12:

> Thank you for forwarding to me the background information on the search for new London accommodation. In order that you are able to get as quickly
D as possible to the point where you submit a proposal, I would suggest the following.
> 1. . . . review Dick's memorandum of the 6th of April 1982 . . .

That was a memorandum prepared early on in Metier's search when certain criteria had been suggested as to what Metier were looking for.

> 2. Review in detail what their requirements will be for other than staff accommodation . . .
> 3. Review on what basis you should calculate space requirements . . .

and so on

> I suggest that you summarise that into a set of requirements which can be provided to our agents, requesting that they submit as rapidly as possible a list
E of all available properties which fit or nearly fit our specifications . . .
> As it will be mid August by the time you receive this letter, and it is my understanding our lease on Metier House expires at the end of December, you should treat the search for alternative accommodation as urgent

and I expect you to do this, that and the other by the end of August. The requirements are all set out. It is a rocket from the American side saying "Get on with finding other accommodation." As a result of that, and as a result of what Mr Roth had told them, Metier, on September 7, went to inspect the premises. They went in the person of Mr Gunn. They went and met Mr McCulloch and they also met Mr Roth and Allan Grossman of Druce's, all on site. Apparently at some stage during the course of that site meeting Mr McCulloch said to Mr Gunn "Oh, you are Metier? Your company has inspected this site
F before." "Has it?" said Mr Gunn, "I did not know." He was not terribly surprised because he knew they had been looking for properties for some time. It certainly did not register with him as a surprise, nor did it register with him as something that he had known.

Very shortly after that inspection and after the letter of September 15 from Phillips Roth to Mr Gunn, there was an executive management meeting of Metier in Houston, Texas, where the London accommodation was discussed.

Mr Spencer was away on holiday and when he returned he had a telephone conversation with Mr McCulloch and was told that Fairview House had been let. I should say that that conversation took place because Mr McCulloch had phoned him for the purpose of informing him that a deal had been done on Fairview House. Mr Spencer's reaction was that he was delighted and would be sending his account for commission. Mr McCulloch said that he had accepted another agent's introduction. It was agreed they would both check their files and get back to each other, which they did. But they did not resolve the question, and that is why we are here today.

Some time in October 1983 Metier signed a lease on Fairview House, and it is said that it is of some significance that Mr McCulloch told the court that he phoned Mr Spencer as a matter of courtesy to let him know the deal was finalised and that he phoned no other agent, although Fairview House had been very widely advertised. He clearly recognised that Mr Spencer certainly had been involved over Fairview House.

On those facts I have to ask the question: what was the effective cause of the letting? Not because at this stage I have come to the conclusion that effective cause is a material question in the course of this contract, but because it is convenient to deal with that question while the facts are before me. If effective cause of the letting is a relevant consideration in this case, I have come to the conclusion that that effective cause was not Mr Spencer but was probably Mr Roth.

Mr Spencer, so far as Metier were concerned, had virtually dropped out of the picture after December 1982. There was a fresh approach and activity by Mr Gunn through Mr Roth from May 1983. Mr Gunn had never heard of the approach of Metier to this building earlier on. True it is that Metier are a corporate entity, and as a corporate entity I suppose Metier were aware of Mr Spencer's introduction, but as a corporate entity I am equally satisfied that Metier had forgotten all about it and that Mr Spencer's introduction did not operate on their mind in coming to sign a lease. Apart from some conversation to which I have already referred with Mr Mr McCulloch in summer of 1983 to say that Metier were not suited, there was no involvement by Mr Spencer since the conversation with Mr Hood after the March re-advertisement. The inspection by Mr Gunn of the completed building was an inspection on the instigation of Mr Roth and the documentary and oral evidence support Mr Roth's detailed negotiations as to the terms. I do not think I need to go into that.

I do not think either that there is any need to try to define the term "effective cause". There is a footnote in *Bowstead on Agency* where the *American Restatement* definition is printed*, but I do not think it is a great deal of help. It is a circular definition in my view. I think the words themselves are clear. In the light of my finding that really Mr Spencer had dropped out of the picture by the time Mr Roth came on the scene, I do not think it can be said he was the effective cause of the sale.

The question I have to decide is this: is the "effective cause" term relevant to this contract? The contract which has to be construed is set out at p 4 of the bundle:

> We confirm that we are pleased to offer a full scale letting fee . . .

It is not, in my view, the contract at p 55 of the bundle, which is in rather different terms. The contract at p 4 was the original contract. Double commission had been agreed between Mr Spencer and Mr McCulloch before we get to March 2 1983, when the second contract came into existence; it had been agreed some time in December 1982. P 55 is a document, so to speak, unilaterally sent by Mr McCulloch to Mr Spencer. It was not the result of any agreement. Mr Spencer had already done work on the basis of p 4 of March 22 1982, and it was not open to Fairview, in these circumstances, unilaterally to impose new terms. They could only be agreed between the parties, as the double commission term had been agreed. In those circumstances, in my view, the question "Was Metier your applicant?", which is the phrase used in the later document of March 2 1983, simply does not arise.

On the authorities, especially the well-known *Luxor* case *[Luxor (Eastbourne) Ltd* v *Cooper* [1941] AC 108], the position appears to be that if the terms of the contract are clear, you cannot introduce such implied terms as "the effective cause". I have been referred to many authorities where the test of effective cause has been imported. I hope I shall not be thought discourteous if I do not review them in the course of this judgment. I think the effect of them is set out in *Bowstead on Agency*, 15th ed, at p 227, at the beginning of Article 59:

> Subject to any special terms in the contract of agency, where the remuneration of an agent is a commission on a transaction to be brought about, he is not entitled to such commission unless his services were the effective cause of the transaction being brought about.

**Bowstead on Agency*, 15th ed., p229, footnote 24: Restatement, Second, Agency (American Law Institute) 1958.

The words at the beginning of the paragraph — "subject to any special terms in the contract of agency" — seem to me to mean no more than "subject to the terms of the contract". That is what the authorities appear to bear out. Generally, the authorities which were cited to me relate to cases where the person who ultimately purchases has been introduced by the agent, and the question is: was the purchase consequent upon the introduction of that person or was it merely subsequent to it? If that is the question to be asked, what it becomes is: what is the effective cause?

In the contract which I have to construe, the word "subsequently" has been expressly used: "should you introduce a tenant who subsequently completes a lease . . . ". The plaintiff's case is that that is really the negation of causation; in other words, what this contract is saying expressly or by necessary implication from the use of the word "subsequently" is: we are not interested here in effective cause; we are interested in the first agent to come along who introduces somebody who subsequently completes a lease.

Mr John M Phillips [senior partner, Henry Butcher & Co] was called by the defendants as an expert. He gave some significant evidence about that, although I think one always has to treat the evidence of an expert cautiously when what he is doing is construing a document and that is the very matter which the court has to do. He described this formulation as being popular and he said that the C reason that it is popular is that it makes it clear that the person entitled to the commission is the one who first introduces the party who finally concludes the lease. He said "It makes for certainty." I, too, think it does.

It looks as though that formulation has been introduced precisely to get rid of arguments about who was the effective cause of the lease. It seems to me that Mr Spencer has satisfied the requirements of the contract. He has introduced a tenant by whom he was unable to be retained (and we know that that means no more than that he was not retained) "with whom we have not been in previous communication" — and Fairview had not been in previous communication with Metier at the time of the introduction by Mr D Spencer. Metier subsequently completed a lease.

Nothing, in my view, could be clearer. I see no necessity, either in terms of business efficacy or otherwise, to import an obligation that Mr Spencer, in order to be entitled to his commission, must be the effective cause of the signing of the lease. As I have already said, I do not think he was, but he satisfies the requirements of the contract as set out in the letter of March 22 1982 and there must be judgment for him accordingly.

For a further case on this subject see p 238

LANDLORD AND TENANT

GENERAL

Court of Appeal

November 20 1985

(Before Lord Justice FOX, Lord Justice MUSTILL and Lord Justice BALCOMBE)

INTERNATIONAL DRILLING FLUIDS LTD v LOUISVILLE INVESTMENTS (UXBRIDGE) LTD

Estates Gazette January 11 1986

277 EG 62-66

Landlord and tenant — Assignment of lease — Landlords' appeal against decision of deputy High Court judge that their refusal of consent to assignment was unreasonable — Lease was for 30 years of a two-storey office building — Proposed assignees intended to use building for provision of "serviced office accommodation" — This was provision of short-term office accommodation in small, fully furnished office suites with services of a receptionist, telephonist and typist and facilities such as word-processing, photocopying, telex and telephone equipment — Landlords' main objection to proposed assignment was that the intended use would adversely affect the value of their reversionary interest — Subsidiary objections were as to the viability of the proposed enterprise and the effects on car-parking facilities — The judge had held that, in the circumstances, there was no ground for a reasonable apprehension of damage to the landlords' interest and he had dismissed the other objections urged by the landlords — In the judge's view there was, on the evidence, no possibility that the use of the building for serviced offices would have a depreciating effect on the letting value of the property at the end of the lease and there was no significant danger that the rent would not be paid throughout the term — Held, after a review of the authorities, that on the facts the judge was entitled to come to the conclusion that he did on the main issue as to damage to the reversion and that he had reached the right decision — He was also justified in rejecting the landlords' other objections — The present case was one of those where it would be unreasonable for the landlord not to consider the detriment to the tenant of the refusal of consent, where the detriment is extreme and disproportionate to the benefit to the landlord — Decision in *Premier Confectionery (London) Co Ltd* v *London Commercial Sale Rooms Ltd* doubted — Appeal dismissed

This was an appeal by the landlord, Louisville Investments (Uxbridge) Ltd, from a decision of Mr Edward Nugee QC, sitting as a deputy judge of the Chancery Division, in favour of the tenant, International Drilling Fluids Ltd, which had sought a declaration that the landlord had unreasonably refused a licence to assign a lease of Colne House, Oxford Road, Uxbridge, to Euro Business Services Ltd. The decision of Mr Edward Nugee QC was reported at [1985] 2 EGLR 74; (1985) 275 EG 802.

Kim Lewison (instructed by Helder Roberts & Co, of Epsom) appeared on behalf of the appellant; Paul Morgan (instructed by Nabarro Nathanson) represented the respondent tenant.

Giving the first judgment at the invitation of Fox LJ, BALCOMBE LJ said: This is an appeal from an order dated June 18 1985 of Mr Edward Nugee QC, sitting as a deputy High Court judge, whereby he F declared that the refusal of the appellant landlord to grant to the respondent ("the tenant") a licence to assign a lease to Euro Business Services Ltd ("Euro") was unreasonable. The lease was made on January 31 1972 between the landlord and Bovis New Homes Southern Ltd ("Bovis"). The property comprised in the lease is a two-storey office building known as Colne House, Highbridge Industrial Estate, Uxbridge. Highbridge Industrial Estate is owned by the landlord.

The term of the lease is for 30 years from December 25 1971, so it now has some 16 years to run. The initial rent was £12,500 per annum, but there is a rent review clause, exercisable only upwards at G five-yearly intervals, and the current rent is £46,000 per annum. The user clause in the lease prohibits the use of the property "for any purpose other than as offices within the meaning of Class II of the Town and Country Planning (Use Classes) Order 1963 with ancillary showrooms . . . ". The relevant part of the lessee's covenant against assignment is in the following terms:

> . . . not at any time during the term hereby granted to assign the whole of the demised premises without the licence in writing of the lessor being previously obtained such licence not to be unreasonably withheld.

The lease was assigned by Bovis to the tenant by an assignment dated March 1 1978. By a deed dated February 28 1978 the landlord granted Bovis licence to assign the lease to the tenant, and the tenant H entered into a direct covenant with the landlord for payment of the rent reserved by the lease. So the landlord now has the benefit of two direct covenants for payment of the rent — that of Bovis and that of the tenant. The tenant is a subsidiary of a well-known public company, and its accounts for the year ended September 30 1983 showed an annual profit (after tax) of £470,000 on a turnover of almost £15m, with fixed assets of £2.16m and net current assets of £963,000.

The tenant occupied Colne House as a single office block for its own use. In July 1983 the tenant instructed Leslie Lintott & Associates ("Lintott"), surveyors and valuers, to try to find someone who would take an assignment of the lease of Colne House, without any premium, since the tenant wished to move to new J premises. The learned judge said he was satisfied that Lintotts had taken the appropriate steps to market the property, but nevertheless, until shortly before the trial of the action, only one serious inquiry had been received. That was from a Mr Brodie and a Mr Gluck, who wished to use the building for the provision of serviced office accommodation. This is a form of business which has developed in recent years to meet the demand for office facilities which can be used on a short-term basis. There are a growing number of businesses which do not wish to take permanent office accommodation but want to have the temporary use of offices which are fully furnished and which provide the services of a receptionist, telephonist and typist, and such facilities as word-processing, photocopying and telex and telephone equipment. K

It was common ground that, if the lease were assigned to Mr Brodie and Mr Gluck, or to a company formed by them, the proposed use of the property would not be in breach of any of the provisions of the lease. Although the landlord's witnesses were not optimistic about the viability of the proposed business, the learned judge held that Mr Brodie had carried out sufficient research into this field of business to entitle him to form the view that it could be successful at Colne House.

Mr Brodie and Mr Gluck first expressed interest in December 1983. An initial application by the tenant for a licence to assign the lease to a particular company owned by them was abandoned, but on August 2 1984 the tenant applied to the landlord for licence to assign the lease

A to Euro, a company owned by Mr Brodie and Mr Gluck, and for whose obligations they were prepared to stand as guarantors. In the meantime, in May 1984, the tenant had vacated Colne House which, at the date of trial, remained empty. Of course the tenant remained, and remains, liable on all its obligations under the lease.

On August 28 1984 the landlord's solicitors wrote saying that their clients were not prepared to grant a licence to assign "on the grounds that the investment value of our clients' interest in the property would be detrimentally affected by the proposed use". These grounds were supplemented by further grounds in a letter from the landlord's solicitors, dated October 26 1984, but the ground of diminution in the value of the reversion remained, and remains, the principal ground
B of the landlord's refusal to consent to a licence to the tenant to assign the lease to Euro. Of the other grounds mentioned in the letter of October 26 1984, two are still relied on by the landlord in its notice of appeal: (i) the viability of the proposed business, and (ii) the effect that the proposed user of Colne House might have on the car-parking facilities on the Highbridge Industrial Estate.

On the principal ground of objection, the learned judge heard evidence from a number of expert witnesses. His relevant findings of fact can be summarised as follows:

(1) By the end of the term of the lease, the site value of the property would be as great as, or greater than, the building value. There was no possibility that the use of the building for serviced offices might have a depreciating effect on the letting value of the property at the end of
C the lease.

(2) In view of the tenant's financial position, there was no significant danger that the rent would not be paid throughout the term.

(3) The rent obtainable on future rent reviews would not be prejudiced by the use of the premises as serviced offices.

(4) There was no prospect of Colne House being placed on the market or mortgaged to the fullest extent possible. Although this finding was attacked by Mr Lewison, who appeared for the landlord before us (as he did below), I am satisfied that it was justified by the evidence of Mr Dibley (a director of the landlord) which the learned judge fully rehearsed in his judgment — if one understands "prospect" in its dictionary meaning of "expectation".

D (5) That reasonable professional men might take the view that, if Colne House were placed on the market, it could fetch less with Euro in occupation of the property carrying on the business of providing serviced offices than with the property having remained vacant for more than a year. The learned judge said that if it were relevant, he would not himself be satisfied that that would be the case, and I can understand his reluctance to reach such a conclusion, since it does seem surprising that the reversion to an empty property, which no one wishes to occupy as a single unit, should be worth more than one occupied by a company providing (with guarantors) a third source from which payment of the rent, and performance of the obligations under the lease, could be secured.

E The learned judge then rehearsed the arguments and the authorities with great care, and his finding on the principal ground was expressed in the following terms:

> I accept that the valuation evidence shows that reasonable professional men might take the view that, if Colne House were placed on the market, it could fetch less with Euro in occupation of the property carrying on the business of providing serviced offices than with the property having remained vacant for more than a year . . . but in the circumstances of this case, in which, so far as the evidence shows, there is no prospect of Colne House being placed on the market or mortgaged to the fullest extent possible, that does not in my judgment constitute a ground for reasonable apprehension of damage to the (landlord's) property interest.

He then dealt quite shortly with the car-parking problem as a ground for the landlord refusing consent to the assignment to Euro,
F commenting that even the landlord's own witnesses considered it a minor point, and said that he was not satisfied that the problem was likely to be substantially greater than if the property were assigned to a company which could use it as its own offices. He concluded that he did not consider that the possibility of a small increase in parking problems was enough, either on its own or in conjunction with the alleged diminution in the value of the reversion, to justify the landlord's refusal to consent to the proposed assignment, or that a reasonable man might consider it enough.

The judge, however, did not accept an argument by Mr Morgan, counsel for the tenant both here and below, that since the proposed use of the property was not forbidden by the lease, the landlord, by refusing consent to the proposed assignment, was trying to secure a collateral advantage in the form of preventing a use which was originally permitted when the lease was granted. This argument has been revived before us by a respondent's notice.

During the course of argument many cases were cited to us, as they were to the learned judge. I do not propose to set them out in detail here; many of the older cases were considered in the full judgment of the Court of Appeal in *Pimms Ltd* v *Tallow Chandlers Co* [1964] 2 QB 547. From the authorities I deduce the following propositions of law:

(1) The purpose of a covenant against assignment without the consent of the landlord, such consent not to be unreasonably withheld, is to protect the lessor from having his premises used or occupied in an undesirable way, or by an undesirable tenant or assignee — per A L Smith LJ in *Bates* v *Donaldson* [1896] 2 QB 241, at p 247, approved by all the members of the Court of Appeal in *Houlder Bros & Co Ltd* v *Gibbs* [1925] Ch 575.

(2) As a corollary to the first proposition, a landlord is not entitled to refuse his consent to an assignment on grounds which have nothing whatever to do with the relationship of landlord and tenant in regard to the subject-matter of the lease (see *Houlder Bros & Co Ltd* v *Gibbs* (*supra*), a decision which (despite some criticism) is binding on this court; *Bickel* v *Duke of Westminster* [1977] QB 517).

A recent example of a case where the landlord's consent was unreasonably withheld because the refusal was designed to achieve a collateral purpose unconnected with the terms of the lease is *Bromley Park Garden Estates Ltd* v *Moss* [1982] 1 WLR 1019.

(3) The onus of proving that consent has been unreasonably withheld is on the tenant — see *Shanly* v *Ward* (1913) 29 TLR 714 and *Pimms* v *Tallow Chandlers* (*supra*) at p 564.

(4) It is not necessary for the landlord to prove that the conclusions which led him to consent were justified, if they were conclusions which might be reached by a reasonable man in the circumstances — *Pimms* v *Tallow Chandlers* (*supra*) at p 564.

(5) It may be reasonable for the landlord to refuse his consent to an assignment on the ground of the purpose for which the proposed assignee intends to use the premises, even though that purpose is not forbidden by the lease — see *Bates* v *Donaldson* (*supra*) at p 244.

(6) There is a divergence of authority on the question, in considering whether the landlord's refusal of consent is reasonable, whether it is permissible to have regard to the consequences to the tenant if consent to the proposed assignment is withheld. In an early case at first instance, *Sheppard* v *Hongkong & Shanghai Banking Corporation* (1872) 20 WR 459, at p 460, Malins V-C said that by withholding their consent the lessors threw a very heavy burden on the lessees, and they therefore ought to show good grounds for refusing it. In *Houlder Bros & Co Ltd* v *Gibbs* (*supra*), Warrington LJ said (at p 584):

> An act must be regarded as reasonable or unreasonable in reference to the circumstances under which it is committed, and when the question arises on the construction of a contract, the outstanding circumstances to be considered are the nature of the contract to be construed and the relations between the parties resulting from it.

In a recent decision of this court — *Leeward Securities Ltd* v *Lilyheath Properties Ltd* (1983) 271 EG 279 — a case concerning a subletting which would attract the protection of the Rent Act — both Oliver and O'Connor LJJ made it clear in their judgments that they could envisage circumstances in which it might be unreasonable to refuse consent to an underletting, if the result would be that there was no way in which the tenant (the sublandlord) could reasonably exploit the premises except by creating a tenancy to which the Rent Act protection would apply, and which inevitably would affect the value of the landlord's reversion. O'Connor LJ said (at p 283):

> It must not be thought that, because the introduction of a Rent Act tenant inevitably has an adverse effect upon the value of the reversion, that that is a sufficient ground for the landlords to say that they can withhold consent and that the court will hold that that is reasonable.

To the opposite effect are the dicta, *obiter* but nevertheless weighty, of Viscount Dunedin and Lord Phillimore in *Viscount Tredegar* v *Harwood* [1929] AC 72 at pp 78 and 82. There are numerous other dicta to the effect that a landlord need consider only his own interests — see eg *West Layton Ltd* v *Ford* [1979] QB 593 at p 605; *Bromley Park Garden Estates Ltd* v *Moss* (*supra*) at p 1027E. Those dicta must be qualified, since a landlord's interests, collateral to the purposes of the lease, are in any event ineligible for consideration — see para (2) above.

But in my judgment a proper reconciliation of those two streams of authority can be achieved by saying that while a landlord need usually consider only his own relevant interests, there may be cases where there is such a disproportion between the benefit to the landlord and the detriment to the tenant if the landlord withholds his consent to an assignment, that it is unreasonable for the landlord to refuse consent.

(7) Subject to the propositions set out above, it is, in each case, a question of fact, depending upon all the circumstances, whether the landlord's consent to an assignment is being unreasonably withheld — see *Bickel* v *Duke of Westminster* (*supra*) at p 524; *West Layton Ltd* v *Ford* (*supra*) at pp 604H and 606-7.

In the present case, the learned judge, having made the findings of specific fact set out above, carefully considered the relevant authorities. He then reached the conclusion that the views of the landlord's expert witnesses about the effect of the proposed assignment on the value of the reversion, although views which could be held by reasonable professional men, did not in the circumstances of this case, where there was no prospect of the landlord's wishing to realise the reversion, constitute a ground for reasonable apprehension of damage to its interests. That was a decision on the facts to which the learned judge was entitled to come. He made no error of law in reaching his decision; he took into account nothing which he ought not to have considered, and he omitted nothing which he ought to have considered. In my judgment, this court ought not to interfere.

But in any event, in my judgment, the learned judge reached the right decision. Although he did not expressly mention the disproportionate harm to the tenant if the landlord were entitled to refuse consent to the assignment, compared with the minimum disadvantage which he clearly considered the landlord would suffer by a diminution in the paper value of the reversion — "paper value" because he was satisfied there was no prospect of the landlord's wishing to realise the reversion — he clearly recognised the curious results to which the landlord's arguments, based solely upon a consideration of his own interests, could lead.

As he said (at p 19C of the transcript)*:

It seems to me that, if Mr Lewison is right, the more substantial the lessee, the more easily the landlord would be able to justify a refusal of consent to an assignment, since unless the proposed assignee's covenant was as strong as the assignor's, a reasonable man might form the view that the market would consider the reversion less attractive if the lease were vested in the proposed assignee than if it were vested in the proposed assignor. To take the matter to extremes, if a lease was made in favour of a government department, it would be unassignable except to another government department; for, as Mr Matthews (one of the expert witnesses) accepted in cross-examination, the market would prefer to have the government as the lessee, whether the premises were being used as serviced offices or not, even if they were standing empty, rather than a company, however strong its covenant.

In my judgment the gross unfairness to the tenant of the example postulated by the learned judge strengthens the arguments in favour, in an appropriate case — of which the instant case is one — of it being unreasonable for the landlord not to consider the detriment to the tenant if consent is refused, where the detriment is extreme and disproportionate to the benefit to the landlord.

I am also satisfied that the learned judge could, and should, have had regard to the fact that the proposed service office user was within the only form of user permitted by the lease. I have already stated the proposition of law, derived from the cases, that it may be reasonable for the landlord to refuse his consent to an assignment on the grounds of the proposed user, even though that proposed user is permitted by the lease. But it does not follow from that that, in all circumstances, it will be reasonable for the landlord to object to a proposed user which is permitted by the lease. In most of the cases cited to us in which it was held reasonable to object to the proposed user, even though not forbidden by the lease, the user clause was, in general terms, merely prohibiting the carrying on of any noxious or offensive trade or business (see eg *Governors of Bridewell Hospital* v *Fawkner* (1892) 8 TLR 637; *Re Spark's Lease* (1905) 1 Ch 456). An exception is the case of *Premier Confectionery (London) Co Ltd* v *London Commercial Sale Rooms Ltd* [1933] Ch 904. This was a decision at first instance which is not binding on us. The facts are that there were two separate tenancies of a shop and a kiosk in the same office building, granted consecutively by the same landlord to the same tenant. Each contained a covenant by the tenant to use the demised premises as a tobacconist's shop only. The tenancies were assigned together to a company which went into creditor's voluntary liquidation. The liquidator applied for consent to assign the tenancy of the kiosk alone to an assignee who wished to carry on the trade of tobacconist in it. The landlord refused consent to the assignment on the grounds that it would prejudice the chances of finding anyone prepared to take an assignment of the tenancy of the shop. Bennett J held that the landlord's consent was not unreasonably withheld.

It may be that the decision can be justified on the particular facts of that case, in that the two tenancies were originally granted to the same person, and that there was no certainty (as there is here) that the landlord had security for the rent of the shop for the residue of the term (see the judgment at p 911); but I find it difficult to reconcile this decision with the second proposition of law set out above, and if it be suggested that it is authority for the proposition that in all circumstances it is reasonable for a landlord to refuse his consent to an assignment on the grounds of the proposed user, even though the proposed user is the only user permitted by the lease, then I am not prepared to follow it. There is all the difference in the world between the case where the user clause prohibits only certain types of use, so that the tenant is free to use the property in any other way, and the case where (as here) only one specific type of use is permitted. In my judgment, in that type of case it is not reasonable for the landlord to refuse consent to an assignment on the grounds of the proposed user (being within the only specific type of use), where the result will be that the property is left vacant and where (as here) the landlord is fully secured for payment of the rent.

I can deal shortly with the remaining grounds of appeal. Mr Lewison did not seriously seek to contend that the learned judge was wrong in his conclusion that the possibility of a small increase in parking problems was enough, or that a reasonable man might consider it enough, to justify the landlord's refusal in the present case. In my judgment that was a conclusion to which the learned judge was entitled to come on the evidence before him, and is unassailable.

Mr Lewison devoted slightly more time to his ground of appeal based on the viability of the proposed business and in particular to certain answers given by Mr Brodie in cross-examination. Euro had submitted to the landlord the draft of a licence which it proposed to grant to the licensees of the serviced offices. This draft provided that among the services which Euro would provide to the licensee for the basic licence fee, there should be included dictating equipment for the use of the licensee, who might use this "to dictate up to four A4-size sheets per day. The Licensor will also provide secretarial staff to transcribe the same on stationery provided by the Licensee".

Under cross-examination, Mr Brodie admitted that this provision was a nonsense. Doubtless this would very soon have been realised by Euro — probably well before any licences were actually issued. Certainly there is no reason to suppose that once the defects of this provision (or indeed any other defects in the form of the draft licence) had been appreciated, they would have been allowed to continue without remedy, or to affect the viability of the business. In my judgment, there is nothing in this point.

Accordingly, I would dismiss this appeal.

FOX and MUSTILL LJJ agreed and did not add anything.

The appeal was dismissed with costs and leave to appeal to the House of Lords was refused.

Court of Appeal

November 12 1985

(Before Lord Justice FOX and Sir Roger ORMROD)

FALGOR COMMERCIAL SA v ALSABAHIA INC

Estates Gazette January 18 1986

277 EG 185-189

Landlord and tenant — Application of covenant in 999 years' lease to company not without lessors' consent to use or occupy flat otherwise than as a single private residence in one occupation only — Lessee company was in fact the lessee of a number of flats in the block and treated them in the same way, but the present proceedings concerned a particular flat — Lessees granted occupational licences to visitors to reside in the flats in return for payments, the flats being equipped and furnished by the lessees and cleaned by employees — Consent to the granting of the licences was never sought or given and no question as to exact status of licensees was raised in the appeal

*[1985] 2 EGLR 74 at p 79; (1985) 275 EG 802 at p 809.

— Appeal was against an order of Scott J, on lessors' application under Order 14, granting an injunction restraining the lessee company from using the subject flat otherwise than as a single private residence without the lessors' consent — Held, affirming the decision of Scott J on this point, that the appellant lessees were not using the flat as the lessees' private residence, but as a residence for such members of the public as were acceptable to the lessees — The case was different from that of a company using property for its directors or staff or its own guests — In the present case the user was not residence user at all but the business user of providing service accommodation — It was, on the other hand, open to the lessee company under the terms of the lease to sublet the flat with the lessors' consent, so that the court's decision did not preclude an exploitation of the lease — However, the lessees raised another point — At the trial a defence of acquiescence or waiver by the lessors was mentioned but not then pursued — It was sought to raise it on appeal and the Court of Appeal in their discretion gave permission — After considering the evidence, the court decided that there was an arguable case that the lessors knew what was going on in the flat and took no action — The facts ought to be fully investigated at a trial and unconditional leave would be given to defend on that issue — Subject to that, appeal dismissed

This was an appeal by Alsabahia Inc, the defendant company, lessees *inter alia* of 4 Carrington House, Hertford Street, London W1, from the order of Scott J, on an Order 14 application by the plaintiffs, the present respondents, Falgor Commercial SA, granting an injunction restraining the appellants from using the flat otherwise than as a private residence without the consent of the respondents.

Derek Wood QC and G Bennett (instructed by Gershon Young Finer & Green) appeared on behalf of the appellants; Gavin Lightman QC, K Farrow and Miss Beverley-Ann Rogers (instructed by Wright Webb Syrett & Co) represented the respondents.

Giving judgment, FOX LJ said: This is an appeal by the defendant from an order of Scott J on an Order 14 application giving summary judgment, by which the learned judge ordered that the defendant be restrained "(whether by its officers servants or agents or any of them or otherwise howsoever) from using the leasehold flat known as Flat 4, Carrington House, Hertford Street, London W1, otherwise than as a single private residence without the consent of the plaintiff".

The plaintiff is the lessor of the block of flats called Carrington House. The defendant is the lessee of a number of those flats, and this action is brought in respect of flat no 4 only, which was leased on December 20 1977 for 999 years. The lessee paid a premium of £23,000, and there is ground rent of £5 per annum plus a service charge.

Between 1976 and 1982 the defendant acquired the leases of some 23 flats in the block — six in 1970 and a total of 17 in 1977. It then assigned away four of them, but retained the rest, including no 4. All the leases are in the same form. Formal consent to the assignment of flat no 4 to the defendant was given in 1979 — although apparently it was purchased earlier than that date, but nothing turns on that and it is not of any consequence.

All the flats owned by the defendant are treated in the same way, which in the main is by granting occupational licences to visitors to reside in the flats in return for a monetary payment. The flats are equipped and furnished by the defendant and are cleaned daily by an employee of the defendant, but consent to the granting of licences was neither sought nor given. The leasehold reversion to the property was acquired by the plaintiff in 1982.

The case is concerned with the interpretation of one of the lessee's covenants in the leases, and the question is whether the defendant is in breach.

The lessee's covenants are contained in the third schedule to the lease. Para 3.8 contains a covenant: "Not to assign transfer underlet or part with or share possession of any part of the flat (as distinct from the whole) in any way whatsoever" and "(ii) Not to assign transfer underlet or part with possession of the flat as a whole without the previous consent in writing of the company such consent not to be unreasonably withheld." The matter has proceeded upon the basis that there have been no sublettings or parting of possession — only the grant of licences.

Mr Fakih in his affidavit on behalf of the defendant sworn on October 22 1984 exhibited a draft form of licence and indicated that the defendant intended to allow persons into occupation of flat no 4 — and presumably other flats belonging to the defendant — on the terms of the licence agreement, and on no other terms. The learned judge took the view that the terms in fact created a licence and, having regard to the way in which the case has proceeded, I need not investigate the correctness of that.

Para 3.26 of the third schedule is in the following terms:

> (a) Not without the company's consent to use or occupy the flat otherwise than as a single private residence in one occupation only so that the total number in residence shall not exceed three persons; (b) Without prejudice to the general application of subclause (a) hereof not to carry on in the flat or any part thereof any business as defined by section 23(2) of the Landlord and Tenant Act 1954 or any statute amending or re-enacting the same.

The defendant no doubt is making use of the flat for a business purpose in the sense of making money out of it; but the defendant is not carrying on any business "in" the flat, or any part of it. Para 3.26(b) is therefore not infringed by the granting of the licences.

Therefore the question turns, as the judge found, on the effect of para 3.26(a), the question being whether the defendant is in breach of that; ie, having regard to the licences, is the defendant occupying or using flat no 4 "as a single private residence in one occupation only"?

The judge held, first, that the defendant is using the flat; he is using the flat for the purpose of carrying on the business of granting licences for occupation and, second, the judge held that the defendant is using the flat otherwise than as a single private dwelling-house in one occupation only.

The test is (and as the learned judge found) whether the lessee is using the flat as the lessee's private residence. The answer to that in my view is that the lessee is not. The defendant is not using the flat as the defendant's private residence in any sense. It is not a case of a company using property for its directors or its staff or its own guests. Each of those classes, it seems to me, might have sufficient nexus with the defendant company as such to justify regarding their occupation as that of the defendant company itself. But what we have here is a situation that the defendant company is using the flat, not for any private residential purpose of its own, but as a residence for such members of the public as are acceptable to the defendant. In my opinion the judge's conclusion is right. This is an undertaking by the lessee not to use the flat otherwise than as a single private residence in one occupation. The lessee in the present case is using the flat. But the lessee is not using it as a private residence, if somebody else (ie the licensee) is using it as a private residence, and the lessee does not reside there at all. The user as a private residence is not the lessee's user; it is that of somebody else altogether. And even if a lessee is actually residing on the premises, but invites members of the public to live there as paying lodgers, he is not keeping the house as a private dwelling-house only — that appears from the decision of the Court of Appeal in *Tendler* v *Sproule* [1947] 1 All ER 193.

In my view, the defendant's user is not residence user at all; it is the business of providing service accommodation, and it seems to me there is no question of the defendant using it as a residence, much less as a private residence. It is true that this was a lease granted to a company in expensive premises, and a company (unlike an individual) may have limited use for such premises unless it can exploit it for value. Licensing in the manner which has been adopted in this case is such a method of exploitation.

But, in my view, the judge's decision does not mean that the property is frozen in the defendant's hands. The defendant can exploit it simply by subletting it, and it seems to me that it is reasonable enough to suppose that a landlord would, in a lease of this sort, stipulate for exploitation by subletting rather than by licensing. The occupancies, in the case of subletting, are likely to be longer periods, with less comings and goings. The subtenants will probably be more carefully chosen than would be the case with licensees, and the creating of subleases requires the consent of the landlord.

The lease in the present case fairly obviously intended a comprehensive system of control, which is understandable in a large block of flats. We were referred to regulation 7(3) of the regulations in the lease, which is in the following terms:

> . . . not to do or permit to do by himself, his family, servants, agents or visitors, any act to the damage or annoyance of the company or the tenants of the company or the occupiers of any part of the building or of any adjoining or neighbouring premises, or any illegal or immoral act.

That indicates a concern about the conduct of occupiers and does not, in its terms, appear to contemplate licensees.

Therefore, in my view, under the lease there is a system of control by consent to subletting and by the provisions which I have just read, and also by the provisions of para 3.26(a) of the third schedule. I conclude that the judge was right on the construction as to para 3.26(a) and I would dismiss the appeal on that aspect of the case.

However, the defendant raises a further point. The defendant's evidence before the judge was directed, in part, to a defence of acquiescence or waiver. It was, so we understand, referred to by counsel before the judge, but the response of the judge was somewhat discouraging and counsel did not, therefore, pursue it. However, it is sought to be raised in this court.

Mr Lightman, for the plaintiff, opposes this, but he does not suggest that his client was prejudiced by the course which the matter took below. He says, however, that it is a matter of discretion for the court whether to allow it to be raised or not, and the court should not exercise that discretion, in the present circumstances, in favour of the defendant. He referred us to the decision of the Court of Appeal in *Hickman* v *Kent or Romney Marsh Sheepbreeders' Association* (1920) 37 TLR 163. The Court of Appeal in that case accepted that it was a matter of discretion whether a point which had been deliberately not raised below should be allowed to be raised in the Court of Appeal. The court, in its discretion, declined to permit it. But that was a case where, as I have said, the matter was deliberately not raised below, and was a case where there had been a full trial. Since in the present case there is no prejudice to the plaintiff in consequence of what happened, for myself I would, as a matter of discretion, permit the point to be raised here. The point was raised in the evidence and it was mentioned to the judge below.

I come, then, to the evidence which has been led by the defendant company on this aspect of the case. In the affidavit of Mr Fakih, sworn on October 22 1984, paras 3 and 4 read thus:

> Alsabahia

that is the defendant

> own the leases of flats

and the affidavit then gives their numbers

> at Carrington House, London W1 that is to say 16 flats out of a total of the 72 flats at the block, which are arranged over seven floors and a ground floor. The lease of each flat is for 999 years with each lease having approximately 990 years to run. The total liability for service charges and rates for these flats is approximately £70,000 per annum.

He then refers to a schedule of flats owned by the defendant and says:

> I do not think it possible that the landlords (whether Alegros Holdings SA or their successors in title the present plaintiffs, who purchased the freehold of Carrington House on 4 February 1982), can at any time have thought in the circumstances that Alsabahia could have wished to acquire so many flats simply for the use of its own employees/directors. 4. To the best of my knowledge and belief there have at all material times been other flat owners who have used their flats for the purpose of occupancies by third parties.

Then he refers to a letter from Brecher & Co of September 24 1982 "written on behalf of the landlords".

He continues:

> It is my belief that the landlords have at all times been aware (and well before the date of this particular letter) that Alsabahia's flats and those of other flat owners have been used as accommodation for third parties. I am informed by Mrs Marcia Green of Central Estates, who has a detailed knowledge of . . .

and he then mentions a number of flats and adds that they

> are all owned by non-residents and are presently used for this purpose. Flats 506 and 508 are owned by Bellview Enterprises Inc (a Panamanian company) and are also so used.

The letter of September 24 1982 from Brecher & Co is in these terms:

> *Re: Carrington House.* We act for the Landlords of the above and we have been asked to write to all the tenants by our clients' Managing Agents, Orbiton Estates, regarding numerous complaints that they have been receiving concerning nuisances, disturbances and difficulties that are being caused by certain of the occupants of the flats in the building. We have been asked to write to all the tenants to remind them of the terms of the Leases under which the flats are held wherein each and every tenant covenants to observe and perform the obligations contained in the Third Schedule to the Lease and in particular to Clause 3.8 of that Schedule which requires the consent of the Landlord prior to any underletting or assignment and to Clause 3.26 which provides for the flats to be occupied as single private residences for not more than three people. We are instructed that these covenants are being disregarded by some tenants which is obviously to the detriment of the building as a whole and it does perhaps go without saying that our clients cannot allow this state of affairs to continue.

Then there is an affidavit of Marcia Green, sworn on October 18 1984, para 2 of which is in the following terms:

> Now produced and shown to me marked "MG1" is a copy of Mr Beuselinck's letter of August 4 1982 addressed to Mr Coates of Orbiton Estates (the Landlords' Agents) and a copy of the letter sent therewith, the original of which was addressed to me. I can confirm that these letters were written following an informal meeting which I had with Mr Beuselinck at my office. Mr Beuselinck had alleged that flats in Carrington House were being sublet without the authority of the Landlords and asked me to confirm that this practice would stop. I did not agree that Mr Beuselinck's allegation was correct and I told Mr Beuselinck that I proposed to continue allowing third parties to occupy the flats as before. I appreciated that Mr Beuselinck was concerned that he should not suffer any disturbance as a result of such occupation and I told him that I would do my very best to ensure that no disturbance to him would result. I can confirm that so far as any flats owned by Alsabahia are concerned, no serious complaint was received by me in respect of any of them since August 1982 until the time when section 146 Notices were served by the Landlords in August 1984. For completeness, I should add that on or about August 21 1984 I received a complaint from Mr Beuselinck about the occupants of Flat 510 Carrington House, which is owned by Alsabahia. Mr Beuselinck's complaint was that the occupants of the flat were creating unacceptable disturbances. On receiving the complaint I immediately took action and made the occupants leave the flat straightaway.

So far as the matter of acquiescence or waiver is concerned, there are two possibilities: the first is that the covenantee had given up his right by acquiescence; second, that if there was no acquiescence in that sense, there is the question whether the covenantee has, by his conduct, acted in such a way that, although there has been no acquiescence in law, the court should not, in its discretion, grant an injunction. It seems to me that, on the evidence, there is an arguable case that the landlords knew what was going on as to the user of the flat, and did not act. I think there is sufficient evidence to say that the facts ought to be fully investigated at a trial to ascertain the full knowledge of the landlords, and the consequent legal position of both parties.

If, as is possible on the evidence, the defendant company was allowed to proceed with the process of licensing, it may well have reasonably acted to its detriment in reliance on that. I express no view as to the outcome of the matter, but I think the defendant has produced sufficient evidence to establish an arguable case on acquiescence or waiver which ought to be allowed to go to trial.

I would, therefore, give unconditional leave to the defendant to defend on that issue.

Subject to that, I would dismiss the appeal.

Agreeing, Sir Roger Ormrod said: The covenant which is numbered 3.26 in the third schedule to the lease in this case is obviously drafted very badly indeed if it were intended that the lease including it would be used where the lessee is a company. It is almost impossible to make sense of that covenant, assuming the lessee to be a company.

It is obvious to my mind that the clause has been lifted out of a lease designed for use with private lessees in the ordinary way, and it would make perfectly good sense in that context, just as regulation 7(3) makes perfectly good sense in that context and makes nonsense in the context of a company lessee.

That fact, coupled with the evidence that this lessee company took the leases of no less than six of these flats simultaneously on December 20 1977, led me to think that perhaps the learned judge had given a much too strict grammatical construction to this clause and that it would be proper to construe it, or to look at it again, in the light of the surrounding circumstances, which were known to both parties. It is obvious that both parties to this lease knew that these premises were being used, or were going to be used, for (to use a neutral word) licensees or tenants (or whatever one likes to call it) and were not intended to be used, in any sense of the word, by the company as residences or for occupation by their staff, directors, or guests.

So at first I thought that the suggested construction put on it by the appellants might be right. The suggestion was that it should be read in such a way as this: that the company, on subletting or licensing other parties to occupy the flat, must so arrange it that those occupants only use the premises as a private residence for themselves. It would be a somewhat forced construction, but it would make sense in the context of the case.

However, my difficulties in this respect were swept away in a sentence by Mr Lightman, who pointed out that it would be perfectly possible (though no doubt inconvenient) for the lessees, the defendant company in this case, to use these flats as they intended to use them, but by the method of subletting rather than licensing, with the landlords' consent to each subletting. That again would make sense, in that it would give the landlords some control over the use of these very expensive and prestigious (if that is the correct word) block of flats.

So my conclusion is the same as my lord's, that this clause 3.26 must be construed in the way the learned judge construed it. That being so, it is obvious that the lessees in this case cannot possibly be using or occupying this flat as a single private residence, but are clearly using it for a completely different purpose, namely, to allow other persons to use it for an appropriate fee. Therefore, I have no doubt in my mind that the lessees are in breach, as the learned judge found, of that covenant.

On the second part of the case, it seems to me that that conclusion, followed as it was by the learned judge granting an injunction, is an extremely serious threat to the defendant company. Obviously it would apply to all the other flats which they own in this block, and it will seriously disrupt their present business methods in connection with it, and that could represent a major loss to them.

In my view, it would, therefore, be wrong to shut out the alternative defence which the defendants have adumbrated in their evidence, namely acquiescence or waiver, because it is, *prima facie* at any rate, clear that the landlords in this case must have been aware of the business practices being carried on by the defendant company for a long time — certainly since the present landlords took over in 1982 and presumably their predecessors in title would equally be well aware of what was going on. Whether that represents an acquiescence or not is a matter for argument — and about that I express no view — but it would be wrong, in my view, to shut out the defendant from advancing that case, and therefore I agree that in this court it was appropriate and right to permit the appellants to raise that point here. And having raised it, I agree with my lord that there is sufficient material before us in the evidence to suggest that there is an arguable case here and consequently, this being Order 14 proceedings, it would follow that leave to defend should be given.

I agree that the appeal should be dismissed but that the appellant should be given leave to defend on the basis indicated.

It was ordered that the respondent company should have the costs of the appeal, the cost below to be in cause.

Chancery Division

July 25 1985

(Before His Honour Judge THOMAS, sitting as a judge of the High Court)

MULTON AND ANOTHER v CORDELL AND ANOTHER

Estates Gazette January 18 1986

277 EG 189-190

Landlord and tenant — Option clauses in leases of a shop and a flat — Tenants claim specific performance of options enabling them to obtain further terms of 21 years after expiry of terms of 35 years — Whether requests effective to exercise options — Whether landlords estopped from disputing exercise of options — Leases provided for the options to be exercised by "the written request the tenant made three months before the expiration of the term hereby created" — Original terms expired in 1984, and three months before the expiry date would have been Christmas 1983 — Requests by tenants to exercise the options were made as early as January 1981 — The first point decided by the judge was whether, in the light of *Biondi* v *Kirklington & Piccadilly Estates Ltd*, such a request in 1981 was made a reasonable time before Christmas 1983 — He held that the option clauses contemplated that the request should be made a reasonably short time before Christmas 1983, especially in view of the proviso in the clauses that there must be no breach or non-observance of covenants at the date of the request — Hence the request made nearly three years before Christmas 1983 was not made at a reasonable time — He also held that some correspondence in 1983 in which the 1981 requests were mentioned did not constitute fresh requests — The second point related to the question of estoppel — One of the landlords had agreed in correspondence with the tenants' solicitors that the phrase in the option clause "request of the tenant made three months before the expiration of the term" meant at any time before the date which was three months before the expiration of the term, but this interpretation was repudiated a few months later by surveyors on behalf of the landlords as incorrect and made without professional advice — It was argued that, despite this repudiation, the original agreement by the joint landlord created an estoppel — The judge, without deciding whether the correspondence would have created an estoppel, held that in any case the repudiation was effective; there was nothing to prevent the landlords from resiling from their earlier position — Claim for specific performance dismissed

In this case Violet Jane Multon and Florence Ivy Clarke, tenants of a shop and of a first-floor flat at 71 St Helen's Gardens, London W10, brought an action for specific performance of options contained in leases of the premises against their landlords, Horace Arthur Cordell and Barry Marshall Cordell.

Robert Pryor QC and Kirk Reynolds (instructed by Nicholas Morris) appeared on behalf of the plaintiff tenants; Ian Lamb (instructed by Pickering & Butters, of Stafford) represented the defendant landlords.

Giving judgment, JUDGE THOMAS said: The plaintiffs are the tenants and the defendants are the landlords under two leases. They are successors in title to the original tenants and landlords. The plaintiffs in this action seek specific performance of options contained in the two leases. By one of the leases, which is dated May 2 1949, the lock-up shop known as 71 St Helen's Gardens, London W10, was demised for the term of 35 years from March 25 1949 at the rent and upon the other terms and conditions set out in the lease. By the other lease, also dated May 2 1949, the first-floor flat at 71 St Helen's Gardens was demised for a like term of 35 years from March 25 1949 at the rent and upon the other terms and conditions set out in that lease.

Each of the leases contained in clause 3(4) a covenant by the landlord in the following terms:

> The landlord will on the written request of the tenant made three months before the expiration of the term hereby created and if there shall not at the time of such request be any existing breach or non-observance of any of the covenants on the part of the tenant hereinbefore contained at the expense of the tenant grant to him a lease of the demised premises for a further term of 21 years from the expiration of the said term at the same rent and containing the like covenants and provisos as are herein contained, with the exception of this covenant for renewal, the tenant on execution of such renewed lease to execute a counterpart thereof.

The term of 35 years from March 25 1949 expired on either March 24 or 25 1984 and three months before the expiration of that term of 35 years brings us to Christmas 1983.

Under cover of a letter dated January 5 1981 the plaintiffs' solicitors wrote to the defendants and enclosed in that letter two formal requests for the grant of a new lease for 21 years in accordance with the provisions of clause 3(4).

The first question I have to answer is whether or not those requests are effective to exercise the option contained in clause 3(4). That depends upon the construction of that clause. A clause in similar terms was considered by Roxburgh J in *Biondi* v *Kirklington & Piccadilly Estates Ltd* [1947] 2 All ER 59. The only difference between the covenant which Roxburgh J was considering and clause 3(4) in the present case was that in the former the request was to be made six calendar months before the expiration of the term granted, which in that case was also 35 years. The term granted by the lease in that case was 35 years from February 8 1911. The lease itself was executed on March 1 1911. The 35 years, therefore, expired on February 7 or 8 1946, and that made the date six months before the expiration of the term August 8 1945. In the *Biondi* case the lessee served a notice on the lessor electing to exercise the option for renewal on March 10 1911; that is just 10 days after the execution of the lease.

Roxburgh J, interpreting the covenant in question, said at p 60-E that it was:

capable of at least four different meanings: (i) made on August 8 1945; (ii) made at any time before August 8 1945; (iii) made on or a reasonable time before August 8 1945; and (iv) given so as to take effect on August 8 1945.

In that case he said that the second possible construction — that is, made at any time — was the only one which would avail the plaintiff, and he considered the case on that footing.

In the present case it is not contended that the same meaning should be ascribed to the covenant I am considering, and it is common ground that the first question I have to answer is: was the request, made in this case on January 5 1981, made a reasonable time before Christmas 1983?

Clause 3(4) in specifying a period of three months before the expiration of the term of 35 years is specifying a brief period before the expiration of the term, and that suggests to me that the request must be made within a reasonably short time before Christmas 1983.

Further, it is to be noted that the clause provides that breaches or non-observance of the tenants' covenants are to be considered at the time that the request is made. There is no suggestion in the present case that there have been any breaches at any time, but it seems to me that it must have been in the contemplation of the parties when the leases were executed in 1949 that consideration of whether or not there were breaches of covenants should be as late in the term of years as was reasonably possible; in other words, that there should not be a long interval between the request and the end of the term.

The request made in this case on January 5 1981 was made three years, or nearly three years, before Christmas 1983 and that does not strike me as a request made at a reasonable time.

The next question arises out of a letter dated May 12 1983 written by the plaintiffs' solicitors to the defendants' solicitors. That letter was written in answer to a letter from the defendants' solicitors to the plaintiffs' solicitors. In the letter dated May 9 1983 the defendants' solicitors wrote:

We note from the correspondence that notices under clauses 3(4) of the two leases held by your clients in relation to the lock-up shop and the first-floor flat have been purported to be served on our clients. We should be pleased if you would let us have copies of such purported notices.

The letter of May 12 in answer said:

As requested we enclose copies of the counterpart underlease dated 10th June, 1968;

which is not relevant

our notice is dated 5th January, 1981 to your clients and of our covering letter. We look forward to hearing from you with drafts for approval of the new leases to which our clients are entitled.

It is urged by the plaintiffs that that letter of May 12 enclosing copies of the notices sent on January 5 1981 constitutes a fresh request for new leases. In my view, it was not a request at all within the meaning of clause 3(4). It did not evince an intention to exercise the option in a clear way, either formally or informally. The last part of the letter of May 12 makes it clear that the author of the letter regarded his clients as being then entitled to new leases; not that they were making a request for them.

The letter of May 12 1983 was written some seven months before Christmas 1983. I do not need to express an opinion as to whether that period of seven months was a reasonable interval before Christmas 1983, but I incline to the view that it probably was a reasonable interval.

The final question with which I am concerned is the question of estoppel. It arises out of some correspondence which I shall read. On November 6 1980 the plaintiffs' solicitors wrote to the solicitors who had been acting for the defendants on other matters, saying:

Clause 3(4) of the leases gives our clients the right to take a further term upon the expiration of the existing term in March 1984 subject to a "written request from the tenant made three months before the expiration of the term hereby created". We have always considered the words "made three months before" to mean made at any time before the date three months before the expiration of the term as otherwise our clients would appear to be limited to a single and specific day upon which to make the request, which can hardly have been the intention of the parties to the lease. First, do you agree with our interpretation? Secondly, if you do, are you prepared to advise your clients to give some suitable form of confirmation?

It seems that at that stage the solicitor who had been acting for the defendants had semi-retired (so it appears from an attendance note) and he passed the correspondence on to the defendants.

On December 27 1980 the second defendant, Mr B M Cordell,
wrote to the plaintiffs' solicitors, saying in the last paragraph of his G
letter:

So far as your second point is concerned I think your interpretation of the words "made three months before" would seem to be reasonable and I concur with this.

That letter having been received, the plaintiffs' solicitors, on January 5 1981, wrote the letter I have already referred to to the defendants enclosing the formal requests for the new leases. In that letter in the second paragraph the plaintiffs' solicitors said this:

Thank you for confirming that you share our interpretation of the words "made three months before" in clause 3(4) of the leases dated May 2 1949. In the circumstances, as our clients obviously wish to exercise the option
contained in the leases, we now enclose the appropriate notices and shall H
appreciate the return of one copy of each of them.

Obviously the plaintiffs' solicitors acted on the written agreement of both parties as to what clause 3(4) meant.

Later on, on March 25 1981, Messrs Willmotts, surveyors and estate agents, wrote to the plaintiffs' solicitors on behalf of the defendants, saying this in the second paragraph:

We have also been forwarded various correspondence passing between yourselves and our clients in connection with the option clause contained under clause 3(4) of the lease dated 2nd May, 1949. We must make it clear that our clients had not sought professional advice on the significance of the wording of this clause prior to agreeing with the contention raised in your letter to him of the 6th November, 1980. We would make it quite clear that our
client now withdraws the statements made in his letter in reply dated 27th J
December and you may also take it that we disagree with your interpretation. The request to exercise the option set out in your letter dated 5th January is not, therefore, acceptable to our clients and we should be obliged if you would convey this to your own clients accordingly.

I have read the correspondence which gives rise to the question of estoppel. What the plaintiffs say is that the defendants are estopped from asserting that the construction of clause 3(4) is other than that referred to in the earlier of the letters I have just quoted. The estoppel they rely on is an estoppel by convention.

In *Spencer Bower & Turner on Estoppel by Representation*, 3rd ed, at p 157, estoppel by convention is described as follows:

This form of estoppel is founded, not on a representation of fact made by a representor and believed by a representee, but on an agreed statement of facts
the truth of which has been assumed by the convention of the parties, as the K
basis for a transaction into which they are about to enter. When the parties have acted in their transaction upon the agreed assumption that a given state of facts is to be accepted between them as true, then as regards that transaction each will be estopped against the other from questioning the truth of the statement of facts so assumed.

That description of estoppel by convention was adopted by Brandon LJ (as he then was) in *Amalgamated Investment & Property Co Ltd* v *Texas Commerce International Bank Ltd* [1982] QB 84, at pp 130 and 131. What Lord Denning MR had to say about the matter is to be found at p 122 of the same report: I shall quote from paras A-B on that page:

The doctrine of estoppel is one of the most flexible and useful in the armoury L
of the law. But it has become overloaded with cases. That is why I have not gone through them all in this judgment. It has evolved during the last 150 years in a sequence of separate developments: proprietary estoppel, estoppel by representation of fact, estoppel by acquiescence, and promissory estoppel. At the same time it has been sought to be limited by a series of maxims: estoppel is only a rule of evidence, estoppel cannot give rise to a cause of action, estoppel cannot do away with the need for consideration, and so forth. All these can now be seen to merge into one general principle shorn of limitations. When the parties to a transaction proceed on the basis of an underlying assumption — either of fact or of law — whether due to misrepresentation or mistake makes no difference — on which they have conducted the dealings between them — neither of them will be allowed to go back on that assumption when it would be unfair or unjust to allow him to do so. If one of them does seek to go back
on it, the courts will give the other such remedy as the equity of the case M
demands.

The letter dated December 27 1980 which I have already read seems to me to be an expression of opinion rather than a concurrence in the truth of facts which are being assumed, but I am not at all certain that there is not, nevertheless, in the correspondence, including the letter of November 6 1980, an underlying assumption by the parties on the basis of which the plaintiffs proceeded: but on March 25 1981 Messrs Willmotts on behalf of the defendants made it clear that the defendants were resiling from the position created by the earlier letters, whatever that position might be.

That date, March 25 1981, is more than two-and-a-half years before Christmas 1983 and at that early stage the defendants were

A making it clear that they were rejecting the position which they had earlier adopted.

It seems to me, looking back at what Lord Denning had to say in *Amalgamated Investment & Property Co Ltd* v *Texas Commerce International Bank Ltd*, that there is nothing unfair or unjust for the defendants to resile from their earlier position. The plaintiffs still had ample time to serve at an appropriate time the relevant requests under clause 3(4), and, indeed, if they had wanted to hold the position, should have done so.

Having reached the conclusions which I have on the three questions which have been posed, I must dismiss the claim for specific performance.

B

Chancery Division

October 23 1985

(Before Mr Justice NICHOLLS)

FIELD v BARKWORTH

Estates Gazette January 18 1986

277 EG 193-194

Landlord and tenant — Construction of covenant in lease —
C **Agricultural holding — Whether a covenant "not to assign or underlet any part of premises without the consent in writing of the landlord such consent not to be unreasonably withheld" prohibited the assignment or underletting of the whole of the premises — Declaration sought by plaintiff tenant that she was entitled to assign or underlet the whole of the demised premises without requiring such consent — Plaintiff contended that, as it was accepted that a covenant against assignment of the whole did not prohibit an assignment of part, the converse should hold good — Defendant submitted that if, after an assignment or subletting of the whole of the demised premises, one asked the question, Has there been an**
D **assignment of any part?, the answer was plainly "yes", because there had been an assignment or underletting of every part — Nicholls J agreed with this submission — He rejected as inconclusive arguments for the plaintiff drawn from different parts of the lease, where the draftsman had been careful to use phrases such as "the farm or any part thereof" when it had been intended to cover the whole — Contrary arguments could be based on other phrases — The words used in the covenant, "not to assign or underlet any part of the premises", were plain and unambiguous and their meaning did not admit of doubt — Held accordingly that the covenant**
E **on its true construction prohibited the plaintiff tenant from assigning or underletting the whole or any part of the premises in question without the consent of the defendant landlord — Declaration refused**

This was an originating summons whereby the plaintiff, Nora Field, sought a declaration that she was entitled to assign or underlet the whole of the agricultural holding known as High Ashes Farm, Abinger, Surrey, of which she was the tenant, without the consent of the landlord, the defendant, Terence Barkworth.

Geoffrey Stephenson (instructed by Downs, of Dorking, Surrey) appeared on behalf of the plaintiff; David Neuberger (instructed by Sharpe Pritchard & Co, agents for Tattersall & Sturt, of
F Bournemouth) represented the defendant.

Giving judgment, NICHOLLS J said: The question raised by this summons is whether a covenant in a lease not to assign or underlet any part of the demised premises is broken by an assignment or underletting of the whole of the demised premises. There is no authority directly in point.

The lease was made on December 13 1948 and by it the landlord demised a farm known as High Ashes Farm at Abinger in Surrey and comprising a farmhouse, three cottages, farmbuildings and some 118 acres of agricultural land. The lease was for a term of 21 years. It was professionally drawn and contained covenants of the type that one would expect in a lease of this nature. I need mention only some of the provisions. A rent was reserved in the usual way and in clause 2 certain exceptions and reservations in favour of the landlord were set out. Among those (in subclause (5)) was power for the landlord to enter upon "the farm or any part thereof" to view its condition, and power to enter upon "the farm or any part thereof" for the purpose of carrying out repairs or improvements. In subclause (6) of the same clause there were reserved to the landlord rights of way across "any parts of the farm" for purposes of access and otherwise as there set out.

The tenant's covenants were set out in clause 3 and the subclause I am concerned to construe is subclause (24). So far as material that subclause reads:

> Not to assign or underlet any part of the premises without the consent in writing of the landlord such consent not to be unreasonably withheld and prior to any such assignment or underletting not to permit any person or persons other than the tenant, his bailiff, servants or workmen to reside in the farm house and cottages hereby demised.

The landlord's covenants were set out in clause 4 and they included in the insurance provision words to the effect that if "the said farmhouse and buildings and cottages or any of them or any part thereof" should be damaged by fire then there was to be reinstatement in the usual way. The forfeiture clause was clause 5(5) and this provided that in certain events the landlord might re-enter into or upon "the farm or any part thereof in the name of the whole" whereupon the term was to determine.

At the end of the lease appeared a schedule listing the fields and areas of the demised property by reference to ordnance survey numbers with a description and a stated acreage.

The plaintiff is the present tenant under the lease and the defendant is entitled now to the reversion. The plaintiff is concerned to know whether an assignment or underletting by her of the whole of the demised premises requires the consent in writing of the defendant. For the plaintiff it was submitted, first, and this was common ground between counsel, that a covenant against assignment or subletting or parting with possession is not a covenant which is implied by law into a lease and that, therefore, such a covenant must be express and the lessor must take care expressly to prohibit all such acts as he may wish to exclude. Further, and this also was not in dispute, a covenant against assignment of the whole of the demised premises simpliciter does not prohibit the assignment of part only of those premises. For the plaintiff it was submitted further that this case is the converse of that. Again, and still in general terms, it was submitted for the plaintiff that a lease is to be construed strictly against a landlord and that a tenant's rights are not to be cut down except by clear words.

In this regard I was referred to the case of *Cook* v *Shoesmith*, a decision of the Court of Appeal [1951] 1 KB 752. There a tenant agreed "not to sublet", and the court held that such an agreement was not broken by a subletting of part of the premises because the words "to sublet" must have an object and the only possible object in that case was the house that was demised. So, thus construed, the agreement fell within the well-established principle, set out in the judgment of Somerville LJ at p 753, that an undertaking not to sublet or to assign the premises is not broken by a sublease or assignment of part only of them. In that case reference was made to the principle that the courts have always held a strict hand over conditions for defeating leases as one of long standing.

Turning to the particular wording of this lease, for the plaintiff it was submitted that clause 3(24) specifically refers only to "part" of the premises. It would have been a simple matter for the draftsman to have followed the common practice of excluding assignment of the whole or any part of the premises or all or any part of the premises, but that he has chosen not to do. As to context it was submitted that from the manner in which the demised premises are described in the list in the schedule to which I have referred it appears that the demised premises were let in sections or parts. Futher, that from the various provisions I have mentioned, and in particular from clause 2(5) with its reference to "the farm or any part thereof", it is to be inferred that the draftsman had a different purpose in mind when in clause 3(24) he used the different language making up the phrase "any part of the premises". It was said that for many years it has been common practice for a prohibition against assignment or underletting to be couched in some such form as "the premises or any part thereof" if the landlord intends to prohibit assignment or underletting of the whole. Reference was made to the manner in which such covenants are referred to in section 19(1) of the Landlord and Tenant Act 1927, where the well-known provision there set out is expressed to apply to leases containing a covenant, condition or agreement against assigning, underletting, charging or parting with possession of "demised premises or any part thereof". I was also shown two extracts from precedents in the 1964 edition of the *Encyclopaedia of Forms and Precedents* where covenants against assignment or underletting were couched in a like form.

In my judgment the relevant words of the covenant in clause 3(24) are clear and unambiguous and their meaning admits of no doubt. I agree with counsel for the defendant that it after an assignment or underletting of the whole of the premises one asked the question: has there been an assignment or underletting of any part of the premises?, the answer plainly would be yes. The answer would be yes, because what had been assigned or underlet would be every part of the premises and this covenant against assignment or underletting of any part of the premises in my view plainly embraces the assignment or underletting of every part. As I see it, that is the beginning and end of this case.

However, in deference to the arguments addressed to me I should deal with the further points raised. For my part I do not think that such pointers as counsel's careful examination of each provision of this lease has produced materially assist on the point of construction. In the phrases "the farm or any part thereof" in clause 2(5), "the said farmhouse and buildings and cottages or any of them or any part thereof" in the insurance provision in clause 4(2), and "the farm or any part thereof" in the forfeiture provision in clause 5(5), part means less than the whole. Against such assistance as the plaintiff may derive from this can be set the use of the phrase appearing in clause 2(6), "any parts of the farm". Some arguments also centred on the meaning and effect of the second limb of clause 3(24) itself. In that limb there is a prohibition against persons other than the tenant and his staff residing in the farmhouse or cottages "prior to any such assignment or underletting". Such assignment or underletting refers back to an assignment or underletting of any part of the premises which, on the plaintiff's argument, does not embrace an assignment or underletting of the whole. Accordingly, for the defendant it was argued that the plaintiff's construction of the first limb of subclause (24) leads to an absurd conclusion with regard to the second limb, namely, that the second limb would operate to prevent any person other than the tenant and his staff residing in the farmhouse and cottages even after there had been an underletting of the whole of the premises. In answer the plaintiff submitted that, as assignments or underlettings of the whole of the premises are permitted and therefore stand outside the subclause altogether, so the prohibition in the second limb would not apply to prevent an assignee or subtenant of the whole taking up residence. I am not convinced that this answer is sound, but I need not pursue this, because even if the plaintiff is correct on this the only result is that the second limb of subclause (24) is of no assistance to the defendant. Overall I do not myself think that these provisions really take the matter any further in the light of what seems to me to be the clear wording of the first limb of subclause (24).

Nor do I derive any assistance from the manner in which the property was listed in the schedule at the end of the lease. The form of the schedule is unexceptional for an agricultural lease.

As to the conclusion to be drawn from the failure to use a common form provision such as "the whole of the demised premises or any part thereof" or "the demised premises or any part thereof", it seems to me that this failure provides a very unsure guide indeed in a case such as this where the wording used is unambiguous and where in practice various phrases are commonly used in covenants against assignment and underletting. The purpose, I apprehend, of including words such as "or any part thereof" in such covenants is to ensure that in conformity with well-settled principle the omission of those words does not leave the tenant free to assign or sublet parts. But it does not follow from this that if the draftsman uses the phrase "any part of the property" he is contemplating that there may be an assignment or underletting of the whole. Had that been intended I would have expected the prohibition to be expressed as applying to assignments or underlettings of "part only" of the premises.

Finally, there is some evidence before me that an assignment or underletting of the whole to a single tenant is not attended by the same disadvantages to a landlord as an assignment or underletting of part or parts only. That may be. But it would be the purest speculation to suggest that that is what the original landlord in this case had in mind. He might equally have had in mind a wish to have some control over who was farming his land, a consideration as applicable to an assignment or subletting of the whole as to an assignment or subletting of part or parts only. I do not think this type G
of speculation assists.

Accordingly, in my judgment, on its true construction clause 3(24) prohibits the plaintiff from assigning or underletting the whole or any part of the premises in question without the consent of the defendant landlord.

Court of Appeal

November 8 1985.

(Before Lord Justice MAY, Lord Justice SLADE and Lord Justice
LLOYD) H

BOLDMARK LTD v COHEN AND ANOTHER

Estates Gazette February 22 1986

277 EG 745-748

**Landlord and tenant — Dispute as to service charge provisions in long lease of a flat — Main issue tenants' liability to pay interest on amount borrowed by landlords to finance provision of services set out in lease — Appeal from decision of county court judge — The lease provided that, in addition to an annual rent and a contribution to insurance, the tenants J
were liable to pay a percentage of the costs, expenses and outgoings incurred by the landlords in carrying out maintenance of the block of flats and in supplying certain services enumerated in a schedule — It was also provided that there should be added to such costs "such sums as the lessors may from time to time expend in respect to the general administration and management of the block" — In proceedings in the county court brought by the landlords to resolve disputes the judge found in favour of the tenants on a number of matters, but gave judgment for the landlords on a claim in respect of interest payments and made no order as to costs — His decision was challenged on these two matters in K
the appeal — The interest point arose because the landlords had from time to time to borrow money temporarily, before receipts from service charges accrued, in order to carry out maintenance or provide services in accordance with their contractual obligations — The judge held that the tenants could be charged their due proportion of the interest payable in respect of such borrowings, as being within the sums expended "in respect to the general administration and management of the block" — On analogy with applications for a new tenancy under the Landlord and Tenant Act 1954, he made no order as to costs — Held by the Court of Appeal that, L
although in some circumstances an express provision in a lease enabling tenants to be charged with such interest might be sensible, the present lease did not authorise such a contribution, whether as a matter of pure construction of the terms or when regard was had to the practical consequences — A subsidiary argument based on a broad interpretation of the word "cost" was also rejected — The judge's order in respect of the costs of the proceedings in the county court was set aside — Appeal by tenants allowed**

This was an appeal by Rachel Cohen and Sally Cohen, defendants in an action brought by the present respondents, M
Boldmark Ltd, in Willesden County Court. The proceedings concerned the lease of a flat in a block, 27 Melvin Hall, Golders Green Road, London NW11. The appeal related to the appellants' liability to pay, in the service charge for their flat, a contribution in respect of interest paid by the landlords on money borrowed by them to finance expenditure on the items covered by the service charge. The appellants also challenged the judge's decision to make no order as to costs.

Derek Wood QC and Miss Joanne Moss (instructed by Winckworth & Pemberton) appeared on behalf of the appellants; J H Weeks QC and D P Friedman (instructed by A Kramer & Co) represented the respondents.

Giving the first judgment at the invitation of May LJ, SLADE LJ

A said: This is an appeal by the defendants in an action, Rachel Cohen and Sally Michal Cohen, who are respectively mother and daughter, from parts of an order of His Honour Judge Hill-Smith made at the trial of the action in the Willesden County Court on October 22 1984. The principal issue raised by the appeal concerns the construction and effect of certain provisions in a lease relating to service charges, though there is a subsidiary issue relating to costs.

The lease, which was dated February 14 1969, comprised a flat numbered 27 ("the flat") in a block of flats known as Melvin Hall, Golders Green Road, London, which consists of 37 flats in all. I understand that most of them are let under long leases in similar form and that the second appellant is chairman of the local residents' B association. The parties to the lease were Charlton Estates Ltd as lessors and the appellants as lessees.

By clause 1 of the lease the lessors, in consideration of a premium of £10,750, demised the flat to the appellants (who were together defined as "the Lessee") for a term of 99 years from January 1 1967 at a yearly rent of £60 and also paying by way of further or additional rent:

> (i) from time to time a sum or sums of money equal to 2.6 per centum of the amount which the Lessors may expend in affecting [sic] or maintaining the insurance of the building and other parts of the Block against loss or damage by fire and such other risks (if any) as the Lessors think fit as hereinafter mentioned such last-mentioned rent to be paid without any deduction on the quarter day for the payment of rent next ensuing after the expenditure thereof C and (ii) on the usual quarter days in advance as aforesaid the sums which shall be payable by the Lessee under the covenant hereinafter contained in Clause 4 hereof.

By clause 4(ii) the lessee covenanted with the lessors and with the owners and lessees of the other flats comprised in the block that the lessee would at all times thereafter:

> (ii) Contribute and pay the yearly sum of SEVENTY EIGHT pounds as a contribution towards the Lessors' yearly costs expenses and outgoings and matters mentioned in the Fourth Schedule hereto PROVIDED that this charge shall be subject to annual review and in the event of the said costs expenses and outgoings and matters rising above the sum of Three thousand pounds in any year expiring the Thirty-first day of December the Lessee shall pay 2.6 per centum of the total amount of increase in such costs expenses outgoings and matters. The said contribution shall be payable by equal quarterly payments D in advance on the usual quarter days in each year and any increases determined after the then current payment or payments shall have been made shall be added on to the payment or payments next falling due after such determination PROVIDED that the Lessors shall be entitled to collect from the Lessee at any time on one month's prior notice in writing such sum as the Lessors may anticipate expending in excess of the annual sum aforesaid in respect to any of the matters mentioned in Clause 1 of the Fourth Schedule hereto. The amount of the increase in any year shall if any lessee shall so require be certified in writing by the Lessors' Accountants whose certificates shall be final and binding upon the parties.

By clause 5(D) the lessors covenanted with the lessee as follows:

> (D) That (subject to contribution and payment as hereinbefore provided) the Lessors will maintain repair redecorate and renew (a) the main structure E and in particular the roof chimney stacks gutters and rainwater pipes of the Block (b) the gas and water pipes drains and electric cables and wires in under and upon the Block and enjoyed or used by the Lessee in common with the owners and lessees of the other flats (c) the entrance passages landings lifts and staircases of the Block so enjoyed or used by the Lessee in common as aforesaid (d) the boundary walls and fences of the Block and (e) the porter's flat and such other premises and/or amenities as may be erected or provided for the common use of the tenants of the Block.

The fourth schedule to the lease set out the lessors' yearly costs, expenses, outgoings and matters referred to in clause 4(ii) as follows:

> 1 The expenses of maintaining repairing redecorating and renewing (A) the main structure and in particular the roof chimney stacks gutters and rainwater pipes of the Block (B) the gas and water pipes drains and electric cables and F wires in under or upon the Block and enjoyed or used by the Lessee in common with the owners and Lessees of the other flats (C) the entrances passages landings lifts and staircases of the Block so enjoyed or used by the Lessee in common as aforesaid and (D) the boundary walls and fences of the Block (E) the porter's flat (F) such other premises and/or amenities as may be erected or provided for the common use of the tenants of the block.
>
> 2 The cost of cleaning and lighting the passages landings lifts staircases and other parts of the Block so enjoyed or used by the Lessee in common as aforesaid and of keeping the forecourt garden way and others parts of the Block in good condition and (as to the garden) in cultivation.
>
> 3 All rates taxes and outgoings (if any) payable in respect of the forecourt garden way and other parts of the Block.
>
> 4 The cost of insurance against third party risks in respect of the Block and any persons employed thereat by the Lessors if such insurance shall in fact be taken out by the Lessors.
>
> 5 The cost of employment of porters and gardeners and others in connection with the aforementioned service.
>
> 6 A reasonable rack rental and all usual outgoings of the porter's flat.
>
> 7 There shall be added to the costs expenses and outgoings and matters referred to in the preceding paragraphs of this Schedule such sums as the Lessors may from time to time expend in respect to the general administration and management of the Block.

On September 29 1980 the reversion of the block Melvin Hall was purchased by Boldmark Ltd, the plaintiffs in the action and respondents to this appeal.

Disputes then arose between the parties, which resulted in proceedings in the Willesden County Court brought by the respondents against the appellants (Case No 8108750). These proceedings were in due course compromised under a consent order made in that court in the "Tomlin" form on October 5 1982 and amended on January 12 1983. The schedule to that order contained a number of provisions by way of variation of the lease, setting out the basis upon which service charges were to be calculated and paid. However, since it has been common ground between counsel appearing before us that these provisions are not material to the issues which we have to decide, I need refer to them no further.

Regrettably, the consent order of 1982 did not put an end to the disputes between the parties. In reliance on clause 4(ii) and the fourth schedule to the lease, the respondents sought to debit the appellants with 2.6% of a number of items in respect of which the appellants claimed they were not entitled to debit them. In particular, the respondents claimed the right to:

> charge as part of the expenditure under Schedule 4 to the lease (of which the defendants are liable to contribute 2.6%) such sums as the plaintiffs may incur during any calendar year in respect of reasonable interest payments reasonably incurred as a result of reasonable borrowing being made by the plaintiffs to finance the provision of the services set out in Schedule 4 to the Lease during a given calendar year pending receipt of reimbursement of such expenditure in accordance with Clause 4(ii) of the Lease.

On May 6 1983 the respondents issued proceedings against the appellants designed to resolve the outstanding disputes. In the prayer to their particulars of claim they sought certain declarations (including a declaration of their entitlement to recover interest in the terms which I have just quoted) and orders for payment of several sums of money. The particulars of claim were followed by a defence and counterclaim in which the appellants denied that the respondents had any such right to recover interest and claimed declarations that a number of specific items were not chargeable to them under the lease.

We have been told that the hearing before the learned judge, who heard oral evidence, lasted for some three and a half days. By that time, as he said in his judgment, the points at issue between the parties had been reduced to five. He dealt with all the outstanding issues in his judgment. On the amended counterclaim he made a number of declarations in favour of the appellant tenants. The correctness of these declarations is not challenged on this present appeal. His decision is challenged only in relation to costs and the first issue, which in his judgment he summarised as being "whether the plaintiff is entitled to recover interest payments on money borrowed to finance the fourth schedule expenditure". His conclusion on this issue and the reasoning which led him to it is to be found in the following passage in his judgment:

> I am satisfied that, on occasion, expenditure in discharge of the landlord's covenants under Paragraphs 2-6 of the Fourth Schedule of advance contribution within the terms of the lease, as amended by the Tomlin Order, can exceed what the landlord is now entitled to demand from the tenants. And I am satisfied the words in Paragraph 7 to which I have referred are of a wide nature; the words "such sums as the Lessors may from time to time expend in respect to the general administration and management of the Block" ought to be construed as including interest on such sums as they are required to find in discharge of its duties.
>
> I say this because the matter was widely explored in the evidence. Of course the landlord is under an obligation to carry out works set out in the lease. There have been in this case disputes relating to the propriety of service charges sought to be made and at times payment or late payment by one or other tenant might arise or be made subsequent to receipt by the landlord of the monies lawfully raised by way of service charge. In the event of such late receipt or default of any tenant, in my judgment, it would not be open to the landlord to charge interest by way of service charge since that is not money expended in the general administration or management of the Block. In my judgment interest can only be recovered on sums spent before the lawful receipt of service charges and can only arise under Paragraphs 2-6 of the Fourth Schedule to the lease.

This reference to "the lawful receipt of service charges", though not perhaps very clearly expressed, must, I think, be intended to refer to the date when service charges become due. Effectively, therefore, the learned judge's decision on the first issue (though I do not think that the wording of the final order drawn up to embody it entirely accurately or completely reflects it) was that the respondents are entitled to charge interest paid by them on money borrowed to finance expenditure on any of the specific items set out in the fourth schedule to the lease as being sums expended "in respect to the general administration and management of the block", but subject to two important qualifications, namely, this entitlement to charge interest does not extend either (1) to interest paid in respect of money borrowed to finance expenditure on the items mentioned in para 1 of the fourth schedule ("the first qualification") or (2) to additional interest which has become payable on any borrowing because of the failure of any tenant or tenants in the block to pay their service charges on the due date ("the second qualification").

In relation to costs, the learned judge concluded his judgment thus:

> This is analogous to the grant of the new lease under the Landlord and Tenant Act 1954 where the court exercises its jurisdiction to adjudicate between disputing parties. In monetary terms the defendant succeeded in defeating the plaintiff's monetary claim; the declaration has gone the other way: I make therefore no order as to costs.

The appellants now appeal both from the part of the learned judge's judgment relating to interest and from his order relating to costs, which they should submit should be varied whether or not their appeal on the first issue is successful.

We have had the benefit of very helpful argument on both sides. On the issue relating to interest, Mr Weeks, on behalf of the respondents, has sought to maintain both the conclusion and the reasoning of the learned judge. Further or alternatively, however, he has submitted that the word "cost" in paras 2, 4 and 5 of the fourth schedule is by itself apt to include reasonable interest on money borrowed by the lessors to finance expenditure of the nature referred to in those respective paragraphs.

I find it convenient to begin by considering this latter submission and in this context to cite a short passage from the judgment of Cairns LJ in the unreported case of *Agavil Investments Ltd* v *Corner* (October 3 1975), where he said this:

> In the court below dictionary meanings were cited as to the word "costs" and, I think, the word "expenses". I do not find such definitions helpful. These are words in common use, but with variable meaning, the meaning depending on the context. For example, if one spoke of the cost of a hat one would not be likely to include in that cost the bus fare for going to the shop to buy it; nor would one be likely to include something in respect of the loss of time while one was visiting the shop. On the other hand, if a man who had served on a jury were asked how much it had cost him, he would be very likely to include, in addition to the actual fare to and from the court, the amount that he had lost by way of losing a day's work. So it is not the dictionary that will help on these matters, but consideration of the context and what the parties could reasonably be taken to have intended by the words in that context.

In the light of these observations, I turn to consider which of their possible variable meanings the words "expenses" and "cost" bear in paras 1 to 6 of the fourth schedule to the lease. In my opinion it is quite clear that in the particular context of para 1 the word "expenses" does not include interest. The second proviso to clause 4(ii) of the lease gives the lessors the right in effect to claim a payment in advance on account of anticipated para 1 expenditure (which clause 5(D) obliges them to incur). The lessors having been given this specific right, it cannot have been the parties' intention that the lessors, in reliance on the word "expenses", should also be entitled to claim interest on money borrowed to effect para 1 expenditure. And indeed Mr Weeks did not so contend. He accepted that the word "expenses" in the context of that paragraph does not include interest. He also accepted that the phrases "all rates taxes and outgoings" in para 3 and "all usual outgoings" in para 6 of the fourth schedule cannot be said to include interest on borrowings effected by the lessors; these phrases plainly refer to the actual moneys paid by the lessor in discharge of these items. However, he submitted that the presence of the word "cost" in paras 2, 4 and 5 gives rise to different considerations. He referred us to *Hinds* v *Buenos Ayres Grand National Tramways Co Ltd* [1906] 2 Ch 654, in which Warrington J, in the course of holding that the company concerned was entitled to charge interest on money borrowed for the purpose of constructing works to capital account, said this (at p 660):

> the money is borrowed for that particular purpose — the £10,000. They have to pay interest on that £10,000 during the period that construction is taking place. In my opinion that asset which they are so constructing costs them not only the £10,000, but the £10,000 plus the amount of interest during that G
> period of construction.

He also referred us to an observation of Lord Upjohn in *Chancery Lane Safe Deposit & Offices Co Ltd* v *IRC* [1966] AC 85 where, in the context of section 170 of the Income Tax Act 1952, he said this at p 124: "The cost of hiring money to rebuild a house is just as much a capital cost as the cost of hiring labour to do the rebuilding."

I do not doubt that in some contexts a reference to the "cost" of doing a particular work will be perfectly apt to include interest on money borrowed to do it. However, I find it impossible to read the word in this broad sense in paras 2, 4, and 5 of the fourth schedule to this lease. These paragraphs have to be read in their own context. They follow para 1, where the reference to "expenses" clearly refers H simply to the moneys actually paid by the lessors in maintaining, repairing, redecorating and renewing the items there mentioned. They are accompanied by paras 3 and 6, which no less clearly refer simply to the moneys actually paid by the lessors in discharge of the rates, taxes and outgoings there mentioned. In the context of this schedule, I find it impossible to read the references to costs in paras 2, 4 and 5 as referring to anything more than the moneys actually paid by the lessors to the persons providing the services in question and for any necessary equipment. I would therefore reject Mr Weeks' alternative submission on the issue of interest.

His principal submission on this issue may, I think, be fairly summarised as follows. When the lessors borrow money for the purpose of effecting expenditure on any fourth schedule items, and J have to pay interest on such borrowings, they are expending sums "in respect to the general administration and management of the Block" within the meaning of para 7 of the fourth schedule. They are therefore entitled to recover from the lessees their proportionate share of interest on such borrowings, subject to the first and second qualifications already mentioned, by which, Mr Weeks accepted, the respondents' rights of recovery must on any footing be limited. He submitted that the whole purpose of the fourth schedule, in conjunction with clause 4(ii) of the lease, must have been to achieve a situation in which the lessors would ultimately make neither a profit nor a loss in managing the block on behalf of all the tenants and in incurring direct and indirect expenses in so doing. The words of para 7 are, as he termed them, "final sweeping up" provisions designed K for this purpose and should be read liberally. In support of this submission he referred to the decision in *Agavil Investments Ltd* v *Corner* (*supra*) where the Court of Appeal, adopting what Goff LJ described as a "liberal meaning", held that, in the context of the lease before it, loss to the landlords by giving up a flat for the occupation of a caretaker fell within the words "costs or expenses incurred by them in carrying out their obligations" under a particular clause of the lease.

I have considerable sympathy with these submissions. The learned judge found as a fact that without default by any of the tenants in payment of the ground rents and service charges, the lessors might be forced to borrow in order to discharge their management functions. L If they are not permitted to debit the tenants with interest on their borrowing, they may find themselves substantially out of pocket. I see no reason in principle why a lease should not provide for a landlord to be compensated for reasonable interest payments of this nature. (Mr Wood, on behalf of the appellants, has not sought to suggest that if the lease in the present case on its true construction does operate to impose a contractual obligation on the appellants to pay an interest, such obligation has been eliminated by the provisions of Schedule 19 to the Housing Act 1980, which include certain statutory restrictions on the imposition of service charges.) It seems to me that in some contexts a provision of this nature might be a sensible one. I would even accept that in some other contexts a reference in general terms to expenditure in respect of the general M administration of a block of flats might perhaps, on a liberal construction, be capable of including an interest element.

However, it has to be accepted that, though the parties to this lease clearly contemplated that the cost to the lessors of the fourth schedule items might in due course come to exceed £3,000, they never provided for the reimbursement of interest on borrowings by clear and unambiguous words. With some regret, I have come to the conclusion that the provisions of para 7 of the fourth schedule to this particular lease, on their proper construction, are not capable of including payments by way of interest.

First, the crucial phrase in para 7 ("such sums as the Lessors may from time to time expend in respect to the general administration and

A management of the Block") is introduced by the words "There shall be added to the costs expenses and outgoings and matters referred to in the preceding paragraphs of this Schedule". Paras 1 to 6 had referred to a number of specific services to be rendered or provided by the lessors. In this context the natural meaning to attribute to the crucial phrase in para 7 is, in my opinion, as referring to sums expended by the lessors in doing *additional* acts of administration and management beyond those specified in paras 1 to 6; if the mention of the items specified in paras 1 to 6 had been intended to give the lessors a right to claim reimbursement for interest in respect of such items, the natural place to provide for this would have been in those paras themselves, not para 7.

B Second, the conclusion that the crucial phrase in para 7 must bear what I have described as its natural meaning is strongly reinforced by the following consideration. If the general wording of that phrase were to be read as capable of including interest paid by the lessors on borrowings for para 7 purposes, two important qualifications, of the nature indicated above, would then have to be imported into the paragraph as a matter of necessary implication. For, as the learned judge recognised, and Mr Weeks accepted, it cannot have been the intention of the parties that the lessors should be entitled to reimbursement from the lessee either in respect of interest paid on borrowings for para 1 purposes (in view of the second proviso to clause 4(ii)) or in respect of additional interest which the lessors find themselves obliged to pay as the result of defaults on the part of other C tenants in the block. The adoption of the wide construction of para 7 urged on behalf of the respondents becomes infinitely more difficult if one has at the same time to import with it restrictions, by necessary implication, so as to qualify the apparently unqualified generality of the crucial phrase.

Third, the correctness or otherwise of this construction may, in my opinion, be fairly tested by looking at its practical consequences. If in any particular year the lessors, without the necessity to borrow, have sufficient liquid resources to pay for the paras 2 to 6 items and choose to pay for them out of such resources, they will be deprived for the time being of the use of the moneys concerned. Yet there will be no question of their being able to claim reimbursement from the lessees in respect of lost interest. Mr Weeks (in my opinion rightly) expressly D conceded that the crucial phrase in para 7 would not on any construction be wide enough to include a claim for interest on notional borrowing. If the construction urged on behalf of the lessors were correct then, as Mr Wood pointed out on behalf of the appellants, the extent of the lessee's liability for service charges would depend on the personal circumstances of the lessors or their policy in arranging their financial affairs. If in any particular year of the tenancy they were impecunious, they would have to borrow, and the lessee's liability for service charges would be correspondingly increased. If, on the other hand, they had sufficient resources to render borrowing unnecessary, the extent of the lessee's liability would depend on whether the lessors chose, as a matter of financial E policy, to borrow. It would, of course, be perfectly possible for parties to a lease by clear words to contract in terms which subjected the tenant to a liability for service charges which would vary according to circumstances of this nature, over which he had no control. For my part, however, I would be slow to construe general phraseology of uncertain import (such as the crucial phrase in para 7) in such a way as to produce these potentially anomalous results.

In the end, the onus must fall on the respondents to show that under this particular lease the appellants have contracted to pay the interest claimed. For the reasons stated, I do not think they have discharged this onus. I would therefore reverse the decision of the learned judge on this issue.

As regards the question of costs, Mr Wood submitted that, even if F the learned judge's decision on the first issue was correct, he still erred in principle in drawing an analogy between this case and applications under the Landlord and Tenant Act 1954, and in making no order for costs. In the light of the decision of this court on the first issue, I think it unnecessary to express any view on this submission. The effect of our decision is that, save in respect of one relatively minor item relating to surveyor's fees, the appellants have successfully resisted all the respondents' claims made in this section. It follows, in my opinion, that the appeal should be allowed with costs in this court and the court below. I would set aside the relevant parts of the learned judge's order. I would grant the appellants a declaration that upon the true construction of the provisions relating to service charges contained in the lease, as varied by the consent order, the respondents as landlords are not entitled to recover from the appellants as tenants any interest in respect of borrowings effected by the respondents. I would further order that the respondents pay to the appellants their costs of this appeal to be taxed and of the hearing in the court below to be taxed upon County Court scale 3.

MAY and LLOYD LJJ agreed and did not add anything.

The appeal was allowed with costs in the Court of Appeal and below. Leave to appeal to the House of Lords was refused.

Court of Appeal

December 16 1985

(Before Lord Justice O'CONNOR and Mr Justice LATEY)

SPITALIOTIS v MORGAN AND OTHERS

Estates Gazette February 22 1986

277 EG 750-753

Landlord and tenant — Landlord's appeal against county court judge's refusal of possession order — There were questions as to the position of the tenant, who had gone to live abroad for a time, and as to the status of persons in occupation of the subject dwelling — At the trial, when the judge had nearly finished giving his judgment, an application was made on behalf of the landlord for an adjournment in order to produce a witness who might be able to give material evidence — The judge refused the application and decided the case against the landlord — On appeal it was submitted that the judge was wrong to refuse the adjournment and that there was fresh evidence available which would satisfy the three requirements laid down in *Ladd* v *Marshall* — These were (1) that the evidence could not have been obtained with reasonable diligence for use at the trial, (2) that the evidence was such that, if given, it would probably have an important influence on the result, and (3) that the evidence is such as is presumably to be believed, or, in other words, it must be apparently credible, though it need not be incontrovertible — Held that the appellant had not shown that the first limb of *Ladd* v *Marshall* was satisfied — Appellant had not been taken by surprise at the trial, the existence of the person wanted as a witness was well known, his part in the events was well known, and the appellant had before the trial been given sufficient information to put him on his guard — It could not, therefore, be said that the evidence in question could not have been obtained with reasonable diligence for use at the trial — Appeal dismissed

This was an appeal by the landlord, Andrew Spitaliotis, from a decision of Judge Stockdale, at Barnet County Court, refusing an order for possession of a dwelling-house at 363a Green Lanes, Palmers Green, North London, occupied by Mr and Mrs Knight, respondents (second defendants in county court action). Mr J Morgan, also a respondent, had been the original tenant and was the first defendant in the action.

R V Bryan (instructed by Alban Gould Baker & Co) appeared on behalf of the appellant; P N Mallender (instructed by Bronkhurst Pallis & Co) represented the respondents, Mr and Mrs Knight; J Morgan appeared in person.

Giving judgment O'CONNOR LJ said: This is a landlord's appeal against an order of His Honour Judge Stockdale made on June 4 1984 refusing to make an order for possession of a dwelling-house, 363a Green Lanes, Palmers Green, North London, on the ground that the second defendants, Mr and Mrs Knight, were in occupation as tenants under the Rent Acts and were protected.

The case is very stale, but what happened was this. In 1972 it seems that a company called James Brooke & Sons Ltd were landlords of this dwelling-house. They let it on a weekly tenancy to the first defendant, Mr Morgan. It seems that he and his wife are joined as first defendants. Be that as it may, he lived there happily and no problems arose until 1981, when Mr Morgan decided that he was going to leave and live in Florida for a year. It seems that he wanted to keep the tenancy of this place in London. What happened was that Mr and Mrs Knight moved in. The rent was paid to the landlords by banker's order drawn on Mr Morgan's bank account. That arrangement went on. There is no information as to when Mr Morgan returned from the United States, but he seems to have been back by 1984 at all events, because he gave evidence.

The plaintiff, Mr Spitaliotis, acquired the freehold or leasehold of the premises (it matters not which), but he was entitled as landlord to possession, subject to the rights of any tenant. He brought these proceedings firstly against Mr Morgan alone. The claim then was that Mr Morgan had abandoned the place, he was not occupying any part of it as a dwelling-house, and he had assigned it or sublet it in contravention of the terms of the agreement without the landlord's permission. He therefore came within the relevant part of the Rent Act 1977.

At some stage it must have been that the plaintiff and his advisers learnt that Mr and Mrs Knight were claiming to be in the dwelling-house as of right, for they joined them in the particulars of claim. The amendment read: "The third and fourth named defendants are the subtenants of the first and second defendants and occupy the said premises without licence or consent of the plaintiff." The defendants put in no defence at all — not even the ordinary formal document issued by the county court enabling a party in person to state very shortly what defence, if any, they allege that they have.

We thus come to June 4 1984. When the case was called on, the plaintiff was there but the defendants were not. It seems as though judgment was given. A little later the Morgans and the Knights turned up. The case was then heard. In his evidence Mr Morgan told the judge that when he was going off to the United States he had been to see a Mr Norman, who was the man at the landlords who was responsible for dealing with letting. He was indeed the secretary of James Brooke & Sons Ltd, and there is no doubt that he was the man responsible for dealing with any letting, assignment and so forth. Mr Morgan's evidence in simple language was that he had cleared the changeover to the Knights, while he was away, with Mr Norman, who had agreed that the rent should go on being paid through Mr Morgan's bank account but that it was all right for the Knights to live in the place. He said that he had a letter from Mr Norman to that effect. The note of his evidence was:

> I wrote to Co Sec and got permission for Knights, who are friends, to live there as tenants whilst I was away. I stopped being the tenant. I went on paying rent by banker's order. The Knights repaid me.

It is said that this came as a great shock and surprise to the plaintiff's camp, that they had no idea that this was to be the defence, and they said the nature of the shock and surprise did not really come home until the judge was giving judgment, at which stage counsel for the plaintiff got up and asked for an adjournment in order to produce Mr Norman, who was away on holiday. The learned judge refused to adjourn the matter further and decided the case in favour of Mr and Mrs Knight.

The nature of the appeal is that the learned judge was wrong to refuse an adjournment and that there is fresh evidence which satisfies the requirements of *Ladd* v *Marshall* [1954] 1 WLR 1489, namely that it was not reasonably available at the time of the trial on the ground that no defence had been filed and the plaintiff was wholly unaware that any such suggestion was going to be made. Therefore he had not got Mr Norman at court.

Of course it is well known that if a party is taken by surprise by something said by the opposing party at the trial, either an adjournment is granted to allow that to be done (if the fresh evidence is recognisably available at that moment) or alternatively, in the absence of any evidence contrary to the surprise, time is taken to discover evidence which, if believed, would have a material effect on the decision (the second limb of *Ladd* v *Marshall*), then normally the appeal will be allowed and the case sent back for a new trial.

The question here really is, "Was the first limb of *Ladd* v *Marshall* really satisfied?" The existence of Mr Norman was well known, and the surprise was the assertion that he had written a letter recognising what Mr Morgan and Mr Knight said was happening. On June 24 he was interviewed, and he swore an affidavit on June 25. He said:

> In 1981 Mr Morgan informed me that he wished to go to America for a year and would like to make arrangements for his rent to be paid by standing order. This I agreed to and arrangements were duly put in hand. . . . Mr Morgan did not make any request of me in writing or verbally to transfer the tenancy of 363a Green Lanes to any other person. At no time since Mr Morgan's request to pay his rent by standing order have I received any sort of request from any person either verbally or in writing to transfer the tenancy.

It will be noted that the affidavit is silent as to whether he knew that Mr and Mrs Knight were living in the place, and if so when, or anything more about it. With that affidavit as the basis of the appeal, notice of appeal was given.

The notice of appeal said that the learned judge was wrong to refuse the application for an adjournment to enable the plaintiff to obtain the evidence of Mr Norman, that he was wrong to act on the oral evidence of Mr Morgan and to receive the hearsay evidence of the letter (because no letter was produced by Mr Morgan), and then that if he had had the evidence of Mr Norman in front of him he would have decided the case the other way. Mr Norman produced a letter dated November 17 1981 to Mr Morgan which refers to a conversation, encloses a banker's order, and says: "Please let me know your son's address and your address in Florida", and he agrees the rent payable at so much a month. There is not a word of any occupation by Mr and Mrs Knight or anybody else while Mr Morgan was away.

That was the state of affairs in June. After the notice of appeal had been served, on October 17 1984 Mrs Knight swore an affidavit in which she said that, when he had bought the flat, Mr Spitaliotis

> came to see the flat where we live, and I met him. We discussed how we came to be living in the flat, and how we came to have taken it over from Mr Morgan. I remember distinctly telling him that Mr Norman had agreed to our staying in the flat after Mr Morgan went to America to set up business there. Having heard the full story Mr Spitaliotis said that that all seemed fine, and there was no reason why we could not stay.

In the final paragraph of her affidavit she said that she had had a conversation with the plaintiff's solicitors about two weeks before the hearing and told them that they would be contesting the case and would be coming to court. She said she had no doubt that he understood that they would be defending the case. She finished up by saying:

> There is not the slightest possibility of the plaintiff's solicitors being taken by surprise in our attending court, nor in our saying in court that we had permission to occupy the premises from Mr Norman. I would add that I was very surprised indeed when we got to court to discover that Mr Norman was not at court.

Note that there has never been any contradiction of any part of that affidavit of Mrs Knight, and the knowledge of the plaintiff is sufficient. It may be that there was a lack of communication between him and his solicitor, and there are good grounds for thinking that his legal representatives were not perhaps fully familiar with the matter with which they were dealing. What happened was that the appeal then went to sleep. The reason for that was that the plaintiff's side did not inform the court how long it was going to last. Until that piece of information was given to the appeal office the appeal was not in for trial. That is how it comes about that it is now December 1985. At the last moment, in a flurry of activity, the plaintiff tried to repair a breach which, as will appear in a moment, is irreparable.

Mr Baker, a representative of the plaintiff's solicitors, swore an affidavit saying that the application for an adjournment had been made by counsel at the end of Mr Morgan's evidence. How he came to swear that I cannot understand, because the fact is that it did not happen that way, as he now recognises. Faced with that, Mr Knight swore an affidavit saying that the application did not come until the judge had nearly finished giving his judgment. Mr Baker has now recognised that that is right. Thus we have got all this happening in December 1985, and nobody took the trouble to ask the judge what had happened. By the time they got around to doing that (some time last week) he had gone on holiday, so we do not know anything from the judge.

Was the first rule of *Ladd* v *Marshall* satisfied? In my judgment it was not. I do not think that on the evidence which is before this court it is right to say that Mr Norman's evidence was not reasonably available on the grounds of surprise. His existence was well known, his part in the affair was well known, and on the affidavit of Mrs Knight, which is unchallenged, she had told Mr Spitaliotis enough to put him on his guard. In any event, as Mr Norman was the man who had done all the letting, I would have thought that any lawyer engaged on behalf of the plaintiff, applying his mind as to what evidence should be brought, would have started with naming Mr Norman as the first necessary witness. After all, the plaintiff had only come on the scene in 1983, and the transactions had all taken place before he had anything to do with the property. That is enough to dispose of this appeal.

During the course of the hearing it has been suggested that we cannot entertain the appeal because the plaintiff has not applied to the county court judge under Order 37, rule 1, for a rehearing. Order 37 and section 77 of the County Court Act 1984 are different from section 108 and the wording of Order 37, rule 1, before the modern statutory changes. We were referred to the case of *Weinbaum* v *Klein* [1950] 1 All ER 353 where, on the old wording of the statute, which was different, and the old wording of the rules, which was different, this court held that where the only relief sought was a new trial the Court of Appeal would not entertain the appeal unless application were being made to the county court judge for a rehearing and had been refused.

In my judgment that case may well require reconsideration in the light of the present statutory wording, but it is unnecessary for me to consider the matter in this case. I will, however, say that as part of the relevant claim was founded on an alleged failure of the judge to grant an adjournment, that would seem to take this case outside the wording of the present Order 37, rule 1, which is limited to cases where there is no error of the court alleged.

As I say, it is unnecessary for us to consider the full effect of the change in the statutory wording for the purposes of this case. I have come to the conclusion that the first requirement of *Ladd* v *Marshall* is not fulfilled, and in the circumstances I would dismiss this appeal.

LATEY J agreed and did not add anything.

The appeal was dismissed with costs; legal aid taxation of the respondents' costs was ordered.

Chancery Division

September 2 1985

(Before Mr Justice WALTON)

SMITH v METROPOLITAN CITY PROPERTIES LTD

Estates Gazette February 22 1986

277 EG 753-754

Landlord and tenant — Forfeiture of lease — Attempt of former tenant, against whom forfeiture proceedings had been completed and landlords had re-entered, to reopen the question of relief against forfeiture by invoking the ancient equitable jurisdiction of the court — Notice of motion by defendant owners, under RSC Order 18, rule 19, or under the inherent jurisdiction of the court, to dismiss the summons as frivolous or vexatious or an abuse of the process of the court — It had been argued that, although the statutory jurisdiction under section 146(2) of the Law of Property Act 1925 could not be relied on, the court still had an equitable jurisdiction to grant relief in any circumstances in which the statutory jurisdiction was not available — Held that it was clear from the speech of Lord Wilberforce in *Shiloh Spinners Ltd* v *Harding*, as explained by Dillon LJ in *Official Custodian for Charities* v *Parway Estates Developments*, that such a submission was not sustainable — Decision of Nicholls J in *Abbey National Building Society* v *Maybeech Ltd* not followed — Lord Wilberforce's reference to equitable jurisdiction not being ousted concerned cases outside the area of leases and underleases covered by s 146 — Originating summons dismissed as an abuse of the process of the court

This was a notice of motion by the defendants, Metropolitan City Properties Ltd, that an originating summons taken out by the plaintiff, Harlock Smith, should be dismissed under RSC Order 18, rule 19, or under the court's inherent jurisdiction. The summons asked for relief from forfeiture under an equitable jurisdiction which, the plaintiff claimed, had not been displaced by the Law of Property Act 1925, section 146.

J Lynch (instructed by Pumfrey & Lythaby, of Orpington, Kent) appeared for the plaintiff in the originating summons; E J Prince (instructed by Broad, Riggall & Godman, of Watford) represented the defendants.

Giving judgment, WALTON J said: This is a notice of motion by the defendant for an order under Order 18, rule 19 of the Rules of the Supreme Court 1965 or under the inherent jurisdiction of the court that the originating summons herein be dismissed on the grounds that it is frivolous or vexatious or an abuse of the process of the court and that judgment be entered for the defendants for the costs of the action including this application.

The originating summons is an originating summons asking for relief from forfeiture of a lease dated May 25 1954 made between the Prudential Insurance Co Ltd of the one part and Fred Harlock of the other part, by which lease the Prudential demised the property known as 13 Romberg Road, London SW17, to the said Fred Harlock for a term of 77 years. That lease contained (and there is no argument about this at all — no dispute between the parties) a proviso for re-entry in the case of breach of covenant. A schedule of dilapidation was served together with a section 146 notice in August 1984. The notice was not complied with and the respondents, the then landlords, re-entered as against the applicant, the tenant, in December 1984.

One would have thought that that would have been the end of the matter because section 146(2) of the Law of Property Act 1925 provides:

> Where a lessor is proceeding, by action or otherwise, to enforce such a right of re-entry or forfeiture, the lessee may, in the lessor's action, if any, or in any action brought by himself, apply to the court for relief, and the court may grant or refuse relief, as the court, having regard to the proceedings and conduct of the parties under the foregoing provisions of this section, and to all the other circumstances, thinks fit; . . .

Again it is common ground that, the landlord having re-entered in December 1984, he is no longer proceeding, by action or otherwise, to enforce the right of re-entering, and therefore section 146(2) is applicable.

However, what has been said by Mr Lynch on behalf of the tenant is simply this. The ancient equitable jurisdiction relating to relief from forfeiture has not been displaced in relation to leases by section 146 of the Law of Property Act 1925 but exists quite independently thereof in full force, vigour and effect. Of course the first comment which must be made upon that startling submission is that if it were to be so it would make a complete nonsense of subsection (2), because instead of the restriction "Where a lessor is proceeding, by action or otherwise", one might just as well write, "Where a lessor is proceeding, by action or otherwise, or has in fact proceeded and obtained possession already". Any reading of a statute in a manner of that nature which makes absolute nonsense of what is quite clearly an essential provision of subsection (2) just cannot be right on that ground alone. Fortunately one does not have to rely merely upon that obvious and logical deduction.

In *Shiloh Spinners Ltd* v *Harding* in the House of Lords in [1973] AC 691 Lord Wilberforce considered the limits of the court's equitable jurisdiction to relieve from forfeiture and the limits upon it; that was a case in which the jurisdiction was involved in a case concerning the assignment of a lease, but outside the area covered by section 146. The very persuasive argument put in front of the House of Lords was that because section 146 dealt with, and, quite clearly, by dealing with, limited, defined and made clear, the equitable jurisdiction in relation to relief from forfeiture of leases and underleases, Parliament had by implication limited the right of the court to grant relief from forfeiture outside the cases of leases and underleases expressly dealt with in the section. What Lord Wilberforce said about that argument was this:

> Secondly, a point of more difficulty arises from the intervention of Parliament in providing specific machinery for the granting of relief against forfeiture of leases:

Pausing there, Lord Wilberforce never uses language inaccurately: when he says "specific machinery for the granting of relief against forfeiture of leases" he means exactly that. That is the specific machinery provided. That and that only. He continues:

> see Law of Property (Amendment) Act 1859 ... Common Law Procedure Act 1852, Law of Property Act 1925, Leasehold Property (Repairs) Act 1938 and other statutes. This, it is said, negatives an intention that any corresponding jurisdiction should exist outside the case of leases. I do not accept this argument. In my opinion where the courts have established a general principle of law or equity, and the legislature steps in with particular legislation in a particular area

here that is precisely and exactly what the legislature has done. It has stepped in with particular legislation in a particular area, namely forfeiture of leases and underleases. By so stepping in it has provided the specific machinery for the granting of relief against the forfeiture of leases. Lord Wilberforce continued:

it must, unless showing a contrary intention, be taken to have left cases outside that area

if I may comment on that, that is to say outside leases and underleases

where they were under the influence of the general law. To suppose otherwise involves the conclusion that an existing jurisdiction has been cut down by implication, by an enactment moreover which is positive in character (for it amplifies the jurisdiction in cases of leases) rather than negative. That legislation did not have this effect was the view of Kay LJ in *Barrow* v *Isaacs & Son* ... when he held that covenants against assigning — excluded from the Conveyancing Act 1881 — were left to be dealt with according to the ordinary law.

What Lord Wilberforce is saying in the plainest possible terms is: "Look, the legislature has interfered with the old equitable jurisdiction to grant relief against forfeiture in the case of leases and underleases, but outside that area it has left the jurisdiction where it was." If, I may add, the situation had been that even by stepping in with the specific legislation, Parliament had not in fact interfered one bit with the equitable jurisdiction, then it is quite clear that Lord Wilberforce would have said: "I do not know what you are talking about. However much legislation there has been, the equitable jurisdiction, even in the case of leases, as well as outside them, is left precisely where it is." That, of course, is precisely what he did not say.

The matter is again assisted by the very helpful observations of Dillon LJ in the case of *Official Custodian for Charities* v *Parway Estates Developments (In Liquidation)* [1985] 1 Ch 151*. The relevant passage from Dillon LJ's judgment is at the top of p 162. I may say that it appears that Mr Lynch, in seeking to rely upon the very clear and, if I may say so, obviously clear words of Lord Wilberforce to which I have already referred, in supporting his argument, was not alone in completely misconstruing them, because apparently that is precisely what Mr Edward Nugee QC did in the case to which I am now referring. If I may start at the bottom of p 164, Dillon LJ said this:

Being unable to rely on the section, Mr Nugee on behalf of the defendant submits that the court in its equitable jurisdiction has, in any appropriate case, power to grant relief against forfeiture of any property interest.

In substance that is really what Mr Lynch has been submitting.

In particular the court can, he submits, under its equitable jurisdiction grant relief to any lessee against forfeiture of a lease in any circumstances in which the court's statutory jurisdiction to grant relief is not available.

That again is in effect Mr Lynch's submission.

Therefore, in Mr Nugee's submission there is no problem about jurisdiction; the only question was whether the jurisdiction should be exercised, and that was a matter for the deputy judge.

These far-reaching submissions are founded on the decision of the House of Lords in *Shiloh Spinners Ltd* v *Harding* [1973] AC 691 (which was not a landlord and tenant case). Mr Nugee relies both on the general principles established in that case as to the width of the jurisdiction and in particular on a passage in the speech of Lord Wilberforce

which I have already quoted. Then Dillon LJ continues with these telling words:

In my judgment this particular passage in Lord Wilberforce's speech is, on a true appreciation, destructive of Mr Nugee's arguments.

Just as, on a true appreciation, it is destructive of Mr Lynch's argument.

Where the legislature steps in with particular legislation in a particular area, effect must be given to that legislation, and in that particular area any wider equitable jurisdiction is ousted. Parliament has stepped in initially by the Conveyancing and Law of Property Act 1892 ... and now by section 146(9) and (10) in the particular area of relief against forfeiture of leases because of the bankruptcy or liquidation of the lessee. The provisions of the subsections make no sense unless they are regarded as comprehensive within that particular area; in cases of certain types of tenancy listed in section 146(9) there is to be no relief at all, and in other cases the insolvent lessee has a year only to apply for relief, but if the leasehold interest is sold within that year, the time-limit does not apply.

Mr Lynch points out, and points out quite accurately, that that case was in fact dealing with section 146(9) and (10) in the particular area of relief against forfeiture of leases, but equally the same argument applies to subsections (1), (2), (3), (4) and so on to (8), as well as (9) and (10). He cannot, and it would be quite impossible to, suggest any rational distinction between them. The only support which Mr Lynch can find for his proposition is the decision of Nicholls J in *Abbey National Building Society* v *Maybeech Ltd* [1985] Ch 190.† It is perfectly true that in that case the learned judge decided that on its true construction section 146(4) of the 1925 Act, G
which again was prefaced by the words "where a lessor is proceeding, by action or otherwise, to enforce a right of re-entering", did not displace the court's inherent equitable jurisdiction to grant relief to a person claiming as underlessee where the lessee failed to pay a sum other than rent due under the lease.

With great respect to the learned judge, first of all he did not have the very helpful and definitive comments of Dillon LJ to which I have referred, and second, in my judgment, he does not appear to have appreciated, perhaps lacking those particular comments of Dillon LJ, the precise effect of what Lord Wilberforce was getting at. Returning really to the point from which I started, it is clear to me that to give effect to the argument of Mr Lynch would be to turn
subsections (2) and (4) completely on their heads. I cannot think that H
where Parliament has clearly laid down that something is to happen, and happen only, where a lessor is proceeding by action or otherwise, that is intended to leave the prior equitable jurisdiction untrammelled in any way so that precisely the same result can be achieved in a situation where the lessor has proceeded and has in fact properly obtained possession. The suggestion really is quite fantastic.

For those reasons it appears to me that the motion is well founded, and I shall therefore give effect to it by dismissing the originating summons on the ground that it is an abuse of the process of the court, and I shall enter judgment for the defendants for the costs of the action including the motion. J

---

*Editor's note: Also reported at (1984) 270 EG 1077.
†Editor's note: Also reported at (1984) 271 EG 995.

Court of Appeal

January 15 1986

(Before Lord Justice DILLON, Mr Justice NICHOLLS and Sir George WALLER)

RAPID RESULTS COLLEGE LTD v ANGELL AND OTHERS K

Estates Gazette March 1 1986

277 EG 856-859

**Landlord and tenant — Appeal by landlords from decision of Mr Scott Baker QC on a preliminary question as to construction of lease — Whether tenants of part of a building were liable to make contributions in the nature of a "services rent" in respect of works done to the parapet on the roof of the building — The demise to the tenants (respondents to the appeal) comprised the second floor of the building (in fact the
top floor) and they had the use of other parts of the building, L
such as entrance hall, staircase, landings and passages — The parapet had to be rebuilt after the bricks had become saturated with damp — Respondent tenants were required to contribute towards the repair works, the contribution being demanded by a certificate signed by the landlords' surveyor, whose decision was stated to be binding on the parties — The claim for a contribution was based mainly on an item, "maintenance of the exterior", in a list of items of expenditure in respect of the offices on the first and second floors — It was argued for the appellant landlords that, as the respondents' second-floor offices were on the top floor,
maintenance of the roof was maintenance of part of the M
exterior of the respondents' offices — Held, distinguishing the decision of *Douglas-Scott* v *Scorgie*, that in the context of the present case the roof could not be regarded as part of the exterior of the second-floor offices — The court also rejected submissions that the claim could succeed under the heading of expenses attributable wholly to the first- and second-floor offices or, alternatively, under the head of repairs and maintenance in respect of parts of the building used in common by the respondents with the landlords and tenants of other portions of the building — The court distinguished the decision in *South West Water Authority* v *Rumble's*, which**

A **had been cited by the appellants in support of this last submission — It was common ground that the landlords' surveyor's certificate was not conclusive as to whether items listed in the certificate did as a matter of law fall within particular provisions in the lease — Appeal dismissed**

This was an appeal by landlords, Rapid Results College Ltd, from a decision of Mr Scott Baker QC, sitting as a deputy judge of the Queen's Bench Division, in favour of tenants on a preliminary issue of construction under a lease of office premises at Brighton House, High Street, Ewell, Surrey. The tenants were Donald Douglas Angell, Terence Michael Thompson, Philip Rodney Gasson and
B Malcolm John Henderson. The decision of Mr Scott Baker QC is reported at [1985] 2 EGLR 66; (1985) 275 EG 247.

K S Munro (instructed by Sowman Pinks & Co) appeared on behalf of the appellants; John Tonna (instructed by Brand & Co, of North Cheam, Surrey) represented the respondents.

Giving judgment, DILLON LJ said: This is an appeal from a decision of Mr Scott Baker QC, sitting as a deputy judge of the Queen's Bench Division on March 21 1985. The appeal raises a question of construction of particular provisions in a lease in relation to what is called a services rent. The lease is a document which the draftsman has attempted to tailor to the particular requirements of a particular building; the question that arises depends on particular
C provisions and not on any general consideration.

The respondents to this appeal, the defendants in the action, are the original lessees, who carried on a profession in partnership in offices on the second floor of the building. The appellants, the plaintiffs in the action, are by assignment the reversioners entitled to the reversion on the lease. The lease is of part of a building known as Brighton House, High Street, Ewell in Surrey. That building was built some 20 years or so ago; it is a three-storey building consisting of shops on the ground floor and offices on the first and second floors. There is a flat roof, and on the roof and going round the perimeter of the building there is a brick parapet wall. Included in the building as I have described it there is a common staircase and landings. The first-floor office and ground-floor shops extend slightly further than the
D second-floor offices; there is a garage space at the back, and built over that, at mezzanine level, there is a further building, to which it seems access can be obtained by a bridge from the first floor of the main building.

The lease was a lease for a term of six years from March 25 1975. That term has of course now expired and the defendants have quit the building; the question that arises is whether they are liable in respect of services rent during the continuance of the term.

The lease was granted at a yearly rent of £4,400, with provision for a review after the expiration of three years of the term granted; the reviewed rent is to represent a fair rack rental market value of the premises. In addition, however, there is reserved by the lease to the landlords by way of additional rent a further sum on account of
E expenses and outgoings incurred by the landlords in compliance with the covenants on the part of the landlords contained in the third schedule to the lease; this further sum is called the "services rent" and is to be calculated in the manner specified in the fourth schedule to the lease.

The covenants on the part of the landlord in the third schedule include the usual covenant for quiet enjoyment and a covenant to insure, and to keep insured, the demised premises; to provide adequate heat during office hours from October to April to panels or radiators fixed in the demised premises; to pay all charges for gas and electricity consumed in the building; and in paragraph 3(4) it provides:

F Whenever reasonably necessary to maintain put in good and substantial repair redecorate and renew:

(a) the external walls and structure and in particular the roof foundations gutters and rainwater pipes of the building

(b) the gas and water pipes tanks drains and electric cables and wires in under and upon or serving the building and enjoyed or used by the Tenants in common with the tenants of other parts of the building

(c) the main entrances common passages landings and staircases and other parts of the building so enjoyed or used by the Tenants in common as aforesaid

and then it says: "together with the boundary walls surrounding the building", which seems to mean the walls around the car park. The demised premises are plainly not the whole of the building. They are defined in the first schedule as follows:

FIRSTLY ALL THAT office accommodation on the Second Floor of the building (hereinafter called "the building") known as Brighton House High Street Ewell as the same is shown on the plan annexed hereto and edged red Together with the use in common with the Landlords and the tenants and occupiers of other portions of the building of the entrance hall staircase, landings and passages leading thereto and the ladies' lavatories on the first floor and the men's lavatories on the second floor of the building shown on the plan edged brown between the hours of 8 am and 7 pm on all days of the said term except Sundays and Good Friday Christmas Day and Bank and public holidays AND SECONDLY ALL THAT the parking space shown and coloured mauve on the said plan

and these are said to be thereinafter together called "the demised premises".

There is also provision for a right of way along a service road and a pedestrian right of way; and there is a grant of the

free passage and running of water soil gas and electricity and other services relating to the Landlords' adjoining premises through the pipes wires and cables in through or under the Landlords' adjoining premises

with a corresponding reservation to the landlords.

The fourth schedule provides in para 1(b) for the calculation of the services rent. Under subpara (i) the amount of the services rent is to be ascertained and certified annually by a certificate signed by the landlords' surveyor, whose decision is to be binding upon the parties; that is to be prepared as soon after the landlords' financial year as may be practicable, and is to relate to that year. "The landlords' financial year" is then defined in subpara (ii), and under subpara (iii) a copy of the certificate for each landlords' financial year is to be supplied to the tenants by the landlords without charge to the tenants. Subpara (iv) reads:

The certificate shall contain a fair summary of expenses and outgoings incurred by the Landlords under the heads of expenditure specified in subparagraph (v) hereof in compliance with the covenants by the Landlords contained in any lease of any part of the building during the Landlords' financial year to which it relates.

Under subpara (v):

The annual amount of the services rent payable by the Tenants as aforesaid shall be calculated as:

(a) Fifty per centum of the aggregate of the following expenses and outgoings incurred by the Landlords in respect of the premises being the offices on the First and Second Floors forming part of the building and shown bounded by the red line on the said plan annexed hereto (hereinafter called "the first and second floor offices"):
1. Insurance
2. Heating
3. Electricity
4. Maintenance of the exterior
5. Any other expenses or outgoings which in the opinion of the Landlords' surveyor (whose decision shall be binding on the parties) are attributable wholly to the first and second floor offices

(b) Twenty-five per centum of the aggregate of the following expenses and outgoings incurred by the Landlords in respect of all other parts of the building specified in the First Schedule hereto of which the Tenants have the use in common with the Landlords and tenants and occupiers of other portions of the building:
1. Cleaning
2. Towels
3. Electricity not falling within subparagraph 3, of paragraph (a) above
4. Repairs and maintenance
5. Any other expenses and outgoings incurred by the Landlords in respect of the said common parts of the building

and

(c) Fifteen per centum of the aggregate of the expenses and outgoings incurred by the Landlords in respect of the car park and boundary walls thereof.

What happened was that there was trouble in relation to the parapet on the roof of the building. It appears that the rendering of the bricks which were on edge had cracked away between the bricks on one side or another, so that the bricks became saturated with damp. On the advice of their architect the landlords took the parapet down and rebuilt it altogether. This cost a considerable sum of money, over £21,000, to which architects' fees of over £2,000 had to be added.

In September of 1980 the landlords submitted a certificate claiming payment, by way of services rent from the defendants as the lessees of the second-floor offices, of a sum of £8,382.70 plus VAT, less £642.18 paid in advance; nothing turns on the payments in advance. The £8,382.70 is calculated for the year ended March 31 1980, first by adding two-thirds of the cost of insurance of the whole building, a sum for gas and maintenance of the boiler, 50% of total paid on electricity, and for the maintenance of the exterior, two-thirds of the amount paid on the parapet and the architects' fees, and small sums for cleaning and towels. This produces £16,765.41, of which the £8,382.70 is the "one-half" — the 50% referred to in subpara (v) of the provisions for the calculation of services rent in the fourth schedule to the lease.

The defendants objected to paying this sum. They took quite a range of points; they said that the reconstruction of the parapet wall was not maintenance; they said that what had been done was not reasonably necessary to maintain and put in good and substantial repair or renew the external walls and structure and roof of the building. But they also took the point that work on the roof and parapet could not fall within any of the factors which the landlords are entitled to take into account under para 1b(v) in the fourth schedule in calculating the services rent.

Since that point did not depend on evidence as to the precise work done or the need for it, or any distinction which might involve expert evidence between maintenance and repair, it was agreed in the court below that that question of construction should be taken as a preliminary question at the trial. The learned judge decided it against the landlords and in favour of the defendants. That put an end to the action against the defendants. If on this appeal the plaintiffs succeed, the action would have to go back for further trial at first instance, to investigate the other issues which were raised by the defendants by way of defence to the claim.

In the court below it was urged that the provisions for the calculation of service rent reflected, and should be construed so as to tie in with, and cover, the matters on which the landlords had covenanted to incur expenditure under the third schedule. Obviously there is a degree of echo in the provisions of the fourth schedule of what the landlords are to do under the third schedule, because the effective charging clause reserving the services rent describes it as a further sum on account of expenses and outgoings incurred by the landlords in compliance with the covenants on the part of the landlords contained in the third schedule; but it does not follow automatically, nor has it been urged in this court, that the services rent must necessarily cover everything that the landlords are to do under the third schedule. The lease is for a term of only six years; one would not automatically expect that a tenant at a rack rent of a suite of offices in part of a building for a term of six years would be undertaking responsibility for the cost of repair of the main structure roof and foundations of the building. It is necessary, while bearing the third schedule in mind, to construe the relevant words in the fourth schedule, where the draftsman has apparently attempted to work out what is to apply, in part 50% of expenses being chargeable and in part 25%.

The main claim by the landlords rests on the words in subpara (v)(a) in the fourth schedule, "4. Maintenance of the exterior". Seeing those words one asks at once, "exterior of what?" It occurred to me in the course of argument that in the light of the provisions in the third schedule, and as the heading "Insurance" in (v)(a)1 in the fourth schedule would probably involve, as a practical matter, an apportionment of an insurance on the building as a whole, there might be a possibility of construing the words "Maintenance of the exterior" as referring to the exterior of the building, and as implying an apportionment in the opening words "50% of the aggregate of the following expenses and outgoings incurred by the landlords in respect of the premises". But I do not think that is possible, because the draftsman of the lease has emphasised that the premises in respect of which these outgoings are to be incurred are the offices on the first and second floors forming part of the building shown bounded by the red line and specifically referred to as "the first and second floor offices". The draftsman has gone on, in item 5, to deal with expenses which are attributable wholly to the first- and second-floor offices, and I do not see that there is scope in that context for saying that two-thirds, or some other fraction, of expenses of maintenance of parts of the exterior, which are not exterior of the first- and second-floor offices, are to be brought within (v)(a)4 in the fourth schedule.

Mr Munro, for the appellants, accepted that "Maintenance of the exterior" meant the exterior of the first- and second-floor offices. His main argument was that as the second-floor offices were on the top floor, maintenance of the roof was maintenance of part of the exterior of the first- and second-floor offices. He supported that argument by reference to the case of *Douglas-Scott* v *Scorgie* [1984] 1 WLR 716*. That is a decision of this court in which it was held that for G
the purposes of section 32 of the Housing Act 1961 the roof of a building might, according to the evidence, form part of the exterior of a top-floor flat in that building. As it seems to me, it is not possible in this case to say that the parapets on the roof were part of the exterior of the second-floor offices. Apart from that, in the context of this case I cannot regard the roof as being part of the exterior of the second-floor offices. The phrase in section 32 of the Housing Act 1961 is "structure and exterior of the dwelling-house" which is a different phrase; and the statutory context is concerned with obligations imposed on a landlord as a matter of law in certain tenancies of dwelling-houses; it is a very different context from that with which we are concerned in the present appeal.

Therefore, as it seems to me, what has been incurred by way of H
rebuilding of the parapet does not fall within the words "maintenance of the exterior" in 4(v)(a) of the fourth schedule.

It is submitted alternatively for the appellants that a proportion of the cost falls within the wording of 5(v)(a):

> Any other expenses or outgoings which in the opinion of the Landlords' surveyor (whose decision shall be binding on the parties) are attributable wholly to the first and second floor offices.

It seems to me that the appellants have a difficulty in attempting to rely on 5, because they are only seeking to charge a proportion, 50% of two-thirds of the cost of the expenditure. They must thereby be recognising that the total expenditure on the parapet is not wholly attributable to the first- and second-floor offices. Therefore, as it J
seems to me, that expenditure cannot fall within item 5, and the wording of item 5 does not permit a prior apportionment by the landlords' surveyor so as to treat the part which he considers appropriate as wholly attributable to the first- and second-floor offices.

There is a further difficulty in the appellants' way in that, as Mr Tonna, for the respondents, has pointed out, the roof covers not only the first- and second-floor offices as shown bounded by the red line on the plan but also the common parts and stairway as shown bounded by a brown line on the plan, and the parapet went round the whole of the roof. That again must indicate that the expenses of rebuilding the parapet cannot be attributable wholly to the first- and second-floor offices. Ordinarily one would regard expenses on a roof K
as attributable to the building as a whole, not merely to the top two floors of the building.

In the alternative, the appellants argue that they can at least claim 25% of their expenses in respect of the parapet under (v)(b) in the fourth schedule, as being expenses and outgoings incurred by the landlords in respect of other parts of the building, of which the tenants have the use in common with the landlords and tenants and occupiers of other portions of the building; they fall under heading 4 in (v)(b), it is submitted, as "repairs and maintenance".

Again I leave aside the question whether the work done is work of repair and maintenance as opposed to work of some other nature, because that is a point which was not argued in the court below. The L
point here is "use in common with the landlords".

Looking at the lease and the parcels demised as set out in the first schedule, it is quite plain that the tenants are given the use, in common with the landlords and other tenants and occupiers, of the entrance hall, staircase, landings and passages leading to the demised premises and various ladies' and men's lavatories; they also have the free passage of water, soil and gas, electricity and other services through various pipes, wires and cables; but they do not, in the sense in which the word "use" is used in the parcels in the grant of the easements, have any use of the roof at all. This is underlined by the wording of para 3(4) in the third schedule.

It is said, none the less, that they do have the use of the roof in the sense in which that phrase was used in the recent decision of the M
House of Lords in *South West Water Authority* v *Rumble's* [1985] AC 609. In that case the position was that occupiers of a ground-floor shop, which had no direct water or drainage facilities, were held to be liable to water charges in respect of water services, because surface water was removed from the roof of the building by way of a downpipe running into a gulley and from there into the water authority's sewers. It is, however, relevant to note that the phrase used in that particular Act was not simply "use of the building" but "use of sewage or drainage facilities". The court was concerned not with use of the building but with the use of facilities which drained to a sewer or drain connected with the local authority's sewer. That is a

*Editor's note: Also reported at (1984) 269 EG 1164.

A different context, and I do not regard it as appropriate language in the context of this lease to say that the occupiers of the first- and second-floor offices have the use in common of the roof and parapet of the building.

I should add that it is common ground that the surveyor's certificate is not conclusive on whether matters listed in the certificate do, as a matter of law, fall within the subheads of the fourth schedule, and in particular in subpara (v) of that schedule.

Therefore, in my judgment the learned deputy judge was right in the conclusion which he reached on the construction of this lease, and I would dismiss this appeal.

Agreeing, NICHOLLS J said: I add only a brief comment on one
B point, namely, the primary way in which the lessors put their case on this appeal. The primary construction contended by the lessors is that in item 4 of clause 1(b)(v)(a) in the fourth schedule, "exterior" means exterior of the premises earlier referred to in that clause, that is, the offices on the first and second floors shown bounded by a red line on the plan.

The next step in the argument is that although the roof and parapet wall do not form part of the second-floor property demised by this lease, none the less they are properly to be regarded as the exterior of the second-floor property. The difficulty that I feel with this argument is that, accepting that the first step is correct, the roof and parapet wall in question extend over not only the part of the building edged in red on the lease plan but also the part of the building edged in
C brown on the plan. Thus at best only part of the roof and parapet wall would be within item 4, because only part of the roof and parapet wall are contiguous to the part of the second floor which is edged in red. Hence, even if one were to accept that the roof and parapet wall are capable of being the "exterior", the end result would be the unattractive one that part, but only part, of the single flat roof and of the one continuous perimeter parapet wall would be within item 4. I find it impossible to believe that such a result was, or is to be taken as having been, the intention of the parties to the lease.

Furthermore, construing item 4 as contended would have the result that although occupying one-third of the building, the tenants under this short lease would be responsible for half the costs of the
D maintenance of the roof and the parapet wall. It may be said that this is not inappropriate, because the tenants occupy the top floor of the building immediately below the roof; but I do not think that this result is so obviously right that it prompts the conclusion that this must be what the parties to the lease intended.

In my view, these considerations point to the conclusion that in this short lease, in item 4 "exterior" does not include the roof of the building or, *a fortiori*, the parapet wall.

SIR GEORGE WALLER agreed with both judgments and did not add anything.

The appeal was dismissed with costs.

E
Chancery Division

December 4 1985

(Before Mr Justice MERVYN DAVIES)

POSNER AND OTHERS v SCOTT-LEWIS AND OTHERS

Estates Gazette March 1 1986

277 EG 859-862

Landlord and tenant — Covenant to employ resident porter — Whether specifically enforceable — Action by tenants of
F block of luxury flats against landlords claiming specific performance of a covenant in the lease to employ a resident porter "(a) to keep clean the common staircases and entrance hall landings and passages and lift, (b) to be responsible for looking after and stoking the central heating and domestic hot-water boilers (c) to carry down rubbish from the properties to the dustbins outside the building every day" — It appeared that there had been a porter in residence in the porter's flat who had performed the services mentioned, but he ceased to be employed by the defendant landlords, vacated the flat and took employment at a nearby block — However, he continued on a part-time basis to carry out his former duties for the defendants while residing elsewhere — By a notice of motion in the action the plaintiffs sought, pursuant to Order 14, specific performance and damages, alternatively interlocutory relief, and, by a second motion, the appointment of a receiver of the rents and profits pending trial — Defendants contended that the duties enumerated in the lease were in fact being performed, although by a non-resident porter, and that consequently there was no breach — The judge held, however, that, even if the particular services mentioned in the lease were performed, the defendants were still in breach of the covenant to provide a resident porter — There was a world of difference between living in a block with a porter in residence and living in a block with no resident porter — The tenant benefited from the feeling of security and the opportunities to ask for help — A resident porter was valued not only for his actual duties but for his very presence — The question then arose as to whether the remedy of specific performance could be granted — *Ryan* v *Mutual Tontine Westminster Chambers Association*, where it was held that a contract to employ a porter was not specifically enforceable, was distinguishable and had been discussed in later authorities, such as *C H Giles & Co Ltd* v *Morris, Shiloh Spinners Ltd* v *Harding* and *Tito* v *Waddell (No2)* — Held that it was, in the light of the authorities, open to the court to make an order for specific performance, and that the order should be made — There would be no need for protracted superintendence — Defendants would merely be doing what they had promised to do and had in fact done for the past 20 years — Order for specific performance granted

These were two motions by the plaintiffs, Mr Posner and eight others, tenants of flats in Danes Court, St Edmund's Terrace, London NW8, and members of the Danes Court Residents' Association, against four defendants, Richard John Scott-Lewis and three others, lessors of the flats. The motions were in an action for specific performance of the defendants' covenant in the lease to provide a resident porter.

Romie Tager (instructed by Beckman & Beckman) appeared on behalf of the plaintiffs; A R Hochhauser (instructed by Malkin Cullis & Sumption) represented the defendants.

Giving judgment, MERVYN DAVIES J said: I have before me two motions by the plaintiffs in this action: the first motion dated October 24 1985 seeks final judgment under Order 14 with an alternative claim for interlocutory relief; the second motion dated November 7 1985 seeks as a further alternative the appointment of a receiver. The writ is dated August 29 1985. There are nine plaintiffs. The plaintiffs are members of the Danes Court Residents' Association. Danes Court is a block of residential flats in St Edmund's Terrace, London NW8. Each of the plaintiffs is a tenant of a flat in the block. Thus the first plaintiff, the chairman of the residents' association, is the lessee of 4 Danes Court, the second plaintiff, Mr Dayan, the secretary, is the lessee of 2 Danes Court, and so on. There are 25 flats in the block together with a flat that has until recently been occupied by a resident porter.

At the issue of the writ the first defendant, Richard John Scott-Lewis, was made sole defendant on the footing that he was the lessor or landlord to the plaintiff tenants. On October 1 1985 the writ was amended to add the second, third and fourth defendants, since it is common ground that the four defendants together are the landlords of the flats in Danes Court. The defendants have been the landlords since about 1982.

Each of the plaintiffs holds his or her tenancy on the terms of a lease in writing. It is common ground that all the Danes Court leases are in terms similar to the lease which is exhibit SN 1 to an affidavit of Mrs S Nagel, the third plaintiff, sworn on September 2 1985. This particular lease is that of Mr Dayan, the second plaintiff. The lease is dated April 6 1964 and is made between Boyton Developments Ltd "(hereinafter called 'the Lessor' which expression where the context so admits includes the estate owner or estate owners for the time being entitled to the immediate reversion to the property hereby demised expectant on the term hereby granted)" of the one part and Mr Dayan "(hereinafter called the 'lessee' which expression where the context so admits includes his assigns and successors in title and the persons deriving title under him)" of the other part. As I have mentioned, the defendants are now the lessors.

The lease witnesses that in consideration of the sum of £8,500 the lessor demises Flat 2 on the ground floor of Danes Court to Mr Dayan (together with and subject to various rights which I need not mention) for 99 years from December 25 1963 at a yearly rent of £100.

There follow various covenants by the lessee including these:

(4) To use the property exclusively as a private dwelling or residence in single occupation only and the garage as a garage incidental to the enjoyment of the property

(6) Not to use or permit or suffer the property to be used for any trade business or profession or for any meeting thereon or for any illegal or immoral purpose or in any manner whereby the character of the properties in the building as high class residential properties may be prejudiced or injured

(18) To observe and perform and conform to the Management Regulations set forth in the Schedule hereto and to such other reasonable rules and regulations as the lessor may from time to time properly make for the orderly and better management of the building and the grounds thereof upon being given notice in writing of such other reasonable rules

There are also covenants by the lessor. These covenants include: clause 3(2), where the lessor promises to maintain etc the main hall, staircases, lift, the central heating system and the internal telephone system (which connects with the porter's flat); clause 3(7), a promise by the lessor to see that the common parts are kept clean and lighted and — see clause 3(1) — to supply and maintain hot water and central heating. There is a lessor's covenant that I must quote in full. Clause 3(11) reads:

(11) To employ (so far as in the Lessor's power lies) a resident porter for the following purposes and for no other purposes: (a) To keep clean the common staircases and entrance hall landings and passages and lift (b) To be responsible for looking after and stoking the central heating and domestic hot water boilers (c) To carry down rubbish from the properties to the dustbins outside the building every day.

Clause 3(12) reads:

(12) To maintain repair and meet all overheads outgoings and expenses in respect of the porter's flat.

Clause 4 of the lease provides for the lessee paying a proportion of the expenditure incurred by the lessor in connection with the fulfilment by the lessor of the obligations undertaken by clause 3 of the lease. I need not set out these detailed provisions, but the effect is, *inter alia*, that the cost to the lessor of providing a resident porter under clause 3(11) is recoverable in part from the lessee. Thus with a similar provision in all the leases the lessor will recover all his clause 3 expenditure. The clause 3 covenants are binding on the lessor only while holding title to the reversion (see clause 5). I need not mention the stipulations in the schedule, ie the Management Regulations, save to quote regulation 1:

The entrance doors of properties shall be kept shut and no owner shall on any account whatsoever leave or permit to be left any invalid carriage or chair or any bicycle perambulator box parcel refuse rubbish or any article whatsoever upon or obstructing any part of the building used in common with other occupiers but lessees shall each morning before nine am have ready within their properties for removal in a proper receptacle the refuse of the previous day.

The writ claims specific performance of the lessor's covenant in clause 3(11) of the plaintiffs' respective leases of their respective flats, that is to say the plaintiffs seek to compel the defendants to employ a resident porter for the purposes itemised in 3(11). There is also a claim for damages by reason of the defendants' failure to employ a resident porter since June 1 1985. The issue of the writ on August 29 1985 was followed by an ex parte order made by Walton J on September 5 1985. The defendants were restrained until judgment or further order from selling or leasing the porter's flat. That order was made, as I understand, because it was then supposed that the defendants were about to dispose of the porter's flat, and if that were done then the relief sought in the writ might not be available.

The next step to mention is that the plaintiffs served a statement of claim on October 2 1985. Particulars of the plaintiffs' and defendants' titles are given, and complaint is made of the defendants' failure to employ or seek to employ a resident porter since June 1 1985.

There followed on October 24 1985 the first notice of motion seeking, pursuant to Order 14, specific performance as mentioned in the writ, with damages. In the alternative there is a claim for an interlocutory mandatory order requiring the defendants to take steps to employ a resident porter. Then, as I have said, there is the second notice of motion dated November 7 1985 whereby there is sought,
pending trial, a receiver of all the rents, profits, service charges and G
other moneys payable by the lessees of all the Danes Court flats.

I will now mention some of the affidavit evidence read to me. The principal affidavit in support of the motion was sworn by the honorary treasurer of the residents' association, Mrs Nagel, the third plaintiff. In para 3 Mrs Nagel describes Danes Court as a modern luxury block of flats ("in the true sense of the term") which was completed in 1964, comprising 26 flats including the porter's flat, on eight floors. Some of the flats in the block have been sublet but all the plaintiffs are residents save that Mrs Nagel, living in no 19 with her husband, has sublet no 18. Mrs Nagel says she receives a rent of £1,300 a month for the subletting of no 18. Mrs Nagel states that Danes Court "has until recently been maintained to the highest
standards with the recognised amenity of a resident porter which H
distinguishes the true luxury flat from other flats in the neighbourhood which do not offer this particular service". In para 8 Mrs Nagel says that there is a flat on the ground floor which has been occupied by the resident porters, who have been employed successively since 1964. The flat ceased to be so occupied on May 31 1985. The porter then in residence, Mr Bewsher, ceased to be employed by the defendants on that date. Mr Bewsher took up employment at a nearby block of flats called Kings Court. That is not in the defendants' ownership. However, it seems that there was some arrangement between Mr Bewsher and the first defendant, because since June 1 1985 Mr Bewsher has, to some extent at any rate,
continued to carry out the works specified in (a), (b) and (c) of clause J
3(11). According to Mrs Nagel the work is done inadequately and clearly it must be done on a part-time basis since Mr Bewsher now has duties at Kings Court, where he resides.

There was a meeting of some of the residents on July 11 1985. The first defendant attended. In para 18 of the affidavit Mrs Nagel says this: "the defendant said that he had made efforts to find a replacement for Mr Bewsher and he said he had interviewed 11 applicants". There was a discussion at the meeting of certain repair works which the defendants said would have to be carried out.

In this connection Mrs Nagel says in para 19:

At this point, the defendant returned to the earlier discussion about the required remedial works and their costs. He said that in order to finance the
building works it would be necessary for him to consider selling the resident K
porter's flat. Mr Bennett intervened once again and asked the defendant why he had bothered to interview 11 applicants for the job if he was proposing to sell off the porter's flat. To everyone's astonishment, the defendant then conceded that he had not in fact interviewed any applicants for the job, had not made any efforts to do so, and had no intention of doing so. The defendant then put forward a proposal that the lessees should pay him in advance the moneys which would be required for the remedial works. He suggested that if an appropriate scheme could be worked out it would not be necessary to sell the porter's flat.

There is some correspondence exhibited as SN 3 to Mrs Nagel's affidavit extending over the period June 6 1985 to August 20 1985. It illustrates the tenants' concern at being left without a resident porter,
and in a letter dated August 16 1985 assurances are sought that the L
porter's flat will not be sold and that a new porter will be put in residence. No assurances being forthcoming, the writ was issued.

There is then a second affidavit of Mrs Nagel sworn on September 4 1985 exhibiting other correspondence. I do not propose to refer to that exhibit (SN 4) since the correspondence relates to old complaints. There is, however, one recent letter there, dated July 2 1985, written by the first defendant, in which he was then saying that he would be interviewing another caretaker on July 15 and that Mrs Nagel had his assurance that it was not the intention to sell the caretaker's flat "at this juncture". That letter was of course written before the meeting dated July 11 1985 that I have mentioned.

A third affidavit of Mrs Nagel sworn on October 24 1985
sufficiently verifies the facts alleged in the statement of claim. M

The first plaintiff, Mr Posner, swore an affidavit on November 7 1985. Therein he seeks to show the inconveniences suffered by reason of there being no resident porter at Danes Court. The inconveniences include: (a) automatic timing of the central-heating system deprives the tenants of the benefits resulting from a resident porter exercising his discretion as to when heating should be put on and off having regard to the prevailing weather conditions, (b) breakdowns in the lift with the consequence that residents outside the lift have to seek help for residents trapped in the lift, (c) there is difficulty in obtaining access to the block, since access has to be obtained by disturbing a resident rather than a porter, (d) recorded mail is sometimes difficult to deliver, and (e) milk and newspapers often have to be left in the

A open outside the main door.

On the defendants' side there are three affidavits, one by Mr Kenneth John Bewsher, who was the resident porter until May 31 1985, and the other two by the first defendant. Mr Bewsher bears out that he is now a resident caretaker at Kings Court, 31 Prince Albert Road, NW8, and that prior to May 1985 he was the caretaker at Danes Court, living in a flat there. He left Danes Court for Kings Court because he was offered better pay. He says he resigned from his position at Danes Court. Mr Bewsher goes on to say this in para 3: "Mr Scott-Lewis asked me to stay on as I was only going to be approximately one minute away in the adjacent block, and I agreed to do so." This means, as I understand, that while Mr Bewsher is B employed as a resident caretaker at Kings Court, he also works for Mr Scott-Lewis, so apparently he is paid by two employers in respect of the same time. He says in paras 5 and 6:

> I am therefore in a position to act as caretaker for both Kings Court and Danes Court. I spend approximately 2 hours in the morning at Danes Court collecting the rubbish and cleaning the common parts, and I also patrol in the afternoon and in the evening. The central heating is on automatic and I order oil when appropriate. I also continue to help out the various tenants from time to time and so far as I am concerned I do exactly the same job now as I was doing when I was actually resident in the premises, with no decline in standards whatsoever.
>
> (6) All the tenants know where I am, which is approximately one minute away, and they either telephone me or pop in to see me to ask me to assist, which of course I am always willing to do.

C Mr Scott-Lewis, in his first affidavit sworn November 15 1985, gives evidence to the same effect as Mr Bewsher, and indeed states that Mr Bewsher was asked to carry out after May 1985 the same duties as he had carried out before that date "albeit on a non resident basis". In his second affidavit sworn on November 20 1985, Mr Scott-Lewis refers to the fact that in his first affidavit he said he had made efforts to obtain an "alternative porter which came to nought". He goes on (see para 2) to say this:

> It is totally incorrect to say that I am not performing the covenant set forth in clause 3(11) of the leases, as Mr Bewsher is doing exactly the same job to perform the covenants as he was doing prior to May 31 1985 albeit that he lives approximately 1 minute away from the block.

D I say now that I do not understand this statement because clause 3(11) obliges the defendants to employ a resident porter and this they are not doing. I need not deal with other matters in the Scott-Lewis affidavit, as it is largely concerned with a denial of any falling-off in standards since Mr Bewsher ceased to reside.

There is reply evidence by the plaintiffs. It is an affidavit by Mr Dayan sworn November 22 1985. Mr Dayan says (see para 7) that it is not the case that Mr Bewsher is now carrying out the same duties as he did when resident. Dealing with the specific duties mentioned in clause 3(11), Mr Dayan complains: (1) of a deterioration in the maintenance of the common parts which "are beginning to look decidedly shabby". He comments on para 3 of the first affidavit of Mr Scott-Lewis where it is suggested that the cleaning of the common E parts and the emptying of rubbish can be done in about two hours. Mr Dayan says that this is an unrealistic estimate, since there are 26 flats on eight floors. He is satisfied that it used to take Mr Bewsher, when he was in residence, considerably longer than two hours to carry out those duties. (2) That the change to automatic time-switching for the central heating means that the central-heating system is operated much less satisfactorily.

As well as making those references to the clause 3(11) tasks, Mr Dayan also refers to the fact that in the past the resident porters at Danes Court have always performed functions and services which have added to "the general amenity of the residents of the block and the efficiency of the day-to-day running thereof".

He itemises some obvious advantages arising from the presence of F a resident porter in the way of giving access to tradesmen, deliveries and a better feeling of security. Mr Dayan takes up Mr Scott-Lewis' statement in para 4 of the first Scott-Lewis affidavit where Mr Scott-Lewis says:

> I am surprised by the allegation that I stated that I had not interviewed people for the post. I instructed my secretary to ask Park Estates to advertise for a new caretaker and as a result 11 people were interviewed. There was not one suitable applicant.

Mr Dayan says there will be no difficulty in finding a resident porter for Danes Court. Mr Dayan advertised in *The Standard* on September 2 1985. A copy of the advertisement is in exhibit LD 2. There were 20 applications received in response to that advertisement, the applications being included in exhibit LD 2.

The plaintiffs' principal claim, pursuant to Order 14, is in these terms:

> (1) Specific performance of the lessors' covenant in clause 3(11) of the plaintiffs' respective leases of their respective flats (and garages in the case of the third and fifth plaintiffs) in Danes Court, nos 1 - 5 St Edmunds Terrace, London NW8 to employ (so far as in the defendant's power lie) a resident porter for the purposes more particularly set forth in the said clause 3(11).

That is to say, the plaintiffs claim in effect that the defendants are now failing to employ, pursuant to clause 3(11), a resident porter for the purposes specified in clause 3(11), and that the defendants ought to be ordered so to employ. It is admitted that the defendants do not now employ a *resident* porter for the purposes specified or for any purpose. Mr Hochhauser for the defendants accepted that no resident porter is employed, and as much is to be inferred from para 2 of Mr Scott-Lewis' second affidavit. However, Mr Hochhauser submitted that the defendants were not in breach of clause 3(11) since, he said, all the duties that are to be performed in (a), (b) and (c) of clause 3(11) are in fact being discharged, and the fact that those duties are being discharged by a non-resident as opposed to a resident porter does not justify the view that there is a breach of clause 3(11). Mr Hochhauser went on to say that at least the defendants should have the opportunity of a trial, so that, in so far as there is in the affidavit evidence a conflict about the present adequacy of the porterage (within (a), (b) and (c) of clause 3(11)), the position can be clarified by oral evidence.

I do not accept Mr Hochhauser's submission. If one assumes, as the defendants allege, that Mr Bewsher is discharging his (a), (b) and (c) duties now as he did before May 31 1985, there is in my view still a clear breach of clause 3(11). Clause 3(11) indicates that: (a) there will be a porter in residence and (b) that he will perform certain functions. To arrange for his functions to be carried out by a non-resident cannot in my view discharge the defendants from their duty to keep a porter in residence. There is, to my mind, a world of difference between living in a block with a porter in residence and living in a block where there is no porter in residence. A tenant in a block of flats understandably attaches great importance to the presence of a resident porter. While the tenant no doubt appreciates the manual work that the porter may perform, he equally appreciates the feeling of security (and the opportunities to ask for help) that arises from the presence of a resident porter. In other words, a resident porter is valued not only for the duties he is expected to perform but also for his very presence.

I am therefore of the view that the defendants are in breach of clause 3(11). I see no reason for sending the matter for trial to discuss that matter.

Since I am of the view that the defendants are in breach of clause 3(11), the question arises whether or not clause 3(11) is a provision susceptible of specific performance.

I was referred to *Ryan* v *Mutual Tontine Westminster Chambers Association* [1893] Ch 116. That is a case where (see p 123) the Court of Appeal considered a contract between a landlord and his tenant by which the landlord undertook to employ a porter to perform certain services for the benefit of the tenant. The contract was held to be not specifically enforceable. One ground of the decision was that the execution of the contract would require "constant superintendence but the court": see p 123 and p 125.

A close examination of the facts in the *Ryan* case as set out at pp 117 to 120 shows the situation in that case differs in some respects from the situation before me. For example, in the *Ryan* case the porter was to "be and act as the servant of the tenants". That is not so at Danes Court. Again the Danes Court lease has, but the *Ryan* lease has not, covenants by the lessor whereby the porter's duties are, at any rate as to (a) and (b) of clause 3(11), elsewhere in the lease seen as direct obligations of the lessor to the lessee. See clause 3(2), (7), (10) and (13). There is also the fact that the scheme of apportioning service charges between the Danes Court tenants involves taking account of the costs of maintaining and repairing the porter's flat: see clause 3(12).

Drawing attention to these differences between *Ryan* and the present case, Mr Tager for the plaintiffs submitted that *Ryan* should be distinguished. In short, he said that since the resident porter's functions at Danes Court were already obligations of the lessors to the lessees, there were no duties on the part of the porter towards the tenants that the tenants were seeking to enforce. All that was required was the appointment of a resident porter, whereas in *Ryan* the plaintiff was in effect seeking to enforce performance of duties said to be owed by the porter to the plaintiff. I do not accept or reject Mr Tager's able argument. I suspect that it is difficult to distinguish the *Ryan* case. However that may be, the *Ryan* case has been remarked upon in many later authorities.

In *C H Giles & Co Ltd* v *Morris* [1972] 1 WLR 307 Megarry J, after referring at p 316 to *Ryan*'s case, says this at p 318:

> One day, perhaps, the courts will look again at the so-called rule that contracts for personal services or involving the continuous performance of services will not be specifically enforced. Such a rule is plainly not absolute and without exception, nor do I think that it can be based on any narrow consideration such as difficulties of constant superintendence by the court. Mandatory injunctions are by no means unknown, and there is normally no question of the court having to send its officers to supervise the performance of the order of the court. Prohibitory injunctions are common, and again there is no direct supervision by the court. Performance of each type of injunction is normally secured by the realisation of the person enjoined that he is liable to be punished for contempt if evidence of his disobedience to the order is put before the court; and if the injunction is prohibitory, actual committal will usually, so long as it continues, make disobedience impossible. If instead the order is for specific performance of a contract for personal services, a similar machinery of enforcement could be employed, again without there being any question of supervision by any officer of the court. The reasons why the court is reluctant to decree specific performance of a contract for personal services (and I would regard it as a strong reluctance rather than a rule) are, I think, more complex and more firmly bottomed on human nature. If a singer contracts to sing, there could no doubt be proceedings for committal if, ordered to sing, the singer remained obstinately dumb. But if instead the singer sang flat, or sharp, or too fast, or too slowly, or too loudly, or too quietly, or resorted to a dozen of the manifestations of temperament traditionally associated with some singers, the threat of committal would reveal itself as a most unsatisfactory weapon: for who could say whether the imperfections of performance were natural or self-induced? To make an order with such possibilities of evasion would be vain; and so the order will not be made. However, not all contracts of personal service or for the continuous performance of services are as dependent as this on matters of opinion and judgment, nor do all such contracts involve the same degree of the daily impact of person upon person. In general, no doubt, the inconvenience and mischief of decreeing specific performance of most of such contracts will greatly outweigh the advantages, and specific performance will be refused. But I do not think that it should be assumed that as soon as any element of personal service or continuous services can be discerned in a contract the court will, without more, refuse specific performance. Of course, a requirement for the continuous performance of services has the disadvantage that repeated breaches may engender repeated applications to the court for enforcement. But so may many injunctions; and the prospects of repetition, although an important consideration, ought not to be allowed to negative a right. As is so often the case in equity, the matter is one of the balance of advantage and disadvantage in relation to the particular obligations in question; and the fact that the balance will usually lie on one side does not turn this probability into a rule. The present case, of course, is *a fortiori*, since the contract of which specific performance has been decreed requires not the performance of personal service or any continuous series of acts, but merely procuring the execution of an agreement which contains a provision for such services or acts.

Those observations do not of themselves enable me to disregard *Ryan*'s case. But then one comes to *Shiloh Spinners Ltd* v *Harding* [1973] AC 691. At p 724 (C-D) Lord Wilberforce seems to say that "the impossibility for the courts to supervise the doing of work" may be rejected as a reason against granting relief. Finally there is *Tito* v *Waddell (No 2)* [1977] Ch 106. At p 321 Megarry V-C says this:

> In cases of this kind it was at one time said that an order for the specific performance of the contract would not be made if there would be difficulty in the court supervising its execution: see, eg *Ryan* v *Mutual Tontine Westminster Chambers Association* [1893] 1 Ch 116, especially at pp 123, 125, 128. Sir Archibald Smith MR subsequently found himself unable to see the force of this objection (see *Wolverhampton Corporation* v *Emmons* [1901] 1 QB 515, 523); and after it had been discussed and questioned in *C H Giles & Co Ltd* v *Morris* [1972] 1 WLR 307, 318, the House of Lords disposed of it (I hope finally) in *Shiloh Spinners Ltd* v *Harding* [1973] AC 691, 724. The real question is whether there is a sufficient definition of what has to be done in order to comply with the order of the court. That definition may be provided by the contract itself, or it may be supplied by the terms of the order, in which case there is the further question whether the court considers that the terms of the contract sufficiently support, by implication or otherwise, the terms of the proposed order.

In the light of those authorities it is, I think, open to me to consider the making of an order for specific performance in this case; particularly since the order contemplated is in the *a fortiori* class referred to by Megarry J (as he then was) in the last sentence of the extract from the *Giles* case quoted above. Damages here could hardly be regarded as an adequate remedy.

Whether or not a specific performance order should be made seems to me to depend on the following considerations: (a) Is there a sufficient definition of what has to be done in order to comply with the order of the court? (b) Will enforcing compliance involve superintendence by the court to an unacceptable degree? (c) What are the respective prejudices or hardships that will be suffered by the parties if the order is made or not made?

As to (a), one may in this case sufficiently define what has to be done by the defendants by ordering the defendants, within, say, two months, to employ a porter to be resident at Danes Court for the purpose of carrying out the clause 3(11) duties. It is to be borne in mind that there is still a vacant flat available for a resident porter. As to (b), I do not see that such an order will occasion any protracted superintendence by the court. If the defendants without good cause fail to comply with the order in due time, then the plaintiffs can take appropriate enforcement proceedings against the defendants. As to (c), I see no hardship or prejudice resulting to the defendants from the order. They will simply be performing what they have promised to do and what has been carried out by the lessors over the past 20 years. On the other hand I see considerable inconvenience, if not exactly hardship, for the plaintiffs if, having bargained for a resident porter and paid a premium and having enjoyed his presence for 20 years, they are to be expected for the future to be content with a porter who simply walks up and down the stairs for two hours only during the day doing his cleaning and refuse collection. It follows that there should be an order for specific performance. I should add that Mr Hochhauser referred to the four *Shelfer* conditions as set out in *Federated Homes Ltd* v *Mill Lodge Ltd* [1980] 1 WLR 594 at p 607E. There Brightman LJ says:

> There remains only the question whether we ought to interfere with the remedy granted by the judge of an injunction against the building of the 32 extra dwellings. *Shelfer* v *City of London Electric Lighting Co* [1895] 1 Ch 287 is authority for the proposition that a person who has the benefit of a restrictive covenant is, as a general rule, entitled to an injunction on the trial of the action as distinct from an award of damages unless (1) the injury to the plaintiff's legal rights is small, (2) it is capable of being estimated in terms of money, (3) it can adequately be compensated for by a small payment, and (4) it would be oppressive to the defendant to grant an injunction. In my view, the first, third and fourth of these conditions have not been shown to be satisfied.

Mr Hochhauser said that all four conditions were satisfied so that the appropriate remedy for breach of clause 3(11) was damages and not the specific relief. If and in so far as it is appropriate to consider the four conditions in the circumstances of this case, I am satisfied that not one of the four conditions can be answered in the defendants' favour.

Mr Tager at my invitation in the course of the hearing drafted a proposed order. The draft is elaborate. I am content to make an order based on that form, but since it is so extremely detailed I think I will give Mr Tager the opportunity of putting it in a minute with a simpler form of order. He may be able to do that with Mr Hochhauser's collaboration. Mr Hochhauser may wish to collaborate, since a simpler order will save the cost of working out the details involved in the original Tager draft. The order will provide for an inquiry as to damages in respect of the time since May 31 1985.

In the light of the order for specific performance, it is not necessary for me to say anything about the other relief claimed in the first notice of motion or about the relief claimed in the second notice of motion.

Court of Appeal

November 28 1985

(Before Lord Justice OLIVER, Lord Justice RALPH GIBSON and Mr Justice ANTHONY LINCOLN)

CIN PROPERTIES LTD v BARCLAYS BANK PLC

Estates Gazette March 8 1986

277 EG 973-976

Landlord and tenant — Dispute, with a complicated conveyancing background, as to a clause in an agreement concerning repairs — The agreement, which was not itself a lease, was between landlords and tenants of parts of a building, the parties having links with other interests in the building — The agreement cast upon the tenants obligations

which were not imposed on them by their lease — It provided for a proportionate contribution by the tenants to the cost of repairs or other works to parts of the structure of the building, subject to a proviso that the landlords would not accept any estimates or place any orders for the repairs without first submitting them to the tenants for approval — Extensive repairs became necessary and the landlords gave instructions to contractors to carry out the works without consulting the tenants or showing them any estimates or tenders as required by the agreement — Landlords sought a contribution from the tenants, who argued that the submission of estimates for their approval was a condition precedent to their liability to contribute — Landlords contended that the proviso was in the nature of mere machinery or a concession to the tenants to enable them to check in advance the bona fides of any estimate or proposed contract — Caulfield J accepted the latter view and held the tenants liable — Held by Court of Appeal, disagreeing with this construction, that the proviso was an unequivocal condition precedent to the tenants' liability — Further submissions by landlords based on alleged waiver or estoppel rejected; the facts did not support either suggestion — Tenants' appeal allowed

This was an appeal by Barclays Bank plc, tenants of parts of the building at Alitalia House, 249-259 Regent Street, London W1, from a decision of Caulfield J in an action by CIN Properties Ltd, their immediate landlords (and the present respondents), for a contribution of £166,254 towards repairs. Caulfield J held that the appellants were liable to pay this amount.

John Colyer QC and Kirk Reynolds (instructed by Durrant Piesse) appeared on behalf of the appellants; B Levy (instructed by Ronald V Cowles) represented the respondents.

Giving judgment, OLIVER LJ said: This is an appeal from an order made on April 12 1984 by Caulfield J on the trial of a preliminary issue in an action by the respondents, CIN Properties Ltd, for the recovery from the appellants, Barclays Bank plc, of a sum of £166,254.99 being three-sevenths of the expenditure by the respondents on external repairs attributable to that part of the building known as Alitalia House 249-259 Regent Street, London, which is occupied by the appellants. The learned judge declared that the appellants were liable to pay three-sevenths of the amount expended by the respondents in carrying out the work referred to in the statement of claim.

The appellants, to whom it will be convenient to refer as "the bank", are the tenants of parts of the building concerned and the respondents (CIN) are their immediate landlords. In order to isolate the issue which arises on the appeal it is necessary to go shortly into the conveyancing background, which is as follows.

In 1948 Jays Ltd were the leaseholders of the material parts of the building 249-259 Regent Street under a long lease, the terms of which do not matter for present purposes. In that year they sold their interest in the whole of 251-259 and in the upper part of 249 to the Representative Body of the Church of Wales, which in turn, on March 25 1948, sublet the major part of the building (but excluding the ground floor of no 251) to Great Universal Stores (GUS) from March 25 to the October 11 1948 and thereafter for a term of 47 years less the last three days. That sublease contained covenants by GUS for the external structural repair of the building. At the same time the Representative Body granted an underlease of the ground floor and basement of 251 Regent Street to Martins Bank Ltd for the same period as the underlease to GUS, the underlease containing no covenant by the tenants to carry out any external or structural repairs but containing a covenant by the landlords to observe the provisions of their headlease.

By an underlease dated May 29 1958 and made between GUS and Martins, GUS let to Martins, from May 29 1958 to October 10 1958 and thereafter for a term of 37 years less the last four days, part of the fourth floor of 249-259 together with the lift shaft connecting the ground floor and fourth floor. That underlease contained covenants by the tenants to repair additions and improvements to the building, to contribute to the cost of repairing party walls used in common with adjoining owners and to paint outside parts and clean stonework on the elevations of the demised premises at stated intervals. Again, however, it contained no covenant to execute any external or structural repairs, the responsibility for which rested with the landlords under the terms of its lease from the Representative Body.

At the same time as the grant of this underlease, however, a further deed was entered into, also dated May 29 1958 and made between the same parties. It is quite clear from this that both sides appreciated the position as regards external repairs under their respective underleases and that this document was entered into to regulate the relationship between them arising out of that position. The first clause dealt with the liabilities of the parties for the reinstatement of certain works of alteration which were either then in progress or had been completed and its terms do not matter for present purposes. Their significance, if any, is only that they contain and were obviously intended to contain a complete and comprehensive code for disposing of those liabilities.

The provision which has given rise to the issues on this appeal is clause 2, which is in the following terms:

> In the event of any repairs or other works being required to be carried out in respect of the roof, main walls, timbers, drains or other parts of the structure of that part of the said building (basement to roof) of which inter alia the ground floor is demised to the Bank (other than any such repairs for which the Bank are responsible to the Lessors under their Lease) including any repairs or works pursuant to any order or requirement made by any local or other authority the responsibility for having such repairs or works carried out shall rest upon GUS and GUS will obtain estimates or tenders therefor on the open market. Provided that GUS shall not accept any tender or estimate or enter into any contract or place any order for the carrying out of any such repairs or works without first submitting the same (together with plans and specifications of the proposed repairs or works if so required by the Bank) to the Bank for approval (such approval not to be unreasonably withheld or delayed).
>
> The Bank will refund GUS three-sevenths of the cost of carrying out such repairs or works immediately the same has been ascertained notwithstanding that the Bank's premises are not directly or indirectly affected thereby.
>
> Provided that as regards any order or requirement made by any local or other authority which calls for repairs or other works to the part of the said building comprised in either of the said Leases which are solely attributable to the nature of the use and occupation to which that part of the said building is put then the responsibility for carrying out such repairs or other works shall be the exclusive responsibility of the party to whom such premises are demised and they shall indemnify the other party in respect thereof accordingly.

It is, as I have said, quite clear that both parties fully appreciated the position as to repairs under their respective leases and this is underlined by the recitals to the deed, the material parts of which are as follows:

> 5. The lease to GUS provides that they will well and substantially repair not only the premises comprised in such lease but also the premises comprised in the lease to the Bank and further that GUS will carry out any works as well to the premises thereby demised as also to the premises comprised in the lease to the Bank and all things necessary in relation to the said building to comply with any order or other requirement of any local or other authority in respect of the premises demised to GUS and the Bank respectively, and GUS covenants to insure the premises demised to GUS in the full value thereof with necessary joinder to the Bank.
>
> The lease to the Bank is framed so as not to cast any liability upon them in respect of the repairs and works aforesaid and contains a covenant by them to insure the premises demised to them in conjunction with GUS.
>
> 7. The underlease to the Bank is framed so as not to cast any liability upon them in respect of the roof, main walls, timbers, drains or other parts of the structure of the said building or in respect of any works or acts necessary to comply with any order or other requirement of any local or other authority in respect of the premises thereby demised but contained a covenant by the Bank to insure the premises demised to them in conjunction with GUS.

The underlease to GUS became at some time prior to 1966 vested in CIN and the appeal bundle contains correspondence in that year between CIN and Martins Bank recording an informal agreement as regards repairs. On April 15 1966 Martins wrote to CIN's property department agreeing a suggestion that "prior notice of large items of repair need only be given where the cost is £150 or over".

The bank are the successors in title to Martins Bank Ltd.

The immediate cause of the dispute giving rise to this appeal occurred in 1977, when a piece of masonry became dislodged on the outside of the fourth floor. The fire brigade was called and the immediate danger removed, but inspection revealed that extensive repairs to the stonework were required and the local authority served a dangerous structure notice. Thereafter CIN entered into contracts (on October 6 1978 with Harveys and on March 7 1980 with Taylor Woodrow) for the carrying out of major works, involving overall as regards the whole building (as opposed to that vertical portion of it containing the bank's premises) upwards of £6m. In no case, however, were the bank consulted about the cost of the work. They

A were not shown or asked to approve any tenders, estimates, orders or contracts as required by clause 2 of the deed of 1958 and it is that omission which gives rise to the first issue on this appeal.

It is common ground that if the bank are under any liability at all for payment for the repairs carried out that liability arises under the 1958 deed and nowhere else and it is common ground that the provisions of that deed have not been observed by CIN, so that in now seeking a contribution under the deed CIN are, in a sense, relying on their own breach of contract.

CIN's contention, however, which found favour with the learned judge, is that the requirement that estimates and tenders for external repairing work should first be submitted for approval to the bank, B albeit it follows immediately on CIN's obligation and precedes that of the bank as a matter of the structure of the relevant clause, is a mere matter of machinery or, as the judge put it, a concession to the bank to enable them to check in advance the bona fides of any proposed estimate or contract. Mr Levy, on CIN's behalf, argues that the operative provision of the clause is that which casts on the bank the obligation to pay three-sevenths of "the cost" of carrying out the repairs, not three-sevenths of an approved estimate, and that the intermediate obligation of submission is merely interposed for CIN's benefit to enable CIN, by obtaining approval in advance, to preclude any argument subsequently about whether the costs were not properly incurred.

That is a view with which I confess to having some sympathy, for C the argument that costs properly incurred for the benefit of all those interested in the building should be irrecoverable only because they have not been scrutinised and approved before the work is carried out is not, on its face, an attractive one.

Nevertheless, Mr Colyer for the bank submits that, as a matter of construction of the clause, there is no escape from the proposition that the submission of estimates etc for approval is and was intended to be a condition precedent to liability to contribute to the cost.

Despite the sympathy that I have with Mr Levy's contention, I have for my part felt compelled to the conclusion that Mr Colyer is right. The question is purely one of construction and there is little prior jurisprudence that is of any assistance. That a provision of this type can constitute a condition precedent to liability is clear from the decision of this court in *Finchbourne Ltd* v *Rodrigues* [1976] 3 All ER 587, where it was held that under a covenant in the lease of a flat for the payment of service charges to be certified by the landlord's managing agent, the issue of a proper and valid certificate was a precondition to any liability, since it was only thus that the extent of any liability could be quantified. That, however, is not precisely on all fours with the instant case. The obligation here is to contribute to the cost and, as the learned judge pointed out, there is a provision that approval cannot be unreasonably withheld. On the other hand, Mr Colyer points, as he is entitled to, to *Barrow* v *Isaacs* [1891] 1 QB 417, where, despite an express finding that there could be no reasonable objection to a proposed undertenant, it was held that the submission of the proposal to the landlord was a necessary prerequisite of a valid sublease under the covenant against assigning or subletting without consent.

In the end one is compelled to look at the structure of the clause and the context in which it was entered into. One starts from the position that it was this provision which was intended to cast upon the tenants an obligation which they did not have under their lease. That was expressly recognised. It starts with a specific obligation to obtain estimates or tenders on the open market. Thus it would, in my judgment, be unarguable that the landlords could recover any part of the expense if they chose, for instance, to carry out the work through their own maintenance department or by direct labour. The obligation assumed to obtain estimates on the open market is expressed unequivocally as a contractual obligation. Equally unequivocal is the proviso attached to that obligation. There is an express contractual provision not to accept any tender or enter into any contract for the repairs without first submitting them to the bank for approval and one must, I think, assume that that contractual provision was intended to be effective and to be enforceable.

But if Mr Levy be right, it is a contractual provision for which no sanction exists and that, as it seems to me, cannot have been the intention as it emerges from the words used. The learned judge regarded this contractual proviso as a not very valuable concession to the tenants, but if, as he held, it was of no effect at all because the tenants would be bound to pay whether it was operated or not, then it seems to me not merely to be of little value but to be completely valueless. G

Equally, if, as Mr Levy argues, it was inserted for the landlords' protection in order to prevent argument in the future as to whether expenditure was properly incurred it (a) would not prevent argument if approval were withheld, and (b) added nothing to what would in any event have been the position without any contractual obligation at all if approval were given.

I start, therefore, from the position that, this being a deed designed to regulate the conditions upon which the tenants were undertaking to assume an obligation that it did not have under the relevant subleases, the proviso must have been intended to have some effect and to be enforceable against the landlords, and I can, for my part, see no other way in which it could be effectively enforced against the landlords than by its being treated, as its position in the clause and its H contractual prohibition would suggest, as a condition precedent to the tenants' obligation to pay that which, apart from the clause, would not have been due in any event. It must also, in my judgment, be read as one clause and as designed to set out as a comprehensive whole the conditions upon which the tenants were to assume the new liability.

I confess to having reached this conclusion with a degree of reluctance, for it means that, subject to Mr Levy's further contentions on waiver and estoppel, CIN have deprived themselves of a recoupment, of which they might have availed themselves, by an unfortunate oversight. Nevertheless, I find myself unable to construe the clause in a way which, for practical purposes, gives no effect at all J to this carefully worded and comprehensive proviso which was clearly intended to ensure that the tenants were fully consulted at every stage from estimate to contract.

That conclusion, however, does not dispose of the matter, for it is still necessary to consider the points raised in the respondents' cross-notice. That raises three contentions, the first of which — that the proviso is simply directed to the lessees' ability to challenge expenditure in fact incurred — has already been dealt with.

The two other points raised are first that, on the facts, the bank either waived compliance with the requirements of the clause or became disentitled to insist on them by reason of estoppel or acquiescence and, second, that CIN are entitled to some payment by way of *quantum meruit*. The latter can be disposed of immediately in K the light of the express contractual provision and it has not been pressed by Mr Levy. As to the former, the learned judge made certain findings of fact and Mr Colyer has devoted some time to a submission that, at any rate as regards his findings that the bank were aware of the necessity for submission of estimates by March 1979, there was no evidence to support them.

For my part I have felt it unnecessary to pursue the inquiry upon which Mr Colyer has invited us to embark, because I am content for present purposes to accept the learned judge's findings. Accepting them at their face value I am, speaking for myself, quite unable to see how they could support the case for waiver or estoppel upon which Mr Levy seeks to rely. L

What the judge found may be summarised in three propositions, viz: (1) that in June 1977, when the damage manifested itself, there was a state of emergency; (2) that CIN were unaware of their rights and obligations under clause 2 of the deed of 1958 until June 1980; and (3) that the bank were likewise unaware of their rights and obligations under the deed until, probably, March 20 1979; but that if that be wrong they were certainly aware of their rights and obligations in December 1979.

I will consider the impact of these findings in a moment, but there is one matter which, in fairness to the bank, ought to be disposed of at this stage. The learned judge drew certain inferences with regard to the stage of knowledge (and by inference to the propriety of the conduct) of a Miss Pleasance who was employed in the bank's M property department and conducted much of the correspondence on the bank's behalf. He appears to have drawn an adverse inference from the fact that Miss Pleasance was not called as a witness. For my part, while I accept that the correspondence is certainly susceptible of an inference that Miss Pleasance may have become aware that the bank were vulnerable if the provisions of the 1959 deed had been observed, I can see nothing sinister or significant in the fact that she was not called to give evidence since, on the pleadings as they stood at the trial, I can see no issue upon which it would have been necessary to call her.

I turn, therefore, to the question of whether, on the facts found by the judge, Mr Levy succeeds in supporting a case of waiver or estoppel. Quite plainly, as it seems to me, no question can arise prior to March 1979, when neither party were, on any analysis, aware of their rights. The bank waived nothing, because they were not aware that there was anything to waive. Mr Levy has sought to make something of the fact that it was the bank who raised the issue of repairs in the first instance and requested CIN to take the necessary action. Nothing, as it seems to me, can turn on that. The disrepair first manifested itself on the exterior face of the bank's premises and it was, as the bank rightly surmised, for CIN and not the bank to deal with it. Thereafter CIN informed the bank, on August 24 1977, that they had instructed Harvey-Damer to carry out repairs and the hope was expressed that this would not occasion too much inconvenience. Then on November 7 1977 the bank wrote to CIN confirming what had been agreed at a meeting on the site including the following:

> 5. It was agreed that should the proposed works result in the moving of computer terminals or other such Bank equipment, the cost of such works will be borne by the Landlords, as will the cost of any other works resulting from the stonework repairs.

However, on September 28 1978 CIN wrote requesting a contribution to repairs and scaffolding, and internal memoranda of the bank reveal that this prompted an inquiry as to the extent (if any) of the bank's liability. The relevant documents seem to have been collected at some time in February 1979, and on March 20 1979 Miss Pleasance wrote to the chief clerk at local head office saying that she (or her department) had been through all the documents and that "there is at least an argument for saying that we are not responsible for the repairs, although we would be responsible for exterior painting and stone cleaning". Again speaking entirely for myself, I cannot see that anything hinges on this. It is, I suppose, possible that Miss Pleasance may have appreciated that the 1958 deed required submission of estimates and contracts and that that had not been done, but certainly nothing was done to suggest to CIN that any such requirement either was being or had been waived. Indeed, on the same day she wrote to CIN disclaiming any responsibility at all for repairs and inviting comments. For my part, I can see nothing whatever wrong with that. It may or may not have been a good point, but its only effect could or should have been to alert CIN to the necessity to look at the relevant documents and formulate the basis for the claim which they knew was being disputed. The response elicited was in fact no more than a quantification in percentage terms of the alleged liability, and on April 5 1979 Miss Pleasance replied stating unequivocally that the bank regarded the repairs as CIN's and not the bank's liability, a stance repeated in a further letter of July 30 1979. On August 10 1979 CIN replied agreeing that the bank would not be liable to pay a proportion of the cost of exterior structural repairs.

It is certainly true that thereafter CIN kept the bank informed from time to time of the probable progress of the work, but I find myself quite unable to see how this correspondence can be said to support a waiver or estoppel.

Mr Levy has submitted that the fact that the bank must be taken, on the judge's findings, to have been aware from March 1979 onwards of the provisions of the 1958 deed enables CIN in some way to invoke the broad equity which I endeavoured to expound in *Taylor's Fashions Ltd* v *Liverpool Victoria Trustees Co Ltd* [1982] QB 133, but I confess that I have been unable to follow the basis for this contention. Mr Levy accepts that there was here no common assumption upon which it could be said that CIN have acted to their detriment. The work of repair was their responsibility and had to be carried out not only so far as concerned the bank but so far also as the superior lessors were concerned.

Can it be said then that either CIN were led by the bank not to submit estimates or that there was a common assumption that the bank would pay their contribution even though no estimates were submitted? Since throughout, rightly or wrongly, the bank were contesting any liability at all, this argument cannot run. There was throughout a disagreement, whether well-founded or not, about the bank's liability to pay and in the absence of some duty on the bank to draw CIN's attention to a legal point on the construction of the deeds which were the subject-matter of the contest, I can see no ground for saying that their failure to do so was unconscionable — assuming that at that stage they had this particular point in mind.

For my part I can see no such duty and I would, therefore, reject the grounds advanced in the respondents' notice. I would accordingly allow the appeal.

RALPH GIBSON LJ and ANTHONY LINCOLN J agreed and did not add anything.

The appeal was allowed with costs in the Court of Appeal and below.

Court of Appeal

October 17 1985

(Before Sir John DONALDSON MR and PARKER LJ)

J WILLIS & SON v WILLIS AND ANOTHER

Estates Gazette March 15 1986

277 EG 1133-1134

Equity — Estoppel — Appeal by occupiers of premises against grant of possession order by county court — It was common ground that the owners were entitled at law to a possession order, but occupiers claimed that they had a right to remain by virtue of the principle of promissory estoppel — They claimed that on the faith of a statement (made on behalf of the owners) that they could live rent free in the subject premises as long as they wanted to, they had spent money on the improvement and repair of the premises — They argued that it would therefore be unconscionable for the owners to insist on their strict legal right to possession — The assistant recorder in the county court had rejected this plea on the ground of conduct by the occupiers which violated the principle of equity that "he who comes to equity must come with clean hands" — There was evidence that the appellants had put forward, in support of particulars of alleged expenditure for the benefit of the premises, a letter from a third party about supposed work done which was wholly fictitious and fraudulent — Held that the assistant recorder was fully justified in deciding that in the circumstances no court could grant equitable relief — Appeal against possession order dismissed

This was an appeal by Simon John Willis and Joanna Willis from a decision of Mr Assistant Recorder Wigmore at Swindon County Court, granting J Willis & Son, the applicants in the county court and the respondents to this appeal, an order for the possession of the Stable Flat, Twatley Farm, Malmesbury, Wiltshire.

Colin Sara (instructed by Stephen Berry & Co, of Bristol) appeared on behalf of the appellants; Graeme Wood (instructed by Goughs, of Chippenham) represented the respondents.

Giving the first judgment at the invitation of the Master of the Rolls, PARKER LJ said: On February 22 1985 Mr Assistant Recorder Wigmore at the Swindon County Court granted to the applicants, who are the respondents in this appeal, an order for possession of premises known as the Stable Flat, Twatley Farm, Malmesbury, Wiltshire. The action had been begun by an originating application dated May 16 1984. The appellants appeal from that order.

It is common ground that the applicants are entitled to possession at law, but it was contended by the respondents that they were entitled to remain on the basis of a promissory estoppel. The estoppel relied on is variously called in the cases "promissory estoppel", "equitable estoppel" or "quasi estoppel". Whatever it is called, however, its origins lie in equity and not in law. The root question is, therefore, whether it would be inequitable to allow the applicants to insist on their strict legal right to possession.

The history of the matter is a long one and the details of it are complicated. They are all dealt with at length in the judgment of the learned judge, which has been recorded in a long and no doubt entirely accurate note by counsel. In the event, it is, however, unnecessary for me to recite the long and detailed history, which is one of family dispute developing over a considerable period.

The basis of the claim for equitable estoppel is that a statement on behalf of the applicants, repeated more than once, was made that the respondents could live in the premises rent free as long as they wanted to, and in reliance upon that promise the respondents expended money on the premises from time to time and in considerable amounts. Hence it is said that it would be inequitable to allow the applicants to rely on their legal right to possession notwithstanding that as long ago as August 18 1983 they had requested the respondents, as they were perfectly entitled at law to do, to move out and had given a formal notice on April 6 1984 that possession was required on May 7 1984.

The learned assistant recorder rejected the defence on a number of grounds, one of which was that the respondents had sought to bolster their claim by putting forward a document which was known by the respondent Simon Willis, at any event, to be wholly fraudulent. He was, in my judgment, entirely right so to do, and I would be content to dismiss the appeal on that ground alone. When a person seeks the aid of the court to obtain the court's assistance, via the principles of equity, to override somebody's strict legal rights, it is clearly a case for the application of the maxim, as indeed is accepted by the appellants, "that he who comes to equity must come with clean hands". This matter is dealt with by the learned assistant recorder, where he says this:

> I have Mr Robins's letter passed [by the respondents] to their solicitor knowing it was fraudulent. Mr Sara says that only if a matter is very serious should it be taken to take away an equity. I take an extremely serious view of this. I am not sure just how far Simon was concerned [in its original procurement]. But one thing is absolutely clear. He nevertheless passed it to his solicitor knowing what it was. It was specifically to bolster a claim to equitable relief. Even if I found he could otherwise have relied on equitable relief I would disqualify him on that ground alone.

I, too, regard it as a very serious matter.

It is submitted on behalf of the appellants that although an equity can be taken away by bad conduct on the part of the person seeking to obtain equitable relief, it must be of a very, very serious character and a balancing exercise must be done to see whether it is sufficiently serious to justify the barring of the relief. He relies on the only case in which, apparently, the matter has arisen in like circumstances to the present, *Williams* v *Staite* [1979] Ch 291. In that case there was a somewhat similar situation. Two cottages were owned by Mrs Staite's mother. She had owned them both from a period in 1946. She had a daughter who lived at home. In 1960 the daughter married Mr Staite, and the mother said to her daughter: "You can live in and have no 2 as a wedding present. You can live there as long as you wish." She said that to the bride and bridegroom on their wedding. The bridegroom, Mr Staite, was not keen to move into no 2 because he had a cottage a mile or two away which went with his job. But the mother, Mrs Moore, wanted the young couple to move into no 2. She wanted to have them next to her so that the daughter could look after her parents as they got older. As a result, Mr Staite, the bridegroom, gave up his other cottage and moved into no 2 so as to look after the father and mother as they got older. On that basis it was held that an equitable licence for life had been established. Subsequently the cottage was sold, and the new owner sought to obtain possession from the Staites. The question arose whether the new owner could be prevented from exercising his legal rights as a result of the equitable licence which had been established to have been granted by the mother. It was said that he could not because of the conduct of the licensees towards the new owner. It was held in that case that the established equitable licence to occupy for life had not been determined by the subsequent misconduct.

This case is somewhat different. Here the question is not whether a subsequent owner can be defeated in exercising his legal rights by the bad conduct of an equitable licensee, but whether there is an equitable licence at all. For my part I consider that there is a difference between the two situations. When the question is whether the equitable licence is established at all, it will normally be the case that the degree of misconduct required is less than would be required if it were sought to cancel a pre-existing equitable licence, if indeed — and there is some ground for saying there is not — such an equitable licence, once established, can be destroyed by such misconduct.

The conduct in the present case was very serious, for this reason. In their original defence the appellants set out the items of expenditure upon which they relied and which they said had been performed in reliance on the promise that they could stay in the flat for as long as they wanted to. The defence and counterclaim appear in the papers, and the items include the following:

> August 1979, £559.90, New bath and sink unit. February 1980, £320.00, Repairs to roof and loft insulation. May 1980, £160.00, Replacement of old hot and cold water pipes with new copper pipes. September 1980, £300.00, Removing of wall in front room, replastering and making a cupboard space.

They were asked for particulars of the allegations of expenditure on August 2 1984, and what appears to be a letter from a Mr Robins to Mr Simon Willis (one of the appellants), dated August 11 1984, was handed to the solicitors for the appellants and was used for the answer to the request for particulars which were delivered on September 6 1984. The answers specified Mr Robins as having been the person who had carried out the items which I have recited. The letter from Mr Robins reads as follows:

> Dear Mr Willis,
> With reference to your request for copies of invoices regarding work carried out at your address in 1979 and 1980, I regret that I no longer have complete details; I can however confirm from entries in my ledgers going back, the following particulars which should help you. In August 1979 there was an entry for £559.90 which would almost certainly be for the new bath and sink unit which I then fitted; again in February 1980 for a sum of £320 involving a number of sheets of foilbacked plasterboard, furring strips, etc for repairs to the roof and insulation. May 1980 you paid me £160 for plumbing work for which I have no details. Then in September removal of wall and reinstatement in lounge area I have an entry of £315. I recall electrical work being requested in early 1981 but have no knowledge of any payment or details because if you remember this was passed on to a man from Sherston whose name I cannot recall. Similarly I have no record of any double glazing. I do hope this helps.

That letter was a complete fiction. Mr Robins had not done any work; he had no ledgers, and he had not been paid any money. The document came into existence apparently at the instance of Mr Simon Willis' father-in-law, who sought Mr Robins' assistance to provide evidence of the expenditure of money. The assistant recorder made no finding as to how far Mr Simon Willis was concerned in the procurement of the original document, but that he knew it was wholly false is accepted and beyond argument.

The applicants' knowledge that it was false arose in this way. The document having been served, I think with the particulars, upon the applicants' solicitors, they made inquiries of Mr Willis for further details, the letter not being very precise. They made three attempts, and eventually went to see Mr Willis. As a result of their visit, he then signed on December 9 1984 a statement in which he admitted that the letter was wholly false and explained the circumstances in which it had been obtained, to which I have already referred.

The document was also included in a list of documents filed by the respondents to the summons, but that was on February 7 1985. It was presumably then known to everybody that it was false because the pleading had by then, if the dates on the document are correct, already been amended in order to remove the Robins item from the defence. I am not entirely clear whether that is correct, for it is strange that the list of documents should have been filed as late as it was. Be that as it may, there is no doubt that when Mr Willis passed the letter to the solicitors he knew that it was false in every single particular. It was used as a formal document in court proceedings, and the inference is in my view inescapable that, if the falsity had not been discovered by the applicants' solicitors, it would have been relied on throughout.

I find it difficult to see how there could be any more serious conduct than that. When a party comes to the court and seeks to obtain from it equitable relief, it is accepted, as I have said, that he must come with clean hands. I accept also, as was submitted on behalf of the appellants, that not every item of misconduct can possibly be sufficient to deprive a party who seeks equity from being granted the relief he seeks. Some misconduct may be trivial. But, when a party acts as these parties have done — and Joanna Willis must be regarded as having been concerned in this, albeit indirectly, inasmuch as the document was put forward on behalf of both the appellants — it seems to be impossible for this court to do other than take the most serious view of it and to decline to grant equitable relief, even if, to which I say nothing because it does not arise on the view I take of this case, they would otherwise have been so entitled.

I accordingly would dismiss this appeal.

Agreeing, SIR JOHN DONALDSON MR said: The conduct of the appellants which has been disclosed in this case was such that no court could, in my judgment, possibly grant equitable relief.

I, too, would dismiss the appeal.

The appeal was dismissed with costs against appellants, not to be enforced without leave of the court; order nisi against legal aid fund on usual terms; legal aid taxation of appellants' costs.

A
Chancery Division

November 20 1985

(Before Mr Justice PETER GIBSON)

BRITISH BAKERIES (MIDLANDS) LTD v MICHAEL TESTLER & CO LTD

Estates Gazette March 22 1986

277 EG 1245-1248

Landlord and tenant — Covenant against assignment
B
without consent of lease of business premises — Tenants challenge landlord's refusal of consent as unreasonable — Landlord refused consent on two grounds, on the proposed user of the premises and on the financial ability of the proposed assignees — The first ground was rejected by the judge in the light of *Killick* v *Second Covent Garden Property Co Ltd* as being an expectation of a future breach of user covenant which did not provide a good reason for a refusal of consent — The main issue was as to the second ground — Tenants put forward six sets of references as to proposed assignees' financial ability, from accountants, bankers, solicitor, estate agents and two trade references, and also
C
submitted accounts relating to the proposed assignees' past trading — The references, however, related to the proposed assignees' previous business and previous premises, whereas they were intending to embark on a further venture in premises with a higher rent; certain property valuations were unreliable; and the unaudited accounts were not such as to remove a reasonable landlord's doubts as to the proposed assignees' financial ability — A suggestion that the proposed assignment would substantially depreciate the landlord's interest was, however, rejected, both on the ground that there was no evidence that this had affected their minds at the date
D
of refusal and for the reasons given by the deputy judge in criticising a similar submission in *International Drilling Fluids Ltd* v *Louisville Investments (Uxbridge) Ltd* — Held that the landlord was not acting unreasonably in refusing consent in view of his real doubts as to the proposed assignees' ability to meet the obligations under the lease — Declarations sought by tenants refused and action dismissed

In this action the plaintiff, British Bakeries (Midlands) Ltd, tenant of premises at 89 Cornwall Street, Plymouth, sought against the defendant, Michael Testler & Co Ltd, the landlord, declarations based on the plaintiff's submission that the defendant had unreasonably refused consent to the assignment of the lease of the
E
premises.

W Poulton (instructed by Metson Cross & Co) appeared on behalf of the plaintiff; David Neuberger (instructed by Hughes Watton & Co) represented the defendant.

Giving judgment, PETER GIBSON J said: British Bakeries (Midlands) Ltd, the tenant under a lease of premises at 89 Cornwall Street, Plymouth, seeks two declarations: (1) that the defendant, Michael Testler & Co Ltd, the landlord under that lease, has unreasonably withheld its consent to an assignment of the lease to two proposed assignees, and (2) that the tenant is entitled, without the landlord's consent, to assign the lease to those assignees.

The lease of those premises is one granted for a term of 35 years from March 25 1962. The premises consist of a building on two
F
floors. I need only refer to the user covenant in the lease relating to the ground floor, the user of which is limited to that of a retail shop for the trade of baker, confectioner and pastry cook. That was varied in 1980 and the present actual and permitted user of the ground floor is a composite use of retail shop, hot food take-away and restaurant/snack bar. Clause 24(a)(ii) contains a covenant by the tenant "not without the previous consent in writing of the landlord, not to be unreasonably withheld, to assign or under-let the premises as a whole". As a result of a rent review in 1983, the rent payable for the remainder of the term of the lease is £14,780 per annum.

On June 26 1984 the tenant's solicitors wrote to the landlord's solicitors and said that the tenant had instructed them in connection with the disposal of their leasehold interest to a Mr P Vasiliou and Mr J Arghyrou, trading as Gregory's Steak House. The landlord's consent was requested. To enable the landlord to consider the proposed assignees, copies of references from the assignees' bankers and accountants, together with a trade reference, were enclosed and the landlord's solicitors were asked to let the tenant's solicitors know whether further information concerning the assignees was required. On July 2 the landlord's solicitors acknowledged that letter and asked whether the proposed assignees were currently trading from any other premises, in which case a landlord's reference was required. They also asked for copies of the last three years' trading accounts and inquired as to the use to which the proposed assignees wished to put the premises. On July 6 a landlord's reference in the form of a letter from Hillier Parker May & Rowden, the managing agents of the lessor, of 94 Cornwall Street, Plymouth, of which the proposed assignees were the tenants, was supplied. Hillier Parker very properly drew attention to the fact that the rent payable was approximately half the rent for 89 Cornwall Street, but said that the proposed assignees had proved to be responsible and respectable tenants, complying with all their covenants as tenants. Two other references were supplied to the landlord's solicitors by the tenant's solicitors, who stated that there would be no change of user. On July 17 accounts were supplied for the years ended March 31 1981 and 1983 respectively. On July 19 the landlord's solicitors wrote to the tenant's solicitors expressing concern that the reference supplied by Hillier Parker referred only to a smaller rent, and they also expressed concern about the proposed user of the premises, saying that it was only reasonable and proper that the landlord should be able to satisfy itself that there would be no breach of the permitted user if the proposed assignment took place. The tenant's solicitors replied on July 20, pointing out that Hillier Parker could only base their reference on the property where the clients of those agents were the lessors, and stated that the assignees would be using the premises as a café/take-away involving both hot and cold foods and, therefore, there would be no breach of user. On July 25 the landlord's solicitors wrote, commenting adversely on figures appearing in the accounts of the proposed assignees, suggesting that the proposed user would be a breach of the lease and stated that they could not advise the landlord to approve the proposed assignment. On September 19 the tenant's solicitors wrote again. They enclosed a letter from the proposed assignees' accountants, who made various comments on the accounts. They also repeated that there would be no change of user, and this was confirmed by the solicitor to the proposed assignees. Finally, on October 19, the landlord's solicitors wrote again. They again commented adversely on the figures in the accounts and concluded that the landlord's opinion was the same as that set out in the letter of July 25 and that the landlord must decline to approve the proposed assignment. It is common ground that that was the relevant refusal by the landlord, which the tenant now claims was an unreasonable withholding of consent to the assignment.

These proceedings were instituted by the tenant on April 10 1985. The only evidence that is relied on by either side is in the form of affidavits. Two have been filed on behalf of each of the tenant and the landlord and there has been no application to cross-examine any of the deponents.

There is a good deal of common ground between Mr Poulton appearing for the tenant and Mr Neuberger appearing for the landlord as to the relevant principles to be applied. I put those principles in my own words as follows: (1) It was for the tenant to satisfy the landlord that its consent to the proposed assignment should not be withheld. Information not known to the landlord when it refused consent is irrelevant to the question whether the refusal was unreasonable. (2) The court should have regard to the reasons which affected the mind of the landlord in refusing its consent. Reasons which were not stated to the tenant before or at the time of the refusal may, if the court is satisfied that they did affect the mind of the landlord at the relevant time, be taken into account, but not reasons which subsequently occurred to the landlord. (3) The onus is on the tenant to establish that consent has been withheld unreasonably. (4) The test to be applied is an objective test: could a reasonable landlord have withheld consent for the reasons actually shown by the landlord? The court cannot substitute its own view for that of the landlord unless on the objective test it can be shown that no reasonable landlord could have withheld consent for the reasons shown by the landlord. (5) If a landlord has a good and a bad reason for withholding consent, consent may nevertheless have been reasonably withheld if the good reason is a sufficient reason and is not otherwise vitiated by the bad reason.

It is apparent from the correspondence that two reasons were advanced on behalf of the landlord for the refusal of consent. The first related to the financial ability of the proposed assignees to meet the obligations that would be imposed on them under the lease, in particular the rent of £14,780 per annum. The second related to the proposed user of the premises. As I read the two affidavits of Mr Testler, a director of the landlord and the person who dealt with the question of consent, the first reason was the substantial reason for the withholding of consent and I shall come back to it. The user of the premises is primarily adverted to by Mr Testler in his affidavits in a way different from that mentioned in the correspondence. In the correspondence the proposed user was said to be a breach of covenant (see in particular the letter of July 25 1984). In the affidavits the primary point taken by Mr Testler on this is put in this way:

> It is my considered opinion

and I note the present tense

> that the proposed assignees in proposing to use the premises as a café/take-away involving both hot and cold foods . . . may amount to an intensification of that part of the actual user which is not recognised by the planning authority and thereby invite action to be taken by the planning authority. Such action would inevitably materially affect the proposed assignees' profitability and thereby their ability to meet their obligations under the lease.

I am not satisfied on this evidence that this was a reason which affected Mr Testler's mind when the landlord refused consent. The point then taken was the alleged breach of covenant, despite the repeated assurances that the assignees would not change the user. I need not, however, take up time on this point as Mr Neuberger accepts that, in the light of the decision of the Court of Appeal in *Killick* v *Second Covent Garden Property Co Ltd* [1973] 1 WLR 658, even if the landlord did expect there to be a breach of the user covenant, it would not be a sufficient reason for withholding consent. Accordingly, I hold that the second reason given by the landlord and based on user is not a good reason.

I turn back to the first reason. The tenant in seeking the consent of the landlord relied first on references and second, when it complied with the request for accounts, on accounts. There were six references: (1) the proposed assignees' accountants, (2) their bank, (3) their solicitor, (4) Hillier Parker, and (5) and (6) two trade references. All suggested that the proposed assignees were satisfactory persons to have had dealings with. But this was said in relation to the assignees' business of which they had experience, that is to say of them trading as Gregory's Steak House. Had the assignees been proposing to transfer their existing business to 89 Cornwall Street, and the obligations under the lease of those premises were not more onerous than they were under the lease of their existing premises, those references would have been of greater weight. But it is common ground that the assignees were proposing to embark on a further business venture by seeking the further premises. Some referees like the accountants, the bank and the solicitor, Mr Daniel, went further in saying that the assignees should prove satisfactory tenants of the new premises, but save in the case of the accountants who prepared the trading accounts of the existing business, it is not apparent — and in my judgment would not be apparent to the reasonable landlord — on what facts that opinion was based. For example, the National Westminster Bank say (and it does not appear who in the bank so said): "Responsible and trustworthy; should prove good. A lease of premises with annual outgoing of £14,780 exclusive." All the referees except the trade referees excluded liability for their statements, and, while this is common enough these days, it nevertheless detracts from the weight to be given to the references.

Mr Poulton submits that I should infer that the references from the accountants, the solicitor and the bank were based on information such as is contained in an affidavit sworn by Mr Brennan, the solicitor of the tenant, on October 28 of this year. He deposes to Mr Vasiliou owning a £60,000 house and Mr Arghyrou owning jointly with his wife two properties each worth £37,000, but one subject to a £12,000 mortgage, and the other, as well as Mr Vasiliou's house, subject to a £50,000 mortgage apparently being to purchase another property in Plymouth. I do not see how I could properly infer that those referees had that information. Indeed, even with the affidavits before me, I do not see how I can infer that such information was available to such referees. None of the information was conveyed to the landlord before its decision to refuse consent and they only appeared in Mr Brennan's late affidavit.

In any event, to my mind it is impossible to place much reliance on valuations which are not stated to have been made by a professional G
valuer; nor do I see that such information would satisfy a reasonable landlord that the rents under the lease would be paid as they fell due.

Of the two trade references, one is wholly unspecific as to the amount of business transacted with the proposed assignees, while the other refers to only doing £300-400 worth of business per month.

Hillier Parker could only speak as to the assignees' behaviour as tenants under a lease requiring payment of a much smaller rent.

Mr Brennan submitted in his affidavit that the various trade references for the proposed assignees which were provided to the landlord were those that might normally be expected for the location and nature of the business under consideration. Mr Brennan appears to be a London solicitor. He does not state what is his experience to H
justify that submission. It is further contradicted unequivocally by Mr Testler in his second affidavit. Mr Testler deposes to the fact that he has been professionally engaged in the commercial property market for 18 years. He draws attention to the fact that the references do not contain much information and that notwithstanding the criticisms made by him of those references in his own first affidavit, no attempt was made to meet the deficiencies. I prefer Mr Testler's evidence on this point in the circumstances. In my judgment, a reasonable landlord would be entitled to regard the references as not justifying the giving of consent to an assignment unless there was other information available to the landlord of the assignees' ability to meet their obligations under the new lease as those obligations fell due, in addition to their existing obligations. J

The accounts, which are unaudited accounts, relate to the business of the assignees trading as Gregory's Steak House. Because they are unaudited, the accountants must have been dependent on what they were told by the proposed assignees. The latest accounts show the position as at March 31 1983, that is to say 18 months before the refusal. No accounts or information were supplied in relation to 1984. The accounts show that in 1983 the gross takings were £138,439, leaving a net profit of £15,654. That profit is struck before the assignees take a penny for themselves. The accountants, in their letter of September 12 1984, point out that the net profits of the business to March 31 1983 were after charging £3,120 for "wife's wages". Mr Poulton suggests that this was relevant because the more K
money that came to the family, the less the assignees themselves needed. I cannot see the relevance of this to a reasonable landlord looking to the ability of the assignees to pay rent. The wives were not offered as guarantors. The accountants further refer to nearly £6,500 of refurbishment expenses and suggest that in normal trading circumstances annual profits of between £18,000 and £20,000 would be expected. The £6,500 appear in the accounts as "repair and maintenance" and the sum is higher by about £4,000 than the previous year. But even if the figures given by the accountants as normal are correct, the profit margins are not large, particularly when the absence of any salary for the assignees themselves is taken into account. Further, the figures should be contrasted with the net profits for the previous year, which were only £8,438, again without the assignees taking a penny. L

A reasonable landlord may well have serious doubt as to the ability of the assignees to pay a rent of £14,780 exclusive of rates, taxes and other outgoings, as well as meeting the full repairing obligations under the lease, if the new venture did not prosper. No profit projections whatsoever were offered to the landlord in relation to the premises which the assignees wish to acquire. Mr Testler's evidence, which was not challenged, was that a generally accepted test of the financial standing of any proposed assignee is that his accounts should show a pre-tax profit of not less than three times the amount payable under the lease in question. It is clear that the assignees cannot satisfy that test.

The accountants in their letter of September 12 also refer to the M
balance sheet. That showed net current assets of only £353. If the 1983 accounts had followed the practice adopted in the 1981 accounts, and shown a loan account belonging to Mrs Arghyrou and Mrs Vasiliou as current liabilities, current liabilities would exceed current assets by £13,000. The accountants suggest that the loans would not have been called in before sale, but a reasonable landlord would understand the position as being that the loans were repayable on demand. The accountants then state that the assets of the business would realise at least £100,000 on the open market. I comment that a reasonable landlord is concerned with the tenant's ability to meet the obligations under the lease as those obligations fall due. Many a debtor is properly made bankrupt, even though he claims he has

assets which when ultimately realised would be sufficient to pay his debts.

In any event, I am not clear how the accounts have arrived at the figure of £100,000. The accounts show that the assets of the business exceed the liabilities by some £7,000, but that ignores the loan account owned by Mrs Arghyrou and Mrs Vasiliou. Thus the true picture is that the liabilities exceed the assets.

The accountants made a general statement about the assets of the business being written down. Mr Poulton accepted that the £100,000 figure must be made up by the value of the lease and the goodwill, the values of neither of which were written down. The accountants do not say that they obtained a professional valuation of the lease and a reasonable landlord might assume that they had not done so because the accountants state that it is their opinion that the value of the business is £100,000. Whether there was any value in the residue of the term of the lease would depend on the relevant provisions of the lease. A major item would appear to be goodwill, but given the low margins on which the business operated, with the two assignees taking no salaries, I find it hard to accept that the goodwill ought realistically to be valued at anything like a figure which would justify the accountants' opinion.

For the sake of completeness I should add that Mr Brennan exhibited a copy of a letter dated August 7 1985 addressed to Mr Arghyrou from a Mr May of May & Trout, who describe themselves as "Estate, Land and Business Agents", but no professional qualifications are given. In the letter, Mr May purports to value the business at 94 Cornwall Street at £110,000, but the only figures on which that valuation appear to be based, that is to say a turnover — no figure is given for profits — of £250,000, as well as a rental of £12,250, contrast with the figures made available to the landlord at the relevant time. Indeed, it does not even accord with Hillier Parker's statement about the rent. This is not relevant evidence, not having been presented to the landlord at the relevant time. In any event, I accept Mr Testler's criticism of the valuation as being one on which no reliance can be placed.

In my judgment, therefore, a reasonable landlord could properly take the view, on the information available to the landlord, that there was a real doubt about the assignees' ability to meet their obligations under the lease as they fell due.

Mr Poulton has criticised the landlord for concentrating attention on the existing business of the proposed assignees. He submitted that the landlord overlooked the fact that the proposed assignees were to take the lease in their own name and so be personally liable. Of course it is right that they would be personally liable up to the full extent of their assets, but the way the tenant chose to present the proposed assignees as suitable persons to receive an assignment was as persons trading as Gregory's Steak House. Even the tenant's letters asking the referees for references were headed with that designation. It was never suggested to the landlord that the assignees had other income or other assets and no profit projections or information about the business to be carried on at 89 Cornwall Street were supplied, save as to user. A reasonable landlord, with such limited information, could properly be concerned lest the proposed assignees were over-extending themselves financially.

Mr Poulton also submitted that the real reason for the refusal of consent was that the landlord was content to have the tenant, a subsidiary of a large public company, as tenant, and did not want the tenant to assign to anybody, unless perhaps another public company. He submitted that this was the inference to be drawn from the penultimate para in the second affidavit of Mr Testler. In that para Mr Testler stated that after careful reconsideration he adhered to the view that the proposed assignees were of inadequate covenant and that consent of an assignment to them would substantially depreciate the landlord's interests. He accepted that the tenant's intermediate covenant would remain, but said that in practice the market would direct its attention to the tenant in occupation and to the quality and nature of its business operations and planning risks. He calculated the damage to the reversion to be that, instead of a yield of 8%, the market would require a yield, with the proposed assignees as tenants, of 9% with a 20% depreciation in the capital value of the reversion.

Various points can be made about this. First, there is no evidence that such diminution in value to the reversion was a matter which affected the landlord's mind at the time of refusal. It was only when Mr Testler on October 29 of this year, in his second affidavit, raised the point, that the matter first came to light. I accept Mr Poulton's submission that it should not be admitted as a reason for the refusal as it is not apparent that Mr Testler had this point in mind when he refused consent on behalf of the landlord. Second, in any event the approach of Mr Testler would appear to be similar to that which was criticised by Mr Nugee QC, sitting as a deputy judge of this division, in *International Drilling Fluids Ltd* v *Louisville Investments (Uxbridge) Ltd* (1985) 275 EG 802 (affirmed today by the Court of Appeal)*. However, Mr Poulton submits that I should infer from the para in question (because it is nowhere stated expressly) that the real reason which motivated the landlord was that the landlord wanted to prevent any assignment. In my judgment, it is illogical to reject the para on the ground that it did not state Mr Testler's reasoning at the relevant time and yet to seek to infer from it what is said to be Mr Testler's true reasoning at the relevant time. In any event, in my judgment, it is quite impossible to draw any such inference from the para in question. The para is directed to a comparison between the value of the reversion with the present tenant as tenant and the value of the reversion if the proposed assignment to the two proposed assignees were allowed. It says nothing about what the value of the reversion would be if another assignee were put forward — indeed, how could it?

In my judgment, therefore, the landlord was not acting unreasonably in refusing consent, because of the real doubt that the landlord had about the proposed assignees' financial ability to meet their obligations under the lease. That reason does not seem to me to be vitiated by the bad reason given by the landlord as to user. It follows that I must refuse the first declaration sought.

Mr Poulton did not suggest that the second declaration could be granted even if the first declaration were refused. Accordingly, I refuse the second declaration as well.

I dismiss this action.

---

*Editor's note: The first instance decision is also reported at [1985] 2 EGLR 74. The Court of Appeal decision is reported at p 39 *ante* and (1985) 277 EG 62.

Chancery Division

December 13 1985

(Before Mr Justice WARNER)

PONDEROSA INTERNATIONAL DEVELOPMENT INC v PENGAP SECURITIES (BRISTOL) LTD

Estates Gazette March 22 1986

277 EG 1252-1260

Landlord and tenant — Claim by tenants that landlords had unreasonably refused consent to assignment of lease — Landlords considered that the financial standing of the proposed assignees and their guarantors was insufficient, not only because of its intrinsic inadequacy but also because the assignment would mean the substitution of a less attractive covenant for that of a group with an international reputation — Tenants argued that, although the market might be influenced by this latter consideration, this attitude was unreasonable as by law the original lessees remained continuously liable, if called upon, despite the assignment — Balcombe LJ's seven propositions in *International Drilling Fluids Ltd* v *Louisville Investments (Uxbridge) Ltd* considered — The actual decision in that case, however, which went in favour of the tenants, was distinguishable from the present case in two respects — In that case there was no prospect or need for the landlords to sell, whereas here the landlords needed to sell, and in that case there was a detriment to the tenants, whereas here there was not as the landlords were willing to consent to a subletting — Consequently, the very principles which in the *International Drilling Fluids* case led to the decision that the landlords had unreasonably refused their consent led in the present case, subject to two questions, to the conclusion that the landlords had not unreasonably withheld it — The first question was whether the landlords were entitled to take into account the attitude of the market mentioned above despite the contingent liability of original lessees after assignment — Held on this point that the

landlords were so entitled, as they lived in the real world and had to take the market as they found it, not as lawyers might wish it to be — The second question was whether the landlords were unreasonable in taking a less than favourable view of the financial standing of the proposed assignees and guarantors — Held on this point that the landlords were not being unreasonable — The assignees' capitalisation and debt position and the references provided as to the guarantors' financial standing justified the landlords' reservations — Plaintiff tenants' motion and action dismissed

This was a motion by plaintiff tenant, Ponderosa International Development Inc, seeking the same relief as claimed in its action by originating summons against the defendant landlord, Pengap Securities (Bristol) Ltd. The action sought declarations that the defendant had unreasonably refused consent to a proposed assignment of a lease and that the plaintiff was entitled to proceed with the assignment without the defendant's consent. The lease was for 25 years from June 24 1985 of a shop unit at 84/90 King Street, Hammersmith, London W6.

Jonathan Brock (instructed by Clifford-Turner, appeared on behalf of the plaintiff; Caroline Hutton (instructed by Manches & Co) represented the defendant.

Giving judgment, WARNER J said: This is a motion in an action brought by an American company called Ponderosa International Development Inc against an English company called Pengap Securities (Bristol) Ltd. I will call them "Ponderosa" and "Pengap" respectively. The action, which is brought by originating summons, is for a declaration that Pengap has unreasonably withheld its consent to the assignment of a certain lease by Ponderosa to another English company whose name is Maturescope Ltd (and which I will call "Maturescope") or alternatively a declaration that Ponderosa is entitled to assign that lease to Maturescope without the licence or consent of Pengap. The motion seeks the very relief claimed in the originating summons, but Pengap raises no objection on that score.

The lease is dated September 23 1985. It is a lease of a shop unit at 84-90 King Street, Hammersmith, for 25 years from June 24 1985. The rent reserved by the lease is £55,000 per annum payable quarterly in advance, subject to a rent review clause. The first quarterly payment is to be made on the 23rd of this month. There is a rent-free period until that date. The reason for that is that 84-90 King Street, Hammersmith, is a new development undertaken by Pengap and the rent-free period was agreed upon to give Ponderosa time to carry out some fitting-out works. The only other provisions of the lease that I think I need mention are clauses 3(21) and 3(30). Clause 3(21) contains a covenant as to user. It forbids the use of the premises otherwise than as a licensed restaurant, with ancillary kitchen and storage facilities, or than for such other use within Class I of the Town and Country Planning (Use Classes) Order 1972 as the landlord may approve (such approval not to be unreasonably withheld). Class I is, I am told, use as retail premises. Clause (30) is the important provision for present purposes. It is a covenant as to alienation. It is a long one. It takes up about two and a half pages in the lease. I do not propose to read it all. The effect of it (so far as material) is to preclude the tenant from assigning or underletting the premises without the previous consent in writing of the landlord, such consent not to be unreasonably withheld. There is an express proviso:

That in respect of any proposed assignee or underlessee being a private limited company the landlord may as a condition of such consent require a guarantor or guarantors of the covenants.

That lease was executed in pursuance of an agreement for a lease which had been entered into between the same parties on April 15 1985. I do not think, having regard to the way in which the argument has proceeded, that I need now trouble with the detailed provisions of that agreement, save to say that they do foreshadow the works of fitting out the shop unit that I have mentioned and that they place the tenant under an obligation to carry out those works. They also make the benefit of the agreement non-assignable.

I must now say something about the parties to the lease. Ponderosa, as I indicated earlier, is an American company. It is a subsidiary of a larger American company, namely Ponderosa Inc, which is well known in the catering field. It specialises in operating restaurants of a distinctive kind, which serve steak and salad meals. Ponderosa Inc has over 650 such restaurants in the United States. Some of those restaurants are managed by Ponderosa Inc itself and some are operated through franchising agreements. Ponderosa (the G plaintiff) was established to promote Ponderosa Inc's interests outside the United States and it has three restaurants already in the UK, in Watford, Croydon and Chester, which it manages itself. Pengap is a small development company which carried out this development at 84-90 King Street. It is, in the jargon of the trade, a "single development" company. From the outset Ponderosa knew that Pengap's object, having developed the property and let its component units, would be to sell the whole development to an investor.

On July 23 1985, that is between the date of the agreement for the lease and the date of the lease itself, Ponderosa entered into a franchise agreement with Maturescope relating to the shop unit that was to be comprised in the lease. It is a very long agreement. It runs to H 30 pages, and I do not propose to read any of it. Unfortunately Ponderosa took the view that that agreement was confidential as between itself and Maturescope and did not disclose its existence to Pengap or its advisers. The result (or so I suspect from what is said in the evidence) was that those acting on behalf of Pengap felt that they were not being treated with candour by Ponderosa and suspicions arose of various kinds which may well have contributed to the relations between the parties becoming soured. However, that is all water under the bridge now. Miss Hutton, on behalf of Pengap, has really given up the suggestion that those suspicions can legitimately form part of the reasons for Pengap refusing the consent sought by Ponderosa for a licence to assign the lease to Maturescope. J

The reasons now relied upon on behalf of Pengap for refusing such a licence are these. The directors and advisers of Pengap consider that the financial standing, as now known to them, of Maturescope and those put forward as guarantors on its behalf, is inadequate. I will go into the evidence about that in a moment. That inadequacy might in itself be a ground for refusing the licence to assign, but the matter goes further than that because, as I have indicated, Ponderosa is a well-known international company, or at all events the Ponderosa group are a well-known international group, and of course one knows, although this is not stated in terms in this evidence, that generally speaking reputable international groups do not allow their subsidiaries to default on their obligations. The view taken by those advising Pengap is that, on a sale of the whole development to an K investor, the fact that the rent provided for in this lease was payable by Ponderosa would be a good selling point. A rent payable by Maturescope, a new and generally unknown company, would not be such an attractive asset. Mr Brock argues that, as a matter of law, that is nonsense because, whatever happens, Ponderosa's covenant will be there to secure the payment of the rent, since Ponderosa is the first lessee and so will remain liable under the covenants in the lease. To someone who understands the law, says Mr Brock, it cannot matter whether he has the benefit of Ponderosa's covenant because Ponderosa is the first lessee or because it is a continuing lessee. There is, however, evidence from experts who have sworn affidavits, Mr Burke for Pengap and Mr Jones for Ponderosa, that it is a widely held view in the property market, and in particular among investors, that, L after the assignment of a lease, the identity of the original lessee is of no interest, so that, in this case, the development as a whole will be more readily saleable by Pengap to an investor if Ponderosa remains as the lessee of the shop unit and sublets it to Maturescope (a course to which Pengap is prepared to consent) than if Ponderosa assigns the lease to Maturescope. Mr Brock argues that under the provisions of clause 3(30) of the lease, as modified by section 19 of the Landlord and Tenant Act 1927, it is not open to Pengap to rely on the attitude of the market in that respect. The attitude of the market is unreasonable, says Mr Brock, and therefore the attitude of Pengap in taking it into account is itself unreasonable. In any case, says Mr Brock, it cannot be legitimate for Pengap to refuse Ponderosa a M licence to assign the lease to Maturescope simply because Ponderosa is a more desirable tenant than Maturescope.

The authority that seems to me most in point on all this is a very recent judgment of Balcombe LJ in the Court of Appeal in a case called *International Drilling Fluids Ltd* v *Louisville Investments (Uxbridge) Ltd* in which judgment was delivered on November 20, that is to say last month. I understand that that judgment is going to be reported.* It is a judgment with which Fox LJ and Mustill LJ expressed their agreement. In it Balcombe LJ distilled from earlier authorities seven propositions which were relevant in that case and

*Editor's note: The Court of Appeal decision is reported at p 39 *ante* and (1985) 277 EG 62.

A most of which are relevant in the present case. Those propositions are these. I quote from Balcombe LJ's judgment omitting his references to, and discussion of, those earlier authorities:

(1) The purpose of a covenant against assignment without the consent of the landlord, such consent not to be unreasonably withheld, is to protect the lessor from having his premises used or occupied in an undesirable way, or by an undesirable tenant or assignee.
(2) As a corollary to the first proposition, a landlord is not entitled to refuse his consent to an assignment on grounds which have nothing whatever to do with the relationship of landlord and tenant in regard to the subject matter of the lease.
(3) The onus of proving that consent has been unreasonably withheld is on the tenant.
B (4) It is not necessary for the landlord to prove that the conclusions which led him to refuse to consent were justified, if they were conclusions which might be reached by a reasonable man in the circumstances.
(5) It may be reasonable for the landlord to refuse his consent to an assignment on the ground of the purpose for which the proposed assignee intends to use the premises, even though that purpose is not forbidden by the lease.

That does not arise here.

(6) While a landlord need usually only consider his own relevant interests, there may be cases where there is such a disproportion between the benefit to the landlord and the detriment to the tenant if the landlord withholds his consent to an assignment that it is unreasonable for the landlord to refuse consent.
(7) Subject to the propositions set out above, it is in each case a question of fact, depending upon all the circumstances, whether the landlord's consent to
C an assignment is being unreasonably withheld.

In that case the facts were these. The property comprised in the lease was an office building, which the tenants had vacated because they no longer wanted to occupy it. They could find no one prepared to take an assignment of their lease except a company which wished to use the property for what is called "serviced office accommodation". There was evidence on which the judge had found that reasonable professional men might take the view that, if the property were placed on the market, it could fetch less with that company in occupation carrying on the business of providing serviced offices than with the property having remained vacant for more than a year. On the other hand, there was no prospect of the landlord wishing to place the property on the market or to mortgage
D it to the fullest extent possible. Nor was there any possibility that the use of the building for serviced offices might have a depreciating effect on the letting value of the property at the end of the lease. The landlord's main ground for refusing to grant a licence to assign was that the investment value of its interest in the property would be detrimentally affected by the proposed use.

The actual decision in the case and the reasons for it (so far as material for present purposes) are to be found in two paragraphs of Balcombe LJ's judgment, which follows the passage in which he enunciates the seven propositions that I have read. He said:

In the present case, the learned judge, having made the findings of specific fact set out above, carefully considered the relevant authorities. He then
E reached the conclusion that the views of the landlord's expert witnesses about the effect of the proposed assignment on the value of the reversion, although views which could be held by reasonable professional men, did not in the circumstances of this case, where there was no prospect of the landlord wishing to realise the reversion, constitute a ground for reasonable apprehension of damage to its interests. That was a decision on the facts to which the learned judge was entitled to come. He made no error of law in reaching his decision; he took into account nothing which he ought not to have considered and he omitted nothing which he ought to have considered. In my judgment, this court ought not to interfere.

But in any event in my judgment the learned judge reached the right decision. Although he did not expressly mention the disproportionate harm to the tenant if the landlord were entitled to refuse consent to the assignment, compared with the minimum disadvantage which he clearly considered the
F landlord would suffer by a diminution in the paper value of the reversion — "paper value" because he was satisfied there was no prospect of the landlord wishing to realise the reversion — he clearly recognised the curious results to which the landlord's arguments, based solely upon a consideration of his own interests, could lead. As he said (at p 19C of the transcript):† "It seems to me that, if Mr Lewison is right, the more substantial the lessee, the more easily the landlord would be able to justify a refusal of consent to an assignment, since unless the proposed assignee's covenant was as strong as the assignor's, a reasonable man might form the view that the market would consider the reversion less attractive if the lease were vested in the proposed assignee than if it were vested in the proposed assignor. To take the matter to extremes, if a lease was made in favour of a government department, it would be unassignable except to another government department; for as Mr Matthews one of the expert witnesses accepted in cross-examination, the market would prefer to have the government as the lessee, whether the premises were being used as serviced offices or not, even if they were standing empty, rather than a company, however strong its covenant." In my judgment the gross unfairness to the tenant of the example postulated by the learned judge strengthens the arguments in favour, in an appropriate case — of which the instant case is one — of it being unreasonable for the landlord not to consider the detriment to the tenant if consent is refused, where the detriment is extreme and disproportionate to the benefit of the landlord.

Mr Brock's submission was that that applies here, in that the case put forward on behalf of Pengap amounted to saying that, because the covenant of Ponderosa was preferable to the covenant of Maturescope, Ponderosa should not be allowed to assign the lease to Maturescope. I think that that is not only an oversimplification of Pengap's case but also a mistaken way of interpreting the judgment of Balcombe LJ. What Balcombe LJ was saying was that, although in general a landlord was entitled to consider only his own interests, there were exceptional cases where the detriment to the tenant if the landlord withheld his consent to an assignment was so disproportionate to the benefit to the landlord from his doing so that it was unreasonable for him to do so. The learned lord justice mentioned the case of the government department as an extreme example of that. There are two respects in which this case is distinguishable from the *International Drilling Fluids* case. One is that in that case there was no prospect of the landlord's wanting or needing to sell. Therefore the diminution that the expert evidence suggested the assignment would cause to what Balcombe LJ referred to as the "paper value" of the reversion did not matter; it was not a real detriment to the landlord. Here in contrast we know that Pengap wants to sell and indeed needs to sell. There is evidence that, because of the nature of the arrangements that it made to finance the development, it will be in difficulties unless it sells fairly soon. Under the general principles enunciated by Balcombe LJ Pengap is entitled to have regard to its interests in that respect. The other difference between this case and that of *International Drilling Fluids* is in the extent of the detriment to the tenant. Ponderosa is not in the sort of difficulty that the tenant was in in the *International Drilling Fluids* case of being unable to part with property it did not want unless it assigned it to the company that the landlord objected to. Pengap has made it perfectly clear to Ponderosa that Pengap will consent to a sublease by Ponderosa to Maturescope and there is no problem about that. The detriment that Mr Brock says will arise to Ponderosa if that course is adopted is that it will lead to administrative complications. Ponderosa will have to collect the rent from Maturescope and pass it on to the landlord. When rent reviews come up there will have to be two rent reviews, one between the landlord and Ponderosa and another between Ponderosa and Maturescope. The same duplication will occur on claims for dilapidations and with any other problems that may arise during the subsistence of the lease. Of course, that is an inconvenience and it may cost some money, but it is nothing like as grave a detriment as was in question in the *International Drilling Fluids* case. Moreover, Miss Hutton mentioned countervailing administrative disadvantages that might in practice arise for the landlord if, instead of keeping in quarterly touch with Ponderosa as the source of the rent, Ponderosa had to be treated merely as a person who would be liable for the rent in the event of the actual tenant failing to pay. So in my view this case is distinguishable for those reasons from the *International Drilling Fluids* case, and the very principles that led the Court of Appeal in that case to hold that the landlord had unreasonably withheld its consent lead in the present case, subject to two questions with which I have yet to deal, to the conclusion that Pengap has not unreasonably withheld its consent.

Of those questions the first is whether Mr Brock is right in saying that Pengap is not entitled to take into account the attitude of the market to which the expert witnesses refer because, to anyone properly instructed as to the law, that attitude would appear unjustified. In my judgment he is not. Bearing in mind the administrative considerations mentioned by Miss Hutton to which I have just referred, I am not even sure that Mr Brock is right in saying that the attitude of the market is unreasonable. Assuming, however, that he is right as to that, the fact is that Pengap has to live in the real world and to take the market as it finds it, not as lawyers might wish it to be. In saying that I do not overlook the evidence of Mr Jones, the expert fielded by Ponderosa, that he believes that in the present case the reluctance of the market to take Maturescope's covenant could be overcome. He says in his affidavit:

I believe that in this instance a strong case could be made to an investor that

†Editor's note: See [1985] 2 EGLR 74 at p 79 A-B.

> any difference that might actually or be perceived to exist between an assignment and a sub-letting would be offset by the fact that the proposed assignee is a UK registered company and is contractually linked by way of the franchise agreement with the assignor.

Mr Jones explains that, to his mind, the franchise agreement:

> demonstrates a close contractual relationship between the parties designed to ensure that Ponderosa, in lending its name and expertise to the development of the business, has proper control over the assignee from a trading and standards point of view.

He goes on:

> Thus, unlike many assignors, whose intention is to rid themselves of any continuing commitment to a property (save their underlying obligations as original tenant), Ponderosa will maintain a continuing and detailed interest in the success of the business and the proper maintenance of the premises for that purpose . . . I would be surprised if, in entering a very competitive market, Ponderosa do not ensure that everything is done by Maturescope Ltd, both operationally and in the conduct of their affairs as tenants, so as not to prejudice the reputation of Ponderosa in the eyes of the public or other potential landlords.

Mr Jones does appear to have seen the franchise agreement, but he seems to have overlooked that Ponderosa and Maturescope are under no obligation to Pengap or to any future landlord to keep that agreement in force. They can, by agreement between them, bring it to an end at any time. There are, moreover, in the franchise agreement provisions enabling Ponderosa to withdraw from it unilaterally in certain events, and that goes for Maturescope too. So there can be no guarantee to any landlord that the relationship of franchisor and franchisee will continue between Ponderosa and Maturescope (or anyone else who may run a restaurant at those premises) for the next 25 years.

So I come to the last question, which is in a sense the first: whether Pengap is being unreasonable in taking the view that it does of the financial standing of Maturescope and of its proposed guarantors. In considering that question I have in mind in particular the fourth proposition enunciated by Balcombe LJ, viz that it is not necessary for a landlord to prove that the conclusions which led him to refuse his consent were justified, if they were conclusions which might be reached by a reasonable man in the circumstances.

I look first at Maturescope itself. Maturescope is what is generally called an "off the peg" company. It was incorporated on February 18 1985 and bought by two gentlemen called Thakkar in April 1985. They are Mr Mahendra Thakkar and Mr Arvind Thakkar. From a company search that was done by Pengap's solicitors it appears that on April 5 1985 a Mr Nadir Lalani and a Dr Manoobai Amin were appointed co-directors with the two Thakkars. Dr Amin is resident in the United States but the other three directors are, it seems, resident here. So far there is no evidence of any issue of any substantial amount of share capital. Mr Henderson, who is the property manager of Ponderosa, has sworn an affidavit to which he exhibits a document called a "funding schedule" showing that what is intended is that the four directors should put up between them share capital of £250,000 and that various institutions should put up in addition £62,500 by way of share capital and a total of £287,500 by way of loan capital. Mr Brock confirmed to me that those loans would be provided on the security of debentures. The document shows that there will be initial expenditure of the order of £600,000, of which we know from other evidence that the expenditure on the fitting-out works that I referred to earlier will absorb between £450,000 and £500,000. So, on the face of it, this company will start life insolvent.

The guarantees that are offered are those of the two Thakkars and of Mr Lalani. Information about their financial standing has been supplied to Pengap in three instalments.

First, in accordance with normal practice, Ponderosa's solicitors, Clifford-Turner, obtained bank and other references for those three gentlemen and, in September, sent them to Pengap's solicitors, Manches & Co. Mr Brock quite rightly concedes that the bank references provided for the two Thakkars are qualified. To my mind they are so qualified that no landlord in his senses would, in the circumstances, rely on the guarantee of either of the Thakkars. In the case of Mr Lalani the bank reference reads: "Respectably introduced customer known to the bank in a private and business capacity and considered good for the amount and purpose of your inquiry. £55,000 rental — guarantor." However, as Miss Hutton points out, we do not know what the terms of the request for that reference were. In particular, there may not have been any mention in it of the fact that Mr Lalani is to subscribe a substantial — albeit, so far as Pengap and this court are concerned, an unspecified — sum to the initial working capital of Maturescope. Then there is a reference from Mr G
Lalani's solicitors, who are a well-known firm. The material part of it is: "In our opinion our above-named client would prove a respectable and trustworthy guarantor of the lease, details of which are mentioned in your letter. We believe that our client would not enter into any commitment which he believed he could not fulfil." There follows a disclaimer of liability on the part of the firm. There again Miss Hutton points out that we have not got the text of the letter written by Clifford-Turner to those solicitors. There are other references given, but they are from people who are certainly unknown to me, solicitors, accountants and traders of one sort and another. There seems to have been no attempt to satisfy Pengap or its advisers as to the standing of those referees. No blame could, I think,
attach to anyone who treated references given by them with some H
reserve.

Manches & Co indicated to Clifford-Turner that those references did not "give great comfort" to Pengap. So, in early October, Clifford-Turner sent to Manches & Co a bundle of documents dated in June and July 1985. I agree with Miss Hutton that the receipt of those documents was calculated to increase Pengap's unease rather than to dispel it. Those documents consisted firstly of "Franchise Applications" completed in June by the two Thakkars and Mr Lalani on printed forms supplied by Ponderosa. There are pages in those forms on which an applicant is supposed to give details of his financial resources and other particulars. On the forms completed by
the two Thakkars and Mr Lalani those pages are simply crossed out. J
The other documents in the bundle consist of letters from Ponderosa to the proposed guarantors' respective banks enclosing authorisations signed by them for the release of personal data and asking specifically for (and I quote) "information on . . . his current assets with your bank, together with any other financial information you may have that would be helpful." In each case the bank's reply conspicuously does not give any specific information. Save that the reply from Mr Lalani's bank states that he is "chairman of a company with 18 supermarket outlets in London", the replies are brief and couched in general terms. A possible inference from them is that each of the gentlemen in question is overdrawn at his bank.

Third, in November, Mr Lalani and Mr Mahendra Thakkar swore K
affidavits as to their own means. Mr Arvind Thakkar did not do so because, I am told, he was abroad. I am bound to say that those affidavits are sketchy in the extreme. Mr Lalani's says that he is the "controlling shareholder in three substantial companies which employ in total 700 people with an overall annual turnover in excess of £50 million", but he does not exhibit the accounts of any of those companies. Then he says: "My house, which is jointly owned with my wife, is worth approximately £1 million on the open market. There is no mortgage over the property, although it is pledged security in respect of company indebtedness." So for aught we know the £1m could be absorbed by company debts. Finally, he says that
his annual income is "in excess of £50,000 per annum". He does not L
say whether that is before or after tax, or what other annual commitments he has.

Mr Mahendra Thakkar mentions his house, which he says is worth about £100,000, and two commercial properties worth together about £250,000, all of those assets being owned jointly with his wife and mortgaged. The mortgages total £90,000 so that, according to the figures he gives, there is a net equity of £260,000, of which half would be £130,000. As against that we do not know what Mr Thakkar's liabilities are in addition to his commitment to contribute to the share capital of Maturescope. Mr Mahendra Thakkar adds that his gross annual earnings are in the region of £60,000 to £70,000. Clearly they are subject to tax and may be subject to other
commitments. M

In the result, I cannot say that Pengap is, in all the circumstances, acting unreasonably in refusing its consent to an assignment of the lease by Ponderosa to Maturescope, while offering its consent to a sublease.

The motion and originating summons were dismissed with costs.

A Chancery Division

December 13 1985
(Before Mr J JEFFS QC, sitting as a deputy judge of the division)

MULLANEY v MAYBOURNE GRANGE (CROYDON) MANAGEMENT CO LTD

Estates Gazette March 29 1986
277 EG 1350-1353

Landlord and tenant — Construction of provisions in a deed of covenant and a lease in regard to repairs and services — B Plaintiff tenant sought by originating summons a determination of the questions whether the defendant landlords were empowered to replace defective old wooden-framed windows in the plaintiff's flat with new double-glazed windows and whether the plaintiff was obliged to pay for the replacement — There was a subsidiary question as to the plaintiff's liability to pay certain other expenses incurred by the defendants — As a result of the joint operation of the deed and the lease the defendant landlords undertook, *inter alia*, to repair and maintain and otherwise provide services and amenities to the structure and common parts of the block — C Plaintiff tenant covenanted to contribute to the costs of so doing and to such further or additional costs as the defendants incurred in "providing and maintaining additional services or amenities" — Defendants claimed contributions from the plaintiff on the ground that the installation of the new windows was a repair, and alternatively that the expenditure was incurred in "providing and maintaining additional services or amenities" to the block — The judge rejected the claim on both grounds — The replacement of the old windows was not a repair but a long-term improvement — Although the new windows had attributes which made them desirable, they could not be regarded in the ordinary sense as an "amenity" D — As regards the subsidiary claim for expenses, the court did not have sufficient information to make a declaration — Judgment for plaintiff tenant on the two main questions raised in the summons.

This was an originating summons by which the plaintiff, Stephen Paul Mullaney, sought determinations on questions of construction raised by a deed of covenant and a lease. Maybourne Grange (Croydon) Management Co Ltd were the defendant.

Gregory Stone (instructed by Symons, of Croydon) appeared on behalf of the plaintiff; Alastair Norris (instructed by Stoneham Langton & Passmore) represented the defendant.

E Giving judgment, MR J JEFFS QC said: By this summons the plaintiff seeks the determination of the court of the following questions:

(1) whether on the true construction of the Deed of Covenant made 4th June 1984 beween the plaintiff and the defendant and the lease made the 31st day of January 1969 between Western Land Holdings Ltd and George William Reed and Olwen Mary Reed, the defendant is empowered to replace the existing windows of the plaintiff's flat at 24 Maybourne Grange aforesaid with double-glazed windows under the provisions of the said Deed of Covenant and the said Lease; (2) whether on the true construction of the said Deed of Covenant and the said Lease the defendant is entitled to seek payment from the plaintiff and the plaintiff is obliged to make payment to the defendant in respect of the replacement of the existing windows in the plaintiff's flat at 24 Maybourne Grange aforesaid by double-glazed windows under the F provisions of the said Deed of Covenant and the said Lease.

The summons goes on to seek further and other relief and costs. By an amendment which was put forward today and which was not opposed by Mr Norris, who appears for the defendant, I am further asked to make a declaration that the plaintiff is not liable to make payment to the defendant in respect of any costs or expenses incurred by the defendant before June 4 1984 under clause 2 of the deed of covenant of the same date.

Maybourne Grange is a modern tower block of flats. It was constructed originally with wooden-framed windows which, it appears, from the very outset have been a source of trouble, leaking water, requiring painting every four years in accordance with the terms of the lease and causing considerable trouble and expense to many of the lessees of the flats in that block. Looking at it in terms of good housekeeping, it may well have been very desirable to get rid of this nuisance and in the interests of everyone to replace the defective and unsatisfactory windows with something more durable and calling for less maintenance.

That, however, is not the question that I am asked to decide in these proceedings. I can summarise very briefly the provisions of the lease and of the covenant in this way. Parts I and II of the first schedule to the lease define the demised premises as excluding external and load-bearing walls of the block of flats, and certain common parts of the Grange are defined as including those external and load-bearing walls. Under para 2 of the fourth schedule to the lease the lessee is required to keep the glass and the windows properly repaired and cleaned.

I now come to the fourth schedule to the lease, which is headed "Covenants by the Lessee in respect of the Demised Premises". By clause 14 of that schedule it was laid down that the lessee "will not assign the demised premises except to a person who has committed himself to the other lessees of flats in Maybourne Grange and Maybourne Grange (Croydon) Management Co Ltd for the performance and observation of the covenants on the part of the lessee contained in clause 7 hereof". Clause 7 laid down *inter alia* that he "will be jointly liable for the upkeep and management of the block", and it refers to the structure and common parts, which I have already mentioned.

Under para 4 of the fifth schedule the lessee is under an obligation to paint the woodwork of the flats every four years, the painting to be executed so that the block of flats looks attractive and harmonious in appearance with other buildings on the estate. It is put forward that the heading of the fifth schedule implies that the outside windows are part of the common parts.

By the deed dated June 4 1984 the defendant covenanted "to perform and observe the covenants in the fifth schedule to the lease". The defendant was accordingly bound to paint the window frames every four years. The plaintiff was responsible, however, personally for damage to the glass.

The defendant's case is put in this way. First that having the maintenance-free windows installed was merely a way of complying with the defendant's repairing and decorating obligations and further or alternatively that the installation of double glazing and maintenance-free windows throughout the block of flats constitutes the provision of a service or amenity to the block and that clause 2(b) of the deed of June 4 1984 entitles the defendant to require from the plaintiff 2.5% of the cost properly incurred in providing and maintaining "additional services or amenities to Maybourne Grange".

At this point it is convenient to refer to the exact terms that were used in that deed of covenant. It is recited in recital 3 that the lease *inter alia* requires the lessee to "repair, maintain, insure, cleanse and otherwise provide services and amenities to the structure and common parts of Maybourne Grange". The operative part of the deed, para 1, states:

In consideration of the premises and the covenant on the part of the Lessee hereinafter contained the Company hereby covenants with the Lessee to observe and perform the covenants on the part of the Lessee contained in the Fifth Schedule to the Lease.

Para 2 states:

The Lessee for himself and his personal representatives hereby covenants with the Company and the other Lessees in Maybourne Grange that so long as the Lease remains vested in him or them he or they will (a) pay to the Company from time to time on demand $2\frac{1}{2}\%$ of the costs and expenses incurred by the Company in the observance and performance of the covenant on its parts hereinbefore contained; and (b) such further or additional costs which the Company shall properly incur in providing and maintaining additional services or amenities

I emphasise those words

additional services or amenities to Maybourne Grange or the said Garage Block and which shall be authorised by resolution at a General Meeting of the Company.

There is therefore provision for the supply of additional amenities.

Let me say here and now that I do not regard the provision of these double-glazed windows as an additional amenity. It is stretching altogether the terms of the lease and the meaning of the word "amenity" so to construe it. "Amenity" is defined in the *Shorter Oxford Dictionary* as *inter alia* "the pleasurable features of an estate". *Chambers Dictionary* puts it rather differently:

"pleasantness as in situation, climate, manners, disposition; a pleasing feature, object, characteristic; civility".

These windows no doubt had certain attributes which may have made them desirable, but I do not regard them as being in any normal sense of the word an amenity. That the time had come to paint and repair them is established beyond doubt. Various estimates were obtained and I have one before me in the affidavit of Mr Fitzpatrick, sworn on April 18 1985. Painting cost £9,641; joinery repairs, £17,333; a supervision fee for the managing agents, £4,046 — making a total of expenditure of £31,019 to which VAT of £4,653 had to be added. So that was what was necessary to pay in terms of the repairs and maintenance.

The new windows cost almost double that, some £70,000. They may well have been a very desirable improvement, but I am quite satisfied that they fell into the category of improvements and not simply of maintenance. I would refer in this connection to *Woodfall's Landlord and Tenant*, vol 1, para 1-1436 at p 598, where the learned author says this:

A covenant to repair does not involve a duty to improve the property by the introduction of something different *in kind*

and the words "in kind" are in italics

from that which was demised, however beneficial or even necessary that improvement may be by modern standards. So a lessor of old premises not constructed with a damp course or with waterproofing for the outside walls was not bound by his repairing covenant to render the place dry by waterproofing the walls. If a defect in design has given rise to dry rot, the elimination of that existing rot and the replacement of affected timber is, within the principle of renewal of subordinate parts which is dealt with below, an obligation imposed by a covenant to repair, but there is no obligation to undertake the structural alteration which is needed to prevent a recurrence of the rot, for that is improvement and not repair.

Thus it appears to me in the context of the replacement of the windows that what was being done here went beyond what was necessary for the purpose of effecting a repair and was by way of a long-term improvement. It was looking to the future not to the present.

Accordingly, I am satisfied that the plaintiff has made out his case on the relief he sought in paras 1 and 2 of the summons which I have just read out.

I am not, however, satisfied that the plaintiff is entitled to the declaration which he seeks in relation to the costs or expenses incurred before June 4 1984, that being the date at which the deed was entered into. It may be useful at this stage to summarise some of the crucial dates. On February 13 1984 the plaintiff took an assignment of the lease. On April 26 there was an extraordinary general meeting of the defendant company which resolved to change the windows. On May 2 there was an invoice for work done. On May 25 the plaintiff became a shareholder in the company. On June 4 the deed of covenant was entered into.

Although the deed of covenant was not entered into until so late a date, the assignment of the lease was taken earlier and was subject to the clause which I have already read out whereby the purchaser of the lease was under an obligation to enter into the arrangement which was finally embodied in the lease. This provision was, I am informed, included in his contract of purchase of the premises, a contract which I have not seen, but it was agreed that it contained such a clause.

I am not for the purpose of this action immediately before me called upon to make any positive finding with regard to any specific costs or expenses incurred by the defendant before June 4 1984, and I do not do so. It may be that if any action is pending relating to such expenses further discovery may throw light on the matter in a way which at the moment I cannot envisage. But I do not think it appropriate on the information at present before me to grant the declaration sought by the amendment.

Court of Appeal G

February 19 1986

(Before Lord Justice SLADE, Lord Justice WOOLF and Sir Denys BUCKLEY)

GORDON AND ANOTHER v SELICO CO LTD AND ANOTHER

Estates Gazette April 5 1986

278 EG 53-62

Landlord and tenant — Appeal from decision of Goulding J in H
favour of long leaseholders of flat in action by them against lessors (Selico Co Ltd) and management agents (Select Managements Ltd) on grounds of fraudulent misrepresentation by lessors and breaches of obligations and trusts under the lease — Goulding J had awarded the plaintiff leaseholders (the present respondents) damages for deceit and he had made an order under section 125 of the Housing Act 1974 by way of specific performance, requiring the lessors to put so much of the block of flats as was in their possession or control into such reasonable condition as not to cause damage to the lessees or their flat by the incursion of water,
propagation of dry rot or otherwise — One of the lessees' main J
complaints had been of an extensive outbreak of dry rot in the flat, indications of which, it was alleged, had been deliberately covered up before the purchase — The lessors and the agents appealed — There was no challenge on the appeal against Goulding J's finding of a deliberate and fraudulent cover-up by a building contractor instructed by the management agents, but it was claimed by the lessors and the agents that neither were answerable for the tort of the contractor on the basis of actual or ostensible authority — Goulding J had held that the lessors were liable for the deceit on the ground that the
dishonest act was done by the contractor in the course of work K
ordered by the management agents within the scope of their authority as the lessors' agents — The Court of Appeal, while upholding the judge's finding of deceit against the lessors, arrived at the same conclusion by a different route — In their view the management company was, through its main shareholder, a party to the deceit and, as the lessors had entrusted the company with sufficiently wide authority, they were vicariously liable for the fraudulent misrepresentation — The Court of Appeal also agreed with the judge that both the lessors and the agents had been guilty of breaches of covenant
and breaches of trust under the lease — As regards relief, L
damages were not by themselves a sufficient remedy and an order for the enforcement of the trusts of the lease was required — In view of the complexity, the Chief Chancery Master wo ıld be charged with the working out of the order — Accordingly, the court ordered the lessors to pay the lessees damages for the tort of deceit and for breaches of contractual obligations under the lease and made orders for the execution of the trusts of various clauses in the lease and for specific performance of the lessors' covenants to keep certain parts of the block in good repair in pursuance of section 125 of the Housing Act 1974 — The Chief Chancery Master was
entrusted with the authority to give all necessary directions for M
the working out of these orders — Appeal dismissed with variations of Goulding J's order as indicated in the judgment of the court

This was an appeal by the defendants, Selico Co Ltd and Select Managements Ltd, from a decision of Goulding J in favour of the plaintiffs (the present respondents), Jack Leonard Gordon and Anya Teixeira. The action and appeal concerned Flat C in a block called Court Mansions, Frognal, Hampstead, London NW3, of which the appellants, Selico Co Ltd, were the lessors and the respondents were the lessees. The decision of Goulding J was reported at [1985] 2 EGLR 79, (1985) 275 EG 899.

A J H G Sunnucks (instructed by Druces & Attlee) appeared on behalf of the appellants; Jonathan Brock (instructed by Sebastian Coleman & Co) represented the respondents.

Giving the judgment of the court, SLADE LJ said: This appeal arises out of a dispute between landlords and tenants concerning premises known as Flat C, Court Mansions, 13-15 Frognal, London NW3. The respondents, who are the tenants and the plaintiffs in the proceedings, are Mr Gordon and Mrs Teixeira. The first defendant and appellant, Selico Co Ltd ("Selico"), has at all material times been the owner of Court Mansions. The second defendant and appellant, Select Managements Ltd ("Select"), is an estate agent, which has at all material times managed Court Mansions for Selico. B By an order made on January 28 1985, after the trial of the action, Goulding J ordered that Selico should pay to the plaintiffs damages for the tort of deceit, such damages to be assessed on inquiry before an official referee. He further ordered that Selico should, by way of specific performance of covenants contained in a lease made between the parties to the proceedings on August 24 1979 in respect of Flat C, on or before September 28 1985 and at its own expense, put so much of Court Mansions as was in the possession or control of Selico "into such reasonable condition as not to cause damage to the plaintiffs or to the premises demised by the said lease by the incursion of water, the propagation of dry rot or otherwise". In addition he ordered that each of Selico and Select should pay to the plaintiffs damages for breach of covenant in the sum of £10 and dismissed their C counterclaim in the proceedings. He directed the two defendants to pay to the plaintiffs their costs of the action and the counterclaim and gave the parties liberty to "apply at chambers for any necessary directions for the purposes of enforcement and clarification of the order for specific performance and generally".

The defendants now appeal from the whole of this judgment. They ask that it should be set aside and that judgment may be entered for them for the relief asked in their counterclaim or that a new trial may be ordered.

The learned judge gave the following description of the block in which Flat C is situated:

> Court Mansions contains five floors symmetrically disposed about a central front door and staircase. On the left, as you look at the building from the D street and raise your eyes from the basement to the roof, there are in succession flats A, C, E, G and K. Similarly, on the right are flats B, D, F, H and J. The basement flat on the left, Flat A, has been subdivided. The anterior part is now known as Flat L — the rear as Flat A. The defendants say that they never consented to the subdivision. So far as material to this judgment, it appears that Flats A, C, F, J and K have been let on long leases, all assumed (though without much evidence) to be in terms similar to the lease held by the plaintiffs. The other flats have been let on short tenancies, in some cases at least with furniture.

The evidence did not establish when or in what condition Selico had acquired Court Mansions, but the learned judge had no doubt that the building was already in a poor state structurally when the plaintiffs first came on the scene in late 1978 and that by then dry rot E organism was already established in some parts of the premises. Furthermore, as the judge found, the defendants "were not unaware of the position". In 1974 the tenant of Flat A had complained of damp due to the blockage of external down-pipes. In January 1976 an occupier of Flat H had drawn attention to a leaking roof. In October 1976 a solicitor, Mr Pringsheim, who was the tenant of Flat F, reported the entry of rainwater from a defective roof and/or gutters and in November 1976 protested at what he described as the appalling state of the outside repairs to the building and common parts. In March 1977 he sent a schedule of dilapidations to Selico's solicitors, alleging, among other things, defective tiling on the roof, perished valley and parapet gutters, loose brickwork, wood rot in and around sashes and window frames and defects of rainwater pipes and gullies, F causing dampness to the external walls. He began proceedings against Selico and appears to have settled his claim and costs for £3,000. Furthermore, in July 1977, Select had received a complaint about dry rot, among many other defects, in Flat H from a Mr MacGillivray, who had occupied that flat for several years.

In or about 1977 a Mr Podolski, who had been the tenant of Flat C, died. The flat then remained unoccupied until the plaintiffs moved into it. During this interim period, however, Mr MacGillivray inspected it because he himself was interested in purchasing it. He works in the timber industry. He gave evidence at the trial on behalf of the plaintiffs. His evidence was that, when he started in the trade, he studied timber technology for four years, that he was an Associate Member of the Institute of Wood Science and that he would know dry rot if he saw it. His evidence in chief was that when he inspected Flat C "it looked a very sorry state indeed. On entering the flat the first thing that you noticed was the smell and on inspection of the front room and also the bathroom you could clearly see evidence of dry rot". The front room in question was a bedroom which, like the adjoining bathroom and lavatory, overlooked the road. Later in his evidence in chief he was asked whether he was satisfied that what he had seen in these two rooms was dry rot rather than some other form of decay. In his reply he accepted that in some cases it is difficult to distinguish between dry rot and wet rot, but expressed the opinion that what he had seen was dry rot. This evidence of Mr MacGillivray was not challenged in cross-examination. The learned judge accepted him as a truthful witness and his evidence is significant for reasons which will appear.

Subsequent to Mr MacGillivray's visit to Flat C, steps were taken by Select, on behalf of Selico, to render the flat more presentable for the purposes of sale. The business of Select, as the judge found, was under the effective direction and control of a surveyor, Mr C J Haskins, who held all but two of its issued shares but was not a member of its board. Under his active control the business was managed by one of Select's directors, Mr P G Colenutt.

As the documentation shows, Mr Haskins and Mr Colenutt were at the material time also both members of a firm of incorporated valuers, surveyors and estate agents known as "Grenadiers". The writing paper of that firm shows the name of Mr Haskins above a line and, below the line, the names of Mr Colenutt, under the heading "Management", and of another gentleman, under the heading "Sales".

The evidence shows that at a time probably in the latter half of 1978 Select, through Mr Haskins and/or Mr Colenutt, instructed a building contractor, Mr Azzam, to do any work to Flat C "to bring it up to a very good standard for the purpose of selling". In his oral evidence Mr Azzam, who was called as a witness by the defendants, substantially exaggerated the extent of the work which he had done to Flat C. However, he undoubtedly did do work to the front bedroom and bathroom.

The first plaintiff, whom the judge accepted as a truthful witness, drew attention in his evidence in chief to a "startling contrast" between the appearance at the beginning of 1979 of the front rooms, which seemed in good condition, and the rear rooms. He said:

> The bathroom and toilet were panelled in shiny new Laconite. The woodwork in the toilet and bathroom appeared perfect, there was perfect painting and the same was true of the front bedroom. The rest of the flat was disfigured by various stains of the type produced by the flooding from somebody's bathroom and the like and certainly the rear, the terrace door, looked as if the woodwork would require replacement.

Mr Azzam's evidence at the trial was that he had worked for Select for about 20 years and that 80% of his turnover was work done for Select. He said his work to Flat C, on which four men had been engaged, had taken 8 to 10 weeks, that it had been supervised by Mr Haskins, who had visited the flat at least weekly while it was proceeding, and that he (Mr Azzam) had found no dry rot. If he had found it, he said, he would have reported it immediately to Mr Haskins. The learned judge, apparently without hesitation, rejected Mr Azzam's evidence that he had not found dry rot. He described Mr Azzam as "a careless witness easily trapped into inconsistencies and demonstrably untrue assertions". The oral testimony of the plaintiffs' surveyor, Mr Wolmark, of the defendants' surveyor, Mr Job, and of Mr Tribbick, an environmental health officer in the service of the local authority, led irresistibly, in his judgment, to the inference that Mr Azzam had deliberately covered up patches of dry rot without any attempt at eradication. On this appeal there is no challenge to this important finding of fact relating to Mr Azzam's activities.

In the latter part of 1978, the plaintiffs obtained particulars of sale relating to Flat C from a firm of estate agents, Hamptons, who were acting as subagents for Grenadiers, the firm in which Mr Haskins and Mr Colenutt were concerned. Grenadiers had apparently been appointed agents for the sale of Flat C by or on behalf of Selico. In or about November 1978 the plaintiffs visited the flat for the purposes of inspection. Mr Azzam gave evidence to the effect that one or both of them inspected the flat while his work was in progress. The judge, however, rejected this evidence, accepting the plaintiffs' evidence that they never saw Mr Azzam before November 1980. It seems clear that Mr Azzam's cover-up operation had been completed before they saw the flat.

Flat C was for sale as a long leasehold and in November 1978 the plaintiffs agreed, subject to contract, to pay £31,000 for a 99-year lease.

The usual preliminary inquiries were then made and answered. In February 1979 the plaintiffs obtained a detailed report relating to the condition of Flat C from surveyors, Spyers & Partners. Among other things they reported that they had found no dry rot, but pointed out that they had only lifted one floorboard in the flat and could not guarantee that dry rot did not exist elsewhere in the building.

Formal contracts providing for the sale of the leasehold interest in Flat C were exchanged between Selico and the plaintiffs in May 1979. The contract incorporated clause 4(2)(a) of the Law Society's General Conditions of Sale (1973 revision) which, so far as material, provides that "the purchaser shall buy with full notice of the actual state and condition of the property and shall take it as it stands . . . ". As the judge found, at the date of the contract certain exterior repair work was in hand and Selico's solicitors promised that the plaintiffs would not be charged a proportion of its cost.

In June 1979 the plaintiffs became aware that dry rot had appeared on the external front elevation between the top of their front bedroom and the bottom of Flat E, above their own flat. They informed their solicitors but, it seems, not the defendants. Their evidence was that they did not take the matter very seriously and were advised, rightly or wrongly, that they must complete their purchase. It seems that they did not realise that the interior of Flat C was also affected.

The contract was completed by the grant of the lease on August 24 1979.

The lease dated August 24 1979

This is an elaborately drafted and lengthy document, but we see no satisfactory alternative to a full summary of its contents coupled with the quotation of a number of its specific provisions. The parties to it were Selico, defined as the lessor, the two plaintiffs, defined as the lessee, and Select, defined as the maintenance trustee. It recited, *inter alia*, an agreement for the grant by the lessor to the lessee of the lease of Flat C for a premium of £30,000. Clause 1(F) defined "the lessor's property" as meaning the land and buildings comprising the block of flats known as Court Mansions and "the building" as meaning the main and ancillary buildings standing on the lessor's property. Clause 1(G) defined "retained property" as including all such parts of the lessor's property not by this or any other lease or tenancy demised or let. Clause 1(H) and the first schedule to the lease contained a lengthy definition of "the demised premises" which, while in general terms including Flat C, excluded (*inter alia*) those parts of the building lying above the surfaces of the ceilings or below the floor surfaces, the main timbers and joists of the building and of the external walls and conduits in the building, not serving exclusively the flat. These wide exclusions had the effect of extending the ambit of the "retained property".

Clause 1(L), so far as material, defined "maintenance year" as meaning every 12-monthly period ending on March 25. Clause 1(N) defined "maintenance contribution" as meaning:

> a sum equal to such percentage proportion of the aggregate annual maintenance provision for the whole of the Building (computed in accordance with the provisions of the Fourth Schedule hereto) as the rateable value of the Demised Premises bears to the aggregate of the rateable values of all the flats in the Building.

It further defined "maintenance account" as meaning:

> the account into which all Maintenance Contributions for the Building are paid and out of which all disbursements relating to the maintenance of the Building are made.

Clause 1(N) contained a definition of a distant "perpetuity date". Clause 1(O) defined "maintenance trustee" as meaning "the trustee or trustees for the time being of the Maintenance Account".

By clause 2 the lessor, in consideration of the premium, demised the demised premises to the lessee for 99 years from June 24 1978 at a yearly rent of £65 for the first 25 years and with certain modest increases thereafter. Clause 3 contained a number of covenants on the part of the lessee, which were expressed to be made with the lessor and the maintenance trustee. These included, *inter alia*, covenants to pay rent and outgoings in respect of the demised premises, a covenant to repair and maintain them and a covenant to permit the maintenance trustee or the lessor entry for the purpose of effecting repairs to other parts of the building.

By clause 4 the lessee further covenanted with the maintenance trustee and with the lessor to pay the maintenance contribution, so G
far as material, in the following terms:

> THAT the Lessee will in respect of every Maintenance Year pay the Maintenance Contribution to the Maintenance Trustee by four equal instalments on the quarter day immediately preceding the Maintenance Year and on the three subsequent quarter days in the Maintenance Year . . .

The fourth schedule to the lease contained provisions for the computation of the annual maintenance provision. Para 1 provided that this provision in respect of any maintenance year, other than the year ending March 25 1980, should be computed not later than the beginning of June immediately following the commencement of the maintenance year and should be computed in accordance with para 2. Paras 2, 3 and 4 provided: H

> 2 The Annual Maintenance Provision shall consist of:
> (a) a sum comprising the expenditure estimated as likely to be incurred in the Maintenance Year by the Maintenance Trustee for the purposes mentioned in Clause 6 of this Lease;
> (b) the remuneration of the Maintenance Trustee which shall be an amount equal to One per centum of the sum calculated in accordance with paragraph (a) hereof after deduction from such sum the remuneration of the Surveyor.
> 3 As soon as practicable after the end of each Maintenance Year the Lessor shall procure that the accounts for that Maintenance Year shall be audited by an accountant in accordance with the provisions of the Housing Finance Act 1972 (as amended) and such audited accounts for each Maintenance Year shall be conclusive of the amount of the Annual Maintenance Provision or the amount of any adjustment thereof for the Maintenance Year.
> 4 Upon receipt of the audited account the Lessor shall deliver a copy thereof to J
> the Lessee and shall notify the Lessee of the amount by which the estimate under 2 above shall have exceeded or fallen short of the actual expenditure in that Maintenance Year and the Lessee shall be allowed or shall pay as the case may be the proportion appropriate to the Demised Premises of the excess or deficiency.

This, then, was the machinery by which the obligations of the lessee to contribute to the maintenance fund were to be quantified and enforced. Clauses 5 and 6 of the lease declared the trusts upon which the maintenance trustee was to hold the maintenance fund. Clause 5 provided:

> The Maintenance Trustee shall pay the Maintenance Contribution into the Maintenance Account and shall pay thereout its remuneration calculated in accordance with sub-paragraph (b) of paragraph 2 of the Fourth Schedule K
> hereto and shall hold the balance (if any) in the Maintenance Account upon trust to apply the same until the Perpetuity Date for the purposes specified in Clause 6 hereof and subject thereto upon trust set forth in Clause 7 hereof.

Clause 6(A) set out the purposes for which the maintenance fund is to be applied. These included such purposes as the employment of a surveyor, the decoration and repair of the structure of the building, the payment of rates in respect of the building, the upkeep of the grounds, insurance, management expenses and the decoration of common parts.

The purposes specified in clause 6(A)(ii) included the following:

> AND ALSO at all times during the said term to keep the walls ceilings and floors of the Retained Property (but excluding those of any flat for the time being not demised by a lease in similar terms to this Lease as envisaged in Clause L
> 10(C) hereof) and the whole of the structure roof balconies foundations and main drains of the Building and the walls rails fences and gates appurtenant thereto in good repair and condition.

Clause 7 declared ultimate trusts of the maintenance fund to take effect on the perpetuity date.

Clause 8 provided that the statutory power of appointing a new maintenance trustee should be vested in the lessor, who was also to have the power by deed to remove from office a maintenance trustee and to appoint a new one.

Clause 9 authorised the maintenance trustee and the lessor to enter into any agreement whereby the maintenance trustee undertook to render any services to the lessor.

Clause 10 contained a number of covenants on the part of the lessor M
which were expressed to be made both with the lessee and with the maintenance trustee. Subclause (A) was a covenant for quiet enjoyment. Subclause (B) placed an obligation upon the lessor, subject to certain provisos, if so required by the lessee, to enforce against the tenant or occupier of any other part of the building the covenants and conditions contained or implied in any lease thereof and enforceable by the lessor, similar to the lessee's covenants in the present lease. By clause 10(C) the lessor covenanted that every lease of a flat in the building thereafter granted for a term of years similar to the term granted by the lease should be substantially in the same form and should contain similar lessee's covenants.

Clauses 10(E), (F) and (G) provided:

A (E) THAT the Lessor will for the period that any flat in the Building is let without the tenant thereof being required to pay a Maintenance Contribution under an obligation similar to that contained in Clause 4(A) hereof or is unlet pay in respect of all such flats to the Maintenance Trustee a sum equal to the total of the Maintenance Contributions that would be payable by the tenants of such flats and every amount so paid by the Lessor shall be dealt with for all purposes as if it were Maintenance Contribution paid by the tenants of such flats.

(F) THAT subject to the payment by the Lessee of the Maintenance Contribution in accordance with the provisions of Clause 4 hereof the Lessor will procure the due performance by the Maintenance Trustee of the obligations contained in Clauses 5 and 6 hereof and in the event of the Maintenance Trustee failing to perform any of its said obligations the Lessor will (subject to the aforesaid) duly perform and observe such obligations.

B (G) AT the Lessor's own expense at all times during the said term to keep the walls and ceilings and floors of such of the Retained Property as comprise any flats for the time being not demised by a lease or leases in similar terms to this Lease as envisaged in Clause 10(C) hereof in good repair and condition.

Of the several other provisions of the lease we need only refer to clause 12, which reads as follows:

NEITHER the Lessor nor the Maintenance Trustee shall be liable or responsible for any damage suffered by the Lessee or any servant or workman of the Lessee or any member of the Lessee's family or any guest of his through any defect in any fixtures conduits staircase lift machinery (if any) or thing in or upon the Lessor's Property or any part thereof (including the flat) or through the neglect default or misconduct of any porter or other servant employed by the Maintenance Trustee or by the Lessor in connection with the Lessor's
C Property except in so far as any such liability may be covered by insurance effected by the Maintenance Trustee pursuant to Clause 6(A)(v) and (vi) hereof.

Events between the grant of the lease and the trial

The plaintiffs moved into Flat C on January 1 1980 and were soon faced with serious problems. Early in the same year a very badly corroded rainwater down-pipe situated immediately adjacent to the external face of the walls of the front bedroom, bathroom and lavatory of Flat C was removed and shortly afterwards extensive dry rot was found to be present in the front bedroom, bathroom and lavatory. The learned judge was satisfied that the probable cause of the disaster was that stated in a subsequent joint report from the
D parties' surveyors, namely "defects to the external plumbing for disposal of rainwater resulting in water penetration of the brickwork". Neither the plaintiffs nor the defendants were willing to do the necessary work at their own expense. The local authority served a dangerous structure notice in respect of the dry rot, dated August 18 1980, and obtained an order of the magistrates' court on January 5 1981. Since it was not complied with, builders instructed by the local authority did the required work in March and April 1981. The plaintiffs told the judge that they had to move into a hotel for 13 days, that Flat C was left in a condition requiring extensive repairs and redecoration and that the dry rot in the flat was not wholly extirpated.

E This outbreak of dry rot, however, was very far from being the only trouble which faced the plaintiffs as Selico's tenants. The learned judge gave this description of the defendants' activities or inactivity during the period between the grant of the lease and the start of these proceedings in November 1982, which has not been challenged on behalf of the defendants:

After completion, more than three years passed before the commencement of this action. During that time things went from bad to worse. The defendants made no attempt to operate the system of maintenance set up by the Lease. Select did not appoint a managing agent as provided by clause 6: it managed (or mismanaged) Court Mansions itself, charging 10% of the Lessor's maintenance expenditure as a management fee, in addition to its 1% as trustee's remuneration. No real attempt was made to budget for estimated
F expenditure in advance. The figures of actual expenditure, audited annually by Mr Schwartz, pitifully increased each year by a conventional £200 called "Reserve Fund", were charged against lessees in proportion to rateable value. No Maintenance Account, as contemplated by the Lease, ever existed, either at a bank or even as a ledger account. Mr Sunnucks for the defendants said that one could be constructed retrospectively. It has not however been done. Not only did the defendants not seek to budget for and raise the money needed to put the building in repair; but they made no serious effort to get the work physically performed. Repeated complaints by the tenants were disregarded. Little was done during these years except under the pressure of statutory notices, of which there were many, from the Camden London Borough Council. Not all such notices were obeyed. More than once the local authority did the work itself, and by the time of the trial it was claiming over £30,000 in respect of Court Mansions. Moreover, such work as was ordered by the defendants was frequently executed by incompetent or ill-instructed builders. On more than one occasion the local authority's environmental health officer required it to be done again.

The judgment of Goulding J

Against this background the plaintiffs issued proceedings against both the defendants on November 18 1982 and in their reamended statement of claim asserted six cumulative or alternative causes of action, namely (1) deceit (against Selico); (2) breach of the defendants' express obligations contained in the lease; (3) breach of terms alleged to be implied by the lease; (4) negligence; (5) nuisance; (6) breach of trust on the part of Select.

The learned judge did not find it necessary to say anything about the last cause of action, which he said had not been seriously developed in the argument. In dealing with the case generally, he said he had found his task made more difficult by what he described as the undue reticence of the defendants. As he observed, constant pressure by the plaintiffs had been required, both in the interlocutory proceedings and during the course of the trial, to obtain adequate discovery of documents. As to *viva voce* evidence, he had not seen any officer or employee of Selico, except a Mr Schwartz, who was called only as the accountant who, for the purpose of service charges and on the instructions of Select, certified Selico's costs. Though three employees of Select were called, the judge described them as having strictly departmental duties. As he said, none of them could speak of the policy or the responsible decisions adopted on behalf of Select.

Neither Mr Haskins nor Mr Colenutt was called to give evidence. We were told that the first time the plaintiffs became aware of the identity of the person who had actually carried out the relevant works was when Mr Azzam came to give his evidence as the last of the defendants' witnesses.

In the event, the learned judge did not find it necessary to make any findings as to the state of knowledge of Mr Haskins or of Select or of Selico in regard to the cover-up because, as we infer from what counsel have told us, of the way in which the case was then argued. The defendants' counsel vigorously submitted that the judge should not make any finding that Mr Azzam had perpetrated the dishonest concealment alleged. As we understand the position, however, while making no explicit concession on this point, he did not seek to argue that if, contrary to his submission, Mr Azzam *had* been guilty of a fraudulent misrepresentation, his tortious act would not be attributable to Selico.

In these circumstances, the learned judge felt able to dispose of the issue of deceit quite shortly. Having found as a fact that Mr Azzam was guilty of the dishonest cover-up alleged, he continued:

Is it, however, actionable at law? The law must be careful not to run ahead of popular morality by stigmatising as fraudulent every trivial act designed to make buildings or goods more readily saleable, even if a highly scrupulous person might consider it dishonest. But it is to my mind quite a different matter for an intending vendor to hide so sinister and menacing a defect as active dry rot. The case is fairly comparable, in my view, with the concealment of cracks indicating the settlement of foundations, considered in *Ridge* v *Crawley* (1958) 172 Estates Gazette 637 and, in the Court of Appeal (1959) 173 Estates Gazette 959. There the plaintiff relied on a combination of words and conduct, but I believe it to be the law that conduct alone can constitute a fraudulent misrepresentation. (See *Horsfall* v *Thomas* (1862) 1 H&C 90 and *Smith* v *Hughes* (1871) LR 6 QB 597.) In my judgment the concealment of dry rot by Mr Azzam was a knowingly false representation that Flat C did not suffer from dry rot, which was intended to deceive purchasers, and did deceive the plaintiffs to their detriment. I am satisfied that the plaintiffs would not have entered into a contract or accepted the lease had they known there was dry rot inside Flat C. I do not know whether the lessor (against whom alone this claim of deceit is made) had by its responsible officers any dishonest state of mind, but the dishonest act was done by Azzam in the course of work ordered by Select within the scope of Select's authority as the lessor's agent. The contrary has not been argued. I accordingly hold that the lessor is liable to the plaintiffs in damages for deceit.

The learned judge next turned to consider the defendants' express obligations under the lease. In this context he expressed his conclusions thus:

Clause 5 of the Lease begins with the words "The Maintenance Trustee shall", and is in my opinion a promise by Select to open and operate a maintenance account and to act as trustee thereof in accordance with clause 6. Then in clause 10(E) the lessor covenants to make its own maintenance contributions and in clause 10(F) to procure Select to carry out its part of the scheme. It seems to me plain that, by failing to carry out the scheme of maintenance provided for by the Lease, Select was in breach of clause 5 and the lessor was in breach of clause 10(E) and 10(F) thereof. On the evidence the Lessor was also in breach of its covenant in clause 10(G) to maintain the walls, ceilings and floors of flats not demised by long leases, the relevant flat being Flat E immediately above the plaintiffs. I am also of opinion that the failure of the lessor to keep the building watertight was a breach of the covenant for quiet enjoyment contained in clause 10 (A).

In expressing these conclusions as to breach of covenant the learned judge rejected four points argued by Mr Sunnucks on behalf of the defendants. We need not refer to three of these points (respectively based on the wording of clauses 10(B), 10(G) and 12 of the lease), since they have not been pursued on this appeal. We will revert to the fourth (which was based on the introductory words of clause 10(F)) later in this judgment.

The learned judge dealt with implied covenants and negligence together. He held that the lease contained an implied covenant by Select to estimate or procure the estimation of each year's likely expenditure by Select for the purposes of the 4th Schedule and that this implied term had been inexcusably broken. There is no appeal against this.

The other implied term alleged by the plaintiffs had been formulated in para 11 of their statement of claim as follows:

> Further or alternatively the First Defendant at all times retained possession and control of the whole of the Building except the Flat and in particular retained possession and control of those parts of the building described in paragraphs 5 and 6 hereof. By reason thereof the First Defendant impliedly covenanted to take reasonable care that the condition of the said retained premises should not cause damage to the Plaintiffs or to the Flat.

The parts of the building described in paras 5 and 6 are the structure, conduits etc. This claim was founded on the general principle established by such cases as *Hargroves, Aronson & Co* v *Hartopp* [1905] 1 KB 472, and *Cockburn* v *Smith* [1924] 2 KB 119 and recently considered by this court in *Duke of Westminster* v *Guild* [1984] 3 WLR 630.

The learned judge accepted this submission in so far as it was based on implied term, saying this:

> So here, the plaintiffs have pleaded negligence as well as the implied term. For my part I think the better characterisation of the alleged duty of the lessor is as an implied term. Where there are gaps in an instrument expressing the reciprocal obligations of landlord and tenant, it is, in my judgment, more natural to fill them by implication, as the House of Lords did in *Liverpool City Council* v *Irwin* [1977] AC 239, than to invoke the law of tort. The distinction may be of more importance since the enactment of section 125 of the Housing Act 1974 to which I shall refer hereafter. On either view, however, the question arises whether any such general duty, be it in contract or be it in tort, is not excluded by the express maintenance scheme now under scrutiny. In my opinion it is a question of difficulty. On the whole I think the particular provisions of the lease are sufficient to exonerate the lessor from the obligation generally arising from a landlord's retention of property whose proper maintenance is necessary for the protection or safe enjoyment of the demised premises. It is clear that the maintenance scheme might not always provide necessary and timely protection, for example, through the continued inability or refusal of other lessees to pay their proper contribution, and in the absence of express provision I do not think it can fairly be taken to exclude altogether a landlord's ordinary duty in this sphere.
>
> Clause 12 of the Lease, however, provides a defence to claims for damages falling within its scope.

Finally, so far as the action was concerned, the learned judge dealt with the plaintiffs' claim in nuisance, which he regarded as precluded by clause 12 of the lease. Since the plaintiffs have not sought to challenge this particular conclusion on this appeal, it need be considered no further.

Having surveyed the several causes of action on which the plaintiffs relied, the learned judge considered and dismissed the defendants' counterclaim for the reasons stated in his judgment. Mr Sunnucks has advanced no argument in relation to the counterclaim on this appeal and we do not think it will require any consideration separate from the other issues.

Finally in his judgment, the learned judge considered the relief to be given to the plaintiffs on their claim. He ruled out the possibility of rescission and pointed out the obvious difficulties which, on the face of it, specific performance would present. However, he drew attention to section 125 of the Housing Act 1974, which empowers the court, in proceedings in which a tenant of a dwelling alleges a breach on the part of his landlord of a "repairing covenant", to order specific performance of the covenant "whether or not the breach relates to a part of the premises let to the tenant and notwithstanding any equitable rules restricting the scope of that remedy . . .". With the support of that section, the learned judge decided in effect to grant specific performance not of the express covenants on the part of the landlord contained in the lease but of the *implied* obligation which he considered to fall upon Selico, arising from its retention of G
property whose proper maintenance was necessary for the protection or safe enjoyment of the demised premises. Having delivered his judgment, he gave the parties the opportunity to address further argument to him as to the precise form of the order on a later occasion. We have already indicated the form which his order eventually took. As he explained in his judgment, the thought behind his awards of nominal damages against both defendants was that the plaintiffs must look to their damages in tort as compensation for their past and present loss, while the order for specific performance would protect them for the immediate future.

The learned judge's order contained a provision by virtue of which the inquiry for damages for deceit has been stayed pending the result H
of an appeal. It contained no similar provision relating to the order for specific performance. The latest date specified for compliance with this order was September 28 1985, but it has not been complied with.

The claim in deceit

Neither in his notice of appeal nor in argument has Mr Sunnucks sought to challenge the learned judge's findings that (a) Mr Azzam deliberately covered up active patches of dry rot in Flat C without any attempt at eradication; (b) the concealment of dry rot by Mr Azzam was a knowingly false representation that Flat C did not suffer from dry rot; (c) this representation was intended to deceive purchasers of Flat C; and (d) it did deceive the plaintiffs to their detriment. These J
findings inevitably mean that this court must proceed on the footing that, in relation to the plaintiffs, at least Mr Azzam personally was by his conduct guilty of the tort of deceit. Mr Sunnucks made it clear that, so far as his clients are concerned, by far the most important issue on this appeal is whether this fraudulent tortious conduct of Mr Azzam is to be imputed to the vendor, Selico, so as to render Selico liable to the plaintiffs for damages for Mr Azzam's fraud.

Mr Sunnucks' principal submissions under this head can, we think, be sufficiently summarised as follows. The capacity of Mr Azzam was no more than that of an independent building contractor employed by the defendants for the renovation and decoration of Flat C. In his submission, neither of the defendants themselves made any representations whatever, whether orally or in writing or by K
conduct, as to the presence or absence of dry rot in Flat C. The plaintiffs, he submitted, did not plead or establish in evidence that either of the defendants had authorised such a representation or held out Mr Azzam as having express or implied authority to make such a representation. Mr Azzam, as a builder, had no ostensible authority to make any representations or to act as agent on behalf of the vendor. In any event, it is said, there is no evidence that either of the defendants knew of the existence of dry rot in Flat C, though they knew of its existence on the other side of the building in Flat H.

The principles according to which one person may be held vicariously liable in tort for the deceit of another have recently been fully considered by this court in *Armagas Ltd* v *Mundogas SA* [1985] L
3 WLR 640 (which, we understand, is now under appeal to the House of Lords). From this decision it appears that in order to render an employee vicariously liable for his employer's deceit, it has to be shown that the employer had conferred on the employees either (i) actual authority or (ii) ostensible authority to make the particular representation complained of (see particularly at p 658 E et seq, per Robert Goff LJ). If the tort of deceit has been committed by the employee outside the scope of the employee's actual or ostensible authority, it cannot be claimed that it has been committed in the course of his employment in such manner as to render the employer vicariously liable. For this purpose the scope of such authority limits the course of the employment. Subject to what the House of Lords may decide, earlier authorities which suggest a wider test of vicarious M
liability, intended for the more liberal protection of innocent parties against the wrongful acts of their employees, can no longer be regarded as good law (see, for example, at p 681 per Stephenson LJ).

The learned judge held Selico liable for Mr Azzam's deceit on the short grounds that the "dishonest act was done by Azzam in the course of work ordered by Select within the scope of Select's authority as the lessor's agent". The only evidence of what instructions were given to Mr Azzam about Flat C is to be found at pp 2 and 3 of the transcript of his evidence. He was, as we have said, to do "any work to bring it up to a very good standard for the purpose of selling". This was clearly the "course of work ordered by Select" to which the learned judge referred. But Azzam's failure to remedy the

A dry rot was, on the evidence before us, not pursuant to any express direction by Select and therefore cannot, in our judgment, be accurately described as done in the course of work ordered by Select. Though, for reasons which will appear, we think that Select must be held responsible in law for the cover-up, we are of the opinion that, on the basis of the *Armagas* decision, the mere fact that the dishonest cover-up was effected by Mr Azzam while he was engaged on work of renovation ordered by Select would not suffice to render his tortious act attributable in law to either of the defendants. Nor would the additional fact that Selico had conferred on Select authority to order the work to be done suffice to saddle Selico with liability. As we have already explained, these points of law do not appear to have been
B canvassed in argument before the learned judge at the trial. If the plaintiffs are to establish liability on the part of Selico for Mr Azzam's deceit, we think they must do so on grounds other than the short grounds relied on by the leaned judge. This they seek to do by a respondents' notice, which they were allowed to submit during the course of this appeal.

Mr Brock, on behalf of the plaintiffs, has not submitted that Mr Azzam himself acted in connection with Flat C in any capacity other than as a building contractor or that he had any ostensible authority to make any representation to any prospective purchaser or lessee of Flat C as to the state of its repair. The essence of his case on deceit, as developed before us, can we think be summarised in the four following propositions:

C (A) On the evidence, the learned judge should have found (and, if the legal argument presented to him on behalf of the defendants had led him to suppose that a finding on this point was necessary, would have found) that Mr Haskins was at all material times fully aware of the cover-up operation which was being effected or had been effected by Mr Azzam and approved of it, whether or not he had expressly directed it.

(B) In view of the relationship between Mr Haskins and Select, such knowledge and approval on the part of Mr Haskins must be attributed to Select.

(C) Select was therefore a party to the fraudulent misrepresentation made by Mr Azzam to prospective purchasers of Flat C.

D (D) Selico had invested Select with ostensible, if not actual, authority wide enough to make representations to prospective purchasers or lessees as to the state of repair of Flat C and was accordingly vicariously liable to the plaintiffs for the fraudulent misrepresentation made by Select.

Mr Sunnucks submitted that the plaintiffs' reamended statement of claim does not adequately cover a claim on these lines, but we think it does. Para 17 begins with an allegation that prior to the execution of the lease, Selico "by its servants or agents" represented that the flat did not suffer from dry rot. The allegation of a fraudulent cover-up of the dry rot by Selico is to be found in subparagraph (4) of the pleading. True it is that subparagraph (4) does not repeat the reference in the opening words of para 17 to Selico's "servants or
E agents". However, Selico, being a corporate body, could only act by servants or agents. The pleading is, in our judgment, wide enough to cover a claim against Selico on the grounds that it is vicariously liable for the fraudulent concealment of the dry rot effected by its servants or agents. It was so treated without objection at the trial. Though it can be said that the plea of deceit is somewhat inadequately particularised, two answers to any such complaint can, in our opinion, fairly be given. First, further and better particulars of this allegation were not sought. Second, the undue reticence of the defendants, which the judge criticised in his judgment, would, in any event, have made it difficult for the plaintiffs to give full particulars until the relevant facts had emerged at the trial. We have already mentioned that until Mr Azzam gave his evidence, the plaintiffs did
F not even know who had done the relevant work.

Apart from this pleading point, Mr Sunnucks' principal opposition to a finding of deceit, based on the lines indicated in (A), (B), (C) and (D) above, was directed to point (A). In the absence of any specific finding to this effect, he submitted (and we accept) that the court should be very slow to infer that Mr Haskins was aware of, or approved, the cover-up operation carried out by Mr Azzam. The evidence, in his submission, does not suffice to justify any such conclusion.

Nevertheless, on the basis of the findings of fact which the learned judge did make and of the other evidence, we are irresistibly driven to make that inference, if only for these principal reasons. It would appear from the evidence of Mr MacGillivray, whom the judge accepted as a truthful witness, that before Mr Azzam began his work on Flat C in 1978, any person reasonably familiar with dry rot would have detected evidence of dry rot on inspecting the front bedroom and bathroom of the flat. Mr Haskins was a qualified surveyor. At the material time he had a very close connection with Court Mansions, and Flat C in particular, in a number of different capacities. He effectively directed and controlled the business of Select, which not only managed Court Mansions for Selico but also in due course rendered to Selico a fee note, which included a claim for £680 "commission upon introducing Mr G L Gordon and Ms A Teixeira to purchase the above leasehold property in the sum of £31,000 and fee re plans and supervision of repairs". The strong probability, therefore, is that Mr Haskins would have seen the flat before Mr Azzam (who had a long working association with Select) began his work and would have been aware of its condition. Significantly, as Mr Gordon's evidence indicates, the work of renovation to Flat C appears to have been concentrated particularly on these two rooms where the dry rot was present. However, matters go further than this. The following questions were put to Mr Azzam and answered thus in the course of his cross-examination.

(Q) When you did the work in Flat C in 1978/beginning 1979, the work about which we have been talking, would Mr Haskins be supervising that?
(A) Yes.
(Q) And how often did he inspect and supervise the work?
(A) Normally as often as he liked and to my recollection he has visited the premises in question several times.
(Q) During that eight to ten week work period when you did the work in Flat C, do you remember how many times he visited to carry out inspection and give advice?
(A) At least once a week.
(Q) At least once a week?
(A) Yes.
(Q) So he was fully aware of what was going on at the flat during that period?
(A) He would be fully aware of what is happening there.

From this colloquy we think it must have been quite obvious to the defendants and their advisers that when the plaintiffs' counsel came to make his final speech he would be submitting, not only that there had been a cover-up by Mr Azzam, but that Mr Haskins was fully aware of what Mr Azzam was doing (as indeed we understand was submitted). Nevertheless, there was no re-examination of Mr Azzam on this evidence and, more importantly, Mr Haskins was not called as a witness on behalf of the defendants to rebut the potentially damaging allegations of a cover-up by Mr Azzam, of which Mr Haskins was said to be aware. It was not suggested to the judge that Mr Haskins was not available to give evidence. No explanation whatever was given of his absence.

We of course accept that the mere failure to call a witness cannot by itself be treated as evidence against that witness or against the parties to the proceedings. As Mr Sunnucks rightly pointed out, the onus has throughout rested on the plaintiffs to establish fraud, not on the defendants to refute it. However, Mr Haskins, as the controlling shareholder in Select, which was a party to the proceedings, must have been fully aware of the course of this action, and we assume that he would have been afforded the opportunity to give evidence by the defendants' legal advisers if he had wished to do so. Mr Azzam said that he himself was giving evidence at the request of Mr Haskins and Mr Colenutt. In default of any other explanation, we can only infer that, for reasons of his own, Mr Haskins did not himself wish to give evidence. In our judgment, in deciding whether or not on the balance of probabilities to draw the suggested inference against Mr Haskins and through him against the defendants, on the basis of all the evidence in this case, in particular the evidence of Mr Azzam, it must be permissible for the court to have regard to the fact that the defendants, without any explanation, failed to call Mr Haskins, who could on any footing have given highly material testimony in relation to the alleged cover-up and, if this represented the truth, could have told the court either that there had been no cover-up by Mr Azzam or that, if there had been, Mr Haskins himself had been unaware of it, because Mr Azzam had succeeded in misleading him, no less than he had deceived the plaintiffs.

In response to Mr Sunnucks' vigorous submissions on point (A) above, we have considered carefully and anxiously whether the evidence as a whole is strong enough to justify a conclusion that Mr Haskins was at all material times aware of the cover-up operation which was being or had been effected by Mr Azzam and approved of it. Taken as a whole, however, the evidence in our judgment inexorably leads to that conclusion on the balance of probabilities.

And, indeed, we believe that the learned judge would have so found if he had considered such a finding relevant, since, significantly, he expressly found that Mr Azzam's cover-up operation was intended to deceive *purchasers* (not Select, with whom he had had such a long working association).

As to point (B) above, we feel no doubt that the relevant knowledge and approval on the part of Mr Haskins must be attributed to Select, having regard to his intimate connection with that company. Indeed, we do not think that Mr Sunnucks submitted to the contrary.

As to point (C) above, Select not only managed Court Mansions but, through Mr Haskins, supervised the relevant works of repair and indeed rendered an account to Selico for them. As we have already concluded, Select, through Mr Haskins, must be treated as having known and approved of Mr Azzam's cover-up operation. Having regard to such knowledge and approval, and to the fact that neither Mr Haskins nor Select took any steps to disabuse the plaintiffs' minds as to the disguised condition of Flat C, we conclude that Select should be held itself to have been a party to that operation and to have knowingly induced the plaintiffs to enter into the lease upon the faith of a fraudulent misrepresentation that the flat was free from dry rot.

As to point (D) above, we think the position is fairly clear. The judge found as a fact that Select had "at all material times managed Court Mansions for the first defendant". The further details of the relationship between Selico and Select were never explored in evidence at the trial, presumably because no one considered them material. The whole of the trial was apparently conducted on the basis that no relevant distinction fell to be drawn between them, and indeed, at least at one stage in his argument before this court, Mr Sunnucks said that he did not seek to draw any distinction between Selico and Select. The authority given by Selico to Select must, in our judgment, have imported a general authority, ostensible if not actual, to do anything relating to the management of the property, including selling a long lease of any flat and making any representations with a view to selling the lease. If, as we consider, on a true view of the facts Select can be held to have knowingly induced the plaintiffs to enter into their lease on the faith of a fraudulent misrepresentation as to the state of repair of Flat C, that representation was within the actual or ostensible authority conferred on Select by Selico, notwithstanding its falsity. In these circumstances, subject to what is said below, Selico, on the principles of the *Armagas* case, must be vicariously liable to the plaintiffs for Select's fraudulent misrepresentation. In this context the following observations of Lord Wright in *Bradford Third Equitable Benefit Building Society* v *Borders* [1941] 2 All ER 205 at p 220 appear to us material:

I do not question that, if a person knowingly and deliberately profits by another's fraud he may be properly held to have participated in the fraud and to be liable for the damage. This may happen where a continuing false representation has been made by a person, on the basis of which the transaction is concluded. I am prepared to assume here that, not only may that person be guilty of fraud (*With* v *O'Flanagan* [1936] Ch 575, at p 584), but so also may any person who, though not a party to the fraudulent original representation, afterwards learns of it and deliberately and knowingly uses the delusion created by the fraud in the injured party's mind in order to profit by the fraud.

On the evidence there seems to be no doubt that both Selico and Select stood to gain and did gain from a sale of this property.

By way of further defence to the claim based on deceit, Mr Sunnucks relied on a number of subsidiary points. Even if any relevant misrepresentation was made by or on behalf of Selico, he submitted, the plaintiffs did not act upon it; they acted on the report of their own surveyors, which had been obtained before the contract was concluded and had warned them of the possibility of dry rot, even though no dry rot had been discovered. He referred us to a letter of April 17 1979 in which, before contracts were exchanged, Selico by its solicitors had expressly refused to give any undertaking as to repairs. He also relied on clause 4(2)(a) of the Law Society's Conditions of Sale (quoted above) under which a purchaser of a property is deemed to purchase "with full notice of its actual state and condition". The plaintiffs, he pointed out, were content to accept this condition and the refusal of the vendor to give any warranty in relation to the condition of the flats. Furthermore, he observed, the plaintiffs and their surveyor had ample opportunity to inspect the flat, an opportunity of which they availed themselves. In these circumstances, decisions such as *Horsfall* v *Thomas* (1826) 1 H&C 90 and *Smith* v *Hughes* (1871) LR 6 QB 597 precluded the plaintiffs from complaining of any misrepresentation.

Both these two cases, however, are distinguishable from the present on their facts. In the former, not only was the defect in the gun patent and discoverable on inspection but the purchaser took no steps to inspect it, so that he did in fact not rely on any misrepresentation as to its condition which might have been made. In the latter case, the vendor did nothing to disguise the character of the oats sold. In the present case, on the learned judge's relevant findings of fact, with which we see no reason to disagree, not only was a fraudulent misrepresentation made, which was intended to mislead prospective purchasers of a lease of the property; the misrepresentation did mislead the purchasers and they acted on it to their detriment. In these circumstances, it is in our judgment no answer in law to the claim in deceit for the defendants to say that the plaintiffs or their surveyor *could have* discovered the dry rot on a closer inspection of Flat C or were content to purchase without any warranty as to the condition of the property; they and their surveyor were in fact misled by the cover-up operation, as they were intended to be. The general principle *caveat emptor* has no application where a purchaser has been induced to enter the contract of purchase by fraud. Nor can clause 4(2)(a) of the Law Society's Conditions of Sale avail a purchaser in these circumstances. These subsidiary submissions made by way of defence to the claim based on deceit are not in our judgment well founded. Even if at the date of completion the plaintiffs had been fully aware of the fraudulent misrepresentation which had been made to them, this would not have precluded them from affirming the contract and seeking appropriate relief, though it would have prevented them from seeking to rescind it: *Campbell* v *Fleming* (1834) 1 Ad & El 40.

For all these reasons, albeit by a much longer route than that which the learned judge followed, we agree with his conclusion that the claim in deceit against Selico must succeed.

The plaintiffs' other claims

We can deal with the plaintiffs' other claims much more shortly, because in this context there is much common ground.

Mr Sunnucks has submitted that, having regard to the comprehensive nature of the provisions of the lease relating to the repair and maintenance of Flat C and of the block, Court Mansions, as a whole, this is not a case where the learned judge was justified in implying any covenants relating to these matters. The lease, he pointed out, is in an unusual form, providing as it does for the repair and maintenance of the building by a system of express covenants and trusts. If the defendants are in breach of any of those express provisions relating to repair and maintenance, the proper remedy, in his submission, is by reference to the agreed system and not by reference to suggested implied covenants to which the parties never agreed.

We think this submission is correct and indeed in this court Mr Brock, on behalf of the plaintiffs, has not sought seriously to challenge it. Where a written tenancy agreement relating to a flat, forming part of a larger building in multiple occupation, manifestly does not embody the complete agreement between the parties the court may well be willing to supplement the written document by implying terms placing obligations on one party or the other. Such a case was *Liverpool City Council* v *Irwin* [1977] AC 239, where the "conditions of tenancy" which tenants were required to sign related only to the obligations on the part of the tenants, not of the lessor council. However, as Lord Wilberforce observed in that case (at p 254), "such obligation should be read into the contract as the nature of the contract itself implicitly requires no more, no less: a test in other words of necessity". Lord Cross in the same case said (p 258) that the court "must be able to say that the insertion of the term is necessary to give — as it is put — 'business efficacy' to the contract and that if its absence had been pointed out at the time both parties — assuming them to have been reasonable men — would have agreed without hesitation to its insertion". The repair and maintenance scheme provided by this lease is a very cumbersome one and we agree with the learned judge that, even if the lessors and their agents were duly to carry out their obligations, the scheme might not always suffice to give the lessees necessary and timely protection — for example, as he pointed out, through the continued inability or refusal of other lessees to pay their proper contribution. Nevertheless, on a reading of the lease, we feel little doubt that it was intended, by all parties, to provide a comprehensive code in regard to repair and maintenance of the block. We are by no means satisfied that the implication of any further terms in this respect is necessary to give the lease business efficacy, or that the lessor, assuming it to have been a reasonable person, would have "agreed without hesitation" to the insertion of the suggested implied additional terms relating to the repair and maintenance of the block.

In agreement with the learned judge, however, we construe clause 5 of the lease as amounting to a covenant by Select with the plaintiffs to open and operate a maintenance account and to act as trustee thereof in accordance with clause 6 — and indeed also to collect all the maintenance contributions which it may be entitled to collect from other tenants. We also agree with his finding of an implied covenant by Select to estimate or procure the estimation of each year's likely expenditure by Select for the purposes of the 4th Schedule. (For brevity we will henceforth refer to this implied covenant as the "implied estimation covenant".)

Subject to one overriding point raised by way of defence in this context, we do not think that Mr Sunnucks sought to challenge the learned judge's findings that Select was in breach of clause 5 of the lease and that Selico was in breach of clauses 10(E), (F) and (G). His one point can be summarised thus. The obligations of the lessee under clause 10(F) are introduced by the words "subject to the payment by the Lessee of the Maintenance Contribution in accordance with the provisions of Clause 4 hereof . . .". Furthermore, there is a general principle that a party cannot claim to take the benefit of provisions in a deed without subscribing to his own obligations thereunder: *Halsall* v *Brizell* [1957] Ch 169 at p 182, per Upjohn J. It appears that in the present case the plaintiffs paid to the defendants sums totalling £129 in April 1980 by way of maintenance contribution but have so far declined to pay any further sums on this account and have intimated that, at least for the time being, they do not intent to do so. Such refusal, it is submitted, disentitles them from claiming any relief against the defendants in respect of alleged breaches of covenant or trust. The learned judge dealt with this point thus:

> The plaintiffs' failure to pay the contribution was justified at first by the demand of sums for which they were not liable and later by the complete and obvious unwillingness or inability of the defendants to carry out the scheme of the Lease in such a way as to keep the building in repair.

Mr Sunnucks has accepted that their initial refusal to pay the contribution was justified for the reasons given by the learned judge; the defendants were asking for contribution in respect of items for which Selico had agreed to pay. As to the later demands, the learned judge's conclusion seems to us clearly correct, if only for the following reason. Para 1 of the 4th Schedule required the annual maintenance provision in respect of any maintenance year after March 25 1980 to be computed in accordance with the provisions of that schedule. No such computation was ever made because Select was at all material times in breach of the implied estimation covenant referred to above. Such computation is an essential preliminary to any demand for maintenance contributions other that the first. No proper demand for further maintenance contributions was ever addressed to the plaintiffs. With due deference, Mr Sunnucks' overriding point is in our judgment without substance.

We thus accept that, as the judge found, the plaintiffs have established breaches by *Selico* of

(1) its covenant under clause 10(E) (to make payments to the maintenance trustee for other flats);

(2) its covenant under clause 10(F) (to procure the due performance by Select of the obligations contained in clauses 5 and 6);

(3) its covenant under clause 10(G) (to keep the walls, ceilings and floors of such of the retained property as comprise any flats for the time being not demised by a lease or leases in similar terms to the lease as envisaged in clause 10(C) thereof in good repair and condition).

We further accept that, as the judge found, the plaintiffs have established breaches by *Select* of its obligation under clause 5 to open and operate a maintenance account and to act as trustee thereof in accordance with clause 6.

The relief to be granted

It is in our judgment clear that, by virtue of clauses 5 and 6 of the lease, Select has covenanted to act as trustee upon the trust purposes for the benefit of the lessee set out in clause 6 of the lease and that it is in breach of this trust. Likewise, in our opinion, Selico itself, by virtue of clause 10(F), has covenanted in the event (which has happened) of Select failing to perform any of its obligations as trustee, itself to assume and perform these obligations. We think that Selico, by virtue of its failure to comply with this undertaking, is itself technically in breach of trust. (Though breach of trust on the part of Selico is not in terms pleaded, a breach by Selico of its obligations under clause 10(F) is, we think, sufficiently pleaded and relief is sought in respect of such breach.) The plaintiffs, as the intended beneficiaries under these trusts and as persons who have given good consideration for the assumption by the defendants of their obligations as trustees, are, in our judgment, entitled to ask the court, as they now do, to make an order for the enforcement of these trusts.

In view of the complicated provisions of this lease, the process of working out any such order would, unfortunately, be likely to be a long, complicated and expensive one. We have therefore considered whether any order for suitable monetary compensation would by itself afford the plaintiffs all the relief for which they are reasonably entitled to ask. But we accept Mr Brock's submission that it would not. Particularly if, as we understand to be the case, the plaintiffs intend to continue living in their flat at least for the time being, we see no reason why, even with the consolation of a monetary payment, they should be expected to tolerate a continuing total failure on the part of Selico and Select to operate the machinery of the lease which was intended to provide for the block of flats being maintained in reasonable condition, but in fact has been operated wholly inadequately, with the result that the condition of the block must be continually deteriorating, to the serious detriment of Flat C itself.

Though we fear that he will be confronted with a difficult and unenviable task, it appears to us, with due respect to all concerned, that a Chancery Master will be better qualified to deal with the working out of an order for the execution of the trusts of this lease than an official referee (by whom the learned judge directed that the inquiry as to damages for deceit should be assessed). It will manifestly be convenient that one tribunal should be charged with the working out of the whole of any order which this court may make. And, in view of the unusual nature and complexity of the matter, we think that this should be the Chief Chancery Master.

While in the course of argument reference has been made to claims against Select and Selico for damages for breach of trust, we think that the obligation of a defaulting trustee is strictly that of accounting by way of restitution to the trust estate, rather than that of paying damages: see *Snell's Equity* (28th ed) at p 282.

In all the circumstances, subject to any further submissions by counsel as to the appropriate form of order, we propose to make an order to the following effect:

(1) An order for payment by Selico to the plaintiffs of (a) damages for the tort of deceit and (b) damages for breach of its contractual obligations under clauses 10(E) and (G) of the lease, all such damages to be assessed on inquiry before the Chief Chancery Master.

(2) An order for execution of the trusts of clauses 5 and 6 of the lease against Select, together with all necessary accounts and inquiries, including an order for payment of all or any sums found to be due from Select to the trust fund (the maintenance fund) in respect of its breaches of those trusts.

(3) An order for execution of the trusts of clause 10(F) of the lease against Selico together with all necessary accounts and inquiries, including an order for payment of all or any sums found to be due from Selico to the trust fund in respect of its breaches of those trusts.

(4) An order for specific performance of the lessor's covenants under clause 10(G) (in exercise of the power conferred on the court by section 125 of the Housing Act 1974).

(5) A provision giving the Chief Chancery Master authority in his discretion to make any necessary or appropriate directions for the working out of all or any of the above orders, including (without prejudice to the generality of the foregoing) the powers to grant a temporary stay in respect of any one or more of such orders and to impose a time-limit within which the order for specific performance must be complied with.

Subject to these variations in the form of the order, we dismiss this appeal.

The appeal was dismissed with costs; the order of Goulding J to be varied as indicated in the judgment of the court.

Court of Appeal

January 16 1986

(Before Lord Justice KERR, Lord Justice NOURSE and Lord Justice BALCOMBE)

MANORLIKE LTD v LE VITAS TRAVEL AGENCY & CONSULTANCY SERVICES LTD

Estates Gazette April 26 1986

278 EG 412-414

Landlord and tenant — Notice to quit business premises — Validity of notice — Date of expiry — Written notice in accordance with provisions of lease — Lease provided for determination "by giving to the tenant not less than three months' previous notice in writing expiring at any time" — Plaintiff landlords gave a notice, stated to be in accordance with the terms of the lease, requiring the tenants "to vacate the said premises within a period of three months from the date of service of this notice" — Tenants contended that the notice was invalid as the period which it allowed was shorter than the lease required, "within a period of three months" being shorter than "not less than three months" — Tenants accepted that a notice requiring the tenants to leave by or at midnight on the relevant date would have been a good notice, but they argued that the words used excluded the stroke of that midnight — "The tenant contends that the import of the word 'within' is that there is to be some outside wall of time, however thin, on the near side of midnight" (per Nourse LJ) — Held, affirming the decision of the deputy judge under appeal, that the notice to quit was valid — The word "within", used in the context of a period of time, was capable of meaning "before or at the expiry of" that period; it is not necessarily shorter than the period itself — To construe the wording of the present notice as connoting a period of less than three months would be to strain the language in a hair-splitting and wholly artificial manner — Appeal dismissed

This was an appeal by tenants in the nature of a test case from a decision of Mr John Peppitt QC, sitting in chambers as a deputy judge of the Queen's Bench Division. The tenants, Le Vitas Travel Agency & Consultancy Services Ltd, appealed against a decision that the respondent landlords, Manorlike Ltd, were entitled to possession of shop premises in Oxford Walk, 150/154 Oxford Street, London W1. The appellants' case was one of 11 which turned upon the same point as to the validity of a notice to quit.

Michael Rich QC and David Smith (instructed by B Dave & Co) appeared on behalf of the appellants; David Oliver (instructed by Jaques & Lewis) represented the respondents.

Giving judgment, KERR LJ said: This is an appeal by the defendant, Le Vitas Travel Agency & Consultancy Services Ltd, against a judgment given by Mr John Peppitt QC, sitting as a deputy judge of the Queen's Bench Division on December 17 1985. The effect of his judgment was that the plaintiff company, Manorlike Ltd, were entitled to possession of a shop unit or business premises off Oxford Street in London.

The action originally concerned, I think, some 17 claims for possession by the plaintiff landlords against various tenants in a shopping arcade or precinct off Oxford Street. In six of those cases there was a point which does not arise in the other 11 concerning the purposes for which the landlords required the tenants to vacate the premises. All 17 cases came before a deputy master for possession under Order 14, and he gave unconditional leave to defend in the six cases to which I have referred. These raised an arguable point as to whether the landlords were entitled to possession under a break clause in the leases having regard to their plans for dealing with these premises and the reasons given in the notices to quit. He made an order for possession in the other 11 cases.

There was an appeal to Mr Peppitt QC sitting as judge in chambers. In all 17 cases a new short point was then raised on behalf of the tenants. He affirmed the order of the master that the plaintiffs were nevertheless entitled to possession in the 11 cases, having dismissed the claims for possession in the other six. G

The present appeal is in the nature of a test case of the remaining 11 cases which turns solely on whether or not the notice to quit was adequate in terms of time, having regard to the provisions of the lease.

The point is as follows. The lease excluded the provisions of section 38(4) of the Landlord and Tenant Act 1954 (as amended) and the tenant agreed, pursuant to an order of the court, that the provisions of sections 24 to 28 of that Act should also be excluded. The tenancy was for a period of three years from May 9 1984, but it contained a break clause, clause 7, on which the validity of the notice to quit turns. Clause 7 was as follows, so far as material:

> If the Landlord requires possession of the demised premises at any time during the said term the Landlord shall have the right to determine this Lease H by giving to the Tenant not less than 3 months' previous notice in writing expiring at any time so as to determine the same . . .

On May 22 1985 the plaintiff landlords gave a notice of which the relevant terms are as follows:

> We . . . hereby give you notice that we require possession of the above-demised premises,

and then I leave out the reasons given, because they are irrelevant. The notice proceeds as follows:

> and in accordance therewith

that is a reference to clause 7 of the lease

> we require you to vacate the said premises within a period of three months J from the date of service of this notice.

Effectively the issue is whether a requirement to vacate the premises "within a period of three months" is a shorter period than the period of notice for which the tenant was entitled under clause 7, which was a period of "not less than 3 months' previous notice". In that connection it is common ground that the notice given on May 22 1985 expired at midnight on August 22-23 1985.

On behalf of the appellant tenant, Mr Rich submitted that there is a material difference between the words "not less than" and the word "within" when used in the context of a period of time, in this case three months. In that connection he referred us to sections 29(3) and 26(6) of the Landlord and Tenant Act 1954 and relied upon those provisions in order to seek to show that these words have a different K meaning in that Act. For myself, I derive no assistance from the true construction of that Act, even after being reminded of the decision of this court in *E J Riley Investments Ltd* v *Eurostile Holdings Ltd* [1985] 1 WLR 1139*, to which Mr Rich also referred. I would add for myself that anything said about the true construction of the present notice to quit in relation to the present lease is not intended to have any bearing on the meaning and effect of any of these provisions in the 1954 Act.

Mr Rich accepted (I think rightly) that if this notice had required the tenants to leave the premises by or at midnight on August 22-23 it would have been a good notice. He referred us to a number of cases, including one on which the judge relied — *Eastaugh* v *Macpherson* L [1954] 3 All ER 214 — in which it was held that a requirement to vacate by a certain date means on or before that date and therefore includes the last day.

It is clear that the tenant was entitled to a period which included the stroke of midnight of August 22-23 because he was entitled to not less than three months' notice. We were referred in that connection to *Lester* v *Garland* (1808) 15 Ves Jun 248 and *Page* v *More* (1850) QB 684. It is also clear, and was not disputed by Mr Rich, that the moment at which the tenants' right to remain in the premises ended must be precisely coterminous with the moment at which the landlords' right to resume possession began. It is for all those reasons that Mr Rich conceded that a notice expiring precisely at midnight on August 22-23 would have been a good notice. M

Accordingly, all that one is left with is the question whether the words used in this notice, to vacate the premises "within a period of three months", excludes the stroke of midnight on August 22-23. To my mind the word "within", used in the context of a period of time, is capable of meaning "before or at the expiry of" that period, as Mr Oliver submits; it is not necessarily shorter than the period itself. Mr Oliver derives some assistance for that submission from one of the definitions of the word "within" in the *Oxford English Dictionary*, where under no 6 the word "within" is defined as follows: "In the limits of a period of time; (most usually) before the end of, after not more than". It is difficult to think of examples where the word

*Editor's Note: Also reported at [1985] 2 EGLR 124; (1985) 275 EG 716.

A "within" would be used synonymously with the much more complex expression "after not more than". However, what emerged from the illustration of the meaning of the word "within" by the expression "after not more than" is that in common parlance "within" is capable of including the final moment of a period of time. If a person is required to do something within a week, or in a week, he has a full week to do it, as it seems to me, including the last moment of that week, and he is not required to complete the task in less than one week. To construe this wording of this notice so that it connotes a period of less than three months, because permission must be given "within" three months with a consequent failure to allow a full period of three months, appears to me to strain the language in a hair-
B splitting and wholly artificial manner. Moreover, as pointed out by Mr Oliver, the present notice referred expressly to clause 7 of the lease, because it required the tenant to vacate the premises in accordance therewith. Any doubt as to whether there could have been any intended difference — which I think as a matter of ordinary English there is not — between "not less that three months" and "within three months" would, in my view, be resolved by that reference.

This is a notice to quit a commercial premises. I am fully aware that over many decades, and perhaps hundreds of years, there have been many cases of a highly technical nature on notices to quit. But if the present appeal were allowed on this point (which had not even occurred to the tenants when these proceedings started), I think that
C the reasoning of the court in cases of this kind would rightly be regarded as wholly artificial, if not absurd. I fully accept the principle that a notice to quit is a highly important legal document which requires to be strictly construed. But, even bearing that in mind, I cannot bring myself to conclude that there is any sensible difference between the terms of this lease and the words used in the notice to quit, either as a matter of ordinary English or as common sense. Accordingly, I would dismiss this appeal.

Agreeing, NOURSE LJ said: The question is a short one. It is also an exceedingly simple one. The landlord has the right to determine the lease by giving to the tenant "not less that 3 months' previous notice in writing expiring at any time". On May 22 the landlord serves a
D notice on the tenants requiring it to vacate the premises "within a period of three months from the date of service of this notice".

It is not in dispute that a period of three months from May 22 expires at midnight on August 22-23. What then is the effect of a requirement that the tenant shall vacate the premises "within" that period? Does it have the effect of giving to the tenant less than three months' previous notice? The tenant answers that question in the affirmative, contending that if someone is required to do something within a period of months he must do it *before* midnight on the last day of the period and that he may not do it *at* midnight on that day. To put it another way, the tenant contends that the import of the word "within" is that there is to be some outside wall of time, however thin, on the near side of midnight.

E Like my lord, I reject the tenant's contentions on grounds both of language and of common sense. The precise meaning of a preposition such as "within" depends on the context in which it is used. Here it is used in a legal document and it is applied to a period of three months' notice. In such a context I see no difference between the meanings of "within" and "during". In my view, if someone is required to vacate premises within or during a specified period, he will comply with the requirement by walking out of the door either before, or on, the stroke of midnight on the last day of that period. So much for language pure and simple. As for common sense, I need say no more than that the tenant's contentions, if correct, would attribute to the landlord the extremely bizarre intention to differentiate between
F midnight on one hand and the nearest measurable point of time before midnight on the other.

In my judgment the landlord gave the tenant three months' previous notice — neither more nor less. The notice was a valid one. I am accordingly of the opinion that the learned deputy judge came to a correct conclusion, and I, too, would dismiss this appeal.

BALCOMBE LJ also agreed that the appeal should be dismissed and did not add anything.

The appeal was dismissed with costs including costs of two previous applications. Leave to appeal to the House of Lords was refused and a further stay of possession order was refused.

Court of Appeal

February 26 1986

(Before Lord Justice PURCHAS, Lord Justice RALPH GIBSON and Lord Justice NICHOLLS)

CRANCOUR LTD v DA SILVAESA AND ANOTHER
SAME v MEROLA AND ANOTHER

Estates Gazette May 10 and 17 1986

278 EG 618-626 and 733-744

Landlord and tenant — Licence or tenancy — *Street* v *Mountford* again considered — Appeals by occupiers allowed but procedural issues influenced the result — Order 113 of Rules of the Supreme Court — Appeals were from decisions of two judges, Hirst J and Boreham J, who, in cases raising the same substantive points, granted the owners possession of the occupiers' rooms and refused remission for trial in the county court or the adjournment of the High Court proceedings to enable further evidence to be produced — Both judges took the view that the occupiers were clearly "lodgers" within the meaning of Lord Templeman's speech in *Street* v *Mountford* — Written agreements in the two cases appeared to be framed to take advantage of the description of a lodger in that speech — Among the terms of the licences were provisions that the licensee was allowed to use the one-room flat only between midnight and 10.30 am and between noon and midnight; that the possession, management and control of the flat were vested in the licensor, who had an absolute right of entry at all times for that purpose and for carrying out repairs, providing attendance and removing or substituting furniture; that attendance would be provided, including a housekeeper, cleaning of flat, removal of rubbish and laundry of bed linen; and that the licensor could require the licensee at any time to move out of his flat into any other flat of comparable size in the building — On the expiry of the agreements, which were for 26 weeks, the appellants failed to leave and the owners applied for possession under the summary procedure of Order 113 — The present appeals were brought against the grant of the possession orders — In allowing the appeals and remitting the cases to the county court, the Court of Appeal did not themselves decide that the agreements were, as the appellants submitted, a sham but that the owners' claim was not sufficiently clear and straightforward to justify the making of possession orders under Order 113 — In so doing the court made some incidental observations that may be of help in subsequent cases — The court agreed with the judges below that if the written agreements represented the true agreement of the parties they created licences only — Two provisions in the agreements had been particularly attacked, the denial of any right to use the rooms between 10.30 am and noon and the licensor's right to remove entirely articles of furniture — Although these provisions were not in themselves evidence that the agreement as a whole was a sham, they were arguably sham provisions and thus a factor to take into account in deciding whether the matter was clear enough to justify a summary order for possession — A third provision, which had been attacked, the right to require an occupier to move to a different flat in the building, was not regarded by the court as unreasonable — It would be for the trial court to decide whether, in the light of all the effective terms of the agreement, exclusive possession was granted to the appellants — In the second appeal, which was from the decision of Boreham J, there was an additional point, an allegation of an overriding oral agreement between the appellants and the respondents' housekeeper from which a grant of exclusive possession might be inferred — This again raised a triable issue — Appeals allowed, possession orders set aside and cases remitted for trial

These were two appeals against orders for possession under Order 113 which, by consent, were heard together. In the first J M da Silvaesa appealed against a decision of Hirst J granting possession of room no 3 at 21 Cranley Gardens, London SW7, to Vassillios Georgiou Markou. In the second, Miss Sophie Merola and Ahmed Kader appealed against a decision of Boreham J granting possession of room no 12 in the same house to Crancour Ltd (which had become the successor to Mr Markou as the owner of the flats).

Kenneth Bagnall QC and Miss Linda Pearce (instructed by Brocklesby & Co) appeared on behalf of the appellant Joao M da Silvaesa; T J Gallivan (instructed by Brocklesby & Co) represented the appellants S Merola and A Kader; Michael Mann QC and Francis Tregear (instructed by Richards Butler & Co) appeared for the respondents in both appeals (V G Markou and Crancour Ltd).

Giving the first judgment at the invitation of Purchas LJ, RALPH GIBSON LJ said: There are two appeals before the court which, with the consent of all parties, have been heard together. In the first appeal Mr J M da Silvaesa appeals against an order for possession of room no 3 in a house at 21 Cranley Gardens, London SW7, made by Hirst J on September 17 1985.

In the second appeal Miss S Merola and Mr A Kader appeal against an order for possession within 28 days of room no 12 in the same house made by Boreham J on November 28 1985.

In each case the order for possession was made under Order 113 of the Rules of the Supreme Court, which makes provision for summary proceedings for possession of land:

Where a person claims possession of land which he alleges is occupied solely by a person or persons (not being a tenant or tenants holding over after the termination of a tenancy) who entered into or remained in occupation without his licence or consent . . .

Accordingly, if the appellants were originally in occupation as tenants and not as licensees, the use of this summary form of proceedings was misconceived and the orders for possession could not be made. Further, the appeals raise questions as to the appropriateness of this form of proceedings in cases where any dispute of law or fact is raised by those against whom the orders are sought.

The issues in the two appeals are not identical, but one major issue is common to both: the proper construction and meaning of the agreements in writing — both in the same form — made by Mr V G Markou, the then owner of the house, with the appellants. In each case the learned judge held as a matter of construction that the agreement granted a licence to occupy only and not a tenancy and that the agreement was a real agreement between the parties and not a "sham". The appellants say that in each case the judge was wrong on the first and, alternatively, on the second point.

In the first appeal Mr da Silvaesa and a Miss Santos signed an agreement in writing dated January 26 1985. It was called a licence and, according to its terms in clause 1, Mr Markou licensed them to use the furnished room no 3 in the house:

on each day between the hours of midnight and 10.30 am and between noon and midnight but at no other times for a period of 26 weeks from 26 June 1985 for the purpose of temporary accommodation for the licensees personal use only . . .

It is to be noted that this paragraph reserves to the landlord, or building owner, the right to treat the licensee as a trespasser if he remains in the building between 10.30 am and noon and, indeed, to require the licensee to remove his possessions from the room during that period of time each day.

Other clauses of the agreement relevant to the issues raised in the appeals include:

2 The possession, management and control of the flat remains vested in the licensor who is the occupier for all purposes including taxation and rating. The licensor will retain keys to the flat and has an absolute right of entry at all times for the purpose of exercising such management and control and . . . for the purpose of effecting any repairs or cleaning to the flat or building or for the purpose of providing the attendance mentioned in Clause 9 hereof or for the purpose of removing or substituting such articles of furniture from the flat as the licensor may see fit.

3 The licensee shall pay the licensor a licence fee of £110 per fortnight for his use of the flat . . .

4 The licensor may for any reason and at any time require the licensee forthwith to vacate the flat and move to any other flat of comparable size in the building which the licensor may offer the licensee provided that in respect of such a flat the licensee shall not be required to make a greater fortnightly payment than the amount payable by Clause 3 . . .

5 This licence agreement confers upon the licensee merely a personal privilege to use the flat . . .

9 The licensor will provide the following attendance for the licensee: (i) Housekeeper; (ii) Lighting of common parts; (iii) Cleaning of common parts; (iv) Window cleaning; (v) Service to front door; (vi) Telephone; (vii) Cleaning of flat; (viii) Collection of rubbish; (ix) Provision and laundering of bed linen; (x) Hot water.

With reference to these provisions it is to be noted that clause 2 on its face appears to permit the landlord not merely to substitute articles of furniture but also to remove any and all of them, although the licensee remains liable to pay for his use of the room for the whole period of 26 weeks. I suspect an error of drafting or typing but that is the effect of the clause.

The right of Mr da Silvaesa and Miss Santos under the agreement expired on July 26 1985. Mrs Liew, the housekeeper, asked them to move out. On September 4 1985 application was made in the High Court in the name of Mr Markou as plaintiff for an order for possession of the room no 3. Various points were taken in the notice of appeal as to the form of the originating summons, which was served on September 9 1985, but these points were abandoned. Notice was given to these appellants by letter dated September 11 and delivered by hand to them at room no 3 to attend before the judge in chambers on Tuesday September 17 1985. An affidavit sworn by Mr da Silvaesa in support of his application for a stay pending appeal states that he received the letter on September 13. An emergency legal aid certificate was issued to Mr da Silvaesa on September 13 1985 limited to representation on the hearing on September 17 1985. On that date these appellants were represented before Hirst J by counsel, Miss Linda Pearce, but no affidavit evidence was tendered. It was submitted on their behalf that the issue — whether they were, on the true construction of the written agreement, licensees or, as they claimed, tenants having exclusive occupation of their room — should not be dealt with summarily but should be remitted to the county court for trial under section 40 of the County Courts Act 1984. In the alternative, Miss Pearce submitted that the summary proceedings should be adjourned to enable these applicants to present evidence that the licence agreement was a "sham".

Hirst J refused to transfer the proceedings for trial; and, having heard what he described as extremely well argued submissions on both sides and directing himself by reference to the principles stated in Lord Templeman's speech in *Street* v *Mountford* [1985] AC 809*, he decided that, upon its true construction, the written agreement created a "lodger type" licence and not a tenancy. The learned judge then considered the alternative submission that an adjournment should be granted for these appellants to adduce evidence to show that the written agreement was a sham. He had invited Miss Pearce to specify what type of evidence she wished to adduce: the answer was that the agreement was renewed over a number of periods since 1983; that no cleaning was provided, although bed linen was provided, and clause 4 was not implemented. The learned judge concluded that such matters, if proved, would not begin to establish that the agreement was a sham and he accordingly refused to grant an adjournment and made the order for possession.

For these appellants Mr Bagnall submitted that the judge was wrong to refuse an adjournment: the question whether the written agreement did or did not create in the appellants an exclusive right of possession, so as to give them a tenancy of the room, could — said Mr Bagnall — only properly be answered by construing the written terms in the "matrix of facts" in which the agreement was made; and the appellants should have been given an opportunity of making a full statement of their evidence. He drew attention to the fact that in *Street* v *Mountford* it was conceded by the landlord that the agreement granted exclusive possession to Mrs Mountford and the issue determined by the House of Lords was that she was therefore a tenant, notwithstanding express terms to the effect that she had a licence only. The issue, for trial of which the adjournment ought to have been granted in this case, was whether this agreement, construed in its factual context, did or did not create a tenancy by the grant of exclusive possession.

On the issue whether the agreement was to be construed according to its terms or was to be treated as a sham Mr Bagnall acknowledged that, if the only matters to be advanced in demonstration that the agreement was a sham were those stated by counsel to the judge, it was difficult to argue that they would by themselves suffice, but the adjournment should have been granted for that purpose as well, in particular having regard to the indications from the terms of the written agreement itself that the agreement or certain terms in it were

*Editor's note: Also reported at [1985] 1 EGLR 128.

A apparently "sham". I refer to these indications later in this judgment. They do not appear to have been relied upon before Hirst J.

A landlord is entitled to use the summary proceedings under Order 113 if he can demonstrate his right to do so, and the court has no discretion to deny such use merely on the grounds that the proceedings are rapid and summary and that the defendants did not enter as squatters: see *GLC* v *Jenkins* [1975] 1 WLR 155. The landlord, however, takes a risk of finding that the evidence put forward by the defendant is such that the landlord's claim is not sufficiently clear and straightforward for these summary proceedings and must be remitted for trial: see *Shah* v *Givert* (1980) 124 SJ 513. In *Henderson* v *Law* (1985) 17 HLR 237 Griffiths LJ said:

B There will obviously be cases in which, although proceedings are started by way of a summary procedure it quickly becomes apparent that a substantial issue has to be tried. If it was apparent to the applicant that a serious issue was bound to arise as to whether a tenancy or a holding ever existed, no doubt the judge would regard the use of the summary procedure as inappropriate, or even in an extreme case as an abuse of the process, and dismiss the application; but I would expect such cases to be rare, because I would not anticipate that solicitors would seek to steal a march by using an inappropriate procedure. From time to time there are bound to be cases such as this where, from the applicant's point of view, an unexpected issue surfaces which raises the question of a tenancy or a holding over. In such cases, the judge must exercise his discretion and decide whether it is wiser to continue the summary hearing, or to adjourn it for a further hearing after the parties have had a chance to C reconsider the position, or possibly to dismiss the application and leave the applicant to have the issues determined in a subsequent action.

It was, accordingly, for the learned judge to decide firstly whether these appellants had had sufficient time to put forward their case. He decided that they had: they had signed the written agreement; their right to remain according to its terms had expired more than seven weeks before; they had had legal advice available to them from Friday September 13 until the hearing on Tuesday September 17; and all that was required of them was to give instructions as to the circumstances in which they had taken the room and signed the agreement. It is not possible, in my judgment, for this court to say that the judge's conclusion on this point was one which he could not properly reach.

D Next, the judge had to decide whether the claim of the plaintiffs was clearly made out on the evidence or whether from the nature of the plaintiffs' claim by itself, or, having regard to evidence which these appellants had put forward, or wished to put forward, it was necessary or appropriate to remit the case for trial or to adjourn the hearing of the summary proceedings for further evidence.

The judge first considered the meaning and effect of the written agreement and concluded that, *if* it represented the agreement between the parties, it was "crystal clear" that, within the principles stated in *Street* v *Mountford* [1985] AC 809, these appellants were lodgers and not tenants. Those principles were stated by Lord Templeman as follows (p 817H):

E In the case of residential accommodation there is no difficulty in deciding whether the grant confers exclusive possession. An occupier of residential accommodation at a rent for a term is either a lodger or a tenant. The occupier is a lodger if the landlord provides attendance or services which require the landlord or his servants to exercise unrestricted access to and use of the premises. A lodger is entitled to live in the premises but cannot call the place his own. In *Allan* v *Liverpool Overseers* (1874) LR 9 QB 180, 191-192 Blackburn J said:

"A lodger in a house, although he has the exclusive use of rooms in the house, in the sense that nobody else is to be there, and though his goods are stowed there, yet he is not in exclusive occupation in that sense, because the landlord is there for the purpose of being able, as landlords commonly do in the case of lodgings, to have his own servants to look after the house and the furniture, and has retained to himself the occupation, though he has agreed to give the exclusive enjoyment of the occupation to the lodger."

F If on the other hand residential accommodation is granted for a term at a rent with exclusive possession, the landlord providing neither attendance nor services, the grant is a tenancy; any express reservation to the landlord of limited rights to enter and view the state of the premises and to repair and maintain the premises only serves to emphasise the fact that the grantee is entitled to exclusive possession and is a tenant. In the present case it is conceded that Mrs Mountford is entitled to exclusive possession and is not a lodger. Mr Street provided neither attendance nor services and only reserved the limited rights of inspection and maintenance and the like set forth in clause 3 of the agreement. On the traditional view of the matter, Mrs Mountford not being a lodger must be a tenant.

Later, at p 823D, Lord Templeman said:

Exclusive possession is of first importance in considering whether an occupier is a tenant; exclusive possession is not decisive because an occupier who enjoys exclusive possession is not necessarily a tenant. The occupier may be a lodger or service occupier or fall within the other exceptional categories mentioned by Denning LJ in *Errington* v *Errington and Woods* [1952] 1 KB 290.

Then, at p 824G:

In *Marchant* v *Charters* [1977] 1 WLR 1181 a bedsitting room was occupied on terms that the landlord cleaned the rooms daily and provided clean linen each week. It was held by the Court of Appeal that the occupier was a licensee and not a tenant. The decision in the case is sustainable on the grounds that the occupier was a lodger and did not enjoy exclusive possession.

Finally, Lord Templeman added (p 827F):

Henceforth the courts which deal with these problems will, save in exceptional circumstances, only be concerned to inquire whether as a result of an agreement relating to residential accommodation the occupier is a lodger or a tenant.

This agreement requires the landlord to provide attendance and services for the licensee but — and I shall return to this point later in this judgment — the obligation to provide attendance and services is not conclusive that the occupier is a licensee: provision of the attendance or services must, as set out in Lord Templeman's speech, require the landlord or his servants to exercise unrestricted access to and use of the premises. Clause 2 of this agreement as set out above provides that the landlord shall "retain keys to the flat" for the purpose of exercising such management and control and for providing the attendance mentioned in clause 9 and for the purpose of removing or substituting such articles of furniture from the flat as the landlord may see fit under clause 2. Clause 9 of the agreement does not specify the extent or frequency of the attendance or services to be provided, and this is an aspect of the written agreement upon which all these appellants have relied in submitting that the agreement by itself could not be shown to require "unrestricted" access to and over the premises. The task of the court in applying the principles laid down in *Street* v *Mountford* is not that of applying the words of a statute. I take the meaning of the word "unrestricted" in this context to be primarily concerned with the landlord's need to go into and out of the lodger's rooms at the convenience of the landlord and without the lodger being there to let the landlord in. The amount and frequency of the attendance and services agreed to be provided are relevant, but the question to be answered is whether, in all the circumstances, having regard to the landlord's obligations, it is clear that the landlord requires unrestricted access and has reserved the right to exercise such access in order to look after the house and the furniture. Where an agreement describes attendance and services, as in this case, without specifying the extent or frequency of them, the court may and, as I think, should treat the agreement as imposing upon the landlord an obligation to provide the attendance and services to an extent and frequency and standard which are in all the circumstances reasonable. Such a standard upon the material before the learned judge in this appeal could not be regarded as either extensive or elaborate; it would be of the order of removing rubbish daily or at least every other day, cleaning once a week and laundering sheets fortnightly. The evidence later put in by Mr da Silvaesa did not show anything different except for the fact that he asserts that these appellants cleaned the flat themselves.

The learned judge, in my opinion, was right in reaching the conclusion that, if the written agreement represented the agreement between the parties, then, upon its true construction, these appellants were lodgers and not tenants. The agreement does require the landlord to provide attendance and services which require the landlord and his servants to exercise unrestricted access to and use of the room. Possession and control of the room are reserved by clause 2 of the agreement to the landlord for the purposes of discharging the obligation to provide attendance and services and that shows, in my judgment, that exclusive possession was not given. The right to require the lodger to move from one room to another could not, as I think, by itself be conclusive to show that there was no grant of exclusive occupation of the room for residential accommodation if the terms otherwise showed that there was such a grant; but the reservation of the right — and I consider it later in this judgment with reference to the submission that the agreement is on its face "a sham" — seems to me wholly consistent with the status of lodger and of an intention on the part of the landlord to look after the house and furniture by himself or his servants and to retain to himself the occupation.

As to matters which the appellants asserted that they wished to prove to show that the agreement was a sham, the judge was of opinion, as I have said, that proof of those matters would not begin to establish that the agreement was a sham. The length of time that a lodger remains could not, he said, convert the agreement to a tenancy if the terms of it constituted a lodging. The fact that services, which the landlord is contractually obliged to provide, are not taken or used by the lodger does not render the agreement a sham, since the lodger was entitled to call for them. The fact that the landlord had not exercised his right to require the lodger to move to another room also, in the view of the learned judge, could not show the agreement to be a sham. The judge was right, in my opinion, that these matters considered by themselves would not demonstrate that the written agreement was a sham agreement. It has been argued in this court, however, with reference to both appeals that, apart from any evidence which the applicants could give to show that the agreement was sham, the terms of the written agreement themselves provide evidence that it was "a sham" in that the parties cannot be supposed to have intended the agreement to operate between the parties. Emphasis was placed upon all the terms, including those which I have mentioned, in particular the denial of the right in the occupier to use the room from 10.30 to noon, and the right of the landlord to remove furniture. It was submitted that, if the written agreement is shown to have contained terms which, upon the evidence before the court, are arguably sham — in the sense that the parties did not intend that the terms should express the legal rights between the parties — that fact is relevant to the decision of the court as to whether the status of the appellants as licensees, remaining in the premises without consent, is shown with sufficient clarity to justify the making of an order under the summary procedure.

It is necessary, in my view, for the court, when faced with this plea of "sham agreement", to require a clear statement of what is said to be "sham" and what the consequence of any finding to that effect is said to be. In *Snook* v *London & West Riding Investments Ltd* [1967] 1 All ER 518, the Court of Appeal was concerned with the title of a hire purchase company to a motor vehicle. Diplock LJ (as he then was) said this (at p 528):

As regards the contention of the plaintiff that the transactions between himself, Auto-Finance Ltd and the defendants were a "sham", it is, I think, necessary to consider what, if any, legal concept is involved in the use of this popular and pejorative word. I apprehend that, if it has any meaning in law, it means acts done or documents executed by the parties to the "sham" which are intended by them to give to third parties or to the court the appearance of creating between the parties legal rights and obligations different from the actual legal rights and obligations (if any) which the parties intend to create. One thing I think, however, is clear in legal principle, morality and the authorities . . . that for acts or documents to be a "sham", with whatever legal consequences follow from this, all the parties thereto must have a common intention that the acts or documents are not to create the legal rights and obligations which they give the appearance of creating. No unexpressed intentions of a "shammer" affect the rights of a party whom he deceived.

In *Street* v *Mountford* Lord Templeman (at p 825D) referred to *Somma* v *Hazelhurst* [1978] 1 WLR 1014. In that case a landlord "licensed" a couple to use a furnished room: the written agreement expressly provided that they were not to have exclusive possession and that the landlord could license another person to share the room. It was contended in the Court of Appeal that the clauses with reference to shared occupancy were illegal and void as contrary to public policy and that, in considering questions arising under the Rent Acts, the court must look at the substance and reality of the situation: if, upon examining the surrounding circumstances, the court finds that a written agreement disguises the reality of the transaction, that disguise should be disregarded even if it cannot be shown that the agreement is fraudulent or such as to attract the label of a mere sham. Cumming-Bruce LJ rejected the argument: he said (at p 1025A):

If that is what both he and the licensee intend and if they can frame any written agreement in such a way as to demonstrate that it is not really an agreement for a lease masquerading as a licence, we can see no reason in law or justice why they should be prevented from achieving that object. Nor can we see why their common intentions should be categorised as bogus or unreal or as sham merely on the ground that the court disapproves of the bargain. This matter was expressed most happily by Buckley LJ in *Shellmex & BP Ltd* v *Manchester Garages Ltd* [1971] 1 WLR 612, 619 as follows:
"It may be that this is a device which has been adopted by the plaintiff company to avoid possible consequences of the Landlord and Tenant Act 1954, which would have affected a transaction being one of landlord and tenant; but, in my judgment, one cannot take that into account in the process of construing such a document to find out what the true nature of the transaction is. One has first to find out what is the true nature of the transaction and then see how the Act operates upon that state of affairs, if it G bites at all. One should not approach the problem with a tendency to attempt to find a tenancy because unless there is a tenancy the case will escape the effects of the statute."

The court held that the agreement created a licence and not a tenancy. In *Street* v *Mountford* Lord Templeman's comment upon *Somma's* case was as follows (at p 825D):

But the Court of Appeal did not ask themselves whether H and S were lodgers or tenants and did not draw the correct conclusion from the fact that H and S enjoyed exclusive possession. The Court of Appeal were diverted from the correct inquiries by the fact that the landlord obliged H and S to enter into separate agreements and reserved power to determine each agreement separately. The landlord also insisted that the room should not in form be let to either H or S or to both H and S but that each should sign an agreement to H share the room in common with such other persons as the landlord might from time to time nominate. The sham nature of this obligation would have been only slightly more obvious if H and S had been married or if the room had been furnished with a double bed instead of two single beds . . . The room was let and taken as residential accommodation with exclusive possession in order that H and S might live together in undisturbed quasi-connubial bliss making weekly payments. The agreements signed by H and S constituted the grant to H and S jointly of exclusive possession at a rent for a term for the purposes for which the room was taken and the agreement therefore created a tenancy. Although the Rent Acts must not be allowed to alter or influence the construction of an agreement, the court should, in my opinion, be astute to detect and frustrate sham devices and artificial transactions whose only object is to disguise the grant of a tenancy and to evade the Rent Acts. I would disapprove of the decision in this case that H and S were only licensees and for J the same reason would disapprove of the decision in *Aldrington Garages Ltd* v *Fielder* (1978) 37 P & CR 461 and *Sturolson & Co* v *Weniz* (1984) 272 EG 326.

As I understand the reference to the "sham nature of the obligation", namely that of sharing the room in common with other persons nominated by the landlord, the House of Lords is there saying, first, that the agreement in that case constituted the grant of exclusive possession; second, that the written obligation to share the room was not effective to alter the true nature of the grant; and, third, that, on the facts of the case, it should have been clear to the Court of Appeal that that landlord cannot have intended the term as to sharing occupation to be a true statement of the nature of the possession intended to be enjoyed by the "licensees". The principle that the Rent Acts must not be allowed to influence the construction K of an agreement, which was re-asserted in *Street* v *Mountford*, means, of course, that it is not open to the court to dismiss as "sham" a term in an agreement merely because it is either effective in demonstrating that the agreement does not grant exclusive occupation or, taken with other terms, contributes to that result. Further, the presence of a term in an agreement, like that which required the "licensee" to share the room in *Somma's* case, and which is seen to be "sham" in the sense described above, is not of itself illegal or contrary to public policy: the term itself is merely ineffective.

As to the denial in the agreement signed by these appellants of the right in them as occupiers to use the room from 10.30 to noon, I agree L that this clause has the appearance of one which might well be "sham" in the sense that the landlord might be shown not to have intended the term to be a true statement of the nature of the occupation to be enjoyed by the licensees. I do not regard it as a clause which must obviously and unarguably be of that nature in all circumstances but, in this case, it appears never to have been enforced and it is astonishingly extreme in apparently extending to the licensee's property as well as his personal presence. The court might be persuaded on the facts that the landlord never intended that clause in the terms used to describe the nature of the possession intended to be enjoyed by the licensee.

Next, the right to remove, as opposed to the right to substitute, furniture appears to me to fall into the same class. It might well be M that the actual intention was to reserve only the right to "remove and substitute", but no explanation of the clause has been offered and it must be construed as it stands. It is not necessary to work out what rights the licensee might have if the landlord had claimed to remove, for example, the bed without replacement. This clause was referred to in argument by Mr Gallivan as "bizarre": in the absence of any explanation I agree with his submission in that it is not consistent with an ordinary agreement with a lodger for a furnished room to which the lodger is tied for six months.

Lastly, on the apparently "sham" nature of the agreement, reference was made to clause 4: the right to require the licensee to move from one room in the building to another of comparable size without increase of rent: this was also described by Mr Gallivan as "bizarre". I do not accept that submission. It seems to me of its nature to be a term which the landlord of a lodging house, providing temporary accommodation, could both sensibly and reasonably impose for the better running of the house. He may want to redecorate a room; or to enable two applicants for rooms to be near to each other; or to separate occupants who have been in dispute.

In the light of these considerations, it seems to me that the question whether, in the first appeal, the terms of the written agreement themselves provide evidence that the agreement was a sham is to be decided, first, by asking whether the presence of clause 1, no right to use the room from 10.30 to noon, and of clause 2, the right to remove furniture, provide any basis for contending that the agreement itself and as a whole must be regarded as a sham. It is plain to me that they do not. In the first appeal the written agreement was the only agreement between the parties and no other is alleged. The house was being run on the basis of service and attendance being provided to the occupants of the several rooms in it. A housekeeper was employed to provide them. The agreement as a whole is not shown to have been arguably a sham.

Second, the agreement must be considered on the basis that these clauses which can fairly be regarded as possibly sham — the terms in clauses 1 and 2 (no right between 10.30 and noon and removal of furniture) — are treated as ineffective. If the written agreement is considered in this light, was it right for the judge to be satisfied that the case was sufficiently clear for him to make an order for possession in summary proceedings, having regard to the request by these appellants for an adjournment to offer evidence on the matters stated by their counsel?

I have reached the conclusion, despite the submissions of Mr Mann for the respondents to which I shall refer later, that, as the case has been argued in this court, it was not right for the order to be made in the summary proceedings, because the claim of the landlord was not sufficiently clear and straightforward for the following reasons. First, clause 1 of the agreement, which defines the right of use and possession of the room or flat, is shown to be arguably not a true statement of the rights intended to be granted, and clause 2, which states the control retained by the landlord and the reasons for retaining it, is shown arguably to contain a term as to removal of furniture of a similar character. The presence of such a term, in the absence of an allegation of an intent to deceive or deliberately to take unfair advantage of the occupier — and none is made — does not, in my opinion, by any process of infection spread some debilitating weakness to other terms of which the genuineness is not in doubt, either on the face of the agreement or according to evidence available or sought to be tendered. The court could, in my opinion, be satisfied that a claim was clearly made out under Order 113 against the lodger who had remained without consent even if there is shown to be in the agreement a term which the court considers should be treated as ineffective because arguably sham. A draftsman might insert such a term out of misguided zeal in an agreement which is clearly a genuine agreement for the grant of accommodation to a lodger. Nevertheless, the presence of arguably sham terms in the agreement is a factor which the court must take into account in considering whether the claim is sufficiently clear for the purposes of these summary proceedings. If clauses of such importance are self-evidently of doubtful effectiveness, the court must consider the possibility that the clauses which impose the obligation to provide attendance and reserve the right of control and access for the purpose of providing it may not accurately reflect the legal relationship intended by the parties with reference to those matters.

Second, the evidence which these appellants wanted to put before the court included, in the note of the judgment of Hirst J, assertions that "no cleaning was provided although bed linen was provided". If it should appear to the court, after consideration of all the evidence, that, from the making of the agreement, and with reference to an earlier agreement in similar terms between Mr da Silvaesa and the landlord, no cleaning of the flat by the landlord had been carried out or offered, and that the laundering of bed linen and removal of rubbish had been carried out without exercise of or need for any right of access to the room, then it might be open to the court to conclude that the true agreement between the parties did not require unrestricted access to and use of the premises by the landlord so as to cause these appellants to be lodgers and not tenants. I do not suggest that that is a probable or likely conclusion. I do not know what the outcome of the testing of the evidence will be. It is sufficient for the purposes of this appeal that these appellants have shown that the claim of the landlord is on the evidence not so clear and straightforward that an order for possession should be made in proceedings under Order 113.

Mr Mann, for the respondents, submitted that the essential test to determine whether an occupier was a tenant with exclusive occupation or a lodger, and hence a licensee only, was whether the landlord had reserved control or "dominion" over the room; and that, even if these appellants were able to raise any point upon separate terms of the agreement, so as to suggest that those terms may not have represented the true relationship intended by the parties to operate between them, nevertheless clause 2, or the essential force of it, clearly remained unimpaired and by itself demonstrated that the appellants did not have exclusive possession and were not lodgers.

He referred to *Bradley* v *Baylis* (1881) 8 QBD 195, heard in the Court of Appeal before Sir George Jessel MR and Baggallay, Brett, Cotton and Lindley LJJ. The case arose under the Parliamentary and Municipal Registration Act 1878 and the Representation of the People Act 1867. It was held that, although, by the 1878 Act, the term "dwelling-house" in the Act of 1867 is provided to mean part of a house separately occupied, yet, in order to be entitled to the borough franchise as the occupier of a dwelling-house, the person must have an occupation in respect of which he can be rated to the relief of the poor and, therefore, he is not entitled to such a dwelling-house franchise by reason of the occupation of part of a house if he occupies such part as a lodger. The tenant of two rooms, which he took unfurnished at a weekly rent, had the exclusive use of such rooms and a key of the outer door of the house. His landlord had also a key of the outer door, and resided in all the rest of the house, but supplied no attendance or service to the tenant. The court held that such a tenant occupied the rooms as a lodger. Mr Mann referred to a passage in the judgment of Sir George Jessel MR at p 218:

> . . . it remains to consider when a man who occupies a rateable tenement is an occupying tenant, and when he occupies or uses it as a lodger only.
> There is, probably, no question on which there has been a greater variety of judicial opinion than this. . . . I think it wiser and safer to say that the question whether a man is a lodger, or whether he is an occupying tenant, must depend on the circumstances of each case. . . . .
> "First of all, take the case of a *lodger*. It seems to me, as to unfurnished lodgings . . . where the owner of the house does not let the whole of it, but retains a part for his own residence, and resides there, and where he does not let out the passages, staircase and outer door, but gives to the "inmates" . . . merely a right of ingress and egress, and retains to himself the general control, with the right of interfering — I do not mean an actual interference, but a right to interfere, a right to turn out trespassers, and so on; there I consider that such owner is the occupying tenant of the house, and the inmate, whether he has or has not the exclusive use of the room, is a *lodger*. That is one extreme case.
> . . . There will be an immense number of intermediate cases . . . Does it make any difference that the inmates have latch-keys to the outer door and also keys to the inner door? I think not. I think they are still lodgers notwithstanding. Does it make any difference that the landlord does not reside there personally, but has resident servants, who occupy, on his behalf, part of the house? I think not. I think that the inmates are still *lodgers*. Does it make any difference that the landlord does or does not repair? I think not; they are still *lodgers*.

Mr Mann contended that upon this authority — which was apparently not cited to the House of Lords in *Street* v *Mountford* — it is clear that a landlord may retain such control over rooms of which the use for residential accommodation has been granted by agreement, that the user is a lodger only, and not a tenant, even though the rooms are not furnished by the landlord and a landlord provides no attendance or service and the user has exclusive use of the rooms. He pointed to clause 2 in this agreement signed by these appellants, which is set out above, as the most important clause. The landlord here had reserved possession and control and he had a servant in the house to exercise it: therefore, the appellants did not have exclusive possession and could not be tenants. In effect, this submission contends that a clause of this nature can effectively prevent the grant of exclusive possession even where the use of rooms is given for residential accommodation to be enjoyed in the way in which such accommodation is ordinarily enjoyed, that is to say with no interruption or interference apart from occasions when the landlord has need to enter the room for his legitimate purposes as landlord, and even though the occasions upon which the landlord would have such need are infrequent because he provides no attendance or services. It is true that, in the passage cited from Lord Templeman's speech at p 818A:

The occupier (at a rent for a term) is a lodger if the landlord provides attendance or services which require . . . unrestricted access . . .

his lordship did not say that the occupier is a lodger if, and only if, such circumstances are proved. Lord Templeman also cited the passage from Blackburn J in *Allan* v *Liverpool Overseers* which refers to the landlord retaining:

to himself the occupation, though he has agreed to give the exclusive enjoyment of the occupation to the lodger.

It is not necessary, in my opinion, for this court to decide for the purposes of this appeal whether the principles laid down in *Street* v *Mountford* mean that, as a matter of law, premises made available for a term at a rent for residential accommodation must imply a grant of exclusive possession, whatever the terms and circumstances of the agreement, unless the landlord provides attendance or services which require the landlord or his servants to exercise unrestricted access to and use of the premises. I incline to the view that *Street* v *Mountford* does lay down such a rule where the accommodation is intended for use as ordinary separate and private residential accommodation and in the absence of any exceptional circumstances. I understand the basis of the rule to be that possession of rooms for occupation as ordinary residential accommodation must, from its nature, be intended by both parties to be exclusive except where, as stated, the landlord requires unrestricted access for the provision of services or attendance. Whatever be the right answer to that question, it is clear to me that clause 2 in this agreement, and the submission founded upon it, cannot prevail so as to justify by itself the making of an order for possession in summary proceedings on the facts of the first appeal. It will be for the court at trial to decide whether, in the light of the effective terms of the agreement as they are found to be, exclusive possession of the room was granted to these appellants. In the event that it is not shown that the landlord:

provide attendance or services which require the landlord or his servants to exercise unrestricted access to and use of the premises

the court would have to consider then whether any term of the agreement, and, in particular, clause 2 to the extent that it survives, is effective to prevent the grant of exclusive possession to these appellants, having regard to the disapproval by the House of Lords in *Street* v *Mountford* of *Somma's* case. In that case also there were terms which, on their face, denied exclusive possession to the licensees but which should have been treated as ineffective to achieve that purpose.

As to the second appeal, the written agreement in the same form was signed by Miss Merola and Mr Kader on April 12 1985 for 26 weeks. It expired on October 11 1985. It was alleged that these appellants failed to pay after August 30 1985. Before the judge there was an issue as to whether these appellants had, by failure to pay, repudiated the agreement as the landlord alleged. That issue has disappeared from the appeal because the application was made on November 11 1985 and the terms of the licence — if such it be — had by then expired.

The application came before Boreham J on November 28 1985. The plaintiffs are Crancour Ltd — successors to Mr Markou. The application was supported by an affidavit by Mrs J E Sandy, managing agent for the plaintiffs. It provided evidence of the making of the agreement, the failure to pay rent and the failure to vacate. These appellants also obtained legal aid — on November 21 1985 — and in this case a joint affidavit was put before the judge sworn by these appellants. It provided evidence of an uncertain nature of an alleged oral agreement preceding the signing of the written agreement and of the provision of services different from those set out in the agreement. The oral agreement was alleged to have been made by Carina, the housekeeper. The plaintiffs put before the judge an affidavit in reply sworn by Mrs Sandy in which it was asserted that Carina had no authority to make any agreement other than in the form of the written agreement signed. Mrs Sandy also described in detail the services rendered by the housekeeper. She acknowledged that the landlord did not enforce clause 1 by insisting that "licensees" vacate the building from 10.30 to 12.00 each day.

On behalf of these appellants Miss Pearce before the learned judge argued a number of points, but the only points now relevant were put thus: first, it was not so clear that the agreement granted a licence, only that it was appropriate to make an order for possession in summary proceedings; second, there was evidence of a collateral agreement which overrode or varied the written agreement; and, third, that the agreement was a sham. The learned judge rejected the submissions. He found that the written agreement was the only agreement between the parties and that, for reasons like those which Hirst J had expressed, this agreement provided for a licence to G
lodgers to occupy the room.

It is not necessary to deal in detail with the arguments before this court by Mr Gallivan upon the construction of the agreement. I have already referred to some of his submissions in dealing with the first appeal. So far as the second appeal turns upon the construction of the written agreement, in my opinion, for reasons which I have already given, the learned judge is not shown to have been in error.

The indications on the face of the agreement that clauses 1 and 2 contain provisions which are arguably sham, which have been considered above, were relied upon also in the second appeal by Mr Gallivan. For reasons set out above it is clear to me that, in this case also, the right of the landlords to an order for possession was not so
clear that an order should be made in summary proceedings. H

Further, there was in this appeal evidence on affidavit which alleged an oral agreement made with the landlord's housekeeper before or at the time of the signing of the form of written agreement. I have reached the conclusion that, in any event, this evidence was such that it raised issues which could not be properly resolved without the process of trial, in particular the cross-examination of the witnesses. Having formed that view, it is better that I say as little as possible in this court as to the nature and apparent quality of the evidence on either side: the issues of fact will have to be tried and resolved in the county court. I must, however, say, out of respect for the learned judge, that I do not find it surprising that he formed the
view that the alleged conversations preceding the signing of the J
agreement could not have been intended upon objective assessment of the actions of the parties to constitute or to survive as legally binding contractual terms. Mr Mann, for the respondents, contended that the question whether any terms were agreed which could vary or affect the written agreement was a matter to be decided according to the probabilities upon the evidence before the court and that the decision of the learned judge upon that issue could not be faulted. In the light of the more extended arguments which this court has heard upon the various clauses of the agreement, and upon the issue of the alleged oral agreement, I have come to the conclusion, however, that it is not possible on the evidence to say that the issue as to what constituted the agreement between the parties was so clear
that it is right to make an order for possession under Order 113. On K
that ground also I would allow the second appeal.

Accordingly, and for these reasons, I would allow both appeals and set aside the possession orders. It would be necessary for directions to be given for the remission of both sets of proceedings to the county court. The court would, of course, hear the submissions of counsel.

Agreeing that the appeals should be allowed, NICHOLLS LJ said: The feature common to these two appeals is the form of the written agreements. Shortly stated, by an agreement dated January 26 1985, Mr V G Markou granted to Miss Santos and Mr da Silvaesa licence to use a furnished room, called Flat 3, in 21 Cranley Gardens, London
SW7, intermittently over a period of 26 weeks from that date, for a L
fortnightly fee of £110. By an agreement in the same form, dated April 12 1985, Mr V G Markou granted to Miss Merola and Mr Kader licence to use another furnished room, Flat 12, in the same building, again intermittently, over a period of 26 weeks from that date, for a fortnightly fee of £108. Both agreements have expired by effluxion of time, and in both sets of proceedings the issue is whether the respective pairs of occupants are in law licensees or, despite the terms of the licence agreements, tenants. If they are tenants, it is not disputed that they have a good defence to the claim for possession made by Crancour Ltd (which has become the leasehold owner of the flats, in succession to Mr Markou). Conversely, if they are licensees, it is common ground that they have no defence to the possession
claim. The appeals are from possession orders made respectively by M
Hirst J on September 17 1985 regarding Room 3 (which I shall call "the first appeal") and by Boreham J on November 28 1985 regarding Room 12 ("the second appeal"). In each case the orders were made on summonses issued under RSC Order 113.

It will be convenient first to consider whether, assuming that the agreements are genuine and took effect according to their tenor, they gave rise to what the law classifies as a tenancy as distinct from a licence. In *Street* v *Mountford* [1985] AC 809 Lord Templeman concluded his speech by saying:

Henceforth the courts which deal with these problems will, save in exceptional circumstances, only be concerned to inquire whether as a result of an

A agreement relating to residential accommodation the occupier is a lodger or a tenant.

The problems referred to by Lord Templeman were problems such as arise on these appeals. Earlier, he gave the following guidance (at p 817):

> In the case of residential accommodation there is no difficulty in deciding whether the grant confers exclusive possession. An occupier of residential accommodation at a rent for a term is either a lodger or a tenant. The occupier is a lodger if the landlord provides attendance or services which require the landlord or his servants to exercise unrestricted access to and use of the premises. A lodger is entitled to live in the premises but cannot call the place his own. In *Allan* v *Liverpool Overseers* (1874) LR 9 QB 180, 191-192 Blackburn J said:
>
> B "A lodger in a house, although he has the exclusive use of rooms in the house, in the sense that nobody else is to be there, and though his goods are stowed there, yet he is not in exclusive occupation in that sense, because the landlord is there for the purpose of being able, as landlords commonly do in the case of lodgings, to have his own servants to look after the house and the furniture, and has retained to himself the occupation, though he has agreed to give the exclusive enjoyment of the occupation to the lodger."
>
> If on the other hand residential accommodation is granted for a term at a rent with exclusive possession, the landlord providing neither attendance nor services, the grant is a tenancy; any express reservation to the landlord of limited rights to enter and view the state of the premises and to repair and maintain the premises only serves to emphasise the fact that the grantee is entitled to exclusive possession and is a tenant.

C Applying those principles to these two cases, I respectfully agree with both Hirst J and Boreham J that this form of agreement did not grant exclusive possession to the occupants, and accordingly that they were lodgers and not tenants. In clause 1, to which I shall return, the occupants were authorised and permitted to use the room in question, not for a period of 26 weeks, but for $22\frac{1}{2}$ hours only in each 24 hours over the 26-week period. By clause 2 the possession, management and control of the room were expressed to remain vested in the licensor, who was the occupier for all purposes. He retained keys to the room and had "absolute right of entry at all times for the purposes of exercising such management and control", and, in particular, for repairs, cleaning, providing the attendance mentioned in clause 9, or "for the purpose of removing or D substituting such articles of furniture from the flat as the licensor may see fit". Under clause 4 the licensor might for any reason and at any time require the licensee forthwith to vacate the flat and move to any other flat of comparable size in the building, but not at an increased licence fee. By clause 9 the licensor was to provide the following services, among others, to the individual flat: window cleaning, cleaning, collection of rubbish, and provision and laundering of bed linen. To my mind, an agreement in that form is firmly on the lodger side of the line drawn by Lord Templeman.

However, it is at this point that difficulties arise. I can best explain these difficulties by considering first the position regarding the second appeal. The originating summons was issued on November 11 E 1985 with a return date of November 28. It was supported by a brief affidavit made by a Mrs Sandy, who is the managing agent of Crancour. In answer Miss Merola and Mr Kader made an affidavit. They described 21 Cranley Gardens as a terraced house on five floors, including the basement. The basement was occupied by a housekeeper named Carina (Mrs Liew) and the rest of the house let out as bedsitting rooms. In April 1985 they met Carina and discussed accommodation. They then stated in their affidavit:

> We were entitled to have our room cleaned as Carina offered to do it. We declined. We were entitled to have the rubbish removed from outside our door. We were entitled to have our bed linen laundered on Mondays. We were entitled to the exclusive occupation of our room and the right to share the common parts with others. The common facilities included bathroom and wc F and a payphone with three extensions for receiving calls. All of this was made clear in our conversation. Neither of us could make head nor tail of the document but believed it to put into effect the oral agreement which we had made. We did not meet Mr Markou until later although we had been told he was the landlord. He lived elsewhere. We were given to understand that the agreement would be renewed each time. All of this accorded with what we had learned from friends to be the arrangement at the house.

They also stated that their time in the house was uneventful and that "things went much as we believed we had agreed". Their sheets were laundered when they put them outside their room door, and their rubbish was removed by Carina. They cleaned their own room and they were left undisturbed in their room. On one occasion Mr Markou entered, coming in to measure up for insurance purposes. In their affidavit they also stated this:

> We now learn that there the written agreement provides that we were to leave our room between 10.30 am and noon each day. This certainly was never proposed nor discussed at any time. It certainly never happened. We would have objected most strongly if it were demanded and would not have taken the room on any such condition.

Finally, they stated that in August 1985 the housekeeper told them that the house had been sold and that they were to pay their rent to Mrs Sandy. Mrs Sandy refused to accept it, saying that they would be getting a letter telling them to leave.

In reply Crancour produced a further affidavit from Mrs Sandy. She said that the resident housekeeper had no authority to enter into any agreement other than the written, standard form agreement. Mrs Sandy explained that the housekeeper had keys to every room in the house for the purposes of the agreements. Each room was cleaned twice per week, rubbish was collected daily from inside each room, and the housekeeper entered each room weekly to change the bed linen. If the occupants did not avail themselves of these services that was a matter for them, but it did not alter the licensor's obligation to provide them. She said that she knew that the right to compel a licensee to change rooms had been exercised during the currency of the licence agreement when necessary. As to clause 1 of the standard form, she said that "it is right the licensor does not insist that the licensees vacate the building during the hours set out in clause 1 of this agreement".

At the hearing on November 28 1985 Boreham J decided that the submission of counsel for the occupants that there was a collateral agreement which overrode the written agreement was not sustainable and that there was no substance in counsel's submission that the agreement was a sham. He said, according to the notes we have of his judgment, "there is really no ground upon which one could say this was a sham".

For my part, with the greatest respect to the learned judge, I have difficulty in accepting these conclusions. It is trite law that the court will be astute to detect a sham where it appears that a provision has been inserted into an agreement for the purpose of depriving the occupant of protection under the Rent Acts which he might otherwise enjoy. By a sham in this context I mean, adopting the well-known words of Diplock LJ in *Snook* v *West Riding Investments Ltd* [1967] 1 All ER 518 at p528, a document which was intended by the parties thereto to give to third parties or the court the appearance of creating between the parties legal rights and obligations different from the actual legal rights and obligations which the parties intended to create. Lord Templeman adverted to this in *Street* v *Mountford* in these terms (at p 825):

> Although the Rent Acts must not be allowed to alter or influence the construction of an agreement, the court should, in my opinion, be astute to detect and frustrate sham devices and artificial transactions whose only object is to disguise the grant of a tenancy and to evade the Rent Acts.

When one looks at this form of agreement, one finds that the first clause, granting the basic right of occupation, is in these terms:

> Subject as is hereinafter provided the Licensor licenses authorises and permits the Licensee to use that furnished flat known as Flat No 12 (hereinafter "The Flat") on the 3rd floor of the building at and known as 21 Cranley Gardens SW7 (hereinafter "The Building") on each day between the hours of midnight and 10.30 am and between noon and midnight, but at no other times, for a period of 26 weeks from the 12th April 1985 for the purpose of temporary accommodation for the Licensee's personal use only and for no other purpose whatsoever, and for so long as the Licensee is entitled to use the Flat as aforesaid the Licensee shall also be entitled to use, in common with all other persons having a like right, the common parts of the Building.

To my mind, that clause is so extraordinary in a grant of a right to use residential accommodation for a period of six months that it calls for an explanation. Was it genuinely intended by Mr Markou that each day over the six-month period the occupants should be unable to use their room for any purpose? Were the occupants intended to remove all their belongings each day by 10.30 am and return them after noon (because, under clause 1, the occupants had no right to use their room at all for any purpose for that period of one and a half hours each day)? That this was ever really intended by the original licensor (and the licensees' evidence is that it was not intended by them) strains my credulity too far, in the absence of some reasonable explanation. Here there is no explanation at all. There is no evidence from Mr Markou, and Mrs Sandy does not suggest that she ever acted for him. She seems to have become concerned with the property only after Crancour acquired the building, which apparently was at the end of October 1985. Moreover, her evidence supports, rather than negatives, the view that clause 1, in so far as it purports to grant a licence which extended only to 22½ hours each day, may not be a genuine record of the transaction being entered into, but that this was a contrivance, a pretence, a sham, inserted to strengthen the landlord's case that the agreement did not create a tenancy.

Thus, I consider that the very terms of the basic clause of the agreement furnish a foundation for the defence of "sham". Of course, even if the omission from the written document of the grant of a licence to use the room each day between 10.30 am and noon is a sham, it does not follow that in law the true agreement between the parties amounted to a tenancy. I have already referred to other clauses in the form which support the landlord's claim that in law the character of the agreement was, as labelled, a "licence agreement". But, with such a dubious starting point as clause 1, I think that the court has to view the remaining clauses with particular circumspection. When, therefore, on top of this there is uncontradicted evidence (for no evidence from the housekeeper has been produced) to the effect that I have mentioned, I, for my part, do not see how that evidence can be rejected out of hand, and with it the occupants' case that a collateral oral agreement was entered into which had the effect of modifying the written agreement and giving the occupants exclusive occupation of their room as they claim was made clear in their conversation with the housekeeper.

The case sought to be advanced by the occupants is open to criticisms: for example, the account of the conversation with the housekeeper in the affidavit evidence is lacking in particularity; it is unlikely that the housekeeper would have, or be thought to have, authority to vary the form of written agreement; and the genuineness of important clauses such as clause 4 has not been challenged by the occupants in their evidence. But, weighing all these matters, I am satisfied that, in the case of the second appeal, the balance comes down on the side of this being a case where there is an issue in dispute which ought to be tried and which cannot be resolved satisfactorily on the evidence as it stands.

I turn to the first appeal. The course which this matter took is as follows. The summons, with no return date specified therein, was supported by a formal affidavit by the plaintiff Mr Markou (Crancour being substituted as plaintiff on December 10 1985, subsequent to the hearing before Hirst J). The summons and the affidavit were served on September 9 1985. On September 11 the plaintiff's solicitors notified the occupants of Room 3, Miss Santos and Mr da Silvaesa, that the summons was to be heard on Tuesday September 17. A legal aid certificate was issued on Friday September 13. On September 17 the first defendant's counsel argued that the question of whether the agreement created a tenancy or a licence should not be dealt with summarily but should go for trial. Alternatively, counsel sought an adjournment to enable evidence to be produced that the agreement was a sham. The second defendant appeared in person and adopted the first defendant's arguments. The learned judge considered various clauses of the form of agreement and, applying the law as laid down in *Street* v *Mountford*, concluded that it was "crystal clear" that this was a lodger type agreement and not a tenancy. On the request for an adjournment, when asked to state what type of evidence she wished to adduce, counsel indicated that this was that the agreement had been renewed over a number of periods since 1983, that no cleaning was provided (although bed linen was), and that clause 4 was not implemented. Hirst J considered that those matters did not begin to establish that the agreement was a sham, and concluded:

I do not consider that there is any potential material to undermine the agreement to show that it is a sham

and he therefore refused to grant any adjournment.

Subsequently, on October 28 Mr da Silvaesa made an affidavit in support of an application for a stay of execution pending an appeal by him (Miss Santos has not appealed). By then, on any view of the matter, he and his advisers had had ample opportunity to consider what evidence they wished to place before the court. The affidavit set out the history of the occupation of himself and Miss Santos of various rooms in the building from 1983. With regard to the agreement dated January 26 1985, he deposed that, as with the previous rooms, they cleaned the room themselves, but that the housekeeper laundered the bed sheets once a fortnight and took the rubbish down when she cleaned the stairs.

It seems to me that the occupants' case in this appeal is much weaker than the occupants' case in the second appeal. In particular, Mr da Silvaesa does not suggest that any oral arrangement was made with the housekeeper varying the terms of the form of written agreement, or that he understood or intended his position to be other G than as provided for in the written agreement. But the sham point was taken in the court below, although (to judge from the absence of any reference to clause 1 in the notes of his judgment) the learned judge's attention does not seem to have been drawn to clause 1 in this context. I have already expressed my view that clause 1 calls for an explanation. Not without some hesitation, I have reached the conclusion that, in the first appeal also, there is an issue needing to be tried with oral evidence. I think that it would be unsatisfactory for summary judgment to be given in favour of a landlord on the strength of a form of written agreement when *prima facie* the basic clause in that agreement is an artificial contrivance intended to mislead and thereby to create or strengthen a claim by the landlord that the agreement was outside the Rent Acts. I think that this consideration H outweighs the force of the argument that, even if clause 1 is a sham and, in reality, the licence granted was for 24 hours each day, the other terms of the written agreement still put the agreement on the lodger side of the line. That may be so, but I think it is right that there should be a proper investigation to see indeed if that is the correct conclusion.

Accordingly, I agree that both appeals should be allowed, the possession orders set aside and directions given for the further conduct of both sets of proceedings.

I add this. I do not consider that this approach to these two appeals is calculated to open the door to landlords' claims against lodgers for summary possession orders being defeated simply by the defendants J swearing affidavits alleging "sham". The court is well able to deal with such situations should they arise. What makes this case a special one is the terms of clause 1. If a landlord includes in his standard form a basic clause which *prima facie* is a sham, and offers no explanation of it in his evidence, he cannot be surprised if he finds that the short-cut summary procedures do not assist him but that he has to prove his case at a trial.

Also agreeing that the appeals should be allowed, PURCHAS LJ said: The relevant history and surrounding circumstances of these appeals have been set out in the judgments of Ralph Gibson and Nicholls LJJ and need not be repeated here. Each appeal raises important issues concerning the status and effect of a document K called "Licence Agreement" which is common to each. To a considerable extent, the appeals involve philosophical questions of social policy which may be said to lie within the especial province of the legislature and are, therefore, an area in which it behoves the court to walk with circumspection. I refer to the balance to be held between a person's right to contract freely as he will and the need for intervention by Parliament in order to protect a person whose bargaining power is handicapped by his personal circumstances. Well-known examples of such intervention are to be found in the Employment Protection Act 1975, section 118(1)(a) and the Matrimonial Causes Act 1973, section 34(2).

Although a tenant cannot contract out of the protection given to L him under the landlord and tenant legislation, Parliament has not seen fit to intervene so as to inhibit the freedom to enter into contracts for the use by one person of another person's premises so long as the use provided for in the contract falls short of creating a tenancy. This is recognised by Lord Templeman in his speech in *Street* v *Mountford* [1985] AC 809 at p 819:

In the present case, the agreement dated 7th March 1983 professed an intention by both parties to create a licence and their belief that they had in fact created a licence. It was submitted on behalf of Mr Street that the court cannot in these circumstances decide that the agreement created a tenancy without interfering with the freedom of contract enjoyed by both parties. My Lords, Mr Street enjoyed freedom to offer Mrs Mountford the right to occupy the rooms comprised in the agreement on such lawful terms as Mr Street M pleased. Mrs Mountford enjoyed freedom to negotiate with Mr Street to obtain different terms. Both parties enjoyed freedom to contract or not to contract and both parties exercised that freedom by contracting on the terms set forth in the written agreement and on no other terms. But the consequences in law of the agreement, once concluded, can only be determined by consideration of the effect of the agreement. If the agreement satisfied all the requirements of a tenancy, then the agreement produced a tenancy and the parties cannot alter the effect of the agreement by insisting that they only created a licence. The manufacture of a five-pronged implement for manual digging results in a fork even if the manufacturer, unfamiliar with the English language, insists that he intended to make and has made a spade.

It was also submitted that in deciding whether the agreement created a tenancy or a licence, the court should ignore the Rent Acts. If Mr Street has succeeded, where owners have failed these past 70 years, in driving a coach and horses

A through the Rent Acts, he must be left to enjoy the benefit of his ingenuity unless and until Parliament intervenes. I accept that the Rent Acts are irrelevant to the problem of determining the legal effect of the rights granted by the agreement. Like the professed intention of the parties, the Rent Acts cannot alter the effect of the agreement.

However, in considering the exercise of construction of the agreement, Lord Templeman referred to the judgment of Jenkins LJ in *Addiscombe Garden Estates Ltd* v *Crabbe* [1958] 1 QB 513 at p 522:

> The whole of the document must be looked at; and if, after it has been examined, the right conclusion appears to be that, whatever label may have been attached to it, it in fact conferred and imposed on the grantee in substance the rights and obligations of a tenant, and on the grantor in
> B substance the rights and obligations of a landlord, then it must be given the appropriate effect, that is to say, it must be treated as a tenancy agreement as distinct from a mere licence.

and (after referring to the observations of Denning LJ in *Errington* v *Errington* [1952] 1 KB 290, "The test of exclusive possession is not decisive") at p 823D:

> I think that wide statement must be treated as qualified by his observations in *Facchini* v *Bryson* [1952] 1 TLR 1386, 1389; and it seems to me that, save in exceptional cases of the kind mentioned by Denning LJ in that case, the law remains that the fact of exclusive possession, if not decisive against the view that there is a mere licence, as distinct from a tenancy, is at all events a consideration of the first importance.

C After further considering the preceding authorities leading up to the judgment of Lord Denning MR in *Marchant* v *Charters* [1977] 1 WLR 1181:

> "What is the test to see whether the occupier of one room in a house is a tenant or a licensee? It does not depend on whether he or she has exclusive possession or not. It does not depend on whether the room is furnished or not. It does not depend on whether the occupation is permanent or temporary. It does not depend on the label which the parties put upon it. All these are factors which may influence the decision but none of them is conclusive. All the circumstances have to be worked out. Eventually the answer depends on the nature and quality of the occupancy. Was it intended that the occupier should have a stake in the room or did he have only permission for himself personally to occupy the room, whether under a contract or not? In which case he is a licensee."
>
> D But in my opinion in order to ascertain the nature and quality of the occupancy and to see whether the occupier has or has not a stake in the room or only permission for himself personally to occupy, the court must decide whether upon its true construction the agreement confers on the occupier exclusive possession. If exclusive possession at a rent for a term does not constitute a tenancy then the distinction between a contractual tenancy and a contractual licence of land becomes wholly unidentifiable.

In considering the case of *Somma* v *Hazelhurst* [1978] 1 WLR 1014, Lord Templeman was able to construe a written agreement under which a young unmarried couple occupied a double bed-sitting room for which they paid a weekly rent as "a sham" or, in other words, a device to avoid the provisions of the Rent Acts in the case of an agreement otherwise providing for the occupier exclusive
E possession in consideration for a weekly payment:

> In *Somma* v *Hazelhurst* [1978] 1 WLR 1014, a young unmarried couple H and S occupied a double bed-sitting room for which they paid a weekly rent. The landlord did not provide services or attendance and the couple were not lodgers but tenants enjoying exclusive possession. But the Court of Appeal did not ask themselves whether H and S were lodgers or tenants and did not draw the correct conclusion from the fact that H and S enjoyed exclusive possession. The Court of Appeal were diverted from the correct inquiries by the fact that the landlord obliged H and S to enter into separate agreements and reserved power to determine each agreement separately. The landlord also insisted that the room should not in form be let to either H or S or to both H and S but that each should sign an agreement to share the room in common with such other persons as the landlord might from time to time nominate. The sham nature of this obligation would have been only slightly more
> F obvious if H and S had been married or if the room had been furnished with a double bed instead of two single beds. If the landlord had served notice on H to leave and had required S to share the room with a strange man, the notice would only have been a disguised notice to quit on both H and S. The room was let and taken as residential accommodation with exclusive possession in order that H and S might live together in undisturbed quasi-connubial bliss making weekly payments. The agreements signed by H and S constituted the grant to H and S jointly of exclusive possession at a rent for a term for the purposes for which the room was taken and the agreement therefore created a tenancy.

In my judgment, the approach which the court should take is to construe the document as a whole in order to determine the nature and quality of the occupancy under the terms of the agreement reached between the parties. To this end the use of the words such as licence or lease in the agreement is not definitive, nor indeed is the *de facto* intention of either or both of the parties. Subject to the agreement on its face appearing to be a sham, the effect in law of the agreement must depend upon its construction in accordance with the normal rules in the context of its factual matrix and genesis.

Being purely a matter of legal construction, it must be rare that triable issues of fact will arise unless these involve the possible existence of another agreement and/or that the agreement under consideration is a sham. Process under Order 113 would, subject to these qualifications, therefore, appear to be appropriate, and the occasions upon which it would be necessary for a judge to remit the matter for trial, unless satisfied that a tenancy had been created, would be rare.

Faced with the exercise of construing the agreement common to both appeals, for what signals should the court look in order to decide whether the occupier thereby achieved the status of tenant or whether he was merely a lodger? The speech of Lord Templeman, with which all the other noble and learned lords concurred, gives helpful indications as to those criteria which the court should ignore or, alternatively, which should not be treated as decisive. These include:
(a) the description of the agreement chosen by the parties, eg "lease" or "licence" — see *per* Lord Templeman at p 822B-C;
(b) the actual subjective intention of the parties, even if they are *ad idem*, see p 819F, unless they do not intend to enter into the agreement at all — see *Isaac* v *Hotel de Paris* [1960] 1 WLR 239, referred to at p 823E;
(c) the effects of the Rent Acts are irrelevant to the construction of the document — p 819H;
(d) the exercise or failure to exercise rights provided by the agreement by one, other or both the parties to the agreement is not of decisive importance;
(e) the court should not draw up a "shopping list" of clauses;
(f) the court should not award marks for drafting — see p 826F.

On the other hand, so long as owner and occupier are permitted by law to reach an agreement under which the latter has the use of the premises merely as a lodger without the protection granted to a tenant, the courts must accept and enforce an agreement which achieves this objective. This proposition merely recognises that a skilful draftsman must, in appropriate circumstances, be allowed to succeed. For my part, I find difficulty in excluding the respective positions of the parties from an analysis of the genesis and factual matrix of the agreement — see *Reardon Smith Line Ltd* v *Yngvar Hansen-Tangen* [1976] 1 WLR 989. I accept Mr Bagnall's submission that this inquiry requires a trial of the issues and is, therefore, inappropriate to process under Order 113. Clearly, it would be no more sufficient for an occupier to cry "sham" than for a landlord to say "bare licence". It is to the document itself that the court will look in the first instance. Where it contains terms which *ex facie* give the appearance of being a "sham", then the court must inquire further and should not make an order under Order 113.

Before Hirst J and Boreham J the appellants each sought to rely upon an allegation that the agreement was a "sham". This defence was based on the views expressed by Lord Templeman of the judgment in the Court of Appeal in *Somma* v *Hazelhurst*, which I have already set out in this judgment. The provisions of the two agreements made with "H" and "S" *ex facie* give rise to such bizarre results that Lord Templeman had little difficulty in stating that both agreements were a sham. When, however, the provisions of the agreement *ex facie* provide for a licence to occupy premises as a lodger, the mere fact that they do this cannot of themselves make the agreement a "sham". The exercise of construction must be carried out on the basis that the provisions of the agreement are genuine unless, from their very terms, they appear to be consistent with a sham agreement as defined in the speech of Lord Diplock in *Snook* v *London & West Riding Investments Ltd* [1967] 1 All ER 518, to which reference has already been made by Ralph Gibson and Nicholls LJJ.

In the present agreement there are two provisions which can be said to invite inquiry: the daily restriction in the use of the premises between 10.30 am and noon and the right to remove furniture at will. On working days there is nothing *ex facie* unacceptable in the first restriction, although on non-working days it would clearly be onerous to enforce such a provision. If the second provision were limited to substitution or to a habitable minimum, there might be no objection. Although these clauses were not relied upon before Hirst J or Boreham J, in my judgment they do provide an indication of the existence of a position in relation to sham which would merit investigation by the court. In the absence of argument upon these points, I can well understand, and sympathise with, the decisions to make the possession orders against which these appeals have been brought.

In the appeal brought by Merola and Kader the appellants go further and allege that they occupied the premises in pursuance of an oral agreement made between them and the caretaker. This is a somewhat different proposition. The appellants seek to place evidence before the court which would establish that, quite apart from its written terms, the agreement was never intended, as a subjective matter, to represent a contract between the parties. It is always open to the party to adduce external evidence to this effect; and this is not a question of construing the document. I agree that, in this appeal, the affidavits disclose the existence of a triable issue of this kind and that, for this reason alone, Boreham J should not have made an order under Order 113.

For these reasons and for the reasons already given in the judgments of Ralph Gibson and Nicholls LJJ I agree that both these appeals should be allowed.

The appeals were allowed with costs in the Court of Appeal and below. Legal aid taxation of both appellants' costs ordered.

Court of Appeal

April 22 1986

(Before Lord Justice PARKER and Sir George WALLER)

DEPARTMENT OF TRANSPORT v EGOROFF

Estates Gazette June 21 1986

278 EG 1361-1363

Landlord and tenant — Whether Crown (the Department of Transport being the Crown in this context) was bound as landlord by sections 32 and 33 of the Housing Act 1961 (now replaced as a result of the consolidation of housing legislation by sections 11-16 of the Landlord and Tenant Act 1985) — The present issue arose out of proceedings brought by the department as plaintiffs against the defendant, the department's tenant, for arrears of rent and possession — On an application for summary judgment the defendant was given leave to defend and the matter transferred to the county court judge — The judge struck out a number of paragraphs of the defendant's defence as disclosing no reasonable cause of action, including material relating to the alleged application to the tenancy of sections 32 and 33 of the 1961 Act — The only issue before the Court of Appeal was whether the Crown as represented by the department was bound by the obligations placed on landlords of short leases by these sections — It was submitted on behalf of the appellant tenant that the general rule that the Crown is not bound by a statute in the absence of express words or necessary implication was subject to an exception mentioned in the *Magdalen College Case* (1615), namely, where the statute was enacted to suppress a wrong — It was suggested that the object of sections 32 and 33 was to suppress a wrong — Held, rejecting this submission, that the principle stated in the *Magdalen College Case* was far too wide to be of general assistance; that its reliability as a guide was disposed of in *Bombay Province* v *Bombay Municipal Corporation*; and that in the comparable context of the Rent Acts it was established (until the position was partly altered by legislation) that the Acts did not bind the Crown — Accordingly the judge acted correctly in striking out the relevant paragraphs in the defence — Appeal dismissed — Judge's comments on documents presented to the court

This was an appeal by Dimitri Egoroff, the defendant, from the decision of Deputy Judge Sumner, at Folkestone County Court, striking out parts of the defendant's defence in an action against him as tenant by the Department of Transport, as landlord, for arrears of rent and possession of Frogholt House, Frogholt, Kent.

Edward Cousins (instructed by S Rutter & Co) appeared on behalf G of the appellant; Nigel Pleming (instructed by the Treasury Solicitor) represented the respondent department.

Giving judgment, PARKER LJ said: This is an appeal by leave of Nourse LJ from a decision of His Honour Deputy Judge Sumner made on July 17 1985 striking out certain parts of the defendant's defence in an action brought against him for arrears of rent and possession in respect of a property known as Frogholt House, Frogholt, in the county of Kent. The premises were let by the plaintiffs, who are the Department of Transport and are for present purposes to be regarded as "the Crown", under an agreement dated February 26 1980. The tenancy was for a term of one month from March 1 1980 and thereafter from month to month, being terminable H on notice. The action was started by a specially endorsed writ in the High Court and an application for Order 14 judgment was made. Leave to defend was given and the matter was then transferred to the county court.

By his defence the defendant pleaded as follows. Para 1 of the particulars of claim (which was the tenancy) is admitted, and it is averred that by virtue of the tenancy being a tenancy from month to month the provisions of sections 32 and 33 of the Housing Act 1961* apply thereto. Further, by clause 3 of the tenancy agreement it is provided that the defendant paying the said rent and performing and observing all his obligations under the agreement may quietly enjoy the property during the tenancy without lawful interruption by the plaintiff or any person rightfully claiming through or under him. J

Paras 3, 4 and 5 of the defence contain admissions of paras 2, 4 and 5 of the writ and a non-admission as to para 3. Para 6 of the defence then proceeds as follows:

> By reason of the matters particularised hereinafter it is averred that the plaintiff is in breach of the said implied covenant pleaded in Paragraph 1 hereof and/or in breach of the said covenant for quiet enjoyment pleaded in Paragraph 2 hereof, whereby the plaintiff has suffered loss and damage.
>
> 7. Further or alternatively, it is averred that the plaintiff has negligently caused and/or suffered and/or permitted the demised premises to remain and/or fall into a state of disrepair whereby the plaintiff has suffered loss and damage.

There follow particulars of breach of covenant and/or negligence all of which relate to alleged disrepairs or failure to repair the K premises which were let under the tenancy. It is then said that the defendant has suffered loss and damage amounting in total to £68,560. The counterclaim is based on the same allegations as are set out in the defence.

The order which was made was that the words "and it is averred that by virtue of the tenancy being a tenancy from month to month, the provisions of sections 32 and 33 of the Housing Act apply thereto" in para 1 of the defence, and the whole of paras 2, 6, 7, 8, 9 and 11 of the defence, be struck out on the ground that they disclose no reasonable defence, and also that the counterclaim be struck out on the ground that it discloses no reasonable cause of action.

It will be seen that that pleading raises three matters. First, an L implied covenant arising under sections 32 and 33 of the Housing Act. Second, an alleged breach by non-repair of a covenant of quiet enjoyment. Third, negligence in failure to repair. There is no allegation in that pleading that there was a contractually implied term to repair the premises. The plaintiffs applied to strike out the paragraphs which were ultimately struck out on the ground that they disclosed no reasonable cause of action or were frivolous and vexatious and an abuse of the process of the court.

On this appeal the only point arising is whether the plaintiffs are bound by sections 32 and 33 of the Housing Act 1961. Mr Cousins included in his skeleton argument contentions concerning the existence of an implied contractual term, but that does not now arise. Since the offending paragraphs were struck out, the plaintiffs have M applied for and obtained leave to deliver an amended defence and counterclaim in which an allegation with regard to an implied contractual term is made, and they are free to argue that in the county court without further relief from us. The matter is therefore confined to the single point, whether the Crown is bound by sections 32 and 33.

Before dealing with that point, however, I should mention that the documents put before this court are some 180 or more, and the copying of those documents appears to me to have been a wholly unnecessary expense. For the purposes of the determination of this appeal, nothing is necessary other than the writ, the defence, the

*Editor's note: These provisions are now to be found in sections 11-16 of the Landlord and Tenant Act 1985, which came into operation on April 1 1986.

A tenancy agreement itself and the judgment below. Quite apart from that, it is to be observed that the bundle does not comply with the practice direction which appears at p 624 of the *Supreme Court Practice*, which requires a master bundle to be prepared where the documents, apart from pleadings, comprise more than 100 pages. Had attention been paid to that practice direction, it would at once have become apparent that the only documents required were those which I have already listed.

Returning to the point in issue, it is common ground that a statute does not bind the Crown save by express words or necessary implication. Mr Cousins concedes that there are no express words binding the Crown and that there are no words in the statute itself B which raise the shadow of an implication, let alone a necessary implication, that the Crown was intended to be bound. He founds himself wholly on two propositions. First, that there are certain classes of statutes which always bind the Crown, and in that connection he relies on the *Magdalen College Case* (1615) 11 Co Rep 66b. The classes of statute which are said to bind the Crown include, according to that case, statutes which are made to suppress a wrong. Mr Cousins submits that this Act was an Act made to suppress a wrong.

As to the first proposition, the classification in the *Magdalen College Case* is, in my view, far too wide to be of general assistance, and is anyhow wholly disposed of as any sort of reliable guide by the decision in *Bombay Province* v *Bombay Municipal Corporation* C [1947] AC 58 at pp 61 and 62. Reference to the unreliability and outdatedness of the *Magdalen College* classification may also be found in *Wade on Administrative Law*, 5th ed, p 715, and also in a number of other places including vol 27 of *Halsbury's Laws of England*. But even if that classification stood, there can, in my view, be no doubt that sections 32 and 33 of the Housing Act do not come within the classification.

It is submitted by Mr Cousins that the sections were to suppress a wrong. In one sense they were. The sense in which they were is that it was recognised by Parliament that a situation which was socially undesirable required correction for the public good. Precisely the same consideration applies to the Rent Acts, which also dealt with a situation which it was considered was socially undesirable and D required correction for the public good. The Rent Acts in no less than five cases and possibly more, starting with *Clark* v *Downes* (1931) 145 LT 20, have consistently been held not to bind the Crown. Mr Cousins submits that there is a distinction between the Rent Acts and sections 32 and 33 of the Housing Act, a distinction which for my part I found difficult to follow. But, as I understood it, it was submitted that the Rent Acts applied *in personam* whereas the sections here under consideration apply to the premises. That submission is, in my view, without foundation. Both Acts are dealing with premises and it is a truism to say that the Rent Acts apply *in rem*.

There being, as I see it, no relevant distinction between the two sets of provisions, and in the light of what is said in the *Bombay* case, to E which I have already referred, it appears to me that the test must be either an examination of the wording of the Act or in certain specific cases a demonstration that the purpose of the sections would be wholly frustrated unless the Crown were bound. That might raise a necessary implication outside the wording of the Act. But unless the matter can be taken that far, and in this case it certainly cannot, it does not seem to me that an Act of this sort can possibly be said to bind the Crown. For those reasons, I am of the view that the learned deputy judge was wholly correct in striking out the paragraphs which he did strike out, and I would dismiss this appeal.

SIR GEORGE WALLER agreed and did not add anything.

The appeal was dismissed with costs.

F

Chancery Division

November 12 1985

(Before Mr Justice WALTON)

Re DISTRIBUTORS & WAREHOUSING LTD

Estates Gazette June 21 1986

278 EG 1363-1370

Landlord and tenant — Guarantee of payment of rent and performance of tenants' obligations under terms of lease — Assignment of freehold reversion but no assignment of benefit of covenant of guarantee — Two peculiarities of guarantee, two guarantors each liable for half only of amounts guaranteed (one of them in any case insolvent) and guarantee effective only so long as lease remained vested in the named company tenant — Company went into liquidation and liquidator in present proceedings sought leave of court to disclaim lease — Landlords, assignees of the original lessors in whose favour the guarantee was given, opposed the application for leave to disclaim — The essential question was whether the landlords were entitled to the benefit of the guarantee, as this affected the attitude of the court to the proposed disclaimer — Where a company in liquidation are the original lessees, as here, leave to disclaim is not normally given if a subsisting guarantee is involved — Landlords accordingly put forward a number of arguments with a view to establishing that, despite the fact that there had been no express assignment of the guarantee when they acquired the reversion, they were nevertheless entitled to the benefit of it — Among these arguments were that the benefit passed without express assignment; that a supplemental deed in which the guarantee was reaffirmed in respect of an enlargement of the premises in fact created a new guarantee; that the deed created an estoppel; that the benefit passed as a result of section 62 of the Law of Property Act 1925; and that section 56 of that Act applied — All these arguments were rejected by the judge — He also rejected a submission that the landlords were entitled to claim a rectification of the transfer of the reversion — Decisions against the landlords' contentions were the Australian case of *Sacher Investments Pty Ltd* v *Forma Stereo Consultants Pty Ltd* and *Pinemain Ltd* v *Welbeck International Ltd* — Finally, even if the landlords were entitled to the benefits of the guarantee, on the peculiar facts of the case the end result would be the same — If disclaimer was not granted, the liquidator could nevertheless proceed to cause the dissolution of the company and with it would go the existence of the guarantee, which was expressly tied to the continuance of the company — Leave given to the liquidator to disclaim the lease

This was an application by originating summons by Ian Franses, the liquidator of Distributors & Warehousing Ltd, to obtain the necessary leave of the court to disclaim the lease vested in the company of warehouse premises situated on the north side of the East Lancashire Road, Haydock, in the County of Lancaster. The respondents were the present landlords, Co-operative Insurance Society Ltd, and two sureties, Peninsular & Oriental Steam Navigation Co Ltd and Yangzekiang (UK) Ltd.

Christopher Pymont (instructed by Crellins, of Cobham, Surrey) appeared on behalf of the applicant liquidator; Oliver Weaver QC and Martin Moore (instructed by the Solicitor's Department, Co-operative Insurance Society Ltd) represented the society (the first respondents); Gabriel Moss (instructed by Masons) represented P & O Steam Navigation Co Ltd (second respondents); Yangzekiang (UK) Ltd (third respondents) were not represented and took no part in the proceedings.

Giving judgment, WALTON J said: On March 6 1970 Townson Developments Ltd ("Townson") demised certain warehouse premises situate on the northerly side of the East Lancashire Road, Haydock in the County of Lancaster, to Distributors & Warehousing Ltd ("the company") for a term of 42 years from December 15 1969. The initial rent payable was the sum of £26,840, which was made subject to, first, alteration and, second, periodic reviews.

The alteration was to take place in the event that the company exercised an option contained in the lease to require the landlord to construct an extension to the demised premises. If it did so, the rent was to be increased from the date when the lessor's surveyors (after consultation with the lessee's surveyors) issued a certificate to the effect that the extension had been duly constructed or from the date when the lessee entered into any part of the extension, whichever should be the earlier. The amount of the increase of rent was to be

A ascertained by the application of a complicated mathematical formula:

$$\frac{(40\,NP + X\,NP) \times S - (S \times 1{,}000 \times 100)}{43{,}560}\,NP$$

where S equalled the ground-floor area covered by the extension expressed in square feet and X equalled the amount expressed in new pence per superficial square foot of rental value representing the increase in building costs arising between November 1 1968 and the date upon which the plans, drawings and specifications relating to the extension should have been agreed by the lessor and the lessee and — surprisingly enough — also Townson (even though this would
B entail that their agreement was required even after they had ceased to be the lessor) and should have been approved for the purposes of the Town and Country Planning Acts 1962 to 1968 or any statutory modification or re-enactment thereof and for the purposes of the local planning bye-laws affecting the same. In default of agreement, there was a provision for determination by an independent surveyor.

There was then in addition a rent review procedure, applicable both to the original reserved rent and any such increase as would be caused by the erection of the extension. The first rent review was destined to take effect from December 15 1983, being triggered off by a notice in writing given by the lessor to the lessee not less than three nor more than 36 months before that date.

The lessor was defined in the usual way as, where the context so
C admitted, including the persons deriving title under it, and the lessee was similarly so defined. Two guarantors were joined in the lease: The Peninsular & Oriental Steam Navigation Co Ltd ("P & O") and Yangzekiang (UK) Ltd ("Yangzekiang"). The covenant of guarantee which P & O and Yangzekiang gave was certainly in a most unusual form and I think I should read it in full.

> The guarantors in consideration of the demise hereinbefore contained being made by the Lessor at their instance and request hereby jointly agree with and guarantee to the Lessor that at all times so long as the term hereby granted is vested in Distributors and Warehousing Ltd that company will pay the rent hereby reserved and all other sums and payments herein covenanted to be paid by it at the respective times and in manner hereinbefore appointed for payment thereof and also duly perform observe and keep the several stipulations herein on the Lessee's part contained and that the guarantors will each pay and make good to the Lessor one-half of all losses costs and expenses sustained by the Lessor through the default of Distributors and Warehousing Ltd in respect of any of the beforementioned matters so long as the term hereby granted is vested in Distributors and Warehousing Ltd provided always that any neglect or forebearance of the Lessor in endeavouring to obtain payment of the said rent and payments as and when the same become due or its delay to take any steps to enforce the performance or observance of the several stipulations herein on the Lessee's part contained and any time which may be given by the Lessor to Distributors and Warehousing Ltd shall not release or in any way lessen or affect the liability of the guarantors or either of them under this guarantee.

It has been accepted by all parties that the effect of this clause is that each of the two guarantors is liable only for one half of the amounts thereby guaranteed. The effect of this in the present situation is that only P & O are, on any view, effective guarantors (assuming that they are still under any liability under this covenant) for the future. For it is beyond all question that Yangzekiang is no longer trading and that its sole asset is a debt of £50,000 from its parent company, which is itself, although still trading, in by no means a satisfactory financial position. But even if the whole of this sum were to be repaid, it would not suffice to meet Yangzekiang's already accrued liabilities under the covenant of guarantee, of course on the same suppositions as before. It is, of course, virtually beyond all belief that its parent company would spend its endeavours in ensuring a prosperous future for Yangzekiang merely for the purpose of meeting the guarantee. And its disinterest in the position is shown by the fact that although made a respondent to the present proceedings it has taken no part therein.

The other curious feature of the guarantee is that it lasts, in any event, only so long as the term granted by the lease remains vested in the company. If the company were to assign the lease — it would of course require the consent of the lessor so to do — the covenant of guarantee would come to an end. Similarly, if the company were to be dissolved, or struck off the register, the covenant of guarantee would again come to an end. The lessor could have no conceivable complaint, because that is exactly and precisely what the covenant of guarantee provides. Mr Weaver, for Co-operative Insurance Society Ltd ("CIS"), one of the parties to this summons, did indeed submit the contrary to me; but it was obvious from the manner in which he made his submission that he had not even convinced himself. G

On March 9 1970 Townson conveyed the freehold reversion expectant upon the lease to CIS. The transfer was an ordinary registered land form, and the curious feature thereof is that no attempt was made to assign to CIS the benefit of the covenant of guarantee. As will be seen hereafter, Mr Weaver, for CIS, has made a number of submissions to the effect that, for various reasons, this omission is of no importance, and that the benefit of such covenant passed automatically to CIS notwithstanding.

Although I do not think that at the end of the day anything turns upon it, it appears to me that neither of the guarantors can, at this stage, have been aware of this omission. Of course, they will soon have learnt of the change of landlords, but probably both of them assumed (of course quite wrongly) that the benefit of their guarantees H had been assigned to the new landlord. Certainly my attention was not called to any piece of evidence which suggested that the two guarantors, at the respective dates of the two documents to which I shall next turn, were in any way aware of the true position.

The first of these documents is a deed dated September 19 1977 made between CIS of the first part, the company of the second part, and the two guarantors of the third part. It is expressed to be supplemental to the lease of March 6 1980; it recites that the reversion immediately expectant on the determination of that lease is vested in CIS; that the extension had been erected by the lessee with the consent and at the expense of the lessor; and that it had been agreed between all parties that the lease should be varied in the manner thereinafter J appearing. The operative parts of the deed, so far as material, then read as follows:

> In consideration of the premises the Lease is hereby varied in manner following, that is to say: (i) The Lease will in addition to allowing substitution for the said yearly rent of £26,840 (hereinafter called "the existing rent") also reserve and make payable from and after the 5th day of July 1976 during the residue of the term granted by the Lease the additional yearly rent of £19,200 (hereinafter called "the additional rent") payable by the like instalments and on the same dates as the existing rent, the payments of the additional rent due from the period from the 5th day of July 1976 to the 29th of September 1977 having been made prior to the date hereof. (ii) Every reference in the Lease to rent including the Lessee's covenant to pay the same and the proviso for re-entry in respect thereof will be read and construed as referring both to the existing rent and to the additional rent. (iii) The Lease will henceforth be read K and construed as though the provisions of the Schedule hereto were substituted for the provisions contained in the Schedule to the Lease. 2. The covenants on the part of the Lessee and the provisions and stipulations contained in the Lease will extend to the extension and in particular the Lease will be read and construed as though the expression "the premises" included the extension as well as the land and buildings thereby demised.

And clause 6:

> The Guarantors hereby consent to the terms of this Deed and acknowledge that their obligations under the Covenant on their part contained in the Lease extends to and includes the payment by Distributors and Warehousing Ltd of the additional rent and all other sums hereby made payable,

and then there is the schedule containing substituted rent review provisions. L

So one can see quite clearly why, on the supposition that the covenant of the guarantors was still effective, they should have been joined in such a deed. First, whereas the lease provided that the building should be erected by the lessor, it had apparently been erected by the lessee, albeit at the expense of the lessor; the formula for calculating the additional rent was a complicated one, and there might possibly be scope for doubt as to whether it had been correctly ascertained; and there was to be a change in the rent review clause. Any of these matters might well have afforded the guarantors scope for saying that there had been a material variation of the terms of the original contract with the lessee, which would have the automatic effect of discharging the guarantors' obligations. It appears to me that this is the reason for this deed, and is indeed a very satisfactory M reason for its having been brought into existence. Mr Weaver has a submission to make upon this deed: first, he says that it contains a fresh guarantee by the guarantors, which in my judgment it manifestly does not, and, more pertinently, that, having joined in this deed in this manner, the guarantors are thereafter estopped from denying that they are bound to CIS in the terms of the guarantee. I shall have to consider this submission later.

The second deed, upon which nobody has suggested that anything material in the present application depends, is a deed of May 3 1978 supplemental to the deed of September 19 1977 and made between exactly the same parties. All it did was to effect a small increase in the additional rent as from September 29 1977.

The company, unfortunately, has fallen upon hard times; and on August 30 1984 it went into creditors' voluntary liquidation, with Mr Franses being the liquidator. The accounts of the liquidator demonstrate that there will be a substantial shortfall of assets as regards creditors: indeed, there will not even be sufficient to pay the preferential creditors in full.

There is one issue in this matter which I do not have to resolve. It is this. Just before the company went into liquidation, the first rent review, reviewing the rent of the demised premises as from December 15 1983, came to a head in the shape of an agreement as to the rent between CIS and the company. Whether in all the circumstances of the case, this is a firm agreement upon which the company (were it necessary for it to do so) and P & O (as a possible guarantor) could rely, I do not determine. CIS evidently took the view that this was not, in that sense, a binding agreement, and invoked the procedure via the appointment of an independent surveyor. Again, there is some possible dispute as to whether notice of his appointment was properly conveyed to the company and the guarantors so that they might put before him such argument as they might wish. In the event, he decided the revised rent without reference to any arguments of any party, as none were put before him. Accordingly, there are three possible views as to the current rent: first, that, there having been no agreement and no valid determination of any fresh rent, whatever might be the position hereafter, the rent remains at its old rent of £19,508 per annum. Second, that the rent is now the "agreed" rent of £60,000 (but due to rise on December 15 1986). Third, that the rent is the rent determined by the independent surveyor of £92,000 per annum.

It would be interesting and instructive to learn precisely how the independent surveyor arrived at this latter figure, for all the professional surveyors on all sides are fairly clearly agreed upon the position that having regard to the facts (i) that the demised premises are situated in an extremely economically depressed area of England (half way between Liverpool and Manchester) and (ii) their considerable size, and the consideration that, except to a limited extent which would still leave one extremely large warehouse, it would be extremely difficult, if not at the moment absolutely impossible, to let the premises as they stand, or even to divide the premises and as so subdivided to let them at all, at any rent. Nobody wants a warehouse of this size, or even one of the scaled-down size which would be produced by such division in this area at the moment.

However, the crucial point is that having regard to the foregoing factors, there is absolutely no premium value in the lease: if the liquidator could find a person willing to take the lease, in the present economic climate he would have to pay him to cart it away.

In these circumstances, the liquidator very naturally wishes to disclaim the lease, and the actual application before me is an originating summons taken out by him on September 28 1984 to obtain the necessary leave of the court to this course of action. Naturally, the landlords, CIS, and the two guarantors are parties thereto. In view of what I have already indicated in relation to their financial position, it comes as no surprise that Yangzekiang has indicated that it does not wish to contest the proceedings.

It makes no difference to the present proceedings, but they are in fact being conducted pursuant to the provisions of the Companies Act 1948 in this respect, and not those of the Companies Act 1985: see Companies Consolidation (Consequential Provisions) Act 1985, section 31(8)(c).

For what purpose was the concept of disclaimer introduced? It was in order to ensure, if at all possible, an orderly winding up of the affairs of the company. It is apparently quite clear that if no leave is given to disclaim a lease, the liquidator will nevertheless be in a position to proceed with the usual procedure for the summoning of a final meeting, leading up to the dissolution of the company pursuant to the Companies Act 1948, section 300: see *Re Katherine et Cie Ltd* [1932] 1 Ch 70. There is, unfortunately, no guidance in any reported case as to the precise manner in which the future claims of the landlord are, in such a case, to be dealt with. That, no doubt, is because in the only case which deals with the matter, the landlord opposed the granting of leave to disclaim because, as a result of such disclaimer, he would have lost the benefit of a full guarantee of the rent and other obligations under the lease. Presumably, he considered that the guarantee would be a complete guarantee; and so no question of future claims by him would arise. He would look to the guarantors. The guarantors, in their turn, would have future claims against the company, for which presumably they might prove in the liquidation. And in the middle of all this, the actual lease itself would be in balk; presumably in evaluating the claims of the guarantors the prospect of the landlord finding it to his advantage to forfeit the lease for non-payment of rent and so releasing the guarantors would have to be taken into consideration. But this would be a future, hypothetical event, extremely difficult to evaluate in monetary terms.

Be all that as it may, the cases have thrown up this curious dichotomy. In a case where the company is not the original lessee, but an assignee, the presence or absence of guarantors makes no difference to the approach of the court to the question of granting the liquidator leave to disclaim. This is because where leave to disclaim is given, the effect is to release the company from all liability of the company "in or in respect of the property disclaimed" (see Companies Act 1948, section 323 (2)) and if the guarantors were not also discharged, they would or might have a right over against the company on the usual indemnity principles. This, being a claim in respect of the property disclaimed, cannot be allowed to continue, and so all liability of the guarantors must be discharged so far as the company is concerned. So that, if the company is the original lessee, there can be no question but that the liability of the guarantors ceases on disclaimer (see *Stacey* v *Hill* [1901] 1 QB 660 CA and *D Morris & Sons* v *Jeffreys* [1932] 148 LT 56). But if on the other hand the company is an assignee of the lease, so that the guarantee is strictly a guarantee of the liability of the original lessee to pay the rent, then no such question of indemnity arises, so that, if the liquidator is given leave to disclaim, both the liability of the original lessee to pay the rent and the guarantee thereof by the guarantors remains.

This leads to the conclusion that, although in a case where the company is the lessee by assignment of the lease, leave to disclaim will readily be given, since such leave in no way affects the rights of the guarantors, in the case where the company is the original lessee, such leave will not normally be given: *Re Katherine et Cie Ltd* (*supra*).

Hence the first crucial question in the present case is whether CIS is entitled to the benefit of the guarantee of P & O and Yangzekiang. If it is, then the court will certainly lean, in the usual case, to a refusal of the leave to disclaim; if it is not, then, unless any particular reason is shown for taking a different course, leave will be granted.

Before turning in more detail to the detailed submissions by which Mr Weaver sought to persuade me that the benefit of this guarantee was now vested in CIS notwithstanding the omission of any express assignment, I must notice an extremely attractive submission which he made to the effect that I need not in any way attempt to decide the point, because all that was necessary was for CIS to show a seriously arguable case that they had the benefit of the guarantee, not to demonstrate that they did in fact have it.

However, I feel that I must reject this submission. As I have already noted it is well established by now, first, that leave to disclaim will normally be given automatically in a case where there are no guarantors involved, and, second, that leave will normally not be given where guarantors are involved. Hence, it must have been obvious from the first to any counsel as astute as Mr Weaver — indeed obvious to any competent counsel skilled in company law — that the crucial or, at any rate, one of the crucial points in issue was going to be whether or not CIS did have the benefit of the guarantee. CIS should therefore have been prepared with the necessary evidence upon which they were going to rely to make good that submission. In this connection, Mr Moss for P & O submitted that I should in any event take the situation as it existed at the time of the hearing of the summons, and that I should not pay any attention to future events — such, for example, as a contemplated assignment of the benefit of the guarantee now from Townson to CIS which was stated by CIS to be a possibility. I think there is a great deal of force in that submission, but I am not prepared to follow him all the way on that. If, for example (which is of course not the case here), CIS had been in the middle of a hotly contested action against Townson for the rectification of the transfer of the reversion to it, I do not think it would be just for any court to say that as at that precise moment no order for rectification had been made, the court would view the situation as if it did not have the benefit of the covenant.

But, on the other hand, I do accept Mr Moss' submission thus far — that in relation to the ingathering of the benefit of the covenant, the onus is heavily upon CIS to show more than an arguable case; in short to show that they have a strong case for saying that they are likely to succeed.

Bearing these submissions in mind, how does Mr Weaver seek to ingather the benefit of the covenant of guarantee for CIS? He boldly first asserts that, notwithstanding there was no mention of the covenant of guarantee in the transfer of March 9 1970, nevertheless the benefit passed in that transfer, and he also submitted that this would give it business efficacy. I do not take this last submission seriously; many leases without covenants of guarantee are assigned every day, and the lease without it is a perfectly viable business document.

Here his sheet anchor was the case of *Griffith* v *Pelton* [1958] Ch 205. This is a case where there was an assignment of a lease containing an option to purchase the freehold in certain well-defined circumstances, followed by, first, an assignment of that lease, without any express mention of the option, followed by an express assignment of the benefit of the option. The Court of Appeal held, affirming the decision of Vaisey J, that the benefit of the option had indeed passed by the assignment of the lease, but that if it had not it certainly had by the express assignment. As Mr Goulding (as he then was) arguing for the plaintiff well put it, the defendant was on the horns of a dilemma.

The reason that the Court of Appeal came to the conclusion they did was that set out with his usual admirable clarity by Jenkins LJ (as he then was) at the top of p 228 of the report:

> We think that upon the true construction of the proviso, including the definition to be read into it of the term lessee as including the lessee's assigns, the original parties to the lease must be taken to have agreed that the option should be exercisable by Miss Blaker herself or by any assignee of the term to whom she might assign the benefit of the option, and that a mere assignment of the term should operate as an assignment of the benefit of the option to the assignment of the term.

That is to say, they treated it as a pure matter of construction of the contract itself. Viewed in that light, it is miles away from the present case. Indeed, Mr Pymont called attention to the fact that that was a case of the assignment of the lease, whereas in the present case what is in issue is the assignment of the reversion, and that in *Woodall* v *Clifton* [1905] 2 Ch 257 CA a different result was reached for that very reason. Indeed, if I may with respect refer to part of Mr Goulding's successful argument in *Griffith's* case, it ran as follows at p 213:

> (2) There is an ambiguity concealed in the use of the words "touching and concerning". The class of covenants which run with the land under a lease is not the same as the class of covenants which pass under the Grantees of Reversions Act 1540. There is no essential identity between the class of covenants where, if the term is assigned, the burden or benefit (as the case may be) of the covenants pass to the assignee of the term, and that in which, if the reversion is assigned, the burden or benefit (as the case may be) of the covenants pass to the assignee of the reversion. These two branches of the law of covenants have a different origin.

Leaving this on one side, Mr Weaver then turned to the deed of variation of September 19 1977. He said that that deed clearly indicated that the transfer had been effective to transfer the benefit of the covenant of guarantee. I do not, however, think that I can accept that submission just like that. It is, of course, obvious that when the guarantors joined in the deed of variation they must have done so under the belief that CIS had indeed obtained the benefit of their guarantees, but this is a vastly different matter. Merely because a person acts on a certain assumption, which is indeed the natural assumption to make, it does not mean that he is accepting the legal consequences of a document which, so far as the evidence goes, he did not see, but was merely informed as to the operation. So, so far from the deed of variation "demonstrating" (as Mr Weaver would have it) that the effect of the transfer was to pass the benefit of the covenant of guarantee, I do not think it did anything of the kind.

However, this is purely negative. Mr Weaver then submitted positively that the terms of clause 6 of the deed of variation which I read again:

> The Guarantors hereby consent to the terms of this Deed and acknowledge that their obligations under the Covenant on their part contained in the Lease extend to and include the payment by Distributors and Warehousing Ltd of the additional rent and all other sums hereby made payable,

constituted a new guarantee, or else provided the materials for estoppel by deed.

I regard it as quite impossible to regard the provisions of clause 6 as producing either result. There is no additional contractual obligation being undertaken by the guarantors: all that they are doing is acknowledging that something is within the terms of their existing obligation. One can well see why it was necessary or, at the lowest, highly desirable to obtain their acknowledgement to the "additional rent and other sums" which the company were undertaking to pay: they would thereby clearly thereafter be estopped from taking the point that due to a variation of the obligations imposed upon the company they were thereby released from their guarantees, on the usual principles. But merely to acknowledge that an obligation which may possibly include x does indeed include x appears to me to be light years away from undertaking an entirely fresh obligation.

As regards the suggestion of estoppel by deed, what is the law? It is in my judgment accurately set out in *Halsbury's Laws of England*, vol 16 at paras 1571 and following:

> Estoppel by deed is based upon the principle that when a person has entered into a solemn engagement by a deed under his hand and seal as to certain facts he will not be permitted to deny any matter which he has so asserted. It is a rule of evidence according to which certain evidence is taken to be of so high and conclusive a nature as to admit of no contradictory proof. The averment relied upon to work an estoppel must be "certain to every intent" without any ambiguity, but may be contained in the recital or in any part of the deed. A mere mistake in a deed on account of which no one has acted to his detriment will not give rise to an estoppel but may give grounds for rectification. A party is estopped from denying any specific facts contained in a recital in the deed to which he is a party provided the recital is certain, precise and unambiguous. He is not bound by inferences which may be drawn from the statements in the deed and a recital, which although incomplete is true so far as it goes, does not prevent the party from averring what is necessary to complete the truth. Nothing is to be taken by a way of "intendment"; there is no such thing as an estoppel by something implied, and the averment relied upon to work an estoppel must be of something particular, not of a generality. A covenant that a man has a thing is not equivalent to a positive statement that he has it, and consequently, in order to work an estoppel, a recital of title must aver precisely that the person in title is seised in fee or has the legal estate, and thus the operative words of an ordinary conveyance by grant create no estoppel.

It could, of course, be said to be an inference from the terms of the deed of variation that the covenant of guarantee had been assigned to CIS, because otherwise, if still in the hands of Townson and therefore incapable of being sued upon, the acknowledgement in clause 6 would be completely without force and effect. But it is, of course, precisely this kind of "inference" which does not found an estoppel. If authority for this proposition is required, a few sentences from Bowen LJ (as he then was) in *Onward Building Society* v *Smithson* [1893] 1 Ch 1 at p 14 will suffice:

> It would be very dangerous to extract a proposition by inference from the statements in a deed, and hold the party estopped from denying it; estoppel can only arise from a clear, definite statement. For instance, it is not enough to say that a vendor is well entitled in fee at law or in equity: he cannot be estopped from denying that he has the legal estate unless there is an express statement in the deed that he has it. In the present case there is an omission in the deed, so that it falls short of saying that the vendors have the legal fee: it may have been an accidental omission: but what right have we to supply it?

Accordingly, it appears to me quite clear that, there being no unequivocal statement upon which it is possible for CIS to latch to the effect that the guarantors are bound to them in the terms of the covenant of guarantee, I do not think that the deed of variation assists CIS in the slightest.

Mr Weaver's next point was that CIS was entitled to enforce the guarantee pursuant to section 62 of the Law of Property Act 1925. This reads:

> (1) A conveyance of land shall be deemed to include and shall by virtue of this Act operate to convey, with the land, all buildings, erections, fixtures, commons, hedges, ditches, fences, ways, waters, water-courses, liberties, privileges, easements, rights, and advantages whatsoever, appertaining or reputed to appertain to the land, or any part thereof, or, at the time of conveyance, demised, occupied, or enjoyed with, or reputed or known as part or parcel or appurtenant to the land or any part thereof.

This is, of course, another way of putting the point that the transfer, as it stood, passed the benefit of the covenant. Mr Weaver referred to the well-known statement in *Rogers* v *Hosegood* [1900] 2 Ch 388 at p 395:

> Covenants which run with the land must have the following characteristics: (1) They must be made with a covenantee who has an interest in the land to which they refer. (2) They must concern or touch the land. It is not contended that the covenants in question in this case have not the first characteristic, but it is said that they fail in the second. I am of opinion that they possess both. Adopting the definition of Bayley J in *Congleton Corporation* v *Pattison* (1808) 10 East 130, 135, the covenant must either affect the land as regards mode of occupation, or it must be such as *per se*, and not merely from collateral circumstances, affects the value of the land.

And he submitted that the covenant of guarantee affected the value of the land just as much as the covenant to pay rent. But this submission was well answered by Mr Pymont for the liquidator by

A the citation of *London & County (A & D) Ltd* v *Wilfred Sportsman Ltd* [1971] Ch 764. The precise facts of that case do not matter, but the question arose whether rent had been paid when, rent not having been paid, the guarantor in fact paid up. The answer returned by the Court of Appeal was in the negative. At pp 780-1 Russell LJ (as he then was) said:

> The plaintiffs, first, say that all Miah's rent from his first direct default from the March quarter day 1964 up to and including the June 1965 quarter day has in fact been paid, inasmuch as the payments by the third party to Miah's then landlord under the guarantee must be considered as made in satisfaction of Miah's rent or on behalf of Miah. Consequently there has never been an occasion of forfeiture. The judge did not accept that proposition, and neither do I. The payments were of sums under the liability assumed by the third party
> B when it guaranteed to the landlord that Miah would fulfil his obligations under the lease. Of course they were payments made because of Miah's failure (or expected failure) to meet his rent obligations, and in that sense were made in respect of Miah's rent. But in law they were nothing but payments under the guarantee in satisfaction of the third party's contractual obligation thereunder and it is not possible to infer that they were made or accepted as agent for Miah so as to release Miah from his obligation to pay rent as such to his landlord. Our attention was drawn by the plaintiffs to phrases in solicitors' letters which, they said, supported the submissions that the payments were made in satisfaction of rent or on behalf of Miah: but I do not consider that the use of such phrases in such correspondence can define the legal characteristics of such payments. We were referred to *Pellat* v *Boosey* (1862) 31 LJCP 281, not as covering this case, but as indicating that there may be circumstances in which payment by a third party stranger to the lease may be taken as payment
> C of rent or in satisfaction of rent under the lease. Maybe so; but the case is of no assistance to the plaintiffs here, and I need not analyse the facts, which were wholly different. Consequently, in my judgment, the rent under Miah's lease was at all relevant times unpaid.

Mr Moss had earlier analysed the position accurately. He said that the lease really embraced two separate contractual arrangements: one between the landlord and the tenant, and the other between the landlord and the guarantors. And the second contract of course had as its subject-matter the performance of the covenants in the first, but that was its subject-matter — namely the performance of covenants — and in no way did it itself touch and concern the land. I suppose one might just say that the covenant of guarantee touched and concerned covenants for the payment of rent etc which themselves
D touched and concerned the land. But Mr Weaver cited no authority for including covenants at this double remove in the terms of section 62(1) of the Law of Property Act 1925.

Mr Weaver's next submission was under section 56 of the Law of Property Act 1925. That reads as follows:

> (1) A person may take an immediate or other interest in land or other property, or the benefit of any condition, right of entry, covenant or agreement over or respecting land or other property, although he may not be named as a party to the conveyance or other instrument.

Mr Weaver here fastened upon some words used by Mr Nugee QC in a case called *Pinemain Ltd* v *Welbeck International Ltd*, to which I shall refer later, and submitted that CIS was an existing and
E identifiable non-party who was therefore in a position to take the benefit of the covenant of guarantee which was a covenant respecting land and who accordingly did so. I am afraid that this submission is extremely wide of the mark. First, as already explained, the covenant in question is not one respecting land; its subject-matter is covenants, not land. But more to the point, it completely fails to understand the scope of section 56. This section is dealing with the ability of parties to confer, by their conveyance or other instrument, rights of certain kinds on third parties. For this purpose, it is essential that the third party, upon whom rights are thus sought to be conferred, should be in existence and identifiable. One cannot confer such rights on the third son of A hereafter to be born, or upon the first counsel who next comes through the swing doors of Court 17. But where section 56 fails
F to connect at all with the present case is that at no stage has anybody expressed a desire to confer a benefit on CIS. Of course, they may always have been potential persons deriving title under Townsons, but the covenant of guarantee as it stood did not confer any rights on them at all; it conferred rights upon Townsons, which Townsons were free to assign (or not to assign) to any person, whoever that person might have been, who derived title under them. The guarantors were covenanting with one person — Townsons — not with two, namely Townsons and CIS. All this is elementary, but if weighty authority be required perhaps the best place to seek it is in the speech of Lord Upjohn in *Beswick* v *Beswick* [1968] AC 58 at p 106:

> First, let me assume for a moment that the agreement in this case is an indenture *inter partes* under seal — does section 56 help B? Plainly not. C did not purport to covenant with or make any grant to B; he only covenanted with A. Had C purported to covenant with B to pay the annuity to B, though B was not a party, then any difficulty B might have had in suing might be saved by section 56. The narrow view which I take of section 56 is, I think, supported by the observations of Simonds J (as he then was) in *White* v *Bijou Mansions* when he said at p 625: "Just as under section 5 of the Act of 1845 only that person could call it in aid who, although not a party, was a grantee or covenantee, so under section 56 of this Act only that person can call it in aid who, although not named as a party to the conveyance or other instrument, purports to grant something or with which some agreement or covenant is purported to be made." So to the same effect Wynn-Parry J in *In re Miller's Agreement* [1947] Ch 615. That was another example of the familiar case where, upon the dissolution of a partnership, the continuing partners covenanted with the retiring partner to pay as from his death annuities to his three daughters. The learned judge said: "In my view, the plaintiffs [the daughters] are not persons to whom the deed purports to grant something, or with whom some agreement or covenant is purported to be made . . . "

Having now dealt with all Mr Weaver's arguments on construction, I need, I think, refer only extremely briefly to two cases which bear out the totality of the views I have already expressed upon his submissions. They are, first, a case from Australia, *Sacher Investments Pty Ltd* v *Forma Stereo Consultants Pty Ltd* [1976] 1 NSWLR 5. The headnote of that case reads in part as follows:

> Two companies were the parties to a lease registered under the Real Property Act 1900. In the lease "the lessor" was defined to mean and include "the lessor, its successors and assigns . . . ". The two directors of the lessee executed the memorandum of lease as guarantors "for the payment of all moneys and performance of all obligations on the lessee's part therein contained". In September 1974 the first company transferred the reversion expectant upon the lease to a third company hereinafter called "the lessor". Both guarantors were aware of this transfer at the time when it took place. There was, however, no express assignment of the benefit of the guarantees. . . . Held . . . (3) In order that the lessor might sue upon the guarantees supposing them to be debts or other legal choses in action, it was necessary pursuant to section 12 of the Conveyancing Act 1919 that there should have been an assignment in writing of the benefit thereof together with appropriate notice to the guarantors and there was no evidence of either. Nor did sections 51 and 52 of the Real Property Act 1900 operate to cure this position. . . . Upon the supposition that the benefit of the guarantees (any assignment of which must in the present case be deemed to have taken place at a time when no rent was owing and when it was completely uncertain whether any rent would, or would not, be owing in the future) was not a legal chose in action within section 12 of the Conveyancing Act, and that the assignee of the benefit of a guarantee may enforce it against a debtor, even though no notice of the assignment had been given, nevertheless, there was no evidence of any assignment of any type however informal. . . . The fact that the guarantors had covenanted with the original lessor "its successors and assigns" (including the transferee of the reversion of the lease) did not of itself without more entitle the latter to enforce the guarantee. *Griffith* v *Pelton* distinguished. 6. Section 117 of the Conveyancing Act

which corresponds to section 62 of the English Act

> did not assist the lessor, because it does not deal with collateral obligations such as that with which the present case was concerned. 7. Section 36(c)

corresponding to section 56

> of the Conveyancing Act did not assist the lessor. *Beswick* v *Beswick* referred to.

I would be perfectly content to adopt the whole of Yeldman J's careful judgment as my own: in substance he comes to precisely the same answers upon precisely the same submissions as Mr Weaver has raised before me. In particular, it will be observed that he also regards *Griffith* v *Pelton* as affording no guidance in this area of the law. The second case is *Pinemain Ltd* v *Welbeck International Ltd* (1984) 272 EG 1168 where exactly the same points as have been raised before me were raised in virtually precisely the same circumstances and rejected by Mr Nugee QC sitting as a deputy judge of this division. My conclusions on this aspect of the case are therefore backed up by the state of existing authority, by which I should in any event be content to be guided, but which accord precisely with my own conclusions.

However, Mr Weaver was by no means finished. His next submission was that CIS was entitled to rectification of the transfer. He submitted, perfectly correctly, that it was no ground for refusing rectification that one of the parties to a transaction, or their solicitor, had been careless. But what is the situation here? In the first place, it must be borne in mind that rectification is a very special remedy, and although views have differed from time to time as to the precise nature of the evidence required to secure rectification, it has been asserted that there is no known case of rectification having been established on oral evidence only. Be that as it may, in the present case clearly Townson and CIS have nothing to lose from rectification. Indeed, by being willing to concede it, Townson may put themselves in the good books of CIS, which, as a body controlling large sums of money, is a very useful contact for a builder to have. It is quite obvious that the only person interested in opposing rectification would be P & O. In these circumstances, the case for rectification would have to be made out with absolutely convincing clarity.

The most obvious way of doing it would be by producing correspondence dealing with the covenant of guarantee, or a contract — presumably there was a contract in the present case — dealing with or referring to the contract of guarantee, and which it can clearly be shown was not carried into effect according to its tenor by the transfer. None of this would be difficult to put in evidence; a trawl through the files of CIS would, if such documents in fact do exist, readily produce them.

Instead, the only evidence that is offered, and it is not really any evidence at all, is a telex from Townson to Mr Weaver himself (he was only used as an addressee as a matter of convenience, and nothing turns upon that, he being definitely in existence and identified). This telex is in the following form:

> We hereby acknowledge that if and insofar that the benefit of the guarantee provided in the lease made the 6th of March 1970 between Townson Developments Ltd as lessor, Distributors and Warehousing Ltd as a lessee, P & O Steam Navigation Company and Yangzekiang (UK) Ltd as guarantors was not assigned to CIS Ltd by reason of the transfer of the freehold reversion dated 9th March 1970. We have always regarded ourselves as holding the benefit of such a guarantee for the benefit of CIS Ltd and will execute any requisite assignment if called upon to do so.

This is really a most confused telex. It says that "if and so far as there was no assignment, we have always regarded ourselves as holding the benefit of the guarantees for CIS". But this is not the way in which people behave in the real world. It would have been logical to have said something like "If the guarantees were not assigned, that was not what we intended, and we would now be willing to assign them forthwith". But if they really did intend to assign the benefit of the guarantees, and thought they had done so, the question of trusteeship would never have crossed their minds, let alone resulted in their "always regarding" themselves as trustees. So, in itself, the terms of the telex are suspicious. But the crucial point is, of course, that no evidence in the form of contracts, memoranda, letters or whatever, of the crucial date are offered in support of the allegation of trusteeship. In the circumstances, it appears to me quite clear that I must proceed on the footing that there is no real prospect of CIS obtaining rectification.

Mr Weaver then submitted that Townson could still assign the benefit of the guarantee to CIS, who would then be in a position to enforce it. It is, of course, obvious from the terms of the telex that such assignment is a possibility which must be taken seriously. However, it appears to me quite clear that, having once duly assigned the lease without the benefit of the covenant of guarantee, it is no longer possible for Townson to act as Mr Weaver suggests with any meaningful result. I think that there are really two strands to this conclusion, which mesh together. One is that it is quite impossible for an assignor to assign anything more than that of which he is himself at the moment seized. Townson at the moment has a covenant of guarantee which is, so far as it is concerned, completely worthless and unenforceable: it is elementary that in such a case it cannot confer upon an assignee any better title than it has itself. I think the simplest way of appreciating this is by imagining that there was now an equitable, instead of a legal, assignment of this legal chose in action. Any action would then have to be brought in the name of the assignor — Townson — and their inability to sue would have been manifest. Second, if Townson are left with anything at all, they are left with a bare right of action: and once again it is elementary that although of course rights of action may be assigned together with the property to which they relate, they cannot be assigned apart therefrom: see *Halsbury's Laws of England*, 4th ed, vol 6, para 86.

The parallel strand of argument is that the guarantee is only taken to make the payment of the rent more secure, a circumstance which would, from the point of view of the lessor considered as a potential assignor of the reversion, make it possible for him to dispose of the lease to better advantage than if the covenants were unsupported. If the lessor then manages to dispose of the reversion without the necessity of assigning the covenants of guarantee, he has no further use for the covenants and they ought not thereafter to be assignable. There appears to be no direct decision in the books, but in the case of *Miles* v *Easter* [1933] Ch 611 it was held that where a vendor had taken G the benefit of a restrictive covenant for the benefit of certain land, but had disposed of the whole of that land without assigning the benefit of the covenant to any purchaser, he could not, thereafter, validly assign the benefit of the restrictive covenant. Romer LJ (as he then was) had this to say at p 632:

> In the third place, the covenant cannot be enforced by the covenantee against an assign of the purchaser after the covenantee has parted with the whole of his land. This last point was decided, and in our opinion rightly decided, by Sargant J in *Chambers* v *Randall* [1923] 1 Ch 149. As pointed out by that learned judge, the covenant having been entered into to enable the covenantee to dispose of his property to advantage, that result will in fact have been obtained when all that property has been disposed of. There is therefore no longer any reason why the Court should extend to him the benefit of the H equitable doctrine of *Tulk* v *Moxhay*. That is only done when it is sought to enforce the covenant in connection with the enjoyment of land that the covenant was intended to protect. But it was also held by Sargant J in the same case, and in our opinion rightly held, that although on a sale of the whole or part of the property intended to be protected by the covenant the right to enforce the covenant may be expressly assigned to the purchaser, such an assignment will be ineffective if made at a later date when the covenantee has parted with the whole of his land. The covenantee must, indeed, be at liberty to include in any sale of the retained property the right to enforce the covenants. He might not otherwise be able to dispose of such property to the best advantage, and the intention with which he obtained the covenant would be defeated. But if he has been able to sell any particular part of his property without assigning to the purchaser the benefit of the covenant, there seems no reason why he should at a later date and as an independent transaction be at J liberty to confer upon the purchaser such benefit. To hold that he could do so would be to treat the covenant as having been obtained, not only for the purpose of enabling the covenantee to dispose of his land to the best advantage, but also for the purpose of enabling him to dispose of the benefit of the covenant to the best advantage. Where, at the date of the assignment of the benefit of the covenant, the covenantee has disposed of the whole of his land, there is an additional reason why the assignee should be unable to enforce it. For at the date of the assignment the covenant had ceased to be enforceable at the instance of the covenantee himself, and he cannot confer any greater rights upon the assignee than he possessed himself.

I think this, *mutatis mutandis*, applies to the present situation.

Accordingly, it seems to me that CIS has not made out a case for the proposition that they are entitled to the benefit of the guarantee of P & O — and, for what it is worth, that of Yangzekiang. I think K therefore that the result which should follow — because I have not had my attention called to any other circumstance which ought to result in the refusal of the relief claimed by the originating summons herein — is that the liquidator should be granted leave to disclaim. I do not think that any consequential directions are required, save and except for this, that on an earlier hearing of this originating summons, on June 19 1985, an adjournment took place at the request of CIS, but upon the terms embodied in draft minutes of order, which read as follows:

> Upon the undertaking by Co-operative Insurance Society Ltd by its counsel that if the court gives leave to disclaim in respect of the lease the subject matter of the liquidator's summons herein dated September 28 1984 Co-operative Insurance Society Ltd will as between itself on the one hand and the above- L named company and the Peninsular & Oriental Steam Navigation Company Ltd on the other hand treat the lease as having been effectively disclaimed on June 19 1985 and the disclaimer as having been duly filed on June 19 1985 and in particular will treat any alleged liability of Peninsular & Oriental Steam Navigation Company as alleged guarantor or part guarantor of the lease as having terminated as at June 19 1985.

I think that that undertaking by CIS should be formally re-embodied in the order which I am now making.

So much for the way in which the matter appears to me at the moment. But let it be supposed that I am wrong on my main supposition, and that CIS ultimately does establish that it is entitled to the benefit of the guarantees in question. Does any different result follow? In my judgment, on the very peculiar facts of the present M case, it does not. In order to demonstrate this, let it be assumed that no order for disclaimer is granted. What follows?

It was quite clearly envisaged by Maugham J (as he then was) in *Re Katherine et Cie Ltd* [1932] 1 Ch 70 at pp 78, 79 that the refusal of an order for disclaimer does not in any way inhibit the liquidator in proceeding with the usual train of events which normally ensue in the case of a liquidation — namely, the calling of a final meeting and ultimate dissolution of the company.

Once that took place, what would the result be? As pointed out by Maugham J in the case cited, having regard to an alteration of the law made by the 1929 Act, the lease would become *bona vacantia*, and, although it is a vesting which has some peculiar incidents, the lease would vest in the Crown. However, for present purposes this is not particularly material: what is the crucial point is that the lease would cease to be vested in the company. Indeed, nothing can be vested in a dissolved company.

Accordingly, having regard to the precise terms of the guarantee, which is expressly tied to the lease remaining vested in the company, it would come to an end, and it would come to an end pursuant to its own express provisions.

Faced with this situation in the course of his argument, Mr Weaver first of all sought to place a construction upon clause 9 of the lease which it patently would not bear; but then, more realistically, he pointed out that in those circumstances CIS could, as a person interested, make an application to the court to declare the dissolution of the company void, under provisions of the Companies Act 1948, section 352. Mr Pymont pointed out that Mr Weaver had cited no authority whatsoever to show that the court would be likely to make such an order, and submitted that Mr Weaver's submission that CIS would have suffered irretrievable loss — a submission which might incline the court to be sympathetic towards the making of such an order — were wide of the mark, because any detriment which CIS suffered as a result of the dissolution of the company would have been caused by its having the benefit of a covenant condition to last only so long as the company lasted. The detriment, in other words, arises from the form of the covenant, which would nevertheless have been performed to the letter down to the date of dissolution.

In other words, the only thing that would suit Mr Weaver's purpose would be the preservation of the rotting hulk of the company upon the broad stream of ordinary commerce for no purpose or benefit involving the company, but merely to enable CIS to take one-sided advantage of a contract of guarantee. I say one-sided advantage, for I think that P & O would equally be in a position to say that artificially to prolong the existence of the company for the purpose of subjecting them to liability under a covenant expressly designed, so far as they were concerned, not to outlast the life of the company would be to subject them to very grave hardship indeed — hardship they have no prospect whatsoever of mitigating at any stage, unlike CIS which could at any stage — and undoubtedly would if the demised premises could be relet at over half whatever the current rental value is — forfeit the lease for non-payment of rent and thus get the property back in hand and relet it.

But there are other objections to the life of the company being artificially prolonged in this way. There would be the continued costs of keeping open the liquidation: where are these to come from? It would be an intolerable burden upon the preferential creditors of the company if the funds available for them had to be milked for the benefit of a non-preferential creditor. And besides such administrative costs, there would be other costs. For although the liquidator would clearly not be personally liable in respect of the covenants in the lease, he would have to ensure that no claims in respect of negligence or nuisance could possibly be brought against him personally. Indeed, he has between November 21 1984 and August 31 1985 spent some £1,697.70 in inspecting and securing the premises and clearing rubbish deposited by third parties away therefrom.

Mr Weaver submitted that this point could be met by my imposing, as a term of the court's order, conditions upon his client which would ensure that all such disbursements by the liquidator would be met at his client's expense. I am not sure if he was really ready on his client's behalf to fund the continued remuneration of the liquidator, but let that pass. However, I do not think that the court has any such power. Subsection (3) of section 323 reads as follows:

> The court before or on granting leave to disclaim, may require such notices to be given to persons interested, and impose such terms as a condition of granting leave, and make such other order in the matter as the court thinks just.

In my view, the concluding words "make such other order in the matter as the court thinks just" are linked to the opening words "The court, before or on granting leave to disclaim". I do not think it can have occurred to the framers of that section that if leave was refused any other order than that would be required.

There is, however, a much more fundamental objection to the court sanctioning the continuance, indefinitely, of a company which ought to be dissolved. It is that it is clearly the policy of the legislature, and indeed most obviously in the public interest, that defunct companies should be removed from the register. Assuming that the company were dissolved, and that Mr Weaver succeeded in his application to have it restored to the register, this would not in any way mean that the company was carrying on business or in operation. Accordingly, it appears to me that, upon discovering this fact, the registrar would be fully entitled to strike it off the register, so that once again the company will be dissolved.

Accordingly, it appears to me that even if I have wholly misapprehended the position in relation to where the benefit of the guarantees now lie, the end result should not be any different. The liquidator must, as I have already indicated, be given leave to disclaim, CIS in any event being ultimately defeated by the circumstances that the guarantee in question is one conditioned to last, and only last, so long as the lease remains vested in the company.

For a further case on this subject see p 240

LANDLORD AND TENANT

BUSINESS TENANCIES

Court of Appeal

November 4 1985

(Before Lord Justice ACKNER, Lord Justice MUSTILL and Lord Justice NOURSE)

PHILPOTS (WOKING) LTD v SURREY CONVEYANCERS LTD AND ANOTHER

Estates Gazette January 11 1986

277 EG 61-62

Landlord and tenant — Rent review clause in lease — Appeal from decision of deputy judge on a difficult point of construction — The initial rent was £8,000 a year — Rent review clause provided that the reviewed rent was to be "a sum equal to the aggregate of the said sum of £8,000 plus the amount (if any) by which the fair rack rental value . . . at the date of the review shall exceed the sum of £8,000 or the yearly rent fixed at the previous date of review as the case may be" — There was a proviso that the rent payable after the date of review should not be less than that payable immediately before that date — The deputy judge decided in favour of the landlord's construction, which gave a normal commercial meaning to the clause, so that the reviewed rent would always be the equivalent of the fair rack rental value at the date of review, or, if higher, the rent payable immediately before the review date — Tenant's construction meant that the rent payable on first review would be equivalent to the fair rack rental value but that payable on every subsequent review would always fall short of such value (assuming that the latter was always on the increase) — On examination of the wording "without any preconceptions" the Court of Appeal preferred the tenant's construction, although it led to admittedly unpredictable and perhaps capricious, but not absurd, results — The words "as the case may be" did not mean "whichever is the higher" — The court did not accept the proposition that the parties to a commercial lease invariably intended that the rent should not in real terms fall below the market rent initially agreed upon — "It is perfectly possible for them to assume that the fair rack rental value will always increase, but none the less to intend that the rent payable from the second review onwards shall always be less" — Landlords had failed to persuade the court that the tenant's construction was absurd — Tenant's appeal allowed

This was an appeal by the tenant, Surrey Conveyancers Ltd, and the surety, Jonathan Howard Roberts, from a decision of Mr Michael Wheeler QC, sitting as a deputy judge of the Chancery Division, in favour of the landlord, Philpots (Woking) Ltd, the present respondents, on the construction of the rent review clause of a lease of office accommodation on the first floor of 1 Guildford Road, Woking, Surrey.

David Neuberger (instructed by Park Nelson & Doyle Devonshire, agents for Mossops, of Woking) appeared on behalf of the appellants; Roger H T Smith (instructed by Jansons, agents for Lester Perring & Co, of Hastings) represented the respondents.

Giving the first judgment at the invitation of Ackner LJ, NOURSE LJ said: This is an appeal from a decision of Mr Michael Wheeler QC, sitting as a deputy judge of the Chancery Division, given on May 7 1985. It raises what the learned judge correctly described as a short, but far from easy, question of construction on a rent review provision in a lease. F

The lease is dated October 20 1981. It was made between Philpots (Woking) Ltd as landlord of the first part, Surrey Conveyancers Ltd as tenant of the second part and Mr J H Roberts as surety of the third part. It is a lease of office accommodation on the first floor of 1 Guildford Road, Woking, Surrey, for a term of 25 years from June 24 1981 at an initial yearly rent of £8,000.

The reddendum itself gives no hint of possible variations in the rent. The review provisions are contained in clause 5 of the lease. G Subclause (1) (i) of that clause provides that the lessor shall have the right to review the yearly rent for the time being payable at the end of each five years of the term ("the date of review") in manner thereinafter provided. Subclause (1) (ii) provides for the exercise of the right by notice in writing. Subclause (1) (iii) is in these terms:

> If the Lessor shall give the Lessee such notice as aforesaid then from and after the date of review this Lease shall be read and construed and shall take effect in all respects as if the yearly rent for the time being payable hereunder (instead of being the sum of Eight thousand pounds) had from the date of review been a sum equal to the aggregate of the said sum of Eight thousand pounds plus the amount (if any) by which the fair rack rental value (as hereinafter defined) of the demised premises at the date of review shall exceed the sum of Eight thousand pounds or the yearly rent fixed at the previous date H of review as the case may be but without prejudice to any of the other terms and conditions contained in this Lease and so that in no event shall the rent payable by the Lessee to the Lessor after the date of review be less than the rent payable by the Lessee to the Lessor immediately before the date of review.

By subclause (1) (iv) it is provided, shortly stated, that the fair rack rental value of the premises shall be the open market rental at the date of review (as agreed between the parties or determined by arbitration), there being disregarded certain matters which are usually disregarded in cases such as this.

The difficulty arises out of the terms of clause 5(1) (iii). In order to explain the rival constructions, I set out the material words again, with alternative number and indentations as put in by Mr Neuberger, who appears for the tenant and the surety. J

The landlord submits that the reviewed rent should be:

> A sum equal to —
>
> (1) The aggregate of the said sum of £8,000 plus the amount (if any) by which the fair rack rental value . . . at the date of review shall exceed the sum of £8,000;
>
> or
>
> (2) The yearly rent fixed at the previous date of review as the case may be . . . and so that in no event shall the rent payable . . . after the date of review be less than the rent payable by the Lessee to the Lessor immediately before the date of review.

The tenant and the surety submit that the reviewed rent should be:

> A sum equal to the aggregate of the said sum of £8,000 plus the amount (if any) K by which the fair rack rental value . . . at the date of review shall exceed the sum of:
>
> (1) £8,000
>
> or
>
> (2) The yearly rent fixed at the previous date of review as the case may be . . . and so that in no event shall the rent payable . . . after the date of review be less than the rent payable by the Lessee to the Lessor immediately before the date of review.

The learned judge preferred the landlord's construction and made a declaration accordingly. Against that decision the tenant and the surety now appeal.

Neither construction gives rise to any arithmetical uncertainty. If the landlord's construction is correct, the reviewed rent will always

A be the equivalent of the fair rack rental value at the date of review or, if it is higher, the rent which was payable immediately before that date. If the tenant's construction is correct then, assuming that the fair rack rental value is always on the increase, the rent payable on the first review will be its equivalent, but that payable on every subsequent review will always fall short of it. An example was given based on the supposition that the fair rack rental value was £11,000 at the end of the fifth year and £17,000 at the end of the 10th year. At the end of the fifth year you take the aggregate of £8,000 and the amount by which the fair rack rental (£11,000) exceeds £8,000 (£3,000) and find that you are left with £11,000. At the end of the 10th year you take the aggregate of £8,000 and the amount by which the fair rack B rental value (£17,000) exceeds the yearly rent of £11,000 fixed at the previous date of review (£6,000) and find that you arrive at £14,000. A second example of the working of the tenant's construction was based on the supposition that the fair rack rental value was £11,000 at the end of the fifth year, but that it had fallen to £9,000 at the end of the 10th year. In that event the fair rack rental value at the latter date would not of course exceed the yearly rent fixed at the former, but the final words of the provision would then take effect and prolong the previous rent of £11,000.

The primary argument of Mr Neuberger on behalf of the tenant was to the following effect. The landlord's construction involves computing "the aggregate of . . . £8,000 plus the amount . . . by which the fair rack rental value . . . shall exceed the sum of £8,000". C You start with the figure of £8,000. You add to it the amount by which the fair rack rental value exceeds it. And then you subtract the same amount. You are left with the fair rack rental value. If that is what you want to say, why do you have to say it in this inexplicable and extraordinarily convoluted way? Mr Smith for the landlord accepted that criticism, but submitted that his construction does produce an intelligible result and is to be preferred to one which produces a capricious result.

Then Mr Neuberger turned to the words "as the case may be". He said that on the landlord's construction those words acknowledge that a choice is to be made between two figures, but give no guidance as to how it is to be made. Indeed, he went further and said that the landlord's construction gives no meaning at all to those words. They D could just as well have been omitted. Mr Smith's answer was that it is obvious, when the commercial object of the clause is borne in mind, that the figure chosen is to be the higher of the two. He said that the words "as the case may be" can be read either as themselves meaning "whichever is the higher" or as meaning that there is a choice to be made, the choice of the higher figure being required by necessary implication. Mr Neuberger replied that if the higher figure is to be chosen then the final part of the provision beginning with the words "and so that . . ." is superfluous. On the other hand the second example above stated demonstrates that on the tenant's construction those words are necessary. Mr Smith rejoined by saying that that showed that the tenant's construction treated the words introduced E by "and so that . . ." as a proviso, which is something which it is not expressed to be. On the landlord's construction they bear their natural meaning, acting as an explanation, albeit a repetition, of what has gone before.

Such were the principal arguments of Mr Neuberger and the answers which Mr Smith gave to them. The principal argument of Mr Smith was more broadly based. He said that the terms of the lease, which, be it remembered, is a commercial lease between a commercial landlord and a commercial tenant, indicate that the object of the rent review provision is the usual one of ensuring that the rent does not in real terms fall below the market rent initially agreed upon. The parties must be taken to have had some intelligible commercial object in mind in inserting those provisions. He said that F a steadily rising market would produce no rise on the second review and the same could, I think, be said of the fourth. Mr Smith added that if the tenant's construction is correct, the level at which rents were fixed after the first review would be a complete lottery, so that the provisions, although not unworkable, would produce absurd results. The tenant's construction should be adopted only if the words point so clearly and inexorably to it that no other reading is reasonably possible.

The arguments below were substantially the same as those which have been advanced in this court. The learned deputy judge, having carefully considered the opposing arguments, stated his conclusion at pp 9E to 10B-C of the transcript as follows:

> But it seems to me that I am forced to rely on relatively slight *indicia* in deciding between producing a result which seems to me to conform to a rent review clause in a commercial lease of this kind on the one hand and, on the other hand, a formula which produces unascertainable and indeed almost capricious results.
>
> I do not think this is a case where I can simply say: "Well there it is, it is quite plain what the words mean and I must give effect to them." This is not such a case in my judgment. I think this is on analysis a difficult provision to construe; but it seems to me both on grounds of probability and of construction that the landlord's contention is the better of the two with which I am faced, and that there is nothing in the clause itself which compels me, if I felt minded to do so (and I do not discount that), to accept the tenant's formula.

I think it very probable that, in accepting the landlord's construction, the learned judge has correctly assessed what the parties did indeed believe and desire to be the effect of clause 5(1) (iii) of the lease. But a court of construction can only hold that they intended it to have that effect if the intention appears from a fair interpretation of the words which they have used against the factual background known to them at or before the date of the lease, including its genesis and objective aim. It is not in this case suggested that there is any material background beyond the fact that the lease is a commercial lease between commercial parties.

Inherent in Mr Smith's principal argument is the proposition that commercial parties to a commercial lease *invariably* intend that the rent should not in real terms fall below the market rent initially agreed upon. I cannot accept that proposition. It is perfectly possible for them to assume that the fair rack rental value will always increase, but none the less to intend that the rent payable from the second review onwards shall always be less. There is nothing in the words of the lease to exclude that possibility, unless it be that the tenant's construction of them leads to results so absurd that the parties cannot be credited with an intention to achieve them. I do not think that it does. The learned judge described the results as unascertainable and almost capricious. By the first of these he meant that they were unpredictable. So indeed they are. But that is not, I think, enough. I am not sure whether I would describe them as capricious. They may be. It is a word which means different things to different people. But for myself I would not think that capriciousness was enough. I think that Mr Smith must show that the results are absurd, and of that he has failed to persuade me.

Once Mr Smith's principal argument has been rejected, so that the words of clause 5(1) (iii) can be examined without any preconceptions as to their intent, I find that the arguments of Mr Neuberger, as I have stated them, must prevail. His primary argument is a formidable one which speaks for itself. The tenant's construction gives full meaning and effect to the words "as the case may be". Those words do not mean "whichever is the higher" and the rejection of Mr Smith's principal argument defeats an implication to that effect. The tenant's construction finds employment for the words introduced by "and so that . . ." and prevents them from being a mere repetition of what has gone before. The only point which is left to sustain the landlord's construction is the explanatory, not restrictive, meaning of those words. I have felt the force of that point, but not strongly enough for it to overgo the points which are against it.

For these reasons, although in a full understanding of how the learned deputy judge came to be of the contrary view, I am of the opinion that the tenant's construction is correct. I would therefore allow this appeal.

ACKNER and MUSTILL LJJ agreed and did not add anything.

The appeal was allowed with costs in the Court of Appeal and below.

Chancery Division

November 5 1985

(Before Mr Justice WARNER)

DATASTREAM INTERNATIONAL LTD v OAKEEP LTD

Estates Gazette January 11 1986

277 EG 66-70

Landlord and tenant — Construction of rent review clause in lease of office block — The issue was whether the hypothetical lease assumed for the purpose of determining the market

rental value at the review date was or was not to include the provisions for rent review contained in the actual lease — This issue, canvassed in a number of authorities with a variety of wording, depended in the present case on the meaning of the words "other than the amount of rent hereby reserved" — Lessors contended that the provisions for rent review were to be excluded in the hypothetical lease, thus enhancing the rental value at the review date, while the lessees argued for the opposite conclusion — Apart from authority the judge said that he would prefer the meaning that did not produce an unreasonable and unfair result, namely one which did not require the lessees to pay, over the next five years, for a benefit which they would not in fact be enjoying, the benefit of a lease which contained no rent review clause — The matter was not, however, free from authority and the judge considered the cases of *Lister Locks Ltd* v *T E I Pensions Trust Ltd*, *Pugh* v *Smiths Industries Ltd*, *Pearl Assurance plc* v *Shaw*, *National Westminster Bank plc* v *Arthur Young McClelland Moores & Co* and *Equity & Law Life Assurance Society plc* v *Bodfield Ltd* — In the end each case turned upon the precise terms of the lease in question and Warner J did not feel constrained by any of the authorities to depart from the view which he would have formed independently of them — Accordingly, he made a declaration in the sense sought by the tenants, to the effect that the provisions for rent review were not excluded from the terms of the hypothetical lease

This was an originating summons by which the plaintiff lessee, Datastream International Ltd, sought a declaration as to the correct construction of the rent review clause in the lease of an office block called Monmouth House, in City Road, Islington, London EC1, of which the defendant, Oakeep Ltd, was the present lessor.

Michael Barnes QC and Kim Lewison (instructed by Herbert Smith & Co) appeared on behalf of the plaintiff; Roger Ellis (instructed by Julian Holy) represented the defendant.

Giving judgment, WARNER J said: This is an originating summons raising a question of construction of a rent review clause in a lease. The lease is dated September 12 1980. It is a lease of a building — an office block I understand — called Monmouth House, in City Road, Islington, for a term of 25¼ years from March 25 1980. The plaintiff is the lessee. The defendant is the successor in title to the lessors.

The provisions of the lease as to rent are these. In clause 2 one finds the words:

> Yielding and paying therefor unto the Lessors during the said term the rents reserved and payable as set out in the Second Schedule hereto payment thereof to be made by equal quarterly payments in advance on the usual quarter days in every year.

The second schedule is headed "Rents". Para 1 of it, which has the side heading "Reddendum", provides:

> The rents reserved and payable are as follows: (1) From the 25th day of March 1980 until the 3rd day of December 1980 a peppercorn (2) From the 4th day of December 1980 until the 23rd day of June 1985 the clear yearly rent of £360,000 (3) During the next five-year period of the term either the said clear yearly rent of £360,000 or a yearly sum equal to the market rental value of the said property at the expiration of the immediately preceding period whichever is the greater (4) During each of the several subsequent consecutive five-year periods of the said term (and during the remainder thereof) either the clear yearly rent payable during the immediately preceding five-year period or a yearly sum equal to the market rental value of the said property at the expiration of the said preceding five-year period whichever in each case is the greater.

Para 2 provides:

> For the purpose of the provisions of this Schedule the expression "market rental value" shall mean the rent at which the said property is worth to be let in the open market as a whole at the best rent reasonably obtainable without taking any fine or premium and on the following bases: (1) Subject to the provisions of this Lease (other than the amount of rent hereby reserved) for a term equal to the residue then unexpired of the term granted by this Lease but having regard to the Lessee's rights under the Landlord and Tenant Act 1954.

There follow subparas (2), (3) and (4) which prescribe further assumptions that are to be made, such as that at the commencement of the relevant period the property is fit for immediate occupation and business use, that the covenants contained in the lease have been fully performed and observed and that the property is used for purposes specified in the lease. Then there is a provision for three matters to be disregarded. They are (1) any effect on rent of the fact G
that the lessee has been in occupation of the property, (2) any goodwill attached to the property by reason of the carrying on there of the business of the lessee or its predecessors in title in that business and (3) any increase in the market rental value of the property attributable to improvements carried out at the lessee's expense. Para 3 confers a special option on the lessors which is not relevant for present purposes. Para 4 provides the machinery for the fixing of the revised rent. It is machinery of the sort that one normally finds in rent review clauses: here, in the absence of agreement between lessors and lessee the revised rent is to be determined at the option of the lessors either by an arbitrator or by an independent chartered surveyor acting as an expert and not as an arbitrator. Finally para 5 lays down what payments the lessors are to make if the revised rent is not H
ascertained at or prior to the relevant review date.

So the first rent review date was June 23 of this year and it is in relation to that that these proceedings have been taken. The parties have not yet gone to arbitration or to the independent surveyor. Before they do they want the court to decide the question of construction that arises.

That question is whether the hypothetical lease by reference to the terms of which "the market rental value" of the property at the review date is to be fixed is or is not to include the provisions for rent reviews themselves that are contained in the actual lease. The answer to that question turns on the meaning of the words in brackets in para 2(1) of the Second Schedule, that is to say the words "other than the J
amount of rent hereby reserved". The defendant contends that the effect of those words is to exclude the provisions for rent reviews from the hypothetical lease, with the result that the lessee must pay a rent enhanced by the fact that, upon this hypothesis, he will be enjoying the privilege of occupying the property for 20 years at that rent. The plaintiff on the other hand contends that the words in brackets are not apt to exclude from the hypothetical lease the provisions for rent review in the actual lease and that therefore it gets the benefit of a lower rent because account must be taken of the fact that in five years' time there will be another rent review and so on.

I have no doubt that if the question were free from authority I would decide it in favour of the plaintiff. The crux is really this. Does the word "amount" in the parenthesis mean only a fixed amount or K
does it include also any amount to be ascertained by means of a formula? Obviously if it includes only a fixed amount it includes for the purposes of the present review only the peppercorn (if one can call that an amount at all) and the £360,000. If it can include an amount to be ascertained by the application of a formula then it will also include so far as para 1(3) is concerned the "yearly sum equal to the market rental value of the said property at the expiration of the immediately preceding period" and so far as para 1(4) is concerned the "yearly sum equal to the market rental value of the said property at the expiration of the said preceding five-year period". Of course, on the narrow construction, according to which "amount" means only a fixed amount, at the next rent review the amount fixed at the present rent review would fall to be treated as an "amount". Faced with that L
choice as to the meaning of the word "amount", I prefer the meaning that does not produce an unreasonable and unfair result, that is to say does not require the lessee to pay over the next five years for a benefit it will not be enjoying, namely the benefit of a lease with no rent review clause in it.

But the matter is not free from authority. In *Bernstein and Reynolds' Handbook of Rent Review* at p S11 of the August 1984 service, there is a paragraph headed "Whether there are reviews during the hypothetical term" in these terms:

> Most modern rent review clauses require the valuer to assume that there is to be a hypothetical letting on the review date, on the same terms and conditions as those contained in the actual lease. However, the actual amount of the rent M
> payable immediately prior to the review date is virtually always expressly excluded from the terms of the hypothetical lease. Even if there is no express provision, it would almost certainly be implied since otherwise the review would be unworkable. It is also usual to provide, in cases where the hypothetical term is longer than the relevant review period, that the hypothetical lease is to contain a provision for review according to the same pattern as the actual lease. This is probably the effect of the usual stipulation that the terms of the hypothetical letting should be the same as the actual lease "other than as to rent" or the other common variant "other than as to the amount of rent". It is, of course, open to the parties expressly to provide that there are to be assumed to be no reviews in the hypothetical lease. This was held to be the effect of the clear words of the leases considered by the court in *Pugh* v *Smiths Industries Ltd* (1982), and *French Kier Property Investments* v *Marconi Co Ltd* (1982).

So I start with this, that the provision which I find in the present lease is of a fairly common kind and that, according to the learned authors of that work, it probably has the effect of leaving the rent review clause to be included in the provisions of the hypothetical lease.

Of the decided cases to which I must refer, the first chronologically is *Pugh* v *Smiths Industries Ltd.* It is reported at (1982) 264 *Estates Gazette* 823 and it is a decision of Goulding J. In that case the relevant words were in a subclause of the rent review clause defining what was there called "the full yearly open market rent". That subclause was in these terms:

> The full yearly open market rent of the property for the purposes of this clause is that rent (exclusive of rates and other payments (if any) to be made by the lessee by virtue of this subunderlease) which would be obtainable upon the day three months before the relevant material date if this subunderlease was not then subsisting upon a letting with vacant possession of the property for the residue of the term hereby granted and on the basis that the lessee would be obliged to perform and observe the covenants and conditions on the part of the lessee contained herein but excluding therefrom the provisions of this clause and disregarding those matters referred to in section 34 of the Landlord and Tenant Act 1954 as amended by section 1 of the Law of Property Act 1969.

The relevant words there "excluding therefrom the provisions of this clause" unambiguously excluded all the provisions of the rent review clause from the hypothetical lease. Goulding J found it easy to see that the lessees felt that to be unfair, but he held the meaning of the words to be quite plain. I do not see how he could have come to any other conclusion, but those words were very different from the relevant words in the present case.

The next case chronologically is *Lister Locks Ltd* v *T E I Pension Trust Ltd*, which is reported in *Estates Gazette* immediately after *Pugh*'s case.* It is there stated to have been a decision of "His Honour Judge Wheeler", but that is a mistake. It was a decision of Mr Michael Wheeler QC, sitting as a deputy High Court judge. In the *Lister Locks* case the lease provided for a "basic initial rent" of £14,982 per annum and an "extra initial rent" of £1,680 per annum. Those rents were payable for the first five years of the term. For every subsequent five-year period the "basic initial rent" was to be replaced by the "fair rack rental market value" and the "extra initial rent" was to be replaced by the "extra reviewed rent". The definition of the "fair rack rental market value" was, so far as material, as follows:

> The fair rack rental market value of the demised premises shall subject as hereinafter provided be the amount which shall be agreed between the landlord and the tenant (or in default of agreement the amount which shall be determined by assessment as hereinafter provided) to be the best rent at which the premises might reasonably be expected to be let for a term of years equivalent to the then unexpired residue of the term hereby granted in the open market by a willing landlord to a willing tenant and subject to similar covenants and conditions (other than the amount of the rent) to those contained in this lease, there being disregarded any of the matters which the court is required to disregard under the provisions of section 34 of the Landlord and Tenant Act 1954 as amended by section 1 of the Law of Property Act 1969.

The provision defining the "extra review rent" was in these terms:

> Upon such fair rack rental market value being agreed or certified as aforesaid the extra reviewed rent for every subsequent period of five years shall be ascertained in the manner following, that is to say by multiplying the extra initial rent by the said fair rack rental market value and dividing the result by the basic initial rent.

In other words, the extra initial rent was to increase in the same proportion as the basic initial rent.

There was no suggestion on the part of the landlord that the words "other than the amount of the rent" in the parentheses in the definition of "the fair rack rental market value" had the effect of excluding the entire rent review clause itself from the hypothetical lease. The contention of the landlord with which Mr Wheeler had to deal was that those words excluded from the hypothetical lease the obligation to pay the "extra review rent". As to that he said:

> I cannot accept this proposition. In my view, the words quoted are merely conveyancing overcaution. "Amount", in my judgment, means what it says, ie initially £14,892 (the "basic initial rent") and £1,680 (the "extra initial rent").

So Mr Wheeler took the word "amount" in that case to mean the fixed amounts, the amounts as to which figures were available. That authority, so far as it goes, is in the plaintiff's favour in the present case.

The next authority is also in the plaintiff's favour. That is the judgment of Vinelott J in *Pearl Assurance plc* v *Shaw* (1984) 274 EG 490*. That was again a case about a rent review clause, but the ambiguity there arose in relation to a provision concerning user. The passage in Vinelott J's judgment that is relevant to this case is as follows:

> The purpose of a rent review clause in general is to enable a landlord to bring the rent originally negotiated up to date and to substitute for it the rent that the parties might have been expected to agree if the rent had been negotiated on the same basis as before, but in the light of market conditions prevailing at the time of the review and, of course, for the shorter term then unexpired. Looked at in that light I think the court should lean against a construction which requires the rent fixed on a revision to be ascertained without regard to the use which, under the lease, the tenant is to be entitled to make of the demised premises, unless, of course, that intention is spelt out in reasonably clear terms. Otherwise, the effect of the review might be to impose on a tenant an obligation to pay a rent appropriate to a very profitable use, but one very obnoxious to the landlord, and one which he had been careful to forbid in the strongest possible terms — the effect, that is, of making the tenant pay for something which he not only has not got, but which he cannot require the landlord to give him.

Transferring that reasoning from a provision about user to the rent review clause itself, it suggests that I should lean against a construction which requires the tenant to pay for something which he has not got, that is a lease which is free from any rent review clause.

The next authority is the decision of Walton J in *National Westminster Bank plc* v *Arthur Young McClelland Moores & Co* (1984) 273 EG 402†, the judgment in which was delivered exactly one month after the judgment of Vinelott J in the *Pearl Assurance* case. That, unlike the last two authorities I have referred to, is in favour of the defendant, firstly because Walton J adopted an approach which was different from that of Vinelott J. Instead of leaning against a construction that would require a tenant to pay for something he was not to get, Walton J said this:

> The first and I think probably the main question which arises is whether, when the arbitrator is deciding what is to be the market rent for the next rent period, he is to do that upon the basis that the lease contains (as, of course, we know it does in fact contain) a rent review clause or whether, on the other hand, he is to fix the fair market rent on the basis that there is no such clause in the lease. I do not think that there is a presumption one way or the other. I think that in every case that must depend upon the precise terms of the lease, because it must be very much borne in mind that there is no such thing as a fair market rent of any premises in the abstract. There is only a fair market rent upon a set of abstractions which may be actual, may be hypothetical and in most cases under rent reviews are a mixture of the one and the other. For example, here we have that the premises are expected to be let with vacant possession, which is, of course, something which you know as a fact just is not the case because the tenants are actually in possession and are certainly not just going to move out for the purpose of a rent review. So the attempt by Mr Morritt to poison my mind in advance to achieve the lower of the two values by defining the fair market rent as something which favours the tenant, as in fact he was attempting to do, I do not think impresses me very much.

The words of the provision that Walton J had to construe were:

> "Fair Market Rent" means such amount as shall represent a yearly rent at which the demised premises might reasonably be expected to be let at the Relevant Review Date in the open market by a landlord to a tenant without a premium with vacant possession and subject to the provisions of this Subunderlease other than the rent hereby reserved there being disregarded any effect on rent of any of the matters set out in paragraphs (a), (b) or (c) of section 34 of the Landlord and Tenant Act 1954 (as amended).

Of that provision Walton J said:

> The words here are unusual and seem to me to be very odd, because, repeating the relevant ones, they are "to a tenant without a premium with vacant possession and subject to the provisions of this Subunderlease other than the rent hereby reserved . . . ". Quite clearly, something is missing from that phrase and why I say something is missing is because it starts off "subject to the provisions of this Subunderlease other than the rent . . .". Quite clearly, "the rent" is not a provision. The rent is something which is payable under, or reservable by, or covenanted to be paid as a result of, a provision in the lease and, therefore, something has got to be put in. The simplest (and it may in fact be the most correct) is to put in the words "other than the provisions relating to the rent hereby reserved" and if that were put in, as Mr Morritt quite correctly and forcefully pointed out, one would have some rather odd suppositions. One might have the supposition that there is no direct covenant to pay the rent. One might have the supposition that there was no power of re-entry for non-payment of the rent. Be it so, if those have any effect on the

*Editor's note: The reference is (1981) 264 EG 827.

*Editor's note: Also reported at [1985] 1 EGLR 92.
†Editor's note: Also reported at [1985] 1 EGLR 61.

valuation of the rent, I should be surprised, but I think that it is much simpler to think that the missing words are something along the lines of "provisions relating to the amount of rent payable". But, whatever it is, it seems to me quite clear that, however unfortunate it may be and whether it was intended or not, in this particular lease the fair market rent is to be ascertained upon the somewhat curious assumption that there is no rent revision clause contained in the hypothetical terms which the arbitrator is considering. In the present case, that is stated by the arbitrator, who told us what the effect of that would be. It would be very considerable because it would put up a not inconsiderable rent by no less than 20.5%; but it is a simple point and it seems to me an inescapable one in view of the way the definition of "fair market rent" is drawn. Mr Morritt did try to draw a distinction between "rent" and "rents" throughout the lease, but it does not seem to me that such a distinction in fact exists and I do not think that even if it did exist, it would really assist me in construing what is a very simple clause defining what is to be taken for this purpose as the fair market rent.

Mr Ellis, on behalf of the defendant, invites me to do two things. The first is to adopt the approach of Walton J rather than that of Vinelott J. However, as I have already indicated, my own preference is for the approach of Vinelott J. Second, Mr Ellis invites me to follow Walton J's example in inserting words into the relevant clause in the lease. He submits that the words in the present lease "subject to the provisions of this lease (other than the amount of rent hereby reserved)" are grammatically as unsatisfactory as were the words in the lease considered by Walton J and that I should therefore insert into the parenthesis the words "provisions relating to", so that it would read "other than provisions relating to the amount of rent hereby reserved". That would clearly, says Mr Ellis, exclude the provisions of the rent review clause itself. I do not think it would, because the crux would still be the meaning of the word "amount". At all events, I do not feel the need to insert words into this clause in the same way that Walton J felt a need to insert words in the clause that he was considering. It seems to me that "subject to the provisions of this lease (other than the amount of rent hereby reserved)" is perfectly intelligible English and that it would be too pernickety to say that the amount of rent is not a provision of the lease. It is obviously something provided for by the lease. From there to say that it may be referred to as a provision of the lease is not a difficult step, and I decline to alter, or possibly to alter, the meaning of this clause by inserting words in it that there is no compelling reason for inserting.

I come to the last of the authorities in this series, which is the judgment of Peter Gibson J delivered on October 18 1985 in the case of *Equity & Law Life Assurance Society plc* v *Bodfield Ltd*, of which I have been supplied with a transcript.* That was a rather special case because, in the rent review clause in the lease there in question, there was an elaborate provision, clause 4(2), the effect of which was to limit the rent payable following any rent review to the greater of £28,500 and 85% of the "net rental value" as ascertained on the rent review. There was no indication at all as to the reason why that provision had been inserted. The reason was unknown.

The presence of clause 4(2) had two consequences as regards the construction of clause 4(4), which was the critical provision. Clause 4(4) was, so far as material, in these terms:

> The "net rental value" means the best rent which the premises hereby demised might reasonably be expected to fetch on the open market upon the following assumptions, that is to say . . . that they are vacant and to let as a whole without a premium or other capital payment for the residue unexpired of the term hereby granted upon the terms of this lease other than as to duration and rent.

So the crucial phrase was "the terms of this lease other than as to . . . rent".

The first consequence of the existence of clause 4(2) was that that phrase had to be given a meaning such that it excluded clause 4(2) itself from the hypothetical lease. The phrase could not therefore be construed as excluding from that lease only those terms of the actual lease that set out in figures the amounts of rents payable. The second consequence was to render inapplicable any general rule that the court should lean against a construction that would require a tenant to pay for something he was not to have. Peter Gibson J, after pointing out that the discount given to the tenant by clause 4(2) might or might not have been intended to compensate the tenant for not including the rent review clause in the terms of the hypothetical letting, went on to say: "In the circumstances the only safe course is to construe the actual words used without regard to such general considerations as those advanced by Mr Barnes." (I infer that the considerations advanced by Mr Barnes in that case were similar to G
those that he advanced on behalf of the plaintiff in the present case.)

Earlier in his judgment Peter Gibson J had referred, with evident approval, to the passage in the judgment of Walton J in the *National Westminster Bank* case where he had said: "I do not think that there is a presumption one way or the other. I think that in every case that must depend upon the precise terms of the lease." I do not know whether Peter Gibson J had had cited to him the judgment of Vinelott J in the *Pearl Assurance* case. At all events Peter Gibson J expressed his reasons for reaching the conclusion that he did in these terms:

> I turn to Mr Lewison's submission. He says that if one faces the question: "What are the terms of the lease as to rent?", the answer that would naturally be given would not be limited to the terms reserving the fixed quantified rents
> but would be bound to include a reference to the terms as to the revised rent H
> payable on a rent review. I agree. There are no difficulties or inconsistencies with other parts of the lease where the words are so construed. Further, support for this construction to my mind can be found in the decision of Walton J in the *National Westminster Bank* case, to which I have already referred. In that case the hypothetical letting for purposes of rent review was to be "subject to the provisions of the subunderlease other than the rent hereby reserved". Those words of exclusion might perhaps more readily than in the present case be taken to refer simply to the quantum of rent. Walton J had no difficulty in construing them as meaning something along the lines of provisions relating to the amount of rent payable, and he reached the clear conclusion that the rent review clause was thereby excluded. Mr Barnes has rightly pointed out that each lease turns on its own wording. I accept that, but my attention has not been drawn to any features of the lease before me that
> differentiates it very significantly from the lease in the *National Westminster* J
> *Bank* case other than clause 4(2), which as I have indicated tends to be a pointer against the construction advanced by the tenant. In these circumstances, therefore, I am assisted by Walton J's decision in reaching the conclusions that I have stated.

So Peter Gibson J regarded the case before him as stronger in favour of the landlord than Walton J's case, and each of them regarded his decision as turning on the precise terms of the lease before him. I do not therefore feel constrained in any way by their decisions, and that all the more so because I am told that in the *National Westminster Bank* case Walton J refused to certify under section 1 of the Arbitration Act 1979 that there was in that case a question of law of general public importance. So, obviously, he did
not think that his decision would be applicable to other cases. In the K
result the authorities do not bring me to depart from the view that I would have formed myself independently of them.

The originating summons seeks a declaration that:

> Upon the true construction of paragraph 2 of the Second Schedule to the above-mentioned lease and in the events which have happened the rent to be paid by the Plaintiff with effect from the 24th June 1985 is the greater of (a) the clear yearly rent of £360,000 or (b) a yearly sum equal to the rent at which the property comprised in the above-mentioned lease is worth to be let in the open market as a whole at the best rent reasonably obtainable without taking any fine or premium subject to the provisions of the above-mentioned lease (other than the amounts of rent specified in paragraphs 1(1) and 1(2) of the said Second Schedule) but including the provisions for the revising of rent
> contained in the said Second Schedule for a term equal to the residue then L
> unexpired of the term granted by the above-mentioned lease but having regard to the Plaintiff's rights under the Landlord and Tenant Act 1954, and otherwise upon the bases set out in paragraph 2 of the said Second Schedule.

I propose to make that declaration.

The plaintiff was awarded costs.

Chancery Division

November 1 1985

(Before Mr Justice KNOX)

STAINES WAREHOUSING CO LTD v MONTAGU M
EXECUTOR & TRUSTEE CO LTD AND ANOTHER

Estates Gazette January 25 1986

277 EG 305-307

Landlord and tenant — Rent review clause in lease of property in trading estate — Whether a letter to the president of the RICS on behalf of the landlords constituted an effective application for the appointment of an expert surveyor in accordance with the clause — In the events which had happened an application for such an appointment had to be

---

*Editor's note: Reported at [1985] 2 EGLR 144; (1985) 276 EG 1157.

made by the landlords before a named date; otherwise the landlords' "trigger" notice became void — It was agreed that time was of the essence in regard to such an application — While negotiations between the parties were in progress, and before the named date, the landlords' surveyors wrote a letter to the president the effect of which was the issue for the court — After referring to the fact that negotiations were still continuing, the surveyors stated that they were writing in accordance with the terms of the rent review clause "to make an in time only" application to the president for the appointment of an expert surveyor — It was submitted by the tenants that this letter did not constitute an effective application and they criticised it on three grounds — They objected that no notice of the letter had been given to the tenants; that no fee had accompanied the letter to the president, contrary to the RICS guidance notes issued to intending applicants; and (the most serious objection) that the letter, expressed to be "an in time only application", was not a genuine application but merely a piece of paper to preserve the landlords' rights — The judge rejected all three objections — The lease did not require notice to the tenants of the application; the absence of the fee which should have accompanied the application did not prevent the application from being actually made and received, although its "processing" would not in fact proceed; and the letter was a genuine application in compliance with the time-limit, although it was hoped that it might not be necessary for the appointment and subsequent procedure to take place — Tenants' originating summons dismissed

This was an originating summons by the plaintiff tenant, Staines Warehousing Co Ltd, seeking a declaration as to the true effect of a letter which the defendant landlord, Montagu Executor & Trustee Co Ltd, claimed to be an application to the president of the Royal Institution of Chartered Surveyors to appoint an expert surveyor under the rent review clause of a lease of a property known as 18 Central Trading Estate, Staines, Middlesex.

Kenneth Bagnall QC and Kim Lewison (instructed by Berrymans) appeared on behalf of the plaintiff; Kirk Reynolds (instructed by Nabarro Nathanson) represented the first defendant, Montagu Executor & Trustee Co Ltd; the second defendant, Mr J E Cullis FRICS FSVA, the surveyor appointed by the president, was not represented and took no part in the proceedings.

Giving judgment, KNOX J said: This is an originating summons in which the plaintiff, Staines Warehousing Co Ltd, is the tenant under a lease and the first defendant, Montagu Executor & Trustee Co Ltd, is the landlord under that lease. The second defendant is a surveyor, who was not represented before me.

The question, which is a short one, arises under the terms of an underlease, dated July 21 1971, and made by different parties, under which the rent was reserved in the following terms: "YIELDING AND PAYING therefor during the first Seven years of the said term yearly and proportionately for any fraction of a year the rent of TWENTY-FIVE THOUSAND POUNDS" — and then there are the usual provisions in relation to the payment on quarter days. Later on the clause proceeds: "and any increase thereof payable under and by virtue of the provisions of the Second Schedule such last mentioned rent also to be due and payable", in the same manner. The term was for 125 years from September 29 1970. The second schedule provides for a rent review clause in these terms. It defines the rent review dates as September 29 1977 and September 29 "in each succeeding seventh year therefrom through the said term". It is common ground that one of the rent review dates was therefore September 29 1984, although the periodicity has subsequently been altered to one of five years rather than seven, but that does not affect the present decision.

There is then a definition of the open market rent in some detail, but nothing, I think, turns on those terms. The clause then proceeds, in clause 2:

> If at any time not more than twelve nor less than nine months before that one of the rent review dates then next ensuing (hereinafter called "the relevant rent review date") the Landlord shall give notice in writing to the Tenant of its desire to vary the yearly rent payable hereunder as from the relevant rent review date such yearly rent shall from and after the relevant rent review date be whichever is the highest of —

and then there are set out effectively the rent then currently payable, and 70% of the market rent which for the future has equally, I understand, been altered, but again nothing turns on that.

Then clause 3 provides:

> If the Landlord and the Tenant shall not have agreed the open market rent six months before the relevant review date the open market rent shall be determined by a Surveyor

defined as "the Appointed Surveyor"

> to be agreed upon in writing by the Landlord and the Tenant not later than four months before the relevant rent review date and in default of such agreement to be nominated by the President for the time being of the Royal Institution of Chartered Surveyors upon the application of the Landlord to be made not later than three months before the relevant rent review date.

Then there is a clause which deals with the effect of the appointed surveyor's decision. Then in clause 5 there is a provision that the yearly rent is to take effect retrospectively once it is determined by the appointed surveyor.

Then we come to the important clause for the purpose of this application, clause 6, which reads as follows:

> If the Landlord and the Tenant shall not have agreed the open market rent six months before the relevant rent review date and the Landlord shall neglect to make an application referred to in paragraph 3 hereof then (unless the parties hereto shall in writing agree otherwise) any notice already given by the Landlord to the Tenant under the provisions of paragraph 2 hereof shall be void and of no effect.

Certain matters are not in dispute. First, it is agreed that that provision which I have just read, clause 6 in the second schedule, does provide for a time-limit in respect of which time is of the essence, there being sufficient expression of an intention in that paragraph to exclude the *prima facie* rule laid down by the House of Lords in *United Scientific Holdings Ltd* v *Burnley Borough Council* [1978] AC 904. Second, as I mentioned, there is no dispute as to the various time-limits in relation to this particular rent review procedure. The timetable has been helpfully set out in the plaintiff's affidavit in support of the originating summons as follows: that the first notice by the landlord was required to be served not later than December 29 1983. In fact, by a letter dated November 15 1983, the landlord's solicitors wrote a letter to the secretary of the tenant, in which they said:

> In accordance with Clause 2 of the Second Schedule to your Underlease of the premises dated June 21 1971 we should be obliged if you would take this letter as notice of our client's intention to review the annual rent payable under the Underlease as from September 29 1984.

There is no dispute but that that is an effective notice to set in motion the rent review procedure. Reverting to the timetable, if no agreement was reached between the landlord and the tenant by March 29 1984, the rent was to be determined by a surveyor, and no such agreement, it is agreed, between the parties was reached.

Third, if the plaintiff and the defendant had not agreed the identity of the surveyor by May 29 1984, he was to be appointed by the president for the time being of the Royal Institution of Chartered Surveyors. There again, it is common ground that there was no such agreement on the identity of a surveyor.

Finally, and critically, the first defendant was required to make an application to the president for the appointment of such a surveyor by June 29 1984. What in fact happened was that there were some negotiations. While those negotiations were still on foot and on May 30 1984, the surveyors to the landlord wrote a letter to the president of the Royal Institution of Chartered Surveyors. That is, of course, just short of a month before the last date for the landlord to comply with the obligation to make an application. As the issue in this action depends entirely on this letter, I propose to read it in full:

> We act for the freeholder of the above factory, warehouse and office premises, Montagu Executor and Trustee Co Ltd, which are held on lease by Staines Warehousing Co Ltd

and their registered offices are set out

> The lease is for a term expiring on September 29 2095, at a current rent of £43,400 per annum, with provision for a rent review at September 29 1984.
>
> The second schedule to the lease requires that application be made to the President of the RICS for the appointment of an Expert Surveyor to determine a revised rent failing agreement between the parties six months prior to the review date. The application to the President is to be made not later than three months prior to the review date.
>
> Whilst no rent has yet been agreed in this case, negotiations are continuing between ourselves and the tenants, and we are, therefore, writing in accordance with the terms of the second schedule to the lease to make an in time only application for the appointment by the President of an Expert

Surveyor to determine a revised rent for the above premises at September 29 1984.

We should be grateful for your acknowledgement of this letter.

The question before me is: was that, or was it not, an application within the meaning of the second schedule to the lease?

No notice was in fact given to the tenant by the landlord or its agents of that letter having been written to the president of the Royal Institution of Chartered Surveyors, and it remained in ignorance of that fact until very much later in the year 1984, when the landlord filled in a form which is issued by the Royal Institution of Chartered Surveyors for use on applications for the appointment of either an arbitrator or an independent expert. That form was in fact signed by, again, the landlord's agents on December 4 1984. It is a form with two alternative initial paragraphs. One alternative is for an actual application to be made on the form: "We (blank) hereby request the President . . . to appoint an Arbitrator or Independent Expert." The other alternative is in relation to an application having been made in the past, and reads as follows: "The application for the appointment of an Arbitrator/Independent Expert to which the following details refer was made in a letter from" — and then there is in this particular form a blank — "dated" — and then in this form there is filled in — "29th May 1984" — and that is a clear reference to the letter from the landlord's agents which I have just read. Then there are various details set out, as one would expect, in relation to where the premises are and what the rent is and the names of the parties, and so on and so forth.

There is provision at the end of the form for the fee to be paid to the president of the RICS. Before I read that, I should refer to *Guidance Notes* which the RICS has issued to persons who are applying to it for the appointment of arbitrators or experts. The particular paras which are relevant to this case are contained at paras 2-1, where it is said: "Application to the President for appointment. Application to the President for the appointment of an Arbitrator or Independent Expert should be in writing and preferably made on the form obtainable on application to the RICS." Second, "Document(s) and fee on application. The application will not be processed until the appropriate non-refundable fee, currently £40 plus VAT, has been received together with a copy of the lease or other document conferring on the President power to make the appointment". Then there are various provisions which set out what the president will do in selecting an arbitrator or an expert as the case may be, and when those things have been done it is stated in the notes that the appointment takes effect from the date on which the president signs the appointment, which he will not do until he has satisfied himself that the person he proposes to appoint is suitable and not in any way disqualified.

So far as the fee was concerned, the facts are that no fee was paid when the letter of May 30 1984 was sent, but a fee appears to have accompanied the document that was signed on December 4 1984.

On those facts, argument has been addressed to me as to the validity of that letter of May 30 1984. Mr Bagnall, for the tenant, has taken the following points. First of all, he says that this has to be construed in a commercial way. That is not, I think, seriously controverted. Second, he says that there are essentially three defects in this letter as an application. The first is that it was not accompanied by the fee. That does not seem to me to be a valid objection. It is of course plain that the lease itself says nothing more than that the landlord must make an application, and there is no description or further elaboration of what is involved in the making of an application.

Second, it is clear that so far as the RICS's Notes have relevance — and they were relied on by Mr Bagnall — they do recognise that the application exists and has an objective existence by the way in which they describe the effect of the payment of a fee. The particular turn of phrase is "the application will not be processed until the appropriate non-returnable fee has been received". That seems to me to recognise the existence of the application. In a sense this is not directly relevant to the decision I have to take, in that the views of the Royal Institution of Chartered Surveyors, though no doubt worthy of serious consideration, cannot be binding upon me, but they do seem to me to accord with the reality of the matter, in that the application is actually made when the president is approached and told that an application is being made. No doubt it may be that the president will not take steps to process (as the expression is) that application until he receives the proper fee, but nevertheless the application is there and, so to say, on the table.

Third, Mr Bagnall relied on the fact that no notice had been given. G Here again I feel that this is not a point of any substance, because there is no requirement for notice to be given in the lease. The provisions of the lease are, of course, detailed, and where a party is required to do something the lease says so. Mr Bagnall argued that there could be difficulties if no notice was given, and a tenant might be prejudiced if he took steps which he would not have taken had he realised that a notice had been given. Those are arguments *in terrorem* as it seems to me. There is a theoretical answer to them, in that the tenant could perfectly safely inquire whether a notice had been given, since, time being of the essence, there would be no risk in his doing so. But equally, of course, there is no obligation in the lease on the tenant, as Mr Bagnall forcefully pointed out, to make any such inquiry. The matter is left entirely neutral. H

Next — and as it seems to me by far and away the most serious point — is the way in which the letter was actually phrased. It will be recalled that the critical expression was "We are . . . writing in accordance with the terms of the second schedule to the lease to make an in time only application for the appointment by the President of an Expert Surveyor", etc. The three words "in time only" have been the subject of argument before me. It was submitted by Mr Reynolds that they are capable of having two meanings: (1) that the agents were asking the president effectively not to take any further steps on the application because it was only a nominal application. The second meaning relied on by Mr Reynolds is that they were making an application now because, as was in fact stated in terms earlier in the J letter, there were negotiations on foot. Mr Bagnall, on the other hand, submits that this is really not a genuine application at all and that it was not calculated to or indeed intended to put the machinery into effective operation. In support of that, he relied on the decision in *Aly* v *Aly*, reported I think only in *The Times* newspaper on December 27 1983, where, in relation to the provisions of Order 12, rule 8 of the Rules of the Supreme Court, there was a requirement that a defendant should apply to the court within a specified time-limit and an application had been made in that case, in the sense that a letter was sent by the litigant to the appropriate court authority, but the summons on that letter was not issued until the time-limit had expired. In those circumstances, the Court of Appeal held that the time-limit had been complied with, and they applied the following K test:

One had to interpret the rule in question in a way that made sense of the whole procedure laid down. It did not make sense to penalise a party who had done everything required of him by the rule, on the basis of something not done in time by the court. Therefore, one could only say that "apply to the court" meant "do all that it is in your power to do to set in motion the procedure necessary for the relief sought".

Mr Bagnall says that in this case the landlord's agents have not done all that it was in their power to do to set in motion the procedure necessary for the relief sought, in that they had written a letter which indicated, perhaps not in very clear terms, that they did not want any action to be taken on the letter and that it was merely a piece of paper L which would preserve their rights. This is a very short point. In my judgment, the true construction of this is that the landlord's agents were taking the step that they understood to be necessary to comply with the time-limit and that it was in that context that they used the words in the phrase "an in time only application". In my judgment, having set out the framework of the clause and made it clear within what limits they were operating, the landlord's agents indicated that they were making an application, but that it was an application on which they hoped it would not be necessary for the president to appoint an expert surveyor who would then determine the rent, simply because there were negotiations on foot. It really comes down to this, in my judgment: whether or not this was a genuine application. In my judgment, it just passes the test. M

In those circumstances I will hear argument on the actual order I should make, but as at present advised it seems to me that I should merely dismiss the originating summons.

The summons was dismissed, the plaintiffs to pay the first defendants' costs.

A

Court of Appeal

November 8 1985

(Before Lord Justice MAY, Lord Justice LLOYD and Sir David CAIRNS)

R J STRATTON LTD v WALLIS TOMLIN & CO LTD AND ANOTHER

Estates Gazette February 1 1986

277 EG 409-412

B **Landlord and Tenant Act 1954, Part II — Appeal from decision of Vinelott J — Main question was whether letters passing between the appellant tenants (the applicants for a new tenancy) and the first respondents constituted an agreement for a new tenancy within the meaning of sections 28 and 69(2) of the 1954 Act — If so, the courts had no jurisdiction to grant a new tenancy under the Act — The difficulties to which this situation gave rise and resulted in this appeal were the following — The first respondents, the reversioners with whom the correspondence in question took place, had transferred their reversionary interest to the second respondents before the contractual date of termination of the**
C **tenancy — It appeared, although the judgments do not go into details on this point, that the correspondence, if it constituted an agreement, had not been registered as an estate contract — The result, unfortunate from the standpoint of the appellants, was that if the correspondence created a binding contract between the appellants and the first respondents, it effectively excluded the jurisdiction of the court to grant a new tenancy under the Act, but at the same time it was not binding as a contract on the second respondents owing to non-registration — It was therefore in the interests of the appellants to argue that the correspondence in question did not constitute a**
D **binding agreement within section 28 of the Act — It was suggested that not all the terms of the new lease had been agreed, that the terms were not sufficiently precise, that the correspondence was not intended to be legally binding until a formal lease had been executed, and that sections 28 and 69(2) envisaged an agreement drawn up with "due formality" — All these objections to the binding force of the agreement were rejected by the Court of Appeal, although the court expressed sympathy with the predicament in which the appellants were placed — Appeal dismissed**

This was an appeal by R J Stratton Ltd from a decision of Vinelott
E J the effect of which was to dismiss the appellants' application for a new tenancy under Part II of the Landlord and Tenant Act 1954 of premises at 15 Broad Street, Lyme Regis. The first respondents, Wallis Tomlin & Co Ltd, were the appellants' landlords at the date when proceedings under the 1954 Act began and when the correspondence which was in issue took place. The second respondents, Ford & Sons (Electrical & Television) Ltd, became the appellants' landlords by assignment of the reversion before the contractual termination of the term.

Grant Crawford (instructed by Richard Weston & Co, of Taunton) appeared on behalf of the appellants; C H Pymont (instructed by Michelmore Davies & Bellamy and Pomeroy & Hunt, of Sidmouth) represented the second respondents; the first respondents were not
F represented.

Giving judgment, MAY L J said: This is an appeal from a judgment and order of Vinelott J of March 16 1984 whereby, in the ultimate, he dismissed the present appellants' application for a new tenancy of premises at 15 Broad Street, Lyme Regis, pursuant to the provisions of the Landlord and Tenant Act 1954.

The case has certain unusual features. By a lease of April 1 1961 a limited company by the name of Lym Syndicate Ltd granted the appellants a term of 21 years of the relevant premises from March 25 1961 on the terms of the lease, a copy of which is in our bundle but into which I think it wholly unnecessary to go.

The conveyancing details that thereafter took place are conveniently and fully set out in the learned judge's judgment below which, with gratitude and respect, I propose to adopt. In the result the first respondents in this matter became entitled to the freehold reversion on the expiry of the term granted by the lease on October 20 1978.

On September 16 1981 the first respondents, who were then, as I have indicated, the landlords of the present appellants, gave notice to terminate the tenancy pursuant to the relevant provisions of the 1954 Act. In that notice they stipulated March 25 1982, which was the contractual date of termination of the term, as the date of termination under the Act. That notice was served on September 19 1981, and it produced very rapidly, on September 23 1981, a counter-notice by the applicants (the appellants, as they are now) stating that they were not prepared to give up possession on the termination of the lease.

At that point, therefore, one had the very common situation which arises between the landlords of business premises on the one hand and the tenants on the other, namely that the only issues which arise between the parties are about the precise terms of the new tenancy to which, under the statute, generally speaking the tenant will become entitled. Such matters as, for instance, the new rent, the new term, and any desired variations that there may be in the covenants that there already are in the existing lease. In this case, in order to reach agreement (and I use that phrase, for the moment, neutrally) on the future tenancy of these premises, both parties instructed their respective solicitors and surveyors to negotiate to try to arrive at an appropriate result.

Those negotiations ended (and I am taking the matter quickly, but if reference is required to the full history of the negotiations they are again, if I may say so, very fully set out in the admirable judgment of the learned judge below) with a letter dated January 4 1982 from the appellants' surveyors to the first respondents', the landlords', surveyors. That letter was headed "without prejudice", as had been all the other letters in the course of the negotiations, and it read as follows:

> I refer to our telephone conversation shortly before Christmas. As arranged I am now writing to confirm the revised terms that have been agreed in respect of the new lease which I understand will commence on March 26 1982.
> (1) The new rent is to be £2,750 pa.
> (2) The lease is to be for a term of 21 years with reviews after the first 5 years and 4 yearly thereafter.
> (3) The restrictions contained in the existing lease relating to the type of business which may be carried on in the demised premises, are to be lifted so long as the business carried out in future does not conflict with the business being carried out by your Client.
> It is also understood that a covenant will be incorporated to prevent the premises being used for an offensive trade, such as the preparation and sale of fish & chips, fried chicken, Chinese food, etc.
> In so far as the existing terms of the lease dated April 1 1961 are not inconsistent with the above, we imagine that the Solicitors will incorporate them into the new lease.
> Upon hearing from you that the contents of this letter are acceptable I will confirm the details to my Client's Solicitor.

That letter was replied to by a letter of January 14 1982 (it is misdated in the bundle, as letters in January frequently are) from the then landlords' surveyors back to the tenants' surveyors, and it is in these terms:

> Thank you for your letter of January 4 and I enclose herewith copy letter received from my client confirming his acceptance of the terms as set out in your letter.
> I am very glad that we were able to reach an amicable agreement in this matter.

The letter which was enclosed with that letter of January 14 1982 was a letter of January 12 1982 from the managing director of the landlords, as they then were, to his surveyors, who wrote the letter of January 14 1982. It read in this way:

> Thank you for your letter of the 5th instant and the copy letter from Humberts.

Humberts were the tenants' surveyors.

> I have sent this letter to Mr Mayo our solicitor for him to deal with the preparation of the new lease and I would confirm as you have asked my acceptance of the terms as set out in their letter with the proviso that our solicitor can prepare the appropriate amendment to the lease as required under Item 3.
> I very much appreciate the help you have given in this matter and I look forward to receiving your account in due course.

The letter of January 14 1982 from the landlords' surveyors enclosing the letter two days earlier to the tenants' surveyors was not headed "without prejudice". It was the first letter in the exchange of negotiatory correspondence which was not so headed.

The matter then took a somewhat unusual turn, and it is from that that these proceedings have developed. Within less than a month, namely on February 8 1982, a receiver was appointed of the first respondents', that is to say the then landlords', undertaking. Fifteen days later, on February 23, the relevant premises were conveyed to the second respondents to the tenants' proceedings below, the respondents to this appeal, who thus became the landlords of the tenants/appellants under the lease, which was then still in existence and had another month to run.

Thereafter there were substantial interlocutory proceedings, the details of which again are referred to in the judgment of the learned judge below and need no repetition here. They led up to an order in the local county court that three preliminary issues should be tried by a judge in the Chancery Division. The order was dated November 30 1982. I think the three issues merit reading. They were the issues which came before the learned judge below for trial. "(1) Whether the applicant" (the appellant) "is estopped or otherwise precluded from denying the agreement set forth in para 2 of the re-amended answer herein." The agreement there referred to is the agreement contended for made by the letters of January 4 and 14 1982 to which I have already referred. To that question the learned judge, for reasons which will appear, found it unnecessary to give any answer.

The second question was: "If the answer to question (1) above is 'no', whether correspondence passing between the respective surveyors of the applicant and the first respondent in the months of December 1981 and January 1982 constituted an agreement for the grant of a new lease of the premises known as ground floor and rear yard of 15 Broad Street, Lyme Regis, Dorset, by the first respondent to the applicant", that is to say by the original landlords who went into receivership and transferred their reversion to the second respondents, the present respondents to this appeal. To that question the learned judge below gave an affirmative answer.

The third issue for him to decide was this: "If the answer to either question (1) or (2) above is 'yes', whether the effect of the agreement mentioned in that question is that the court has no jurisdiction to entertain the applicant's application for a new tenancy of the said premises." That was the original application to the local county court by the appellants/tenants for a new lease upon receipt of the notice of termination to which I have earlier referred. To that question the learned judge below also gave an affirmative answer and, as will immediately be apparent from the matters to which I refer hereafter, the only result was that the originating application had to be dismissed and this was what occurred.

The learned judge's answers to the various questions were vital because of the provisions of sections 28 and 69 of the Landlord and Tenant Act 1954. Section 28 is in these terms:

> Where the landlord and tenant agree for the grant to the tenant of a future tenancy of the holding, or of the holding with other land, on terms and from a date specified in the agreement, the current tenancy shall continue until that date but no longer, and shall not be a tenancy to which this Part of this Act applies.

By section 69(2) it is provided:

> References in this Act to an agreement between the landlord and the tenant

except in immaterial sections

> shall be construed as references to an agreement in writing between them.

Thus, if those letters of January 1982 to which I have referred did form an agreement within those provisions of the Landlord and Tenant Act, as the learned judge below found by his answers to the preliminary questions, and if there was no registration of the agreement thereby arrived at between the parties, the result was that because of the timing of the transfer of the reversion the appellants lost their rights both under the Landlord and Tenant Act and under the agreement of January 1982 to a new tenancy of the relevant premises. It is perhaps not surprising that in those circumstances the appellants now appeal the decision of the learned judge below in this court.

It was originally contended that the surveyors (and, for that matter, solicitors, but principally the surveyors) to the parties, who negotiated at the end of 1981 and through to 1982, had no sufficient authority to conclude any legally binding agreement. However, that point is no longer persisted in. It is accepted that they did have the appropriate authority. Mr Crawford, who has appeared on behalf of the appellants, who has put before us succinctly and forcefully everything that can be said on their behalf, and for whose assistance we are grateful, in effect has taken four points. He rhetorically asks, in the first, whether all the terms of the new lease were indeed agreed by the exchange of correspondence in January 1982? He bases that G
submission upon the suggestion that, as appears in the correspondence, after January 1982 there did appear to be some points which remained to be agreed between the parties which had played some part in the negotiations, although agreement upon them had not been ultimately reached.

However, when one just considers that submission, the answer must be that if that exchange of correspondence in January 1982 did amount to a binding contract between the parties, then the point raised by Mr Crawford to which I have referred is indeed irrelevant. It may be that there were further matters to be considered and possibly negotiated between the parties, but those facts and matters cannot, as I think, undo the the contract which, *ex hypothesi*, had already been concluded. H

The second point which Mr Crawford takes on behalf of the appellants is to query whether, and to submit that, the terms of the exchanged correspondence in January 1982 were not sufficiently precise to constitute a binding contractual arrangement between the parties. He relies upon two points referred to in the letter of January 4 which I have already read. The first was the terms of the proposed user covenant in the new lease; and the second was not only the terms of "the competing business covenant", if I can so describe, which it was proposed should be contained in the new lease, but also of the precise user covenant which, it will be remembered, referred to such matters as fish and chips and Chinese food.

In so far as the covenant restricting future competition is J
concerned, although from a conveyancing point of view the terms of the letter of January 4 1982 may need filling out in order to give full precision and clarity to what was intended, the position was I think made sufficiently clear in that letter to enable it to constitute a firm offer.

In my judgment the only real doubt that there could be on the letter of January 4 1982 is as to the scope of the user clause. It might be said that there are two opposing lines of judicial authority on this particular point. One is that an agreement may nevertheless be complete and enforceable, although it has not been worked out in meticulous detail; that is the line of authority referred to in para 103 of vol 1 of the 25th edition of *Chitty on Contracts* to which we were referred. The second line of authority is to the effect that the terms of K
an agreement must be sufficiently certain for there to have been a binding contract arrived at between the parties; that line of authority is the one referred to in paras 117 and 118 of the same volume of *Chitty* to which again we were referred.

For my part, in general, and not in relation to the circumstances of this particular case, I have more sympathy with the second of the two approaches than with the first, but that matters not. The point is dealt with particularly in the very well-known dictum of Lord Wright from his speech in the case of *Hillas & Co Ltd* v *Arcos Ltd* (1932) 147 LT 503, 514, which is printed in para 118 of *Chitty* and was to this effect:

> Businessmen often record the most important agreements in crude and summary fashion; modes of expression sufficient and clear to them in the L
> course of their business may appear to those unfamiliar with the business far from complete or precise. It is accordingly the duty of the court to construe such documents fairly and broadly, without being too astute or subtle in finding defects, but on the contrary the court should seek to apply the old maxim of English law *verba ita sunt intelligenda ut res magis valeat quam pereat*. That maxim, however, does not mean that the court is to make a contract for the parties or to go outside the words they have used, except in so far as there are appropriate implications of law.

The contrast to which the learned law lord referred in that dictum between the two approaches leads to the difficulty that often arises in cases such as the present.

However, in the instant case, like the learned judge below, I for my part have no doubt at all that a court would on a construction summons or on an application for specific performance, for M
instance, adequately resolve the scope of the user clause on very much the same principles as the learned Lord Chancellor resolved a similar problem in the case of *Parker* v *Taswell* (1858) 27 LJ Ch (NS) 812. I do not think that I need refer to that judgment in detail.

The references to the type of food in which the intended premises should not deal are, I think, merely examples on the proper construction of that letter, and for my part I deprecate any too meticulous analysis of the second paragraph under head 3 in the letter. Thus again I cannot accept Mr Crawford's argument on this point.

The third argument that Mr Crawford puts forward is that in the circumstances of the instant case the agreement, if agreement it was,

A reached between the parties in January 1982 was no binding agreement in law because it was as it were, in the usual phrase, "subject to the execution of a formal lease". In *Rossiter* v *Miller* (1878) LR 3 App Cas 1124 there is the very well-known dictum of Lord Blackburn at p 1152 where, after referring to an opinion of James LJ, he said:

> I infer, though of this I am not quite sure, that he wishes it to be a canon of construction that, wherever there is a stipulation for a farther and more formal agreement, the previous arrangements should be held to be only of this preliminary nature. I doubt whether such a canon of construction would not often defeat the intention of the parties; but I think it is too late now to introduce it. I think the decisions settle that it is a question of construction whether the parties finally agreed to be bound by the terms, though they were
> B subsequently to have a formal agreement drawn up.

Of course in this case it is common ground that the parties did expect and did intend that there should be a formal lease drawn up and executed by them. But again I am satisfied that the learned judge below was correct in the view which he reached that the agreement which was come to between the surveyors in 1982 was indeed an agreement *per se* and not one expressed to be subject to a formal lease and not to be legally binding until such a formal lease had been drawn up and executed. As the learned law lord said, "it is a question of construction" of the negotiations that had theretofore taken place between the parties. Without going through the correspondence in detail, as I have said the whole tenor of it seems to me quite clearly to have reflected the parties' intention that the surveyors should come
C to an agreement which would bind them, the parties. The surveyors indeed referred back to the parties and to their solicitors on a number of occasions in the course of the correspondence. When they did reach the agreement in January 1982 this was intended to cover and to be binding about, as I think, all the matters which were then within their remit. It must not be forgotten that there was the current lease there as a potential model needing only amendment to take account of the negotiations which led to that agreement in January 1982.

Mr Crawford took a point (and I deal with it, I trust, not disrespectfully but swiftly) on the proviso in the letter of January 12 1982 which was sent on in the letter of January 14 1982. I think it is sufficient to say that that point is, in my judgment, adequately
D covered by the reasoning in the case of *Chipperfield* v *Carter* (1895) 72 LT 487 at p488.

Finally, although this is not decisive, as Sir David Cairns said in the course of argument it is a pointer, it did so happen that the rubric "without prejudice" which had been apparent on all the prior correspondence ceased to appear on the, as I think, acceptance of the firm offer constituted by the letter of January 12 1982. For those reasons therefore I again respectfully cannot accept the submissions by Mr Crawford on this third point.

He will, I trust, forgive me if I say that his fourth and final point was probably a fall-back position. It was based upon what he submitted was the proper construction of and approach to sections 28 and 69 of the Landlord and Tenant Act 1954 having regard to the
E intention of the Act, namely that in general a tenant should be entitled to obtain a new tenancy of business premises, other things being equal. The submission in effect was that when one construes the word "agreement" in section 28, or "the agreement in writing" in section 69, one must construe those words with, as he put it, due formality, or as a reference to an agreement made with due formality. Understandably, and as he realised, Mr Crawford found it a little difficult to answer the question from the court as to what he intended by the phrase "due formality". In my opinion, an agreement as referred to in sections 28 and 69 of the Landlord and Tenant Act 1954 is an agreement, that is to say a binding contractual arrangement enforceable by the parties at law. Again, I cannot accede to Mr Crawford's fourth submission.

F I have, as did the learned judge below in the circumstances, considerable sympathy with the appellants in the predicament in which they find themselves. It may be that it results from the non-registration of the contract which was come to in January 1982. Whether in such circumstances reasonable solicitors would have registered is not a matter for us, and of course it arises only, as was pointed out in the course of the argument, because the relevant land is unregistered. For the reasons that I have given I think that the appeal should be dismissed, and in the circumstances it is unnecessary to consider the interesting issue, as it appears on the documents, of an estoppel, with which the learned judge below also, as I have indicated, found it unnecessary to deal. In my judgment this appeal should be dismissed.

Agreeing, LLOYD LJ said: The letter of January 4 1982 and the reply of January 14 1982 which have been read by my lord show to my mind as clearly as any letters could that there was here a binding agreement for the grant of a new lease between the appellants and the first respondents. Accordingly, section 28 of the Landlord and Tenant Act 1954 took effect. Unfortunately for the appellants, the second respondents are not bound by that agreement. If I could have found a way of holding that the appellants' statutory rights to a new tenancy thereupon revived, I should have done so. But that is not a possible construction of section 28 of the Act. Indeed Mr Crawford did not feel able to argue that it was. It follows that section 28 applies in this case according to its terms and this appeal must be dismissed.

SIR DAVID CAIRNS also agreed and did not add anything

The appeal was dismissed with costs.

Chancery Division

November 28 1985

(Before Sir Nicolas BROWNE-WILKINSON V-C)

NUNES AND ANOTHER v DAVIES LAING & DICK LTD

Estates Gazette February 1 1986

277 EG 416-418

Landlord and tenant — Rent review clause in lease — Whether letter from tenants in response to landlords' trigger notice was a valid counternotice — Whether counternotice must be "clear" as to tenants' intention or whether it must be "unequivocal" — In the present case, where the stipulations as to time were stated to be of the essence of the contract, the landlords served a trigger notice within the time allowed specifying an open market rental figure from the review date of £23,000 per annum — Tenants, also within the time allowed, replied by their agent, in a letter headed "In the matter of rent review as at July 1 1985", that "I am instructed by the tenants to give you hereby formal notice that the open market rental is £12,000 per annum and call on you under the terms of the above lease to agree this. Please confirm that this is accepted as due notice" — The Vice-Chancellor adopted the test applied by the Court of Appeal in *Amalgamated Estates Ltd* v *Joystretch Manufacturing Ltd* that the counternotice should give a "clear indication" to the landlord rather than the apparently more stringent test accepted in *Edlingham Ltd* v *MFI Furniture Centres Ltd* and *Sheridan* v *Blanecourt Investments Ltd* that it should be "unequivocal" — It was established by such cases as *Bellinger* v *South London Stationers Ltd* and *Oldschool* v *Johns* that a mere expression of disagreement with the rent proposed by the landlord was not sufficient — If the test required the notice to be "unequivocal" it could have been argued that it failed the test, because the draftsman might have been under the impression that he had to give formal notice of disagreement — However, applying the "clear indication" test, there were several points in favour of the validity of the letter as a counternotice, eg the heading, the description as a "formal" and as a "due" notice and the fact that the only provision for a formal notice in the context was a counternotice — Held, taking these and other indications into account, that the letter was a valid counternotice by the tenants within the meaning of the rent review clause, so that the rental value would be determined, in accordance with the lease, by an independent surveyor

By this originating summons the plaintiffs, Jacqueline Flavia Nunes and Michael Christopher Kinkead Allen, sought declarations in regard to the application, in the events which had happened, of provisions in a rent review clause of a lease of premises at 19 Pembridge Villas, London W11. The plaintiffs were the landlords and the defendants, Davies Laing & Dick Ltd, were the tenants.

P W Birts (instructed by Penningtons) appeared on behalf of the plaintiffs; Leolin Price QC and Miss Siân Thomas (instructed by Gouldens) represented the defendants.

Giving judgment, SIR NICOLAS BROWNE-WILKINSON V-C said: This is yet another case raising the question whether or not, under a rent review clause, a tenant has duly exercised his right to have the rent determined by an independent surveyor.

The facts are very simple. Under a lease dated July 16 1973 the landlords demised the premises 19 Pembridge Villas, London W11, for a term of 18 years from July 1 1973. The rent reserved was for the first six years, £6,000, for the next six years either £6,000 or the open market rental as defined in the lease at the review date which was also defined, whichever was the higher, and for the remaining six years either the yearly rental reserved as at the first review date or the open market rental value at the second review date, whichever was the higher. The first review date was July 1 1979 and the rent as at that date was agreed at £10,000. The second review date was July 1 1985.

Clause 2 contains the provisions which define the open market rental value and the review date, but this case turns on proviso (iii) to clause 2 which regulates the determination of what is the open market rental value. It reads as follows:

> The open market rental value shall be determined in manner following that is to say it shall be such annual sum as shall be
> (a) specified in a notice in writing served on the Lessee at the demised premises at any time before the beginning of a clear period of two quarters of a year . . . immediately preceding the review date; or
> (b) agreed between the parties before the expiration of three months immediately after the service of such notice; or
> (c) determined at the election of the Lessee (to be made by counternotice in writing served by the Lessee on the Lessors not later than the expiration of the said period of three months) by an independent Surveyor . . .

The lease expressly provided that the stipulations as to time in proviso (iii) were to be of the essence of the contract.

The solicitors for the landlords by a letter dated November 29 1984 headed up "In the matter of the lease" gave the trigger notice under para (a) in these terms:

> On behalf of the present Lessors we hereby give you notice, pursuant to Clause 2 of the Lease that the market rent from the review date (that is to say the expiration of the twelfth year of the term) shall be the annual sum of £23,000.

The response to that letter came from agents instructed by the tenants, Scott Enterprise. In a letter dated December 6 1984 and headed up "In the matter of the lease" and also "In the matter of Rent Review as at July 1 1985" they wrote as follows:

> Dear Sirs, I am instructed by the tenants to give you hereby formal notice that the open market rental is £12,000 per annum and call on you under the terms of the above Lease to agree this. Please confirm that this is accepted as due notice. Faithfully, J S M Scott

who was the principal of the agents.

The only response to that was a letter from the landlords' agents dated December 18 1984 acknowledging receipt of the letter of December 6:

> in which you state that the open market rental value amounts to £12,000 per annum. Regrettably we cannot recommend our clients to accept this rental and look forward to discussing this matter further with you in due course.

It will be noted that that letter does not give the confirmation asked for in the letter of December 6, namely that there was due notice. Nothing further happened. No further notice was given by the tenants.

The short question raised by the originating summons is whether the letter of December 6 1984 from the tenants' agents constitutes a valid counternotice for the purposes of para (c) of proviso (iii) of clause 2 of the lease. If it does, the rent for the final six years of the term will be such rent as is determined by an independent surveyor. If it is not a valid counternotice, the rent will be the £23,000 per annum specified in the landlords' trigger notice.

The same question has arisen on rent review clauses in identical or very similar terms in no less than five reported cases to which I have been referred. It appears that the clause is based on a precedent in the *Encyclopaedia of Forms and Precedents*, 4th ed. In a number of those decided cases the question for decision has been characterised as a question of construction of the letter from the tenants which is alleged to be an election under para (c); but, as the argument in this case proceeded, it emerged that there are in reality two questions:

(a) what in law is required to constitute a valid election under para (c)?

(b) does the letter of December 6 1984 satisfy those requirements?

I will in due course consider those separate points in turn but, before doing so, I should explain why, in my judgment, those are the two relevant questions. In an ordinary case of construction the court G is bound either to give the document under construction some meaning or to hold that it is void for uncertainty, against which the court always leans. Often, the true construction of a document in court is a marginal decision which the court reaches with some doubt. In other types of case — for example, notices to quit, and, in the employment field, notices of dismissal — the law requires not simply that a document on its true construction has a particular meaning but that the document should clearly have that meaning so as to bring home to the mind of the recipient of the notice what is intended by it. So, in cases of the kind before me, the first question is whether the law does require of a notice under para (c) not merely that on its true construction it communicates an intention to have the rent fixed by an independent surveyor but that it should satisfy some higher H standard such as that it should be clear or unequivocal.

(1) *What are the legal requirements of a valid notice under para (c)?*

In the reported cases this question has not been addressed separately. It has been treated as rolled up in the question of the construction of the alleged counternotice. But, in my judgment, the authorities do disclose that the law requires the alleged counternotice to have a certain degree of clarity in its phraseology. It is not merely a question of the judge construing the document; the question is whether the document brings home to the mind of the ordinary landlord the fact that the tenant is exercising his rights under para (c).

But the authorities disclose two possible tests. The first is that the alleged counternotice must constitute "a clear indication" to the J landlord as to the tenant's intention, or must make it "clear to the landlord that he proposes to go to arbitration" (see per Lawton LJ and Templeman LJ in *Amalgamated Estates Ltd* v *Joystretch Manufacturing Ltd* (1980) 257 Estates Gazette 489). However, in two later cases the test has been expressed to be that the letter alleged to constitute the counternotice must be "unequivocal" (see per McNeill J in *Edlingham Ltd* v *MFI Furniture Centres Ltd* (1981) 259 Estates Gazette 421 and per Nicholls J in *Sheridan* v *Blanecourt Investments Ltd* (1984) 270 Estates Gazette 1290). It may be that in those two cases the learned judges were not meaning to apply any more stringent test than that of clarity laid down by the Court of Appeal in the *Amalgamated Estates* case.

This case discloses that there may be a difference between what is K clear and what is unequivocal. If the expression "unequivocal" means "incapable of bearing any other meaning" then the way is open for arguments based on remote possibilities of construction, however unlikely they may be, in construing the counternotice.

In my judgment, the test is that applied by the Court of Appeal in the *Amalgamated Estates* case, namely that the counternotice should be in terms which are sufficiently clear to bring home to the ordinary landlord that the tenant is purporting to exercise his right under para (c), and that is the test which I propose to apply.

(2) *Does the letter of December 6 1984 constitute a valid counterclaim?* L

It is clearly established that a letter which merely disagrees with the rent proposed by the landlord, with or without a request for additional information as to the basis on which it has been fixed, does not constitute a valid counternotice (see *Bellinger* v *South London Stationers Ltd* (1979) 252 Estates Gazette 699, *Oldschool* v *Johns* (1980) 256 Estates Gazette 381 and the Court of Appeal decision in *Amalgamated Estates*). Such a letter is equally consistent with a mere attempt to agree the new rent operating the provisions of para (b).

In the present case the letter of December 6 1984 goes much further than that. It was a letter written in response to the landlords' trigger notice, which expressly referred to clause 2 of the lease. The letter of December 6 is itself headed up "In the matter of the Rent Review". Most importantly, to my mind, the letter is described in itself as M being a "formal" notice and asks for confirmation that it is "due" notice. That is to say, the letter on its face is a document which purports to be a notice which is formally required by and is due under some legal requirement. Now, proviso (iii) envisages only two notices: the landlords' trigger notice under para (a) and the tenants' counternotice under para (c).

In my judgment, it must have been clear to anyone receiving the letter of December 6 that the tenants' agent was giving a formal notice and, knowing the terms of the lease, must have known that the only provision for a formal notice under proviso (iii) was a counternotice under para (c). Mr Birts, in his very able argument, while accepting the force of this point, relied on the fact that, although the tenants'

A agent in the letter describes it as a formal notice, the only thing of which the letter gives notice is that the tenant alleges the proper rent is £12,000 per annum, ie it does not in terms exercise any right of election to go to the independent surveyor. Therefore, says Mr Birts, this is not notice of any election under para (c) but merely notice of a disagreement as to what the rent should be — which, on the authorities, is insufficient to constitute a valid notice. He relies on the decision in the *Edlingham* case, where the letter under consideration as being a counternotice describes itself as being a counternotice, and yet the judge held that it was not sufficient to constitute a valid counternotice within para (c).

There is much force in those submissions, but they do not persuade me. B As Mr Price for the tenants pointed out, the period for negotiation under para (b) and the period for a valid counternotice under para (c) is the same period of three months. Under the machinery of the clause, negotiations and notice to go to arbitration may well coincide. Therefore, a reference in any letter to a disagreement as to price is in no way inconsistent with an intention on the part of a tenant to go to arbitration.

Moreover, the background of the case is that, if there is no agreement as to the rental, in default of election to go to the independent surveyor, the landlords' rent of £23,000 per annum will be the rental. The letter of December 6 is formal notice that the tenants do not accept the rental of £23,000. In my judgment, in the surrounding circumstances, a letter giving formal notice of C disagreement which is said to be due notice clearly brings home to the landlord that the tenant is not accepting the landlord's figure and is giving the only notice which proviso (iii) envisages, namely a counternotice under para (c).

In the *Edlingham* case the learned judge does not explain what force he gave to the words "counternotice" in the letter there under consideration; there may well have been special features in that case which forced him to a contrary conclusion. If not, I must respectfully decline to follow that decision, which may have been influenced by his formulation of the test as a requirement that the letter should be unequivocal. It is certainly arguable in the present case that the letter of December 6 is, in one sense, equivocal. There is a theoretical possibility that the draftsman was under the misapprehension that he D had to give formal notice of disagreement. Therefore, if the test was that the notice must be unequivocal, it could be said that the letter of December 6 was not a good notice. But, in my judgment, that is not the test. Any ordinary landlord receiving the letter of December 6 would know that he was in receipt of formal notice under proviso (iii) and that the only provision for such notice in the event of disagreement with the rent was a notice under para (c).

For those reasons I hold that the letter of December 6 was a valid counternotice under para (c).

Declaration was made accordingly. Plaintiffs were ordered to pay defendants' costs.

E

Chancery Division

November 14 1985

(Before Mr Justice SCOTT)

LINDEN AND ANOTHER v DEPARTMENT OF HEALTH AND SOCIAL SECURITY

Estates Gazette February 8 1986

277 EG 543-550

Landlord and Tenant Act 1954, Part II — Whether Secretary F of State for Social Services was entitled to claim a new tenancy under Part II of property belonging to plaintiffs which was used for the accommodation of staff employed at National Health Service hospitals — Premises were let to secretary of state under a lease which expired on June 24 1985, were managed by a district health authority, and were physically occupied by the health service staff concerned — The plaintiff freehold owners of the property, which had been converted into eight self-contained flats, sought a declaration as to their rights in relation to the secretary of state's claim — The staff who occupied the flats did not have exclusive possession and the district health authority maintained a large measure of control over the property — The staff accommodated changed frequently and usually one or more of the flats stood vacant — Provisions of the National Health Service Act 1977 as amended by the Health Services Act 1980 and of the Landlord and Tenant Act 1954 considered — Section 56(3) of the 1954 Act provided that Part II of the Act applied where a tenancy was held by a government department if the premises "were occupied for any purposes of a government department", even if the condition of section 23(1) as to occupation by the tenant was not fulfilled — There was no requirement that the department holding the tenancy should be the same as that for whose purposes the premises were occupied — In the present case the district health authority could be regarded as in occupation of the premises — Features which justified this conclusion were that vacant flats were under the control of the authority, that the authority retained keys to the flats and visited them regularly, that exclusive possession was not given to the staff accommodated in them, and that the authority provided furniture, crockery and blankets and carried out repairs and decorations — Furthermore, the authority was in occupation "for any purposes of a government department" within section 56(3), as the occupation was in furtherance of the functions of the secretary of state under the National Health Service Act, one of which, delegated to the authority, was to provide residential accommodation for employees of the service — Among decisions considered were *Lee-Verhulst (Investments) Ltd* v *Harwood Trust, William Boyer & Son Ltd* v *Adams, Groveside Properties Ltd* v *Westminster Medical School* and *Pfizer Corporation* v *Minister of Health* — Plaintiffs' action dismissed

In this action Bennie Linden and Helen Goldie Linden, freehold owners of 40 Nottingham Place, London W1, plaintiffs, sought a declaration against the tenant, the Secretary of State for Social Services, defendant, in regard to his claim to be entitled to a new tenancy of the premises under Part II of the Landlord and Tenant Act 1954.

Michael Barnes QC and R J Furber (instructed by Berwin Leighton) appeared on behalf of the plaintiffs; Kim Lewison (instructed by Beachcrofts) represented the tenant.

Giving judgment, SCOTT J said: The plaintiffs in this case are the owners of premises known as 40 Nottingham Place, London W1. The premises were, by a lease dated October 25 1971, demised by the plaintiffs' predecessors in title, Rickatson Estates Ltd, to the Secretary of State for Social Services for a term of 14 years from June 24 1971. The contractual term expired, therefore, on June 24 1985. The question in issue is whether, pursuant to the provisions of Part II of the Landlord and Tenant Act 1954, the secretary of state is entitled to a new tenancy of the premises comprised in the lease. The defendant in the action is the Department of Health and Social Services although, as I understand it, the secretary of state remained the lessee throughout the contractual term. Nothing turns on this.

The plaintiffs acquired the freehold reversion to the lease in May 1978.

The case raises questions both of fact and of law. I will first set out the relevant facts and then turn to the law.

No 40 Nottingham Place is a terraced property with a basement, a ground floor and three upper floors. It has been converted into eight self-contained flats, one in the basement, one on the ground floor and two on each of the upper floors. This conversion was carried out pursuant to a licence dated June 19 1975, granted by Rickatson Estates Ltd to the secretary of state. Clause 3(b) of the lease contains the lessee's user covenant: "To use the said premises only as a hostel for nurses whilst in employment or under the control of the lessee." Clause 1(b) of the licence authorised the secretary of state:

> to use the premises as eight self-contained flats provided . . . that the lessee shall not permit the same to be occupied other than by persons in his employment or control and then only on fully furnished service occupancies.

Pursuant to the licence, the conversion of 40 Nottingham Place, into eight self-contained flats was carried out. Thereafter the flats have been used in the manner authorised by clause 1(b) of the licence. They have been occupied by persons employed in National Health

Service hospitals in the area. The nature of the employment of these occupants has varied widely. Some have been engineers; some accountants; some have been doctors or otherwise employed in a medical capacity; some have been administrators. All past and present occupants have been or are employed for the purposes of one or other of the local National Health Service hospitals.

The flats are let fully furnished and equipped with cutlery and crockery. Blankets are provided, but no bed linen. They are managed by the Paddington and North Kensington Health Authority (which I shall call "the authority") to which powers of management of the provision of health services in the area have been delegated. I will later in this judgment describe more particularly the statutory relationship between the secretary of state and the authority. The flats are used by the authority as a means of providing low-cost furnished accommodation to hospital employees.

The occupants of the flats occupy their respective flats under the terms of written agreements with the authority. The agreements are all in the same form. No rent is reserved. Instead an agreed deduction is authorised to be made from the salary of the occupant in question. The occupants do not have exclusive possession. Para 4 of each written agreement provides that:

The possession, management and control of the premises shall remain vested in the employer, who shall be the occupier for all purposes and shall have the right of entry at all times for the purpose of exercising the management and control.

Consistently with that provision, the authority retains keys to each flat and causes each flat to be visited regularly. Naturally, the authority is responsible for carrying out all repairs and decorations. Even trivial repairs are, it seems from the evidence, seen to by the authority. The agreements impose very light obligations on the occupants. There is an obligation to vacate the flat on ceasing to be employed by the authority or when so required by the authority. There is an agreement not to commit any nuisance; an agreement not to attach any fixtures or fitments to any part of the flat; and an agreement not to use the flat for any purpose other than as a residence for the employee and his family. As to the condition of the flat, there is simply an agreement:

To take reasonable care of the interior of the premises including all fixtures and furniture therein and keep the same in good and clean condition, reasonable wear and tear allowed, and to replace all cracked or broken glass and make good all damage not attributable to reasonable wear and tear.

There is no evidence before me as to the arrangements made for the cleaning of the common parts of the building, that is to say, the hall, the staircases and the landings. In view, however, of the content of the written agreements, it seems likely that the authority is responsible for making appropriate arrangements for these common parts to be cleaned.

Mr High, the acting district administrator of the authority, deposed in his affidavit sworn on May 31 1985 that "the authority maintains a high degree of control over both the occupation and the maintenance of the premises".

In general, the persons who occupy the flats do so for short periods. They tend to be persons who, having taken up employment in one or other of the National Health Service hospitals in the area, need immediate accommodation. They are likely to remain at 40 Nottingham Place for no longer than is necessary to enable them to find suitable permanent accommodation. The occupants of the flats are, therefore, constantly changing and usually one or more of the flats will be standing vacant.

The authority regards the flats as a highly desirable facility for the purpose of attracting staff. Mr T M Higgins, a chartered surveyor and estate agent who gave evidence before me, estimated that on average it took about three months for someone to find and complete the purchase of a house or flat in central London. It seems obvious that the attractions of an offer of employment in central London will be greatly enhanced if the offer can be accompanied by the offer of immediate accommodation. In his affidavit Mr High pointed out the difficulties of obtaining staff for employment in central London. He said this: "This Health Authority, because of its geographical area, has to compete with an exceptionally high proportion of private companies offering considerably higher salaries with better fringe benefits and general terms and conditions of service." A little later in his affidavit he said this: " It is of paramount importance to the efficient running of this authority for the authority to be able to offer accommodation for a reasonable charge to para-medical, nursing and junior medical staff especially where the staff have to be available at short notice for work."

These are views expressed by a senior officer of the authority. He was not challenged on them in cross-examination. Moreover, these views seem to me self-evidently reasonable.

It is an important feature of this case that the tenancy of 40 Nottingham Place was vested in the secretary of state, whereas management of the property was carried out by the authority and the occupants of the flats were all employed by the authority and not, directly at least, by the secretary of state.

The relationship between the secretary of state and the authority is provided for by the National Health Service Act 1977 as amended by the Health Services Act 1980. Section 1 of the 1977 Act imposes on the secretary of state the statutory duty of providing a health service for England and Wales. Section 2 of the Act authorises him "(a) to provide such services as he considers appropriate for the purpose of discharging any duty imposed on him by this Act and (b) to do any other thing whatsoever which is calculated to facilitate, or is conducive or incidental to, the discharge of such a duty". Section 3 of the Act imposes on the secretary of state a duty "to provide throughout England and Wales, to such extent as he considers necessary to meet all reasonable requirements (a) hospital accommodation; (b) other accommodation for the purpose of any service provided under this Act". Section 87(1) of the Act provides:

The Secretary of State may acquire (a) any land, either by agreement or compulsorily; (b) any other property, required by him for the purposes of this Act; and (without prejudice to the generality of paragraph (a) above) land may be so acquired to provide residential accommodation for persons employed for any of those purposes. (2) The Secretary of State may use for the purposes of any of the functions conferred on him by this Act any property belonging to him by virtue of this Act, and he has power to maintain all such property.

Section 8 of the 1977 Act, as amended by the Health Services Act 1980, imposes on the secretary of state the duty to establish regional health authorities, area health authorities and district health authorities. The Paddington and North Kensington Health Authority is a district authority established by the secretary of state pursuant to section 8, as amended. Section 13 of the Act, as amended, gives the secretary of state power to "direct a Regional Health Authority . . . to exercise on his behalf such of his functions relating to the health service as are specified in the directions . . .". The section goes on to provide that "(subject to section 14 below) it shall be the duty of the body in question to comply with the directions". Section 14 of the Act, as amended, gives a Regional Health Authority power to:

direct any Area or District Health Authority of which the area or district is included in its region to exercise such of the functions exercisable by the Regional Health Authority by virtue of section 13 above as are specified in the directions . . .

Here, too, the section provides that "it is the Area or District Health Authority's duty to comply with the directions."

In paras 3 and 4 of his affidavit, Mr High deposes as follows:

3. The individual Regional Health Authorities pursuant to powers contained in the National Health Service Acts 1946 to 1973 and directions from the Secretary of State delegate powers of management to the District Health Authorities in their individual Regions. 4. The North West Thames Regional Health Authority has delegated to the Paddington and North Kensington Health Authority (the Authority) the responsibility for the management of property within the District of the Authority. The property at 40 Nottingham Place, London W1 (the premises) is geographically outside but administratively within the District of the Authority and consequently the Authority manage the premises under the terms of the lease dated 25th October 1971 made between Rickatson Estates Ltd and the Secretary.

The delegations to which Mr High refers, although made under previous Acts, were made pursuant to statutory provisions in all respects the same as the statutory provisions in the 1977 Act, as amended, which I have read.

The authority, by virtue of these delegations, manages 40 Nottingham Place. It also, by virtue of these delegations, employs the various employees who work at the National Health Service hospitals for which it is responsible. These employees, as I have said, include the occupants of the flats. In doing these various things and generally in discharging its duties pursuant to the delegations the authority is exercising, on behalf of the secretary of state, his functions relating to the Health Service. The authorities, whether regional, area or district, are, by para 8 of the Fifth Schedule to the 1977 Act, declared to be bodies corporate. Para 15(1) of the Fifth Schedule is relevant. It provides that:

An authority shall, notwithstanding that it is exercising any function on

> A behalf of the Secretary of State or another authority, be entitled to enforce any rights acquired in the exercise of that function, and be liable in respect of any liabilities incurred (including liabilities in tort) in the exercise of that function, in all respects as if it were acting as a principal.

Then the subpara goes on:

> Proceedings for the enforcement of such rights and liabilities shall be brought, and brought only, by or, as the case may be, against the authority in question in its own name.

I now turn to the Landlord and Tenant Act 1954 and its relevant sections. Section 23(1) of the Act is in these terms:

> Subject to the provisions of this Act, this Part of this Act applies to any tenancy where the property comprised in the tenancy is or includes premises
> B which are occupied by the tenant and are so occupied for the purposes of a business carried on by him or for those and other purposes.

This subsection expresses two criteria. First, the premises comprised in the tenancy must be occupied by the tenant. Second, the premises must be occupied for the purposes of the business carried on by the tenant. If these criteria are satisfied, Part II of the Act applies to the tenancy. Section 23(1), however, does not have to be satisfied in the case of a tenancy held by a government department. That case is specially dealt with under section 56 of the Act. Section 56(1) provides that:

> Subject to the provisions of this and the four next following sections, Part II of this Act shall apply where there is an interest belonging to Her Majesty in right of the Crown or the Duchy of Lancaster or belonging to the Duchy of
> C Cornwall, or belonging to a government department or held on behalf of Her Majesty for the purposes of a government department, in like manner as if that interest were an interest not so belonging or held.

Subsections (3) and (4) of section 56 are important:

> (3) Where a tenancy is held by or on behalf of a government department and the property comprised therein is or includes premises occupied for any purposes of a government department, the tenancy shall be one to which Part II of this Act applies . . . (4) The last foregoing subsection shall apply in relation to any premises provided by a government department without any rent being payable to the department therefor as if the premises were occupied for the purposes of a government department.

Accordingly, where a tenancy is held by a government department, Part II of the Act will apply, if the requirements of subsection (3) or
D of subsection (3) supplemented by subsection (4) are satisfied, regardless of whether or not the case can be brought within section 23(1) of the Act. It is accepted that a minister of state is, for the purpose of these provisions in section 56, a government department.

I should, however, mention at this point a preliminary submission made by Mr Barnes, counsel for the plaintiffs. He drew my attention to the decision of the House of Lords in *Town Investments Ltd* v *Department of the Environment* [1978] AC 359. This case concerned a tenancy which had been granted to the Ministry of Works. A question arose as to the applicability to the tenancy of the Counter-Inflation (Business Rents) Order 1972 and it became necessary to decide whether the demised premises were occupied by the tenant for the purposes of that order, notwithstanding that they were being used
E to accommodate civil servants in other government departments. The House of Lords, reversing a unanimous Court of Appeal, held that tenancies granted to ministers of state or to government departments were, in reality, tenancies granted to the Crown, and that the demised premises were being occupied by the tenant, the Crown, if occupied or used by any government servants, irrespective of the department to which they belonged. The point that the tenancy in question was held not by the Ministry of Works but by the Crown was succinctly made by Lord Diplock at p 381:

> In my opinion, the tenant was the government acting through its appropriate member or, expressed in the term of art in public law, the tenant was the Crown.

So, submitted Mr Barnes, there is in law no such creature as a
F tenancy held by or on behalf of a government department. It follows, he argued, that subsections (3) and (4) of section 56 were drafted on an erroneous view of the law and are devoid of any effect whatever.

I find it difficult to take this submission seriously. It is obvious that the subsections were intended to apply to tenancies which had been expressed to be granted or assigned to a minister of the Crown or to a government department and which were still so held. The legal misdescription established in 1978, 22 years after the 1954 Act was enacted, cannot change the true construction of the subsections. I hold that they apply to the lease of 40 Nottingham Place, which was expressed to be granted to the secretary of state.

Subsection (3) applies Part II of the Act to a tenancy held by a government department if "the property comprised therein is or includes premises occupied for any purposes of a government department". There is, therefore, apparently one criterion only to be satisfied. The subsection, unlike section 23(1), does not in terms add the requirement that the premises must be occupied by the tenant.

There are two other significant features of the wording of subsection (3). First, there is no requirement that the government department that holds the tenancy should be the same government department as that for whose purpose the premises are occupied. Second, the requisite occupation must be "for *any* purposes of a government department": contrast the language of section 23(1), where the occupation is required to be "for the purposes of the business carried on by" the tenant.

In my judgment, subsection (3) brings within Part II of the Act any tenancy held by a government department if the premises are being occupied for any purposes of any government department.

In the present case the only government department involved is the secretary of state and the only relevant purposes of a government department are his purposes under the National Health Services Acts. So, Part II of the Act applies to the secretary of state's tenancy of 40 Nottingham Place if it can be said that the property comprised in the lease is or includes premises occupied for any purpose of his under those Acts.

I turn to subsection (4) of section 56. If that subsection applies, then, for the purposes of subsection (3), the premises are to be assumed to be occupied for the purposes of a government department, whether or not that is in fact the case.

Mr Lewison, counsel for the defendant, submitted that subsection (4) applied to the facts of this case. No 40 Nottingham Place has, he said, been provided by the secretary of state to the authority. No rent is payable to the secretary of state for the property. So, he submitted, the statutory hypothesis contained in subsection (4) comes into effect.

Mr Lewison's approach appears to follow from a literal reading of subsection (4). It has, therefore, much to recommend it. But I am not satisfied that it is right. If his approach is right, it would seem that every tenancy held by every government department would be entitled to the protection of Part II of the 1954 Act unless rent for the premises was being paid to the government department under, presumably, a subletting. In the absence of a subletting, no rent would be payable, the statutory hypothesis of subsection (4) would bring the case within subsection (3) and Part II of the Act would apply, whether or not the premises were being occupied for any purpose of a government department and, perhaps, even if the premises were vacant. This seems a very odd result. I do not think it can be right.

I must, I think, look again at subsection (4) for its true construction and effect. The subsection provides:

> the last foregoing sub-section shall apply in relation to any premises provided by a government department without any rent being payable to the department therefor as if the premises were occupied for the purpose of a government department.

The reference there to "premises" is obviously a reference to the "premises" mentioned in subsection (3).

If subsection (4) is, then, incorporated into subsection (3), so that the two subsections are read together, subsection (3) would read as follows:

> Where a tenancy is held by or on behalf of a government department and the property comprised therein is or includes premises provided by a government department without any rent being payable to the department therefor, the tenancy shall be one to which Part II of this Act applies.

If the two subsections are read together in that way, the reference to the premises being provided by a government department cannot sensibly be taken to be a reference to the premises being provided by the tenant/government department. It can only, in my view, be read as a reference to the premises being provided by the landlord. Read thus, subsection (4) seems to be dealing with the case where one government department has granted a tenancy to another government department, but without any rent being payable thereunder, and to be applying Part II of the Act to the tenancy whether or not the premises are actually occupied for the purposes of a government department.

I would accept that there are difficulties in this construction. It seems to contemplate some sort of dispute between two government departments as to whether or not a tenancy purported to have been granted by one and appearing to be held by another qualifies for protection under Part II of the Act. It may be said that there is little reality in such a dispute. But I prefer this construction to that which would read the reference to premises being "provided" as a reference to the premises being provided by the tenant/government department. That latter construction seems to me to lead to such nonsensical consequences as to be impossible to be accepted as representing the true parliamentary intention.

It is also to be noted that the construction I have preferred is shown by the House of Lords decision in *Town Investments Ltd* v *Department of the Environment* to be based on an erroneous view of the nature of a tenancy or other interest in land held by a minister of state or government department. All such tenancies or interests in land are, held the House of Lords, simply held by the Crown. The Crown cannot very well grant a tenancy to itself or hold a tenancy from itself as landlord. So what is to be made of the construction of subsection (4) that I have adopted? That criticism is a formidable one, but the 1954 Act was, as I have already remarked, enacted some 22 years before the House of Lords clarified the law in the *Town Investments* case. In doing so it reversed a unanimous Court of Appeal. It is not clear to me that the legislators in 1954 may not have had in mind the possibility of one government department granting a tenancy to another government department. At all events, a construction of subsection (4) which proceeds on that footing produces, in my opinion, fewer anomalies and is less unsatisfactory than any other construction. It is the one that I adopt.

Accordingly, I hold that subsection (4) has no application to this case. If Part II of the Act is to apply, the secretary of state must bring his tenancy within subsection (3) unaided by subsection (4).

I must, therefore, now consider how the case stands under subsection (3) of section 56. Can the secretary of state satisfy the criterion that the property, 40 Nottingham Place, is or includes premises occupied for a purpose of his under the National Health Service Acts?

Although section 56(3), unlike section 23(1), imposes only one criterion, that one criterion cannot be satisfied unless there is someone in occupation of the premises whose occupation can be said to be for a purpose of a government department.

The occupation in question cannot, in my judgment, be that of the employee occupants themselves. Their occupation is not, so far as they are concerned, for the purposes of the National Health Service Acts. They occupy their respective flats for ordinary personal residential purposes.

Mr Lewison submitted that, in addition to the employee occupants occupying their respective flats, the authority, as managers of the flats, was, for the purpose of section 56(3) of the Act, in occupation of the whole property. Mr Barnes submitted that the connection between the authority and the property was insufficiently close to justify a finding that the authority was in occupation thereof.

The question of whether a manager of premises which are occupied by others can claim, by virtue of his management function, to be in occupation of the premises was considered in two cases cited to me. They were *Lee-Verhulst (Investments) Ltd* v *Harwood Trust* [1973] QB 204 in the Court of Appeal and *William Boyer & Sons Ltd* v *Adams* (1975) 32 P & CR 89, a decision of Templeman J (as he then was).

In *Lee-Verhulst (Investments) Ltd* v *Harwood Trust* the property comprised in the tenancy was a house which had been subdivided into 20 separate fully furnished apartments. The tenant lived in the basement and carried on the business of letting the apartments. There was a considerable degree of control which the tenant exercised over the whole property. The Court of Appeal upheld the conclusion of the judge at first instance that the tenant occupied the whole property for the purpose of his business. Sachs LJ said this at p215:

> The court must look at the substance of the position as a whole, take into account the various elements which have been discussed and then come to a common sense conclusion as to whether the tenant "occupies" the premises for the purpose of his business . . .

He went on:

> The county court judge was right in the course of his helpful judgment in attaching importance, *inter alia,* to the time and attention devoted by Mr Lee to running the business, to the services rendered to occupants, and to the control exercised over the occupancies. They were all relevant elements and he came to the correct overall conclusion.

In *William Boyer & Sons Ltd* v *Adams* the demised property had been divided into 11 units used for the purposes of light industry. Templeman J set out the facts of the case, referred to the *Lee-Verhulst* case and then said this at p92:

> In the *Lee-Verhulst* case the time devoted by the landlord to the activities of the business, the services rendered and the control exercised were extensive; the landlord's presence in the flats was marked by his chattels. In the present case all these factors are only present on a reduced scale or are missing. But in my judgment the activities of the defendants are sufficient to show that he is not so much acting as a landlord passively receiving rent but as the manager of a business actively earning profits by providing accommodation, facilities and services and by devoting time for this purpose. Some degree of presence by the defendant in the units is provided by his central heating and other fixtures, by the need for him to enter the units to maintain the services and the facilities which he provides for an inclusive rent and by his nightly visits. The organisation of the complex which enables the tenants of the units to carry on their work requires a manager and not only a landlord. G

He concluded that the defendant tenant was:

> in occupation of premises including the units for the purpose of his business of providing accommodation, power, facilities and services for the carrying on of light industries. H

I think the present case is on the borderline, but, on balance, in my opinion, the authority is in occupation of 40 Nottingham Place for the purposes of section 56(3). The features of the case which lead me to that conclusion are, particularly, that one or more of the flats will usually be found to be vacant and under the control only of the authority; that the authority retains keys to all flats and visits them regularly; that under the agreements exclusive possession is not given to the occupants; that the authority provides not simply furniture but crockery and blankets at the premises; and that the authority's maintenance of the flats includes decorating and the carrying out of J quite trivial repairs. The evidence shows, in my view, that 40 Nottingham Place, like the units with which Templeman J was concerned, needs not just a landlord but a manager. On the evidence that is the function discharged by the authority. The conclusion is, in my view, justified that the authority is in occupation of 40 Nottingham Place for the purposes of section 56(3).

The next question is whether or not the authority is in occupation of 40 Nottingham Place "for any purposes of a government department". That requirement will be satisfied if it can be shown that the authority is in occupation for the purposes of the secretary of state in carrying out one or other of his functions under the National Health Service Acts. Mr Barnes submitted that where the property in question was used for the residential purpose of housing employees K in the relevant business, the property could not be said to be used for the purposes of the business unless it could be said that it was necessary for the purposes of that business that the employees should reside there. Convenience was not enough. On the facts of the present case, however convenient it might be for the running of the local hospitals to have 40 Nottingham Place available for accommodation, it could not, submitted Mr Barnes, be said to be necessary.

Mr Barnes relied on *Chapman* v *Freeman* [1978] 1 WLR 1298, a decision of the Court of Appeal, as authority for his submission. In that case the defendant tenant owned a hotel. He rented a cottage for the accommodation of hotel staff. It was accepted that he was in L occupation of it. The question, however, was whether he occupied the cottage for the purpose of his hotel business. At p1300 Lord Denning MR said:

> Speaking generally, the test is whether it is *necessary* for the individual to live in the house in order to perform his own particular duties properly; or whether it is just for *convenience* that he should live there in connection with his duties (Lord Denning's emphasis).

Then a little later on Lord Denning expressed his conclusion thus:

> The truth is that it is a dwelling-house simply for the convenience of the person carrying on the business. It is not a business tenancy.

Geoffrey Lane LJ (as he then was) said, however, at p1301:

> It was merely a matter of convenience that the Andrews should be housed in M this particular cottage. That, to my mind, is not enough. The tenant must go beyond mere convenience and show that the occupation was in furtherance of his business activities — in other words, that it was for business reasons that he was occupying the cottage and not merely for reasons of convenience.

Geoffrey Lane LJ, therefore, while agreeing with Lord Denning that mere convenience was not enough, was not proposing quite the same test as was Lord Denning. Lord Denning was proposing the test of necessity. The test proposed in this passage by Geoffrey Lane LJ is that the "occupation was in furtherance of his business activities".

The third member of the court, Eveleigh LJ, said at p1302:

> We must look to see what the servant is doing there. Why is he there? The servant is not occupying for the purpose of the hotel business but simply as a resident. He is not there for anything to do with the hotel, although his

A employment at the hotel provided the reason for his being permitted to live there.

This approach is concentrating on the servant's reason for occupying rather than on the master's reason for putting the servant in occupation. Lord Denning and Geoffrey Lane LJ had both approached the matter on the latter footing which must, if I may respectfully say so, be right.

These dicta do not, in my view, establish a common ratio. I agree with Mr Lewison that the case is not authority for the proposition that the test of necessity must be passed before tenant's occupation of premises in order to provide residential accommodation for his employees can be taken to be occupation for the purposes of his B business.

The point was further considered by the Court of Appeal in *Groveside Properties Ltd* v *Westminster Medical School* (1983) 47 P & CR 507.* In this case, in my view, the Court of Appeal in effect adopted Geoffrey Lane LJ's test that the occupation must be "in furtherance of" the tenant's business. The case concerned a flat leased by the Westminster Medical School and used for providing residential accommodation for four medical students. It was held, first, that the school's overall control of the flat justified the conclusion that it was in occupation of the flat and, second, that its occupation was in furtherance of its educational purposes. Fox LJ referred to *Chapman* v *Freeman* at p511, cited the passages from the judgments of Geoffrey Lane LJ and Eveleigh LJ that I have cited and C continued:

> That decision does not lead me to any different view of the present case. The user by the students is not simply for convenience, and it is not simply as a residence; it is to achieve an educational purpose for the advancement of their medical training, and it is to assist the achievement of the same purpose and to the same end that the medical school occupies the flat.

Stephenson LJ at p515 said this of *Chapman* v *Freeman*:

> I do not think that that decision requires us to apply any kind of "tied cottage" or "service tenancy or occupancy" test for the tenancy of such a body as this school, or to treat desirable arrangements of this kind as mere matters of convenience.

In my judgment, the question in the present case is whether the D occupation of 40 Nottingham Place by the authority is an occupation in furtherance of one of the functions of the secretary of state under the National Health Service Act. In my judgment, it clearly is. The secretary of state has power under section 2 of the 1977 Act "to do any . . . thing whatsoever which is calculated to facilitate, or is conducive or incidental to, the discharge of any duty imposed on him" under the Act. He has power under section 87(1) of the 1977 Act to acquire property "to provide residential accommodation for persons employed for any of the purposes of this Act".

The exercise of these functions of the secretary of state has been delegated to the authority. Pursuant to that delegation and to the duty cast on it by section 14 of the 1977 Act, as amended, the authority is managing 40 Nottingham Place. In so doing it is E discharging that duty and exercising one of the functions of the secretary of state. Accordingly, in my judgment, the authority's occupation of 40 Nottingham Place is an occupation for a purpose of a governing department. The criterion of section 56(3) is, in my judgment, satisfied. Part II of the Act applies.

That is unfortunately not an end to the difficulties of this case. Subsection (3) of section 23 must now be considered. It is in these terms:

> In the following provisions of this Part of this Act the expression "the holding" in relation to a tenancy to which this Part of this Act applies, means the property comprised in the tenancy, there being excluded any part thereof which is occupied neither by the tenant nor by a person employed by the tenant and so employed for the purposes of a business by reason of which the tenancy F is one to which this Part of this Act applies.

That subsection determines the extent of the holding in respect of which a new tenancy can be claimed by the secretary of state under the Act.

Mr Barnes submits that the secretary of state is the tenant — so he is; that at best the authority is in occupation — I have held that it is — and that the secretary of state is not in occupation of any part of 40 Nottingham Place. Moreover, he submits, the employees who live in the flats at 40 Nottingham Place, are not the employees of the tenant, the secretary of state: they are, he says, employees of the authority. So, he submits, the effect of section 23(3) on the present case is that, even if under section 56(3), Part II of the Act applies to the tenancy, the whole of the premises comprised in the tenancy are withdrawn from "the holding" in respect of which a new tenancy can be sought. There is, it would follow, "no holding", in respect of which the secretary of state can claim a new tenancy. Mr Lewison's answer is that, for the purposes of section 23(3), the authority's occupation of 40 Nottingham Place must be taken to be the occupation of the secretary of state, the tenant. He submits that the authority, in carrying out the management function which has, in my view, justified the conclusion that it is in occupation of 40 Nottingham Place, is acting as the agent of the secretary of state.

I have already referred to sections 13 and 14 of the National Health Service Act 1977, as amended. Under section 13 a regional health authority on which a delegated function devolves exercises the function on behalf of the secretary of state. Under section 14 a district health authority is directed to exercise "such of the functions exercisable by the regional health authority as are specified . . .". So a district health authority, too, acts on behalf of the secretary of state. Mr Lewison is, in my view, right in submitting that the authority is in occupation of 40 Nottingham Place on behalf of the secretary of state.

Mr Barnes submitted, however, that a statutorily authorised delegation of the sort provided for by sections 13 and 14 did not constitute the authority the agent of the secretary of state. The authority, he submitted, discharged its duty under the delegated powers as principal. This submission is, however, in my view, inconsistent with authority. In *Pfizer Corporation* v *Ministry of Health* [1964] Ch 614 in the Court of Appeal and [1965] AC 512 in the House of Lords, the question was whether the Minister of Health was entitled to supply a patented drug to National Health Service hospitals for use in those hospitals. He was entitled under the relevant statutory powers to authorise the use of the drug "for the services of the Crown". The National Health Service hospitals in question were managed by regional health boards, which then discharged the functions now discharged by the regional, area and district health authorities. Regional health boards, too, were statutory corporations. The contention of the plaintiff in the case was that the regional health boards to whom the drug was supplied by the minister for use in the hospitals were using the drug for their own purposes and that their use of it in the hospitals could not be properly described as use "for the services of the Crown". In the Court of Appeal Diplock LJ (as he then was) said this at p 651:

> The Ministry of Health, of which the minister is the head, was established by the Ministry of Health Act 1919 and is plainly a "government department"; but it is contended by counsel on behalf of the plaintiffs (1) that the use with which this appeal is concerned is not use by a government department or alternatively (2) that it is not use for the services of the Crown.

Diplock LJ then referred to various statutory provisions and continued at p 652 with this:

> The scheme of the Act is thus plain. The duty to provide hospital and specialist services is imposed upon the minister. It is in its nature a duty which he can only perform vicariously through agents acting on his behalf. The Act requires him to do so through the immediate agency of the Regional Hospital Boards. The Regional Hospital Boards, being corporations, can themselves only do the physical acts involved in the provision of the services on behalf of the minister, vicariously through their officers and servants. Any act done by an officer or servant of a Regional Hospital Board for the purpose of providing hospital or specialist services is accordingly done on behalf of the minister in performance of the statutory duty which is imposed upon him. Their acts are acts of a government department. Counsel for the plaintiffs placed considerable reliance upon section 13 of the Act, as showing that the Regional Hospital Boards exercised their functions as principals and not as agents of the Minister. But the section seems to me to be directed to quite a different matter. The National Health Service Act 1946 was passed before the Crown Proceedings Act 1947. At that date the fact that a Regional Hospital Board was acting on behalf of a minister of the Crown would have entitled it to shelter behind the immunity of the Crown from suit except by petition of right: see *Feather* v *The Queen; Raleigh* v *Goschen*. All that section 13 does is to re-assert that a Regional Hospital Board is acting on behalf of the minister (ie on behalf of the Crown) but to provide that for the purpose of legal proceedings for the enforcement of rights and liabilities it shall be treated, notwithstanding that it is an agent of the Crown, as if it were acting as a principal and shall sue and be sued in its own name. Such a provision would have been unnecessary unless a Regional Hospital Board in exercising its functions under the Act, was acting on behalf of the Crown.

Section 13 to which Diplock LJ was referring is to the same effect for all relevant purposes as para 15 of Schedule 5 to the 1977 Act which I have read.

* Editor's note: Also reported at (1983) 267 EG 593.

Willmer LJ at p 641 said this:

The section

and he, too, was referring to section 13

does not do anything to alter the fact that Regional Hospital Boards in exercising their functions, by themselves and through the Hospital Management Committees in their area, are discharging the duties laid upon the minister.

The judgment of the Court of Appeal was upheld by the House of Lords. No doubt was expressed as to the accuracy of the analysis of the relationship between the minister and the regional hospital board as expressed, in particular, by Diplock LJ.

In my judgment, the authority, in acting on behalf of the secretary of state in discharge of the functions laid on the secretary of state by the National Health Service Acts, is acting as agent for the secretary of state. The relationship justifies, in my view, a conclusion that, for the purposes of section 23(3) of the Landlord and Tenant Act 1954, the authority's occupation of 40 Nottingham Place is the occupation of the secretary of state, the tenant. In my opinion, for the same reason, the persons employed by the authority in discharge of its delegated functions are, for the purposes of section 23(3), to be treated as the employees of the secretary of state.

Accordingly, I conclude that the holding in the present case extends to the whole of 40 Nottingham Place and that section 23(3) does not exclude any part of the premises comprised in the lease from the holding in respect of which the secretary of state is entitled to claim a new lease.

I therefore dismiss the action.

The action was dismissed with costs.

Court of Appeal

December 4 1985

(Before Lord Justice O'CONNOR and Mr Justice LATEY)

JONES v JENKINS

Estates Gazette February 15 1986

277 EG 644-646

Landlord and Tenant Act 1954, Part II — Appeal by tenant from decision of county court judge refusing a grant of a new tenancy — Appellant occupied a house from which she carried on a nature cure business and also held tenancies of two flats in an adjoining house of which the respondent was the landlord — It was the use made of these flats in connection with the appellant's business which gave rise to the present proceedings — The respondent served a 1954 Act notice of termination in respect of both parts and they were treated by the judge, and by the Court of Appeal, as constituting the holding in respect of which a new tenancy was sought — The respondent's grounds of opposition for the purpose of the appeal were confined to those under paras (c) and (g) of section 30(1) — The county court judge held that the respondent succeeded under both heads, but the Court of Appeal decided that the judge had fallen into error in each case — As regards para (g), the judge was wrong, in the light of the decisions in *Lee-Verhulst (Investments) Ltd* v *Harwood Trust* and *Bagettes Ltd* v *G P Estates Ltd*, in holding that para (g) could apply — There was no evidence that the respondent was actually going to occupy the premises for the purpose of a business to be carried on by her therein — As regards para (c) there was no evidence to support the judge's finding that three rooms in the basement flat had been converted into a laundry in breach of the tenancy obligations — There was evidence to support a finding of a breach in the use of the top flat as a massage parlour, but, as the judge had been partly in error in regard to para (c), the exercise of his discretion to refuse a new tenancy might have been vitiated — Unfortunately for the parties, the case would have to go back for retrial by another judge, as it would not be right for the Court of Appeal simply to allow the appeal and order a new tenancy — Order for rehearing accordingly

This was an appeal by the tenant, Miss W E Jones, from a decision G at Malvern County Court in favour of the landlord, Mrs M I Jenkins, whereby the appellant was refused a new tenancy of a basement and top-floor flat at 7 College Grove, Malvern. The appellant tenant was herself the occupier of 5 College Grove and conducted a business known as "The Malvern Nature Cure Centre".

J R West (instructed by Gaynor Smith Owen & Co, of Malvern) appeared on behalf of the appellant; J K Quirke (instructed by Parkinson Wright, of Worcester) represented the respondent.

Giving judgment, O'CONNOR LJ said: This is a tenant's appeal against the refusal of the learned recorder who tried the case in October 1984 to grant her a new lease of certain premises in Malvern under Part II of the Landlord and Tenant Act 1954. The facts giving rise to the dispute were domestic and simple. There are two houses, 5 H and 7 College Grove. The tenant, Miss Jones, occupies no 5 and runs a business there as "The Malvern Nature Cure Centre". The landlord of no 7 (Mrs Jenkins at the time of the trial — now Mrs Buckland, by which name I will refer to her for the rest of this judgment) owns no 7, and that is divided into four flats at the moment. She lives on the ground floor and has the garden, which is substantial, as we can see from the photographs. The basement flat, naturally called the garden floor now — part of it at least gives on to the garden — is one. The first floor, with which we are not concerned, is let to a teacher, and the top floor formed part of what has emerged as the business tenancy.

What appears to have happened was that Miss Jones was being J helped in her business by a man named Foreman, who was also a friend of Mrs Buckland at that time, and Mrs Buckland agreed, first of all, to put some money into the business and, second, to let the basement flat for the purposes of putting up clients of the business, as she thought, and to let the top flat to be occupied by Mr Foreman while he worked there. At the time when all these things were being agreed it seems that there was an intention that Mr Foreman and Mrs Buckland should depart for Australia. I need say no more about it because it never happened. The relationship became unhappy for a variety of reasons, as the evidence showed before the judge. To start with, while the kitchen in no 5 was being redone, it seems that Mrs Buckland agreed that the laundry could be brought over to no 7, the basement, and done there, and a washing machine was put in. But K apparently that arrangement went on for keeps, and we have photographs of baskets of laundry being carried through the garden and brought over to no 7. There was a dispute with Foreman about an old ambulance which he had brought there, about knocking down fences and coming across from no 5 and so forth. In addition, Mrs Buckland did not take kindly to the fact that, so far as the basement being used for guests was concerned, the cook was put in there, who sometimes had her child, but, much more importantly, she did not take kindly to the fact that the top-floor flat, instead of being used as accommodation for Mr Foreman while he worked next door, was turned into a massage parlour to which clients were brought across and introduced.

L

Both these flats were let at a weekly tenancy. The rent was calculated as at £30 for the basement and £20 for the top. It may well be that there was never any intention to have a business letting of the top at all, but that is beside the mark now, because a 1954 Act notice was served in respect of both parts and, for the purposes of the Landlord and Tenant Act, they were treated by the judge, and must be treated by us, as the holding in respect of which a new tenancy was sought. Notice to quit was given, and the relevant notices were served. Application for a new tenancy was made on October 21 1983. The tenancies were determined as at April 30 1984. In her answer Mrs Buckland opposed the grant of a new tenancy on a number of grounds which are available under section 30 of the Act. In the end we are concerned only with two, namely, section 30(1) paras (c) and (g). M

As to (c) she said:

The tenant ought not to be granted a new tenancy in view of other substantial breaches by her of her obligations under the current tenancy or for some other reason connected with the tenant's use or management of the holding,

and remember that the holding was the two flats: one in the basement, which has got six rooms in it; and one upstairs, which has got four or five rooms in it.

Under (g) she said in her answer:

That on the termination of the current tenancy I intend to occupy the building for the purposes, or partly for the purposes, of a business to be carried on by me therein or as my residence.

In that paragraph of her answer for "building" one has to read, in

A order to comply with the section, "holding", because para (g) provides that:

> Subject as hereinafter provided, that on the termination of the current tenancy the landlord intends to occupy the holding for the purposes, or partly for the purposes, of a business to be carried on by him therein, or as his residence.

Mrs Buckland, perfectly properly and legitimately, said that the business that she intended to carry on was to let the two flats for residential purposes and not for business purposes. She also said that she proposed to divide the basement flat at least into two; that there was plenty of room to make two flats, and therefore she would be able to make more money. That was the broad outline of the case that
B went before the learned recorder.

There were, therefore, two issues. Was she entitled to resist the application for the grant of a new tenancy, which *prima facie* the tenant was entitled to, on either of those two grounds? So that the recorder had to apply his mind to whether there had been substantial breaches of the existing agreement or of the tenant's obligations under the current tenancy, or for any other reason connected with the tenant's use or management of the holding. It will be seen that the wording of para (c) inevitably gives a discretion to the trial judge as to whether, even if there is a substantial breach, nevertheless he is entitled to order the grant of a new tenancy. It is obvious that it is desirable that there should be a discretion where para (c) is concerned; this is not so when one comes to para (g). There, if the
C landlord satisfies the court that he intends to occupy the holding for the purposes, or partly for the purposes, of a business to be carried on by him therein or as his residence, he is entitled to have the application for a new lease refused, subject of course to the period that he has been landlord in the proviso to the section.

Unfortunately, I have come to the conclusion that, although the learned judge appreciated the difference between the two sections (both counsel who appeared before him have appeared in this court and we have been told that he was referred to the relevant law), he has fallen into error in two substantial ways. First of all, he came to the conclusion that para (g) had been satisfied by the landlord. He did it very shortly. He said:

> D I am satisfied therefore that the premises are within the Act. Working backwards the landlord relies on the [section] 30(1)(g). Her residence doesn't apply; she says that she intends to let the rooms to tenants; it cannot be doubted that letting rooms to a tenant is a business. Mr West

who appeared for the tenant

> says that if she lets she doesn't occupy but I am satisfied particularly in view of the case of *[Lee-Verhulst (Investments) Ltd* v *Harwood Trust]* [1973] QB 204 where a landlord let fully furnished rooms. Mrs Jenkins

that is, Mrs Buckland

> says that she would alter these rooms and let them and I am satisfied that that would be her occupation. Mr Quirke says that if I found that it would be conclusive, but I will deal with the other matters.

E He then rejected the plea under section 30(1)(f). There has been no respondent's notice on that topic, because it had been supported by the contention that Mrs Buckland proposed to convert the basement at all events into two flats.

As far as section 30(1)(c) was concerned, the judge found:

> Mrs Jenkins said she let the ground floor as a residential annexe and thought it would be occupied by middle class middle-aged people. She permitted the use of the kitchen as a laundry whilst work was being done at number 5, now she says the basement is being used for a laundry. The photograph shows staff with bundles of laundry. She also says that two members of staff occupy the rooms there and that a quiet room is occupied by someone who is rowdy. I am satisfied that to use a residential annexe as a place where there are now three rooms used as a laundry is a substantial breach of the terms agreed. I am quite satisfied that the agreement on the top floor was for Mr Foreman to occupy
> F and not for long for a residence for himself rent being paid by the Nature Cure Centre. Now he sleeps there and other people constantly go there for treatment. An entirely different use from that contemplated and therefore the landlord succeeds in her contention under ground (c).

Both those matters are challenged on behalf of the tenant by Mr West, who submits that, as far as para (g) is concerned, in law the learned judge, on the evidence before him, simply could not find that the requirements of para (g) had been satisfied. Our attention has been drawn to two decisions of this court. *Lee-Verhulst*, which was cited to the learned judge, is a case where the business of the tenant company was letting furnished rooms with services. The effective owners of the company lived in one of the flatlets or flats in the premises. They had a kitchen, a boilerhouse and storage places, and there were 20 rooms which were service rooms: furnished; serviced by the landlords, who had keys so that the chambermaids could go in and make the beds and keep the place clean and so forth; food was available to be brought up — there were no cooking facilities in the rooms, except for a gas ring for boiling a kettle. The question was as to whether, when the lease of the whole building fell in, the tenants were entitled to a new lease on the basis of section 23. The argument in that case was that they were not within the wording of section 23(1):

> The property comprised in the tenancy is or includes premises which are occupied by the tenant and are so occupied for purposes of a business carried on by him for those and other purposes.

Sachs LJ said at p 213:

> Is there anything in the Act of 1954 which precludes the court from giving to the word "occupied" in section 23 its natural and ordinary meaning in the context of the subject matter of that Act — a meaning which would in the set of circumstances above described clearly lead to it being held that the tenant did occupy the premises for the purpose of the business? Being unable to find anything in the Act which so precludes the court, I have come to the conclusion that this tenant did so occupy the whole of the premises.
>
> For reaching that conclusion it is neither necessary nor desirable to provide a definition of that word which would deal with all the greatly varying sets of circumstances that can exist. As a number of elements have been taken into account, each of a physical nature and each involving a degree of presence on the part of the tenant personally or by goods under his ownership, it is however as well to observe that it could be proper in some other case to reach the same conclusion even if one or more of those elements were subtracted. For instance if the furniture was that of the occupants or if some of the services were not rendered or if the occupancies were not so much controlled, there could still be an occupation by the tenant of the premises as a whole. Much depends on questions of degree.

Earlier in his judgment Sachs LJ had considered the case of *Wheat* v *E Lacon & Co Ltd* [1966] AC 552. That was an occupier case. Then, on p 213, in a passage before that which I have cited, he said:

> That a person who by himself or his agents resides in an apartment house, "lets" the apartments on terms of a month's notice, and in return for the "rent" provides the chamber-maid and other services described above, is carrying on a business is plain, and in that behalf it makes no difference whether his contact with the occupant results in law in a tenancy or in a licence. If it helps to point to statutory recognition of that fact, then it can be found in section 17(3)(b) of the Landlord and Tenant Act 1927 where reference is made to "the business of subletting the premises as residential flats".

In *Lee-Verhulst* the court had been referred to the decision of this court in *Bagettes Ltd* v *G P Estates Ltd* [1956] Ch 290. In that case once again the tenants of a block of flats were asking for a new lease under section 23. At the time of the hearing they contended that they brought themselves within section 23 for these reasons: that one flat was occupied by, I think, a caretaker, that there were the common parts of the building — staircases, passages and boilerhouse, and that they in fact had three vacant flats in hand. They submitted that therefore they were in occupation of part of the premises for the purposes of their business of letting flats. Jenkins LJ (as he then was), who gave the principal judgment, found that the property comprised in the tenancy included premises which were occupied by the tenant and occupied for the purposes of a business carried on by him for those and other purposes. He held that nevertheless a new tenancy could not be granted because of the definition of the holding in section 23(3) which provides:

> In the following provisions of this Part of this Act the expression "the holding", in relation to a tenancy to which this Part of this Act applies, means the property comprised in the tenancy, there being excluded any part thereof which is occupied neither by the tenant nor by a person employed by the tenant and so employed for the purpose of a business by reason of which the tenancy is one to which this Part of this Act applies.

The court held that, as it was only possible to grant a new lease of "the holding", the holding on the facts of that case comprised only the caretaker's flat and common staircases and passages and boilerhouse, all of which of course were not being used for any business once you disconnected them from the flats in the building.

Both of those cases are binding on us. I have no difficulty in seeing the difference between *Lee-Verhulst* and *Bagettes*. As far as the present case is concerned, I am quite satisfied that the business, for so it was, and is, of Mrs Buckland of letting flats (it matters not that it is on a small scale), the holdings, which are the flats — and the holding here was the two flats — has got to be given the same meaning when we are considering section 30(1)(g) as it has elsewhere. When I ask myself was there any evidence that she was proposing to occupy the holding for the purposes of a business to be carried on by her therein, I am satisfied that she was not. It was submitted by Mr Quirke on her behalf that it was sufficient that there was evidence which the judge appears to have accepted that she intended to convert a substantial part of the holding, namely, the bottom flat, and that she would be occupying it during that period of conversion and therefore that she fell within para (g). He placed reliance on the case of *Cam Gears Ltd* v *Cunningham* [1981] 1 WLR 1011, where Oliver LJ gave the first judgment. That was a case where the holding comprised a vacant lot which had been concreted and was used as a car park. The landlords, who apparently had neighbouring premises, wanted to build a new building on it for the purposes of their business, which I think was a garage or filling station of some kind — a workshop and a waiting room. Although the case would have fallen fairly and squarely under para (f), the notice had been given under para (g) only, and the court held that para (g) applied; that the holding was going to be occupied for the purposes of a business to be carried on therein, although a new building was going to be put up on the holding; the whole of it was to be used for the new premises. Mr Quirke has submitted, therefore, that the court would be entitled to apply para (g) to the facts of the present case. For my part, I cannot agree with that. It seems to me that *Cam Gears* v *Cunningham* is an exceptional case, and we are not concerned with the whole of the holding being taken over and used in the sense which that one was going to be used. The present case differs from it, because the use which Mrs Buckland proposes to put the converted flat or flats is to relet them; and that immediately takes them outside the holding. She is not going to occupy them once they are relet; that is clear. I do not think that temporary occupation can possibly be sufficient to bring the case within (g). If it were, holding any commercial premises in hand for any purpose, such as redecoration or otherwise, would be sufficient occupation to provide an answer to a tenant's request for a new lease or *prima facie* rights to a new lease.

I am satisfied that that is not the proper construction of para (g) and I have come to the conclusion that the learned judge fell into error in holding that the landlord had satisfied para (g).

Now I turn to the finding on para (c). I can deal with that very shortly. I am quite clear that he made a mistake. There was no evidence to support the finding that three rooms had been converted into a laundry in the basement. There simply was no evidence to that effect. I of course take into account that he was not asked to agree the note of his judgment until many months after it had been given. I of course take into account that the judgment was given at the end of a long hearing and that it may be that he did not mean what he said, but we have got to work on the basis that he did make that finding, and I am satisfied that it is quite plainly wrong.

As far as the top flat is concerned, the breach which he found was that he held that it was a term of the letting that it should be occupied by Mr Foreman and not used as a massage parlour. Mr West has submitted that the learned judge has wholly misconceived what he was doing. He made the finding (and I repeat it):

> I am quite satisfied that the agreement on the top floor was for Mr Foreman to occupy and not for long for a residence for himself rent being paid by the Nature Cure Centre. Now he sleeps there and other people constantly go there for treatment. An entirely different use from that contemplated.

First of all, he submits that there is no sufficient finding as to what the terms of the agreement were; that it is not good enough to talk about compensation. I cannot read that finding in that fashion. It seems to me that the learned judge was finding that there was a substantial breach and that it was a term that the top floor was to be used as a residence for Mr Foreman; that was a restriction on the use of the top floor and that it was broken when it was used as a massage parlour. Mr West submitted that by adding the "not for long" it shows that the learned judge was not dealing with the question properly because he was harking back to the evidence that at the time of the letting Mrs Buckland and Mr Foreman were planning to go off to Australia and that it is to be concluded that the judge was reading into the agreement that that would put an end to the tenancy so far as it concerned the top floor, whereas that cannot be right because that would not be lawful under the Act.

I do not read the passage in the judgment as making a finding of that kind. I think that, when the passage is read as a whole, that was only an aside and that the reality is that he was finding that it was a term that the top floor was to be used only as a residence for Mr Foreman and that the term was broken when it was used as a massage parlour by others.

So, as far as para (c) is concerned, one half of the breach which the judge has found is not to be supported on the evidence; the other half, in my judgment, is. The difficulty about that is that the judge has got G
to exercise a discretion, and it seems to me inevitable that in this situation, where he has gone hopelessly wrong on one half, it is impossible to say that the exercise of his discretion to refuse a new lease and support the landlord's answer has not been vitiated. This is because, although he makes no reference to the exercise of his discretion, he was referred to the relevant authority, and the wording of para (c) shows that it is discretionary, it is inevitable that the discretion has not been properly exercised because in part it has been exercised on a false basis. Mr West has submitted that in those circumstances the appeal should be allowed and a new lease ordered. I do not think that would be just. It seems to me that there has been a mistrial of these proceedings (I need say no more about it) and that the learned judge has simply not tackled the case with proper clarity. H
He has made a mistake on the basement, but there was plenty of evidence to support the finding that the top flat was to be for the occupation of Mr Foreman and nothing else; there was a breach and it requires the exercise of discretion as to whether that satisfies para (c).

We invited the parties to consider whether this court, in order to save expense, should try to do the necessary exercise, but it was pointed out that that would require bringing the facts up to date, as the case is a year old already, and we cannot do that.

Unfortunate though it is for the parties (and I am deeply conscious that it will be putting them to further expense unless they can come to terms), I see no alternative but to allow this appeal and to send the J
case back for retrial by another judge.

LATEY J agreed with the order proposed and the reasoning of O'Connor LJ and did not add anything of his own.

The appeal was allowed to the extent that the case should be remitted for rehearing before another judge; the costs below and of the appeal to be determined by the judge who rehears the case. Legal aid and taxation of appellant's costs ordered.

Chancery Division

January 31 1986

(Before Mr Justice HOFFMANN) K

MFI PROPERTIES LTD v BICC GROUP PENSION TRUST LTD

Estates Gazette March 1 1986

277 EG 862-872

**Landlord and tenant — Construction of rent review clause in underleases of commercial premises — This was another case in which the issue was whether the hypothetical lease assumed for the purpose of determining the market rental value at the L
review date was or was not to include the rent review provisions in the actual tenancy — This depended on the correct construction of the words "other than those relating to rent" — Landlords contended that, on the plain meaning of the words, the rent review clause should be excluded as a term relating to rent — Tenants submitted that such a construction was so contrary to commercial common sense as to require the words to be construed more narrowly — The judge, in some general observations, said that it was not disputed that the particular language of the review clause must be construed in the context of the document as a whole and against the commercial background of the transaction — The purpose of M
such clauses points to a *prima facie* presumption that the parties intend the hypothetical letting to be for the residue of the actual lease on the same terms and in the same circumstances as actually exist at the relevant date, although in some cases there is a clearly expressed intention to depart from reality — Where, as here, the language is capable of more than one meaning the court is entitled to select that meaning which serves the apparent commercial purpose of the clause — A literal interpretation, as put forward by the landlords, would exclude from the hypothetical tenancy not only the rent review provisions but other terms relating to rent,**

A **such as the provision for payment of the amount of an insurance premium by way of rent, payment of a service charge by way of further and additional rent and a proviso for re-entry for non-payment of the reserved rents — These exclusions would flout business common sense and face the valuer with extraordinary assumptions — The hypothetical lease would be a different animal from anything existing in real life — The narrower construction propounded by the tenant was therefore to be preferred — The judge considered, agreeing with the views of Warner J in *Datastream International Ltd* v *Oakeep Ltd*, that the cases of *National*
B *Westminster Bank plc* v *Arthur Young McClelland Moores & Co* and *Equity & Law Life Assurance Society* v *Bodfield Ltd* "gave less weight than I would have done to the commercial purpose of the clauses there in question" — Declaration in favour of tenants' construction**

This was a summons by which the plaintiff tenants, MFI Properties Ltd, sought the determination of the court as to the true construction of the rent review clause in an underlease and reversionary underlease of commercial premises known as Unit A, Norwich City Trading Estate. The defendants were the landlords, BICC Group Pension Trust Ltd.

Nigel Hague QC and Nicholas Patten (instructed by Theodore
C Goddard & Co, agents for Andrew M Jackson & Co, of Hull) appeared on behalf of the plaintiffs; Nicholas Dowding (instructed by Stanleys & Simpson North) represented the defendants.

Giving judgment, HOFFMANN J said: This summons raises a short question of construction on two identical rent review clauses in an underlease and reversionary underlease of commercial premises in Norwich. The initial underlease is for a term of 15 years from December 3 1974 and the reversionary underlease for a term of 35 years commencing upon the expiry of the initial underlease. Both underleases provide for rent reviews at intervals of five years, at which dates:

> the rent hereinbefore first reserved shall be revised so as to equal the rent at which having regard to the terms of this sub-underlease (other than those
> D relating to rent) the demised premises might then reasonably be expected to be let in the open market by a willing landlord to a willing tenant for a term of twenty years with vacant possession and on the assumption (if not a fact) that the demised premises are equipped with all necessary and usual heating, lighting and fire-fighting equipment for a building of the nature of the demised premises which if installed shall be regarded as landlord's fixtures for the purposes hereof there being disregarded the matters set out in section 34 of the Landlord and Tenant Act 1954 as amended by section 1 of the Law of Property Act 1969 ...

The issue in these proceedings is whether the words in parenthesis "other than those relating to rent" require an assumption that the hypothetical letting contains no rent review. The landlord says that upon the plain meaning of the words, the rent review clause is a term
E relating to rent. It must therefore be assumed to be excluded. The tenant says that while this may be the literal meaning, the consequences of such a construction are so contrary to commercial common sense that the words should be construed more narrowly. It is argued that in the context in which they appear, they are capable of meaning and should be held to mean only those terms concerning rent which the rent review is intended to revise.

The lease contains numerous terms which can upon a literal construction be said to relate to rent. Thus the reddendum deals not only with the amount of the initial yearly rent but also the manner in which it is to be paid. It includes provisions for payment by way of "further rent" of a sum equal to the premium which the landlord has to pay for insurance which elsewhere in the lease he covenants to
F provide. The tenant covenants to pay the rents reserved in the reddendum and to pay "by way of further and additional rent" a proportion of the cost of various services which elsewhere in the lease the landlord covenants to provide. There is the usual proviso for re-entry for non-payment of the rents reserved and a term for suspension of the liability to pay rents if the premises are destroyed by fire.

It is not disputed that the language of the parties must be construed in the context of the document as a whole and against the commercial background to the transaction. A rent review clause is designed to deal with a particular commercial problem, namely, that of the tenant who wants security of tenure for a lengthy term, and a landlord who, in times of inflation or a rapidly changing property market, does not want to commit himself to a fixed rent for the whole of that term. The clause therefore permits the rent to be periodically revised to an amount calculated on the assumption of a fresh letting at the relevant date.

The terms upon which such hypothetical letting must be assumed to take place will naturally depend in each case upon the particular language of the rent review clause. In general, the purpose of such clauses would point to a *prima facie* assumption that the parties intended the hypothetical letting to be for the residue of the actual lease held by the tenant on the same terms and in circumstances which actually exist at the relevant date. Thereby, the tenant would be required to pay on the assumption that he was being granted the interest which he actually held on the rent review date. But the language of the clause may show clearly that a departure from reality was intended. In some cases it will be easy, even for an outsider who was not privy to the negotiations between the parties, to see why such an assumption should have been made. For example, fairness to the landlord might explain an assumption that the tenant has complied with his repairing covenants even if he has not in fact done so. Similarly, fairness to the tenant will explain an assumption which excludes from consideration any improvements which the tenant has made at his own expense or his acquisition of a goodwill which he would pay a higher rent to protect. But there will also be cases in which the language used by the parties shows beyond doubt that they intended an assumption for which, to a third party who knows nothing of the negotiations, no commercial purpose can be discerned. In such circumstances the court has no option but to assume that it was a *quid pro quo* for some other concession in the course of negotiations. The court cannot reject it as absurd merely because it is counter-factual and has no outward commercial justification. On the other hand, if the language is capable of more than one meaning, I think the court is entitled to select the meaning which accords with the apparent commercial purpose of the clause rather than one which appears commercially irrational.

The present clause exhibits assumptions in both the categories I have mentioned. The assumption of a letting with vacant possession and the disregard of the matters set out in section 34 of the Landlord and Tenant Act 1954 are easily explicable. The assumption of a 20-year term and the existence of heating, lighting and fire-fighting equipment have no explanation readily accessible to a person who was not privy to the negotiations. It may be that the assumption of a 20-year term is partly connected with the existence of the reversionary underlease and partly in order to simplify the task of the valuer. The reference to the heating, lighting and fire-fighting equipment may be connected with an obligation assumed by the tenant under some other agreement. All this is pure speculation, but the assumptions would present a valuer with no practical difficulty of application and language of the clause leaves the intention of the parties in no doubt. The question in this case is whether the same can be said about the words "other than those relating to rent".

I do not think that these words can be said to be incapable of bearing more than one meaning. They must take their colour from the context in which they are used. Precisely the same words are used in section 34 of the Landlord and Tenant Act 1954, and it seems clear that in that context they have been treated as bearing the meaning contended for by the tenant rather than the wider meaning suggested by the landlord. It may be that the statutory context is somewhat different from that of the clause in this case, but that is not the point. What is, I think, significant is that one cannot, in my view, treat the generally accepted meaning of the words in the statute as one which, whatever the context, they could not reasonably bear.

Given that the words then are reasonably capable of bearing both the wider and the narrower meanings, I think that I am entitled in choosing between them to have regard to the fact that they appear in the context of a rent review clause and to have regard to the normal purpose of such a clause. It seems to me that a literal construction must exclude from the hypothetical tenancy not only the provision for rent review but all the other terms relating to rent which I have mentioned.

Mr Dowding for the landlord suggested some refinements upon the literal construction which might save some of these other provisions from exclusion. The argument seemed to me to qualify for the description of "detailed semantic and syntactical analysis" which Lord Diplock in *The Antaios* [1985]AC 191 at p 210E said should not be allowed to flout business common sense. In my view these were arbitrary qualifications propounded to save the literal meaning from total absurdity rather than founded upon some rational commercial purpose. Without these refinements, the literal construction seems to me to face the valuer with so extraordinary a set of assumptions that I would not assume the parties to have chosen them unless their language left me with no alternative. It is not so much that they would favour one side or the other. This, as I have said, could be the result of concessions in the course of negotiations. It is that the hypothetical lease would be an animal so different from anything which ever existed in real life that its valuation would be outside the experience and expertise of the ordinary valuer. No doubt a valuer would be able to produce some kind of figure, but what I find wholly improbable is that the parties intended to make such assumptions specifically for the purpose of submitting them to the process of valuation.

Even on the construction suggested by Mr Dowding, the exclusion of the rent review clause and the covenant to pay rent from the hypothetical lease is an assumption which has no apparent commercial justification and gives rise to practical problems. It is common knowledge that landlords do not nowadays grant long leases of commercial premises at rack rents without rent reviews. In this case the animal is not so much fabulous as extinct, but the problem is the same. A valuer asked to determine the rent at which premises could be let in the open market for a lengthy term without a rent review therefore has no comparables in the real world to which he can refer. All that he can do is to make some more or less arbitrary adjustment to the rent which would be commanded on a letting of the same premises with the conventional provisions for rent review. It seems to me improbable that the parties intended the rent to be determined in this way.

It seems to me, therefore, that the only plausible alternative to the literal construction is the narrower construction advanced by the tenant. This allows the language to be qualified by the rent review context in which the words appear and thereby to confine them to those rent provisions which would be affected by the operation of the rent review. Since I think that the words are capable of bearing this meaning and that it is in accordance with the context and commercial purpose of the clause, it is the construction which I prefer.

I have been referred to a number of cases in which words having some similarity to those of the present lease have been construed by various judges. Most of them were considered by Warner J in the recently reported case of *Datastream International Ltd* v *Oakeep Ltd* (1985) 277 EG 66. All of them turned upon the language of the particular clause, and, having regard to Lord Diplock's remarks in *The Antaios* deprecating a generous citation of judicial authority on questions of construction, I do not think that it would be helpful for me to analyse them. I shall therefore say no more than that, like Warner J, I think that the cases of *National Westminster Bank plc* v *Arthur Young McClelland Moores & Co* (1984) 273 EG 402 and *Equity & Law Life Assurance Society plc* v *Bodfield Ltd* (1985) 276 EG 1157 gave less weight than I would have done to the commercial purpose of the clauses there in question.

The result is that in my judgment the true construction of the rent review clauses is that for which the tenants contend and I will so declare.

Chancery Division

October 30 1985

(Before Mr Justice SCOTT)

JAMES AND ANOTHER v BRITISH CRAFTS CENTRE

Estates Gazette March 8 1986

277 EG 976-980

Landlord and tenant — Construction of rent review clause in lease — The difficulty in this case arose in determining what exactly were to be the terms of the hypothetical lease which had to be assumed in arriving at the "commercial yearly rent" for the purpose of the rent review clause — After providing that such rent should be the open market rent of a lease for a term equal to the residue of the actual term of the current lease, the review clause stated that the hypothetical lease was to be "in the same terms in all other respects as these presents" — Thus the tenant's covenants in the hypothetical lease were to be the same as those in the existing lease — The construction problem was due to the fact that both the user covenant and the assignment covenant contained a specific reference to the lessee by name, "the British Crafts Centre" — The user covenant permitted the use of the premises for storage, sale and display of craftsmen's work while the lessee was the British Crafts Centre — The assignment covenant permitted the sharing of occupation with a holding or subsidiary company "but only whilst the lessee is the British Crafts Centre" — The relevance of the point was that the broader the class of potential lessees the higher the rent that might be expected to be obtained — The tenants accordingly contended that the provisions should be construed narrowly as directed to the particular personal position of the British Crafts Centre, whereas the landlords argued that they should be taken to refer to the lessee who took the hypothetical lease, whoever that lessee might turn out to be — In the event the judge decided that the two covenants should be applied differently in this respect — Held that in the case of the user covenant it was intended to grant a personal privilege to the Centre and that the hypothetical lease should contain a similar restriction — Held, however, that in the case of the assignment covenant the hypothetical lease should leave blank the name of the lessee, the intention being to prevent the privilege of sharing occupation to be enjoyed by anyone other than the original lessee, whoever the original lessee might be — *The Law Land Co Ltd* v *Consumers' Association Ltd* considered

This was an originating summons by which the plaintiff landlords, Robert Hedley James and Christopher Brian Carr, sought declarations as to the true construction of provisions in a rent review clause in a lease to the defendant tenants, British Crafts Centre, of premises at 43 Earlham Street, Covent Garden, London WC2.

J P Whittaker (instructed by Hempsons) appeared on behalf of the plaintiffs; Paul de la Piquerie (instructed by Sacker & Partners) represented the defendants.

Giving judgment, SCOTT J said: I have before me a short question of construction of a lease. The plaintiffs, Robert Hedley James and Christopher Brian Carr, are the landlords in respect of the lease — not the original landlords, but the landlords under a subsequent transfer. The defendant, British Crafts Centre, is the original lessee under the lease. The lease was dated August 12 1977. The premises comprised therein are 43 Earlham Street, Covent Garden, WC2. Those premises comprise four floors. There is a basement, a ground floor, a first floor and a second floor. At the date of the lease it appears that the defendant was already in occupation of the basement and ground floor. At the date of the lease the first floor was let to a company called High Vision Ltd. There are in evidence planning permissions relating to the first floor and to the basement and ground floors. The planning permission relating to the first floor is dated September 16 1975. It authorises the use of the first floor by High Vision Ltd as a studio and ancillary offices and it contains the condition that the permission granted should be personal to High Vision Ltd and that upon cessation of the use permitted for High Vision Ltd the premises should not, without some further permission, be used for any other purpose than the established use. There is no evidence of the established use, but I think it must be the case that the established use, varied by that planning permission, is use for office purposes. The other planning permission was dated September 8 1971. It was granted to Craft Centre of Great Britain, the same organisation as is now known as British Crafts Centre, the defendant in this action. That planning permission was expressed to authorise the continued use of the ground floor and basement of the premises for the display and selling of craftsmen's work. That was the position on the date on which this lease was granted. The parties to the lease were the predecessors in title of the plaintiffs of the one part and British Crafts Centre of the other part. It seems from the description of British Crafts Centre given in the lease that it is a corporation. Since the word "limited" is missing, I imagine it must be established for charitable purposes, although there is no evidence of that. After the name of the defendant, British Crafts Centre, its registered office is given, 43 Earlham Street. There is then in parenthesis the usual provision "(hereinafter called 'the lessee' which expression where the context so admits includes the persons or corporate body in whom the term hereby granted may from time to time be vested)''.

Under this lease a term of 14 years from September 29 1976 was granted to British Crafts Centre. The lease reserved an initial yearly rent of £10,500. That was subject to a rent review clause which has given rise to the difficulties leading to this construction summons.

The rent review clause is to be found in clause 5 of the lease. It provides for a rent review to be carried out in order that a new rent may be fixed as from review dates. The first review date for this purpose was the date on which the fifth anniversary of the grant of the lease expired, that is to say on September 29 1981. That first rent review brought to light the difficulty of construction that I must resolve.

The rent review provision in clause 5 of the lease provided for the lessors to give notice of their wish to have a rent review at any time during the year preceding the review date and contained a definition of ''the commercial yearly rent'', which was to be the new rent as from the review date provided that it exceeded the previous rent. The review provisions in this lease, as is commonly the case, provided only for an upward review.

The definition reads thus:

> ''The commercial yearly rent'' means the clear yearly rent at which the demised premises, assuming the due performance and observation of the covenants on the part of the lessee and conditions contained in these presents, might reasonably be expected to be let at the review date by a willing landlord in the open market with vacant possession and without premium or any other consideration than that evidenced by execution of a lease thereof to a willing tenant for a term equal to the residue then unexpired of the term hereby granted by a lease in the same terms in all other respects as these presents (including this subclause), there being disregarded (i) any effect on rent of the fact that the lessee or any other person may have been in occupation of the demised premises and (ii) any goodwill attached to the demised premises by reason of the user of the demised premises by the lessee or any other person or the purposes for which the demised premises may have been used and (iii) any effect upon the rents of any improvement carried out by the lessee after the commencement of the term hereby granted otherwise than in pursuance of an obligation under the terms of this lease.

Provision was made in the clause for the new rent, assessed in accordance with that definition, to be determined by surveyors acting for the respective parties. The surveyors who have been instructed by the respective parties to determine the commercial yearly rent as at September 29 1981 have met difficulties as to the approach that they should adopt.

It is common ground that for the purpose of fixing the commercial yearly rent there must be assumed to be a hypothetical lease being offered to potential lessees. The object of the notional exercise is to determine what the market rent for a lease on those terms would be. The difficulty in the present case arises in deciding exactly what would be the terms of that hypothetical lease. The definition of ''the commercial yearly rent'' that I have read contemplates that the hypothetical lease should be ''a lease in the same terms in all other respects as these presents''. That contemplates that the hypothetical lease will contain the same tenant's covenants as the tenant's covenants contained in the existing lease.

I turn to the tenant's covenants which have caused problems. Clause 2(16) contains the user covenants in the lease. By that clause the lessee covenants, so far as material to the point now in issue:

> not to use or permit or suffer the demised premises or any part thereof to be used for any purpose other than (i) for high class business, commercial or professional offices (subject to compliance with the provisions of clause 2(13) hereof) or (ii) in respect of such part of the demised premises as shall for the time being be occupied and used by the lessee (here meaning the British Crafts Centre party hereto) for storage, sale and display of craftsmen's work and ancillary offices and in respect of the first floor of the demised premises (whilst not occupied and used by British Crafts Centre) as an office and studio for the trade or business of designers, advertising and press agents.

Clause 2(13), to which reference was made under subclause (i) of this user covenant, is a covenant by the tenant ''To comply in all respects with the provision of every enactment'', which, in this subclause, means ''(every Act of Parliament already or hereafter to be passed, and any byelaws, statutory instruments, orders, rules and regulations and any notices, orders, directions, licences, consents or permissions made or given under or by virtue thereof) so far as the same shall affect the demised premises''; in effect, a covenant to comply with current planning permissions.

Subclause (17) of clause 2 of the lease contains covenants by the lessee in respect of assignments and parting with possession. Subclause (A)(i) is a covenant ''Not to assign or (subject to the provisions of paragraph (v) of this subclause (17)(A)) demise, underlet or otherwise part with possession of any part of the demised premises, here meaning a portion only and not the whole thereof, or, subject to the provisions of paragraph (ii) of this subclause (17)(A), share occupation of the whole or any part thereof for all or any part of the said term''. Paragraph (ii) there referred to is in these terms: ''Notwithstanding the provisions of paragraph (i) of this subclause (17)(A) but only whilst the lessee is the British Crafts Centre the lessee may share occupation of the demised premises or any part thereof with a holding or subsidiary company of it or a subsidiary of such holding company.'' There are then various provisos that I need not, I think, read.

The problem is this — and I will explain it by reference, first, to subclause (16)(ii), the user clause: should the hypothetical lease on the basis on which the market rent is to be ascertained specify the British Crafts Centre as the party whose use of the premises for storage, sale and display of craftsmen's work and ancillary offices is to be permitted, or should the name of the person whose use of the premises for those purposes is to be permitted be left blank on the footing that it will be the new lessee whose use of the premises for those purposes will be permitted under this clause?

The relevance of the point goes to the class of persons who might be expected to be interested in becoming lessees. Plainly enough, I suppose, the broader the class of potential lessees, the higher the rent that might be expected to be obtained. The larger the class of potential lessees, the greater will be the competition; and supply and demand no doubt influence rent levels as it does the price of most commodities.

The definition in the lease of ''the commercial yearly rent'' requires, it will be recalled, that the hypothetical lease be in the same terms in all other respects — the ''other'' can, for this purpose, be disregarded — as the instant lease. Accordingly it is submitted by Mr de la Piquerie for the tenant that the express reference to the British Crafts Centre which is to be found in the existing lease should be retained in the hypothetical lease. If that is so, it would not be possible for any person or company, other than the British Crafts Centre, who wanted to use the premises for storage, sale and display of craftsmen's work, to take a lease. At any rate, it would be a foolish thing for any such person or company to do because they would not be able to use the premises for those purposes. So, if Mr de la Piquerie is right, the potential class of lessees would be reduced accordingly. On the other hand, Mr Whittaker for the landlords contends that the clause in question should simply be regarded as a clause restricting the use of the premises for storage, sale and display of craftsmen's work to use for those purposes by the original lessee of the lease. He points out, correctly, that the British Crafts Centre was the original lessee in the actual lease and accordingly the hypothetical lease, which is in all respects, so far as can be achieved, to be in the same terms as the actual lease, should leave a blank for the name of the lessee whose use of the premises for storage, sale and display of craftsmen's work is to be permitted.

As it seems to me, both approaches require to some extent a rewriting of this clause. It does not seem to me possible for the hypothetical lease to follow, *ipsissima verba*, the words of this subclause. Mr de la Piquerie's approach would require the words ''used by the lessee (here meaning the British Crafts Centre party hereto)'' to be deleted and for there to be substituted simply the words ''used by the British Crafts Centre''. It could not be assumed that the British Crafts Centre would be the lessee or would be a party to the hypothetical lease. Accordingly, the clause could do no more than simply reproduce the name of the party whose use of the premises for those purposes was intended to be permitted. Mr Whittaker, on the other hand, would simply delete from the provision the words ''the British Crafts Centre''. He would leave a blank in place of those words on the footing that that blank could not be completed until it were known who would be the lessee.

Neither approach permits literal effect to be given to the requirement in the definition of ''commercial yearly rent'' that the hypothetical lease be ''a lease in the same terms in all other respects'' as the existing lease.

In order to decide which approach is the right one to adopt, it seems to me that I must direct my attention to the substance of the user provision contained in subclause (16)(ii).

It was said by Mr Whittaker that, in order to determine the market rent of the premises, one could not cut down or ought not to cut down the number of potential lessees who might be interested in taking a lease. The proposition, so broadly stated, cannot, in my view, be right. The hypothetical lease is intended to contain (*inter alia*) the tenant's covenants in the original lease. If the effect of those covenants, properly construed, is to limit the class of persons who would be interested in the lease, which, to some extent, must, I would think, be true of virtually every lease there has ever been of business premises, then it follows that, to that extent, the number of persons interested is going *pro tanto* to be cut down. The question then is to determine the real nature of the provision agreed upon by the parties to the lease and incorporated therein as subclause (16)(ii) of clause 2.

Mr de la Piquerie has submitted that the subclause was directed to the particular personal position of the British Crafts Centre. It was intended, he submitted, to allow use of the premises by the British Crafts Centre for its business, which must have been well known to the lessors, namely, storage, sale and display of craftsmen's work. It was not intended, he said, thereby to permit any use of the premises for those purposes other than use by the British Crafts Centre. The nature of the user restriction embodied in clause 2(16) was, he submitted, that use by anyone else for storage, sale and display of craftsmen's work was prohibited.

In my view, that is a correct analysis of clause 2(16). It would seem to me to make very little sense to argue that any original lessee would have been permitted to use the premises for storage, sale and display of craftsmen's work. The primary use authorised by this user covenant is the use authorised in subclause (i), namely, use "for high class business, commercial or professional offices". There are likely to be a number — probably a very large number indeed — of entirely suitable and desirable lessees who would wish to use the premises for those purposes and to whom use for storage, sale and display of craftsmen's work would be a complete irrelevance. In the course of argument — and I may have been responsible for this particular direction of argument myself — the position of solicitors was considered. Use of these premises by solicitors would no doubt be regarded as use "for high class . . . professional offices". But a provision in the lease under which use of the premises by those solicitors for storage, sale and display of craftsmen's work was permitted would seem silly. On the other hand, given that the British Crafts Centre was in occupation of the premises on the date of the lease, it makes sense that there should have been an exception to permit the British Crafts Centre, the particular party in occupation, to continue to use the premises for the same business purposes for which it was then using those premises. Hence, I think, the drafting and incorporation in the lease of subclause (ii) of this user covenant.

I was referred by Mr Whittaker to *The Law Land Co Ltd* v *Consumers' Association Ltd* (1980) 255 EG 617. This case, too, involved a rent review. In this case, too, the lease contemplated the determination on the rent review date of the open market rent of the premises. Here, too, the open market rent was to be decided by considering what rent would be obtainable on a lease of the same terms as the lease in question. Under the lease in question the tenant, the original lessee, was the Consumers' Association. The lease contained a clause under which the tenant covenanted not, without the prior written consent of the landlord, to use or permit the demised premises or any part thereof to be used other than as offices of the Consumers' Association and its associated organisations. It was argued in this case by the tenant that the hypothetical lease on the basis of which the open market rent was to be assessed should contain a similar clause, restricting use of the premises to any use other than as offices of the Consumers' Association and its associated organisations. If that had been the right approach, there could have been only one possible lessee, only one possible person who would be interested in taking this hypothetical lease, namely, the Consumers' Association. Nobody else would have been interested because nobody else, otherwise than by breaking the user covenant, could have used the premises. The Court of Appeal rejected the notion that a hypothetical lease with such a restriction could be used as the basis for determining an open market rent. It was plainly the intention of the parties to the lease that an open market rent should be ascertained for the purpose of the rent review. That could be done only if some modification were made to the user covenant. Accordingly, it was concluded that, for the purpose of the rent review, the user covenant should be regarded as imposing a restriction on use of the premises otherwise than as offices for the lessee who took the hypothetical lease, whoever that might turn out to be. That approach opened up the area of potential lessees and would permit an open market rental G to be assessed.

Mr Whittaker urged me to adopt the same approach in this case. He referred me in particular to a passage in the judgment of Buckley LJ in which Buckley LJ indicated that the hypothetical lease, for the purposes of the case with which he was concerned, should be a lease which contemplated that the name of the lessee would be specified in the user covenant. On the footing that the lessee had not yet been identified no name, Buckley LJ held, should be incorporated into the user covenant. For the purpose of the rent review the name should be left blank. However, the whole purpose of the approach in that case was to enable sense to be made of the requirement in the rent review clause that an open market rent be ascertained. That could not be H done if the lease in its actual form were regarded as the hypothetical lease.

In the present case that approach is not necessary. There is bound to be a market — and, I would think, a substantial market — for a lease of premises which can be used "for high-class business, commercial or professional offices". The fact that a provision limiting use for storage, sale and display of craftsmen's work to use by the British Crafts Centre is included in the lease does not, to my mind, advance at all the proposition that a lease so restricted cannot be used for the purpose of assessing an open market rent.

I do not, therefore, regard the reasoning and approach in *The Law Land Co Ltd* v *Consumers' Association* case as relevant to the problem before me. J

As I have said, I think the nature of the user restriction here under review is that it was intended to grant a personal privilege to the British Crafts Centre to use the premises for storage, sale and display of craftsmen's work. It was not intended that anyone else should use the premises for that purpose.

Accordingly, I think the hypothetical lease should contain a similar restriction.

I now turn to subclause (17)(A)(ii), the assignment clause. Mr de la Piquerie argued that the same approach as I have found correct in the case of the user covenant should be applied here. The point is, in my view, a fine one; but I do not feel confident when reading subclause (17)(A)(ii) that the privilege of sharing occupation with subsidiary companies or holding companies was intended to be personal to the K British Crafts Centre. A landlord can, of course, accept or reject a tenant as he or it wishes. When faced with a prospective tenant a landlord will take up references and make inquiries about that prospective tenant. If a landlord concludes that a tenant is not suitable to be granted the benefit of being allowed to share occupation with associated companies, then, if the lease is to include such a provision, that prospective tenant will not be granted the lease. The argument that the privilege of sharing occupation with associated companies granted under (17)(A)(ii) should be regarded as a personal privilege intended for the British Crafts Centre does not seem to me to be on all fours with the comparable argument regarding (16)(ii). It seems to me that the intention of the parties lying behind (17)(A)(ii) was to restrict the privilege of sharing occupation there L granted from being enjoyed by anyone other than the original lessee, whoever the original lessee might be.

Accordingly, it seems to me that the hypothetical lease, so far as subclause (17)(A)(ii) is concerned, ought not to specify the British Crafts Centre but ought to leave blank the reference to the name of the lessee. I am not satisfied, as I am in regard to subclause (16)(ii), that the privilege was intended to be personal and restricted to the British Crafts Centre as such, as opposed to restricted to the person who happened to be the first lessee under the lease.

I would accordingly make declarations for the purposes of assessment of the open market rent on the rent review in accordance with this judgment, that is to say, that the hypothetical lease should M contain a provision as to user on the lines of or in the words of subclause (ii) of clause (16), but substituting simply a reference to the British Crafts Centre for the extended description in the actual lease, namely, "the lessee (here meaning the British Crafts Centre party hereto)", and that the hypothetical lease should contain, so far as subclause (ii) of (17)(A) is concerned, simply a reference to the lessee with the name left blank.

I will leave counsel to draft declarations, if they think it appropriate to do it, in more formal form.

There was no order as to costs.

A
Chancery Division

February 6 1986

(Before Sir Nicolas BROWNE-WILKINSON V-C)

BRITISH GAS CORPORATION v UNIVERSITIES SUPERANNUATION SCHEME LTD

Estates Gazette March 8 1986

277 EG 980-984

Landlord and tenant — Construction of rent review clause in lease of office premises — Conflicting decisions considered by B Vice-Chancellor — Another case on the question whether, in determining the rack rental value for the purpose of the clause, the hypothetical lease should be assumed to include or exclude the rent review provisions contained in the actual lease — Construction of the "rent exclusion provision" — The material words in the present lease, which provided for five-yearly reviews, were that the hypothetical lease should be a lease "containing the same provisions (other than as to the yearly rent)" as in the actual lease — Considering the matter first without reference to authority, the Vice-Chancellor rejected as offending against common sense a wholly literal C construction to the effect that all provisions relating to rent (such as the proviso for re-entry on non-payment) should be ignored — Further, a construction requiring the valuer to ignore both the actual rent payable before the review and the provisions for future reviews conferred a windfall on the landlord going beyond the adjustment of rental values in line with the market which was the underlying purpose of a rent review clause — Accordingly, apart from authority, and in the absence of any clear words indicating the parties' intention to enter into an unusual bargain, the correct construction in the present case was that the hypothetical letting was to be on the terms of the actual lease, excluding only the rent payable D before the review date but including the provisions for five-yearly rent reviews — Examining the authorities, there was a conflict between, on the one hand, the *Pearl Assurance* case, the *Datastream* case and the *MFI* case, which approached the construction on the basis of the underlying commercial purpose of the clauses, and, on the other hand, the *National Westminster Bank*, *Pugh*, and *Equity & Law Life* cases, which did not adopt that approach, although in all these cases there were differences in the particular wording of the clauses — Subject to a clearly expressed intention (as in the *Pugh* case) that the rent review provisions were to be disregarded, it was E proper to give effect to the underlying commercial purpose of a rent review clause and to require future rent reviews to be taken into account in the hypothetical letting — There was an urgent need for certainty in this field and it was most desirable that there should be an early decision of the Court of Appeal to resolve the conflict of judicial approach — Declaration in favour of tenants

This was an originating summons in which the plaintiffs, the British Gas Corporation, sought a declaration as to the correct construction of the rent review clause in a lease of office premises at 152 Grosvenor Road, London SW1, of which they were the tenants, the defendant landlords being the Universities Superannuation F Scheme Ltd.

John Colyer QC and Kirk Reynolds (instructed by the Director of Legal Services, British Gas Corporation) appeared on behalf of the plaintiffs; Michael Barnes QC and David Elvin (instructed by Coward Chance) represented the defendants.

Giving judgment, SIR NICOLAS BROWNE-WILKINSON V-C said: This is yet another case raising the question whether or not a valuer, in fixing the new rent under a rent review clause in a lease, should take into account the fact that the lease in question contains provisions for further rent reviews in the future.

Shortly stated, the problem arises in this way. For very many years now, in the overwhelming majority of cases landlords granting long terms of years have insisted on the insertion of a rent review clause in the lease whereby the rent payable is periodically adjusted to reflect changes in the value of money and property. The most common machinery is to provide for the new rent to be fixed by a valuer as being the rack rental which would be obtainable in the market at the review date if the demised premises were to be let on that date. Such rent review clauses usually lay down the formula by reference to which such rack rental is to be fixed by the valuer. The formula commonly takes the form that the valuer is to assume a hypothetical letting of the premises on the open market on the terms of the actual lease but subject to certain artificial variations. A very common variation is that the valuer is to assume that the hypothetical letting is on the terms of a lease containing the same provisions as the actual lease "other than as to rent" or "other than those relating to rent" or "other than the amount of rent hereby reserved". I will call such words "the rent exclusion provision". The question is whether in any particular case the effect of the rent exclusion provision is to require the valuer to ignore the fact that the actual lease contains provisions for future rent review, on the footing that the rent review clause is itself a provision relating to rent. The answer to that question may be of great financial importance. In the present case I was told that, if the valuer had to disregard future reviews of rent, the new rent might be as much as 20% greater than the rent which would be fixed if the future reviews are taken into account.

The present case concerns a lease of very substantial office premises granted by the predecessor in title of the defendant to the plaintiff for a term of 35 years from March 25 1980. The rent reserved (defined in the lease as "the Yearly Rent") was a peppercorn until July 1 1980 and "thereafter throughout the remainder of the term the rent of £1,150,000.00 per annum (and so in proportion for any lesser period) or such other rent as may from time to time be substituted therefor pursuant to the provisions of the Second Schedule . . . ". The second schedule, para 1(a), provides that the landlord may give the tenant a review notice in which event the yearly rent payable as from the review date "shall be the higher of (i) the Yearly Rent payable immediately before such Review Date and (ii) the rack rental value of the Demised Premises at the relevant Review Date". "The Review Dates" are defined as the dates of commencement of the 6th, 11th, 16th, 21st, 26th and 31st years of the term, ie at five-yearly intervals. Para 3 of the second schedule provides as follows:

> Rack rental value of the Demised Premises means such rent as may be agreed or determined as hereinafter provided to be the best yearly rent at which the Demised Premises could reasonably be expected to let in the open market by a willing landlord to a willing tenant for a term equal to the term hereby granted by means of a lease containing the same provisions (other than as to the yearly rent) as are herein contained on the following assumptions . . .

A review having been required by the landlord at the first review date (March 25 1985) and the parties having failed to agree the new rent, under the terms of the lease the new rent has to be fixed by an independent valuer. The question is whether such valuer in assessing the "rack rental value" should assume a hypothetical letting on terms which include provisions for review of the rent every five years.

The question is one of construction. Accordingly I will first consider the matter apart from authority and then turn to see whether I am precluded by the decided cases from giving effect to my own views.

Despite the definition of "the Yearly Rent" to which I have referred, the draftsman of the lease was not careful in his use of the terminology. In various clauses he uses the phrases "yearly rent" (ie without capital letters) and "the rents hereby reserved" (in the plural). It is common ground that such variations in terminology disclose no method and can be ignored. The question therefore turns entirely on the words "containing the same provisions (other than as to the yearly rent)" in para 3 of Schedule 2 construed in the context of the lease as a whole and the surrounding circumstances. There is no evidence of any special circumstances applying to this case which might affect the question of construction.

Counsel agreed that there were three possible constructions of the rent exclusion provision in this case, viz:

(1) that it requires the valuer to ignore *all* provisions relating to rent in the lease;

(2) that it requires the valuer to ignore those provisions which relate to the quantification of rent, ie the rent payable immediately before the relevant review date *and* the provisions for future rent reviews; and

(3) that it requires the valuer to ignore the rent actually payable before the review date only, ie he must take into account the provisions for future reviews of the rent.

The first of these constructions is the only one that the words literally bear. But the result of such a literal construction is so surprising as to offend common sense. The covenant for payment of rent is a "provision . . . as to the yearly rent". So is the right of re-entry reserved for non-payment of rent. Accordingly, if the literal construction were right, the valuer would be faced with the wholly unrealistic (and I would have thought impossible) task of fixing the open market rental for a large office block on terms which contained no covenant for payment of rent nor any right of re-entry in the event of non-payment of rent. Mr Barnes, for the landlord, to my mind rightly did not contend for this fully literal construction.

Once one departs from the literal meaning of the words, one is necessarily implying some limitation into the words actually used in the lease. The only question is whether such implication should be that inherent in construction (2) or construction (3) above. At this stage, in my judgment, one is necessarily forced to seek to discover the underlying purpose of the provisions for rent review so as to give effect to that purpose.

There is really no dispute that the general purpose of a provision for rent review is to enable the landlord to obtain from time to time the market rental which the premises would command if let on the same terms on the open market at the review dates. The purpose is to reflect the changes in the value of money and real increases in the value of the property during a long term. Such being the purpose, in the absence of special circumstances it would in my judgment be wayward to impute to the parties an intention that the landlord should get a rent which was additionally inflated by a factor which has no reference either to changes in the value of money or in the value of the property but is referable to a factor which has no existence as between the actual landlord and the actual tenant, ie the additional rent which could be obtained if there were no provisions for rent review. Of course, the lease may be expressed in words so clear that there is no room for giving effect to such underlying purpose. Again, there may be special surrounding circumstances which indicate that the parties did intend to reach such an unusual bargain. But in the absence of such clear words or surrounding circumstances, in my judgment the lease should be construed so as to give effect to the basic purpose of the rent review clause and not so as to confer on the landlord a windfall benefit which he could never obtain on the market if he were actually letting the premises at the review date, viz a letting on terms which contain provisions for rent review at a rent appropriate to a letting which did not contain such a provision.

Therefore, apart from authority I would hold that on the true construction of para 3 the rack rental value has to be fixed on the basis that the hypothetical letting is on the terms of the actual lease excluding only the rent actually quantified and payable before the review date but including the provisions for five-yearly rent review.

Does authority preclude me from reaching this result? In my judgment it does not. The authorities were recently reviewed by Warner J in *Datastream International Ltd* v *Oakeep* (1985) 277 EG 66*; I gratefully adopt his analysis. I add to the cases there considered the recent judgment of Hoffmann J in *MFI Properties Ltd* v *BICC Group Pension Trust Ltd* (as yet unreported)†, decided since the conclusion of the argument in this case. In the latter case, the words of the rent exclusion provision were "other than those relating to rent": Hoffmann J held that in fixing the hypothetical rent the provision for rent review had to be taken into account by the valuer. His reasoning closely accords with my own and I gratefully adopt it.

Although each of the decided cases is a decision on the construction of the lease in question (and therefore not directly authority on the meaning of the lease I have to construe), they do disclose a marked difference of approach to the construction of these clauses. On the one side are Vinelott J (in *Pearl Assurance plc* v *Shaw* (1984) 274 EG 490), Warner J (in the *Datastream* case) and Hoffmann J (in the *MFI* case) who, like myself, treat the literal construction or any intermediate construction as offending commercial common sense and therefore give effect to the underlying commercial purpose. On the other side are those cases in which either the words were given their literal meaning (see *National Westminster Bank plc* v *Arthur Young McClelland Moores & Co* (1984) 273 EG 402) or the judge rejected the view that one should approach the construction of a rent review clause on the basis that it is intended only to give effect to the normal commercial reason for

*Editor's note: See p 98 *ante*.

†Editor's note: Now reported at p 115 *ante* and (1986) 277 EG 862.

G

including a rent review clause: *ibid*; *Pugh* v *Smiths Industries Ltd* (1982) 264 EG 823; *Equity & Law Life Assurance Society plc* v *Bodfield Ltd* (1985) 276 EG 1157*. I am far from suggesting that the *Pugh* case and the *Equity & Law Life* case were wrongly decided. In the former, the words used in the rent exclusion provision were so clear that imputed intention could not override them and the literal construction did not produce an absurdity. In the latter, the other terms of the lease in question were very unusual. But neither Goulding J nor Peter Gibson J thought it right to give any effect to the underlying commercial purpose of such clauses.

H

In these circumstances, there are in my judgment conflicting decisions as to the correct approach to the construction of these clauses. I am accordingly free to adopt the approach I prefer. In my judgment the correct approach is as follows:

(a) words in a rent exclusion provision which require *all* provisions as to rent to be disregarded produce a result so manifestly contrary to commercial common sense that they cannot be given literal effect;

(b) other clear words which require the rent review provision (as opposed to all provisions as to rent) to be disregarded (such as those in the *Pugh* case) must be given effect to, however wayward the result; and

(c) subject to (b), in the absence of special circumstances it is proper to give effect to the underlying commercial purpose of a rent review clause and to construe the words so as to give effect to that purpose by requiring future rent reviews to be taken into account in fixing the open market rental under the hypothetical letting.

J

I am conscious that such an approach is perilously close to seeking to lay down mechanistic rules of construction as opposed to principles of construction. But there is an urgent need to produce certainty in this field. Every year thousands of rents are coming up for review on the basis of clauses such as the one before me: witness the growing tide of litigation raising the point. Landlords, tenants and their valuers need to know what is the right basis of valuation without recourse to lawyers let alone to the courts. The question cannot be left to turn on the terms of each lease without the basic approach being certain. It is in my judgment most desirable that this, or some other case, should at an early stage be taken to the Court of Appeal so as to resolve the conflicting judicial approaches that have emerged.

K

Finally, I must deal specifically with the submissions which Mr Barnes made on the basis of the existing authorities. He submitted that a distinction should be drawn between those cases where the rent exclusion provision (as in this case) excluded *provisions* as to rent on the one hand and, on the other, cases where the rent exclusion provision excluded *the amount* of rent (as in the *Datastream* case). He submitted that in the former case the rent review provisions had to be disregarded whereas in the latter they did not. In my judgment such narrow semantic distinctions are both legally and commercially undesirable. Moreover, since Mr Barnes made the submission, it has become inconsistent with the decision of Hoffmann J in the *MFI Properties* case. Once one departs from the truly literal construction (which Mr Barnes did not contend for), in my judgment there is no alternative, save in special circumstances, but to give effect to the underlying commercial purpose of the clause.

L

I will therefore make the declaration asked for in the originating summons.

*Editor's note: Also reported at [1985] 2 EGLR 144.

M

Chancery Division

January 30 1986

(Before Mr Justice HOFFMANN)

MY KINDA TOWN LTD v CASTLEBROOK PROPERTIES LTD

Estates Gazette March 15 1986

277 EG 1144-1146

Landlord and tenant — Arbitration Act 1979 — Challenge to award of arbitrators determining the market rent under a rent review clause of a lease of a "destination restaurant" in

Knightsbridge — Award increased rent from £24,000 a year agreed at previous review date to £98,885 a year five years later — Restaurant, described by tenants as "idiosyncratic" and referred to in the award as belonging to the class of "destination restaurants", was situated in the comparative "backwater" of Raphael Street — It was said to be the kind of restaurant which customers chose on account of its particular food and atmosphere rather than because of its location nearby or where they might happen to be passing — Comparables put forward by tenants included one in Knightsbridge, while the landlords put forward three in the West End, one in Covent Garden and one near Hyde Park Corner — Tenants attacked the award of the arbitrators as erroneous in law on the ground that, in the words of Lord Radcliffe in *Edwards* v *Bairstow*, "no person acting judicially and properly instructed as to the relevant law could have come to the determination" in question — The error of law alleged was said to lie in comparing the tenants' premises, which had been a successful destination restaurant for a short period, only with other established destination restaurants — Held, rejecting the tenants' submissions, that the arbitrators' decision was based on two findings of fact from which there was no appeal and which justified the use of the "comparables" cited by the landlords — These two findings were that the premises were marketable as suitable for use as a destination restaurant and that the market for such premises covered the whole of the restaurant districts in central London — Appeal dismissed

This was an appeal by leave, pursuant to section 1 of the Arbitration Act 1979, against an award by Mr David Widdicombe QC and Mr David Hyams FRICS as arbitrators under the terms of a rent review clause in a lease of premises in Raphael Street, Knightsbridge, London SW1. The appellants were My Kinda Town Ltd and the respondents Castlebrook Properties Ltd.

Leolin Price QC and J N Arbuthnot (instructed by Gouldens) appeared on behalf of the appellants; Terence Cullen QC, and Miss Hazel Williamson (instructed by Hughes Watton & Co) represented the respondents.

Giving judgment, HOFFMANN J said: This is an appeal by leave pursuant to section 1 of the Arbitration Act 1979 against an award dated September 23 1985 by Mr David Widdicombe QC and Mr David Hyams FRICS acting as arbitrators under a rent review clause in a lease of premises in Raphael Street, Knightsbridge, London SW1. The lease was granted for a term of 25½ years from March 25 1975 until September 28 2000. The initial rent was £12,500 a year rising to £16,000 a year in 1977 and with rent reviews in 1979 and at five-yearly intervals thereafter. At the 1979 review the new rent was agreed at £24,000 a year. In 1984 the parties were unable to agree and in accordance with clause 2(e) of the lease the matter was referred to arbitration. The arbitrators had to determine the market rent on January 30 1984, which clause 2(c) of the lease defined as:

> the best rent that would be expected to be obtained in the open market for the said demised premises if let at [the relevant date] with vacant possession without a premium by a then willing lessor on a lease for a term equivalent to the residue then unexpired of the term granted by this lease and containing the like conditions and provisions as are contained in this present lease.

The present tenants took an assignment of the lease in 1981 and since February 1982 have carried on business there as a restaurant called The Chicago Rib Shack. The terms of the lease and applicable planning permission permit such user. Mr Price for the tenants described the restaurant as "idiosyncratic" and said that it was the kind of restaurant which customers choose on account of its particular food or atmosphere rather than because it was nearby or they happened to be passing. Restaurants of the former kind were conveniently labelled in the award as "destination restaurants". The tenants' restaurant has been very successful.

Both parties called valuation experts who put forward schedules of lettings claimed to be sufficiently comparable to afford guidance in determining the market rent of these premises. Of these, the arbitrators said that they found seven "helpful". They included one restaurant in Knightsbridge put forward by the tenants, but also five restaurants outside the Knightsbridge area which had been put forward by the landlords. Three of these were in the West End, one in Covent Garden and one near Hyde Park Corner. The arbitrators justified having regard to these restaurants by saying that, like The Chicago Rib Shack, they were "destination restaurants". They said:

> We accept the tenant's evidence that Raphael Street is, comparatively speaking, a backwater. However, for a destination restaurant, that is not important. The landlord's evidence was that it does not matter much where a destination restaurant is located within Central London. Knightsbridge is a restaurant area, as is Covent Garden, Piccadilly etc. Restaurant society is mobile. We think this is correct, and thus it would not appear inappropriate to compare rents of different destination and theme restaurants in central London, as the landlord has done, making adjustments where necessary.

Having regard to these comparables, the arbitrators determined the market rent of the premises on the relevant date to have been £98,885 a year.

Mr Price contends that the arbitrators' reasoning involves a logical error which means that they must have erred in law because they had come to a conclusion which "no person acting judicially and properly instructed as to the relevant law could have come to" (see Lord Radcliffe in *Edwards* v *Bairstow* [1956] AC 14 at p 36). The error is said to lie in having regard to the fact that the tenants' premises have for a relatively short period been used as a successful destination restaurant and that the comparables are successful destination restaurants. That, says Mr Price, is to give the landlords the benefit of the tenants' commercial success rather than the value of the premises as such. The premises must in accordance with clause 2(c) of the lease be assumed to be let with vacant possession, and that entails the absence of everything which gives the premises the attributes of a destination restaurant.

I think that this analysis distorts the reasoning of the arbitrators, which appears to me to rest upon two findings. The first is that the premises are marketable as suitable for use as a destination restaurant and therefore that for the purpose of fixing a rent the appropriate market to be considered is prospective tenants who require premises for such a use. This seems to me no more than common sense. The most obvious class of potential tenants to be taken into account when evaluating the market for premises is those who would want to carry on a business similar to that for which the premises have already been shown to be suited. The relevance of the fact that the tenants in this case have carried on the business of a destination restaurant upon the premises is no more than to show that the premises could reasonably be let as suitable for such use.

The other relevant finding by the arbitrators is that the market in the restaurant districts of central London for premises suitable for use as destination restaurants is a single market within which the general location of the premises (eg in Covent Garden, the West End or Knightsbridge) does not appreciably affect the rent which a prospective tenant would be willing to pay. Of course the specific location of the premises (in an attractive street or an interesting building or in a prominent position which can attract some passing trade as well as "destination" customers) may affect the rent, and these factors will call for the "adjustments" to which the arbitrators refer. But the unity of the market means that one need not treat the evidence of lettings in the West End as irrelevant simply because they were in the West End rather than Knightsbridge.

These two findings, that the premises are marketable as suitable for use as a destination restaurant and that the market for such premises covers the whole of the restaurant districts of central London are, in my judgment, findings of fact from which there is no appeal. Given these findings, I can see no illogicality in the arbitrators' use of comparables. There is no inconsistency between taking into account lettings of destination restaurants and the assumption that the instant premises are let with vacant possession. In applying the comparables, the arbitrators were comparing like with like, because all the comparables were either actual lettings with vacant possession or rents fixed at rent reviews on the hypothesis that there was a letting with vacant possession. It follows that the award contains no error of law and that the appeal must be dismissed.

The appeal was dismissed with costs. Leave to appeal to the Court of Appeal was refused.

Chancery Division

November 21 1985

(Before Mr Justice Peter GIBSON)

STANDARD LIFE ASSURANCE CO v OXOID LTD; OXOID LTD v STANDARD LIFE ASSURANCE CO

Estates Gazette March 22 1986

277 EG 1248-1252

Landlord and tenant — Construction of unusual rent review clause in lease of industrial premises — Instead of usual rent review provisions based on a hypothetical letting of the actual premises on the open market, the review rent was to be determined by applying a mathematical formula to "standard accommodation" — The reason for this unusual method appears to have been the difficulty of finding satisfactory comparables — The main dispute giving rise to the present summons was as to the meaning of "standard accommodation" in the formula — The characteristics mentioned were that it must be a single-storey industrial building, in the same locality as the subject premises and of a like age, but there was no mention of size, although both parties agreed that size must be a necessary element in the conception of the standard — At the date of the lease there were, in addition to the subject premises, six other units in existence on the industrial estate — All were single-storey industrial buildings of about the same age as the subject premises, but whereas their sizes varied from 10,000 to 15,000 sq ft, with most of them about 10,000 or 11,000 sq ft, the size of the subject premises was 100,000 sq ft — Landlord argued that for the purpose of the standard accommodation in the review formula the size should be taken as 10,000 or 11,000 sq ft — Tenant submitted that it would be an unfair distortion of reality to apply such a standard size to premises in fact ten times larger (in view of the variations of rent per sq ft in accordance with the size of the premises) and that, if necessary, a wider locality than the industrial estate should be regarded — Held that as a matter of construction the landlord's view was correct — The tenant's construction would involve writing into the formula a requirement which was not there, namely, that the size must be the same as that of the subject premises — Declaration accordingly

In these proceedings there were two originating summonses, one taken out by the landlord, the Standard Life Assurance Co, and the other by the tenant, Oxoid Ltd, but both summonses raised the same question as to the construction of the rent review provisions in a lease of a unit on the Kingsland Industrial Estate in Basingstoke.

Edward Nugee QC and Kim Lewison (instructed by Herbert Smith & Co) appeared on behalf of the Standard Life Assurance Co; Peter Langan QC and R Wakefield (instructed by Alsop Stevens) represented Oxoid Ltd.

Giving judgment, PETER GIBSON J said: I have before me two originating summonses, each raising what is in effect the same question. That question is the construction of a particular term in the rent review clause under a lease. The lease is one dated November 4 1974 whereby Stan-Bilt Ltd ("Stan-Bilt") demised to Oxoid Ltd ("the tenant") certain premises part of the Kingsland Industrial Estate in Basingstoke in Hampshire. Those premises are now known as Unit 11 on that estate. The term of the lease was 35 years from September 29 1974 and the initial rent was £95,300 per annum, but the lease provided for five-year rent reviews. Stan-Bilt was itself the lessee of the estate under a lease from the Standard Life Assurance Co ("the landlord") but it surrendered its interest to the landlord in 1984.

The estate prior to 1969 consisted of only three completed units. Between 1969 and 1977 Stan-Bilt built a number of other units. At the date of the lease the estate was still in the course of development, but seven units built by Stan-Bilt were in existence at that date. All were single-storey industrial buildings containing warehouse or factory space and some office accommodation. One, Unit 11, differed from the other six in size and the amount of office accommodation. Of the other six, five were about 10,000 or 11,000 sq ft and the sixth, Unit 4, was nearly 15,000 sq ft. But Unit 11 was 100,000 sq ft. In all the other six not more than about 20% of the available space was office accommodation. In Unit 11, however, 49% of the available space was office accommodation.

The first rent review of Unit 11 resulted in an increase in the rent to £263,000, but the premises were at that time let to a company other than the tenant. On the second rent review the question of construction that is the subject of these proceedings has emerged between the landlord and the tenant. To understand the question I must read clause 4(i) of the lease. This provides:

> (i) for the purposes of this clause the following expressions shall have the following meanings: "initial standard rent" means the fair annual rack rent calculated on a per square foot basis at which a standard single storey industrial building in the same locality as the demised premises and of like age (hereinafter called "standard accommodation") if vacant could now be let in the open market by a willing lessor to a willing lessee on the same terms and conditions including period of demise as are contained in this Lease which the parties hereby agree is at the rate of £0.65 per square foot; "review standard rent" means the fair annual rack rent calculated on a per square foot basis at which standard accommodation if vacant at the rent review referred to in paragraph (ii) of this clause could be let at that time in the open market by a willing lessor to a willing lessee on the same terms and conditions including period of demise as are contained in this Lease; "Lease rent" means the rent reserved by this Lease in respect of the demised premises; and "initial Lease rent" means the Lease rent at the date hereof which the parties hereby agree at the rate of £0.95 per square foot; and "review Lease rent" means the Lease rent which shall be payable following each relevant review of the Lease rent hereinafter provided.

Clause 4(ii) then provides for the rent review. If the landlord sets the review in motion, and if on such review the review standard rent is found to exceed the initial standard rent, the lease rent is to be increased so that the review lease rent bears the same proportion to the review standard rent as the initial lease rent bears to the initial standard rent. In other words, having ascertained the rent for the standard accommodation at the relevant date, one then increases that rent by multiplying by 95 and dividing by 65. If the parties fail to agree, then the matter is to be determined by a surveyor.

The dispute between the parties turns on the meaning of standard accommodation as defined. The landlord attributes to it a meaning which it asks, by its originating summons dated July 9 1985, the court to declare, that is to say, that it means a standard single-storey industrial building situated on the Kingsland Industrial Estate, Basingstoke, having a floor area of 10,000 sq ft or thereabouts of which 20% or less was designed or adapted for use as offices. The tenant attributes to "standard accommodation" a different meaning; the meaning it asks the court, by its originating summons dated July 10 1985, to declare is that it means "a building having the same floor area as the building on the demised premises". Each of the parties has put in an affidavit on which it relies, but there has been no cross-examination or other oral evidence and there is no dispute of fact.

Nor is there any dispute on the principles to be applied in construing the lease. Mr Nugee, appearing for the landlord, has reminded me of the much cited passage from the speech of Lord Wilberforce in *Reardon Smith Line Ltd* v *Yngvar Hansen-Tangen* [1976] 1 WLR 989 at pp 996-997. From that it is clear that the proper test is an objective one, to ascertain the intention which reasonable persons would have had in agreeing the rent review clause if placed in the position of the parties to the lease at the relevant date. Mr Nugee submitted that that date was the date when the parties agreed the terms and that was likely to be a date preceding that of the lease. He submitted that there was a real possibility that an unusual building like Unit 11 might well have been the subject of a prior agreement between Stan-Bilt and the tenant and the terms of the lease might therefore have been agreed much earlier than the date of the lease. But that seems to me to be much too speculative. All that I have to go on is the fact that the lease was dated November 4 1974, and that I must take to be the relevant date when the parties committed themselves to the terms of the lease.

The rent review clause is in a very unusual form. Instead of the usual rent review provision whereby the updated rent of the demised premises is ascertained on the basis of a hypothetical letting of the actual premises on the open market, the rent to be paid, in the present case, is to be determined by applying a mathematical formula to the rent of premises other than Unit 11, that is to say, standard accommodation as defined. The parties to the lease must be taken to have anticipated some difficulty in ascertaining the market rent of

A Unit 11 in the conventional way, and I would infer that difficulty to have been that of finding comparables to Unit 11. That of course would be the ordinary way of ascertaining the market rent. The parties must, I think, also be taken to have adopted the definition of standard accommodation on the footing that the building therein defined avoided that difficulty. One would therefore expect that there would be sufficient comparables known to the parties to exist at the date of the lease which would ensure that that description to be found in the lease made the task of ascertaining the review standard rent a relatively easy one. One would also expect that all the information necessary to enable the rent review to be carried out would be provided in the definition.

The identifying characteristics of standard accommodation as
B defined are (1) that it is a single-storey industrial building which answers the description "standard", (2) in the same locality as Unit 11, and (3) of like age. There is little difficulty over the third characteristic. It is common ground that a building built within a few years of the date of the building of Unit 11 would answer that description. However, in determining whether the parties must be taken to have had in mind particular buildings, one should I think look only at buildings existing at the date of the lease, unless there is some evidence, which there is not, of other buildings which were in the course of being built or about to be built and which were known to the parties. The difficulties of construction relate to the word "standard" and "in the same locality".

C The approach of the landlord is quite simple. If one looks at the facts in order to find standard accommodation as defined, there in the immediate area of Unit 11 on the estate itself are to be found five single-storey industrial buildings, which have the common features of being approximately the same size, that is to say, about 10,000 or 11,000 sq ft, and having not more than about 20% office accommodation in the available space, and which are all of about the same age as Unit 11. The rents per sq ft obtained for those units prior to the date of the lease were 57p on September 8 1971 for Unit 5, 62.5p on November 20 1972 for Unit 6, 70p on February 8 1973 for Unit 7, and 77p in September 1974 for Unit 9. Unit 12 was let after the lease of Unit 11 on December 20 1974 for £1.08. The rent of 65p, agreed to be the rent of standard accommodation at the date of the lease, must,
D the landlord says, have been derived from the rents of Units 5, 6, 7 and 9.

The tenant's approach is different. Mr Colin Yates, a chartered surveyor, has sworn an affidavit on which the tenant relies. He looked at an area wider than the estate on the footing that "locality" would naturally be understood as having a wider connotation. He has looked at an area up to three-quarters of a mile from Unit 11. His purpose, as he deposes in his affidavit, was to determine the situation, size and rentals of other industrial buildings, which might be relevant to the issue in the proceedings. He listed 19 buildings built between 1965 and 1978, but it appears that he excluded those buildings on the estate which are of between 10-15,000 sq ft and any others that there may be outside the estate of that size. The buildings
E he lists vary from 19,000 to 113,000 sq ft. Mr Yates concludes that in what he has taken to be the locality there appears to be no standard industrial building in terms of size. He then proceeds to look at the rents of units on the estate, but takes only the rents of units 9 and 12 because they were determined close to the date of the lease of Unit 11, and regards those units as the only two relevant units on the estate which are truly comparable for the purpose of finding the level of rentals for industrial buildings. Because 65p is well below those rentals, Mr Yates rejects the landlord's view that the standard accommodation is a building of about 10,000 sq ft. He asserts that the initial standard rent should be fixed by reference to a building having the size of Unit 11. The result is that on his own evidence there was no example in the locality of standard accommodation except possibly
F one building of approximately 113,000 sq ft built in two stages between 1969 and 1971, a half mile away from Unit 11, and another of 90,000 sq ft which Mr Langan, appearing for the tenant, accepts should be excluded as it was built in 1965. That is a remarkable result given the obvious business purpose of the unusual rent review clause, to provide an easily ascertainable rent by reference, one must assume, to comparables. With all due respect to Mr Yates, to my mind he has approached the matter from the wrong end. Instead of surveying the area round Unit 11, to look at all industrial buildings to see whether there were some which had the defined characteristics of standard accommodation, he has assumed what he intended to prove, namely, that those buildings on the estate which do have common characteristics should be excluded because they are too small. Yet he accepts that two of the units of the estate are relevant for the purposes of comparability of rents for standard accommodation. I cannot accept that Mr Yates' approach is what the parties to the lease can be taken to have intended.

Mr Nugee and Mr Langan agree that standard accommodation has this characteristic, that it is of a building of which up to about 20% of the available space is given to office accommodation. Mr Langan says that it is therefore not necessary to look further for the meaning of "standard" in that a sensible meaning can be given to it. But it is common ground between Mr Nugee and Mr Langan that there must be a reference to size in the definition of standard accommodation. That must be right, given the fact that the rent per sq ft of premises will vary according to the size of the premises. Mr Langan cannot point to any words in the definition that procure for him the attribute of standard accommodation which he asserts, that is to say, that it should be the same size as Unit 11. In effect he asks the court to write in those words. That is a very bold submission, the more so having regard to the actual language of the definition. There one finds the parties have expressly agreed two matters which standard accommodation has to have in common with Unit 11, namely locality and age. It would be remarkable if the parties should be taken to have intended a third characteristic, size (and a particular size), but had not expressed it. Moreover, it is so obviously important a characteristic to be mentioned for the guidance of those who have to ascertain the rent on a rent review. In contrast Mr Nugee, on his construction, can point to the word "standard" as including a reference to size. To my mind "standard" connotes a norm, the first, the most obvious and natural characteristic of which is size.

Mr Langan, however, urges on me two arguments against size being comprehended in the word "standard" in this lease. The first is that, if standard accommodation relates to premises of only 10,000 sq ft, that is to say, one-tenth of the size of Unit 11, that would be a great distortion of reality and, he submits, it is to be assumed that the parties intended to stick as close as possible to the actual facts. With respect, this seems to me to be an impossible argument. Standard accommodation is accommodation other than Unit 11. Save for the two characteristics to which I have already referred, locality and age, it cannot be assumed that the parties intended Unit 11 to be in any other way like standard accommodation. The second and more attractive argument of Mr Langan is one to which Mr Yates adverted, namely, that in the light of the rents obtained for Units 9 and 12 the agreed figure of 65p for standard accommodation does not relate to a building of 10,000 sq ft. Mr Langan argues that otherwise the landlord would obtain a discount on the true rental value plus the application of the $\frac{95}{65}$ multiplier. The basis of that submission is that the 65p rental was intended by the parties to be a genuine estimate of the market rental for standard accommodation and that a rental of not less than 77p, and as much as £1.08, was the actual market rental known to the parties for standard industrial buildings of about 10,000 sq ft.

As for the actual market rent in November 1974, I do not think it permissible to take account of the £1.08 figure which appeared in the subsequent lease dated December 20 1974, that is to say, after the date of the lease of Unit 11. I do accept that the rents agreed for other units on the estate are likely to have been known to the parties given the practice of surveyors to obtain comparables, and it is a fact that 77p, the rent obtained in September, is higher than 65p. But I am not thereby persuaded that 65p was necessarily not the actual market rent, as the parties perceived it to be in November 1974, for a building of about 10,000 sq ft. I can take judicial note of the fact that in November 1974 the financial and property markets were in considerable disarray. The true market rental must have been a matter of some uncertainty at that time and it is possible that the parties would have regarded 65p as the true market rent, looking at the rents which had been obtained in the previous three years for other units on the estate. But in any event the primary purpose of the reference in the lease to 65p appears to be to provide the denominator in the $\frac{95}{65}$ multiplier. The parties had agreed 95p per sq ft as the actual rental for Unit 11. Whether the 65p figure played any part in the negotiations for that initial rent is not known. The landlord, of course, would have wanted the denominator in the multiplier fraction to be as low as possible. In all the circumstances I cannot give the reference to 65p the significance that Mr Langan has urged upon me. In my judgment Mr Langan's argument is not sufficient to overcome what to my mind is the clear indication from the language of the definition and in particular the use of the word "standard".

Mr Langan also submitted that the landlord was wrong to treat the locality as limited to the estate. Mr Langan rightly points out that the parties have not used the words "the estate" although that term is to be found in another clause of the lease. He submits that a wider meaning must have been intended by the parties. I see great force in that argument. But "locality" is a word of some imprecision; there is no particular logic in going as little or as far as three-quarters of a mile away from Unit 11, and in this case to go that far crosses local authority boundary lines. On the facts, it has not been shown that there are industrial buildings of the like size and otherwise possessing the characteristics required by the definition in an area wider than the estate. The words "the locality" are capable of being limited to the estate on which Unit 11 stands, and in the circumstances I am prepared to hold that that is its meaning.

In the result I reach the clear conclusion that the landlord's construction is to be preferred as according better with what I take to be the intentions of the parties, viewed objectively. Accordingly I shall make the declaration asked for by the landlord by its originating summons and I shall dismiss the tenant's originating summons.

Court of Appeal

February 21 1986

(Before Lord Justice SLADE, Lord Justice NICHOLLS and Mr Justice SHELDON)

METROLANDS INVESTMENTS LTD v J H DEWHURST LTD

Estates Gazette March 29 1986

277 EG 1343-1350

Landlord and tenant — Rent review clause — Principles restated by Court of Appeal — Decision of Peter Gibson J reversed — A simple correlation between rent review clause and break clause not by itself a sufficient contra-indication to displace the *prima facie* rule laid down in *United Scientific Holdings* v *Burnley Borough Council* that time was not of the essence — In the present case there was "the clearest possible interrelation between the timetable embodied in the rent review clause and that embodied in the break clause" and Peter Gibson J followed the decision in *William Hill (Southern) Ltd* v *Govier & Govier* in holding time was of the essence, although he did "not find the simple correlation approach completely satisfying logically" — After reviewing authorities, the Court of Appeal, in a judgment of the court, formulated some underlying principles — The reason for the *prima facie* rule stated in the *United Scientific* case was that it was likely that the detriment to the landlord of losing his review by failure to adhere strictly to the stipulated time-limit would be wholly disproportionate to the disadvantage to the tenant of a delay in the assessment of rent — It was therefore not right, in the absence of sufficient contra-indications, to impute to the parties the intention that time should be of the essence for the purpose of a rent review clause — Where the timetables of a rent review clause and a break clause are closely interlocked, this may constitute a sufficient contra-indication, but everything depends on the structure and wording of the particular lease — Good examples of such a contra-indication were *C Richards & Son Ltd* v *Karenita Ltd* and *Coventry City Council* v *J Hepworth & Sons Ltd* — These two cases had one most important feature in common, namely, that the event as to which time was held to be of the essence was one in the landlord's full control (service of his trigger notice) — In the present case, on the other hand, the event as to which time was said to be of the essence was the actual obtaining of the arbitrator's decision — No matter how diligent the landlord was, there could be many reasons why the arbitrator's decision might not be obtained until after the material date — In the *William Hill* case the tenant, no less than the landlord, had the right to refer the matter to the arbitrator, but the Court of Appeal did not express any conclusion on a submission that the case was wrongly decided — Held that in the present case the interrelation between the break clause and the rent review clause was not a sufficient contra-indication to rebut the *prima facie* rule that time was not of the essence — The two main reasons were (1) that the date of the relevant event, the actual obtaining of the arbitrator's award, was to a substantial degree outside the landlord's control, and (2) that any potential hardship to the tenant through the landlord's delay in initiating the rent review procedure could be eliminated, or substantially mitigated, by the tenant initiating such action itself — Appeal allowed

This was an appeal by the plaintiff landlord, Metrolands Investments Ltd, from a decision of Peter Gibson J on the trial of a preliminary issue in the plaintiff's action against J H Dewhurst Ltd, the defendant and present respondent. The case related to the lease of a butcher's shop at 18 Mercian Way, Ashton-under-Lyne, Lancashire. The decision of Peter Gibson J was reported at [1985] 1 EGLR 105; (1985) 274 EG 1388-1396.

Michael Mark (instructed by Willey Hargrave & Co, of Harrogate) appeared on behalf of the appellant; J A F Thom (instructed by R A Roberts) represented the respondent.

Giving judgment, SLADE LJ said: This is the judgment of the court on an appeal by the plaintiff in certain landlord and tenant proceedings from an order of Peter Gibson J made on December 17 1984 on the trial of a preliminary issue in the action. The appellant landlord is Metrolands Investments Ltd ("Metrolands"). The respondent tenant is J H Dewhurst Ltd ("Dewhurst"). The dispute between the parties concerns a rent review clause.

The decision of Peter Gibson J contains a very clear statement of the relevant facts and is reported at [1985] 3 All ER 206*. The action related to a butcher's shop which was let by Metrolands to Dewhurst by a lease dated January 30 1968 for a term of 21 years from February 19 1968. The reddendum was in this form:

> YIELDING AND PAYING therefor during the first three years of the said term the yearly rent of [£1,500] during the next eleven years of the said term the yearly rent of [£1,800] and during the remaining seven years of the said term the yearly rent at which the demised premises might on [February 19 1982] reasonably be expected to let without premium in the open market between willing landlord and willing tenant on terms similar to those contained in this present Lease and assuming that the Lessee has observed and performed all the covenants and conditions to be observed and performed hereunder but disregarding any goodwill attaching to the demised premises by reason of the carrying on thereat of the Lessees business such yearly rent to be agreed between the Landlord and the Lessee and failing agreement to be determined by arbitration as hereinafter provided Provided always that the decision of such arbitrator shall be obtained before the expiration of the first half of the Fourteenth year of the term hereby created BUT such yearly rent shall not in any event be less than [£1,800]...

The learned judge found it convenient to divide the relevant part of the reddendum relating to the last seven years of the term into four limbs: limb 1 commencing with "during the remaining seven years" and ending with "the Lessee's business"; limb 2 commencing with "such yearly rent" and ending with "as hereinafter provided"; limb 3 commencing with "Provided always" and ending with "term hereby created"; and limb 4 commencing with "but such yearly rent" and ending with "£1,800". I will adopt the same convenient division.

There are two other material provisions of the lease. Clause 4(4) provides:

> If any dispute question difference or controversy shall arise which under the terms of this Lease is to be referred to arbitration the same shall be referred to a single arbitrator to be agreed between the parties or failing agreement to two arbitrators (one to be appointed by each party to the reference) or their umpire pursuant to and so as with regard to the mode and consequences of any such reference and in all other respects to be in conformity with the provisions of the Arbitration Act 1950 or any statutory modification or re-enactment thereof.

Clause 5 ("the break clause") provides:

> PROVIDED LASTLY AND IT IS HEREBY AGREED AND DECLARED that if the Lessee shall be desirous of determining this present lease at the end of the fourteenth year of the term hereby granted and of such its desire shall deliver to the Landlord or leave for the Landlord or send by registered post to the Landlord at its registered office within the six months previous to the end of the said

*Editor's note: See also [1985] 1 EGLR 105; (1985) 274 EG 1388.

A fourteenth year of the said term not less than three months notice in writing and shall pay all rent and perform and observe all the covenants and conditions hereinbefore contained and on the part of the Lessee to be performed and observed up to such determination then and in such case immediately after the expiration of the said period of fourteen years this present Lease shall cease and be void without prejudice to any claim by the Landlord against the Lessee in respect of any antecedent breach of any covenant or condition herein contained.

The timetable envisaged by the express provisions of the reddendum and the break clause was thus as follows:

(1) The date by reference to which the open market yearly rent payable by Dewhurst for the last seven years of the term was to be calculated was to be February 19 1982 (being the first day of that
B seven-year period).

(2) The open market yearly rent payable by Dewhurst for this seven-year period was to be fixed according to the formula set out in the reddendum, either by agreement between the two parties or in default of agreement by arbitration. If arbitration was necessary the decision of the arbitrator was to be obtained before August 19 1981.

(3) Dewhurst would be free to operate the break clause by serving written notice on Metrolands at any time between August 19 1981 and November 18 1981 with the effect of determining the lease on February 18 1982.

Neither Metrolands nor Dewhurst made any attempt to agree a rent or to go to arbitration before August 19 1981. Dewhurst did not attempt to operate the break clause before November 19 1981. On
C December 2 1981, for the first time, Metrolands by its surveyors wrote to Dewhurst indicating that it wanted the rent to be reviewed, although no figure was then suggested. On December 17 1981, Dewhurst replied to the effect that the rent review notice was invalid as being too late and pointed out that it was by then also too late for it to serve a notice under the break clause. Dewhurst, however, does not suggest that Metrolands deliberately delayed serving notice of its intention to have the rent reviewed.

On October 24 1983 Metrolands issued proceedings against Dewhurst. By para 1 of the prayer for relief in its statement of claim it claimed a declaration, in effect, that the rent payable in the last seven years of the term is that set out in the formula embodied in limb 1 of
D the reddendum.

The master made a consent order that the question raised by this para, and the question whether the amount of the yearly rent was to be determined by the court or by arbitration or in some other way, should be tried as a preliminary issue, on the basis of an agreed statement of facts. The learned judge declined to grant any of the relief sought. He upheld Dewhurst's contention that time was of the essence of the rent review procedure embodied in the lease and that Metrolands' attempt to initiate that procedure in December 1981 onwards was accordingly too late.

As the learned judge said, the starting point for any consideration of the question whether time is of the essence of rent review provisions is the decision of the House of Lords in *United Scientific*
E *Holdings Ltd* v *Burnley Borough Council* [1978] AC 904, where Lord Diplock (at p 930) stated the principle that:

In the absence of any contra-indications in the express words of the lease or in the interrelation of the rent review clause itself and other clauses or in the surrounding circumstances, the presumption is that the timetable specified in a rent review clause for completion of the various steps for determining the rent payable in respect of the period following the review date is not of the essence of the contract.

Neither side in argument has placed any reliance on the surrounding circumstances as an aid to the construction of this lease. Mr Thom, however, on behalf of Dewhurst, has submitted to us, as he submitted to Peter Gibson J, that this particular lease contains two contra-indications sufficient to rebut the normal presumption that
F the timetable specified in a rent review clause is not of the essence of the contract. The first suggested contra-indication is the express terms of limb 4 of the reddendum. The learned judge (at p 210C-210E) rejected this submission and Dewhurst challenges this rejection in a respondent's notice. However, he accepted the sufficiency and full force of the second contra-indication relied on by Dewhurst, which was the interrelation between the time-limit in limb 3 of the reddendum and the break clause. This is the part of his decision (set out at pp 210C to 214J of the report) which Metrolands seeks to challenge on this appeal. It has not sought to pursue certain alternative submissions which were advanced in the court below and are referred to at pp 215A to 215G of the report. The effect of the learned judge's decision (see pp 215H to 216A) was that the lease had made no express provision for the amount of the rent to be payable for the last seven years of the term if it was not ascertained by agreement or arbitration within the specified time-limits, that the court itself had no power to determine a rent if those time-limits were not complied with, and that, in the event of non-compliance, the annual rent payable for the last seven years must continue to be the same as that payable for the preceding period, namely £1,800 (see *Weller* v *Akehurst* [1981]3 All ER 411).

It will be convenient to begin by dealing with the first of the suggested contra-indications relied on by Dewhurst, since it raises a short, albeit not entirely easy, question of construction arising on the reddendum, when read in isolation.

Mr Thom, on behalf of Dewhurst, pointed out that the words "but such yearly rent shall not in any event be less than £1,800" in limb 4 of the reddendum immediately follow the proviso at the end of limb 3, which provides (in apparently mandatory form, in view of the use of the word "shall") that "the decision of such arbitrator shall be obtained" before the stated date. In his submission, the parties to the lease must have contemplated the possible contingency of a failure to obtain such a decision by the stated date. The words "in any event", as he put it, do "double duty" by covering the possible contingency of the arbitrator fixing an annual rent of less than £1,800 *and* the possible contingency of the failure to obtain a timeous decision from the arbitrator. Thus, in his submission, one of the two purposes of limb 4, when properly construed, is to specify the rent that is to be payable in default of compliance with the time-limits laid down in limb 3. The coupling of this "in default" provision with the final time-limit in limb 3 shows, in his submission, that this time-limit is intended to be a strict one.

The learned judge (at p 210D-E) rejected these submissions in these words:

Limb 4 relates to "such yearly rent". That, as counsel for Metrolands pointed out, can only refer to the yearly rent on the hypothetical basis set out in limb 1 or on that basis and agreed or determined in limb 2. All that limb 4 does is to put a minimum figure on the yearly rent on that basis or on that basis and agreed or determined. It does not refer to the rent not ascertained by reference to limb 1 or limb 2. In my judgment, therefore, limb 4 cannot be construed as a default clause to operate on the failure to go to arbitration at all or to obtain the decision in time.

The phrase "such yearly rent" in limb 4 echoes the same phrase which is used in limb 2. In limb 2 it clearly refers to a rent to be agreed or determined according to the formula specified in limb 1. *Prima facie* the phrase must bear the same meaning in limb 4 as it bears in limb 2. There is, however, a further point which supports the learned judge's construction, to which Mr Mark on behalf of Metrolands drew our attention. If limb 4 were intended to operate as a provision fixing the rent in default of its being otherwise determined by agreement or arbitration within the specified time-limits, the words "not . . . less than" in limb 4 would be quite inappropriate.

For all these reasons, we conclude that limb 4 cannot properly be read as an "in default" provision. We respectfully agree with the learned judge's conclusion on this point, and that the reddendum, read by itself, does not contain any clear indication that time is to be of the essence of the rent review provisions. While the phrase "in any event" in limb 4 is not a very clear one, we read it as meaning "whatever may be the arbitrator's decision", so as to ensure that on any footing the rent review will be incapable of resulting in a drop in rent.

We now turn to the other suggested contra-indication relied on by Dewhurst, based on the correlation of the rent review clause with the break clause. Let it be said at once that there is the clearest possible correlation of this nature. Earlier in this judgment we have set out the timetable envisaged by the express provisions of the lease. The relevant dates were clearly fixed by the draftsman in the contemplation that any necessary decision of the arbitrator would be obtained at the very latest before August 19 1981, which was to be the start of the three-month period (August 19 1981 — November 18 1981) during which Dewhurst was to be free, if it thought fit, to serve a notice to determine the lease. The draftsman manifestly envisaged a timetable by virtue of which the lessee would know the rent which would be payable for the last seven years of the term by the time when it came to make its decision whether or not to exercise its right to determine the lease.

One other important point is common ground, as it appears to have been before the learned judge. It is not disputed that time is of the essence for the purpose of applying the time-limits specified in the break clause itself. The reasoning sufficiently appears from the following passage in the speech of Lord Diplock in the *United Scientific* case (at p 929):

. . . there is a practical business reason for treating time as of the essence of such a clause, which is similar to that applicable to an option to acquire property. The exercise of this option by the tenant will have the effect of depriving the landlord of the existing source of income from his property and the evident purpose of the stipulation as to notice is to leave him free thereafter to enter into a contract with a new tenant for a tenancy commencing at the date of surrender provided for in the break clause.

It seems clear that, in referring in the *United Scientific* case to "the interrelation of the rent review clause itself and other clauses", Lord Diplock himself had particularly in mind the interrelation of rent review clauses and clauses giving the tenant the right to determine the lease. In his decision in *Coventry City Council* v *J Hepworth & Sons Ltd* (1981) 261 EG 566 (upheld by this court on appeal at (1982) 46 P & CR 170),* Warner J helpfully identified and cited the following passages from the speeches in the *United Scientific* case which concerned this interrelation.

Lord Diplock ([1978] AC 904 at pp 935-936) said this:

Samuel Properties (Developments) Ltd v *Hayek* [1972] 1 WLR 1296 may be regarded as the origin of the dichotomy between "option" on the one hand and "obligation" or "machinery" on the other: the word option having been used in the lease itself to describe the landlord's right to require the rent to be reviewed. It should be treated as overruled. There was a complication in that the rent review clause was associated with a break clause which gave to the tenant the right to surrender the residue of the term on any rent review day by giving prior notice. The time-table in the rent review clause for the determination of the new rent was obviously correlated with the time by which the tenant had to give notice of his intention to surrender, so as to enable him to make his decision whether or not to exercise that right in the knowledge of what the new rent would be if he continued in possession after the review date. Had that been all, as it had been in the previous and rightly decided case of *C Richards & Son Ltd* v *Karenita Ltd* (1971) 221 EG 25, it would, I think, have been sufficient by necessary implication to make time of the essence of the rent review clause because of its interrelation with the time by which notice was to be given under the break clause — a time which, for reasons I have given earlier, I consider to be of the essence of the contract.

In *Samuel Properties (Developments) Ltd* v *Hayek,* however, the break clause itself contained a provision under which the period during which the tenant could exercise his right to surrender would be extended in the event of the reviewed rent not having been ascertained within the time stipulated in the rent review clause. So the implication that would otherwise have arisen from the association of the rent review clause with a break clause was negatived.

Lord Dilhorne (at p 940) agreed with what Lord Diplock had said about these cases.

Lord Simon of Glaisdale (at p 946) said:

However, where a rent review clause is associated with a true option (a "break" clause, for example), it is a strong indication that time is intended to be of the essence of the rent review clause — if not absolutely, at least to the extent that the tenant will reasonably expect to know what new rent he will have to pay before the time comes for him to elect whether to terminate or renew the tenancy (cf *Samuel Properties (Developments) Ltd* v *Hayek* [1972] 1 WLR 1296). That situation stands in significant contrast with those in the instant appeals.

Lord Salmon (at pp 951-952) said:

In *Samuel Properties (Developments) Ltd* v *Hayek* the rent revision clause which laid down the procedure for having the open market rent value ascertained at the end of the seventh and fourteenth years of the term and the rent then being raised to that level, was dressed up to look like an option. Indeed the word "option" appeared in the clause. But, for the reasons I have already stated, I do not think that it was a real option in the sense that any option to renew or determine a lease is an option. The clause required the landlord to give notice to the tenants six months prior to the expiry of the seventh year if he required the rent to be raised to the open market rental value. If within one month of the notice, the parties failed to agree the open market rental value this figure was to be determined by a valuer appointed by the President of the Royal Institution of Chartered Surveyors. But the date by which this determination was to be made was not specified. The landlord gave his notice about one month late. The Court of Appeal held that time was of the essence and that the landlord was precluded from putting the rent revision clause into operation. Clause 5 of the lease so far as relevant gave the tenant a true option to determine the lease at the end of the seventh year of the term by giving the landlords at least one quarter's notice in writing. This break clause was obviously inserted to protect the tenant should he not wish to pay the increased rent during the next seven year period of the term.

The proviso to the break clause strongly suggests however that the time provisions relating to rent revision were not of the essence of the contract. It reads: "Provided always that if one quarter before the expiration of the first seven . . . years of the term . . . the reviewed rent . . . shall not have been reviewed then the right of the lessee to terminate as herein provided shall be G
extended until the expiration of one month from the date of the notification of the reviewed rent to the lessee."

There is nothing in the proviso nor in any other part of the lease to suggest that the new rent may not be determined by the valuer and notification of this rent may not reach the lessee until after the expiration of the first seven year period. In my view *Samuel Properties (Developments) Ltd* v *Hayek* [1972] 1 WLR 1296 was wrongly decided and should be overruled.

Lastly, Lord Fraser (at pp 962-963) said:

For these reasons I am of the opinion that the equitable rule against treating time as of the essence of a contract is applicable to rent review clauses unless there is some special reason for excluding its application to a particular clause. The rule would of course be excluded if the review clause expressly stated that time was to be of the essence. It would also be excluded if the context clearly H
indicated that that was the intention of the parties — as for instance where the tenant had a right to break the lease by notice given by a specified date which was later than the last date for serving the landlord's trigger notice. The tenant's notice to terminate the contract would be one where the time-limit was mandatory, and the necessary implication is that the time-limit for giving the landlord's notice of review must also be mandatory. An example of such interlocked provisions is to be found in *C Richards & Son Ltd* v *Karenita Ltd* (1971) 221 EG 25 where the decision that time was of the essence of the landlord's notice could be supported on this ground, although not, as I think, on the ground on which it was actually tested. The case of *Samuel Properties (Developments) Ltd* v *Hayek* [1972] 1 WLR 1296 is not in this class because, although there was a tenant's break clause, the time allowed to the tenant for giving notice was automatically extended until one month after the notification of the reviewed rent to the lessee. J

The relevant facts of the *Hayek* case appear sufficiently from these passages which I have cited from their lordships' speeches. The relevant facts of the *Richards* case appear sufficiently from this passage from the judgment of Goulding J at (1971) 221 EG 25:

by an underlease of June 5 1963, Nobbs (Cleaners) Ltd demised the second floor of 52a Blackstock Road, London N, to the defendants for a term of 14 years at a yearly rent of £700. Clause 2(11) provided that "if the landlords shall by giving notice in writing to the tenants at any time during the first three months of the seventh year of the term hereby created require a review of the rent payable hereunder, such rent shall be revised with effect from the expiration of the seventh year . . ." Clause 4(4) provided that if the tenants desired to determine the term at the expiration of the first seven years and gave to the landlords three months' previous notice in writing, then K
immediately on the expiration of the seven years the demise should cease. The period during which the landlords' notice might have been given began on June 6, 1969, and ended on September 5, 1969. No notice under clause 2(11) was given during that period, but the plaintiffs, who were now landlords, purported to give such a notice on February 26, 1970, when nearly nine months of the seventh year had elapsed. The question before the court was whether that late notice was effective or not.

The *Coventry City Council* case (*supra*) likewise concerned a rent review clause, coupled with a break clause. The rent review clause itself laid down what Warner J described as an elaborate and precise timetable. For present purposes it will suffice to quote the summary of the relevant provisions given in his judgment ((1981) 261 EG 566 at L
p 567):

On or before December 31 1973 the corporation might give to the tenants notice of its desire to increase the rent in respect of the period from April 1 1975 until the expiration of the term created by the lease, that is, roughly speaking, in respect of the second 21 years of it. I shall call that a "rent review notice". If the corporation gave such a notice and no agreement had been reached by February 28 1974 (that is, within two months), there was to be a reference to arbitration. Clearly, what was envisaged was that, in the two months between December 31 1973 and February 28 1974, there would be negotiations between the parties, which would either produce agreement on a new rent or be abortive. Then, if the negotiations proved to be abortive, the parties were given one month in which to agree upon a single arbitrator. If they failed to do so, each must appoint an arbitrator. The last day for doing that was April 30 1974. The provisions of sections 7 and 8 of the Arbitration M
Act 1950 would then apply. On or before September 30 1974 the tenants might give notice of their intention to determine the lease as from March 31 1975, that is the day before any increase of rent might take effect.

The ultimate conclusion of Warner J and its *ratio decidendi* was stated by him as follows (at p 569):

Mr Cripps submitted that it was not in every case where a rent review clause was associated with a tenant's option to break that time was of the essence of the rent review clause. That is manifestly right, as the House of Lords' treatment of the *Hayek* case illustrates. I do not think, however, that the features of the present case to which Mr Cripps pointed are sufficient to negative the presumption that, where a rent review clause is linked to a tenant's option to break, time is of the essence of the rent review clause. The length of the term and the fact that the lease provided for only one break at 21

*Editor's note: See also (1982) 265 EG 608 for CA decision.

A years are, I think, sufficiently explained by the circumstance that it was granted in 1953, at a time when inflation was at a much lower rate than now. The fact that substantial subletting by Hepworths was envisaged seems to me neutral. The period of nine months between December 31 1973 and September 30 1974 was, I think, intended, not to give the tenant an opportunity to serve a notice making time of the essence, but to enable any necessary arbitration to have been completed by September 30 1974 or at least to be sufficiently advanced by that date to give the tenant a good idea of what the revised rent was likely to be. Lastly, I do not think that it matters that there could be no certainty that the tenant would know the precise amount of the revised rent by that date. So far as the report of *C Richards & Son Ltd* v *Karenita Ltd* shows, there was no such certainty in that case either.

The Court of Appeal affirmed Warner J's judgment in the B *Coventry City Council* case (1983) 46 P & CR 170.* Lawton LJ, referring to the *United Scientific* decision, said (at page 176):

It comes to this, that their Lordships . . . were all of the opinion that, where you have a triggering off of a rent review provision started by the landlord, followed by an option given to the tenant to break the lease if he so wishes, then time is to be presumed to be of the essence of the agreement unless there are contra-indications.

He added (at p 176) that it made no difference that in the case of a lease which provides for triggering action followed by a break clause, but also for an interval of time between the triggering action and the date when a decision has to be made about breaking the lease, the tenant can always serve upon the landlord a notice making time of the essence of the agreement. As he observed: "No landlord and tenant, C in my judgment, when making a lease of this kind, would have contemplated such a possibility."

Griffiths LJ (as he then was) agreed (at p 176), saying that he found the reasoning of Warner J so compelling that he did not wish to give any separate reasons of his own. Fox LJ (at p 176) also agreed without adding any further observations.

The *Coventry City Council* decision was followed by Mr Evans-Lombe QC, sitting as a deputy judge of the Chancery Division in *William Hill (Southern) Ltd* v *Govier & Govier* (1984) 269 EG 1168. In that case, by a subclause beginning "the rent shall be reviewed at the 25th December 1982 and at the 25th December 1987", the lease provided for rent reviews at the end of the fifth and tenth years of the term. The wording of the lease thus appeared to make the reviews D mandatory. If the landlord and tenant failed to agree the best rack rent, obtainable in the open market, nine months before the review date, then either party might within a further period of one month refer the question of the amount of rent to a chartered surveyor nominated in default of agreement by the president of the Royal Institution of Chartered Surveyors. There was a break clause, which entitled the tenant to determine the lease at the end of the fifth or tenth year by giving the lessor three months' written notice to expire on the relevant review date.

The lessor, in argument, sought to distinguish the *Coventry City Council* case, particularly on the grounds that the provisions for rent review in the lease in the *William Hill* case were mandatory and took E effect without any requirement of a "trigger notice" on the lessor's part. The learned deputy judge rejected this suggested distinction. He described (at p 1170) Warner J and the Court of Appeal in the *Coventry City Council* case as having held that "where such provisions are linked with provisions for rent review, *prima facie* the court will construe the provisions to make time of the essence of those relating to the review". He could find nothing sufficient to lead him to a conclusion contrary to "Warner J's *prima facie* rule".

In the present case, Peter Gibson J (at p 214A) said that the main submission put before him by counsel for Metrolands was that "in the *United Scientific* case itself the House of Lords had reached the conclusion that time was not of the essence, *inter alia*, because the tenant as well as the landlord had it in his power to initiate the rent F review". However, as he said at p 213F, there was a major point of similarity between the case before him and the *William Hill* case in that "the tenant, no less than the landlord, had the right to refer the matter to the arbitrator". Nevertheless, in the *William Hill* case the deputy judge clearly had considered, but rejected, the suggested distinction drawn on that account, between a landlord's "trigger notice" case and a case such as that before him where the tenant, as well as the landlord, could refer the matter to an arbitrator. Peter Gibson J, after a very careful analysis of the *William Hill* decision, observed (at p 213G):

The decision in the *William Hill* case is of a judge, albeit a deputy, of equal jurisdiction on a point on which there is no other direct authority by way of a decided case; and whilst it is not binding on me I would not wish to introduce further uncertainty into this difficult area of the law by a decision inconsistent with his unless I was convinced that the deputy judge reached the wrong conclusion.

He summarised his own ultimate conclusions thus (at p 214H-J):

In other words, in many landlord's trigger notice cases it would be open to the tenant to find out the likely rent, just as much as in a case where he has an express right to refer to arbitration; and yet that was not treated by Lord Diplock, Lord Simon and Lord Fraser as relevant. The simple correlation between the rent review clause and the break clause was enough, save where there was a specific inconsistent provision such as the proviso in the *Samuel Properties* case. Moreover, I think that Lord Diplock in referring, as he did, to the necessary implication as to time being of the essence where there was an interrelation between a rent review clause and a break clause, and Lord Simon to the like effect, were intending to lay down general guidelines. I confess that, for the reasons so attractively advanced by counsel for Metrolands, I do not find the simple correlation approach completely satisfying logically. But I am unable to say that Warner J was wrong to state the applicable presumption in the way that he did, nor am I able to say that the deputy judge in the *William Hill* case was wrong to follow that statement.

Accordingly, I shall follow the decision in the *William Hill* case in holding that here, too, time is of the essence of the reddendum by reason of the correlation with the break clause.

In the course of argument in this court, in support of the learned judge's judgment, Mr Thom relied strongly on the applicable presumption as stated by Warner J in the *Coventry City Council* case. He submitted that the judgments of the Court of Appeal in that case support the existence of that presumption and that, where it is applicable, it wholly displaces the presumption ordinarily applicable to the construction of rent review clauses as stated by Lord Diplock in the *United Scientific* case (at p 930). Mr Mark, on the other hand, submitted that in so far as this court in the *Coventry City Council* case had supported the existence of any such presumption as was referred to by Warner J, the judgment of Lawton LJ makes it clear that the presumption applies only in a case of a lease which provides for the triggering off of a rent review provision started by the landlord alone, followed by an option given to the tenant to break the lease; in the present case the tenant, no less than the landlord, has the right to initiate any review. He challenged the correctness of the principle which Peter Gibson J derived from the speeches in the *United Scientific* case that "the simple correlation between the rent review clause and the break clause was enough" to make time of the essence of the rent review clause "save where there was a specific inconsistent provision . . .".

Throughout this debate, we think it must be borne in mind that the ultimate object of the court in construing a rent review clause, like any other contractual provision, must be to ascertain the parties' intentions from the particular words which they have used to express those intentions, read, of course, in the light of any admissible evidence as to surrounding circumstances — albeit with the assistance of the guidelines as to construction afforded by earlier authorities. In many, perhaps most, cases, of which the present is one, the rent review (if any) can result only in an increase, and thus is only for the landlord's benefit. Essentially, therefore, the question to which the court has to direct its mind is this: Is the proper intention to impute to the parties, from the words which they have used, the intention that the landlord shall lose his right to a review if the stipulated timetable is not strictly adhered to in the relevant respects?

In practice it is ordinarily likely that "the detriment to the landlord of losing his review altogether by failure to adhere strictly to the stipulated time-limit will be wholly disproportionate to the disadvantage to the tenant of a delay in the assessment of the rent". This , as Dillon LJ pointed out in *Touche Ross & Co* v *Secretary of State for the Environment* (1983) 46 P & CR 187 (at p 190),* appears to have been the reason why the House of Lords in the *United Scientific* case concluded that *prima facie*, and in the absence of sufficient contra-indications, it is not right to impute to the parties to a lease the intention that time is to be of the essence for the purpose of a rent review clause. This, we think, should be the initial starting point on the consideration of any such clause.

However, as their lordships recognised, in a case where a lease contains a break clause as well as a rent review clause and the timetables of the two clauses are closely interlocked, the interrelation of the two clauses is *likely* to suffice as a contra-indication sufficient to rebut the ordinary presumption — though everything must depend on the wording of the particular lease. This, we believe, was the sense

*Editor's note: See also (1982) 265 EG 608.

*Editor's note: See also (1982) 265 EG 982 at p 983.

in which Warner J referred to a "presumption" applicable to a case of this nature.

A good example of a case where the interrelation of the two clauses did constitute a contra-indication of this nature is *Richards* itself. There the latest date specified for the giving of the landlord's trigger notice was September 5 1969 and the latest date specified for the giving of the tenant's notice to determine was six months later, namely March 5 1970. Lord Diplock, Lord Dilhorne and Lord Fraser, in approving that decision in the result, did not find it necessary to spell out in any detail the reasons why they considered the ordinary presumption to be rebutted, but we infer that their reasoning could properly be elaborated as follows: The intention of the parties to the lease in providing for this six-month gap must have been to allow adequate time for the process of the review of the rent to have been completed, or at least sufficiently advanced, by the latest date for service of the tenant's notice to determine, so as to give the tenant the advantages of knowing before that latest date (a) that the lessor intended to invoke the rent review procedure, and (b) what the revised rent was likely to be. The tenant himself must on any footing be held strictly to the time-limits laid down for the service of any notice by him to determine the lease. If time was not regarded as being of the essence for the purpose of the lessor's trigger notice, these potential advantages would have been subject to at least erosion and possible elimination by tardy service of any such notice. In these circumstances, despite the severe potential loss which the landlord would suffer if that notice was served even one day late, it could not reasonably be supposed that it was in the contemplation of the parties that the tenant should be exposed to the loss of these advantages by tardy action on the part of the landlord in initiating the rent review procedure.

Precisely the same reasoning, *mutatis mutandis*, was applicable in the *Coventry City Council* case, where the timetable specified in the lease provided for a five months' gap between the latest date for the giving of the landlord's trigger notice and the latest date for the giving of the tenant's notice to determine. We respectfully agree with that decision, which in any event is binding on us.

The lease under consideration in the *William Hill* case contained a number of features not present in the lease now before us and, since that decision does not bind us, we do not find it necessary to embark on any detailed consideration of it or to express any conclusion on Mr Mark's submission that it was wrongly decided. We have doubts as to the usefulness of a minute comparison of the different provisions of different leases in a case such as this. However, as Mr Mark's argument in this court developed, it became apparent that by far his most important point (at least to our minds) was one which, though touched on by the learned judge in his judgment (at p 213D-F), does not appear to have been placed in the forefront of Metrolands' case in the court below. There is a basic difference in the provisions of the lease in the present case from those under consideration in *Richards* and in *Coventry City Council*. In those two cases, the leases had one most important feature in common, that is to say, *the event as to which time was held to be of the essence was one in the landlord's full control* (namely, service of his trigger notice). In a case where an event included as part of the timetable in a rent review clause is one within the landlord's full control, it is by no means inconceivable that he should have been willing and should have intended to commit himself to a rigid adherence to that part of the timetable, even though the consequence of a failure to do so would be that of losing the entire benefit of the relevant rent review and even though the lease does not expressly state that time is to be of the essence.

In contrast, the lease under consideration in the present case has this special and, we would think, unusual feature: the event as to which time is said to be of the essence is the *actual obtaining of the arbitrator's decision*. As Mr Mark pointed out, it was readily conceivable that the landlord might have acted with irreproachable promptness and diligence in setting in motion the arbitration machinery provided for by the lease, for the purpose of a rent review, but that, without any fault whatever on the landlord's part and due to events entirely beyond his control, the arbitrator's decision might still not have been obtained until after August 19 1981. The arbitrator himself, for example, might have been ill or unduly dilatory; or the final conclusion of the arbitration proceedings might have been delayed by the submission of a number of perfectly legitimate points on the part of the tenant or by other circumstances quite beyond the lessor's control. All these are matters to be taken into account when considering the intention to be imputed to the parties regarding the significance of the date of the obtaining of the arbitrator's decision.

If time was to be treated as being of the essence in respect of this G
date, this would have meant that (subject to any powers of the court to extend time under the Arbitration Acts — as to which see below) a delay in the obtaining of the arbitrator's decision by only one day beyond August 19 1981 would have entirely deprived the lessor of its right to a review in respect of the last seven years of this 21-year term and would have left it saddled with a continuing annual rental as low as £1,800, which had been the rental payable for the preceding 11 years. In these circumstances, we think that the following observations of Lord Dilhorne in the *United Scientific* case [1978] AC 904 at pp 938H-939A are apposite, spoken as they were in relation to a lease which, apart from the absence of a break clause, bore a marked similarity to that under consideration in the present case:

H

> It is most unlikely in these circumstances that the lessors, if they had been asked at the time the leases were entered into to agree that time should be of the essence, would ever have agreed to that, and I see no reason for imputing to them an intention which no reasonable landlord would have had.

In saying this, we do not overlook the importance to the lessee of knowing before August 19 1981 what the revised rent was likely to be, so as to assist in deciding whether or not to operate the break clause. Nevertheless, as it is reasonable to assume, the parties would have been aware at the time when the lease was executed that the lessee would by no means have been left at the lessor's mercy if the lessor did not choose to set the rent review machinery in operation promptly; for the lessee itself had the same right as the lessor to set the arbitration procedure in motion. If, therefore, the lessee foresaw a J
risk of the arbitrator's decision not being obtained by August 19 1981 and that risk caused it concern, an obvious remedy lay in its own hands. For these reasons there can, in our judgment, be no doubt that the potential detriment to which the lessor under this particular lease would have exposed itself by agreeing that time should be of the essence as regards the stipulated date for the obtaining of the arbitrator's decision would have far outweighed any potential detriment to which the lessee would have exposed itself by agreeing that it should not be.

Mr Thom referred us to section 27 of the Arbitration Act 1950, which gives the court power to extend the time for commencing arbitration proceedings, and to section 13 of the Arbitration Act
1950, which gives it power to enlarge the time for the making of the K
award. He submitted that, in appropriate circumstances, it would have been open to the lessor to make an application under those sections for the purpose of avoiding the rigours of any strict time-limit imposed by the lease in relation to the obtaining of the arbitrator's decision. With due deference to this argument, however, we would find it fanciful to attribute to the parties the intention that, on the one hand, time should be strictly of the essence as regards the latest date for the obtaining of the arbitrator's decision, but that, on the other hand, such date should be capable of extension under the Arbitration Act 1950.

This case has been very well argued on both sides. We hope it will be
forgiven if we do not refer to all the points raised in the course of L
argument. It will suffice to summarise our conclusions thus: Though there is the clearest possible interrelation between the timetable embodied in the rent review clause and that embodied in the break clause, such interrelation in the context of this particular lease is not, in our judgment, a "contra-indication" sufficient to rebut the initial presumption of construction (from which we think it right to start) that the timetable specified in the rent review clause for the obtaining of the arbitrator's decision is not of the essence of the contract. The principal reasons why it does not suffice for this purpose are two, which must be considered in conjunction with one another and have not been applicable in conjunction with any previous case which has been cited to us.

First, the date of the relevant event, namely the actual obtaining of M
such decision, is one which is to a substantial degree outside the lessor's control. Second, any potential hardship to the tenant which might otherwise arise through tardy action by the landlord in initiating the rent review procedure can be eliminated or, at least, substantially mitigated, by the tenant initiating such action itself. The ultimate object of the court in construing clauses such as this, as we have already indicated, must be to ascertain the parties' intentions from the words which they have used. Weighing in the balance all the competing considerations that have been urged on us, we have come to the clear conclusion that the intention properly to be attributed to the parties to this particular lease is the intention that time should not be of the essence as regards the obtaining of the arbitrator's decision.

A We therefore allow this appeal, but would welcome submissions as to the precise form of relief which should be granted.

The appeal was allowed with costs in the Court of Appeal and below. The judgment below was set aside and a declaration made to the effect that the rent was to be determined, in default of agreement, by arbitration on the footing that time was not of the essence for the purpose of the material part of the rent review provisions.

Court of Appeal

February 27 1986

B (Before Lord Justice FOX, Lord Justice DILLON and Lord Justice NEILL)

PITTALIS v SHEREFETTIN

Estates Gazette April 12 1986

278 EG 153-162

Arbitration Act 1950, section 27 — Whether tenant could invoke section 27 to obtain an extension of time to give notice of election under rent review clause in lease to have open market rental determined by arbitration — The rent review
C **provisions in the lease gave the lessors the right to notify to the lessee the amount of the open market rental, but with the right of the lessee not later than three months after such notification (time to be deemed to be of the essence) to require the open market rental to be determined by arbitration — Lessors notified an increase (after seven years) from £850 to £6,000 per annum — Lessee did not serve a notice of election for arbitration within the time-limit — On lessee's failure to pay the increased rent the lessors brought proceedings for possession, arrears of rent and mesne profits — Lessee applied for extension of time under section 27 — After giving judgment in favour of the plaintiff lessors the county court**
D **judge had second thoughts, recalled his first judgment and gave judgment in favour of the lessee, extending the time for giving notice of election under section 27 — Lessors appealed — Before the Court of Appeal several issues arose for decision — It was submitted by the appellant lessors that section 27 of the Arbitration Act 1950 did not apply because in the present case there was no arbitration agreement at all — It had been decided in *Tote Bookmakers Ltd* v *Development & Property Holding Co Ltd*, in reliance on a dictum by Davies LJ in *Baron* v *Sunderland Corporation*, that an arbitration clause must give bilateral rights of reference — In the present case, as in the**
E ***Tote* case, the lessee had a unilateral right to refer, so that the Arbitration Act had no application — Held on this point, rejecting the appellants' submission, preferring the case of *Woolf* v *Collis Removal Service* to the *Tote* case, and pronouncing the latter wrongly decided, that parties are entitled, if they choose, to agree to confer a unilateral right to require arbitration — The court therefore had jurisdiction to extend the time-limit for election under the power given by section 27 — The appellants also submitted that the judge was not entitled to recall his first judgment and substitute a different one; they relied on *Re Barrell Enterprises* — Held that in the exceptional circumstances of this case, where the**
F **judge decided almost immediately that his first thoughts were wrong, and there was no detriment to the plaintiffs, he had acted correctly in the sensible administration of justice — The remaining question was, however, whether he had exercised his discretion properly — The court could interfere only if the judge had misdirected himself — In fact his conclusion, that the rent claimed by the appellants was "manifestly grossly inflated", did constitute a misdirection — £850 in 1968 was not a valid comparison with £6,000 in 1982 — It was then for the Court of Appeal to exercise the discretion themselves — Taking everything into account, including the lessee's delay, his failure to establish undue hardship, and the lessors' position, the court's discretion ought not to be exercised under section 27 in favour of the lessee — Lessors' appeal allowed**

This was an appeal by Mr Errato Pittalis and three other persons, lessors of a shop and premises at 355 Caledonian Road, Islington, London N7, from a decision of Judge Martin QC at Westminster County Court, allowing the respondent lessee, Mr Ali Sherefettin, an extension of time to elect for arbitration under a rent review clause in a lease of the property.

J Littman (instructed by Bernard Oberman & Co) appeared on behalf of the appellants; J Sarch (instructed by Miles Griffiths Piercy & Co) represented the respondent.

Giving judgment, Fox LJ said: This is an appeal from a decision of Judge Martin QC at the Westminster County Court. It concerns the operation of a rent review clause in a lease. It gives rise to three main questions, namely:

(i) Whether in the circumstances section 27 of the Arbitration Act 1950 is applicable so as to empower the court to extend a time-limit in the lease for an election by the tenant that the rent be determined by an independent surveyor.

(ii) If section 27 is so applicable, whether the judge having, in an oral judgment, decided that he should decline to exercise the discretion conferred by the section in favour of the tenant, was entitled to recall that judgment, which had not been entered, and by a further judgment to review the matter afresh and reverse his previous decision.

(iii) Whether, if the judge was so entitled, he exercised his discretion properly. And if he did not, the question then arises: How should this court exercise the discretion?

The lease is dated August 30 1968, and was made between Errato Pittalis (the first plaintiff) and Maritsa Georgiou of the one part and Hassan Ahmet of the other part. The lease demised to Mrs Georgiou and her permitted assigns the shop and the premises known as 355 Caledonian Road, Islington ("the property"), for 21 years from June 24 1968.

The rent payable under the lease was £800 per annum for the first seven years, and £850 per annum for the next period of seven years (ie until June 24 1982). From June 24 1982 for the remainder of the term of the lease (the review period) the rent should be whichever was the higher of £850 and the open market rental value of the property for the review period.

The lease further provided that the open market rental value of the property for the review period meant the sum in relation to that period

. . . determined as hereinafter provided as being at the time of such determination the rental value of the [property] in the open market for a lease for a term of seven years certain consisting of the review period with vacant possession at the commencement of the term and on the same terms and conditions other than as to the amount of the rent and the length of the term as are herein contained without the payment of any premium and disregarding any goodwill attached to the holding by reason of the carrying on thereat of the business of the Lessee . . . .

The lease further provided:

(2) The aforesaid open market rental value shall be determined in relation to the review period in the following manner that is to say either

(i) it shall be such a sum as shall be notified in writing by the Lessors to the Lessee not earlier than the twenty fourth day of June one thousand nine hundred and eighty one or as shall within three months after such notification be agreed between the Lessors and the Lessee in writing in substitution for the sum so notified or

(ii) at the election of the Lessee by notice in writing to the Lessors not later than three months after the Lessors' notification in writing mentioned in sub-clause (i) above (time in this respect to be deemed to be of the essence hereof) it shall be determined (in accordance so far as not inconsistent herewith with the provisions of the Arbitration Act 1950 or any statutory modification or re-enactment thereof for the time being in force) by an independent surveyor appointed for that purpose by the Lessors and Lessee by agreement in writing or failing such agreement as to such appointment within one month after the Lessee's said notice of election then by an independent surveyor appointed for that purpose by the President for the time being of the Royal Institution of Chartered Surveyors . . .

Some time prior to 1981 the plaintiffs acquired the freehold reversion to the lease.

On February 24 1972 the lease was assigned to the defendant, Mr Sherefettin (and I will refer to him hereafter as the tenant).

On September 29 1981 the plaintiffs gave notice to the tenant that the open market rental value as from June 24 1982 should be £6,000 per annum. It is common ground that the letter was a valid notice for the purposes of proviso (2)(i) of clause 1 of the lease.

The tenant claimed that he had, by notice in writing to the plaintiffs within three months after the notification by the plaintiffs in the letter of September 29 1981 elected to have the open market rental value determined by an independent surveyor pursuant to proviso (2)(ii) of the lease. Judge Toyn found that no such notice had been served; there is no appeal from that finding. The result is that no election has been made under proviso (2)(ii).

Stopping there, unless the tenant can obtain an extension of time for electing, the position under the provisions of the lease is that as from June 24 1982 the rent for the premises is £6,000. The tenant therefore invokes the provisions of section 27 of the Arbitration Act 1950. That section is as follows:

> Where the terms of an agreement to refer future disputes to arbitration provide that any claims to which the agreement applies shall be barred unless notice to appoint an arbitrator is given or an arbitrator is appointed or some other step to commence arbitration proceedings is taken within a time fixed by the agreement, and a dispute arises to which the agreement applies, the High Court, if it is of opinion that in the circumstances of the case undue hardship would otherwise be caused, and notwithstanding that the time so fixed has expired, may, on such terms, if any, as the justice of the case may require, but without prejudice to the provisions of any enactment limiting the time for the commencement of arbitration proceedings, extend the time for such period as it thinks proper.

The first issue which has arisen is whether section 27 applies at all. It is said on behalf of the plaintiffs that the parties were not contemplating arbitration; they merely required the amount of the rent to be determined by an expert. It is contended, therefore, that the case is not one of arbitration but of valuation (see *Re Dawdy* (1885) 15 QBD 426). I do not think that is correct. The lease provides that the determination should be ". . . in accordance so far as not inconsistent herewith with the provisions of the Arbitration Act 1950 or any statutory modification or re-enactment thereof for the time being in force". I see no reason why the parties should expressly incorporate the provisions of the Arbitration Act in relation to this matter unless they were intending that the proceedings should indeed be an arbitration.

Then it is said that, since notice of election by the tenant was not given in due time, there is no subsisting agreement to arbitrate. But section 27 is applicable to "an agreement to refer future disputes to arbitration". Leaving aside for the moment the question which I mention later, based upon *Baron* v *Sunderland Corporation* [1966] 2 QB 56, in my opinion the lease did contain an agreement to refer a further dispute to arbitration. That agreement was in no sense an agreement to agree. It was contractual. It is true that there would be no reference to the independent surveyor unless the tenant elected. But an agreement to arbitrate in future if a party so elects can, in my opinion, correctly be described as an agreement to refer a future dispute to arbitration; if there is an election, both parties are bound. Looking at the matter at the point of time when the lease was made, there was an agreement to refer a future dispute to arbitration, and not the less so because the reference was upon a contingency (ie election).

The next point taken by the plaintiffs is based upon the decision in *Tote Bookmakers Ltd* v *Development & Property Holding Co Ltd* [1985] 2 WLR 603*. In that case the provisions of the lease were very similar to those with which we are concerned here. The open market rental was to be determined by an independent surveyor at the election of the lessee by counternotice in writing to the lessor not more than three months after the lessor's notice (time to be of the essence). The lessee failed to serve a counternotice within the three months. The lessee made an application to the court under section 27 of the Arbitration Act 1950 for the extension of the period of three months. Peter Gibson J held that it was an essential attribute of an arbitration clause that it gave either party the right to refer the dispute to arbitration, and that since the lessee had a unilateral right to refer, there was no "agreement to refer future disputes to arbitration".

That decision was based upon the statement by Davies LJ in *Baron* v *Sunderland Corporation* (*supra*) at p 64 as follows:

> It is necessary in an arbitration clause that each party shall agree to refer disputes to arbitration; and it is an essential ingredient of an arbitration clause that either party may, in the event of a dispute arising, refer it, in the provided manner, to arbitration. In other words, the clause must give bilateral rights of reference.

The judgment of Davies LJ was concurred in by the other two members of the court.

*Editor's note: See also [1985] 1 EGLR 94; (1984) 274 EG 585.

G Peter Gibson J, after observing that the statement of the law by Davies LJ had been subjected to powerful criticism in both *Russell on Arbitration* (2nd ed) and *Mustill and Boyd on Commercial Arbitration*, came to the conclusion that, whatever his personal views on the question, he was compelled as a matter of authority to follow the statement in *Baron* v *Sunderland Corporation*.

Looking at the matter apart from authority, I can see no reason why, if an agreement between two persons confers on one of them alone the right to refer the matter to arbitration, the reference should not constitute an arbitration. There is a fully bilateral agreement, which constitutes a contract to refer. The fact that the option is exercisable by one of the parties only seems to me to be irrelevant. The arrangement suits both parties. The reason why that is so in cases H such as the present and in the *Tote Bookmakers* case is because the landlord is protected, if there is no arbitration, by his own assessment of the rent as stated in his notice; and the tenant is protected, if he is dissatisfied with the landlord's assessment of the rent, by his right to refer the matter to arbitration. Both sides, therefore, have accepted the arrangement and there is no question of any lack of mutuality. The authorities prior to *Baron* v *Sunderland Corporation* (*supra*) lead me to the same conclusion.

In *Woolf* v *Collis Removal Service* [1948] 1 KB 11, the plaintiff contracted with the defendants to store his goods in their warehouse. The contract contained an arbitration clause requiring any claim by the customer (but no claims by the warehouseman) to be submitted to arbitration. Asquith LJ, giving the judgment of the Court of Appeal, J said at p 17:

> There is nothing in its
>
> the arbitration clause's
>
> unequal operation to divest it, in our view, of the character attributed to arbitration clauses in general in the speeches of Lords Macmillan and Porter in *Heyman* v *Darwins Ltd* as distinguishing them from exception clauses.

The court granted a stay of proceedings pending a reference to arbitration.

In *Heyman* v *Darwins Ltd* [1942] AC 356 Lord Macmillan, at p 370, said:

> Arbitration clauses in contracts vary widely in their language for there is no limitation upon the liberty of contracting parties to define as they please the K matters which they desire to submit to arbitration.

In *Baron* v *Sunderland Corporation* the plaintiff was a school teacher employed by the Sunderland Corporation (the education authority) who were required to pay him a salary in accordance with the Burnham Report. A dispute arose between the parties whether the plaintiff was entitled to a special increment to his salary under the Burnham scales. The Burnham Report established a joint committee to determine questions of interpretation of the report brought forward by a local education authority, or by an association of teachers or by the chairman of the Burnham Committee.

The plaintiff sued the corporation for the increment. The judge stayed the action on the basis that the provisions of the report for the determination of questions constituted an arbitration agreement. L The Court of Appeal allowed the plaintiff's appeal, and held that there was no arbitration agreement.

That decision was quite plainly right. As Davies LJ said at p 64, the position under the report was about as unlike an arbitration clause as anything one could imagine. There was no agreement between the plaintiff and the corporation to refer any issue for determination. All that the plaintiff could do was to ask his association of teachers to bring the matter forward. That circumstance was enough to dispose of the case. I do not think it is correct to say that "the clause must be given bilateral rights of reference". All that is necessary is that there shall be a contract which gives right of reference (whether unilateral or bilateral). That is present in this case.

M I feel some doubt whether Davies LJ, in disposing in an extempore judgment of a perfectly plain case, was intending to do anything more than to make it clear that there could be no arbitration unless both parties agreed to arbitration. But his language is very widely expressed, and I do not think that Peter Gibson J could do other than take it at its face value. It seems to me, however, that so read, the observations of Davies LJ are inconsistent with *Woolf* v *Collis Removal Service*. I prefer the approach in *Woolf* v *Collis Removal Service* which, in my opinion, is correct in principle; that decision, it seems, was not cited in *Baron* v *Sunderland Corporation* or in *Tote Bookmakers Ltd* v *Development & Property Holding Co Ltd*.

The result in my view is that the parties are entitled, if they so choose, to confer a unilateral right to insist upon arbitration. It follows, in my opinion, that the *Tote Bookmakers Ltd* case was wrongly decided. We are informed that there was no appeal in the latter case.

The next question is the ambit of the words (in section 27): ". . . or some other step to commence arbitration proceedings is taken within a time fixed by the agreement". Do they include the election to arbitrate required by proviso (2)(ii) of the lease? It can be said that the election is simply the conclusion of the agreement and is not part of the arbitration proceedings. In reality, however, it is effectively the first step in the arbitration proceedings, and I think it would be an unduly narrow construction of section 27 to exclude it from the operation of the section. Although section 27 was originally construed rather restrictively, the courts have, since the decision in "*The Pegasus*" [1967] 2 QB 86, construed it more liberally (see *The American Sioux* [1980] 3 All ER 154 at p 158 (per Lord Denning MR)).

In *Babanaft International Co SA* v *Avant Petroleum Inc* [1982] 1 WLR 871, Donaldson LJ giving the leading judgment (with which Sir David Cairns and Stephenson LJ agreed) in dealing with the case of *Nestlé Co Ltd* v *E Biggins & Co Ltd* [1958] 1 Lloyds Rep 398, considered the position which arose under an agreement which required arbitration "to be claimed" within 14 days from weighing and/or warehousing the goods. Donaldson LJ said (at p 885) that the concept of "claiming arbitration" was well known in the commodity trades; the arbitration rules of the trade then provide the steps to be taken by each party. He said that in such cases he would accept that "claiming arbitration" may be regarded as a step to commence arbitration proceedings within the meaning of section 27.

I conclude that, in the present case, election under proviso (2)(ii) would be a step to commence arbitration proceedings.

There is one further matter to which I should refer in relation to the construction of section 27. The time limitation in proviso (2)(ii) is expressed to be a provision of which time is of the essence. If section 27 is applicable to this case, as I think it is, it is applicable notwithstanding the provision in the lease that time is of the essence. The lease, it is true, states that the provisions of the Arbitration Act 1950 are applicable "so far as not inconsistent herewith". There are provisions of the Act which the parties can exclude (eg sections 8(1) and (2), and sections 12(1) and (2)). Section 27, however, is mandatory.

I conclude, therefore, that the court had jurisdiction under section 27 to extend the three months' time-limit for election. The jurisdiction is conferred on the High Court, and this case was heard in the county court; but it is common ground between the parties that, the case having been transferred from the High Court, the judge had High Court jurisdiction for the purpose.

The course of the present action was as follows:

The writ was issued on July 20 1982. The tenant had failed to pay the increased rental of £6,000 per annum. The writ sought possession of the premises and arrears of rent and mesne profits. The defence pleaded was that by a letter of October 8 1981 the tenant had, within the three months' period specified in the lease, notified his unwillingness to agree the increased rent, and elected to have the rent matter determined by an independent surveyor. The defence was served on December 1 1983. It contained no reference to section 27.

The hearing started before Judge Martin on February 15 1985. It lasted the whole day, by which time the hearing on the question whether the tenant had served his notice of election in due time had been completed. The judge indicated that he would give a full judgment at a later date, but his finding would be that no valid notice had been served in the three-month period. It seems that the judge's reasons for giving this indication of his decision was to enable the tenant's advisers to prepare an application for an extension of time under section 27. On February 19 1985 the tenant served a notice of application for extension of time under section 27.

On February 25 Judge Martin gave judgment and held that the tenant had not served a notice of election in due time. The judge then heard evidence and counsel's speeches on the section 27 application. The judge then retired for 20 minutes or so, after which he returned to court and gave judgment dismissing the application. Judgment was then given for the plaintiffs for arrears of rent, mesne profits and interest; the judge also made an order for possession if the judgment sum was not paid in three months. The day's proceedings ended at 5.30 pm.

On February 26 the county court wrote to the parties' solicitors, stating that the judgment had not been entered and that the judge had had second thoughts about the case, and had now proposed to allow the section 27 application and to extend the time for serving a notice of election. The letter stated that the judge would hear further argument, if desired.

The case was listed for further hearing on March 18. Previously, in reply to an inquiry by the plaintiffs' solicitors, the county court stated that there were no specific issues upon which the judge desired to be addressed.

On March 18 the judge indicated in general terms the grounds upon which, as he said: "I would and shall, if not persuaded otherwise, alter the judgment and orders which I pronounced on Monday February 25."

The grounds were, according to the note taken by the plaintiffs' solicitors (and approved by the judge), as follows:

1. The very great hardship which the defendant would suffer if I were to refuse the application for an extension of time to serve his counternotice under the lease, ie the loss of the benefit of the £25,000 he has spent on repairs to the demised premises, or of a part of it, and the loss of the £4,000-odd that he paid for the lease;
2. The rent that the plaintiffs are now claiming would be almost precisely seven times the rent that the defendant paid for the period ending June 24 1982: in the absence of evidence as to the open market rental value of the premises, this is manifestly grossly inflated;
3. Following receipt of the plaintiffs' notice of Review Rent dated September 29 1981, the defendant did take the advice of an architect, and on his advice the letter of October 8 1981 was written. It was not the defendant's fault that the advice was bad, he being a person who cannot read or write English. In his letter of October 8 1981 the defendant said he was asking for their surveyor's figures, the plaintiffs having intimated by letter of September 29 1981 that a surveyor instructed by them would carry out a survey in the very near future; the relevant passage is, "I will wait to see . . ." to ". . . pay this rent" and so on. The plaintiffs never indicated whether or not they would instruct their surveyors to value the property, nor, if so, whether they would let the defendant know; they never replied to that letter, and the defendant was waiting for a reply;
4. Mr Pittalis gave the defendant the impression that he was willing to negotiate over the rent — I have in mind there the meetings which were arranged and the telephone calls — but in fact never did so. The plaintiffs could have started proceedings around the end of 1981 or beginning of 1982, though in fact they issued their Writ on July 25 1983. I think that the defendant was lulled into a sense of false security. I placed too much weight on the fact that the defendant did not read the plaintiffs' solicitors' letter dated July 25 1982; the defendant had in fact taken advice from the architect on more than one occasion, and had followed it.

The plaintiffs' council did not ask for an adjournment to consider these matters. After addresses by counsel, the judge gave a further judgment, exercising his jurisdiction under section 27 in favour of the tenant, and extending time for service of a notice of election. In the course of that judgment the judge said that, within minutes of delivering his first judgment, he was sure that his decision was wrong and that he remained of that view. From that decision the plaintiffs appeal.

Was the judge entitled to recall his first judgment? It is clear that the jurisdiction to recall a judgment which has not been passed and entered exists (see, for example, *Re Harrison's Share* [1955] Ch 260 and *Bastow* v *Bagley & Co* [1961] 1 WLR 1494). And there is no difference between the jurisdiction of the county court and the High Court (see *Moon's Motors Ltd* v *Kiuan Wou* [1952] 2 Lloyds Rep 80).

The plaintiffs, however, rely upon *Re Barrell Enterprises* [1973] 1 WLR 19. In that case the Court of Appeal (at pp 23-24) stated that:

> When oral judgments have been given either in a court of first instance or on appeal, the successful party ought save in the most exceptional circumstances to be able to assume that the judgment is a valid and effective one.

Re Barrell Enterprises does not seek to particularise the cases where an order may or may not be recalled. The case itself is, on its facts, far removed from the present. Miss Barrell was, in May 1972, attempting to set aside a committal order of January 1971. In the present case, the judge recalled the order the day after he made it. It is not suggested that the plaintiffs in any way, in proper reliance upon the order, acted to their detriment. We are dealing with a case where the judge, practically as soon as he gave the judgment, decided that he was wrong. As a matter of the sensible administration of justice and fairness between parties, it seems to me proper in the circumstances that the judge should be at liberty to recall his order. The position can properly be called exceptional.

In *Millensted* v *Grosvenor House (Park Lane) Ltd* [1937] 1 KB 717 (which was not cited in *Re Barrell Enterprises*) the judge gave judgment for £50 plus costs. The following day the judge said that he thought £50 was excessive, and he reduced it to £35. The Court of Appeal accepted that a judgment can be recalled at any time until it has been drawn up (see, in particular, per Slesser LJ at p 722). Further, the court upheld the decision of the judge to reduce the damages to £35. In my opinion, Judge Martin in the present case was entitled to review the matter.

I come then to the manner in which the judge exercised his discretion. In the circumstances of this case, we can interfere only if the judge misdirected himself. In considering the matter, we have to bear in mind that section 27 requires that the court be of opinion that the tenant will suffer "undue hardship" if the time-limit is not extended. The onus of establishing that must be on the tenant.

In his second judgment, the judge said that in his judgment on February 25 he had given insufficient weight to certain matters, one of which was:

> The rent that the plaintiffs are now claiming would be almost precisely seven times the rent that the (tenant) paid in the period ending June 24 1982; in the absence of evidence as to the open market rental value of the premises, this is manifestly grossly inflated.

In my opinion that was a misdirection. £850 per annum is not a valid comparison with £6,000 per annum unless it was the market rental in 1982. In fact, however, it had been fixed in 1968, and that must have been by comparison with 1968 rentals. It was for the second seven years of the term, and the rent for the first seven years (commencing in 1968) was £800. We do not even know if £800 per annum was a rack rent in 1968; it may not have been because the premises were then in a bad condition and the lessee was under an obligation to put them into repair. The likelihood is that by 1982 the open market rental was much more than £850; it appears that the Retail Price Index had, between 1968 and 1982, gone up by five times. The onus was on the tenant to prove hardship, but he called no evidence as to market rental in 1982. He referred to some other premises in the neighbourhood, but did not prove sufficient detail about them to establish any comparable rentals.

In my view the judge did not have before him any evidence which justified his conclusion that £6,000 per annum was a "grossly inflated" rental.

I think, therefore, that the judge erred in a material matter in his approach to the case. I think that invalidates the exercise of the discretion and that we, therefore, are entitled to exercise the discretion ourselves.

The central question is whether the tenant would suffer hardship if an extension of time were refused. The grounds which he advances in support of his contention that he would indeed suffer undue hardship are as follows:

First, that he would be paying an excessively high rent. I have already dealt with this in part. It may well be that £6,000 per annum is higher than would be the open market rent as fixed by a surveyor, but we have no means of judging to what extent that is a hardship, because there is no evidence about it. As the judge quite rightly observed in his judgment on February 25, there was no evidence one way or the other whether £6,000 was excessive or not. The absence of evidence may be due to the fact that the section 27 point was taken only at the last moment. But the fact remains that the tenant has wholly failed to establish his case on this issue. I have already commented on the unreliability of any comparison with the £850 per annum payable in the second seven years of the term of the lease.

Second, the tenant says not merely is a rent of £6,000 per annum excessive, but it will prevent him from continuing to conduct his café business on the premises. This, he says, is a modest business producing a profit of about £6,000 per annum. He tendered some accounts in support of that. They are unsigned and relate to the year ended February 1983 only. I do not think that such a document advances the matter.

The basement of the premises is used by a mini-cab business. The tenant says that he receives about £10 a week from that.

It may be that the tenant could not continue to occupy the premises at a rental of £6,000 per annum. It may equally be that he could not do so on the open market rental as determined by an independent surveyor. His business may simply not be profitable enough.

Third, the tenant says that if the plaintiffs obtain possession of the premises, he will lose the sum of £4,500 which he paid by way of premium when the lease was assigned to him in 1972 and the sum of £25,000 he has spent on repairing the premises. But so far as the £25,000 is concerned, the lease was a full repairing lease. If the expenditure was on repairs, it was in discharge of his obligations. It is suggested that part, at any rate, of the expenditure was on G
improvements. There is, however, no evidence to that effect.

Fourth, it is said, in effect, that the tenant's troubles arise from the fact that he is a simple man who had no legal advice, and could not read or write English. But he has been in England for a long time and has been in business in London since 1972. The judge thought he was by no means stupid. He had previously instructed solicitors in relation to the purchase of the property. On this occasion he chose not to seek legal advice.

The tenant's troubles, it seems to me, are largely of his own making. The judge appears to have taken the view that the plaintiffs were at fault in not replying to the tenant's letter of October 8 1981, which said that he would wait for their surveyor's figure — the
plaintiffs having said that their surveyor would carry out a survey H
shortly. But the judge had already found that the letter of October 8 had not been served on the plaintiffs. There is no evidence that they received it. The plaintiffs' counsel, on instructions, says that they did not.

I do not, therefore, think that the tenant has established a case of undue hardship. Further, one must, I think, take into account the plaintiffs' position. Section 27 is a valuable power enabling the court to prevent the hardship that may result from time provisions. But the delay in the present case is wholly excessive. The tenant was three and a half years out of time before he sought to rely on section 27 at all. It will be four years next June since the increased rent became payable,
and the plaintiffs are still not in receipt of the proper rent — whether J
it be £6,000 or an arbitrator's award. An arbitration will delay the matter further.

I have come to the conclusion that the tenant has not demonstrated a case of undue hardship and I would, therefore, allow the appeal.

Agreeing, DILLON LJ said: The plaintiff landlords appeal against the ultimate order of Judge Martin QC of March 18 1985, on three grounds, viz:

(1) It was not proper for the judge to recall his earlier order of February 25, although it had not been drawn up, since there were no "exceptional circumstances" to warrant his doing so, and the
procedure he followed, in only formulating on March 18 when K
counsel were before him the reasons which were leading him to change his mind, was unfair.

(2) The judge had no jurisdiction to extend the tenant's time for serving a notice of election under proviso (2)(ii) in clause 1 of the lease because in truth section 27 of the Arbitration Act 1950 was not applicable, and

(3) In any event, the judge misdirected himself and erred when he re-exercised his discretion under section 27 in the tenant's favour on March 18.

Ground (1) can be disposed of quite shortly. It is submitted for the plaintiffs that the decision of this court in *Re Barrell Enterprises* [1973] 1 WLR 19 is to be regarded as a comprehensive exposition of
the exceptional circumstances in which it is proper for a court or a L
judge to recall an order which has been pronounced orally but has not yet been drawn up, registered, or otherwise perfected. But I cannot regard *Re Barrell Enterprises* as overruling or qualifying the earlier decision of this court in *Millensted* v *Grosvenor House (Park Lane) Ltd* [1937] 1 KB 717 (which apparently was not cited in *Re Barrell Enterprises*). In *Millensted* this court approved statements of the law in earlier authorities to the effect that a judge can always reconsider his decision until his order has been drawn up or perfected, and more importantly, this court upheld the action of a High Court judge, who had in an oral judgment awarded a certain sum by way of damages but then withdrew that judgment and substituted judgment for a lower figure, because, before his order was drawn up, he was satisfied
after serious further consideration that the sum he had originally M
awarded was excessive. In *Moon's Motors Ltd* v *Kiuan Wou* [1952] 2 Lloyd's Rep 80 it was held that the position in the county court is, in this respect, exactly the same as in the High Court.

Millensted is, in my judgment, a close parallel to the present case. It is indeed exceptional for a judge who has pronounced an order in court to be completely satisfied, before the order has been drawn up, registered or perfected, that the order was wrong. That happened, however, in the present case, and accordingly the judge was entitled, taking the view he did, to recall his earlier order. I see nothing unfair in the procedure he followed to do so.

When counsel came before him on March 18 he told them the four points which, on an overall view of the evidence, had led him to want

A (unless persuaded otherwise) to change his order. Counsel were familiar with the case and had a full opportunity to address him on the merits of the case as much as on the powers of a judge to recall an order. If — which seems to me unlikely with experienced counsel — either counsel felt embarrassed at the absence of time to prepare to argue the judge's four points, he could have applied for an adjournment. Ground (1) therefore, in my judgment, fails.

Ground (2) raises three points in relation to section 27. First, it is pointed out that section 28 can apply only where there is an agreement to refer future disputes to arbitration, and it is said that as the tenant never gave any notice of election there was never any agreement to refer the assessment of the open market rental value of these B premises, or any dispute over that value, to arbitration.

Second, it is said that the giving of such a notice of election is not the taking of "a step to commence arbitration proceedings" within the meaning of section 27.

Third, and most importantly, it is submitted that it is established by the decision of this court in *Baron* v *Sunderland Corporation* [1966] 1 QB 56, and especially by the judgment of Davies LJ, that it is an essential ingredient of an arbitration clause or arbitration agreement within the meaning of the 1950 Act that it must confer bilateral rights, and not merely (as in the present case) a unilateral right of reference of any dispute arising between the parties to an arbitrator.

There are, of course, many forms of rent review clause, involving C references to an independent surveyor, or to two surveyors or their umpire, to determine the rent for premises on a review which are cases of mere valuation, and not of arbitration at all. The distinction between cases of arbitration and cases of mere valuation is discussed in such cases as *Re Dawdy* (1885) 15 QBD 426 and *Re Carus-Wilson & Greene* (1887) 18 QBD 7. In the present case, however, the clause expressly provides that, if the tenant serves a notice of election, the open market rental value of the premises is to be determined by the independent surveyor referred to "in accordance so far as not inconsistent" with the lease "with the provisions of the Arbitration Act 1950". Subject only, therefore, to the three points referred to in the previous paragraph, this is a case of arbitration and not of mere valuation.

D The rent review clause is not, in my judgment, a mere agreement to agree. There is in the lease itself an agreement, binding on the landlords, that there shall be arbitration over the open market rental value of the premises if the tenant, by notice, so elects. That is a binding contract to refer disputes over the open market rental value if the tenant so elects, and it must be an agreement for arbitration within section 27 unless the third point set out above is valid, that an arbitration clause must confer bilateral rights and not merely a unilateral right to reference.

Subject again to that third point, a notice of election by the tenant under this lease is clearly, in my judgment, the taking of a step to commence arbitration proceedings within the meaning of section 27. E See *Babanaft International Co SA* v *Avant Petroleum Inc* [1982] 1 WLR 871 at 885 D-E, where Donaldson LJ said:

> The concept of "claiming arbitration" is well known in the commodity trades. Telex messages fly to and fro and at some stage one party or the other says "we claim arbitration". The arbitration rules of the trade concerned then provide the steps to be taken by each party. I would therefore accept that in such cases "claiming arbitration" may be regarded as a step to commence arbitration proceedings within the meaning of section 27.

I turn therefore to the third point, and the extent to which *Baron* v *Sunderland Corporation* is binding on this court.

The actual decision in that case was, if I may respectfully say so, plainly right. The plaintiff, a teacher employed by the defendant corporation, was entitled by statutory instrument to remuneration in F accordance with the scales set out in a report called the Burnham Report. He sued to recover extra remuneration to which he claimed to be entitled under the Burnham Report. The corporation applied to stay the action, pursuant to section 4 of the 1950 Act, on the ground that section V of the Burnham Report, which established a joint committee of reference for the determination of "any question relating to the interpretation of the provisions of (the report) brought forward by a local education authority acting through the authorities' panel or by any association of teachers acting through the teachers' panel" or by the chairman of the Burnham Committee, was an arbitration clause which had been incorporated into the plaintiff's contract of employment with the corporation, and that it constituted a valid submission to arbitration.

This court held that there was no submission to arbitration. The main basis of the decision, as I read the report, was that there could have been no submission to arbitration because no question could fall to be considered by the joint committee of reference unless it had been brought forward by a local education authority acting through the authorities' panel, or by the chairman, and that had never happened in respect of the plaintiff's claim. That seems unanswerable, but in giving his judgment, Davies LJ (with whose reasons Salmon LJ agreed) said, at the top of p 64 of the report:

> It seems to me that this is about as unlike an arbitration clause as anything one could imagine. It is necessary in an arbitration clause that each party shall agree to refer disputes to arbitration; and it is an essential ingredient of an arbitration clause that either party may, in the event of a dispute arising, refer it, in the provided manner, to arbitration. In other words, the clause must give bilateral rights of reference. The present clause, as I see it, does nothing of the kind.

It may be that all Davies LJ was intending to cover by these words was the point which Salmon LJ dealt with at p 65G when he said, in relation to the judge in the court below whose decision this court reversed:

> The judge approached this case as if section V of the Burnham Report were an arbitration clause which is written into the contract of service between the teacher and the local authority by the relevant statutory instrument. For my part . . . I think that this is the wrong approach . . .

In other words, there was a complete lack of mutuality because the teacher was not a party to section V of the Burnham Report, and section V was not a part of his contract of service. If that is all that Davies LJ meant, what he said is not applicable to the present case, where the provisions of the lease as to arbitration are part of the contract between the landlords and the tenant which binds their successors in title, the parties to this action.

But the more natural reading of the judgment of Davies LJ is, I apprehend, that the words quoted above from his judgment are to be taken at their face value as part of the ratio of the decision, and would cover any case, such as the present, where only one party to the agreement can refer a disputed question to arbitration. If that is the true interpretation of the judgment of Davies LJ then the question arises whether this court is bound to apply it in the present case.

There is, of course, no want of mutuality in the present case. The landlords and the tenant mutually agreed by the lease that the tenant should have the right, within a certain time, to elect to refer the determination of the open market rental value of the premises to arbitration. From the nature of the transaction, which was perfectly legitimate, there was no occasion to give the landlords a corresponding right of election, because the rental value proposed by the landlords would be applicable automatically unless some other rent was agreed, or the tenant elected for arbitration. I can see no reason in principle why such an agreement under which one party only has the right to call for arbitration should not be a valid agreement for arbitration within the meaning of the 1950 Act; indeed, if it is not valid, the reference to that Act in the lease becomes meaningless.

We have not been referred to any earlier authority which suggests the existence of such a doctrine as has apparently been enunciated by Davies LJ; no relevant authority was cited in *Baron* v *Sunderland Corporation*. On the contrary, in *Heyman* v *Darwins Ltd* [1942] AC 356, Lord Macmillan, in a comprehensive analysis of the function of an arbitration clause in the context of the application of the court's power to stay proceedings pending arbitration under what is now section 4 of the 1950 Act, says categorically at p 370:

> Arbitration clauses in contracts vary widely in their language for there is no limitation on the liberty of contracting parties to define as they please the matters which they desire to submit to arbitration.

More importantly, in *Woolf* v *Collis Removal Service* [1948] 1 KB 11, where a contract provided that claims by one party, though not claims by the other, were to be referred to arbitration, another division of this court held, in a reserved judgment, that there was nothing in the unequal operation of the clause to divest it of the character attributed to arbitration clauses in general in the speech of Lord Macmillan in *Heyman* v *Darwins*; see the judgment of Asquith LJ at the foot of p 17. Consequently, it was appropriate to grant a stay of proceedings pending arbitration under section 4, and that necessarily imports that, despite its unilateral operation, the agreement was an arbitration agreement within the meaning of the Act.

I prefer the ratio in *Woolf* v *Collis Removal Service* to the ratio of Davies LJ in *Baron* v *Sunderland Corporation* in the passage quoted above from his judgment, if that passage is understood according to the more natural meaning. Accordingly this court is not bound by *Baron* v *Sunderland Corporation* to hold that because of its unilateral nature there is no arbitration clause or agreement for arbitration in the present case. On the contrary there is, in my judgment in the lease, such an agreement within the meaning of the 1950 Act in general, and section 27 in particular.

Accordingly I would reject the appellants' ground (2).

It follows that the case at first instance of *Tote Bookmakers Ltd* v *Development & Property Holding Co Ltd* [1985] 2 WLR 603 was, in my judgment, wrongly decided and should be overruled.

It remains to consider the appellants' ground (3), which relates to the judge's re-exercise of his discretion on March 18. This court could interfere on this ground only if the judge has misdirected himself.

Under the terms of section 27, the court can extend time, so that an arbitration may take place, only if the court is of opinion that in the circumstances of the case undue hardship would otherwise be caused. The onus must be on the applicant for the extension to show a *prima facie* case of undue hardship. But in the context of the present case, where a rent review has to take place under the terms of the lease, the hardship must stem, not from the fact that the rent is being increased to the £6,000 per annum proposed by the landlords from the £850 per annum payable under the lease in previous years, but from the fact that the tenant will have to pay the £6,000 per annum instead of what he would reasonably have put forward as a rack rent for the premises in an arbitration under the rent review clause.

But the tenant — probably because the possibility of obtaining an extension of time under section 27 occurred to him or his advisers only at such a very late stage in the proceedings — has never put forward any suggestion of what the open market rental value of the premises ought to be fixed at if there is an arbitration. He did indeed, at a very early stage in October or November 1981, suggest to Mr Pittalis that the rent ought to be only £1,000 a year, but that suggestion was rejected as derisory and has never been resuscitated.

The tenant claimed, in the somewhat brief affidavit he swore in support of his application under section 27, that though he and his wife worked hard in running the café at the premises, with very little assistance, the return was modest, and it would be an undue hardship if the rent of £6,000 were to stand without being considered by an arbitrator. He produced unsigned draft accounts which purported to show an annual profit from the café business of £6,259 after allowing for wages to his wife of £1,196, and he admitted receiving a further £520 per annum (£10 per week) from a mini-cab business carried on from the basement of the premises. On these figures it might be said that any substantial increase in the rent from the premises would be difficult for the tenant to meet and might cause him hardship, but that would not be the sort of undue hardship referred to in section 27.

The judge stated correctly, in his second judgment of February 25 1985, that he had heard no evidence one way or the other as to whether the rent of £6,000 proposed by the landlords was excessive or not. He said that he therefore had to use his own common sense, and he expressed the view, from the description of the premises, that it was a very substantial jump from £850 to £6,000 in June 1982. By March 18 this had developed into a belief that the £6,000 per annum rent claimed by the landlords, being seven times the rent paid just before June 24 1984 was, in the absence of evidence as to the open market rental value of the premises, manifestly grossly inflated, but the judge does not appear to have had clearly in mind on either occasion that the £850 per annum figure had been fixed as long ago as 1968. The £850 per annum figure is by itself no guide to what the open market rental value of the premises ought to be in 1982. Indeed, it is not even clear that the rents of £800 for the first seven years and £850 for the second seven years reserved by the lease were intended to be rack rents, since the premises were plainly in a poor state in 1968, with a closure order on the basement, and an obligation on the tenant to spend money, and probably quite a lot of money, in repairing the premises.

The judge additionally took into account on March 18 as indicative of the very great hardship which the tenant would suffer if the judge refused his application for an extension of time under section 27 that the tenant would lose the benefit, or part of it, of the premium of £4,500 which the tenant had paid for the lease when he took the assignment in early 1972, and of the £25,000 he spent on "repairs" to the premises. But the lease was a full repairing lease, and the rent for the last seven years of the term was to be a rack rent; accordingly, I cannot see that expenditure by the tenant on repairs in 1972 can have any relevance to hardship to the tenant in relation to the fixing of the G
rent in 1982.

Mr Sarch, for the tenant, submitted that the expenditure by the tenant of the £25,000 may have been on improvements rather than repairs, and he pointed out that under the lease the effect on rent of improvements carried out by the tenant was to be disregarded on the rent review. But if that is a relevant consideration, it is for the tenant to give evidence to show what the works were on which he spent his £25,000.

I would expect that any rent fixed by an arbitrator would be lower than the £6,000 per annum put forward by the landlords, since (a) on an arbitration over rent it is the common experience that an arbitrator will fix the rent at a figure somewhere between the rent levels *bona* H
fide contended for by the opposing parties; (b) under a clause of this type it is to be expected that the landlords' opening proposal of a rent will be on the high side; and (c) at about the time when they put forward the rent of £6,000 per annum, the landlords overestimated the insurance value of the premises. But I can see no material on which it can properly be concluded that the proposed rent of £6,000 per annum was, as the judge put it, "manifestly grossly inflated".

On a separate point, in the third of the grounds which he put to counsel on March 18, the judge appears to hold it against the landlords that they had not answered the tenant's letter of October 8 1981. But he had earlier found that that letter had not been served on the landlords. Counsel for the landlords states on instructions that the landlords had never received the letter, but there was no positive J
evidence to that effect. There was, however, equally no evidence that the landlords *had* received the letter and, in view of his finding that it had not been served, no basis on which the judge could, in the context of this case, infer that the landlords had received the letter. The judge, therefore, was not entitled to hold it against the landlords that they did not answer the letter.

I accordingly conclude that the learned judge misdirected himself in re-exercising his discretion under section 27 on March 18. This court is, accordingly, entitled to form its own view as to how the discretion should be exercised.

I would accept that there is a measure of hardship to the tenant, who on any view is in a small way of business, if he has to pay a rent of £6,000 per annum which is higher than would probably be fixed by K
arbitration. That is, however, a hardship which the tenant has largely brought on himself by the stupid way in which he dealt with the landlords' letter of September 29 1981, which put forward that rent, and by his failure to acquaint himself with the relevant terms of the lease, although he had that document to hand. There was then enormous delay before any claim for an extension of time under section 27 was made. That delay operates adversely to the landlords in that, if, after such delay, there is then an arbitration, the date will be yet further deferred when the landlords will be able to recover rent, above the previous £850 per annum, accruing for the premises since June 1982, and that rent will then seemingly be recoverable without interest. Much, if not most, of that delay might well, however, have L
been avoided if the tenant had treated the landlords' solicitors' letter of July 15 1982 sensibly, instead of ignoring it, and had then gone to a solicitor.

It is urged on behalf of the tenant that he is a simple man who suffers from the disadvantage that he cannot read or write English. The finding of the judge was, however, that he is a fairly simple man, but not stupid by any means, and he can read and write other languages. He has lived in England for very many years, has bought and sold property in England, having solicitors acting for him, and has been in business in London since 1972. His inability to read and write English must be largely a matter of his own choice, in that he has never troubled to learn to do so. I cannot regard it as justifying his unreasonable and unbusinesslike conduct over the letters from the M
landlords and their solicitors which I have mentioned.

For the foregoing reasons, I am of the view that the discretion of the court under section 27 ought not to have been exercised in favour of the tenant, and on that ground I would allow this appeal.

Also agreeing, NEILL LJ said: I have had the advantage of reading in draft the judgments of Fox LJ and Dillon LJ. I agree with them, and I can therefore restrict my own judgment so as to deal expressly with two matters only.

The first matter to which I propose to refer is the question whether an agreement which gives one party alone the right to claim arbitration can constitute a valid "agreement to refer future disputes to arbitration" within section 27 of the Arbitration Act 1950.

Looking at the question apart from authority, I see no reason why the parties to an agreement should not include a provision that only one party to the agreement should be entitled to refer a dispute to arbitration. Indeed, where one party is entitled to put forward a suggested variation of the terms of the agreement such as a variation of a price or a rent, it may be sensible that the other party should be the one who is given the right, under the agreement, to elect to refer the matter to arbitration if he does not concur with the suggested variation. In these circumstances the right to claim arbitration is a unilateral right but it has been mutually agreed. Moreover, such an approach appears to be wholly consistent with the words of Lord Macmillan in *Heyman* v *Darwins Ltd* [1942] AC 356, where he said at p 370:

> Arbitration clauses in contracts vary widely in their language for there is no limitation on the liberty of contracting parties to define as they please the matters which they desire to submit to arbitration.

I turn therefore to consider the authorities. My lords have already referred to the decision in *Woolf* v *Collis Removal Service* [1948]1 QB 11, where this court granted a stay of the proceedings pending a reference to arbitration in a case where the storage contract between the parties provided that claims by one of the parties (the customer) should be submitted to arbitration but included no similar provision in respect of claims by the other party (the warehousemen). The court held that the fact that the arbitration clause operated unequally did not divest the agreement of "the character attributed to arbitration clauses in general in the speeches of Lords Macmillan and Porter in *Heyman* v *Darwins Ltd* (*supra*) as distinguishing them from exception clauses"; see per Asquith LJ at p 17.

Furthermore, in the earlier case of *Ronaasen & Son* v *Metsanomistajain* (1931) 40 Ll L Rep 267, the Divisional Court upheld part of the award of an arbitrator in a case where the arbitration clause in a contract for the sale of timber was restricted to disputes arising from complaints by the buyers only; the court set aside the second part of the award on the basis that there was no provision in the contract for an arbitration to be claimed at the instance of the sellers.

In the present case, however, it was argued that an agreement whereby the tenant alone can elect arbitration is not an agreement to refer disputes to arbitration within the meaning of section 27 of the 1950 Act. In support of this submission, reliance was placed on a passage in the judgment of Davies LJ in *Baron* v *Sunderland Corporation* [1966] 2 QB 56, where he said at p 64:

> It is necessary in an arbitration clause that each party shall agree to refer disputes to arbitration; and it is an essential ingredient of an arbitration clause that either party may, in the event of a dispute arising, refer it, in the provided manner, to arbitration. In other words, the clause must give bilateral rights of reference.

Our attention was also drawn to a recent decision of Peter Gibson J in *Tote Bookmakers Ltd* v *Development & Property Holding Co Ltd* [1985] 2 WLR 603 where, considering himself to be bound by the view expressed by the Court of Appeal in *Baron*'s case, the learned judge held that a clause giving a unilateral right to a tenant to elect arbitration did not constitute an agreement to which section 27 of the 1950 Act applied.

In these circumstances it is necessary for this court to come to a conclusion as to the meaning and effect of this passage in the judgment of Davies LJ in *Baron*'s case.

For my part I find it impossible to regard the words used by Davies LJ as ambiguous. They seem to me to amount to a plain statement that unless bilateral rights to refer are given by the relevant clause, no arbitration agreement within the 1950 Act is concluded. But I further consider that this view of the law is inconsistent with the speeches in *Heyman* v *Darwins Ltd* (*supra*) and does not accord with the decision in *Woolf*'s case in 1948, to which the court were not referred in *Baron*'s case. Accordingly, I have no hesitation in saying that in my judgment the court should follow the decision in *Woolf*'s case and should decline to adopt the dictum of Davies LJ in *Baron*'s case.

It follows, therefore, that I agree with my lords that there is before the court in the present case an agreement to refer future disputes to arbitration within the meaning of section 27 of the 1950 Act.

I turn therefore to the second matter to which I propose to make express reference. I agree with my lords that there was no adequate material before the learned judge for him to come to the conclusion that the proposed rent of £6,000 a year was "manifestly grossly inflated". The judge appears to have overlooked the fact that the previous rents of £800 a year and £850 a year were fixed in 1968 and that there was no satisfactory evidence before the court as to whether either of these rents was intended to be a rack rent.

It follows, therefore, that this court is entitled to, and should, exercise a fresh discretion as to how the power conferred by section 27 of the 1950 Act should be applied to the facts of the instant case. Some guidelines as to the way in which a court should exercise its jurisdiction under section 27 of the 1950 Act were formulated by Brandon J (as he then was) in *The Jocelyne* [1977] 2 Lloyds Rep 121 at p 129, and were later approved and applied by this court in *The Aspen Trader* [1981] 1 Lloyds Rep 273 at p 279.

In addition, a convenient though more elaborate summary of the relevant principles is to be found in the current edition of *Mustill and Boyd on Commercial Arbitration* at p 178.

For the purpose of the present case it is sufficient to refer to Lord Brandon's summary as set out in *The Aspen Trader* (*supra*):

> (1) The words "undue hardship" in section 27 should not be construed too narrowly.
> (2) "Undue hardship" means excessive hardship and, where the hardship is due to the fault of the claimant, it means hardship the consequences of which are out of proportion to such fault.
> (3) In deciding whether to extend time or not, the court should look at all the relevant circumstances of the case.
> (4) In particular, the following matters should be considered:
> (a) The length of delay;
> (b) The amount at stake;
> (c) Whether the delay was due to the fault of the claimant or to circumstances outside his control;
> (d) If it was due to the fault of the claimant, the degree of such fault;
> (e) Whether the claimant was misled by the other party;
> (f) Whether the other party has been prejudiced by the delay, and, if so, the degree of such prejudice.

The critical question to be resolved, therefore, is whether in the circumstances of the particular case undue hardship would be caused to the applicant if the time limited for the commencement of arbitration proceedings were not extended.

I turn to the facts of the present case. It seems to me that if one applies the factors listed by the Court of Appeal in *The Aspen Trader* only one conclusion is possible. The delay, which in no sense is the responsibility of the landlords, exceeds three years. There is no satisfactory evidence as to what reduction in the suggested rent the tenant would achieve if the matter went to arbitration, nor is there any satisfactory evidence as to whether the tenant would be able to pay such a reduced rent and thus remain in occupation even if he succeeded in achieving a reduction. It is true that this is not a case where the passage of time will affect the recollection of witnesses and thus cause prejudice to the other party, but delay of this magnitude is by itself prejudicial.

It seems to me therefore that this court is driven to the conclusion that the tenant has failed to make out any sufficient grounds to justify the grant of an extension of time.

On these grounds, as well as those set out in the judgments of my lords, I, too, would allow this appeal.

The appeal was allowed with costs, together with costs of an application before Kerr LJ, such costs to be limited to £400. No order was made against the Law Society.

Chancery Division

December 13 1985

(Before Mr Justice HOFFMANN)

NORWICH UNION LIFE INSURANCE SOCIETY v TRUSTEE SAVINGS BANKS CENTRAL BOARD

Estates Gazette April 12 1986

278 EG 162-168

Landlord and tenant — Construction of rent review clause — Arbitration — Application for leave to appeal from arbitrator's award — Arbitration Act 1979 — *The Nema* guidelines — Lease for 22 years, with rent reviews at seventh and each subsequent fifth anniversary of commencement — Dispute arose at the first of the five-year reviews — The rent review clause provided that the new rent was to be assessed on the assumption that the premises were let in the open market

by a willing lessor to a willing lessee taking a lease otherwise on the terms and conditions of the present lease — Issue was whether it should be assumed that the hypothetical lease was for a period of 22 years or only for the unexpired residue of the actual lease, ie for 10 years — Arbitrator decided that the assumption should be for a grant equal only to the unexpired residue — Landlords sought leave to appeal — Held that there was a presumption that the hypothesis on which the rent should be determined ought to bear as close a resemblance to reality as possible — The purpose of rent review was to enable adjustments to be made to take account of inflation and market changes, not to enable the landlords to benefit from an assumption that a brand new lease of 22 years was being granted where in fact there was a residue of 10 years only — In the present case the judge had no doubt that the arbitrator was right — The court was not, therefore, concerned with such questions as whether there was a strong *prima facie* case or an arguable case that the arbitrator was wrong or with possible adjustments of the guidelines given by Lord Diplock in *The Nema* — Leave to appeal from the arbitrator's award refused

This was an application by the landlords, Norwich Union Life Insurance Society, for leave to appeal under the Arbitration Act 1979 from the award of an arbitrator in a rent review arbitration in respect of premises at 25 Milk Street, London EC2, the respondents being the tenants, Trustee Savings Banks Central Board.

Paul Morgan (instructed by the solicitor to the Norwich Union Insurance Group) appeared on behalf of the applicants; David Neuberger (instructed by J F Willcocks, Legal Division, Trustee Savings Banks Central Board) represented the respondents.

Giving judgment, HOFFMANN J said: This is an application for leave to appeal from the award of an arbitrator in a rent review arbitration. The rent review clause provided that the new rent was to be assessed on the review date on the assumption that the premises were being let in the open market by a willing lessor to a willing lessee taking a lease otherwise on the terms and conditions of this lease.

The lease in question had been granted for a term of 22 years from September 29 1972. Rent reviews were provided at the seventh and on each subsequent fifth anniversary of the commencement of the term. On the first rent review date the parties were able to agree upon a rent and this dispute, therefore, arose for the first time when the lease had run for 12 years. The issue between the parties was whether the hypothesis required by the lease was the assumption that a new lease for 22 years was being granted at the rent review date or a lease only for the unexpired period of the existing lease, which would have been 10 years. Another way of putting the same question would be to ask whether it was to be assumed that a lease of 22 years commencing upon the date of the original lease had been granted or such a lease commencing on the rent review date.

The arbitrator came to the conclusion that upon its true construction the clause required the assumption of a grant for the unexpired period alone. Mr Morgan, for the landlords, asks for leave to appeal and has referred me to the guidelines laid down by the House of Lords in *The Nema* [1982] AC 724. These indicate that I should give leave to appeal only if I consider that there is a strong *prima facie* case that the arbitrator was wrong.

Mr Morgan has also submitted that *The Nema* guidelines are no doubt appropriate for international commercial arbitrations but do not apply with the same force to a domestic rent review arbitration. In the case of a domestic rent review arbitration, he suggested that the proper test should be whether the applicant for leave has an arguable case.

There is nothing in the judgment of their lordships to suggest that the guidelines are not of general application to all arbitrations subject to the Arbitration Act 1979. It is no doubt true that the prospect of adjustment to the guidelines is contemplated on a reading of the subsequent decision in *Antaios Compania Naviera SA* v *Salen Rederierna AB* [1985] AC 191, but those adjustments, I think, are a matter for their lordships rather than for me.

In any event, in a case such as this, which is very much a matter of impression, I do not think that the distinctions between *"prima facie* case", "arguable case" and various other grades of persuasion which might exist really have very much meaning. One thinks either that the arbitrator was right or that he was wrong, and for my part, so far from thinking that there is a *prima facie* case that he was wrong, I have no doubt that he was right. This is not something upon which it is necessary, particularly in an application such as this, to elaborate. There is, I think, a presumption that the hypothesis upon which the rent should be fixed upon a review should bear as close a resemblance to reality as possible. In this case the reality was that at the date of the rent review the tenant's interest was an unexpired period of 10 years. He had been paying in the earlier part of the term a rent calculated, at any rate for the first five years, according to the market rent for what was then being granted, namely a lease for a period of 22 years. The purpose of the rent review is to enable that rent to be adjusted at a subsequent date in order to take into account the effects of inflation and changes in the market since the original grant. But I think that the landlord would be having it both ways if he was entitled not only to an adjustment for changes in the market and changes in inflation but also to the assumption that what was being granted on the rent review date was a brand new lease rather than what was in fact the case, a lease which by then was 12 years expired. That construction also seems to me to be in accordance with what Viscount Dilhorne in *Ponsford* v *HMS Aerosols Ltd* [1979] AC 63 at p 76H and Vinelott J in *Pearl Assurance plc* v *Shaw* (1984) 274 EG 490 regarded as being commonsense.

For those reasons, I refuse leave to appeal.

The application was dismissed with costs. Leave to appeal to the Court of Appeal was refused.

Court of Appeal

November 15 1985

(Before Lord Justice O'CONNOR, Lord Justice BALCOMBE and Sir Denys BUCKLEY)

BASILDON DEVELOPMENT CORPORATION v MACTRO LTD

Estates Gazette April 26 1986

278 EG 406-412

Landlord and tenant — Construction and application of user covenants in lease of shop intended to be a small supermarket in a parade of shops — Appeal by lessors (a development corporation, plaintiffs below) and cross-appeal by lessees from a decision of Whitford J — Dispute centred on two sub-clauses in lease — The first contained a positive covenant to use the demised premises as in effect a "food supermarket", selling groceries and provisions, including meat and vegetables, bread and other bakers' items, with an off-licence for beer, wines and spirits — The second contained a complementary negative covenant not to carry on any trade or business other than that specified in the positive covenant, but also contained more particular restrictions — These were not to sell goods "not usually sold or dealt with by any other trade or business of the like nature"; not to sell cigarettes, tobacco and confectionery from a sales area exceeding a certain percentage of the net retail area; and not to sell any newspapers, magazines, stationery, toys, haberdashery or wools — Disputes arose when the respondent lessees began to sell certain electrical goods and some items usually sold in chemists' shops and the plaintiffs (appellants) sought and obtained an interlocutory injunction — At the trial of the action Whitford J construed the relevant clauses as entitling the lessees to deal in articles which at the date of the lease were in fact being dealt with in supermarkets for the sale of groceries and provisions and that comparison would have to be made with a business predominantly concerned with the sale of groceries and provisions — Whitford J granted the appellants an injunction and awarded them the whole costs of the action — The Court of Appeal agreed with the judge as to the construction of the covenants, but held that he was in error in directing himself as to the burden of proof and as to what had been established by the evidence — The court also criticised the terms of the injunction granted, while pointing out that the earlier interlocutory injunction had been too wide

A **and had caused the appellants to become liable on their cross-undertaking in damages — In view of the unsatisfactory way in which the action had evolved the appellants were asking for a new trial — In the result the appellants had failed to establish that the respondents were carrying on a business which they were not entitled to carry on and, except in regard to certain items where breaches were conceded, had failed to establish a breach of covenant — Held, rejecting the plea for a new trial, that the appeal should be dismissed, the cross-appeal allowed and a substituted injunction granted restraining the respondents from selling those items as to which they had** B **admitted a breach of covenant — The court also ordered an inquiry into damages**

This was an appeal by Basildon Development Corporation, lessors, from a decision of Whitford J in an action in which as plaintiffs they had claimed relief by way of injunction against Mactro Ltd, lessees of a lock-up shop at 1 Felmores End, Basildon, Essex, the defendants and present respondents. The plaintiffs had obtained an interlocutory injunction from Dillon J subject to the usual undertaking as to damages.

A F B Scrivener QC and Brian Leech (instructed by Wortley Byers & Co, of Brentwood) appeared on behalf of the appellants; Gavin Lightman QC and Jonathan Sofer (instructed by Kingsford Dorman C & Co, agents for Barrett & Boxall, of Westcliff-on-Sea) represented the respondents.

Giving the first judgment at the invitation of O'Connor LJ, BALCOMBE LJ said: This is an appeal from an order of Whitford J made on March 25 1983. The plaintiffs in the action (the appellants in this court), the Basildon Development Corporation, were at all material times the freehold owners of a block of lock-up shops, numbered 1 to 9 Felmores End, Basildon, Essex. They asserted in their statement of claim (and the learned judge so found) that in the summer of 1980 they were negotiating with intended lessees of the various shop premisees, and they were taking the view that what they wanted to see was some feature which would provide for the inhabitants of the area a collection of shops which would meet the D majority of their needs and would be within easy walking distance of quite a large number of residents. They also took the view that, if this aim were to be achieved, what was necessary was to grant leases to a variety of businesses which would not be competing with each other, as no doubt, with a limited public available, it would not be possible effectively for the various persons who took leases to sustain their business if competition of this kind within such a restricted field of activity were taking place. Mr Scrivener, who has appeared for the plaintiffs before us, submits that this is the matrix within which the relevant covenants of the lease with which we are concerned ought to be construed.

The lease in question is dated November 21 1980 and is made E between the Basildon Development Corporation of the first part and the defendants, Mactro Ltd, of the second part, and a gentleman as a surety of the third part. The lease was of what appears from the plan to be the largest unit in this parade of nine shops, no 1. We are told that it has a floor area of some 2,400 sq ft. The term is 15 years from November 21 1980 and the lease contains the usual lessee's covenants. The two clauses of the lessee's covenants with which this court is concerned are subclauses (20) and (21) of clause 2.

By subclause (20) the defendants covenanted:

> To use the demised premises for carrying on the trade or business of a supermarket for the sale of groceries and provisions (including fresh and frozen meat and fresh and frozen vegetables, bread and other items sold at a bakers) and as incidental to this main use (subject to the necessary Justices' F Licence being first obtained) for the sale of beer, wines and spirits for consumption off the demised premises.

Pausing there, I read that as a positive covenant to use the demised premises for what I shall, for the purposes of shorthand, call a "food supermarket". It continues:

> and at all times of the year during the usual business hours of the locality to keep the demised premises open as a first class shop for carrying on the said business in its several branches

and that again is a positive covenant; and then

> and not to do or permit or suffer to be done anything to injure the connection or goodwill of such business.

Subclause (21) is negative in its terms, and by it the defendant covenanted:

> Not to carry on or permit or suffer to be carried on in or upon the demised premises or on any part thereof any trade or business other than that specified in clause 2(20) hereof

(in other words, the trade or business of what I have called a "food supermarket"). That has been described as the first limb of subclause (21) and in my judgment — I will refer to the authorities in a moment — it is clear that that refers to the question of the trade or business, namely, that what the defendants were not allowed to do under that limb of the covenant was to carry on something which could be properly described as another trade or business. It is not now alleged (though in the statement of claim it was so alleged) that the defendants have broken that limb of the covenant.

The second limb of subclause (21) is in these terms:

> And not to sell or deal in articles commodities or goods of any description not usually sold or dealt in by another trade or business of the like nature.

That limb, the second limb, of the covenant deals with a restriction upon the goods which the defendants may sell at their premises, even though they are carrying on a business of a food supermarket, because it is probably well-known today that most food supermarkets also sell some non-food items of goods. Indeed that construction of the second limb of the clause is made plain by the third limb of the subclause, which is in the following terms:

> For the avoidance of doubt and without prejudice to the generality of the restriction contained in this sub-clause it is hereby agreed that the Lessee shall not be entitled to sell or otherwise deal in any of the following commodities or goods namely cigarettes, tobacco and confectionery where the total sales area devoted to such commodities exceeds two per centum of the net retail area of the demised premises and shall not be entitled to sell any newspapers, magazines, stationery, toys, haberdashery or wools whatsoever.

Without looking at any of the authorities, it seems to me that it is clear from that third limb of the subclause that what the second limb of the subclause was intended to do was to enable the defendants to sell goods, even though they were not items of food or drink, provided that they were usually sold or dealt in by other food supermarkets, possibly (although this was an issue which was not dealt with at the trial) other food supermarkets of a size similar to that of the defendants' premises. Because if that was not the intention of the second limb of the clause, there was no point at all in having a total prohibition on the sale of newspapers, magazines, stationery, toys, haberdashery or wools (none of which are items of food or drink) nor the qualified prohibition on the sale of cigarettes and tobacco.

That that construction of the clause is the correct one is apparent from the three cases to which we have been referred, namely *Stuart* v *Diplock* (1889) 43 Ch D 343, *Lewis & Co (Westminster)* v *Bell Property Trust* [1940] Ch 345, and *Labone* v *Litherland Urban District Council* [1956] 1 WLR 522, all of which make the distinction between the carrying on of a trade or business and the sale of particular items within that trade or business, even though those items might themselves be properly sold by another trade or business. Thus, for example, in the *Lewis* v *Bell Property Trust* case the trade or business which the tenant was entitled to carry on was that of a tea shop. In the course of that trade or business they sold some cigarettes and tobacco. It was held that that was not of itself a breach of the covenant to carry on business only as a tea shop.

What happened in this case was that quite soon after the grant of the lease the defendants started carrying on business and included in their business — which it is conceded has always been that of a food supermarket — certain items of electrical goods and certain items which might more normally be expected to be found on sale in a chemist's shop.

Some of the other local traders complained, and in the event, on May 29 1981, the plaintiffs issued a writ claiming against the defendants injunctive relief. It is significant to see what the form of the relief originally claimed was. First:

> An order that the defendants do use the premises for carrying on the trade or business of a supermarket for the sale of those goods only and none other specifically referred to in Covenant 2(20) of the lease — namely for the sale of groceries and provisions (including fresh and frozen meat and fresh and frozen vegetables, bread and items sold at a bakers) and as incidental to the main use for the sale of beer wines and spirits for the consumption off the demised premises

That relief claimed in that way suggested that the plaintiffs thought that the defendants could be stopped from selling any kinds of electrical items and any kinds of chemist's items, and that was complemented by clauses (2) and (3) of the relief claimed in the writ, (2) being

For an injunction restraining the defendants whether by themselves their servants or agents or otherwise howsoever from selling in and from the demised premises any items usually and normally sold in chemist shops including: toiletries, cosmetics, sanitary towels, photographic equipment, drugs, medicines, accessories for babies, hair brushes, scissors, cotton wool and other prophylactic items.

Clause (3) claims:

An injunction restraining the defendants whether by themselves, their servants or agents or otherwise howsoever from selling in and from the demised premises any electrical goods including: plugs, adaptors, batteries and light bulbs.

The plaintiffs then applied on motion for interlocutory relief, and the matter came before Dillon J on June 4 1981 when, after the plaintiffs had given their usual cross-undertaking as to damages, he ordered that "the defendants be restrained from selling in their premises . . . until after judgment in this action or . . . further order" certain items listed in a schedule, being an exhibit to an affidavit filed before him, a copy of which was annexed. Those were almost entirely chemists' items. They run to 16 in all and I need not enumerate them here. They include items such as hair colourant, surgical hosiery, surgical dressing, babies' bottles, teats and dummies, cotton wool and scissors. In addition the defendants were injuncted from selling electrical plugs, adaptors, light bulbs and batteries. The costs of the motion were reserved.

The plaintiffs served their statement of claim in February 1982 which set out both the subclauses (20) and (21) of clause 2 of the lease. In para 6 it is pleaded that "In breach of the said express terms set out in paragraph 3 above", that is, the two subclauses, "the defendants have between March 1981 and June 1981" (those were the relevant dates)

sold from the shop premises at No 1 Felmores End goods not usually sold in a supermarket for the sale of groceries and provisions and not in a grocery and provision supermarket, including: cosmetics, toiletries, sanitary towels, baby lines, proprietary medicines; goods usually sold at electrical shops, namely plugs, batteries, adaptors and light bulbs.

The claim for relief in the statement of claim repeated the claims made in the writ.

By their defence the defendants admitted that they sold the items listed in para 6 of the statement of claim between March and June 1981, but then denied that they had, whether by such sales or otherwise, acted in breach of the terms.

The matter came before Whitford J in March 1983, and both sides gave what was termed expert evidence before him. For the plaintiffs there was called a Mr I C Hayward [FRICS], a surveyor from a well-known firm of estate agents, who put in as part of his evidence-in-chief a proof of evidence upon which he was cross-examined. I will say no more about his evidence than this: it seems to be reasonably clear from both the proof and the transcript of his oral evidence that he was not putting himself forward as an expert in the grocery or food trade, and when he gave evidence of what other similar supermarkets within the area concerned sold he was speaking as the result of an inspection he had made in the summer of 1982.

The expert who gave evidence on behalf of the defendants was a Mr A Toft, who could fairly be described as an expert in this particular field, because he had been concerned in the grocery trade as a marketing manager for a number of good companies and then he had launched a journal called *The Independent Grocer*, of which he was editor, and had been researching for that journal for some months before January 1981. He was in a position to speak (and his qualifications for so doing are not challenged) as to what goods were usually sold (and I quote from the lease itself) "or dealt in by any other trade or business of the like nature" to that which the defendants were permitted to carry on at their premises.

It appears that the plaintiffs were maintaining, up till quite a late stage, that there was a breach of the covenant against carrying on any other business, because much of their expert evidence was concerned with what was the business they were carrying on and the extent of the shelf space devoted to the particular goods in question and so on, and of course the defendants had to be able to meet that and their evidence also went to a great degree to that question, which it was conceded before us was not the relevant one.

The first part of the learned judge's judgment dealt with the question of the construction of the covenants, and although certain of the reasons which the learned judge put forward for reaching the conclusion to which he eventually came are not reasons which appeal to me, nevertheless, in my judgment, he came to the right answer, although possibly by the wrong road, when on p 7 of the transcript at G letter C he says this:

The result of it is that to my mind, on the true construction of these two clauses, the defendants are in a situation where they are entitled, under the terms of their lease, to deal in articles of a kind which as of the date of the lease — and, indeed, that is accepted as the relevant date — were in fact being dealt in in supermarkets for the sale of groceries and provisions, and Mr Caplan

who was leading counsel for the defendants in the court below

accepts that in considering a business of that kind one would have to make a comparison with a business which was predominantly concerned with the sale of groceries and provisions.

I do not wish to express an opinion as to whether the date of the lease is, in the construction of a covenant of this nature, the relevant H date, but, as it was conceded by both parties below that it was, I say no more about it. So having, as I have said, come to the right result on the question of construction, the learned judge then went on to consider what he termed "the rather difficult question of the breaches". He then said this:

The fact is that I have got no evidence at all as to what in fact was being sold in supermarkets for the sale of groceries and provisions as at the relevant date other than what can be inferred from certain material which is dealing with events subsequent to 1980.

Mr Scrivener for the plaintiffs (the appellants before us) concedes that that is not correct; that there was indeed before the learned judge the evidence of Mr Toft and, in particular, that part of his evidence which is set out in an addendum to his proof of evidence, and in order J to understand this one has to look first at para 7 in his original report where he says:

7. I have been instructed: (i) to express an opinion as to which of the items listed on p 11 of Mr Hayward's report

and they were these chemists' items to which I have already referred briefly — the items scheduled to the order made by Dillon J

are habitually sold in "supermarkets for the sale of groceries and provisions".

A little later in his report he says this:

17. Those chemist lines and cosmetics which need expert handling and advice from the other side of the counter I would not expect to find in a supermarket or in a small convenience store. By this I mean such items as cough remedies which are only legally handled by pharmacists — and not proprietary brands K which are available generally, diabetic foods but not health foods. There is a clear distinction here.

In para 18 he says this:

I would not expect to find in a grocery store or supermarket dermatological creams for skin problems, but I would expect to see a ladies' hand cream.

Then, when one comes to the addendum to his report, he says this:

In summary form the answers are as follows to the three questions, as set out in para 7 of my report, upon which I was asked to express an opinion:
1. Subject to the qualifications expressed in paras 17 and 18 of my report, and non-self-service cosmetics and perfumes, all the items are, and were in 1980, habitually sold in "supermarkets for the sale of groceries and provisions" which is a description commonly understood to mean premises in which goods which are predominantly groceries and provisions are offered for sale L by predominantly self-service methods.

That evidence was before the trial judge. It does not appear that Mr Toft was ever challenged about that, and so the learned judge was in error when he said that he had no evidence as to what in fact was being sold in such supermarkets in 1980. He then goes on to consider the evidence, first, of Mr Hayward, but discounted that because Mr Hayward had said he did not know what was being sold in 1980. That appears to be correct. The judge then considered Mr Toft's evidence and, having gone through that in some detail, he said:

In the end I think Mr Caplan conceded that it could not be suggested even as of today that plugs and adaptors, I think babies' bottles and teats for babies' bottles and scissors were in fact among the items which are usually sold or M dealt in in supermarkets for the sale of groceries and provisions. The only thing which follows from this is that on the face of it, it would seem undoubtedly to be the case that there has been a breach of covenant.

Then he turned finally to the question of the lease:

Against this assertion that in breach of the terms there has been a sale of some goods not usually sold in supermarkets for the sale of groceries and provisions what we get in the defence is an admission of the sale the subject of complaint and an assertion that the commodities pleaded in para 6 of the statement of claim, and listed more fully in a schedule to one of the orders made on the interlocutory proceedings "are usually sold in supermarkets for the sale of groceries and provisions and grocery and provision supermarkets conducted as a first class shop for carrying on the said business in its several branches".

Then he says this:

A the defendants have not in fact adduced any evidence which in any way establishes their case that any of these goods were usually sold in supermarkets for the sale of groceries and provisions as at the relevant date.

That sentence contains two errors. Firstly, it shows a misconception of the burden of proof because it was for the plaintiffs to establish both that the goods in question of which complaint was made were being sold (that was admitted by the defence) and also that their sale constituted a breach of the covenant, which in turn threw upon the plaintiffs the onus of proof that these were goods which were not of a nature which were usually sold in similar types of supermarkets. It was not for the defendants to take upon themselves that burden, but, even if it had been, there lies the second error, because by the evidence of Mr Toft the defendants had established, with the exception of B those goods in respect of which Mr Caplan made a concession and one or two other items which were not clear, that many of the goods of which complaint was made in fact were goods which were sold in businesses of a similar type. From that error the learned judge went on to say that, because the defendants had not established their case, "the plaintiffs are, in my view, accordingly entitled to some form of relief, but I will hear counsel on the appropriate form", because the concession was correct. He then went on to award to the plaintiffs the whole of the costs of the action.

The order, in the form in which it was eventually drawn up (that is the order of Whitford J), was an injunction

that the defendant be restrained from selling (whether by its directors or by its C servants or agents or any of them or otherwise howsoever) in and from the premises known as 1 Felmores End, Basildon, Essex, any items not as of November 21 1980 usually sold in supermarkets for the sale of groceries and provisions.

An injunction in those terms was certainly something to which the plaintiffs were entitled, because it merely repeated, in slightly different words and introducing the concessionary date, the covenant by which the defendants were already bound, but it was not a particularly helpful kind of injunction, because on any question of enforcement (committal for contempt or something of that nature) it would leave wide open the question what goods were sold in supermarkets on that date. But, secondly, it also introduced problems into the inquiry as to damages, because the second branch D of the order was that the following inquiry be made, that is to say:

an inquiry as to what damages (if any) the defendant has sustained by reason of the injunction contained in the Order dated June 4 1981 which the plaintiff ought to pay by reason of its undertaking in damages in the said Order contained.

Of course in the order of June 4 1981 the defendants had been injuncted from selling a very large number of chemists' items as well as the limited number of electrical items to which I have already referred. The plaintiffs have failed to establish, save by Mr Caplan's limited concession, that the defendant was in breach of his covenant at all in respect of the sale of those items. Therefore, except in respect of what I will call the concessionary or conceded items, the injunction granted in June 1981 was wider than that to which the plaintiffs E became entitled as a result of the trial, and accordingly the plaintiffs became liable on their cross-undertaking in damages because the defendants had been prevented from selling goods which the plaintiffs had failed at the trial to establish that they should not have sold. Again, the injunction in its original form leaves open the issue to what the inquiry as to damages should have related.

Before us we have an appeal by the Basildon Corporation and a cross-appeal; though it is contained in a document headed "Respondents' Notice", it is in fact a cross-appeal by the defendants.

The appellants at the end of the day are asking for a new trial. They concede that the judgment was not a satisfactory one. Initially Mr Scrivener asserted that the clause, that is the second limb of subclause F (21), was entitled to a wider construction than that which the learned judge gave it, but, as I have already said, in my judgment his construction was correct, though he reached it by an unusual route. So they finally came down to submitting that the case evolved in such an unsatisfactory way — in fact, if I can use Mr Scrivener's own words, it was such a mess — that the only satisfactory solution was to discharge the order in its entirety and send the case back for a new trial.

Mr Lightman, on the other hand, submitted that that would be unduly hard on the defendants, who have had this case hanging over them now for some four years. In my judgment, Mr Lightman's submission is justified. Whatever the reason, the position is that the plaintiffs brought this case; they brought it, as I have said, under two heads — one asserting that the defendants were carrying on a business which they were not entitled to carry on, and on that they have failed and do not now maintain that they should have succeeded; and they also ask for relief stopping the defendants selling certain specified items of goods, but even in respect of those items they failed to adduce any evidence at the trial which would have established that the sale of such items was a breach of the covenant in question, except those items which eventually the defendants conceded were sold in breach of covenant.

In the result, in my judgment the appeal should be dismissed, the cross-appeal allowed, and for the order which the learned judge made there should be substituted an injunction restraining

the defendants, whether by its directors or by its servants or agents or any of them howsoever, from selling in and from the premises known as no 1 Felmores End, Basildon, Essex, the following items or any of them,

and the following items would be those which the defendants conceded represented a breach of covenant, namely, plugs, adaptors, babies' bottles, teats, scissors, non-self-service cosmetics and non-self-service perfumes. For my part, I think that the words "non-self-service cosmetics and non-self-service perfumes" may not be sufficiently precise to be the subject of an order and, if there is any doubt about that, I think the appropriate course would be for junior counsel for each of the parties to settle a form of minute which would accurately reflect the substance of what I believe should be the right order, namely, the injunction should be limited to the sale of the conceded items. Then there should be an inquiry as to damages in the form in which it was contained in the original judgment, but I make it clear that the case on which the plaintiffs eventually succeeded was only in relation to the conceded items. Accordingly the interlocutory injunction granted on June 4 1981, in so far as it prohibited the sale of items beyond the conceded items was beyond that to which the plaintiffs were entitled by way of relief, and accordingly the inquiry will be directed to the damages which the defendants have sustained by being precluded from selling items other than the conceded items from the date of the interlocutory injunction.

That leaves the question of costs below. We have not yet heard any submissions as to the costs of this appeal, but, so far as costs below are concerned, in my judgment the learned judge, because of his error as to the burden of proof and to the question of what evidence there had been before him about the type of goods that were in November 1980 sold in food supermarkets, took the view that the plaintiffs had won. They had won to a limited extent because the concession about the conceded goods appears to have been made at a fairly late stage, but, as against that, they had lost, both on their initial claim to restrain the defendants from carrying on a particular type of business, and also by failing to adduce any relevant evidence about a large number of the particular items of goods, the sale of which they were complaining of. It is very difficult of course for this court to place itself in the position of the court below when it comes to dealing with a question of costs, but I am satisfied that, as a result of the view I take upon the true meaning of the covenant and the evidence which was adduced before the learned judge, the order for costs below cannot stand, and, doing the best I can in the circumstances, I would substitute for the order for costs below an order for no order for costs below, leaving each side to pay their costs at first instance, because each succeeded to an extent and failed to an extent.

Agreeing, SIR DENYS BUCKLEY said: I only desire to add, I hope, quite a short word upon the analysis of covenants (20) and (21) contained in the relevant lease.

Covenant (20) contains what can be described as a positive obligation to use the demised premises for a particular kind of business, and it identifies the type of business to be carried on as the

business of a supermarket for the sale of groceries and provisions (including fresh and frozen meat and fresh and frozen vegetables bread and other items sold at a bakers) and as incidental to this main use (subject to the [availability of a suitable licence]) for the sale of beer wines and spirits for consumption off the . . . premises

That covenant identifies the type of business which the tenants are to carry on upon the premises, but it does not identify in any way the particular classes of goods which might be dealt with in the course of that business.

Covenant (21) has been referred to as a restrictive clause, but in my view it is restrictive in only a very limited sense or extent. For the purpose of analysing it, I divide it into three parts, which I call subparas (1), (2) and (3). Subpara (1) starts with the first word and goes down to the words "in clause 2(20) hereof and". Subpara (2) starts with the words "not to sell or deal in articles" and goes to the end of that sentence. Subpara (3) starts with the words "For the avoidance of doubt" and goes down to the end of the covenant.

Subpara (1) excludes the carrying on of any business other than a business of the type identified in covenant (20). It does not cut down covenant (20) in any way; it merely makes it clear that the defendants are to confine their activities to the carrying on of a business within the type of business identified in covenant (20). Subpara (2) of (21) identifies the articles to be dealt in in that business. It does so by an exclusive mechanism. It excludes goods not usually sold or dealt in in businesses of the permitted class, that is to say, the business of a supermarket for the sale of groceries and provisions. It is precisely complementary with covenant (20); it does not, to my mind, restrict what may be done under covenant (20). It merely identifies the classes of goods which may be properly dealt in in that business. It is not, in my view, a restrictive clause. Subpara (3) is to some extent restrictive, and it is inserted in the covenant, as the opening words indicate, for the avoidance of doubt. It does more than avoid doubt, because it does in fact to some extent restrict the classes of goods which can be dealt in in the course of the business. It restricts the sales of cigarettes, tobacco and confectionery by reference to the area or areas devoted to goods of those classes, and it contains a positive embargo on the sale of certain classes of goods, namely, any newspapers, magazines, stationery, toys, haberdashery or wool whatsoever.

In my judgment, there is nothing in either clause 2(20) or clause 2(21) to indicate that the tenants are in any way restricted in their business to goods that are food or drink. I would not wish to define precisely what falls within the term "groceries"; I am not sure that upon its true interpretation it should be confined to consumables in the sense of foods and drink, but in my judgment, upon their true construction, the two covenants clearly leave it open to the tenants, in the course of their business of a supermarket for the sale of groceries and provisions, to deal in any goods which would in the usual course of businesses of this character be dealt in in addition to the sale of those articles which precisely fall within the definition of groceries and provisions.

So, although I would arrive at the conclusion by a different course of reasoning from that which was adopted by the learned judge in his judgment, I think that he was right in his conclusions at p 7 of the transcript of his judgment in saying that the result, on the true construction of the two clauses, is that "the defendants are in a situation where they are entitled, under the terms of their lease, to deal in articles of a kind which as of the date of the lease", and I say a word about that in a moment, "were in fact being dealt in in supermarkets for the sale of groceries and provisions". Before the learned judge it was common ground that the lease should be read, construed and applied by reference to the state of affairs at the date when the lease was executed. We have not heard argument about that because it was common ground before the learned judge. I do not think it would have been right for either party to attempt in this court to recede from that standing, and I think that for the purposes of this appeal we must also treat the matter as to be decided in the light of the facts as they existed at the date of the lease. That does not mean that I would necessarily take the view that, upon the true construction of the clauses, one has to have regard exclusively to what was the state of affairs when the lease was executed.

For these reasons I agree that the result of this appeal should be that which has been formulated by Balcombe LJ, and I agree with the views which he has expressed about what should be done about the costs both here and below.

Also agreeing, O'CONNOR LJ said: Subclause (20) of clause 2 in this lease imposes a positive covenant upon the lessee to carry on a particular type of business. The premises demised were a shop of 2,400 sq ft and the lessee was required to use it for the sale of groceries and provisions. He was also required to sell fresh and frozen meat and fresh and frozen vegetables and bread and other items sold at a bakers. So that, effectively, what the lessors, who perfectly properly were setting up a small shopping parade, intended (and in my judgment achieved) by their positive covenant was to see to it that there was a small supermarket which was going to deal with groceries and provisions and was to be the equivalent of a local butcher, greengrocer and baker. In addition, it is quite plain that the covenant required the lessee to apply for a justices' licence and operate an off-licence.

When one comes to consider that covenant, the business of a supermarket for sale of groceries and provisions, including all these other matters, must be limited by the size of the premises; that is within the matrix of the covenant. Quite plainly, the lines of goods G carried in a supermarket of that size would not necessarily be the same as those carried by one twice the size, 10 times the size or somewhat smaller.

Subclause (21) was in three parts, as my lords have said. The first part adds nothing to the positive covenant in subclause (20); it is the other half of it. Then the rest of subclause (21), in my judgment, operates to restrict that which the lessee can sell in his shop. Some things he is barred from selling altogether, such as newspapers, magazines, stationery, toys, haberdashery or wool. The reason why that is necessary is because it might well be that the sale of those items in small quantities would not change the nature of the business carried on and required to be carried on by subclause (20). H

The other matters were that he was "not to sell or deal in articles, commodities or goods of any description not usually sold or dealt in by any other trade or business of the like nature". I would construe that as saying that, where you have got effectively a food market with its ancillary matters which may properly be included with groceries (and, without defining them, it seems to me that there will be a good many household articles customarily bought at a small supermarket properly described as selling groceries and provisions), in so far as the lessee deals in other goods, they must not be such as are not commonly found also dealt in by other supermarkets of equivalent size.

Within those limits he is free to sell anything that he likes, as long as it does not change the nature of the business which he is required to J carry on by subclause (20).

I do not wish to say anything beyond what my lords have said as to the judgment of the learned judge. I agree that unfortunately a number of errors occurred. I agree that the onus of proof on a breach of subclause (21) was mistakenly put upon the defendants and, for the reasons given by my lords, I agree that the injunction granted cannot stand, and I also agree with the order proposed by Balcombe LJ.

The appeal was dismissed and the cross-appeal allowed. There should be substituted for the order below an injunction and inquiry into damages in the terms stated in the judgment of Balcombe LJ. For the order as to costs below there should be substituted no order for costs below; and the respondents to have the costs in the Court of K *Appeal.*

Court of Appeal

February 10 1986

(Before Lord Justice DILLON and Mr Justice SHELDON)

THOMAS v HAMMOND-LAWRENCE

Estates Gazette April 26 1986

278 EG 414-416 L

Landlord and Tenant Act 1954, Part II — Application by landlord to fix interim rent under section 24A — Question as to the date from which the interim rent was to run — Section 24A(2) provides that such rent shall be payable as from the date on which the proceedings were commenced or the date specified in the landlord's notice or the tenant's request, whichever is the later — The proceedings referred to are the proceedings by the landlord to fix the interim rent — In the present case the landlord, in her answer to the tenant's originating application for a new tenancy, after setting out various matters such as her grounds of opposition added a M **paragraph in which she applied under section 24A for the determination of an interim rent — In doing so the landlord was following a paragraph in Practice Form N400 in the *County Court Practice* — This was in May 1983 — After nearly two years, during which negotiations and discussions continued, in April 1985 the landlord issued a notice that she intended to apply to the county court to fix an interim rent — The judge decided that the date on which proceedings were commenced, for the purpose of the interim rent, was the notice in April 1985, not the date of the answer in May 1983 — He held that the para in the landlord's answer could not**

A **constitute an application, which took place only when some notice of appointment was taken, as in the present case, under Order 13, rule 1 — On appeal by the landlord the Court of Appeal held that the judge was wrong on this point — Order 13, rule 1 was a convenient practical rule dealing with interlocutory applications but it did not cover the whole ground — Where there is a dispute as to rent and terms of the new tenancy as well as the interim rent, and the tenant's application is before the court on a single substantive appointment at the trial of the action, the court can determine the terms of the new tenancy, and the interim rent, because the**
B **application for the latter has been made in the answer — If the application for the interim rent is granted it will date back to the date of the answer — Here the interim rent would run from May 1983 — Appeal allowed**

This was an appeal by the landlord, Mrs Joan Hammond-Lawrence, from a decision of Judge Motta Singh at Clerkenwell County Court as to the date from which an interim rent under section 24A of the Landlord and Tenant Act 1954 was payable in respect of business premises at 288 Camden Road, London N7. The respondent to the appeal, the applicant for the grant of a new tenancy, was Mrs Dorothy Gwinfell Thomas.

R J Moshi (instructed by Hurd & Haigh) appeared on behalf of the appellant; Sir Ashley Bramall (instructed by Davidsons) represented
C the respondent.

Giving judgment, DILLON LJ said: This is an appeal by a landlord against a decision of Judge Motta Singh QC given in the Clerkenwell County Court on July 25 1985. The appeal raises a short point of practice in relation to proceedings in the county court under Part II of the Landlord and Tenant Act 1954 and, in particular, in relation to an application by a landlord for the fixing of an interim rent for premises in respect of which there is an application for a new tenancy under Part II of that Act.

Section 24A of the Landlord and Tenant Act 1954 was introduced into the Act by the Law of Property Act 1969. Subsection (1) of section 24A provides:

D The landlord of a tenancy to which [Part II] of this Act applies may
in certain events
apply to the court to determine a rent which it would be reasonable for the tenant to pay while the tenancy continues by virtue of section 24 of this Act, and the court may determine a rent accordingly

Subsection (2) provides:

A rent determined in proceedings under this section shall be deemed to be the rent payable under the tenancy from the date on which the proceedings were commenced or the date specified in the landlord's notice or the tenant's request, whichever is the later.

In the case of *Stream Properties Ltd* v *Davis* [1972] 1 WLR 645, it was held that the "proceedings under this section", referred to in subsection (2), were the proceedings by the landlord to fix an interim
E rent and not the proceedings by the tenant to obtain a new tenancy. Therefore, the reference in subsection (2) to "the date on which the proceedings were commenced" is a reference to the date on which the landlord's proceedings for fixing an interim rent were commenced. That is the date which has to be ascertained to find out what the date is from which the interim rent, when granted, will run. *Stream Properties Ltd* v *Davis* was a High Court case, and in the High Court there is express provision in Order 97, rule 9A of the Rules of the Supreme Court that

An application under section 24A of the Act of 1954 to determine an interim rent shall (a) if the tenant has begun proceedings for a new tenancy under section 24 of the Act, be made by summons in those proceedings, and (b) in any other case, be made by originating summons.

F It was accordingly held in *Stream Properties Ltd* v *Davis* in the context of High Court proceedings that the date from which the interim rent was to run was the date of the issue of the summons in the tenant's proceedings which was issued by the landlord by way of application to determine the interim rent under section 24A. There is no parallel in the County Court Rules to Order 97, rule 9A of the Rules of the Supreme Court.

The facts of the present case are as follows. The appellant is the present landlord, and the tenant, who is respondent to this appeal, is the tenant of business premises known as 288 Camden Road, London N7. The tenant was granted in 1975 a tenancy of those premises for a term of seven years from March 26 1976 at a rent of £1,000 per annum. It is common ground that that tenancy was within the protection of Part II of the 1954 Act.

On November 23 1982 the landlord served notice on the tenant determining the tenancy under section 25 of the 1954 Act. The notice gave May 25 1983 as the date for the termination of the tenancy; it also indicated that the landlord would oppose the grant of a new tenancy. Accordingly, on January 18 1983 the tenant made application in the county court for the grant of a new tenancy. The application is in the normal form, Form N397, for an originating application for a new tenancy under Part II of the 1954 Act, and it sets out the various matters which such an application is required by the rules to set out. It accordingly became the landlord's duty to serve an answer to that application. The answer was served on May 27 1983 by the landlord's solicitors, and it is addressed to the registrar of the county court and to the applicant and her solicitors. This set out that the landlord opposed the grant of a new tenancy on certain grounds which are indicated, and it also objected to the new tanancy, if granted, being granted on the terms proposed by the applicant. It set out various other matters which an answer has to contain, and then in para 5 it said this:

The Respondent hereby applies under Section 24A of the Landlord and Tenant Act 1954 that the Court should determine the rent which it would be reasonable for the Applicant to pay while the tenancy continues by virtue of Section 24 of the said Act.

As is often the case with applications for a new tenancy, there were negotiations and discussions and delays for various reasons. There was an abortive appointment to decide certain preliminary issues in November 1984, the preliminary issues concerned being whether the landlord made out the grounds for objecting to the grant of a new tenancy at all. Ultimately, on April 11 1985, the landlord issued a notice, which was served on the tenant a few days later with a covering letter, giving notice that the landlord intended to apply to the judge of the court on July 10 1985 for an interim rent to be fixed for the premises and for costs. The notice intimated that a day had been allowed for the hearing. The hearing commenced on July 10 1985, but for various reasons the hearing was not then concluded; it was adjourned to the 25th and then concluded, the order under appeal then being made.

The judge decided that the date on which the proceedings were commenced for the purposes of subsection (2) of section 24A was not the date of the answer which contained the para which I have read but the date of the issue of the notice in April 1985 of the hearing to fix the interim rent.

The question is: Was he right in so concluding, or was the date when the proceedings were commenced the date of the answer?

In the *Green Book* it is set out in the notes, at p 1506 of the 1985 *County Court Practice,* under section 24A against the word "Procedure":

If the tenant has already applied to a county court for a new tenancy, an application under this section may be made in the course of those proceedings in accordance with Ord 13, r 1 . . . or in the landlord's answer. If the tenant has not already applied for a new tenancy, an application to a county court under this section should be made by originating application in accordance with Ord 3, r 4.

A similar note has appeared in the corresponding part of the *Green Book* for many years past, even before the introduction of the County Court Rules 1981.

The practice forms issued in respect of county court proceedings include, as Form N400, a form of answer to originating application for new tenancy under Part II of the Landlord and Tenant Act 1954. This practice form includes, as something to be added if required, a para 6 in these words:

I hereby apply to the court under section 24A of the Act to determine a rent which it would be reasonable for the applicant to pay while the tenancy continues by virtue of section 24 of the Act.

It will be seen that the form of the para in this landlord's answer in this case was closely parallel to that in the practice form N400. Before the 1981 County Court Rules came in, the form of answer in the county court forms was a mandatory form, but it did not include any reference to application to determine an interim rent. The present form is not mandatory; it is a practice form issued with the approval of the Lord Chancellor but not prescribed by rule. Such forms are issued with the two-fold object of assisting litigants to take steps in the proceedings and promoting uniformity of practice among the courts. Notwithstanding that, the learned county court judge held that the para in the landlord's answer in the present case and the para in the practice form are in truth of no effect at all, because they do not constitute an application to the court to determine a rent; they do not commence proceedings for the determination of the interim rent; that happens only when some notice of appointment is taken, as in the present case, by a notice or application under Order 13, rule 1.

The learned judge held that there was nothing to make it mandatory or to give official support to the inclusion of the application for a new tenancy in the answer, and, seeing no official support in the way of a mandatory provision, he felt that he should not hold it effective.

It has been submitted by Sir Ashley Bramall for the respondent in this court that, in truth, Order 13, rule 1 of the County Court Rules makes it mandatory to make the application for an interim rent by summons under Order 13, rule 1 if proceedings have been commenced by the tenant, and therefore makes the issue of the summons (in this case in April 1985) the date of the commencement of the proceedings for the interim rent.

Order 13, rule 1 provides:

(1) Except as otherwise provided, the following paragraphs of this rule shall have effect in relation to any application authorised by or under any Act or rule to be made in the course of an action or matter before or after judgment.
(2) Unless allowed or authorised to be made ex parte, the application shall be made on notice, which shall be filed and served on the opposite party not less than two days before the hearing of the application.

That is dealing with interloctory applications and is an obviously convenient practical rule. Sir Ashley says that the application by a landlord is an application authorised by statute, and so under Order 13, rule 1 it must be made by notice and is effectively so made only when the notice is issued. But I do not see that that covers the whole ground. One might have a case in which there was no dispute that the tenant was entitled to a new tenancy and therefore no preliminary issue to be tried. But there was dispute between the parties over what the rent under a new tenancy should be, or what some other terms should be, and over a claim by the landlord to an interim rent. In such a case, as it seems to me, if the landlord has in his answer, in accordance with the practice form, put forth a claim for an interim rent, then the tenant's application could come before the court on a single substantive appointment at the trial of the action and the court would then be able to determine the terms of the new tenancy and the interim rent because the application for the interim rent had been made in the answer. I do not see that it makes any difference that there is an issue as to whether the tenant should have a new tenancy at all; the answer is the pleading in which the claim for an interim rent is made. It is then a matter of sensible practice how the various issues raised in the tenant's application and the answer can best be dealt with, but, as the answer has raised the application for an interim rent, if one is granted by the court, that will date back to the date of the answer. Equally, whereas it may often happen that the parties will wish to negotiate over the amount of an interim rent as over the amount of the final rent under an order for a new tenancy, they will be able to negotiate with the knowledge that the application had been raised and is properly included in the proceedings so that time is not running against the landlord in the sense that he would be precluded from claiming interim rent over the period of the negotiations.

With every respect, therefore, to the learned judge, I think he was wrong on this point and I would accordingly allow this appeal.

SHELDON J agreed, for the reasons given by Dillon LJ, that the appeal should be allowed and he did not add anything further.

The appeal was allowed with costs and it was declared that the landlord's application was validly made in her answer of May 27 1983, the interim rent to run from that date.

Chancery Division

February 24 1986

(Before Mr Terence CULLEN QC, sitting as a deputy judge of the division)

ELECTRICITY SUPPLY NOMINEES v F M INSURANCE CO LTD

Estates Gazette May 3 1986

278 EG 523-528

Landlord and tenant — Rent review clause in lease — Question as to whether the hypothetical lease should be assumed to exclude the provision for rent reviews — Rent was to be determined on review "upon the terms of this lease except as regards rent" — Plaintiff landlords sought a declaration that the effect of the words "except as regards rent" was that a schedule to the lease which contained the rent review provisions should be treated as omitted from the hypothetical lease — In argument, however, the landlords submitted primarily that the effect of the words quoted was to excise every term which could be said to be a term "as regards rent"; if that argument failed, then at least the schedule containing the rent review provisions should be excised — Terms "except as regards rent" were said to cover nine terms, including the covenant to pay rent, the reservation of insurance and service rents and the proviso for entry on non-payment of rent, as well as the rent review provisions — Tenants submitted that the words meant only that the rent actually passing at the review date should be disregarded — In considering the respective arguments, the judge referred to the restatement by Stephenson LJ in *Bonalumi* v *Home Secretary* of Lord Wensleydale's "golden rule" in *Grey* v *Pearson* as "giving the words their natural and ordinary significance unless there is some compelling reason to the contrary" — The usual purpose of a rent review clause was to ensure that the rent at which a property was let did not fall in real terms at rent reviews while the tenant remained in occupation — That purpose required certain hypotheses to be made which were contrary to reality, eg the hypothesis of vacant possession, and the parties might make it clear that they intended the hypothetical lease to depart still further from reality — But if the wording was not clear, and a literal interpretation resulted in an alteration of rental values which went beyond a reflection of changes in the value of money or in the property market generally, that could be a compelling reason for not applying the "golden rule" — The words in dispute in the present case could be given a literal meaning which would be inconsistent with the commercial purpose of rent review provisions, or which would be plainly futile or absurd, or they could be interpreted as only excluding the passing rent — The last might merely produce the result which would have been produced without these express words, but that was no reason to reject this interpretation, which the judge held to be the correct one — A number of authorities were considered, including the Vice-Chancellor's judgment in *British Gas Corporation* v *Universities Superannuation Scheme Ltd*, which did not enunciate special principles of construction applicable to rent review provisions, but set out the result which flowed from the application of the general principles of construction to such provisions — Held that the tenants were right in submitting that the words referred only to the reservation of rent, which in this particular case resulted only in the exclusion of the passing rent — Landlords' summons dismissed.

This was a summons by the plaintiff, Electricity Supply Nominees, seeking a declaration that, on the true construction of a lease of accommodation at 97-105 Victoria Street, London SW1, the rental value on rent reviews was to be assessed on the basis that the lease contained no provision for rent reviews. The defendant, the tenant, was F M Insurance Co Ltd.

Paul Morgan (instructed by Jaques & Lewis) appeared on behalf of the plaintiff; R J Furber (instructed by Linklaters & Paines) represented the defendant.

Giving judgment, MR TERENCE CULLEN QC said: This summons raises a point on the construction of a lease dated June 13 1980 and made between the plaintiff as landlord and the defendant as tenant, together with a guarantor. The demised property was the fifth floor and part of the ground floor of 97-105 Victoria Street, London SW1. The plaintiff seeks a declaration that on the true construction of the lease the rental value on rent reviews under the lease is to be assessed on the basis that the hypothetical lease contains no provisions for rent reviews. The lease is for 25 years from March 25 1980.

The present dispute arises on the occasion of the first rent review, which is operative as at March 25 1985. The rent review provisions are contained in the fourth schedule to the lease. By para 1 the expression "rental value" is defined, so far as material, as:

> the best yearly rent at which the demised premises are reasonably worth to be let as offices

then I omit provisions as to area

> in the open market with vacant possession and as a whole for a term of twenty-five years at a relevant date hereunder (that is to say a date as at which rental value falls to be agreed or determined in accordance with the following provisions of this Schedule) without taking a fine or premium and upon the terms of this Lease except as regards rent.

There then follow certain immaterial assumptions and common form provisions for disregarding certain matters.

The schedule goes on to provide for five-yearly reviews with a reference to arbitration if the rental value cannot be agreed. Para 5 provides for the attachment to the lease of a memorandum of any agreement or determination of matters under the schedule.

The declaration sought by the plaintiff is that under para 1 of the fourth schedule the hypothetical lease is on the same terms as the actual lease save that it is for 25 years from the date of review (this is not disputed) and by reason of the words "except as regards rent" the hypothetical lease omits the fourth schedule.

However, in argument, while confining himself to that relief, Mr Morgan for the landlord submitted primarily that the words "except as regards rent" had the effect of excising all terms which can be said to be a term "as regards rent". It was only if that argument failed that he submitted that, at the narrowest, the words extended to the fourth schedule.

Mr Furber for the tenant submitted that the words should be confined to meaning the rent actually payable immediately prior to the relevant review date, which I shall call "the passing rent". I was told that the parties had agreed that if the landlord were right, the rental value as at March 25 1985 was £18.50 per sq ft, while if the tenant were right it was £17 per sq ft. In view of that agreement, there was no factual dispute between the parties, so that by consent they came before the court on this construction issue rather than having it resolved by arbitration under the lease.

It is common ground that the words "terms except as regards rent" could include:

(i) the reservation of rent. This is in the following terms:

> YIELDING AND PAYING therefor during the term hereby granted FIRST the yearly rent of FOUR HUNDRED AND FOUR THOUSAND ONE HUNDRED AND NINETY-THREE POUNDS (£404,193) or (after the end of the fifth year) such greater yearly rent as shall be agreed or determined in accordance with the provisions of the Fourth Schedule hereto.

Thus at the first review date the reservation of rent will reserve a rent of £404,193, and at each subsequent review date it will reserve that sum or such other sum as is shown on any memorandum as is attached to the lease under para 5 of the fourth schedule. In other words, the reservation will, at each review date, simply set out the passing rent. It is therefore the defendant's contention that the material words are limited to the reservation of rent, and it is only those words that are omitted from the hypothetical lease.

(ii) the reservation of an additional yearly rent in respect of service charges, as to which there are elaborate provisions for quantification and administration in the fifth schedule.

(iii) the reservation of an additional yearly rent in respect of a proportion of insurance premium payable by the landlord.

(iv) the covenant to pay rent in clause 2(1) of the lease.

(v) the proviso for interest on arrears of rent in clause 4(1).

(vi) the proviso for re-entry for non-payment of rent in clause 4(2).

(vii) the proviso for rent suspension in clause 4(3).

(viii) the proviso that receipt of rent should not be a waiver of the right of re-entry in clause 4(9), and

(ix), of course, the rent review provisions in the fourth schedule.

Mr Morgan submitted that the words "terms except as regards rent", in their literal sense, covered all nine of the terms or provisions I have mentioned. He then directed me to *Halsbury's Laws of England*, 4th ed, vol 12, para 1463, and submitted that I must apply the literal meaning and excise all nine terms unless to do so would be absurd. He further submitted that, even if I found that there would be an absurdity in the case of the excision of some of the terms, I was still bound to give the words the widest meaning short of absurdity, and that meant that at least the rent review provisions must be excised from the hypothetical lease, because it could not be said that to exclude them would be an absurd result.

In view of three authorities: *French Kier Property Investment Ltd* v *Marconi Co Ltd*, an unreported decision of McNeill J delivered on March 16 1982; *Pugh* v *Smiths Industries Ltd* (1982) 264 EG 823; and *Safeway Food Stores Ltd* v *Banderway Ltd* (1983) 267 EG 850, in which tenants entered into leases which plainly excluded rent review provisions from the hypothetical lease, I agree with Mr Morgan that I could not conclude that the excision of the rent review provisions could be said to be absurd.

However, the canon of construction set out in para 1463 is not confined to the rejection of the absurd. The golden rule, as it is called, which applies to wills, statutes, and all written instruments, is usually taken from the speech of Lord Wensleydale in *Grey* v *Pearson* (1857) 6 HL Cas 61 at p 106, and covers not only absurdity but also an inconsistency with the rest of the instrument, in which case the ordinary sense of the words can be modified so as to avoid that inconsistency. In the case of a contractual transaction, in concluding whether there is an inconsistency with the rest of the instrument, the court takes into account the whole of the instrument, together with the commercial purpose of the transaction, viewed objectively (see *Reardon Smith Line* v *Hansen-Tangen* [1976] 1 WLR 989 at p 996).

The golden rule itself has been restated in modern terms by the Court of Appeal in *Bonalumi* v *Secretary of State for the Home Department* [1985] 1 QB 675, where Stephenson LJ said at p 682F:

> the court . . . must start by apply Lord Wensleydale's golden rule in *Grey* v *Pearson* . . . (put bluntly) of giving the words their natural and ordinary significance unless there is some compelling reason to the contrary.

So I turn to this lease, looking at it as a whole, and in the context of the commercial purpose of rent review provisions viewed objectively, as to which there was no doubt in 1980, although such purpose was expressed in various ways. To take one example, known to the parties at the date of the lease, I refer to *Ponsford* v *HMS Aerosols Ltd* [1979] AC 63, where Viscount Dilhorne, at p 76G, stated: "Their object is to secure that in real terms the rent payable does not fall below that initially agreed on."

That is a limited purpose. The rent was initially agreed on the basis of the value of the property when let on the terms of the actual lease. To achieve the purpose of securing that that rent does not fall in real terms at rent reviews while the tenant remains in occupation, so the premises cannot actually be let at the then current rental, requires certain hypotheses to be made which are contrary to reality; for example, the hypothesis of vacant possession to be found in this and in most other rent review provisions. Of course, the parties may have a particular commercial or other reason which causes them to agree to depart still further from reality, but a hypothesis that would result in an alteration in rental value other than one merely reflecting an alteration in the value of money or in the property market generally is, in my judgment, inconsistent with the limited commercial purpose of a rent review provision. If the wording of the lease makes it clear that the parties intended to affect the rental value on review by using a basis other than the terms of the actual lease, then I must give effect to that intention. It is not for me to speculate what particular commercial or other reasons prompted such alteration, nor to rewrite the lease so as to make it conform to the usual limited commercial purpose. Thus in the present case the term of the hypothetical lease is to be 25 years at each review date, although the actual lease will have a shorter residue at each such date.

However, if the wording of the lease is not clear, and if the words, used literally, result in altering rental values other than by merely reflecting an alteration in the value of money or the property market generally, and if the words can be applied in a more limited sense which is consistent with the commercial purpose I have described, then in my judgment there is a compelling reason for not applying the golden rule. In the lease before me the literal approach clearly results in alterations in rental value which have nothing to do with inflation or the property market. If the words "terms except as regards rent" apply to the service charge rent and the insurance rent so that these are to be excised from the hypothetical lease, the rental value will obviously be different from that which would result from a lease on the terms of the actual lease at the rent review date. So with the rent review provisions of the fourth schedule.

It follows that, if there is a narrower meaning that could be applied to the words, I would restrict them to such narrower meaning.

Then there are the terms relating to actual payment of rent and interest, to the securing of such payment by the right of re-entry, and to the suspension of the payment of rent. I do not know if these would affect the rental value or not. If they did, then they would be inconsistent with the commercial purpose as in the case of the other terms I have dealt with and would not be within the relevant words for the same reason. If they would have no effect on rental value, then they have no discernible effect at all, save to emasculate the lease, and it would be an exercise in futility for the parties to have provided that the hypothetical lease should not include such provisions.

In the latter event, the question of exclusion or inclusion is only of academic interest. However, I should deal with the point. In such event, the terms would not be inconsistent with the commercial purpose, but futility or absurdity seem to me to be a like compelling reason for not extending the words to include these terms.

I am therefore left with the reservation of rent, which, as I have said, effectively merely sets out the passing rent. To exclude the passing rent from the actual lease in order to arrive at the hypothetical lease in which the new rent was to be inserted is something which has to be done. Indeed, Mr Morgan points out that the words would be impliedly excluded anyway. I agree.

To sum up, the relevant words can be given a literal meaning that includes terms that are inconsistent with the commercial purpose of the rent review provisions. They can be given a narrower meaning which includes terms which it would be futile or absurd to include, or they can be given a meaning narrowed down to the passing rent, which meaning would be implicit and therefore, Mr Morgan says, unnecessary. However, if the words so construed would be implied anyway, it follows that they are neither inconsistent nor futile, and therefore in my judgment I must give that narrowest meaning to them. I see no compelling reason in the context of this lease to reject the possible meaning of the words simply because the parties would be held to have intended that result even if the lease had been silent on the point.

It follows that I consider that the tenant is right in submitting that the words refer only to the reservation of rent, which in this particular lease results only in the exclusion of the passing rent.

I was referred to a number of authorities. However, I am construing this particular lease, so that, apart from one authority which I shall come to, other decisions on other leases are not of great assistance. I merely mention those in the landlord's favour. In *Safeway Food Stores Ltd* v *Banderway Ltd* (1983) 267 EG 850 and the two earlier decisions to which I have already referred, the terms of a lease were such that they could have no meaning other than that the rent review provisions were to be excluded from the hypothetical lease.

In *National Westminster Bank plc* v *Arthur Young, McClelland Moores & Co* (1984) 273 EG 402* the wording of the particular lease was such that the relevant words could not be confined to the passing rent. Even with adjustments for grammatical reasons, the words still included "rent payable from time to time", and that in turn, under the reddendum, included the rent review provisions, which therefore had to be excluded from the hypothetical lease.

In *Equity & Law Life Assurance Society plc* v *Bodfield Ltd* (1985) 276 EG 1157†, the relevant words again could not be confined to the passing rent as they can in this case.

I was also referred by Mr Furber to section 34 of the Landlord and Tenant Act 1954 and by Mr Morgan to section 8 of the Agricultural Holdings Act 1948 and to the reversal of approach occasioned by the amendment of that section by virtue of section 1 of the Agricultural Holdings Act 1984. However, I get no assistance from such statutory provisions.

As I have said, there is one authority I should mention more fully. That is the recent unreported decision of the Vice-Chancellor in *British Gas Corporation* v *Universities Superannuation Scheme Ltd*.‡ The judgment was delivered on February 6 1986, and I was provided with a transcript. Mr Furber submitted that in that case the Vice-Chancellor laid down new principles of construction that are applicable to rent review provisions in leases, and he further submitted that I am bound by that decision under the principles referred to by Nourse J in *Colchester Estates (Cardiff) Ltd* v *Carlton Industries plc* [1984] 3 WLR 693.§ However, I do not consider that the Vice-Chancellor was enunciating special principles of construction which apply only to rent review provisions in leases. It seems to me that the Vice-Chancellor was merely setting out what would be the usual result that would flow from the application of the general principles of construction to rent review clauses in leases. As I have indicated, in my judgment, that result flows in this case. I therefore dismiss the summons.

The plaintiff's summons was dismissed with costs.

*Editor's note: Also reported at [1985] 1 EGLR 61.
†Editor's note: Also reported at [1985] 2 EGLR 144.
‡Editor's note: Reported at p 120 *ante* and (1986) 277 EG 980.
§ Editor's note: Also reported at (1984) 271 EG 778.

Court of Appeal

March 24 1986

(Before Lord Justice MAY, Lord Justice LLOYD and Mr Justice CAULFIELD)

PLEASURAMA PROPERTIES LTD v LEISURE INVESTMENTS (WEST END) LTD

Estates Gazette May 17 1986
278 EG 732-733

Landlord and tenant — Rent review clause — Whether an obligation contained in a licence to reinstate premises at the end of a subtenancy to their former condition should be taken into account in assessing the rent on a rent review — In this case a sublessee was granted a licence by the sublessor and the head landlord to convert part of premises into a dolphinarium, with a dolphin pool and auditorium — Licence contained a covenant by sublessee to reinstate the premises for normal shopping use at the expiration of the underlease — Licence contained no provision for monetary consideration either by increased rent or a premium, the sublessor and head landlord apparently being content with the sublessee's obligation to reinstate at his own expense — When the first rent review approached, the parties were unable to agree as to whether the sublessee's obligation to reinstate should be taken into account or disregarded, the sublessee contending that it should be taken into account (thus depressing the rent) and the sublessor arguing that it should be disregarded (with the opposite result) — Nourse J (as he then was) decided in favour of the sublessor — Before the Court of Appeal the sublessee argued that the covenant in the licence to reinstate should in effect be regarded as one of the covenants contained in the underlease to which the hypothetical term was to be subject — The court rejected this submission and held, affirming the decision of Nourse J, that it could not have been intended that the sublessee, for whose sole benefit the licence was given, should reap an additional benefit in the form of a reduction in the rent which would otherwise become payable on review — In addition, as the judge pointed out, the sublessee's contention was irreconcilable with the provision in the underlease that there should be disregarded any effect on rent by reason of any improvement carried out by the sublessee in order to put the premises in any better condition — Appeal dismissed

This was an appeal by Pleasurama Properties Ltd, the plaintiff company in the court below, the sublessees of premises in Oxford Street and Soho Square, London W1, from a decision of Nourse J in favour of the respondents, Leisure Investments (West End) Ltd, the defendant company below, the sublessor of the premises. The decision of Nourse J was reported at [1985] 1 EGLR 54; (1985) 273 EG 67.

Derek Wood QC and E J Prince (instructed by Barnett Alexander Chart) appeared on behalf of the appellants; Leolin Price QC and Isaac Jacob (instructed by Jerrard Saunders Donn) represented the respondents.

Giving the first judgment at the invitation of May LJ, LLOYD LJ said: This is a plaintiffs' appeal from a decision of Nourse J, as he then was, given on October 3 1984, whereby he gave judgment in favour of the defendants on a short point of construction arising on a rent review clause in an underlease dated May 15 1969. The facts and relevant clauses of the lease are fully set out in the judgment. I do not intend to repeat them in any detail.

The underlease relates to commercial premises in Oxford Street and Soho Square. The plaintiffs, Pleasurama Properties Ltd, are the sublessees. The defendants, Leisure Investments (West End) Ltd, are the sublessors.

There was a provision in the lease for additions to the rent at the end of the 14th and 28th years, based on any increase in the then yearly rack rent value of the premises on the open market. The only clause in the lease which I need set out is clause 2.

Clause 2(1) provides:

> The additional rent to be paid for the period commencing at the beginning of the fifteenth year of the said term and continuing for the remainder of the said term shall be a yearly rent equal to the amount (if any) by which the yearly rack rent value as hereinafter defined of the demised premises (hereinafter called "the yearly rack rent value") at the end of the fourteenth year of the said term shall exceed the initial rent.

Clause 2(3) provides:

> In this clause the expression "yearly rack rent value" means such an amount as shall be agreed by the parties hereto to represent the yearly rent which ought reasonably to be expected to be obtainable for the demised premises let as a whole in the open market as between a willing Landlord and a willing Tenant without taking any fine or premium for a lease for a term equal to the residue unexpired of the term hereby demised at the end of the fourteenth and twenty-eighth years of the said term (as the case may be) and subject to covenants on the part of the Tenant and conditions similar to those herein contained and for the purpose of carrying on therein the same trade or trades as that or those being carried on therein by the then occupier but disregarding . . .
>
> (c) any effect on rent by reason of any improvements or other work carried out by the Tenant or any person deriving title under it after the date of this Underlease in order to put the demised premises in any better condition than they are now in.

The permitted user under clause 3(13) of the underlease was in wide terms and included use for the purposes of "a high class entertainment and amusement centre and/or shopping arcade". At the date of the underlease the premises were, according to the judge, being used as retail shopping premises.

By a deed of licence dated September 7 1970, the head landlord and the sublessor gave permission for the sublessee to convert part of the premises into a dolphinarium, on terms, *inter alia*, that the sublessee would reconvert the premises to shop premises at the expiry of the underlease. The only clause in the deed of licence to which I need refer is clause 2, which sets out the covenants entered into by the sublessee. By clause 2(vi) the sublessee covenanted:

> at the expiration or sooner determination of the said Underlease unless released from compliance with this stipulation by the Society and the Tenant at its own cost and to the reasonable satisfaction of the Society and the Tenant to reinstate and make good the dolphin pool and the auditorium only to such a state (which shall be to the reasonable approval of the Society's and the Tenant's respective Surveyors) as shall be consistent with normal shopping standards that is to say to construct a level basement floor and a ground floor on the same level as Oxford Street together with such a connecting staircase as may be reasonably required and in such a manner as shall be to the reasonable satisfaction of the local authority's town planning and bye law requirements to enable the demised premises (as so reinstated) to be utilised for the retail shopping purposes thereafter.

It is important to notice that the sublessor did not stipulate for any payment in return for its consent.

There was a subunderlease dated January 16 1973, which was entered into after the conversion had been completed. The judge did not find it necessary to refer to the terms of the subunderlease: nor do I.

The question which arises is whether the obligation to reinstate the premises contained in clause 2(vi) of the deed of licence should be taken into account in determining the notional yearly rack rental value under clause 2(3) of the lease. It is in the plaintiffs' interest to argue that it should, since it will reduce the notional yearly rack rent, and thereby reduce the additional rent payable by the plaintiffs after the 15th and 29th years. It is in the interests of the sublessors to argue the contrary.

The learned judge decided in favour of the defendants. He did so on two grounds. Clause 2(3) of the lease provides, as I have already said, that the notional yearly rack rental shall be calculated on the basis of "covenants on the part of the tenant and conditions similar to those herein contained". It was common ground before the judge as before us that there is no specific provision in the lease, or the licence, that the covenant to reinstate contained in the licence should be taken into consideration under clause 2(3) of the lease. The judge held that a reference to that covenant could only be imported or implied in clause 2(3) if it was necessary to do so in order to give business efficacy to the agreement between the parties. The judge held that there was no such necessity. He referred in that connection to *Liverpool City Council* v *Irwin* [1977] AC 239.

The second reason given by the judge was that to decide in favour of the plaintiffs would offend against common sense. The defendants were doing a favour to the plaintiffs by allowing them to convert part of the premises into a dolphinarium. The defendants got no sort of benefit from the conversion themselves. It was, in the judge's view, contrary to common sense that the defendants should, as a consequence of giving their consent, suffer an increased burden by reason of the operation of the rent review clause.

Mr Wood for the plaintiffs has put the emphasis before us somewhat differently. He does not rely on necessary implication. He submits that he is entitled to succeed on construction. He puts the argument in two stages. If the lease had been varied by a deed of variation, then the words "herein contained" would be taken to refer not just to the covenants contained in the lease at the date of the lease but also to the covenants contained in the lease as varied at the time of the rent review. So that if, for instance, the permitted user had been widened by a deed of variation, then the widened user would be reflected in the notional yearly rack rent.

The second stage of the argument was that the deed of licence was on its true construction a variation of the lease. It is more than a mere personal licence. The covenants imposed on the underlessee, including the covenant to reinstate, run with the lease. The covenant to reinstate must therefore be added to clause 2(3) of the lease. By reason of the variation of the lease the covenant became a covenant "herein contained" on the true construction of that clause.

That, in outline, is Mr Wood's argument on construction. But in my judgment it contains a fallacy. I accept that the words "herein contained" are capable of applying to the lease as varied. They are not necessarily confined to the covenants at the date of the lease. I will assume — though I certainly do not accept — that the deed of licence is to be regarded as a variation of the lease. I prefer the view that the licence is what it purports to be, namely, a licence granted pursuant to the lease. But I will assume, as I say, that it was a variation of the lease. Even so, it does not follow that the covenant contained in the licence is to be read into clause 2 of the lease. The two steps of Mr Wood's argument do not carry him the whole way. There is a third step which he has to take. We have to see not only whether there has been a variation, which I am assuming, but what is the nature and extent of the variation.

In the case of a deed of variation the effect of which was to widen the permitted user, then I could well understand the courts holding that the words "herein contained" should be read as referring to the covenants as varied. But the same does not follow in the present case. I accept, of course, that the two documents must be read together. But they must be read together so as to produce a sensible and business-like result. The question we have to ask ourselves is whether, on a fair reading of the deed of licence, the parties intended the covenant in clause 2(vi) to be brought within clause 2 of the lease. I am quite clear that they did not. It cannot have been intended that the plaintiffs, for whose sole benefit the consent was given, should, as the result, reap an additional benefit in the form of a reduction in the rent which would otherwise become payable on review. I agree with the judge when he says that so to read the documents offends against common sense.

Mr Wood argued that where a tenant applies for consent to carry out improvements or alterations, there are three possibilities. The first is that the landlord will refuse consent on reasonable grounds. The second is that he will give unconditional consent. The third is that he will give conditional consent as contemplated by section 19(2) of the Landlord and Tenant Act 1927. The first alternative will not affect the value of the demised premises. The second alternative will increase the value of the demised premises. So will the third alternative, but not by so much as the second. Mr Wood argues that the difference in the increase in value between the second and third alternatives should be reflected in the rent payable on review. Since under the third alternative the landlord has given away less than he might, it is reasonable that there should be a reduction in the rent otherwise payable. I cannot agree with Mr Wood's argument. There is still a net benefit to the tenant under either the second or the third alternative. It does not make sense that the landlord should have to pay for that benefit.

The argument which I have just mentioned leads on to the last point. To hold in favour of the defendants would in my view be inconsistent with the express provisions of clause 2(3)(c) of the lease. The point is put very clearly in a single sentence on the last page of the judge's judgment:

> That provision — which requires that there should be disregarded any effect on rent by reason of any improvements carried out by the tenant in order to put the demised premises in any better condition — is simply not reconcilable with the notion that regard should be had to an obligation to eliminate such improvements and reinstate the premises to their former condition.

I can find no answer to that point. Mr Wood argued that the covenant to reinstate is still there. The burden on the tenant should be reflected at the rent review. But the covenant to reinstate is only the obverse of the permission to improve. If the benefit from the temporary improvement is to be disregarded and excluded from account, it is wholly illogical that the burden of reinstating should be included.

For the reasons I have given I would dismiss the appeal.

MAY LJ and CAULFIELD J agreed and did not add anything.

The appeal was dismissed with costs. Leave to appeal to the House of Lords was refused.

Chancery Division

March 21 1986

(Before Sir Nicolas BROWNE-WILKINSON, Vice-Chancellor)

LUCAS INDUSTRIES PLC v WELSH DEVELOPMENT AGENCY

Estates Gazette May 24 1986

278 EG 878-880

Landlord and tenant — Rent review clause in lease — Arbitration Act 1979, section 1(3) — Application for leave to appeal on a question of law arising out of arbitrator's award — Guidelines given by House of Lords in *The Nema* and *The Antaios* as to the correct judicial approach to the exercise of discretion in regard to leave to appeal under the 1979 Act — Extent to which such guidelines, concerned with appeals against decisions of commercial arbitrators, apply to appeals from arbitrators dealing with rent review clauses — Vice-Chancellor's reasons for concluding that such guidelines are not capable of direct application to the latter and his formulation as to the correct approach to them — In the case of disputes concerned with the standard terms of commercial contracts, the House of Lords had laid it down that leave to appeal should be given if the judge is satisfied that there is a strong *prima facie* case that the arbitrator is wrong in law; in the "one-off" type of case the judge should only give leave if satisfied that the arbitrator is obviously wrong — Such criteria are difficult, if not impossible, to apply to arbitrations on rent review clauses — Such clauses in no way correspond to the standard printed terms in commercial contracts; they give rise to questions of law of general application; and the arbitrator's decision can constitute an issue estoppel affecting future reviews — The approach should be that if the judge is left in real doubt as to whether the arbitrator was right in law he should grant leave to appeal — In the present case the review clause did not specify the terms of the hypothetical lease and the question was whether such terms should embody the very strict provisions as to alienation and user in the actual lease or the much less restrictive provisions on which the premises would have been let by the lessors in accordance with standard forms which they had introduced by the revision date — The arbitrator had determined that the hypothetical lease should embody the latter, less restrictive, terms — Applying his own test, the Vice-Chancellor concluded that he was left in real doubt as to whether the arbitrator was right — He was satisfied that, as required by section 1(4) of the 1979 Act, the determination of the question could substantially affect the rights of the parties — Leave to appeal accordingly granted

This was an application by Lucas Industries plc under section 1(3) of the Arbitration Act 1979 for leave to appeal on a question of law G
arising out of an award made by Mr D W Marsh FRICS. The arbitration concerned a lease for 25 years from May 11 1978 granted by the Welsh Development Agency in respect of Unit 4, Tafarnaubach Industrial Estate, Tafarnaubach, Mid Glamorgan, and the question arose under the rent review clause in the lease.

D J Pearce-Higgins (instructed by Evershed & Tomkinson, of Birmingham) appeared on behalf of the applicants; Kirk Reynolds (instructed by J E Bradwick, Welsh Development Agency) represented the Agency.

Giving judgment, SIR NICOLAS BROWNE-WILKINSON V-C said: This is an application under section 1(3) of the Arbitration Act 1979 for leave to appeal on a question of law arising out of an award made by Mr D W Marsh as arbitrator. The arbitration concerned the rent H
payable under a rent review clause contained in a lease. As is now usual in such cases, the application for leave to appeal has been referred by a judge of the Commercial Court for determination by a judge of the Chancery Division.

The rent review clause is contained in a lease dated December 21 1978 granted by the Welsh Development Agency ("the Agency"). The term of 25 years from May 11 1978 granted by the lease is now vested in the applicant, Lucas Industries plc. The rent reserved by the lease during the first five years of the term was £32,500 pa and for the remainder of the term a yearly rent to be agreed or determined from time to time as provided in clause 4 of the lease. Clause 4 provides for five-yearly rent reviews. The machinery is that the Agency gives to the J
lessee notice of the proposed revised rent and the lessee can then give a counternotice requiring the rent to be fixed by arbitration. Such a counternotice was given at the first review date and Mr Marsh (a surveyor) was appointed the arbitrator. Clause 4(b) provides as follows:

> The amount of the yearly rent payable from the appropriate revision date shall be the sum equivalent to the amount which represents the rental value in the open market on a date being three calendar months immediately before the appropriate revision date of the demised premises properly maintained in accordance with the covenants on the part of the Lessee herein contained . . . .

It should be noted that that formula does not specify the terms of the lease on which the property is to be deemed to be let to the K
hypothetical tenant.

The lease dated December 21 1978 contained covenants imposing restrictions as to user and alienation. The user covenant is as follows:

> . . . not to use the demised premises . . . for any other business than the . . . manufacture of vehicle and industrial braking and suspension systems and purposes ancillary thereto and the manufacture of such other products to which the Agency may give prior approval in writing.

The covenant against alienation provides an absolute bar on alienation subject to a very limited concession permitting assignment or sharing possession with a subsidiary or holding company of the lessee.

The lease dated December 21 1978 was in the standard form of L
lease then adopted by the Agency (Type A). In 1980 the Agency introduced a new standard form (Type B) which permits change of use for any light and general industrial purposes and permits assignment of the whole of the demised premises, in both cases subject to the landlord's consent such consent not to be unreasonably withheld. If the demised premises were being let for the first time by the Agency as at the revision date, the Agency would have let on the terms of the Type B lease, ie on terms much less restrictive as to user and alienation than those contained in the 1978 lease.

One of the main issues on the arbitration was whether in fixing the rent obtainable on the open market the arbitrator should assume a letting to the hypothetical tenant on terms as to user and alienation being those in the actual lease or those in a lease of Type B. In a M
carefully reasoned award, the arbitrator adopted the Agency's submission that the hypothetical letting should be on the terms of a Type B lease. The question which Lucas wish to raise on appeal is whether the arbitrator erred in law in so holding.

Under section 1(3) an appeal against the decision of the arbitrator can only be brought with the consent of all parties or with the leave of the court. Under section 1(4) the court is directed not to grant leave unless it considers that, having regard to all the circumstances, the determination of the question of law concerned could substantially affect the rights of one or more of the parties to the arbitration agreement. There are two decisions of the House of Lords laying down guidelines as to how a judge should approach the exercise of the

A discretion to give leave to appeal, viz *Pioneer Shipping Ltd* v *B T B Tioxide Ltd (The Nema)* [1982] AC 724 and *Antaios Compania Naviera SA* v *Salen Rederierna AB (The Antaios)* [1985] AC 191. Those cases (and many others in lower courts) have all been concerned with appeals against decisions of commercial arbitrators. Difficulty has been experienced in determining the extent to which those guidelines are applicable to appeals from arbitrators under rent review clauses. It is for that reason that in this case I have, with some trepidation, decided to give my reasons, a practice which Lord Diplock in *The Antaios* (at pp 205-6) characterised as not "normally" desirable for a "commercial" judge to adopt.

As the speeches in the House of Lords make clear, the purpose B behind the changes introduced by the 1979 Act was to get rid of the use made of so-called points of law to cause delay and expense in arbitrations arising out of commercial and shipping contracts. In that context the House of Lords has drawn a distinction between those cases which involve terms contained in a standard form commercial contract widely adopted in a particular field (eg The New York Produce Exchange time-charter form) and those which involve a one-off contract or event. In the case of disputes concerned with standard terms of commercial contracts, leave should be given if the judge is satisfied that there is a strong *prima facie* case that the arbitrator was wrong in law. In the one-off type of case, the judge should only give leave if satisfied that the arbitrator was obviously wrong.

C This approach is difficult, if not impossible, to apply to arbitrations on rent review clauses. First, although there are precedents of such clauses in the books, they are frequently modified in practice: they in no way correspond to the standard printed terms of contract commonly incorporated into commercial contracts. Second, although the exact words used in rent review clauses vary from lease to lease, there are questions of law which apply generally to rent review clauses of a particular type, eg the question in this case whether in the absence of express provision the hypothetical lease is to be treated as having the same terms as the actual lease between the parties. Third (and to my mind the most important), save in the case of the last review, the proper legal effect of a rent review clause will be material on future reviews between the same parties under the same D lease. Thus in the present case exactly the same point will arise between these parties (or their successors in title) on the reviews which will have to take place in the 10th, 15th and 20th years of the term. There are strong dicta which suggest that the decision by the arbitrator on this point of law will constitute an issue estoppel which will prevent a different arbitrator taking a different view of the point of law on future reviews: see per Diplock LJ in *Fidelitas Shipping Co Ltd* v *V/O Exportchleb* [1966] 1 QB 630 at p643. In all the commercial decisions I have seen, the point of law in question has always been "one-off" as between the parties to the arbitration and did not regulate their future contractual relationships.

For these reasons, I do not think that the House of Lords guidelines E are capable of direct application to arbitrations on rent review clauses. I take some comfort from the fact that in *The Antaios* (at p200B) Lord Diplock said that the guidelines are not immutable but subject to adaptation to meet problems of kinds that were not foreseen and are not covered by what was said in *The Nema*. In my judgment, in neither of the cases in the House of Lords were they considering problems raised by rent review clauses.

What then should be the correct approach? All that the House of Lords has said about the need to avoid delay and expense in commercial arbitrations is applicable to rent review arbitrations, although possibly to a lesser extent. Accordingly the need to avoid appeals on over-ingenious points of law which have no real prospect of success is equally applicable to arbitrations on rent review clauses. F But, in my judgment, in cases where the same point of law will regulate future rent reviews a lower standard than a strong *prima facie* case is appropriate. If, after hearing submissions, I am left in real doubt whether the arbitrator was right in law, it seems to me appropriate to grant leave so that the law regulating the future relationship of the parties can be authoritatively determined by the court. This is especially so in a case such as the present where the decision on the point of law is likely to be of importance in relation to all rent review clauses which do not specify what are to be the terms of the hypothetical lease. There are many such clauses, especially those drafted some years ago.

Adopting that test, I am left in real doubt whether the arbitrator was right in the conclusion which he reached. I am also satisfied (as required by section 1(4) of the 1979 Act) that the determination of that question could substantially affect the rights of parties. Following the House of Lords guidelines I do not think it appropriate to give my reasons: to do so would merely embarrass the judge who will have to determine the question after full argument.

Finally I should mention that the Agency has granted other leases of Type A and the same point will arise on the rent review clauses in those leases. I would have granted leave to appeal in the absence of this factor; but it provides an additional reason for so doing.

Leave to appeal granted.

Court of Appeal

February 28 1986

(Before Lord Justice FOX and Lord Justice MUSTILL)

BEARD (FORMERLY COLEMAN) v WILLIAMS

Estates Gazette June 7 1986

278 EG 1087-1090

Landlord and tenant — Landlord and Tenant Act 1954, Part II — In this case there had been a "tangled history" of previous disputes and litigation, but the issues in the present appeal arose under section 30(1)(c) of the 1954 Act — Appellant was a trainer or breeder of greyhounds who carried on this enterprise on land belonging to the respondent — There had been a dispute as to his exact status on the land, but it had been decided that he had a business tenancy subject to an express term prohibiting him from living on the land — He had been ordered to remove himself and a van from the land and he had eventually removed the van but had placed it outside the respondent's premises about 100 yds away from the kennels, near enough to enable him to carry on his breeding activities — The van 's precise location was not clear but was said to be on a grass verge on or near the highway and its position might have been unlawful — Respondent served a section 25 notice on the appellant terminating his tenancy and notifying opposition to the grant of a new tenancy on the ground stated in section 30(1)(c) of the 1954 Act — Appellant made a cross-application for a new tenancy — County Court judge held that no breach of obligations under the tenancy had been proved, but that the appellant's action in keeping his van unlawfully on the verge was an act which brought the second limb of section 30(1)(c) into operation, as a reason "connected with his use or management of the holding" for not granting him a new tenancy — The Court of Appeal upheld the judge's decision on this point, although arriving at it on the basis of a different process of reasoning, but then went on to consider whether the judge had exercised his discretion correctly in deciding that in all the circumstances the appellant should not be granted a new tenancy — Held that, in exercising his discretion, the judge should have asked himself whether the respondent's interest as landlord was likely to be prejudiced by the matters relied on under section 30(1)(c) — The judge had given no indication of the matters which he took into account, apart from the question of illegality, which was not material to the exercise — In these circumstances the case would have to go back to the judge to look at the matter again, as the Court of Appeal did not have sufficient information to exercise the discretion themselves — Appeal allowed to this extent

This was an appeal by William Beard from a decision of Judge Taylor, at Bishop's Stortford County Court, in favour of the respondent, Mrs Hilda Daphne Williams, the owner of land at Great Hallingbury, Bishop's Stortford, in proceedings under the Landlord and Tenant Act 1954, in which the judge decided that the appellant should not be granted a new tenancy of certain fields and stables which he had been using for breeding and training greyhounds.

A E M Cooper (instructed by Pellys, of Bishop's Stortford) appeared on behalf of the appellant; J Brooke-Smith (instructed by Nockolds, of Bishop's Stortford) represented the respondent.

Giving the first judgment at the invitation of Fox LJ, MUSTILL LJ said: This is an appeal against the judgment of His Honour Judge Taylor given at the Bishop's Stortford County Court on February 21 1985. The appeal raises a short question on the law relating to landlord and tenant which has been well and succinctly argued on both sides.

The history of the matter is as follows. The appellant, William Beard, has at all material times been either a trainer or a breeder of greyhounds. The respondent, Mrs Hilda Daphne Williams, was the proprietor of land and premises at Great Hallingbury, Bishop's Stortford. During 1971 the appellant moved on to the land of the respondent and set up a business of training greyhounds in certain stables and on certain fields belonging to the respondent. Three months later he took over an additional 5 acres of the respondent's land at a rent of £20 per week, which had been agreed should be paid for the occupation of the premises, but the rent was then increased to £25 per week.

Disputes arose between the parties, and during March 1983 the respondent gave the appellant an informal notice to quit. During June of that same year a summary application for possession was made, essentially on the ground that the appellant had been allowed on to the land as an act of kindness; that he was, accordingly, a licensee, and that proper notice had been given to terminate the licence. That application was supported by various complaints about the conduct of the appellant — including the placing of a decrepit van on the land in question.

The application elicited an affidavit in response, asserting that the arrangement between the parties amounted not to a licence but to a business tenancy. Thereafter the dispute proceeded in the shape of a possession action, with pleadings being delivered, and the matter eventually came to judgment before His Honour Judge Taylor on April 12 1984.

In essence, the learned judge determined in favour of the present appellant, that the nature of the relationship was that of a business tenancy, comprising the use of certain fields, but that the tenancy was subject to an express term prohibiting the appellant from living on the land.

There was an appeal by the present appellant to another division of this court against that part of the judgment of Judge Taylor which had deprived the appellant of the right to live on the land — and in particular to keep his van there. That appeal was dismissed in June 1984.

Immediately thereafter the respondent served on the appellant a notice under section 25 of the Landlord and Tenant Act 1954 to terminate the tenancy with effect from January 31 1985, and the grant of any further tenancy was opposed on the basis of the grounds set out in section 30(1)(c) of the 1954 Act.

In addition there were proceedings brought by the respondent in the nature of an application for an injunction to cause the appellant to move himself and his van from the respondent's land. Final orders were made to that effect during August and October 1984. In response to those orders, the appellant did move his van from the respondent's land. However, this apparently caused him to lose his licence to train greyhounds, because it is a condition of the grant of such a licence that the trainer shall reside on the premises where the training is carried out. But he still carries on the business of breeding dogs — he has some 30 animals still on the respondent's land.

His van was removed from the respondent's land and was placed outside her premises at a location which is apparently some 100 yds away from the stables. At that point he is sufficiently near (so he says) to the place where the dogs are kept for him to be able to carry on his breeding activities satisfactorily.

It appears to be his contention that he cannot move substantially further away without imperilling his business. However, where exactly the van is located has not emerged clearly from the evidence. It is said to be on the grass verge on or near the highway; and it is also said that the appellant has been told by a police officer that the van should not be there. In spite of this warning, the van remains on the grass verge, and apparently no steps have been taken, either by the police authorities or by other public authorities or private persons, to have it removed. There the matter rests, at least for the time being.

In response to the application by the respondent for an order under section 25, the appellant on his part made a cross-application to the court for a new tenancy for seven years at a rent of £25 per week. G

It will be convenient at this stage if I read the relevant parts of the statute. Section 29 provides:

(1) Subject to the provisions of this Act, on an application under subsection (1) of section 24 of this Act for a new tenancy the court shall make an order for the grant of a tenancy comprising such property . . .

By section 30(1) it is provided:

The grounds on which a landlord may oppose an application under subsection (1) of section 24 of this Act are such of the following grounds as may be stated in the landlord's notice under section 25 of this Act or, as the case may be, under subsection (6) of section 26 thereof, that is to say . . .

then omitting subparas (a) and (b):

(c) that the tenant ought not to be granted a new tenancy in view of other substantial breaches by him of his obligations under the current tenancy, or for any other reason connected with the tenant's use or management of the holding. H

In response to the application by the appellant for a new tenancy, the respondent answered with grounds of opposition which consisted in the main of allegations that the conduct of the appellant amounted to breaches of express or implied terms of the tenancy agreement. If such breaches had been established, they would have fallen within the first part of subpara (c) of section 30(1).

In addition, by ground (v) of the answer to the application, the respondent alleged as follows:

By the express terms of the tenancy the applicant is not entitled to reside on the land and it is therefore impracticable for the said business to be managed properly or at all and for the applicant to be granted a new tenancy in the terms sought by the originating application herein. J

That contention was prefaced by the factual assertion that the appellant was precluded by the previous judgment of Judge Taylor from living on the land and that there was nowhere in practice where he could live within such proximity to the land as to enable his business to be carried on in a proper and efficient manner.

That was the position when the matter came again before Judge Taylor. Evidence was given on both sides, on the lines which I have briefly summarised, and the respondent also sought to support by evidence the allegations of breaches of the tenancy agreement, to which I have already referred. K

At the conclusion of his judgment, the learned judge dismissed all the allegations of breach, saying that he was not satisfied that they had been proved by the landlord. Nevertheless, he did decide the case in favour of the respondent on grounds which are set out in an agreed note of his judgment, in the following terms:

Mr Beard says that this (ie the present arrangement) is an illegal arrangement and has been told as much by the police, but that the police are keeping a blind eye and he wishes to continue living there. Thus if I grant his application he will be carrying on his business in a manner which he says is illegal and has been told so by the police.

I now look at section 30(1)(c), where the landlord's right of opposition to a new tenancy is on the grounds "that the tenant ought not to be granted a new tenancy in view of other substantial breaches by him of his obligations under the current tenancy, or for any other reason connected with the tenant's use or management of the holding". I disregard the first limb of that paragraph because there are no substantial breaches. However this case comes four square within the terms of the second limb of this paragraph. This applicant is enjoying an illegal residence on the grass verge and in my view this is well within terms of the Act. In my discretion I hold that paragraph 1(v) is proved and I do not grant the tenant's application. L

It is against that decision that the present appeal is now brought. It is common ground that, for the purposes both of the exercise to be performed by the learned judge at first instance and for this appeal, two distinct questions must successively be considered: First, whether or not the manner in which the appellant chose to arrange his living accommodation and hence chose to supervise the carrying on of the business on the holding was capable of being a reason, in the words of section 30(1)(c), "connected with the tenant's use or management of the holding". M

Second, if the matter complained of is capable of being such a reason, whether, in all the circumstances, the tenant ought not to be granted a new tenancy, this matter (it being acknowleged on both sides) being a question for the discretion of the trial judge. I shall consider these two questions in turn.

As to the first question, two distinct ways of approaching the respondent's case must be identified. The first is the ground relied on by the learned judge. Essentially, what underlay the decision now under appeal was the proposition that the keeping of the van on the verge was an illegal act and that, accordingly, there is a sufficient

A nexus between the performance of the tenancy agreement and an act of illegality to require the court, *ipso facto*, to decline any suggestion that a new tenancy should be ordered.

The only reported case brought to our attention which comes anywhere near the present matter is the decision of another division of this court in *Turner & Bell (trading as Avro Luxury Coaches)* v *Searles (Stanford-le-Hope) Ltd* (1977) 33 P & CR 208.* In that case the premises in question were used by the tenant as a depot in connection with the business of coach transporters. The local planning authority served on the tenant an enforcement notice requiring them to discontinue the use of the premises for the operation of the coach business. An appeal against the notice was unsuccessful, and an extension of time which had been granted for
B compliance was allowed to expire without any steps being taken to discontinue the business. The continued operation of the business was a criminal offence, albeit no prosecution was ever brought in respect of it. The landlords served a notice under section 25 and the tenant served a counternotice. In answer to the application for the new tenancy, the landlord relied on section 30(1)(c).

The trial judge held that no new tenancy should be ordered, and that decision was upheld in this court. Two grounds may be identified in the judgment; the first is exemplified by the following passage from the judgment of Bridge LJ at p 211:

> It seems to me that the short answer to this appeal is that, quite independently of the express provisions of the statute, this is a case where the court would be
> C bound to refuse the relief claimed on the simple ground that if the court were to order a new tenancy in the circumstances indicated it would be ordering the parties to enter into an illegal contract which the court could not enforce because the illegal purpose of the tenant was clearly known to both parties. That would be an absurdity.

The second ground, which was closely allied to the first, was simply that the matter fell within the words of section 30(1)(c) because the fact that the premises were being used and were intended to be used for an unlawful purpose was manifestly an "other reason connected with the tenant's use . . . of the holding".

In my judgment, this decision provides little guidance in the circumstances of the present case. There the illegality question was directly concerned with the use of the land. It would have been
D impossible for the court to enforce the business tenancy without enforcing the user of the land, which was illegal. The illegality was implicit in the proposed new tenancy. But nothing of that kind exists here. The user of the land is not illegal; the appellant has done nothing to suggest that the manner in which he conducts his business of breeding greyhounds will be carried on otherwise than in accordance with the law. Any unlawful character of the arrangements which he makes to put himself into the position of performing the subject-matter of the tenancy agreement is, in my judgment, in a quite different category from the kind of illegality which was under consideration in *Turner & Bell* v *Searles*.

So much for the ground upon which the learned judge principally relied.

E Mr Brooke-Smith, who appears on behalf of the respondent, has, however, been able to draw attention to his original grounds of opposition to the new tenancy, from which I have already quoted. He naturally concedes that, in the ordinary course of events, the living arrangements of the tenant would not be connected with the use or management of the holding; but he maintains that this particular business demands a particular type of living arrangement — namely, one which is adjacent to the place where the dogs are kept. Accordingly, so the argument runs, since the possibility of the tenant's living on the land has already been eliminated by the prior judgment of Judge Taylor, the business on the holding cannot be conducted in a proper manner. Hence (so the argument concludes), in
F the exceptional circumstances of this case, the appellant's living arrangements do constitute a matter "connected with the tenant's use or management of the holding".

In my judgment, this argument is correct in principle. The learned judge in my opinion would have been entitled to hold that the precarious nature of the tenant's living arrangements, quite aside from any direct impact of illegality on the proposed tenancy agreement, would have constituted a reason why the tenant ought not to be granted a tenancy; for, if the tenant was driven away from his existing place of abode, the business would (so the evidence suggests) rapidly deteriorate, with the obvious risk of prejudice to the respondent.

Accordingly, I would hold that the matters complained of are capable of forming reasons "connected with the tenant's use or management of the holding"; and that the first stage in the process of argument is established.

I should mention briefly that Mr Brooke-Smith also put forward an alternative approach, to the effect that one can find in the authorities support for the proposition that a wider meaning of section 30(1)(c) than the one contended for by Mr Cooper should be adopted. Mr Brooke-Smith referred to certain passages in the judgment of *Turner & Bell* v *Searles*, and in particular to the judgment of Roskill LJ, suggesting that any fact which could be said to be relevant to the use or management may legitimately be taken into account.

I find it unnecessary to express any concluded view on this argument. As at present advised, it seems to me that the members of the court were doing no more than rebutting an argument which had been put forward on behalf of the appellant in that case, to the effect that section 30(1)(c) was confined to matters which were directly connected with the relationship of the parties, *qua* landlord and tenant. That narrow construction was negatived and I do not at present find any warrant for giving the judgments a wider meaning than what they actually say.

Be that as it may, I conclude that the respondent does bring herself within the words of the latter part of section 30(1)(c).

The next question is whether, in the exercise of his discretion, the learned judge should have concluded that the tenant ought not to be granted a new tenancy. As I understand the position, it was common ground in the argument before us, that, as part of the exercise of his discretion, the learned judge should have asked himself whether the landlord's interest was likely to be prejudiced by the occurrence of the matters relied upon as constituting reasons within section 30(1). Authority for the proposition that such matters should be taken into account can be found in *John Kay Ltd* v *Kay* [1952] 2 QB 258; *Lyons* v *Central Commercial Properties (London) Ltd* [1958] 1 WLR 869 and *Eichner* v *Midland Bank Executor and Trustee Co Ltd* [1970] 1 WLR 1120.

In the present case, the notes of the judge's decision give no indication of the matters which he took into account besides the question of the illegality, to which I have already referred. For the reasons which I have stated, I regard this question of illegality as immaterial to the exercise which the learned judge had to perform; and I am therefore driven to the conclusion that when exercising his discretion, he took into account material which was not germane. Conversely, it is impossible to be sure, from the note of his judgment, that he did take into account other matters which were germane.

In these circumstances, it appears to me that the discretion was not exercised in the light of material considerations and no others. It would be open to this court, in those circumstances, to exercise for itself the discretion which originally fell to the learned judge below. But we are not confident that we know sufficient of the facts to enable us to exercise that discretion with any degree of reliability. We should wish to know more of the nature of the supposed illegality — including the question whether what was described as "illegality" was perhaps no more than a mere unlawfulness constituted by a trespass. We should need to know exactly how precarious the appellant's present place of abode may be said to be. We should need to know what alternative accommodation may be available if the appellant is really forced to look for it. These are matters which would have to be investigated by evidence more extensive than is to be found in the papers before us.

I therefore feel constrained to hold that, although the discretion may well have been exercised in an appropriate manner, this case must be remitted to the county court judge so that he can look at the matter again in the light of the opinions expressed by this court. He is already well familiar with the tangled history of this matter; he has a "feel" for the case which this court cannot obtain merely by reading the notes of the evidence. He will be able to hear further evidence on the matter and, in the light of all the considerations then before him, decide whether his original conclusion should be upheld or whether, on the contrary, in all the circumstances, a new tenancy ought to be granted.

Accordingly it is to that extent, but no further, that I would allow this appeal.

FOX LJ agreed and did not add anything.

The appeal was allowed and the case remitted to the county court judge to reconsider his decision in the light of the observations of the

---

*Editor's note: See also (1977) 244 EG 1023.

Court of Appeal. No order was made as to costs; legal aid taxation of the appellant's costs ordered.

Court of Appeal

April 10 1986

(Before Lord Justice SLADE and Mr Justice CAULFIELD)

TRANS-BRITANNIA PROPERTIES LTD v DARBY PROPERTIES LTD

Estates Gazette June 14 1986

278 EG 1254-1260

Landlord and Tenant Act 1954, Part II — Whether tenants were in occupation of the relevant premises for the purpose of a business carried on by them within the meaning of section 23(1) of the Act — Landlords' appeal from decision of county court judge on a preliminary point in favour of tenants — Land used by tenants for the carrying on of the business of proprietors of lock-up garages — On application for a new tenancy landlords had opposed the grant on the ground *inter alia* that tenants did not satisfy the occupation requirement of section 23(1) — The facts were that some garages were at any time vacant but the majority were actually sublet — One garage was used by the tenants as a store room — Tenants were in rateable occupation of the whole premises but each subtenant had exclusive possession of his garage — Tenants, although not under any contractual obligation to do so, in fact carried out some maintenance and cleaning of the sites and sometimes, with the consent of the subtenant, would effect internal repairs to a garage — The manager made fortnightly visits to the sites — On the evidence before him the county court judge found that the tenants retained sufficient occupation to support an application for a new tenancy — Court of Appeal considered *Bagettes Ltd* v *GP Estates*, *Lee-Verhulst (Investments) Ltd* v *Harwood Trust*, *William Boyer & Sons Ltd* v *Adams*, *Hancock & Willis* v *GMS Syndicate Ltd*, and *Linden* v *Department of Health and Social Security* — Held that, although the question raised by this kind of case must always be one of degree, the judge had misdirected himself in purporting to apply the authorities and had placed the present case on the wrong side of the line — There was here no office or living accommodation on the site, no water, no electricity, no security guard, and regular visits by staff were only fortnightly — In the authorities where sufficient occupation had been established the extent of the physical presence, the amount of services rendered to subtenants or licensees, the degree of control exercised and the time spent in connection with the relevant premises were all much greater — Tenants here failed to qualify — Landlords' appeal allowed

This was an appeal by the landlords, Darby Properties Ltd, from the decision of Judge Davison at Northampton County Court on a preliminary point in proceedings under Part II of the Landlord and Tenant Act 1954. The tenants, in whose favour the judge had decided the point, plaintiffs in the court below and respondents to the appeal, were Trans-Britannia Properties Ltd. The premises consisted of land at Far Cotton, Northampton, on which 46 lock-up garages were at all material times available for letting.

Jeremy Carey (instructed by Ray & Vials, of Northampton) appeared on behalf of the appellants; A Geddes (instructed by Wigram & Co) represented the respondents.

Giving judgment, SLADE LJ said: This is an appeal by Darby Properties Ltd, pursuant to the leave of the judge, from an order of His Honour Judge Davison made in the Northampton County Court on January 3 1985, whereby he decided a preliminary point against the appellant landlords in landlord and tenant proceedings.

The history of the matter is briefly as follows. On October 1 1962, a lease of a piece of land at Far Cotton, Northampton, was granted by the appellants ("Darby") to Banbury Buildings (Lock-Ups) Ltd for a G
term of 21 years, expiring on September 30 1983. Clause 3 of the lease contained, *inter alia*, the following tenant's covenants:

(4) Not to carry on or permit or suffer to be carried on upon the said land nor to use or occupy the same for the purposes of any trade or business other than the business of proprietors of lock-up garages and to use the land only for the garaging of private vehicles.

(6) Not to form any refuse dump or scrap heap on the land and to keep the same clear of all refuse rubbish and scrap and in a neat and tidy condition.

(9) Not at any time during the said term to assign or underlet or part with possession of the demised premises or any part thereof or of this lease without the written consent of the landlords, such consent not to be unreasonably withheld Provided Always that the tenants may without such consent grant periodic tenancies (determinable within the term hereby granted) of any lock-up garages erected by the tenants upon the said land but so that any such H
tenancies shall be granted upon and subject to the conditions that the same shall not be used or occupied for the purposes of business.

At the date of the lease there were no buildings on the land, but, as the lease itself indicates, it was always understood that the land would be used by the tenants for the carrying on of the business of proprietors of lock-up garages. The then tenants erected a number of such garages. It appears that at all material times there have been 46 of them available for letting.

On June 18 1981 the original lessees assigned the lease to Trans-Britannia Properties Ltd ("Trans-Britannia"), who are the respondents to this appeal. On September 30 1983 the lease expired. On October 3 1983 Trans-Britannia made a request for a new tenancy J
under the provisions of section 26 of the Landlord and Tenant Act 1954. On December 20 1983 they issued an originating application in the Northampton County Court seeking a new tenancy under the 1954 Act, describing the nature of the business carried on at the premises as being "Lettings of Lock-up Garages". This was followed by an answer to the originating application, dated January 9 1984, in which Darby stated that they opposed the grant of a new tenancy on three grounds. The first of these grounds was:

You do not satisfy the conditions of the Landlord and Tenant Act 1954 part II, section 23 (1) (2) and (3) because neither you nor any person employed by you occupies the premises for the purpose of a business carried on by you.

Having regard to this contention, the registrar ordered the trial of the preliminary issue as to whether Trans-Britannia were entitled to K
make an application for a new lease under the 1954 Act.

The relevant subsections of section 23 of the 1954 Act state:

(1) Subject to the provisions of this Act, this Part of this Act applies to any tenancy where the property comprised in the tenancy is or includes premises which are occupied by the tenant and are so occupied for the purposes of a business carried on by him or for those and other purposes.

(2) In this Part of this Act the expression "business" includes a trade, profession or employment and includes any activity carried on by a body of persons, whether corporate or unincorporate.

(3) In the following provisions of this Part of this Act the expression "the holding", in relation to a tenancy to which this Part of this Act applies, means the property comprised in the tenancy, there being excluded any part thereof which is occupied neither by the tenant nor by a person employed by the tenant L
and so employed for the purposes of a business by reason of which the tenancy is one to which this Part of this Act applies.

At the trial the learned judge heard oral evidence from one witness on behalf of Darby, namely Mr Roger Darby, one of its directors, who had sworn an affidavit which was also before the judge. He also heard evidence from Mr Frederick Potter, who was employed by Trans-Britannia as manager of their lock-up garages department. There were in addition before him two affidavits sworn on behalf of Trans-Britannia, one by their solicitor, Mr Hansom, and one by Mr Dewan, their managing director.

The learned judge in his judgment did not refer specifically to the evidence of any of these witnesses, other than that of Mr Potter, on whose evidence he heavily relied. Counsel in argument before us did M
not rely specifically on any such other evidence, beyond a very few features which are, I think, common ground. First, Mr Darby, in his affidavit, verified by his oral evidence, stated that the premises comprise 46 lock-up garages which are available for letting, that there is no living accommodation on the site and there is no office. Mr Hansom exhibited to his affidavit three sample agreements providing for occupation of the garages by sublessees on a monthly basis determinable on one month's written notice by either party. They also contain a clause which reads as follows:

The tenant to use the premises as a lock-up garage for a motor vehicle only and to be responsible for repair and maintenance of door and locks.

However, the evidence of Mr Potter, to which I am about to refer, makes it clear that he and his team of workmen do certain maintenance work to the doors and locks of garages on behalf of Trans-Britannia.

We have the benefit of a very clear and careful note of the oral evidence, taken by the learned judge. Since the record of Mr Potter's evidence is important and not very lengthy, I propose to quote it quite fully. In chief, having explained his position with Trans-Britannia, he continued:

> We have about 100-120 sites. Nationwide. My work involves a visit twice a month to the site and we have a permanent store there with materials and vans call there to collect and deposit what they need — to look into any complaints we receive from the agents Horne & Co. We actually maintain the roofs and doors. We spent over £1,000 last year on up and over doors. Gutters are replaced but they are often broken by vandals overnight. The tarmac forecourts are cleared constantly by local skips. We employ four men to carry out this work but I hire local labour as necessary. The doors are maintained by the Company and all the padlocks are bought and supplied by the Company. I have never received any instructions that we should not pay for the doors and locks. We have spare keys to all the garages and when a complaint is received we get the key from the agents and then return it to the agents on completion of the repair. We have never had a tenant trying to prevent us getting in to do repairs. We keep the rubbish down and also the weeds. People do use it as a dumping ground — we clear it up and make the place look tidy constantly. People use the site to park and turn. We go to the garages when there is a changeover and do repairs and clean as necessary the agents write to us and say that garages 1, 2 or 3 are void and to go in and prepare for reletting. I have been involved with this site for over twenty years. I worked for Banbury Buildings.

From the notes of Mr Potter's cross-examination, I need only refer to the following extracts:

> I don't give this site special priority. I visit twice a month. It's only given priority because we have stores there and it happens to be eight miles from my home — but it has no particular attention. I do over a thousand miles every week. I don't only go to a site when there is a complaint. I always check the sites in the area where I am travelling. The stores are used for materials on any site. We have six storage sites I know of now . . . The agent has a key to keep control of the sites. They know which is void and let. We respect the tenant's property. We never go in unless for maintenance purposes. This is an area where there are a lot of break ins.

Mr Potter was not re-examined.

On a first reading, section 23(1) might be thought to render Part II of the 1954 Act applicable when the property comprised in the tenancy merely "includes" premises which are occupied by the tenant for the purposes of a business carried on by him. However, having regard to the provisions of sections 23(3) and 32 of the 1954 Act, Mr Geddes, on behalf of Trans-Britannia, has accepted that they must be found to occupy the entire premises for the purposes of a business carried on by them, if they are to be entitled to apply to the court for the new lease which they claim.

One obvious possible hurdle in their way is that while some of the lock-up garages are vacant, awaiting subletting, many others are actually sublet. At first sight it might appear difficult for lessees to assert that they are in occupation of the whole of a site when substantial parts of it are sublet. However, as the authorities show, the position is not quite as simple as that. Mr Geddes has pointed out that the business of Trans-Britannia is one of subletting lock-up garages and, on the basis of the authorities, has submitted that there was sufficient evidence to justify the learned judge in finding that they were occupying the whole of the premises for the purposes of that business within the meaning of section 23(1) of the 1954 Act. I now turn to consider the authorities which have been cited to us in their chronological order.

The earliest of them affords no assistance to Trans-Britannia's case. In *Bagettes Ltd* v *GP Estates Ltd* [1956] Ch 290, the defendants carried on the business of holders, managers and landlords of real property. One of their assets was a leasehold interest in premises comprising 13 self-contained flats, 10 of which were sublet at the material time, halls, passages, landings and stairs in common use, together with further accommodation in the basement for a caretaker, boiler rooms, storerooms and fuel stores. They provided hot water for their tenants and cleaned the common parts of the premises, though not obliged by the terms of the subleases to do so. They also stored articles for the tenants in the basement storage rooms. The Court of Appeal held that the defendants' tenancy was not one to which Part II of the 1954 Act applied. Jenkins LJ, with whose judgment Lord Evershed MR and Birkett LJ agreed, accepted (at p 299) that the subletting of unfurnished residential flats is a "business" within the meaning of section 23 and that the purposes for which the entire premises there in question (including the flats sublet) were *used* by the defendants for the purposes of a business carried on by them within the meaning of section 23(1). However, he expressed the view (at p 300) that a building wholly sublet into flats from top to bottom could not qualify for protection under Part II of the Act of 1954 "because the tenant would *ex hypothesi* not be in occupation of any part of the premises". Consistently with this reasoning he found (at p 302) that the ascertainment of the holding inevitably involved the extraction from the property involved in the tenancy of the 10 sublet flats. He continued:

> There thus remain as the potential subject-matter of the new tenancy the caretaker's rooms, storerooms and boiler rooms in the basement, the common parts of the premises, and the three unlet flats. Divorced from the 10 sublet flats, can any part of these premises be said to be occupied by the tenant for the purposes of any business carried on by him? That question must, I think, be answered in the negative. There can be no business carried on by the tenant of providing hot water for the sublet flats, or of cleaning the common parts of the premises for the benefit of the tenants of those flats, or of providing those tenants with means of access to and from those flats, once the sublet flats are ruled out of consideration as disqualified from inclusion in the holding to which any new tenancy granted must be confined. Nor, in my view, can there be any business carried on by the tenant of serving in any of the ways above mentioned the three unlet flats so long as they remain unlet, while if let at any time before the ordering of a new tenancy, they would be subject to the same disqualification as the 10 flats now let. Finally, I do not think that the three unlet flats (all of which I treat for the present purposes as in the occupation of the defendants) can properly be treated as occupied by the defendants for the purposes of any business carried on by them. The defendants no doubt hope in due course to turn these three flats to account in their business by subletting them to tenants. But this, in my view, is not occupation of these three flats by the defendants for the purposes of a business carried on by them. On the contrary, it is occupation until such time as the defendants may be able, for reward in the shape of rent, to exclude themselves from occupation by subletting.

On facts superficially somewhat similar to those in the *Bagettes* case, this court reached a different conclusion in *Lee-Verhulst (Investments) Ltd* v *Harwood Trust* [1973] QB 204. I take the facts from the headnote:

> The tenant company of a house carried on there the business of letting furnished rooms with services. L, the sole director of the company, lived in the basement and devoted all his time to running the business. In the basement there was a boiler which provided hot water throughout the house, a store for luggage and a telephone on which L received incoming calls which would be taken by the room occupants on extensions on the landings. In the hall there was a coin operated telephone on which the room occupants could make calls. L was the subscriber and liable for rental and all calls made. There were 20 rooms. Most had bathrooms and gas rings but L controlled the extent to which cooking took place and was willing to provide light meals if required. The rooms were supplied with linen and towels. Maids with pass keys cleaned the rooms and changed the linen. Each occupant had a rent book and paid a weekly rent.

On these facts, having regard to the degree of control and extent of the services provided by the tenant company, this court concluded that the tenant occupied the entire premises for the purposes of its business within the meaning of section 23 of the 1954 Act. Sachs LJ, with whose judgment Karminski LJ agreed, said this (at p 213):

> Thus one comes to the critical issue — were the premises "occupied" by the tenant for the purpose of that business? The tenant company must of course be taken, when considering this question, as being represented on the premises by Mr Lee and the staff he engaged: the position in practice being no different than if Mr Lee had himself been the tenant. By Mr Lee and the staff the tenant company were present in the premises for the purpose of the business day and night; in the course of their services to the occupants they pervaded every room there; control was exercised by Mr Lee over the manner in which the occupancies were conducted — a control (eg by limitations over the cooking that was permitted and over who could stay in each apartment) of a degree much beyond that usual when a flat is let to a tenant on a normal lease: and in addition the tenant company's furniture was in every room. Is there anything in the Act of 1954 which precludes the court from giving to the word "occupied" in section 23 its natural and ordinary meaning in the context of the subject matter of that Act — a meaning which would in the set of circumstances above described clearly lead to it being held that the tenant did occupy the premises for the purpose of the business? Being unable to find anything in the Act which so precludes the court, I have come to the conclusion that this tenant did so occupy the whole of the premises.

Sachs LJ (at p 214) went on to explain that his approach to the issue, looking at the substance of the position, involved:

> much diminishing — in most cases eliminating — the weight to be attached to the form of the agreement between the tenant and an occupant (eg whether in law it constitutes a subtenancy or is only a licence . . .).

At pp 214 and 215 he distinguished the *Bagettes* case on the grounds that:

> it was conceded by the tenant that the subtenancies of the flats put the premises outside the Act of 1954 — so the relevant point was never argued,

and on the further ground that:

> nothing in the report of that case shows the length of the subtenancies and in addition it is to be noted that the tenant was apparently under no obligation to provide the subtenants with any services.

Summing up the position (at p 215) Sachs LJ said:

> the court must look at the substance of the position as a whole, taking into account the various elements which have been discussed and then come to a common sense conclusion as to whether the tenant "occupies" the premises for the purposes of his business . . .

Stamp LJ, agreeing in the result, distinguished the *Bagettes* case on its facts, pointing out (at p 217) that in that case:

> the tenant did not by his servants enter the flats which had been let to carry on the business which one finds here of providing services to the tenants.

In *William Boyer & Sons Ltd* v *Adams* [1975] 32 P&CR 89 the defendant, who owned a farmhouse and outbuildings, lived in the farmhouse and sublet the outbuildings in a number of units for the purposes of a business. He claimed that he occupied the whole of the premises, including the units, for the purpose of carrying on the business of subletting and providing facilities and services to enable his subtenants to engage in light industry. Though he did not devote the whole of his time to the business, he was always on the premises between 10 and 11 in the morning, his wife was generally there when he was not and when they went away on holiday they arranged for someone to be there. The defendant provided a number of services for the occupants of the units, for which he had to enter, and did enter, the units. Templeman J upheld his claim, saying (at pp 92-93):

> The activities of the defendant are sufficient to show that he is not so much acting as a landlord passively receiving rent but as the manager of a business actively earning profits by providing accommodation, facilities and services and by devoting time for this purpose. Some degree of presence by the defendant in the units is provided by his central heating and other fixtures, by the need for him to enter the units to maintain the services and facilities which he provides for an inclusive rent and by his nightly visits. The organisation of the complex which enables the tenants of the units to carry on their work requires a manager and not only a landlord.

Neither of the two last-mentioned decisions appears to have been cited to this court in *Hancock & Willis* v *GMS Syndicate Ltd* (1982) 265 Estates Gazette 473. It is not necessary to refer to the facts of this case, which were far removed from those of the present case. It will suffice to refer to the following statement of principle by Eveleigh LJ, who delivered the leading judgment. Having referred to the relevant words of section 23(1) of the 1954 Act, he said (at p 474):

> The words with which we are concerned import, in my judgment, an element of control and user and they involve the notion of physical occupation. That does not mean physical occupation every minute of the day, provided the right to occupy continues. But it is necessary for the judge trying the case to assess the whole situation where the element of control and use may exist in variable degrees. At the end of the day it is a question of fact for the tribunal to decide, treating the words as ordinary words in the way in which I have referred to them.

Finally, we were referred to *Linden* v *Department of Health and Social Security* [1986] 1 WLR 164*. In that case Scott J, though he thought it was "on the borderline", held that the flats sublet to employees in the Health Service were "occupied" by the health authority for "purposes of a Government department" within the meaning of section 56(3) of the 1954 Act. He said (at pp 174-175):

> The features of the case which lead me to that conclusion are, particularly, that one or more of the flats will usually be found to be vacant and under the control of the authority; that the authority retains keys to all flats and visits them regularly; that under the agreements exclusive possession is not given to the occupants; that the authority provides not simply furniture, but crockery and blankets at the premises; that the authority's maintenance of the flats includes decorating and the carrying out of quite trivial repairs. The evidence shows, in my view, that 40, Nottingham Place, like the units with which Templeman J was concerned, needs not just a landlord but a manager. On the evidence that is the function discharged by the authority.

In the present case the learned judge, having summarised the issues and having stated that some of the garages were let out and some were not, said this:

> I have no evidence of the turnover of the tenant's clientele but it is clear that the applicants have from time to time experienced letting problems. The applicants are undoubtedly in rateable occupation of the whole site and I find that not totally irrelevant. They are responsible for the whole site. They are factually in occupation of one garage which they use as a store room for this and some 100 to 120 similar sites around the country. It is one of six such store G
> rooms. That is the tenant's business: the hiring out of lock-up garage space. I accept the evidence of Mr Potter called on behalf of the applicants. He seemed to me to be an honest, sincere and efficient manager. I accept that the services he and his labour force provide are as he said. There is no doubt that the applicants are carrying on a business on the site, but are they in occupation within the meaning of section 23 of the Landlord and Tenant Act 1954?

The learned judge then referred to the four decisions of this court mentioned earlier in this judgment, and concluded:

> All those cases are binding on me. It seems to me that I have to put myself in the position of a jury, and say "Here are the facts. What is the proper conclusion from the facts?" The facts here are very different from the *Bagettes* case. That was a residential property with all sorts of covenants affecting the situation. This is a straightforward case where the applicants let H
> out the garage space and they undertake to clean, maintain and service the site as a whole. I have to look at the totality of the operation. They do what is required by the nature of the business. They have access keys to all the garages. They appear to maintain a fairly close watch on the garages through themselves and their agents whilst seeking to maintain a measure of privacy for the occupants. It seems to me that when you stand back and look at this this is garage space with a measure of security. The applicants maintain an active maintenance and control as Mr Potter has said; not a passive role, the mere collection of rent. On the totality of the evidence, I would be wrong to say that they have parted with possession of the site. Bearing in mind the purpose for which the land was let — it is a matter of fact and degree — and trying to balance, as I must, the case, I find the applicants have retained a sufficient measure of occupation to justify an application under the Landlord and Tenant Act 1954. J

The learned judge thus accepted the evidence of Mr Potter in its entirety.

If the *Bagettes* decision had stood alone, I think that Trans-Britannia's claim that it was in occupation of the whole premises would have been almost unarguable. However, in the light of the subsequent decisions in the *Lee-Verhulst* and *William Boyer* cases, I accept that the mere fact that a majority of the garages were sublet would not by itself inevitably prevent Trans-Britannia from being in occupation of the whole of the premises for the purposes of their business. I further accept Mr Geddes' submission, on behalf of Trans-Britannia, that in the light of the guidance given by this court in the *Lee-Verhulst* case, the court has to look at all the facts, K
including the time and attention devoted by Trans-Britannia to running their business, so far as it affects those premises, to the services rendered to occupants and to the control exercised over the occupancies (see *ibid* at p 215 F, per Sachs LJ), and then has to come to a "commonsense" conclusion, having regard to the natural and ordinary meaning of the word "occupied" in section 23, in the context of the subject-matter of the 1954 Act (see *ibid* at pp 213 E and 215 E).

In the forefront of his argument, Mr Geddes contended that the nature of the services provided by a tenant to his subtenants or licensees must be looked at in the light of the nature of the particular business which he conducts. True it is that the nature of the services provided by Trans-Britannia to their subtenants are less extensive L
than those which were provided by the tenants to the subtenants in the *William Boyer* case and much less extensive than those provided to the subtenants in the *Lee-Verhulst* case. However, in his submission, this is inherent in the nature of the relevant business. All reasonable services, he contended, are provided by Trans-Britannia to the subtenants of their garages, most particularly by way of repair and maintenance, to which work, on the evidence, none of the tenants ever objects. In his submission, there is no reason to suppose that there was insufficient evidence to support the learned judge's decision or that he misdirected himself either as to law or as to fact.

This court would be slow to interfere with the finding of the learned judge on a pure finding of fact. However, his conclusion in this case was one of mixed fact and law. With all respect to him and to Mr M
Geddes' well-presented argument in support of it, I think this conclusion was erroneous.

I turn to consider the facts upon which it was based. The learned judge made the following express findings of primary fact:

(1) Trans-Britannia are in rateable occupation of the whole site.

(2) They are factually in occupation of one garage which they use as a store.

(3) They let out the garage space and undertake to clean, maintain and service the site.

(4) They have access keys to all the garages.

(5) They appear to maintain "a fairly close watch" on the garages through themselves and their agents.

*Editor's note: Also reported at p 108 *ante* and (1985) 277 EG 543.

A Since he also accepted the evidence of Mr Potter in totality, he must also be taken to have accepted the further facts stated by Mr Potter in the material passages of his evidence quoted above.

Occupation for the purposes of rating is a legal concept all on its own, so that I do not think that (1) above lends Trans-Britannia's case any significant support. As to (3) above, the terms of their lease (clause 2 (6)) place Trans-Britannia under an obligation to Darby to keep the site "neat and tidy". Mr Potter's evidence showed that they clean and maintain the area around the garages, including the access roads. However, I have been able to find no evidence that they ever expressly agreed with their subtenants to clean, maintain and service the site. Mr Potter's evidence showed that from time to time they repair and maintain garages following complaints from subtenants B and that they clean out, and if necessary repair, garages before they are relet. However, so far as can be discovered from the notes of his evidence, he gave no evidence, or no clear evidence, that Trans-Britannia ever claimed or exercised any right to enter lock-up garages during the subsistence of subtenancies, save with the consent of the subtenants, either for the purpose of effecting repairs or for any other purpose. In other words, it would appear from the evidence that each subtenant enjoyed exclusive possession of his garage. As to (5) above, the extent of any regular "close watch" on the site by Trans-Britannia appears from Mr Potter's evidence to have been limited to a visit by their employees about once a fortnight.

I believe I have mentioned all the primary facts relied on by Trans- C Britannia in supporting the learned judge's conclusion. They certainly support the conclusion that if the premises are occupied by Trans-Britannia, they are occupied "for the purposes of their business". This much is accepted by Mr Carey on behalf of Darby. The burden of his argument has, of course, been that they do not "occupy" the premises in the relevant sense.

Against these facts, which are said to support Trans-Britannia's case, have to be set (*inter alia*) the following. There is no living accommodation on the site. So far as the evidence shows, there is no office, no water and no electricity. There is no security guard. Regular visits to the premises occur no more than once a fortnight. Mr Geddes stressed that Trans-Britannia provide all the services for their subtenants which it is reasonable for them to provide, having D regard to the nature of their business. This may well be so. However, respectfully differing from his submissions and the views of the learned judge on this point, I do not think it lends much assistance to Trans-Britannia's case to say, in the judge's words, "they do what is required by the nature of their business". The very nature of some businesses involves the owner parting with the occupation of his land (see the *Bagettes* case (*supra*) at p 302, per Jenkins LJ). Though the learned judge referred to Trans-Britannia as maintaining "active control" of the premises, I think that the very nature of their business and the manner in which they conducted it meant that this control was of a very limited nature.

I have referred above to the facts of the *Lee-Verhulst* and *William* E *Boyer* cases in some detail if only, in the end, to show how far removed they were from those of the present case. Without thinking it necessary to list the many points of distinction, I merely say that in both cases the physical presence of the tenants on the premises, the services they rendered to their subtenants or licensees, the degree of control which they exercised over their occupancies and the time which they spent in connection with the relevant premises were each incomparably more extensive than the equivalent factors in the present case. In both those cases, the tenants were regularly and frequently entering upon every part of the premises in question. In the present case, as I have already indicated, there is no clear evidence that Trans-Britannia ever enter lock-up garages during the substance of subtenancies, save with the consent of the subtenants. For by far F the greater part of any year, they themselves maintain no physical presence on the premises at all.

Mr Geddes, as I have indicated, has accepted that Trans-Britannia must establish "occupation" of the whole premises in order to succeed in their claim.

In all the circumstances, with all respect to the learned judge, I think he misdirected himself in the course of purporting to apply the principles established by the four decisions to which he referred. The question raised by cases such as this must always be one of degree, but I for my part feel no doubt that Trans-Britannia's case falls on the wrong side of the line. In my judgment, applying the natural and ordinary meaning of the word "occupied" to the construction of section 23(1), the evidence points inexorably to one conclusion. This conclusion is that Trans-Britannia are not in occupation of the relevant land and garages within the meaning of section 23(1) of the 1954 Act and that accordingly they are not entitled to apply for a new business tenancy thereof. I would therefore allow this appeal and make a declaration to that effect.

CAULFIELD J agreed and did not add anything.

The appeal was allowed with costs in the Court of Appeal and below and declaration made as mentioned at the end of Slade LJ's judgment.

Court of Appeal

May 7 1986

(Before Lord Justice SLADE, Lord Justice CROOM-JOHNSON and Sir John MEGAW)

BRETT v BRETT ESSEX GOLF CLUB LTD AND ANOTHER

Estates Gazette June 28 1986

278 EG 1476-1483

Landlord and tenant — Construction of rent review clause in relation to question whether works carried out by tenants should be disregarded in ascertaining the open market rent — Appeal by landlord from decision of Judge Finlay QC, sitting as a High Court judge, who held, in favour of tenants, that the works in question were to be disregarded — Tenants were a golf club company — During the course of their original lease, granted in 1973, they had built a clubhouse on the land leased and, in addition to laying out an 18-hole golf course on that land, had laid out a 9-hole course on nearby land not then included in their demise — Subsequently, in 1978, this lease was surrendered and a new lease granted to the tenants which included the 9-hole course — This new lease was rectified by consent during the trial to cure the unintentional omission of provision for further rent reviews after the first — The review clause in the 1978 lease provided for the disregard, in ascertaining the open market rent, of "those matters set out in paragraphs (a), (b) and (c) of section 34 of the Landlord and Tenant Act 1954" — Judge Finlay decided that, as a matter of construction, the reference to section 34 in the lease was intended to be to that section before it was amended by section 1 of the Law of Property Act 1969, although the lease was granted after the amendment had taken effect — He held, however, that it did not follow from this that the principle of the House of Lords decision in *East Coast Amusement Co Ltd* v *British Transport Board* (the *Re "Wonderland", Cleethorpes,* case) applied so as to require the clubhouse and the 9-hole course to be taken into account (ie not to be disregarded) in determining the relevant open market rent — He reached this conclusion by construing the reference to section 34 in the context of the review clause in the lease, which differed from the statutory context of the application for a new tenancy under the 1954 Act, and by considering the factual matrix against which the lease was executed — Held by the Court of Appeal, allowing the landlord's appeal, that the judge was correct in construing the reference to section 34 as being to that section as originally enacted, but that he had not addressed himself to the two critical questions of construction — These were (a) whether the erection of the clubhouse and the creation of the 9-hole course were "improvements" within the meaning of section 34 and incorporated in the 1978 lease, and (b) whether they were carried out by the "tenant" within that meaning — The answer to the first question, which really disposed of the second also, was "no" — The word "improvements" in the context must mean improvements made to the premises demised by the 1978 lease and could not therefore include the clubhouse or the 9-hole course as these

were already part of the premises demised by the lease — "Tenant" meant the tenant of the premises so demised — The "matrix of facts" supported this view rather than the view taken by the judge — Appeal allowed

This was an appeal by the plaintiff landlord, John Dudley Brett, against the decision of Judge Finlay QC, sitting as a High Court Judge in the Chancery Division, in favour of the tenant company, Brett Essex Golf Club Ltd, on the construction of the rent review provisions of a lease granted by the landlord to the tenant company in 1978. The second defendant before Judge Finlay was National Westminster Bank plc to whom the premises had been charged.

Robert Pryor QC and Timothy RF Jennings (instructed by Bazley White & Co) appeared on behalf of the appellant landlord; Alan Steinfeld (instructed by Leonard Kasler & Co) represented the respondent tenants: the bank was not represented on the appeal.

The proceedings before Judge Finlay were reported at [1985] 1 EGLR 63.

Giving judgment, SLADE LJ said: This is an appeal by Mr John Dudley Brett, the plaintiff in an action, from a judgment of His Honour Judge Finlay, sitting as a judge of the Chancery Division, which was delivered on October 19 1984. This judgment concerned the construction of a rent review clause in a lease granted in 1978, and in particular the construction and effect of a reference in that clause to section 34 of the Landlord and Tenant Act 1954.

I think it will be convenient to begin this judgment by explaining briefly the statutory background to the two leases which will fall to be considered. Section 34 of the Landlord and Tenant Act 1954 is comprised in Part II of that Act, which gives security of tenure to business tenants. In its original form, so far as material, it read as follows:

The rent payable under a tenancy granted by order of the court under this Part of this Act shall be such as may be agreed between the landlord and the tenant or as, in default of such agreement, may be determined by the court to be that at which, having regard to the terms of the tenancy (other than those relating to rent), the holding might reasonably be expected to be let in the open market by a willing lessor, there being disregarded —
(a) any effect on rent of the fact that the tenant has or his predecessors in title have been in occupation of the holding,
(b) any goodwill attaching to the holding by reason of the carrying on thereat of the business of the tenant (whether by him or by a predecessor of his in that business),
(c) any effect on rent of any improvement carried out by the tenant or a predecessor in title of his otherwise than in pursuance of an obligation to his immediate landlord . . .

There is a para (d), but I do not think I need read it.

Para (c) of this section, in its original form, gave rise to certain problems of interpretation, as was illustrated by the decision of the House of Lords in the case of *In re "Wonderland", Cleethorpes*, which is reported under the name of *East Coast Amusement Co Ltd* v *British Transport Board* [1965] AC 58. In that case the appellant company had carried out certain works of improvement in 1926 under a lease granted in 1912 and subsequent leases. In 1938 the landlord granted to the company a new lease for 21 years from a date in 1939. Shortly before this lease expired the company served on the landlord a request for a new tenancy under section 26 of the 1954 Act. There was no counternotice. The question arose whether, by virtue of section 34(c) of the 1954 Act, the work done in 1926 should be disregarded for the purpose of determining the rent payable under a new tenancy. The House of Lords held that it should not. It held that section 34(c) referred only to improvements carried out by the tenant who was making the application for the tenancy, and effected during the term of the tenancy current when the application was made, or by a predecessor in title of his to the same tenancy.

A few years later, presumably having regard to this decision, section 34 of the 1954 Act was amended by the Law of Property Act 1969. Section 1(1) of that Act substituted an entirely new paragraph for para (c) and added to section 34 an entirely new subsection (2), by providing:

In section 34 of the Act of 1954 (rent under new tenancy) the following paragraph shall be substituted for paragraph (c) (improvements to be disregarded):-
"(c) Any effect on rent of an improvement to which this paragraph applies" and the following subsection shall be added (the present section, as amended by the foregoing provisions, becoming subsection (1)):-
"(2) Paragraph (c) of the foregoing subsection applies to any improvement carried out by a person who at the time it was carried out was the tenant, but only if it was carried out otherwise than in pursuance of an obligation of his immediate landlord and either it was carried out during the current tenancy or G
the following conditions are satisfied, that is to say, —
(a) that it was completed not more than twenty-one years before the application for the new tenancy was made; and
(b) that the holding or any part of it affected by the improvement has at all times since the completion of the improvement been comprised in tenancies of the description specified in section 23(1) of this Act; and
(c) that at the termination of each of those tenancies the tenant did not quit.

Against this statutory background, I now revert to the facts of the present case. By a lease dated August 13 1973 the plaintiff let certain parcels of land forming part of Clapgate Farm, Little Warley, Essex, which were defined in the lease as "the demised premises" to The Brett Essex Golf Club Ltd, the first defendant in these proceedings (which I shall call "the company"), for a term of 50 years from July 9 H
1973. It appears that at the time when the lease was granted the plaintiff held 750 out of the 1,000 issued £1 shares in the company, while a Mr Greene held the remaining 250 shares. The yearly rent payable under the lease was expressed by clause 1 to be £2,000 for the first two years of the term, £5,000 for the next five years, and for each successive five years, up to and including the 46th year, a rent to be agreed by the parties, or in default of agreement to be determined in accordance with the provisions of clause 2. The first period during which a reviewed rent would be payable would thus begin in July 1980.

Clause 2, so far as material, provided for such rent to be determined in accordance with the following formula, that is to say: J

. . . such rent shall be the rent (but not less than the rent reserved in the period of five years immediately expired) at which the demised premises might reasonably be expected to be let in the open market by a willing Landlord by a lease for a term equal to the residue of the term unexpired at the end of the period of five years immediately before expired with vacant possession on the same terms and subject to the same incidents in all other respects as this present demise and upon the supposition (if not a fact) that the Tenant had complied with all the repairing covenants herein imposed on the Tenant (but without prejudice to the rights and remedies of the Landlord in respect thereto) and there being disregarded (if applicable) those matters set out in paragraphs (a), (b) and (c) of Section 34 of the Landlord and Tenant Act 1954.

Clause 3(4) of the lease contained a covenant by the company

To lay out the demised premises as a golf course and not to erect at any time thereon any dwellinghouse or other building except such buildings as are K
authorised by Clause 4(2) (hereinafter referred to as "the permitted buildings") for the use of the club and the members and servants thereof and to make no other alteration in the character and general arrangement of the demised premises save as may be necessary or desirable for laying out and maintaining the same as a golf course and as may be approved for that purpose by the Landlord or his surveyor.

Clause 4(2) contained a covenant by the landlord with the tenant

that the Tenant shall be at liberty to erect at its own expense on the demised premises at the point marked 'A' on the said plan a clubhouse with all necessary offices and outbuildings for the use of the members and servants of the club.

I think there is nothing in the 1973 lease to suggest that there were any buildings on the land at the date when it was granted. L

After this grant the company not only laid out an 18-hole golf course on the demised land but built a clubhouse which was completed in about July of 1975. It also laid out a 9-hole golf course on some adjoining land, not included in the 1973 demise. That was completed at about the end of 1977.

In January 1978 or thereabouts, negotiations took place between the parties with regard to the proposed sale by the plaintiff of his 750 shares in the company to Mr Greene. At about the same time there was also negotiated a project for the prolongation of the term granted to the company and for the enlargement of the premises comprised in its demise by the inclusion of the 9-hole golf course. Eventually the sale of the shares was completed by a transfer of the plaintiff's M
shareholding to Mr Greene on February 17 1978. In the meantime, on February 14 1978, the plaintiff duly granted to the company an entirely new lease.

In that lease the company was defined as "the Tenant", which expression was stated to include, where the context so admitted, its successors in title. Clause 1 defined the subject-matter of the lease as:

ALL THOSE parcels of land (hereinafter called "the Demised Premises") forming part of Clapgate Farm aforesaid as the same are for the purpose of identification only delineated on the plan annexed hereto and thereon edged red TOGETHER WITH the buildings erected on parts thereof but excluding the lakes hatched blue.

The plan referred to made it plain that the 9-hole golf course was included in the property demised, as of course was the, by then

A already existing, clubhouse.

The lease was granted for a term of 50 years from February 14 1978, at a rent of £12,000 for the first three years, with provision for review at the end of the third year, that is to say in February 1981. It also provided for the tenant from time to time to pay to the landlord the amount which the landlord might expend in effectuating or maintaining the insurance of the demised premises. Clause 2(8) contained a covenant by the tenant

> to use the demised premises and such buildings as may be erected thereon as a private golf course and/or a country and sporting club with the requisite ancillary buildings for the convenience of members of the club and its staff only.

B The earlier clause 2(4), which was in somewhat different form from the corresponding covenant contained in clause 3(4) of the 1973 lease, contained a covenant by the tenant:

> Not to erect at any time thereon any other buildings and to make no other alteration in the character and general arrangement of the demised premises save as may be necessary or desirable for laying out and maintaining the same as a golf course and/or a country and sporting club with associated buildings and as may be approved for that purpose by the Landlord or his surveyor which approval shall not be unreasonably withheld.

The lease also contained, in clause 3, a covenant by the landlord to insure and keep insured the demised premises. It further provided that, in the event of any part of the demised premises at any time during the tenancy being damaged or destroyed by fire so as to be C unfit for habitation and use, then the rent thereby reserved, or a fair proportion thereof, should be suspended until the premises should again be rendered fit for habitation and use. It further provided that:

> . . . any dispute concerning this clause shall be determined by a single arbitrator in accordance with the Arbitration Act 1950 or any statutory modification or re-enactment thereof for the time being in force.

Clause 7 contained a general arbitration clause, which made a reference in similar terms to "the Arbitration Act 1950 or any statutory modification or re-enactment thereof for the time being in force".

Clause 6 provided:

> The provisions of Section 196 of the Law of Property Act 1925 as amended by the Recorded Delivery Service Act 1962 shall apply to any notice under this D lease.

When the 1978 lease was executed, it was apparently the intention of both the company and the plaintiff that it should provide for the rent to be reviewed, not only at the end of the third year but at each successive seventh year thereafter. But the 1978 lease, on its face, provided for review only at the end of the third year.

By a writ issued on June 3 1983 the plaintiff claimed rectification of the 1978 lease as against the company so as to provide for a review of the rent at the end of the third and each successive seventh year thereafter. The proceedings were subsequently amended so as to join as second defendant the National Westminster Bank plc, in as much as the company had charged the property to that bank by way of E mortgage. The bank has not been represented on this appeal.

By the amended statement of claim, rectification of the rent review clause in the 1978 lease was claimed against both defendants. The pleading also claimed a declaration as against the company that the clubhouse and the laying out of the 9-hole golf course were not to be disregarded in determining the relevant open market rental value for the purposes of clause 4 of the 1978 lease.

At the trial before the learned judge the rectification claim was disposed of by consent as against both defendants. By consent the court declared that the 1978 lease ought to be rectified in the manner specified in the schedule to the order and ordered that the lease should be rectified accordingly. It further declared that:

> F . . . the said lease as so rectified as aforesaid ought to take effect in like manner as if the said rectification had been made in the said lease at the time of the execution thereof by the parties thereto.

Clause 1 of the 1978 lease as rectified provided for the payment by the company for the first three years of the term of the yearly rent of £12,000 and

> for each next successive seven years of the said term and for the last five years of the said term (each of which successive periods is hereinafter respectively referred to as "the Review Period") a rent to be determined in accordance with the provision in that behalf contained in clause 4 hereof (hereinafter when specifically referred to, called "the reviewed rent").

The first period during which a reviewed rent was to be payable under the 1978 lease was thus to begin in February 1981, a little later than the start of the first review period provided for by the 1978 lease.

Clause 4 as amended provided for the reviewed rent to be the higher of the first reserved rent and the open market rental value of the demised premises for the relevant review period and continued:

> (1) The expression the open market rental value as aforesaid means a sum in relation to the review period determined in manner hereinafter provided as being at the time of such determination the annual rental value of the demised premises in the open market on a lease for a term of seven years certain save in the case of the last review period, in which case the appropriate term shall be a term of five years certain with vacant possession at the commencement of the term but upon the supposition (if not a fact) that the tenant has complied with the obligations as to repair and decoration herein imposed on the tenant such lease being on the same terms and conditions other than as to the amount of rent and the length of the term as are herein contained without the payment of any fine or premium and disregarding (if applicable) those matters set out in paragraphs (a), (b) and (c) of section 34 of the Landlord and Tenant Act 1954.

I pause to comment that by the time the 1973 and 1978 leases were executed the 1954 Act had of course been amended.

Clause 4(2) of the 1978 lease as rectified contained provisions for the ascertainment of the open market rental value, which I do not think I need read. Henceforth in this judgment all references to the 1978 lease are intended as references to that lease as rectified.

Despite the compromise of the claim for rectification, the learned judge still had to decide the question of construction raised by the statement of claim. This turned on the construction and effect of the reference in clause 4(1) of the 1978 lease to paras (a), (b) and (c) of section 34 of the Landlord and Tenant Act 1954.

The trial proceeded, I understand, without oral evidence, though a number of the background facts to which I have referred, or to which I shall refer hereafter, appear to have been common ground on the basis of the documents and the pleadings. The learned judge in his judgment in effect reached two main conclusions. First, he accepted a submission made on behalf of the plaintiff that clause 4(1) of the 1978 lease, in referring to the three paras (a), (b) and (c), was referring to the unamended version of section 34 and not to the version as amended by the 1969 Act. Second, however, he accepted a submission made on behalf of the company that, when the references in clause 4(1) to paras (a), (b) and (c) of the 1954 Act are properly construed in their context, having regard to the factual matrix against which the 1978 lease was executed, they require the clubhouse to be disregarded in determining the relevant open market rental value for the purposes of the 1978 lease, even though the clubhouse was erected before the 1978 lease was executed. Though I understand that submissions had been made to him on behalf of the company to the effect that the 9-hole golf course likewise fell to be disregarded for this purpose, the learned judge did not specifically advert to this argument in his judgment; indeed, I think he did not refer to the 9-hole golf course at all. However, we have been told that when this point was drawn to his attention after he had delivered his judgment, he indicated that he thought that the 9-hole golf course fell to be treated in the same way as the clubhouse. He accordingly made a formal declaration:

> That upon the true construction of the said lease (1) The clubhouse and (2) the laying out of the nine-hole golf course in the said Writ of Summons [sic] are to be disregarded in determining the relevant open market rental value for the purposes of the said lease dated 14th February 1978 mentioned therein.

By his notice of appeal the plaintiff in effect challenges the correctness of this declaration. The company, by a respondents' notice, disputes the learned judge's holding that the reference in clause 1 of the 1978 lease to section 34 of the Landlord and Tenant Act 1954 constituted a reference to that section in its unamended form.

Logically, I think the first point which falls to be considered is that raised by the respondents' notice, so I propose to consider it first. As the learned judge said, there is no direct authority on the question whether a reference in a lease, granted after the passing of the amending Act of 1969, to section 34 of the Landlord and Tenant Act 1954 or to provisions contained in that section, is to be construed as a reference to the provisions contained in the 1954 Act as originally passed, or to that Act as amended in 1969. In one case, *Euston Centre Properties Ltd* v *H & J Wilson Ltd*, reported in (1981) 262 EG 1079, Cantley J had to consider the effect of a rent review clause which required to be disregarded "the matters referred to in section 34(a), (b) and (c) of the Landlord and Tenant Act 1954". He seems to have assumed that this provision referred to the Act of 1954 as amended. However, since the point does not appear to have been argued before him, and he gave no reasons for his assumption, I agree with Judge Finlay that the decision cannot be regarded as guiding authority for present purposes; and in any event, of course, it is not binding upon this court.

The reasons why the learned judge himself concluded that the reference in clause 4(1) of the 1978 lease is to the unamended Act of 1954 are to be found at p 9 of his judgment:

> It will be noted that the paragraphs are referred to as paragraphs of section 34, not as paragraphs of section 34(1), which is what they would be if it were the 1969 Act that was referred to. Subsection (2) of the amended Act also contains paragraphs (a), (b) and (c), but it is the case the paragraphs so designated in the second subsection of the amended section 34 are in the nature of conditions rather than matters. But there is this further significant difficulty, that paragraph (c) in section 34(1) of the amended Act is in these terms: "any effect on rent of an improvement to which this paragraph applies, and when one looks at subsection (2) to determine to what paragraph (c) is referring, one finds the improvements to which the paragraph applies must be improvements in relation to which the three conditions designated (a), (b) and (c) in subsection (2) are satisfied, and the first of those conditions is "(a) that it" — that is the improvement — "was completed not more than 21 years before the application for the new tenancy was made". It appears to me to involve a considerable degree of imaginative construction to ascertain how that first condition can be made applicable to a provision in a lease which is not concerned in any way with applications made under statute but is simply concerned with a review of the rent. On that ground alone, fortified by the further ground that there is no reference to subsection (1) in the wording of the lease, I come to the conclusion that it is to the unamended Act that the lease is referring, when it refers to paragraphs (a), (b) and (c) of section 34 of the Landlord and Tenant Act 1954.

Mr Steinfeld, in support of his respondents' notice, submits that the more natural construction of the reference to section 34 in clause 4(1) of the 1978 lease, is a reference to that section as it stood at the date of the grant of that lease, February 14 1978, and thus to the section as expressly substituted by section 1 of the 1969 amending Act. In his submission the reference in clause 4(1) to paras (a), (b) and (c) of section 34 is, by necessary inference, a reference to the corresponding paragraphs of subsection (1) of section 34 as amended. In his submission, therefore, no significance need, or should, be attached to the omission of any specific reference to subsection (1) in clause 4(1) of the 1978 lease. Condition (a) of the amended subsection (2), he suggests, is not incapable of being applied to a rent review clause in a lease; if the parties incorporate in such a clause paras (a), (b) and (c) of subsection (1) of the amended section 34, as he says they did here, then they must necessarily have intended them to be subject to such modification of the statutory language as would enable them to be applied in the context of such a clause. On this basis he suggests that the learned judge ought to have held that para (c) of the amended subsection (1) was to be treated as incorporated in clause 4(1) of the 1978 lease subject to the modification that the reference in para (a) of subsection (2) to the date when the application for the new tenancy was made must be treated as a reference to the date which, for the purposes of the rent review provision in the lease, most nearly equates with the date of application by a tenant for a new tenancy under the 1954 Act. In his respondents' notice he suggested two alternative candidates for this date, that is to say, first, the date from which the reviewed rent was to take effect or, second, the date of the tenant's counternotice under clause 4 of the lease. In argument, however, he submitted that the second date was the correct one.

For my part, I agree with Mr Steinfeld that in many, perhaps the majority of, cases where parties to a written contract have incorporated in it a reference to a statute which has been amended, it may be reasonable to impute to them an intention to refer to the statute in its amended form. However, no authority has been cited to us which suggests that under the ordinary law there is any presumption of construction to this effect.

Mr Pryor, on behalf of the plaintiff, has drawn our attention to section 20(2) of the Interpretation Act 1978, which expressly provides that:

> Where an Act refers to an enactment, the reference, unless the contrary intention appears, is a reference to that enactment as amended . . .

However, as he pointed out, the very fact that Parliament thought it necessary to enact this new subsection seems to lend some support to the view that, under the ordinary law, there is no presumption of construction that a bare reference in a contract to an Act of Parliament which happens to have been amended is a reference to the Act in its subsequently amended form. Whether it is to be read in this sense, or as a reference to the Act in its original form, must, in my judgment, depend entirely on the context. In the present case the following points, most of which were referred to by the learned judge, seem to me to lend compelling support to his conclusion that G
the intended reference is to the unamended Act of 1954.

First, the reference in clause 4(1) of the amended 1978 lease to "those matters set out in paras (a), (b) and (c) of section 34 of the Landlord and Tenant Act 1954" *simpliciter* would be entirely accurate if intended as a reference to the matters set out in paras (a), (b) and (c) of the unamended section. It would not, however, be accurate if intended as a reference to the matters set out in those paragraphs of the *amended* section, since the amended section contains two sets of paras (a), (b) and (c), and furthermore the reference should have been an express reference to subsection (1) of section 34. Second, the unamended three paragraphs can be readily incorporated into a rent review clause; on the other hand the wording H
of the amended subsections (1) and (2), as the learned judge pointed out, could not be incorporated into the 1978 lease without "a considerable degree of imaginative construction". In the course of any attempt at such incorporation, one particular difficulty would lie in determining the date on which the 21-year period referred to in subsection (2)(a) was to begin. In my view, Mr Steinfeld's ingenuity has not been able to overcome this difficulty.

Third — and this point follows from the second — I think that, if the draftsman of the 1978 lease had known of the amendments to section 34 of the 1954 Act and had intended to refer to the section as amended, he could reasonably have been expected to specify in greater detail the manner in which the relevant provisions of the amended section were to be applied in the operation of the rent review J
clause. Finally, it is to be observed that in clause 6 of the 1978 lease the draftsman took care specifically to refer to the amended version of section 196 of the Law of Property Act and, in clauses 3(3), 4(2) and 7, to any statutory modification of the Arbitration Act 1950 for the time being in force. I think that if he had had the knowledge and intention to which I have referred he could, in relation to section 34, reasonably have been expected to take a similar course by referring expressly to the amended version of that section.

Taken together, these points lead me to agree with the learned judge that, on the true construction of clause 4(1) of the 1978 lease, the determination of the "open market rental value" of the premises has to be carried out "disregarding if applicable those matters set out in paragraphs (a), (b) and (c)" of section 34 of the 1954 Act *as* K
originally enacted. On this basis, what is the true construction of the clause when applied to the facts of this case?

Mr Steinfeld, who made it clear that the submissions made in his respondents' notice were not his principal submissions in resisting this appeal, has advanced an attractively simple alternative argument in support of the learned judge's judgment. I hope and think that it can be fairly summarised as follows: the relevant words in clause 4(1) constitute simply a shorthand method of incorporating in the subclause, as matters to be disregarded, the matters set out in paras (a), (b) and (c) of section 34. That subclause can and should accordingly be read as though it contained an express provision directing disregard of, *inter alia*, "any effect on rent of any L
improvement carried out by the tenant or a predecessor in title of his otherwise than in pursuance of an obligation to his immediate landlord". It is these words as incorporated in the subclause that fall to be construed and not the same words in the context of the statutory scheme from which they have been lifted. In the *"Wonderland", Cleethorpes* case, he reminded us, the House of Lords was concerned with the construction of para (c) of section 34 in the context of an application by a tenant for a new tenancy under the 1954 Act, following the expiry of the previous tenancy. In his submission it does not follow that the same words must be given the same construction in the different context of the 1978 lease. As will appear a little later, I think that the speeches in the *"Wonderland", Cleethorpes* case give us helpful guidance by analogy in deciding the present issues. Subject M
to this qualification, I would accept in their entirety these submissions of Mr Steinfeld up to this point, as indeed I think they were accepted by the learned judge.

His argument proceeded on the following lines. The company is "the tenant", both in fact and by definition in the 1978 lease. The improvements in question were carried out by the company. It is common ground that the company did not erect the clubhouse or the 9-hole golf course in pursuance of any obligation to its landlord. In these circumstances, Mr Steinfeld submitted, there is no escape from the conclusion that clause 4(1), on its literal wording, requires the two improvements in question to be disregarded. He sought to derive some support for this conclusion from the recent decision of Scott J in the case of *Hambros Bank Executor & Trustee Co Ltd* v *Superdrug Stores Ltd*, reported in [1985] 1 EGLR 99,* in which that learned judge, in the course of construing a rent review clause in a lease, attached importance to the fact that the lease in question defined the defendants as "the tenant".

Mr Steinfeld accepted that the construction of the word "improvements" is also important in the construction of clause 4(1) of the 1978 lease. In his submission, however, there is no sufficient reason to confine the operation of this word to improvements carried out by the company during the subsistence of the 1978 lease. They are likewise, in his submission, perfectly apt to include improvements carried out during the subsistence of the 1973 lease. Support for this conclusion is, in his submission, to be derived when one considers the construction of the 1978 lease in the light of the factual matrix of the transaction and of the principles expounded by Lord Wilberforce in *Prenn* v *Simmonds* [1971] 1 WLR 1381. The 1978 lease was not being granted on the expiry of an old lease; as the facts showed, in Mr Steinfeld's submission, it was being granted substantially as a replacement of the 1973 lease on altered terms. This makes it all the more likely, he suggested, that improvements carried out during the subsistence of the 1973 lease would fall to be disregarded.

As I have already indicated, I accept that the decision of the House of Lords in the *"Wonderland", Cleethorpes* case as to the true construction of section 34 of the original 1954 Act is not decisive of the question: what is meant by words taken from section 34 and incorporated in the 1978 lease by reference? Nevertheless, I think it is clearly relevant to consider the reasoning which led the House of Lords to their conclusion, because so many arguments rather similar to those advanced on behalf of the company in the present case were rejected by them on that occasion.

Lord Simonds, with whose speech Lord Reid, Lord Hodson and Lord Pearce concurred, pointed out ([1965] AC 58 at p 70) that the immediately relevant clause of section 34 was (c) and that the question was whether the relevant works were "improvements" carried out by "the tenant" or a "predecessor in title of his" within the meaning of that clause. He referred to an argument on behalf of the appellant company in these words:

> It is urged on the one hand by the appellant company that they were at the date of their application under section 24 the tenants of the premises and that it was they who had in fact carried out the improvements in question: therefore, they say, the language of the section is unambiguously apt to cover their case.

However, he did not accept this argument. Later, at p 70, he said:

> For me, the critical question is what is meant by "the tenant"? and I reject the premise upon which the appellants must rely that it includes the individual tenant in any other capacity than that in which he makes his application to the court, ie as tenant under the current tenancy which he seeks to have renewed. Part II of the Act (within which section 34 falls) begins and ends with the tenancy then current which is to be extended or renewed. When that tenancy was created, a new relation was established between landlord and tenant. It is to be assumed that, when it was established and the old chapter, if there was one, was closed, the parties took into account what were then their respective rights and liabilities and founded on them accordingly. A new chapter then began and it is only with what thereafter happened that the court is concerned. This is perhaps only another way of saying that in the context the words "carried out by the tenant" import carried out during the current tenancy, but I arrive at that conclusion upon a consideration of the meaning which must be given to "the tenant".

Lord Simonds further referred to an argument advanced on behalf of the respondent that the same result could be reached by having regard to the meaning of the word "improvement", which it was suggested could only refer to something done after the start of the current tenancy. As to this, he said at p 71:

> It may well be that in the context of section 34 (c) this is a valid argument, but I prefer not to rest on it, for it is easy to see what difficulties might arise if it was sought to adopt such a meaning generally in an interpretation of the Act.

Lord Morris, however, appears to have found the last mentioned argument compelling. He said at p 74:

> . . . the tenancy that is now continuing is the current tenancy which was created by the lease of November 4 1938, and which began on March 25 1939. The "holding" (see section 23) is the property comprised in that tenancy. That must be the property as it existed at the beginning of the current tenancy together with any physical improvements carried out during the tenancy. The "tenant" who is referred to in section 34(c) is the tenant of that property. If what is to be regarded as the effect from the rent point of view of any improvements carried out by the tenant, that, I consider, denotes any improvements carried out by him in respect of the property comprised in the current tenancy. That involves a consideration as to whether between the start and the end of the period of the current tenancy improvements have been carried out by him. It seems to me, therefore, that the reference in section 34, sub-section (c), is a reference only to works carried out during the currency of the current lease.

As in the *"Wonderland", Cleethorpes* case, the critical questions of construction seem to me to be (a) whether the relevant works — that is to say, the erection of the clubhouse and the creation of the 9-hole golf course — were "improvements" within the meaning of the words taken from section 34 and incorporated by reference in clause 4(1) and, if so, (b) whether they were carried out "by the tenant" within that meaning.

In my judgment the first of the two questions can only be answered in the negative for these short reasons. In the context of the rectified clause 4(1) (including the incorporated words), the word "improvements" must mean improvements to the demised premises. Yet, in the context of that subclause the reference to "the demised premises" clearly includes the clubhouse and the 9-hole golf course, which were part of the subject-matter comprised in the original grant under the 1978 lease. To construe clause 4(1) as requiring the disregard of improvements to the demised premises, consisting of the erection of the clubhouse and the creation of the 9-hole golf course, when the original demised premises themselves included the already erected clubhouse and the already created 9-hole golf course, would not seem to me to make sense. Even though the rest of their lordships may not have found the reasoning of Lord Morris entirely compelling in the construction of section 34, it seems to me that, *mutatis mutandis*, it is compelling in the construction of clause 4(1) when the wording of paras (a), (b) and (c) of section 34 are incorporated by reference in that subclause. The word "holding" in those paragraphs has to be treated as a reference to the demised premises. The demised premises are the property comprised in the 1978 lease. That must be the property comprised in the 1978 lease as it existed at the beginning of the term granted by that lease. The "tenant" who is referred to is the tenant of that property, ie all the property comprised in that lease, including the clubhouse and the 9-hole golf course. If what is to be disregarded is the effect, from the rent point of view, of any improvements carried out by "the tenant", that must denote any improvements carried out by him in respect of the property comprised in the current tenancy (ie the demised premises including the clubhouse and the 9-hole golf course). That involves a consideration as to whether, between the start of the term granted by the 1978 lease and the time of the review, improvements have been carried out by the tenant. It seems to me, therefore, that unless the matrix compels a contrary conclusion, the reference to para (c) of section 34 when incorporated in clause 4(1) is in its context apt to refer only to works carried out during the currency of the 1978 lease and not before.

This makes it unnecessary to say much more about the meaning of the word "tenant" in the context of clause 4(1) when the wording of para (c) is incorporated in it. Suffice it to say that I do not regard the word "tenant" as unambiguous in this context any more than Lord Simonds regarded it as unambiguous in the context of section 34. In my opinion, as in that case, the context imperatively demands that some qualification be placed on the words, the obvious qualification in the present context being that the reference to "the tenant" means the tenant in his capacity as tenant of the demised premises under the 1978 lease. The decision of Scott J in the *Hambros Bank* case is, in my view, plainly distinguishable on its facts.

Lord Wilberforce in *Prenn* v *Simmonds* [1971] 1 WLR 1381 reminded us, at p1383, that:

> . . . the time was long passed when agreements, even those under seal, were isolated from the matrix of facts in which they were set and interpreted purely on internal linguistic considerations.

In the light of this reminder, I end by considering briefly whether the matrix of facts leads to a conclusion as to the true construction of clause 4(1) different from that which I would reach on the basis of the words used.

The learned judge himself considered that the matrix of facts reinforced the conclusion that the relevant works fell to be disregarded in the application of the rent review clause. He commented:

> Here the improvement was carried out by the tenant holding the premises of

*Editor's note: first reported at (1985) 274 EG 590.

the same landlord; it was carried out during the currency of a term which at the time when the improvements were effected had some 48 years to run. That term was within a few years surrendered by operation of law when the same landlord granted to the same tenant a new lease at a different rent and with some slight changes of the terms of the lease particularly as to rent review, for another term of 50 years, thus extending the original term in effect for a further five years. In that matrix of fact I find support for and nothing repugnant to the adoption of the construction which I place upon paragraph (c) as incorporated in the lease . . .

Mr Steinfeld, as I have already indicated, in support of the reasoning of the learned judge, stressed that the 1978 lease was executed in the special context of a surrender of the 1973 lease. In my opinion, however, these points, which are said to favour the company's construction of the clause, are counterbalanced by a number of factors pointing in the opposite direction. As Mr Pryor pointed out, the 1978 lease conferred a number of benefits on the company which it had not previously enjoyed under the 1973 lease, and which could be said to have made it not unreasonable that the relevant improvements should be taken into account when a reviewed rent came to be fixed.

The term granted by the 1973 lease was extended by nearly five years. The 1978 lease provided for longer review periods (seven years, as opposed to five years). The hypothetical term to be valued on a review of rent under the 1978 lease was shorter than that provided for by the 1973 lease (seven years, as opposed to the residue of the term). The 1978 lease contained new provisions for the insurance of the demised premises, and in the event of damage to them, suspension of the rent until they should be rendered fit for habitation. Finally, the execution of the 1978 lease was part and parcel of an arrangement under which the plaintiff landlord agreed to give up his controlling interest in the tenant company.

In all the circumstances, I think that the matrix of fact, so far as known to us, is at the very least equally consistent with the intention that the clubhouse and the 9-hole golf course should be taken into account in fixing the reviewed rent as with the intention that they should be disregarded.

For my part, I would go a little further. I think that with all respect **D**
to him, the learned judge, when delivering his judgment, either overlooked or attached too little weight to the particular fact that the 1978 lease included a substantial area of property which had not been included in the earlier demise (the 9-hole golf course). *Mutatis mutandis*, some of the observations of Lord Simonds, which I have quoted, appear to me to apply in the present case. Looking at the matter broadly, it seems to me that when the 1978 lease was granted, a new relation was established between the plaintiff and the company. The intention reasonably to be imputed to the parties is that when it was established the old chapter, constituted by the 1973 lease, was closed, and the parties, in entering into the 1978 lease, took into account what were then their respective rights and liabilities, **E**
contractual and statutory, and proceeded accordingly. A new chapter then began, and in my judgment the intention properly to be imputed to them is that only improvements effected by the tenant to the demised premises during the course of this new chapter would fall to be disregarded for the purpose of a futurc rent review.

For all these reasons, based primarily upon the linguistic interpretation of the 1978 lease, but, in my view, supported by the matrix, I would allow this appeal. I would set aside that part of the judgment of the learned judge which dealt witih the construction of the lease. In lieu thereof, I would order and declare that, upon the true construction of the lease, the clubhouse and the laying out of the 9-hole golf course mentioned in the pleadings are not to be disregarded in determining the relevant open market rental value for **F**
the purposes of the 1978 lease.

CROOM-JOHNSON LJ and SIR JOHN MEGAW agreed and did not add anything.

The appeal was allowed with costs in the Court of Appeal and below. It was declared that on the true construction of the lease the clubhouse and the laying out of the 9-hole golf course were not to be disregarded in determining the rental value for the purpose of the review clause.

For a further case on this subject see p 244

LANDLORD AND TENANT

LEASEHOLD REFORM

Court of Appeal

March 4 1986

(Before Lord Justice FOX, Lord Justice MUSTILL and Lord Justice STOCKER)

MOSLEY AND OTHERS v HICKMAN
SAME v HAGAN
SAME v FRANCIS

Estates Gazette May 17 1986

278 EG 728-732

Leasehold enfranchisement — Leasehold Reform Act 1967 as amended — Appeals by lessors by case stated from decision of Lands Tribunal on three references — Question in each case was as to whether the price payable for the freehold was to be calculated subject to an extended lease granted under section 14 of the 1967 Act — Lands Tribunal (Mr W H Rees FRICS) decided, in accordance with the lessees' submissions, that the price should be so calculated — Rateable values on April 1 1973 were over £1,000 and extended leases had been granted in each case expiring in the year 2047 — The main issue was as to the meaning of the words "subject to the tenancy" in section 9(1A)(a) of the 1967 Act as amended by section 118(4) of the Housing Act 1974 — It was argued on behalf of the lessors that if the words referred to the existing tenancies, that is the new terms created by the extended leases which were substituted for the original tenancies, then an unacceptable anomaly arose — It meant that a lessee who chose to take an extended lease before making his request for the freehold would pay a substantially lower price than the lessee who had not done so — In the present cases the prices would be £45,000, £56,000 and £49,500, as compared with £98,000, £137,500 and £111,000 — Court of Appeal, however, upholding the decision of the Lands Tribunal, held that the words "the tenancy" in section 9(1A)(a) could have no other meaning than the existing tenancy held by the tenant — The original tenancy had been wholly extinguished and the tenancy as extended, which was a new tenancy, in substitution for the old, was the only tenancy to which the words could refer — Appeals dismissed

These were appeals by case stated by S J Mosley and others, trustees of the Phillimore Kensington Estate, against the decision of the Lands Tribunal (Mr W H Rees FRICS) in favour of the submissions of three lessees, John Hickman, David Lloyd Hagan and Clare Mary Francis, lessees of, respectively, 27, 46 and 13 Argyll Road, London W8, as to the basis on which the valuation of the freeholds for enfranchisement should be made. The decision of the Lands Tribunal is reported at [1985] 1 EGLR 205; (1985) 274 EG 1261 *sub nom Hickman* v *Phillimore Kensington Estate Trustees*.

Derek Wood QC and David Neuberger (instructed by Frere Cholmeley) appeared on behalf of the appellants; Nigel Hague QC and Miss Judith Jackson (instructed by Michael Freeman & Co) represented the respondents.

Giving judgment, FOX LJ said: This is an appeal from a decision of the Lands Tribunal, Mr W H Rees. The case concerns three references (which were heard together) to determine the amounts to be paid by the tenants for the freehold in three houses in Kensington, F London W8, upon enfranchisement under the Leasehold Reform Act 1967, as amended by the Housing Acts of 1974 and 1980. In each case prior to the notice of enfranchisement, the tenant for the time being had been granted an extended lease under section 14 of the 1967 Act.

It is the contention of the tenants that the valuation should be made subject to the extended lease; the landlords contend that the extension should be ignored.

In each case the property is a dwelling-house in Argyll Road. Each had a rateable value on April 1 1973 of £1,492. The three houses in question are nos 27, 46 and 13 Argyll Road. The tenant at no 27 served notice of his desire to have an extended lease under section 14 of the 1967 Act on December 18 1980. The extended lease was granted G on September 29 1981. The tenant of no 46 served notice of a desire to have an extended lease on October 22 1981; the extended lease was granted on January 17 1983. The tenant of no 13 served notice of a desire to have an extended lease on July 10 1980; the extended lease was granted on March 13 1982.

All three extended leases expire on March 25 2047, which is 50 years after the term date of the lease running at the date of the service of the notice for an extended lease. The original leases were granted in 1936 (no 27) and 1947 (nos 46 and 13).

It is accepted that the new leases, which were granted in pursuance of the notices, comply with the provisions of section 15 of the 1967 Act. Each of the tenants has served a notice under section 8 of the H 1967 Act of his or her desire to have the freehold. In each case the notices were accepted. These notices were on February 9 1982, October 3 1983 and July 23 1982 — in relation, respectively, to nos 27, 46 and 13 Argyll Road.

The respective amounts payable by the tenants would be as follows — if the contentions of the tenants are correct — no 27, £45,000; no 46, £56,000, and no 13, £49,500. If the contentions of the landlords are correct, no 27, £98,000; no 46, £137,500, and no 13, £111,000.

Section 1 (1) of the 1967 Act, as amended, provides:

> This Part of this Act shall have effect to confer on a tenant of a leasehold house, occupying the house as his residence, a right to acquire on fair terms the freehold or an extended lease of the house and premises where —
>
> (a) his tenancy is a long tenancy at a low rent and (subject to subsections (5) J and (6) below) the rateable value of the house and premises on the appropriate day is not (or was not) more than £200 or, if it is in Greater London, than £400;
>
> (b) (I can omit this provision)
>
> and to confer the like right in the other cases for which provision is made in this Part of this Act.

Subsections (5) and (6) were added by the Housing Act 1974 and extend the provisions of the Act to higher rated properties. It is not in dispute that the houses in the present case fall within the relevant extended provisions.

At first sight the form of section 1 suggests that the tenant can have either the freehold or an extended lease. It is common ground, however (and indeed it appears from later provisions of the Act, for K example, section 16), that the tenant is entitled both to have an extended lease and to acquire the freehold. The rights are cumulative.

Section 8(1) of the 1967 Act provides:

> Where a tenant of a house has under this Part of this Act a right to acquire the freehold, and gives to the landlord written notice of his desire to have the freehold, then except as provided by this Part of this Act the landlord shall be bound to make to the tenant, and the tenant to accept (at the price and on the conditions so provided), a grant of the house and premises for an estate in fee simple absolute, subject to the tenancy and to tenant's incumbrances, but otherwise free of incumbrances.

Section 14(1) of the 1967 Act provides:

> A Where a tenant of a house has under this Part of this Act a right to an extended lease, and gives to the landlord written notice of his desire to have it, then except as provided by this Part of this Act the landlord shall be bound to grant to the tenant, and the tenant to accept, in substitution for the existing tenancy a new tenancy of the house and premises for a term expiring fifty years after the term date of the existing tenancy.

Section 15 deals with the terms of the tenancy to be granted on extension. Subsection (2) (a) and (b) provide:

> The new tenancy shall provide that as from the original term date the rent payable for the house and premises shall be a rent ascertained or to be ascertained as follows:
>
> (a) the rent shall be a ground rent in the sense that it shall represent the letting value of the site (without including anything for the value of B buildings on the site) for the uses to which the house and premises have been put since the commencement of the existing tenancy, other than uses which by the terms of the new tenancy are not permitted or are permitted only with the landlord's consent;
>
> (b) the letting value for this purpose shall be in the first instance the letting value at the date from which the rent based on it is to commence, but as from the expiration of twenty-five years from the original term date the letting value at the expiration of those twenty-five years shall be substituted, if the landlord so requires, and a revised rent become payable accordingly.

No premium is payable on the extension of the lease.

The purchase price to be paid on enfranchisement is dealt with by section 9 of the 1967 Act; subsection (1) (a) of that section is as C follows:

> Subject to subsection (2) below, the price payable for a house and premises on a conveyance under section 8 above shall be the amount which at the relevant time the house and premises, if sold in the open market by a willing seller . . . might be expected to realise on the following assumptions:
>
> (a) on the assumption that the vendor was selling for an estate in fee simple, subject to the tenancy but on the assumption that this Part of this Act conferred no right to acquire the freehold; and if the tenancy has not been extended under this Part of this Act, on the assumption that (subject to the landlord's rights under section 17 below) it was to be so extended.

Section 9(1) (a) was amended by the Housing Act 1969, which inserted, after the words "willing seller", words which excluded the tenant and members of his family residing with him as possible purchasers of the freehold.

D The 1967 Act, as amended by the Housing Act 1969, applied to all houses of the specified rateable value in Greater London.

Section 9(1) provided a single formula to determine the price, and in particular it provided that where the tenant had not obtained an extended tenancy under section 14, the price was to be calculated on the assumption that such an extension was to be made.

Section 118 of the Housing Act 1974 altered that position; houses of higher rateable value were brought within the provisions, and a different method of assessment of the price payable for the freehold of such houses was also provided. That provision is dealt with in the new subsection (1A) as follows:

> Notwithstanding, the foregoing subsection, the price payable for a house and E premises, the rateable value of which is above £1,000 in Greater London and £500 elsewhere, on a conveyance under section 8 above, shall be the amount which at the relevant time the house and premises, if sold in the open market by a willing seller, might be expected to realise on the following assumptions:
>
> (a) on the assumption that the vendor was selling for an estate in fee simple, subject to the tenancy, but on the assumption that this Part of this Act conferred no right to acquire the freehold;
>
> (b) on the assumption that at the end of the tenancy the tenant has the right to remain in possession of the house and premises under the provisions of Part I of the Landlord and Tenant Act 1954;
>
> (c) on the assumption that the tenant has no liability to carry out any repairs, maintenance or redecorations under the terms of the tenancy or Part I of the Landlord and Tenant Act 1954;
>
> (d) on the assumption that the price be diminished by the extent to which the value of the house and premises has been increased by any F improvement carried out by the tenant or his predecessors in title at their own expense;

There then follow further subparagraphs which it is not necessary to set out.

It is common ground that if the tenants of the three houses in Argyll Road with which we are concerned had not obtained the extended leases, the purchase price which they would have had to pay for the freehold would have been calculated on the assumption that the freehold was subject to the original lease. However, it is the contention of the tenants (accepted by the Lands Tribunal) that in cases within section 9(1A), if the tenant has obtained an extended lease under section 14, the price payable for the freehold must be assessed subject to the extended lease.

The crucial provision is to be found in section 9(1A) (a); it is the words:

> on the assumption that the vendor was selling for an estate in fee simple, subject to the tenancy, but on the assumption that this Part of this Act conferred no right to acquire the freehold.

The question to be decided is the meaning of the words "subject to the tenancy". Do they mean the original tenancy — as the landlords submit? Or do they mean the existing tenancy, as the tenants submit?

In approaching those questions, there are a number of circumstances to which I should refer. First, by section 1 of the 1967 Act, the right to enfranchisement is conferred upon the tenant of a leasehold house, and his tenancy must be a long tenancy at a low rent. His tenancy is the basis of his right to enfranchisement and must, when he gives notice of his desire to enfranchise, be a subsisting tenancy.

Second, the 1967 Act provides expressly, in section 14(1), that an extended lease is "in substitution for the existing tenancy". As a matter of property law, if a landlord and tenant agree that the tenant's existing lease should be extended, the extension can be effected either by the grant of a lease taking effect in reversion expectant upon the determination of the existing lease or by the surrender of the existing lease and the grant of an entirely new lease. One cannot convert an existing lease into a different estate by adding more years to that lease (see *Jenkin R Lewis & Son Ltd* v *Kerman* [1971] Ch 477 at p 496). It is clear that the procedure under the 1967 Act involves the grant of a new lease — indeed, section 14 (1) uses the expression "a new tenancy". It is as much a new tenancy as if the parties negotiated a new tenancy themselves.

The extended leases granted in the present case complied with that; they created new terms entirely. They are clearly new terms, notwithstanding that the expiry date of those extended leases is 50 years after the expiry date of the original term, in accordance with section 14 of the 1967 Act.

Third, in section 9(1) of the 1967 Act (which deals with the purchase price payable on enfranchisement by a leaseholder of the lower rateable properties) there are the words, "on the assumption that the vendor was selling for an estate in fee simple, subject to the tenancy". The "tenancy" referred to must be the existing tenancy, otherwise there is no point in the words which follow, namely "and if the tenancy has not been extended under this Part of this Act".

Fourth, in determining the value of a freehold reversion for the purposes of a purchase by a sitting tenant, the length and provisions of the existing tenancy are, apart from any express statutory provision, ordinarily of central importance to the valuation.

Fifth, section 9(1A), like section 9(1), is dealing with the price payable by the tenant for a house and premises "on a conveyance under section 8 above". Section 8(1) requires the landlord to make to the tenant, and the tenant to accept ". . . a grant of the house and premises for an estate in fee simple absolute, subject to the tenancy". That is dealing with the content of the actual conveyance to the tenant; the tenancy referred to cannot, it seems to me, be other than the existing tenancy.

Against that background, I come to the meaning of the words "the tenancy" in section 9 (1A)(a). I do not feel able to give the words any other meaning than the existing tenancy. In my opinion, in this provision dealing with the determination of the price to be paid by a sitting tenant for the acquisition of the freehold reversion, the words "the tenancy" must, *prima facie*, refer to the existing tenancy held by the tenant.

There is no other tenancy affecting the property; the original tenancy has been wholly extinguished. The extended tenancy is a new tenancy and by section 14 of the Act it is expressly provided that it is in substitution for the original tenancy.

The words "the tenancy" are, in a proper context, capable of referring to an extinguished tenancy. Thus in section 16(1)(a) by the words ". . . given not later than the original term date of the tenancy" I would agree that the tenancy there refers to the original tenancy. But that does not alter the fact that, for example, in subpara (c) of the same subsection, the words "the tenancy" quite clearly refer to the extended tenancy.

In section 9 (1A) it seems to me that, as a matter of the ordinary use of the English language, the words "the tenancy" refer to the existing tenancy and that there is no context which compels any different conclusion. It has to be borne in mind that the subsection directs that the price shall be the amount which, at the relevant time, "the premises, if sold in the open market . . ., might be expected to realise" on the assumptions set out in the subsection. The relevant time is the date of the tenant's notice that he wishes to have the freehold (see section 37). At the "relevant time" in the present cases the original tenancies had ceased to exist.

That Parliament, having directed that the assumed sale should be at the relevant time, and that it would be subject to the tenancy, must have intended that the words "the tenancy" should refer to something which has already ceased to exist — and even though there is a subsisting tenancy at the relevant date — is a conclusion which, in my view, the language of the statute will not bear.

Mr Wood, for the landlords, says that if "the tenancy" is read as "the existing tenancy" an unacceptable anomaly arises in that the tenant who chooses to take an extended lease prior to making his request for the freehold will pay a substantially lower price than the tenant who has failed to do so.

But that anomaly does not, in my view, justify the court in departing from the clear language of the statute. The practical results of which the landlords complain is really the consequence not so much of section 9 (1A) but of the fact that Parliament chose to confer on the tenants of these houses a right to extend their leases for a substantial period at ground rents, and without premiums.

Mr Hague for the tenants is, I think, correct when he says that if we adopt the landlords' construction, we are introducing another anomaly, namely, that Parliament having conferred on a tenant the right to extend the tenancy (which is obviously a valuable right) should, when it comes to a valuation of the reversion, disregard the exercise of that right by language which is quite inappropriate to such a course.

Any anomalies which result from the construction which I have adopted of the words "the tenancy" can, I think, only be corrected by Parliament.

I accept that we must construe section 9 (1A) in the light of the statute as a whole; but looked at as a whole I do not see compelling indications in support of the landlords' case. It is true that section 9 (1A) is introducing a new provision for a different type of property from that originally dealt with by the 1967 statute, but the court must construe it according to its actual language. The landlords are really saying that for the words "the tenancy", we must substitute the words "the tenancy or, if it is an extended tenancy, the previous tenancy". That is a very substantial rewriting which in my view wholly displaces the ordinary meaning of the words "the tenancy" in their context in this statute; and which involve the requirement of valuation by reference to a tenancy which has wholly ceased to exist.

The fact that section 9(1)(a) provides that, if the tenancy has not been extended, it shall be assumed that it is to be extended is not a reason for reading into section 9 (1A) a provision that, if the tenancy has been extended, it shall be assumed that it has not.

It is true that section 9 (1A)(b) requires the assumption of a right to remain in the house at the end of the tenancy under Part I of the Landlord and Tenant Act 1954, which, in the case of an extended lease, is excluded by section 16 (1)(c) of the 1967 Act. But that is only an assumption in favour of the tenant and does not displace the meaning which the words "the tenancy" would otherwise bear.

In my view the Lands Tribunal came to the correct decision, and I would therefore dismiss these appeals.

MUSTILL and STOCKER LJJ agreed and did not add anything.

The appeals were dismissed with costs. Leave to appeal to the House of Lords was refused.

Court of Appeal

April 17 1986

(Before Sir John DONALDSON MR, Lord Justice PARKER and Sir George WALLER)

R v VALUATION OFFICER FOR WESTMINSTER AND DISTRICT, ex parte RENDALL

Estates Gazette June 7 1986

278 EG 1090-1094

Leasehold Reform Act 1967 — Housing Act 1974, Schedule 8, para 3(2) — Adjustment of rateable value in respect of improvements — Lessee's entitlement to enfranchisement depended on establishing that the rateable value of his property on April 1 1973 did not exceed £1,500 — The value in the list was £1,597 and the applicant sought to show that if certain improvements were taken into account the value would be reduced to the qualifying figure — The valuation officer, however, certified a reduced rateable value of £1,509 — As there was no right of appeal, the applicant challenged the valuation officer's certificate by proceedings for judicial review — Taylor J dismissed the application and the applicant appealed to the Court of Appeal — Appellant's main complaint was that the valuation officer failed to make any reduction for the conversion of a bedroom into a bathroom, although there was a subsidiary point about the replacement of an old coke boiler by a new Potterton gas boiler — In giving judgment the Master of the Rolls emphasised the importance of the distinction between an ordinary appeal (which was not available in the present case) and judicial review — *CCSU* v *Minister for the Civil Service* and *Associated Provincial Picture Houses Ltd* v *Wednesbury Corporation* referred to — A failure to take account of a relevant consideration is a possible ground for judicial review, but a complaint about the assessment of relevant considerations, all of which had been taken into account, was not — One point had been raised which was possibly relevant, namely, the failure of the valuation officer to obtain and consult files in the Valuation Office which might have shown what precisely had been taken into account in the 1973 valuation — The Court of Appeal had itself made inquiries into the existence of such files and, after considerable difficulty, had discovered a pre-1973 file relating to the appellant's house, but it proved to be of no significance — If the valuation officer had found it there would have been no cause to revise his view — Even if there was a failure on his part to look for such a file, this could not form a basis for judicial review — There was nothing in the point about the replacement of an old boiler — Appeal dismissed

This was an appeal by Mr J L Rendall from a decision of Taylor J dismissing his application for judicial review by which he sought to challenge the certificate of the valuation officer under Schedule 8 to the Housing Act 1974. The valuation officer had certified that the amount by which the rateable value of £1,597 in respect of Mr Rendall's leasehold house at 74 Eaton Terrace, London SW1, would have been reduced if certain improvements had not been made was £88. The freehold owners of the house were the Grosvenor Estate.

The appellant, Mr J L Rendall, appeared in person; D Mole (instructed by the Solicitor of Inland Revenue) represented the valuation officer for Westminster and District, Mr Hardy; Nigel Hague QC (instructed by Boodle Hatfield & Co) represented the Grosvenor Estate.

Giving judgment, SIR JOHN DONALDSON MR said: Mr Rendall is the assignee of a long lease of 74 Eaton Terrace, London, of which the freehold owner is the Duke of Westminster. Understandably he would like to enfranchise his interest in the property and it is conceded that, subject to one crucial point, he is entitled to do so. That crucial point turns on the rateable value of the property on April 1 1973. The relevant rateable value is not necessarily that which appeared in the valuation list on that date, since, for this purpose, that value is subject to downward adjustment in so far as it reflected improvements made by the tenant or a previous tenant.

The target rateable value, from Mr Rendall's point of view, was £1,500 or less. The value appearing in the valuation list was £1,597 and Mr Rendall has therefore applied his mind to getting rid of £97 or more. The procedure is governed by the Leasehold Reform Act 1967 as amended by the Housing Act 1974. The first stage in this operation was an application to the county court for a determination of whether, and to what extent, the improvements specified by Mr Rendall qualified for consideration in reduction of the rateable value. He succeeded in respect of four improvements, namely:

(1) The addition of a room on the ground floor;
(2) The creation of a patio area on the mezzanine floor;
(3) The conversion of a bedroom on the third floor into a bathroom; and

(4) The installation of a gas boiler in place of a coke boiler to supply a hot-water system and one radiator.

The second stage consisted of an application to the valuation officer for a certificate under para 3 (2) of Schedule 8 to the Housing Act 1974:

(a) whether or not the improvement has affected the rateable value on 1st April 1973 . . . of the hereditament of which the premises consist . . . and (b) if it has, the amount by which the rateable value would have been less if the improvement had not been made.

It was for Mr Rendall to specify the improvements to which the certificate would relate, provided that they were agreed by the ground landlord as a qualifying improvement or were held to be such by a county court. Mr Rendall in fact specified only improvements (1), (3) and (4) of those so determined by the county court.

On April 21 1981 Mr Hardy, the local valuation officer, certified that the amount by which the rateable value would have been less if those improvements had not been made was £88. This was, of course, £9 short of Mr Rendall's target and there is no right of appeal.

In these circumstances Mr Rendall decided to seek to challenge the valuation officer's certificate by judicial review. It was reasonably clear, and was subsequently confirmed, that the £88 reduction related solely to the increase in floor area consequent upon the addition of a room on the ground floor and that the valuation officer had made no reduction on account of the conversion of the third-floor bedroom into a bathroom or of the replacement of the boiler. In his application Mr Rendall concentrated upon the failure by the valuation officer to make any allowance in respect of the bathroom conversion, stating, in the grounds upon which relief was sought, that "the valuation officer failed to perform his statutory duty or proceeded on wrong or unlawful principles in excluding (as he admits he did) from his certificate the said bathroom which (if included) is likely to have caused a further reduction of at least £25 in the gross value and thus substantially more than £9 in the rateable value". No mention was made of the replacement of the boiler, although in argument, both here and below, it made fleeting and somewhat insubstantial appearances.

Taylor J dismissed the application and Mr Rendall has appealed to this court.

Mr Rendall is a solicitor of very considerable experience in rating matters and he has conducted his own case with great skill. Indeed such was his skill that at times he succeeded in blurring the lines between the jurisdiction of the High Court, and of this court on appeal, to entertain an appeal from the decision of the valuation officer, which is non-existent, and the jurisdiction of the High Court, and of this court on appeal, to judicially review that decision. The distinction is of fundamental importance. The valuation officer's statutory duty was to inquire whether the specified improvement or improvements had in fact affected the rateable value of the property on April 1 1973 and, if it had, the amount by which the rateable value would have been less if the improvements had not been made. The issues were essentially issues of fact. If there had been a right of appeal, it would have been relevant to consider, in the light of the inquisitorial nature of his duties, whether the valuation officer had sought for and found all relevant evidence and whether he had correctly assessed that evidence. It is quite otherwise if the duty of the court is to consider an application for judicial review. In *CCSU* v *Minister for the Civil Service* [1985] 1 AC 374 at p 410, Lord Diplock classified the grounds for judicial review under three main headings, namely, "illegality", "irrationality" and "procedural impropriety". Only "irrationality" is relevant in the instant case.

"Irrationality" was defined by Lord Diplock as "what can now be succinctly referred to as 'Wednesbury unreasonableness' " (*Associated Provincial Picture Houses Ltd* v *Wednesbury Corporation* [1948] 1 KB 223). It applies to a decision which is so outrageous in its defiance of logic or of accepted moral standards that no reasonable person who had applied his mind to the question to be decided could have arrived at it. Mr Hardy, in an affidavit upon which he was cross-examined, stated that "Provided that the house had at least one bathroom, the practice was to assess bathrooms at the same price per square metre as other space in such houses. There is no evidence to show that the conversion of bedrooms to extra bathrooms increased rental value of such houses as at April 1 1973: some hypothetical tenants would prefer the extra bathroom, while others would prefer to retain the bedroom." For my part, I can see nothing irrational in such a conclusion, whether or not I should have reached it.

However, all concerned seem to have accepted that a wider inquiry by the court was permissible, based upon the *Wednesbury* decision itself. There the court was concerned with a discretionary decision and Lord Greene MR said that the decision-maker "must call his own attention to the matters which he is bound to consider. He must exclude from his consideration matters which are irrelevant to what he has to consider". I do not suggest that different considerations apply where the decision is not discretionary, but a word of caution must be uttered. It is one thing to fail to take account of a relevant consideration. It is quite another to take account of all relevant considerations but to assess them in a way which, though not irrational, is not a way which would have appealed to everyone. The former is a ground for judicial review. The latter is simply a ground for appeal.

It was, I think, a failure to appreciate this distinction which expanded the scope of the evidence and arguments in this case to the point at which it was indistinguishable from an appeal against the valuation officer's finding of fact. Thus we had Mr Hardy explaining not only the basis of his decision, which I have already quoted, but exhibiting the rating précis sheets for the 1973 valuation to which he had referred. Taylor J was treated to the cross-examination of Mr Hardy on this evidence. In addition he and we were enlightened by an affidavit from Mr Rendall exhibiting a letter from a distinguished rating surveyor expressing the view as an expert that "the addition of a third floor bathroom must have added to the value of any house in this district", and an affidavit from another distinguished rating surveyor expressing his expert opinion that while the addition of a second bathroom would increase the rateable value, bathrooms in excess of two would not do so. We even had evidence relating to remarks about the effect of additional bathrooms alleged to have been made by Mr Hardy's successor, not in the context of a 1973 valuation but in that of a dispute before the valuation court in 1984. Little of this evidence is appropriate to an application for judicial review.

Standing back and looking at the mass of material available to us, and indeed thrust upon us, there are really only two points which are deserving of serious consideration. The first is one taken against Mr Rendall by the Duke of Westminster and upheld by Taylor J. This is that the rating précis relating to the 1973 valuation stated that no 74 had two bathrooms, not three as was the fact. Accordingly it is to be assumed that the 1973 valuation took no account of the third bathroom which had been converted from a bedroom. If it was not taken into account, the rateable value would manifestly have been no lower if the improvement had not been made. I see the force of this argument, but it is relevant only if (a) Mr Hardy's decision is *prima facie* fit to be quashed on review and (b) the court is then considering whether to refuse relief upon the basis that any fresh certification must lead to the same result. In any event, if we had got to that point, it would have been necessary to consider whether the evidence of the précis sheets was conclusive or might be displaced by other evidence.

The second point requires slightly more elaboration. Mr Hardy has sought to check the correctness of his conclusion that the additional bathroom in no 74 would have no effect upon its 1973 value by referring to entries in the précis sheets relating to other houses which he considered to be comparable. Mr Rendall contended that some of these houses were not comparable. So be it. They may not have provided Mr Hardy with the confirmation which he sought, but equally, on this hypothesis, they could not cast doubt upon his conclusion. For the rest, Mr Rendall devoted much of his argument to seeking to persuade us that Mr Hardy has misinterpreted the précis sheets and failed to make inquiries as to what certain notations meant, all of which is really only appropriate to an appeal, as contrasting with judicial review. But he did suggest that the rating authority must have had files relating to no 74 and other comparable houses which would have revealed what precisely was taken into account on the 1973 valuation. Alternatively he suggested that the 1973 valuation was merely a revision (to take account of changed values) of a full valuation in 1963 and that the files would show what was taken into account in 1963 and thus, on revision, in 1973. This point about the files does not seem to have been developed before Taylor J. It certainly should have been and Mr Rendall could hardly have complained if he had been told that it was now too late. However, it did seem to us that if it was a fact that such files existed, this was the obvious source to which any reasonable surveyor would turn when asked to give a certificate of this nature and that a failure to do so could come within the *Wednesbury* principle. How he assessed that material, having referred to it, would be a quite different matter and *prima facie* not susceptible of forming a ground for judicial review.

It was clear that Mr Hardy did not consult any such files and we asked that inquiries be made as to whether they existed. The first such did indeed reveal a file for no 74, but it had been opened long after 1973 and was irrelevant. Similar files were produced for the other "comparable" houses, but, with two exceptions, they contained no 1973 or earlier material. A second and even more thorough search revealed another file for no 74 which had been misfiled by being in the wrong coloured folder, thereby indicating that it was a commercial and not a residential property, and being filed under Eaton Square instead of Eaton Terrace. It did indeed include a survey sheet relating to the 1963 valuation and showed that at some unspecified point of time before 1984 the rating authority had become aware that a new bathroom had been created on the third floor. But that is all that it did show. It may well have been available to those who undertook the 1973 valuation, but it casts no light upon whether at that time it showed the conversion and, if so, whether this had any effect upon the value.

I will go as far with Mr Rendall as to say that a valuation surveyor who is required to give a certificate relating to the effect of improvements on the value of a property at April 1 1973 should make all reasonable efforts to refer to any working papers still extant which relate to that valuation. Whether or not Mr Hardy did make such efforts, I do not know. All I know is that, if he did so, he did not succeed. The matter either was first raised or at least first achieved significance at the hearing in this court and at that time Mr Hardy was ill and could not be asked. However, it is clear that if he had attempted to find the pre-1973 file relating to no 74 it is most unlikely that he would have succeeded, for the extent of the search mounted in consequence of these proceedings was wholly exceptional and more than could have been expected from Mr Hardy. Furthermore, it is clear that had he found this file, there would have been nothing to cause him to revise his view of what the 1973 valuers must have done, and this failure to look for such a file, if such failure there was, cannot form a basis for judicial review.

The suggestion that any valuer would necessarily have increased the value of the property to take account of the substitution of a new Potterton gas boiler for an old coke boiler is quite insupportable.

I, like the learned judge, cannot fault the valuation officer's actions, other than in respect of a possible failure to seek the 1973 file relating to no 74. This failure, if such it was, could not have affected the form of his certificate and I would dismiss the appeal.

Agreeing, PARKER LJ said: In the *CCSU* case, although Lord Diplock classified under the three heads "illegality", "irrationality" and "procedural impropriety" the grounds upon which administrative action is subject to control by judicial review, he said also ([1985] 1 AC at p 410):

> That is not to say that further development on a case by case basis may not in course of time add further grounds.

He also regarded (see p 411A) "procedural impropriety" as covering "failure to observe basic rules of natural justice or failure to act with procedural fairness towards the person who will be affected by the decision".

On an application for a certificate under para 3 (2) of Schedule 8 to the Housing Act 1974 the valuation officer has two questions to determine:

(a) whether or not the improvement has affected the rateable value on April 1 1971, and

(b) if it has, the amount by which the rateable value would have been less if the improvement had not been made.

The first is a question of historical fact. The second involves an assessment of the consequences which would have taken place upon a hypothesis.

If the valuation officer carried out the 1973 valuation himself and has a good enough memory he can no doubt properly answer both questions without more ado. The valuation officer in the present case did not, however, make the 1973 valuation; he therefore had to obtain information as to what happened in 1973. Mr Rendall's only possible sustainable complaint is, in my view, that Mr Hardy failed either (i) to act with procedural fairness in that he failed to obtain or at least make reasonable efforts to obtain documents which would be the obvious primary source of information relating to the pure question of historical fact, or (ii) to take into account a relevant consideration, namely information in such documents.

I say "only possible sustainable complaint" because any suggestion that Mr Hardy's decision was "irrational" is without foundation.

In my view a valuation officer with no personal knowledge of what happened in 1973 would be, at least *prima facie*, guilty of procedural impropriety if he purported to determine the question of historical fact without making reasonable efforts (i) to ascertain whether documents relating to the 1973 valuation still existed and (ii) to obtain and inspect them in so far as they did exist. The matter may be tested in this way. Suppose a VO, when asked to certify whether or not three improvements had affected the rateable value in 1973, certified that none of them had done so on the simple basis that, having inspected the property, he was confident from his general experience that none of them would have done so and it was therefore unnecessary to look further. Suppose further that, after the certificate had been given, it emerged that within the valuation office there was, readily available, a file relating to the property and that, in that file, there was a detailed breakdown of the 1973 valuation showing clearly that two of the improvements *had* affected the 1973 valuation. In such circumstances it appears to me that there would have been a clear failure on the part of the VO and that whether it were regarded as a failure to act with procedural fairness or as a failure to take into account a relevant consideration it would be right for the court to grant relief by way of judicial review. If not a procedural impropriety such a failure is clearly analogous thereto.

In the present case, however, it will not avail Mr Rendall, even if it be the case that Mr Hardy did fail to take all proper steps to find and obtain papers relating to the 1973 valuation. Searches have now been made and I agree that nothing which has emerged is of any significance. It would therefore be wrong as a matter of discretion to grant relief.

I, too, would dismiss the appeal.

SIR GEORGE WALLER agreed that the appeal should be dismissed for the reasons given by the Master of the Rolls and did not add anything further.

The appeal was dismissed. The appellant was ordered to pay the costs of the valuation officer, but not the costs of the Grosvenor Estate, in the Court of Appeal. Leave to appeal to the House of Lords was refused.

For further cases on this subject see pp 211 and 231

LANDLORD AND TENANT

RENT ACTS

Queen's Bench Division

September 23 1985

(Before Mr Justice McNEILL)

DAEJAN PROPERTIES LTD v CHAMBERS

Estates Gazette January 25 1986

277 EG 308

Rent Act 1977 — Appeal by landlords against decision of rent assessment committee — Rent officer determined a fair rent of flat at £2,325 per annum inclusive of £826.58 per annum in respect of services — Landlords had asked for £2,555 per annum inclusive of £898 for services — Rent assessment committee fixed fair rent at £2,365 per annum inclusive of £865.97 for services — Committee gave reasons for their decision, as requested, and the landlords on appeal to the court put forward two main grounds of complaint — The first was that the reasons contained a statement that "In view of the fact that the committee were unable to gain access to the flat, they made an adjustment in the total rent to reflect the updated service schedule" — This, it was submitted, indicated that the committee considered that inability to obtain access justified them in merely updating the service schedule and not reviewing the rent in the light of all the material before them and their knowledge and experience —The second complaint was that the committee did not take into account evidence of valuation which was submitted to them on behalf of the appellants or at least did not show in their reasons that they had taken it into account — Held that the case must be remitted to the committee for reconsideration in view of their apparent failure to consider the whole of the evidence presented to them, or at any rate their failure to give reasons for rejecting it, and in view of the fallacious ground mentioned in their reasons that they were unable to obtain access to the premises — Committee's decision remitted for reconsideration in the light of the judgment

This was an appeal by landlords Daejan Properties Ltd from a decision of a rent assessment committee for the London area, in respect of the committee's determination of a fair rent of a flat at 17 Conduit House, Hyde Vale, London SE10, of which the tenant was Miss Doris Chambers.

P Wulwik (instructed by Clifford Watts, Compton & Co) appeared on behalf of the appellant landlords. The respondent, Miss Chambers, did not appear and was not represented.

Giving judgment, McNEILL J said: This is an appeal from a decision of a London rent assessment committee, dated October 5 1984, determining the fair rent of a dwelling, 17 Conduit House, Hyde Vale, London SE10. The appellants are the landlords of the premises and by the notice of motion ask that the decision be set aside and that the matter be remitted to the committee for rehearing and determination. Neither the committee nor the respondent, who is the tenant of the premises, has appeared before me.

The premises, no 17, are a flat in a building of which the appellants are the owners. From time to time, over the years, fair rent for the premises has been determined under the provisions of the Rent Act 1977. The matter fell to be reconsidered in 1984 on what I assume was the normal biennial reconsideration. On July 11 1984 the rent officer made a determination to which the appellants objected and the F
objection therefore came before a London rent assessment committee. The position was this. The rent officer had determined the fair rent at £2,325 per annum inclusive of £826.58 per annum for services. I need say little about the service element because the committee were persuaded that the figure for services was too low in the light of material which was more up to date than that which was available to the rent officer. The result of that readjustment, however, meant that the fair rent was determined by the committee at £2,365 per annum, that is to say an increase of £40. It now appears that the appellants were contending for a total figure, by way of fair rent, of £2,555 per annum, including £898 for services. G

The committee, as they are required to do, gave their reasons. They are at p 31 of the bundle. The chairman provided her notes and they are to be found at p 38. The tenant, the present respondent, did not attend the hearing or communicate with the committee. The first paragraph of the reasons at p 31 reads as follows: "At the hearing the landlord company was represented by Mrs Carlyle, an employee of the Freshwater Group of Companies. Mr P Miles of May & Philpot who was appearing in the following case was able to assist the committee by answering some questions."

They then deal with the updated service schedule, and continue: "The Committee inspected the subject block but were unable to gain access to Flat 17." They deal with the general description of the block H
and were "inclined to agree with Mr Miles that Flat 17 was probably the most attractive situation in the block with a southerly view across the park".

Then, omitting the next paragraph, which is not material for these purposes, they said this:

In view of the fact that the committee were unable to gain access to the flat they made an adjustment in the total rent to reflect the updated service schedule. Having regard to all the evidence put before us, to our inspection, to our knowledge and experience, and to the provisions of section 70 of the Rent Act 1977 we determined the fair rent, to be registered exclusive of rates, at £2,365 per annum.

The chairman's notes, which I need not set out in detail, refer in an abbreviation to written representations by the appellants: "Mrs J
Carlyle: from Freshwaters. Written proof of evidence." Then there are three short descriptions of the contents of that proof, which in fact adopted the rent officer's description of the premises. There then appears this line: "She could not answer Mr Emeny's questions." Mr Emeny, it appears from the documents, was the surveyor member of the committee.

That is the whole of the material which the committee have disclosed, and upon that three points are taken. First of all it is said in relation to the penultimate paragraph of the reasons, which I have read, that the committee were determining that in view of the fact that they were unable to gain access to the flat they should only make an adjustment to reflect the updated service schedule, that they were applying the wrong test in law and that their inability to gain access K
was not a reason of itself for failing to apply their experience and knowledge and such material as was before them to reviewing the rent. Of course, at the end of the day they were entitled to say that the rent should remain as it was, which is what they did, as long as they applied themselves to the material that was before them.

In the second ground of appeal it is said that the committee failed to consider adequately or at all the evidence of the appellants' surveyor that there should be such an increase, that is to say the increase to which I have already referred of some £200 per year.

Third, it is said they failed to give any adequate reasons or explanation in respect of their decision and in particular for not accepting the evidence of valuations submitted by the appellants' surveyor. The appellants' surveyor, Mr Miles, has sworn an affidavit which is before me. At para 8 he says:

> The appellants' evidence before the Rent Assessment Committee consisted of written proofs of evidence from Mrs Carlyle and from myself, together with an updated service schedule. . . True copies of the said proofs of evidence and service schedule which were before the Rent Assessment Committee are now produced . . .

He refers to his proposal as to the fair rent.

Whatever be the force, if there is force in the first of those submissions, it seems to me entirely clear from what the committee have recorded in their reasons and from what appears in the notes of the chairman that they do not appear to have regarded Mr Miles as a witness whose evidence was before them, or to have taken into account in any way the proof of evidence which is before me and which plainly refers to no 17. In that respect at the very least the committee were in error in disregarding that material or, if they took it into account, in failing to deal with it in their reasons. It might be said that the penultimate phrase is badly worded. It may well be the committee felt they should record the fact that they were unable to obtain access to the flat, but that is wholly insufficient when set against the fact that they also seem wholly to have disregarded the fact that Mr Miles provided evidence in relation to no 17. It also makes even more mysterious the chairman's note that Mrs Carlyle could not answer Mr Emeny's questions when, if Mr Miles was there, as he clearly was (although the committee seem to have thought he was there for a different case), Mr Emeny should have put his questions to Mr Miles, who no doubt would have been able to deal with them.

For those reasons it seems to me that this matter must go back to the committee for reconsideration. They may come to the same conclusion having considered the whole of the evidence, but it is I think plain that they failed to consider it, or, if they did, they failed entirely to give any reasons for rejecting it or for coming to the conclusion which they did other than, if I may say so, the fallacious reason that they were unable to get access to the premises. Accordingly, the matter will be remitted.

There was no order for costs.

Court of Appeal

January 30 1986

(Before Lord Justice NOURSE, Lord Justice WOOLF and Sir George WALLER)

R v CHIEF RENT OFFICER FOR ROYAL BOROUGH OF KENSINGTON AND CHELSEA, EX PARTE MOBERLEY

Estates Gazette April 19 1986

278 EG 305-308

Rent Act 1977, section 67 — Question as to validity of an application by a tenant for the registration of an existing registered rent — Appeal by tenant from the decision of McCullough J dismissing her application for judicial review — Tenant had sought an order of *certiorari* to quash a registration by the rent officer increasing her rent — The proceedings originated in a mistaken belief by the tenant in 1983 that her landlords would be able to increase her rent of £14 per week, registered in 1980, without any fresh registration — In order, as she thought, to forestall such an increase she applied to the rent officer to re-register her existing rent of £14 per week, and the landlords were, as the legislation required, given notice of her application — The rent officer explained to the tenant that her rent could not have been increased without an application to him by the landlords — He informed the tenant that it was open to her to request a withdrawal of her application, which, if the landlords consented to it, would put an end to the proceedings — The tenant claimed later to have withdrawn the application, but the rent officer stated that no request to withdraw had been received — He proceeded to make his determination and registered a rent of £16 per week — In her appeal the tenant submitted that section 67(1) and (3) allowed only two kinds of application, an application for registration of an initial rent and an application to register a rent different from the registered rent; there was no warrant for the view that an application could be made to re-register at the same figure an existing registered rent — Paras 4 and 5 of Schedule 11 to the Act confirmed this construction — Held that this submission was correct — The opposite view would lead to strange results; as section 67(3) restricted only applications for a different rent, it would be possible to make several applications to register the same rent, thus effectively prolonging the period during which a landlord could not obtain an increase of rent — The consequence was that the application to the rent officer to re-register the existing registered rent was invalid and the determination of the new rent which he made as a result of that invalid application must be quashed — The court in the exercise of their discretion would grant the order of *certiorari* sought — In view of this decision on the point of construction, the court did not consider certain other grounds advanced by the appellant, including a contention that an application to the rent officer could be withdrawn without the other party's consent at any time before registration

This was an appeal by Miss Sheila Moberley, a tenant of Room 13 at 38 Penywern Road, London SW5, from a decision of McCullough J dismissing an application for judicial review for the purpose of quashing a registration by the rent officer of a rent of £16 per week for the appellant's accommodation.

The appellant appeared in person; Duncan Ouseley (instructed by the Treasury Solicitor) represented the respondent, the chief rent officer for the Royal Borough of Kensington and Chelsea.

Giving judgment, NOURSE LJ said: This is an appeal from a decision of McCullough J given on November 29 1984 in proceedings for judicial review. The applicant, Miss Sheila Moberley, is the tenant of Room 13, 38 Penywern Road, London SW5. That property is situated in the Royal Borough of Kensington and Chelsea. By her application Miss Moberley seeks the quashing of a determination of the rent of those premises by a rent officer for the Royal Borough on December 7 1983. The chief rent officer for the Royal Borough is accordingly the respondent to these proceedings.

By his determination the rent officer assessed Miss Moberley's rent at £16 per week inclusive of services and use of furniture. Her rent had last been determined in 1980, the relevant application having been made on July 25 of that year. It was then fixed at £14 per week inclusive of services and the use of furniture. That earlier rent was registered on October 15 1980. Under the relevant statutory provisions it took effect as from the date of the application on July 25.

The material facts, which I take almost verbatim from the judgment of the learned judge, are these. In 1983 Miss Moberley was fearful that after the end of the three-year period her landlords would be free to increase her rent without further reference to the rent officer. That in fact was a mistake. There could of course have been no increase without a further determination. However, acting under her misapprehension, Miss Moberley, on October 13 1983, applied for a fresh registration of the existing rent of £14 per week. The landlords were duly given notice of that application. They replied, saying that they thought the rent should be increased.

On November 22 the rent officer, Mr Souster, went to 38 Penywern Road, where he met Miss Moberley. He asked her why she had asked for the registration when her rent could not be increased unless the landlords first applied to the rent officer themselves. She explained her belief and, when Mr Souster had explained the true position to her, she said that had she known that she would not have made her application. Mr Souster then, in my view very properly, informed her that it was open to her to make a request in writing seeking to withdraw her application. He said that, provided the landlords consented to the withdrawal, he would take no further action. So far as Mr Souster and anyone else in his office were concerned, no written request asking for the application to be withdrawn was ever received. Accordingly, on December 7 1983, Mr Souster made his determination and registered the new rent of £16 per week.

On December 22 Miss Moberley wrote to the rent officer saying, among other things, that she had withdrawn her application to register the rent and that no notice had been taken of that. On January 16 1984 she wrote again, sending a copy of a letter which she said was the letter of withdrawal and which she said she had delivered personally to the rent offices in Cromwell Road on November 22 1983. That letter said this:

> Dear Sir, re Re-registration of Rent. This is to inform you that I wish to withdraw my application to have the rent registered again for the above premises.

On February 8 1984 the chief rent officer wrote to Miss Moberley and said this to her:

> There is no procedure in law for any retrospective withdrawal after the date of registration, so there the matter must rest.

Miss Moberley, not being satisfied with the position, on February 11 1984 applied for leave to move for judicial review on a number of grounds. On June 28 of that year Glidewell J gave her leave to move on one ground, and when the matter came effectively before McCullough J on November 28 1984 he gave her leave to add three further grounds.

The four grounds were summarised by the learned judge as follows. First, that the rent officer had ignored the letter of withdrawal which Miss Moberley said she had delivered on November 22 1983. Before the judge there was an issue of fact as to whether that letter had been delivered or not.

Second, that the registration on December 7 1983 was of no effect because it was founded on an application from Miss Moberley which in itself was of no effect, since the relevant legislation, namely the Rent Act 1977, does not allow a tenant to apply for registration of a rent which is the same as that currently registered. That ground can conveniently be described as raising the construction point.

The third ground is that Miss Moberley was wrongly told in the letter of February 8 1984 that there was no process of law whereby a registered rent could be removed. There she contends that she ought to have been informed that she could apply by way of judicial review for an order of *certiorari*.

The fourth ground is that Mr Souster wrongly told Miss Moberley on November 22 1983 that she could withdraw her application only if the landlord consented to that withdrawal. She says that an application can be withdrawn unilaterally at any time before the rent is registered.

The convenient course is for me to start with the construction point. The Rent Act 1977 was amended in 1980, but the amendments are not material to this case. I therefore refer to the Act in its unamended form. Section 67(1) is in these terms:

> An application for the registration of a rent for a dwelling-house may be made to the rent officer by the landlord or the tenant, or jointly by the landlord and the tenant, under a regulated tenancy of the dwelling-house.

I need not refer to subsection (2) at this stage.

Subsection (3) is in these terms:

> Subject to subsection (4) below, where a rent for a dwelling-house has been registered under this Part of this Act, no application by the tenant alone or by the landlord alone for the registration of a different rent for that dwelling-house shall be entertained before the expiry of 3 years from the relevant date (as defined in subsection (5) below) except on the ground that, since that date, there has been such a change in

certain specified circumstances

> as to make the registered rent no longer a fair rent.

Subsection (4) is immaterial. Subsection (5) defines the relevant date. By the combined effect of that subsection and section 72 of the Act, the relevant date in the present case was the date of the application which led to the registration of the 1980 rent, namely July 25 1980.

Accordingly the effect of section 67(3) in this case was that no application by Miss Moberley alone or by the landlord alone for the registration of a different rent for the property, except on the ground of a change in any of the specified circumstances, could be entertained before July 26 1983.

McCullough J dealt with the construction point in this way at p 6D of the transcript of his judgment:

> An application to register is made under section 67(1). There is nothing in section 67(1) which restricts its operation to an application for registration for the first time; therefore an application by either landlord or tenant, whether or not made for the first registration, is made under section 67(1), and there is nothing in section 67(1) which prevents an application being made for registration of a rent which is the same as that already upon the register.
>
> Section 67(3) gives protection to both landlord and tenant. The tenant benefits because, until the expiry of three years from what I may loosely call G
> the last registration, the landlord cannot even apply for an increase unless there has been a material change in the circumstances. Similarly, the landlord benefits because until the expiry of the same three-year period the tenant cannot even make such an application. It may of course be advantageous for either to apply within the three-year period for registration of the existing rent, because if this results in a fresh registration of that rent, then the appropriate period, which used to be three and is now two years, starts to run again. There is nothing in section 67(3) which prevents such an application being made.
>
> Miss Moberley also referred me to Schedule 11 to the Act, paragraph 5(b), but I see nothing in that provision either which supports the construction for which she contends.

For those reasons the learned judge rejected Miss Moberley's arguments on the construction point. He also rejected her arguments H in relation to the other three grounds and accordingly dismissed her application. Miss Moberley now appeals to this court. She has conducted her own case both here and below.

Miss Moberley's argument on the construction point is essentially a simple one. She says that section 67(1) is concerned with only two kinds of application: first, an application for the registration of an initial rent; and, second an application to register a rent different from the registered rent for the time being. She says that there is no warrant, either in section 67 or in the Act as a whole, for the view that section 67(1) enables an application to be made for the re-registration of an existing registered rent.

As to that argument, I would observe first of all that it certainly is J rather an odd notion that the Act should permit an application to re-register a rent which is already on the register. It seems to me that there are at least three solid reasons for concluding that it does not. First, if an application of the kind suggested were permissible, it might lead to many more applications being made than Parliament can reasonably be taken to have intended. Section 67(3) only restricts applications for a different rent. Accordingly, if applications to re-register existing rents were permitted under section 67(1), they could be made frequently and, in theory at least, they might be made several times during the three-year period. Bearing in mind the administrative problems involved, I cannot believe that Parliament intended that applications should, in the absence of special circumstances, be made more frequently than once every three years. K

Second, it would in certain circumstances be open to either a landlord or a tenant to manipulate applications in such a way as to defeat the purpose of section 67(3). To give an example which is the opposite of the rather more compelling one given by Woolf LJ during the course of the argument, a tenant who opined that the rate of inflation was likely to increase rapidly in about three years' time could, one year after the registration of one rent, apply for its re-registration and, assuming that there had been no significant increase during the first year, effectively prolong from three to four years the period during which his landlord was unable to secure a higher rent. Again, I cannot believe that the Act was intended to have that sort of result. L

Third, Miss Moberley has referred us to paras 4 and 5 of the Eleventh Schedule to the 1977 Act, which contains detailed provisions relating to applications for registration of rent. I will start by reading the first part of para 5:

> After considering, in accordance with paragraph 4 above, what rent ought to be registered or, as the case may be, whether a different rent ought to be registered, the rent officer shall, as the case may require, — (a) determine a fair rent and register it as the rent for the dwelling-house; or (b) confirm the rent for the time being registered and note the confirmation in the register.

It appears to me, on a fair reading of that provision, that what is to be considered is either an application for the registration of an initial rent or one for the registration of a rent different from that which is for the time being suggested. There seems to be no third category. M Moreover, it will be noted that para 5 refers to a consideration in accordance with para 4. Turning to para 4(1), I find that it is expressed in these terms:

> Where representations are made as mentioned in paragraph 2 above or the rent officer is not satisfied that the rent specified in the application is a fair rent or, as the case may be, that the rent for the time being registered is no longer a fair rent, he shall serve a notice under this paragraph.

Once again, it seems to me that that provision recognises, and recognises only, the two kinds of application to which I have referred. I find it very difficult to think, as Mr Ouseley for the chief rent officer submits, that it is intended also to embrace an application for the re-registration of an existing registered rent. If that was what

A was intended, para 4(1) could have been drawn in these simple terms: "Where representations are made as mentioned in paragraph 2 above or the rent officer is not satisfied that the rent specified in the application is a fair rent, he shall serve a notice under this paragraph." I say that because on Mr Ouseley's argument the existing rent would, *ex hypothesi* and just like an initial rent or a different rent, be specified in the application itself.

In the circumstances it seems to me that the provisions of paras 4 and 5 of the Eleventh Schedule serve only to confirm the view at which I would have been prepared to arrive merely on the construction of section 67(1) and (3) and the purpose of the Act as a whole.

B It seems very doubtful whether Miss Moberley's argument on the construction point was pushed home in front of the learned judge as clearly and fully as it has been in this court. However, having heard her argument, to whose quality I would like to pay the same tribute as did the judge, I am left in no doubt that it is correct. It does not seem to me that the countervailing considerations to which the judge referred in the passage which I have quoted from his judgment can fairly displace that view. I would therefore decide the construction point in favour of Miss Moberley and hold that section 67(1) does not enable an application to be made for the re-registration of an existing registered rent.

What then is the effect of that disablement? Does it invalidate the subsequent determination or is there some way in which it can be
C upheld? As to that question, Mr Ouseley has referred us to the decision of the Divisional Court in *Chapman* v *Earl* [1968] 1 WLR 1315 and also to the later decision of this court in *Druid Development Co (Bingley) Ltd* v *Kay* (1982) 44 P & CR 76,* in which the earlier decision was distinguished.

Chapman v *Earl* was a case under the predecessor legislation to the 1977 Act, where two tenants of the same landlord had completed on the prescribed forms, and submitted to the rent officer, applications for the registration of fair rents for their flats. However, neither tenant stated the rent which he sought to register. At first, that was not thought to be an objection. The matter proceeded in the normal way, rents were determined by the rent officer and the landlord appealed to the rent assessment committee, who slightly increased
D them. Then the landlord applied to the Divisional Court, and it was held, granting orders of *certiorari*, that the statement of the proposed rent was an essential condition to the exercise by the rent officer of his powers. Accordingly, a failure to specify a rent was a fatal defect which could not be waived. In the result, the court, in the exercise of their discretion, quashed the determinations of the rent assessment committee.

I now return to section 67(2) of the Rent Act 1977, which I have not yet read. That subsection provides: "Any such application" (ie any application under section 67(1)) "must be in the prescribed form and contain the prescribed particulars in addition to the rent which it is sought to register". That is a provision comparable with those which
E were under consideration in the two earlier cases. Mr Ouseley submits that this is a case where the three-year period has now expired and an application specifying even a marginally different rent could have been entertained. Therefore, he says, there was adequate compliance with the requirements of section 67(2) and the case is covered by *Druid Development Co (Bingley) Ltd* v *Kay*.

I cannot accede to that submission. If, as I have held, the application is not permitted by section 67(1), it never becomes necessary to consider section 67(2). If there is no power to make the application, there is no power for the rent officer to act on it. Alternatively, if the true view be that it is necessary to consider section 67(2), an application specifying a rent which cannot be re-registered must, at the least, stand on the same footing as one which
F does not specify a rent at all. The defect is a fundamental one. For these reasons I am of the opinion that the present case is either *a fortiori* to, or at the least on the same footing as, *Chapman* v *Earl*. The determination of the new rent is liable to be quashed.

In all the circumstances, and subject to one final point, it would seem to me that Miss Moberley has made out her case on the construction point and its consequences, and that this appeal should be allowed on that ground. The final submission made by Mr Ouseley is that this court should, in the exercise of its discretion, refuse to make an order of *certiorari* in the present case. I have taken into account everything which Mr Ouseley has said on that point, including the fact that some hardship may indeed be caused to the landlord if he thought that the matter was properly in issue and that he had been given the opportunity to contend for a higher rent as from October 13 1985. I have also taken into account the fact that this is a case where the order is sought by the person who was herself responsible for making an application which she had no power to make.

Having considered those points, I am not persuaded that there is any sound reason for our declining to make the order. That makes it unnecessary to consider any of the other three grounds of Miss Moberley's application and we have not heard argument from Mr Ouseley on them. I propose to say nothing at all about them.

I would allow this appeal.

WOOLF LJ agreed and did not add anything.

Also agreeing, SIR GEORGE WALLER said: I would very briefly say what my view was on the question of construction. In my opinion section 67, with subsections (1), (2) and (5), contemplates two possibilities, namely the registration of rent and the confirmation of a rent already registered. There is no provision in the section for re-registering a rent which has already been registered. Confirmation can take place only where an application has been made for a different rent. As my lord said, this view of the section is strengthened when one looks at Schedule 11, paras 4 and 5, and particularly para 5, which says:

> After considering, in accordance with paragraph 4 above, what rent ought to be registered or, as the case may be, whether a different rent ought to be registered, the rent officer shall, as the case may require, — (a) determine a fair rent and register it as the rent for the dwelling-house; or (b) confirm the rent for the time being registered and note the confirmation in the register.

There is no provision there for re-registering a rent for which an application has been made to re-register the existing rent. The only opportunity of perpetuating for a further time the existing rent is where an application has been made for a different rent and the existing rent is confirmed.

The appeal was allowed with costs in the Court of Appeal and below.

Court of Appeal

December 20 1985

(Before Lord Justice FOX and Sir Roualeyn CUMMING-BRUCE)

HEWITT v LEWIS

Estates Gazette May 3 1986

278 EG 515-516

Rent Act 1977, Schedule 15, Case 11 and Rent (Amendment) Act 1985 — Retrospective effect of latter Act — Application to pending actions — Appeal by owner from decision of county court judge dismissing claim for possession of dwelling-house — Appellant or a member of his family lived in the house from 1965 until October 1982, but the house was unoccupied from October 1982 until February 1983, when it was let on a regulated tenancy to the present respondent for a term of 13 weeks — Appellant gave respondent the notice required by Case 11 that possession might be recovered under this Case, but at the end of the 13 weeks the respondent remained in the house — Appellant brought an action in the county court for possession — Judge, relying on the decision of the Court of Appeal in *Pocock* v *Steel*, dismissed the action on the ground that the appellant did not occupy the house immediately before the grant of the regulated tenancy to the respondent — The judge's decision was given before the Rent (Amendment) Act 1985, which was passed to reverse the effect of *Pocock* v *Steel*, became law — Appellant now claimed that his appeal should be allowed as a result of the 1985 Act, which provided that Case 11 applied where the owner-occupier had occupied the house as his residence "at any time before the letting" — It was clear that the 1985 Act was intended to have some retrospective effect, as section 1(4) provided that Case 11 in Schedule 15 to the 1977 Act should apply to tenancies

*Editor's note: Also reported at (1982) 264 EG 1080.

granted and notices given before, as well as after, the commencement of the 1985 Act — The question was, however, whether the retrospective operation applied to pending actions — The fact that the decision in the court below was given before the 1985 Act became law was immaterial because the present appeal was by way of a rehearing — Held that the only realistic interpretation of the 1985 Act was that it was intended to apply to pending actions such as the present one — *National Real Estate & Finance Co* v *Hassan* distinguished

This was an appeal by the plaintiff, Mr Hewitt, owner of a freehold dwelling-house in Horsham, Sussex, from a decision of Judge MacManus at Horsham County Court dismissing his action against the respondent, Mr Lewis, defendant in the action, for possession of the house.

J C Harper (instructed by Robbins Olivey & Blake Lapthorn) appeared on behalf of the appellant; Miss C Jones (instructed by Capel Cure Peachey & Bates, of Brighton, Sussex) represented the respondent.

Giving judgment, FOX LJ said: This is an appeal by the plaintiff, Mr Hewitt, from an order of Judge MacManus at Horsham County Court, dismissing his application for possession of a dwelling-house in Horsham of which he is the freehold owner.

Mr Hewitt first occupied the house in 1965, when it had just been built. He lived there with his wife and children until June 1977, when he went to live in Holland. He was accompanied by his wife and three of the children. The other child, David, who was then aged 17, was completing his education in England. He continued to live in the house until October 1982. Mr Hewitt paid all the outgoings (including household bills and the mortgage payments). From time to time he went back to the house — sometimes by himself and sometimes with his whole family.

In October 1982 David finished his education. Mr Hewitt decided to let the house and contacted estate agents, King & Chasemore; we do not know precisely when this decision was made. The property, however, was unoccupied from October 1982 to February 1983, when it was let furnished to the defendant, Mr Lewis, for a term of 13 weeks. Prior to the commencement of the tenancy, Mr Hewitt gave to Mr Lewis a notice under Case 11 of Part II of Schedule 15 to the Rent Act 1977. That is recorded in clause 25 of the tenancy agreement, which is dated February 24 1983. Mr Lewis went into occupation under that tenancy in February 1983. The tenancy expired on May 20 1983. Mr Lewis remained in possession.

In June 1983 a further tenancy agreement seems to have been prepared (but was not signed) by Mr Hewitt. However, Mr Lewis remained in occupation as a tenant. The tenancy was at all times a regulated tenancy for the purposes of the Rent Acts.

On September 27 1984 Mr Hewitt sought an order for possession on the ground that he had previously occupied the premises as a residence and that they were now required as a residence for Mr Hewitt's son Richard, who resided with Mr Hewitt when he last occupied the house as a residence. Notice to quit was served on Mr Lewis on September 15 1983. The Rent Act 1977 specifies certain "Cases" in which an order for possession may be made. One of these is Case 11 in Part II of Schedule 15 to the Act. The Case (as amended) so far as material was in the following terms prior to the Rent (Amendment) Act 1985:

> Where a person who occupied the dwelling-house as his residence (in this Case referred to as "the owner-occupier") let it on a regulated tenancy and — (a) not later than the relevant date the landlord gave notice in writing to the tenant that possession be recovered under this Case, and (b) the dwelling-house has not, since ...(ii) 14th August 1974, in the case of a regulated furnished tenancy ... been let by the owner-occupier on a protected tenancy with respect to which the condition mentioned in paragraph (a) above was not satisfied, and (c) the court is of the opinion that of the conditions set out in Part V of this Schedule one of those in paragraphs (a) and (c) to (f) is satisfied.

The conditions set out in Part V of Schedule 15 include:

> (a) the dwelling-house is required as a residence for the owner or any member of his family who resided with the owner when he last occupied the dwelling-house as a residence....

If the case comes within Case 11, the court would be obliged, in the present case, to make an order for possession (Rent Act 1977, section 98).

The relevant date is the date of the commencement of the tenancy in question (Schedule 15 Part III para 2). Para (a) of Case 11 is therefore satisfied. Para (b) is also satisfied. As regards para (c), the question before us was the applicability of condition (a) in Part V of Schedule 15, which I have already set out. G

In *Pocock* v *Steel* [1985] 1 WLR 229*, the Court of Appeal held that, upon the true construction of Case 11, a landlord claiming to be the owner-occupier had to have occupied the dwelling-house as his residence immediately before he let it to the statutory tenant. On the basis of that decision, Judge MacManus, in the present case, held that Mr Hewitt did not succeed in bringing the case within the provisions of Case 11, because immediately before the grant of the tenancy in February 1983, Mr Hewitt was not, himself, personally residing in the property — nor did he do so immediately prior to the separate grant (if any) of a subsequent letting. Accordingly, the judge dismissed the claim for possession. H

Pocock v *Steel* (*supra*) was decided on November 16 1984. Parliament intervened very quickly. On May 23 1985 the Rent (Amendment) Act 1985 was enacted. Section 1(1) provides:

> In Case 11 in Schedule 15 of the Rent Act 1977 ... for the words from the beginning to "tenancy" (where it first appears) there shall be substituted the words "Where a person (in this Case referred to as "the owner-occupier") who let the dwelling-house on a regulated tenancy had, at any time before the letting, occupied it as his residence".

Section 1(4) provides:

> Case 11 in Schedule 15 to the Rent Act 1977 and Case 11 in Schedule 2 to the Rent (Scotland) Act 1984, as those cases have effect by virtue of this section, apply to tenancies granted and notices given before, as well as after, the commencement of this Act. J

Judgment in the present case was given on February 15 1985. The notice of appeal was dated March 12 1985; we gave leave, with consent, to Mr Hewitt to amend it, adding the contention that the appeal should be allowed as a result of the enactment of the Rent (Amendment) Act 1985.

The first question is whether the retrospective provision of section 1(4) of the 1985 Act applies to this case, notwithstanding that judgment in the court below was given before the statute became law. If it does, then that disposes of the case, because Mr Hewitt had undoubtedly at some time previous to the letting occupied the house as his residence. It is not in dispute that any necessary notice under Case 11 was given at the proper time. K

The general rule is that statutes are construed as operating only in cases or on facts which came into existence after the statute was enacted, unless a retrospective effect is clearly intended. In the present case there is no doubt that a retrospective effect was intended; that is evident from section 1(4). The question is whether the 1985 statute was intended to apply to pending actions. I do not think that, in this case, the fact that the hearing at first instance was before the passing of the statute determines the matter. This appeal is by way of rehearing (see RSC Order 59, rule 3), and the statute plainly has a retrospective effect so the matter resolves itself into the issue whether, if the hearing at first instance had been immediately after the statute was enacted, the provisions of section 1(1) would have applied to it. L

I do not think that the decision at first instance gave to the defendant anything in the nature of a vested right. And the appeal (as I have said) is a rehearing. In *A-G* v *Vernazza* [1960] AC 965 Lord Denning, at p 978, said that it was:

> clear that in the ordinary way the Court of Appeal cannot take into account a statute which has been passed in the interval since the case was decided at first instance, because the rights of litigants are generally to be determined according to the law in force at the date of the earlier proceedings. ... But it is different when the statute is retrospective either because it contains clear words to that effect, or because it deals with matters of procedure only, for then Parliament has shown an intention that the Act should operate on pending proceedings, and the Court of Appeal are entitled to give effect to this retrospective intent as well as a court of first instance. M

I appreciate that a statute may be intended to have some retrospective effect without necessarily being intended to affect pending actions. But in my view the only realistic interpretation of the 1985 statute is that Parliament did intend it to have such effect. Thus, the Act expressly applies to tenancies granted, and Case 11 notices given, before as well as after the Act. As a matter of language it embraces, therefore, the relevant facts of a pending, just as much as a future, action. It is difficult to see any reason why pending actions should be unaffected. This is not a situation where the pre-existing law had been allowed to continue for a long period before the

---

*Editor's note: Also reported at [1985] 1 EGLR 115.

intervention by Parliament. The 1985 statute was enacted very quickly to displace the effect of *Pocock* v *Steel*, a decision which, as I understand, was thought in the profession to have overturned the law as it was then believed to be.

To construe the 1985 statute as not extending to pending actions will, it seems to me, only cause expense and inconvenience because it would immediately be open to the unsuccessful landlord to start a new action for possession in reliance on the new statute. There could be no question of his being defeated by a plea of *res judicata* because the issue raised in consequence of the provisions of section 1(1) of the new statute would never have been litigated in the previous action. The issues would be quite different. I can see no reason why Parliament, having acted so speedily, should be taken as intending to force upon the landlord a new action on an issue which more easily could be determined in the existing proceedings. If the action had been heard at first instance a few days after the statute became law, I cannot see any reason of policy which could conceivably have excluded the full operation of the statute.

We were referred to the decision of the Court of Appeal in *National Real Estate & Finance Co Ltd* v *Hassan* [1939] 2 KB 61. Under the Law of Property Act 1925, section 146, a landlord's right to re-entry or forfeiture for failure to repair could not be enforced by action or otherwise until the lessor served a repair notice on the tenant, and the tenant failed to remedy the breach. By the Leasehold Property (Repairs) Act 1938 the tenant was authorised to serve a counternotice, claiming the benefit of the Act. Where a counternotice was served, the landlord was not permitted to take any proceedings to enforce the right of re-entry without the leave of the court. The Act of 1938 was expressed to apply to leases created, and to breaches occurring, before the commencement of the Act. The plaintiff's notice under the Law of Property Act 1925, section 146, was served on January 12 1937. The plaintiff's writ claiming possession was issued on July 13 1937. The Act of 1938 became law on June 23 1938. Croom-Johnson J decided that, in view of the Act of 1938, the claim was bad. The Court of Appeal allowed the landlord's appeal. Goddard LJ at p 79 said:

> The argument here is to the effect that an action which had been brought but which did not come on for trial until after the Act of 1938 came into force is defeated because the condition precedent in the serving of the notice did not when served in January 1937 contain a reference to an Act passed in 1938

And earlier, on p 79, he said:

> But how can it be supposed Parliament meant to say that notices properly served under an Act which was then in force

ie section 146 of the Law of Property Act 1925

> and which gave a property owner the right to bring an action, ought to be set aside because they did not contain a reference to an Act which had not then been passed?

The *National Real Estate* case seems to me to be far removed from the present. The requirements of the Act of 1938 were simply incapable of application to an action already in existence when the Act was passed. There could be no question of a retrospective effect in the manner contended for by the defendant.

We were also referred to *Zainal bin Hashim* v *Government of Malaysia* [1980] AC 734, which was a Privy Council decision. That again was a very different case from the present. The retrospective clause in the statute took the form of a deeming provision. The case is material, however, for its emphasis upon a clear indication of an intention that the retrospective effect of the statute should extend to pending actions. For the reasons which I have indicated, it is, in my view, clear that the retrospective operation of the Rent (Amendment) Act 1985 does so extend.

On the view which I take of the effect of the 1985 Act, it is not necessary to deal with the alternative contention advanced on behalf of Mr Hewitt that he was in fact occupying the house as his residence immediately before the letting.

I would allow the appeal.

SIR ROUALEYN CUMMING-BRUCE agreed and did not add anything.

The appeal was allowed, the respondent to pay three-quarters of the costs of the appeal; the order for costs below to stand.

Court of Appeal

February 25 1986

(Before Lord Justice MAY and Sir George WALLER)

BRETHERTON v PATON

Estates Gazette May 10 1986

278 EG 615-618

Landlord and tenant — Rent Act — Application of principle of *Street* v *Mountford* — Parties entered into an arrangement by which the appellant was allowed into occupation of a dwelling-house with the ultimate intention that she should purchase the property — It was contemplated that she should do some repairs and later obtain a mortgage which would enable her to purchase — The respondent owner did not wish to grant a tenancy — No contract of sale was made and no purchase price was agreed — Appellant paid the rates and made a small weekly payment to the respondent, described as a contribution towards insurance — It was common ground that she had exclusive possession — After a time difficulties arose, appellant's enthusiasm for the scheme waned, but she wished to continue in occupation — Respondent gave her notice to quit and, on her refusal to leave, brought proceedings for possession — Judge in the county court treated the transaction as an ineffective arrangement for the sale and purchase of land, within the exceptions to the creation of a tenancy mentioned by Lord Templeman in *Street* v *Mountford*, and granted the respondent possession — On appeal the appellant submitted that all the indicia of a tenancy identified in *Street* v *Mountford* were present, exclusive possession for a term at a weekly payment in the nature of rent, and it was conceded that if the appellant was a tenant she was protected by the Rent Act 1977 — Respondent argued that, although there was here no concluded contract of sale, there was an intention to enter into such a contract; that the whole flavour of the arrangement was one of sale and purchase; and that the occupation should be regarded as referable to a sale and so within a specific exception mentioned by Lord Templeman — Held, rejecting the respondent's argument, that although a possible sale was ultimately in contemplation when the appellant went into occupation, the only intention which was relevant was that demonstrated by the facts of the transaction, namely, the grant of exclusive possession at a rent for a periodic term — That was, in this case, a tenancy protected by the Rent Act — Appeal allowed

This was an appeal by Karen Lorraine Paton, defendant in an action for possession in Preston County Court brought by William Bretherton, the plaintiff and present respondent, in relation to a dwelling-house in Freckleton, near Preston.

The appeal was from a decision of Mr Assistant Recorder Hegarty, giving judgment in favour of the respondent for possession and mesne profits.

Jeremy Mark Nicholson (instructed by David Coupe & Co, of Kirkham, Lancs) appeared on behalf of the appellant; Mark Blackett-Ord (instructed by David Blank & Co, of Manchester) represented the respondent.

Giving judgment, MAY LJ said: This is an appeal from the order of Mr Assistant Recorder Hegarty, made in the Preston County Court on July 9 1985 in possession proceedings between the plaintiff (the respondent in this court) and the defendant (the appellant). The assistant recorder gave judgment for the respondent for possession of a dwelling-house in Freckleton, near Preston, with certain mesne profits. The appellant now appeals against that order, contending that at the material time she had a tenancy which was protected by the provisions of the Rent Act and that accordingly the assistant recorder had no power to make a possession order.

Briefly the facts of the case are as follows. The respondent has owned the dwelling-house for 35 years. For many years he let it to tenants. In so far as the most recent letting to a Mrs Foreman was concerned, not only did he have rent arrear problems with his tenant but also during her occupation of the property the condition of the premises deteriorated to a serious degree. At that stage, as the result of those experiences, the respondent decided that he did not in the future wish to lease the house, but instead determined to sell it.

In about March 1982 the appellant, Miss Paton, heard that the property was about to become vacant. She got in touch with the respondent and pressed him to let the house to her. But for the reasons which I have given the respondent resisted her proposal and told her that he was not prepared to do so but he was willing to sell it to her in its then condition at a price of £17,000.

For a period each of them maintained their respective stances. There then emerged another proposal, namely that Miss Paton should be given time to put the property into good condition, occupying it the while, so that once it was in good condition she would be able to obtain a mortgage and then buy it from the respondent.

The learned judge held that in broad terms that was the arrangement which was made between the parties. He then went on to say:

> it is clear that for the purposes of this scheme [the appellant] was to be permitted to have possession of the property and, it would seem, to reside there. It was accepted on behalf of [the respondent] that as a result of the arrangement [Miss Paton] obtained the exclusive possession of the property.

However, there were other aspects of the arrangement to which the parties came. It was common ground that, during the period of her occupation, Miss Paton made certain payments which were referable to her occupation of the premises. She paid the rates direct to the relevant authorities and in addition she paid the sum of £1.20 each week to the respondent. There was an issue of fact about the weekly payments of £1.20. The learned judge preferred the respondent's evidence with regard to it, namely that they were contributions towards the insurance premium for insuring the property until the completion of the anticipated sale by the respondent to the appellant when the repairs had been done, as I have indicated. There was included in the £1.20 a small sum which, apparently at the behest of the respondent's accountant for some reason which is unclear, represented the respondent's costs of collecting the premium instalments. There was a dispute, or uncertainty, as to how much had from time to time been paid; thus an old rent book was used to record the payments. In preferring the respondent's account of the arrangement in relation to these payments, the learned judge reached the conclusion that the respondent did not in the course of the discussions refer to them in any way as rent.

Miss Paton went into possession. Some work was carried out to the property thereafter, but within a short space of time she began to realise that there might well be substantial practical problems, particularly from the point of view of obtaining the necessary finance in implementing the loose arrangement, especially in so far as doing the proper repairs was concerned. The learned judge, having so held, went on to find that in his view Miss Paton was quite happy to remain in occupation of the property but that her enthusiasm for the original scheme had waned after she had gained possession. The original period during which it was envisaged that Miss Paton would have to carry out the necessary repairs was about two years. Before the end of that time Miss Paton and her boyfriend, or fiancé, had the property valued. It would seem that she was advised that subject to her occupation (I use for the moment a neutral phrase) the house was valued at £8,000 or thereabouts and at £14,500 with vacant possession. In consequence the appellant made various offers to the respondent to buy the property at sums lower than £17,000, but throughout the respondent was adamant that that was his price and not a penny less.

Eventually, at the behest of the local authority — no doubt under the Housing Act or similar public health legislation — the respondent himself had to carry out certain repairs. There was some inconclusive discussion between the parties about an extension of time. Ultimately the respondent's patience gave out and he served a notice to quit by his solicitors on December 4 1984, giving the appellant four weeks' notice. That was a document in standard printed form, which is in our papers. It is to be noted that the first reference to a tenancy was deleted, as was the prescribed information in the notes to the form which the relevant legislation requires in relation to lettings and tenancies. The appellant did not vacate the premises pursuant to that notice to quit and accordingly the respondent brought his successful claim below for possession and mesne profits.

As I indicated at the start of this judgment, it is common ground between the parties that, if the appellant is a tenant of the property, G she is protected by the provisions of the Rent Act 1977. If that is the situation the claim for possession had to fail because no statutory grounds for possession were relied upon or shown. Thus, as the learned assistant recorder indicated below, the legal issue between the parties was whether, in the light of the facts to which I have referred, the appellant was a tenant of the premises.

The learned judge considered that question in the light of the recent decision of the House of Lords in *Street* v *Mountford* [1985] 1 AC 809*. In that case Lord Templeman, who delivered the single speech with which the rest of their Lordships' House agreed, discussed the legal effect of a document which regulated the relationship between the two parties in that case and in respect of which the issue was whether it created the licence it expressed to create, or whether in the H circumstances a tenancy had arisen. In the event their lordships decided that a tenancy had been created. In the course of his speech Lord Templeman, at p 816, began by saying "there is no doubt that the traditional distinction between a tenancy and a licence of land lay in the grant of land for a term at a rent with exclusive possession". He then considered and compared two cases in which the purpose of the transaction had been to cut timber and whether exclusive possession had been granted or not for that purpose. He then turned to the case of residential accommodation and said this:

> In the case of residential accommodation there is no difficulty in deciding whether the grant confers exclusive possession. An occupier of residential accommodation at a rent for a term is either a lodger or a tenant. The occupier J is a lodger if the landlord provides attendance or services which require the landlord or his servants to exercise unrestricted access to and use of the premises. A lodger is entitled to live in the premises but cannot call the place his own.

Then, after a reference to a judgment of Blackburn J in the earlier case of *Allan* v *Liverpool Overseers*, the learned law lord continued with a passage which, as I read the speech as a whole, contains the ratio of their lordships' decision. The passage reads in this way:

> If on the other hand residential accommodation is granted for a term at a rent with exclusive possession, the landlord providing neither attendance nor services, the grant is a tenancy; any express reservation to the landlord of limited rights to enter and view the state of the premises and to repair and maintain the premises only serves to emphasise the fact that the grantee is K entitled to exclusive possession and is a tenant. In the present case it is conceded that Mrs Mountford is entitled to exclusive possession and is not a lodger. Mr Street provided neither attendance nor services and only reserved the limited rights of inspection and maintenance and the like set forth in clause 3 of the agreement. On the traditional view of the matter, Mrs Mountford not being a lodger must be a tenant.
>
> There can be no tenancy unless the occupier enjoys exclusive possession; but an occupier who enjoys exclusive possession is not necessarily a tenant. He may be owner in fee simple, a trespasser, a mortgagee in possession, an object of charity or a service occupier. To constitute a tenancy the occupier must be granted exclusive possession for a fixed or periodic term certain in consideration of a premium or periodical payments. The grant may be express, or may be inferred where the owner accepts weekly or other periodical payments from the occupier. L

Thereafter the learned law lord considered in detail a substantial number of the earlier cases in which the issue whether the relationship between the parties was a licence or a tenancy had arisen, in respect of some of which at least it is I think permissible to say that, having regard to the incidence of landlord and tenant legislation, the merits of certain cases may have had rather more effect upon the ultimate result than was perhaps warranted. In the course of his analysis of those earlier decisions Lord Templeman, at p 821, in referring to the case of *Errington* v *Errington and Woods* [1952] 1 KB 290, spoke of ways in which the intention to create a tenancy could be negatived in these terms:

> The intention to create a tenancy was negatived if the parties did not intend to enter into legal relationships at all, or where the relationship between the M parties was that of vendor and purchaser, master and service occupier, or where the owner, a requisitioning authority, had no power to grant a tenancy.

Finally, after further consideration of other cases, the learned law lord, towards the end of his speech, said:

> My Lords, the only intention which is relevant is the intention demonstrated by the agreement to grant exclusive possession for a term at a rent. Sometimes it may be difficult to discover whether, on the true construction of an agreement, exclusive possession is conferred. Sometimes it may appear from the surrounding circumstances that there was no intention to create legal relationships. Sometimes it may appear from the surrounding circumstances that the right to exclusive possession is referable to a legal relationship other

*Editor's note: Also reported at [1985] 1 EGLR 128.

> A than a tenancy. Legal relationships to which the grant of exclusive possession might be referable and which would or might negative the grant of an estate or interest in the land include occupancy under a contract for the sale of the land, occupancy pursuant to a contract of employment or occupancy referable to the holding of an office. But where as in the present case the only circumstances are that residential accommodation is offered and accepted with exclusive possession for a term at a rent, the result is a tenancy.

Having considered that decision, in the instant case the learned judge concluded that the three indicia of exclusivity of possession, a term fixed or periodic, and the payment of rent, were satisfied on the facts of the case. He took the view that, subject to a point with which he then dealt, the necessary consequence was that the nature of the appellant's occupation was a tenancy. But he then went on to B consider the effect upon that *prima facie* result, applying the ratio in the case of *Street* v *Mountford*, of the loose arrangement for occupation by the appellant pending completion of repairs and thereafter the sale at an unspecified figure to which I have already referred. In connection with that, the learned judge said first:

> Miss Paton's occupation of the property is referable to and forms part of an ineffective arrangement for the sale and purchase of land. It therefore seems to fall into or across two of the categories of exceptional circumstances recognised by Lord Templeman.

That was a reference to the last passage from the speech of Lord Templeman which I have just quoted. Then on the following page in his judgment the learned judge said:

> C In my judgment, therefore, the original arrangement between the parties did not make Miss Paton a tenant of the premises. Her occupation of the property is referable to an ineffective arrangement for the sale and purchase of land and the inference of a tenancy that would otherwise arise from the exclusive occupation of premises for a term at a rent is thereby rebutted.

It was on that basis that he concluded that the appellant's occupation of the premises was not a tenancy and not a tenancy protected by the Rent Act, and consequently made the order now appealed against.

Before this court the argument for the appellant has been based fairly and squarely upon the decision in *Street* v *Mountford*. Counsel has submitted that, when one analyses the learned judge's findings, there was a contract between the appellant and the respondent. There was clearly an intent to create legal relations, and indeed Mr Blackett-D Ord, on the respondent's behalf, accepted that that was the situation. In the course of his judgment the learned judge spoke expressly of "the legal relationship between the parties in the instant case", and there seems little doubt that he found (as I think rightly on the material that was before him) that there was a legally enforceable relationship between the two parties. Clearly it was not a contract which embodied the somewhat vague scheme that I have outlined, because a number of the necessary incidents of that scheme at no time were agreed between the parties. The ultimate purchase price, to take only one example, was never agreed between Miss Paton on the one hand and the respondent on the other. That being so, counsel for the appellant submitted that that was a contract pursuant to which the appellant had, as the learned judge found, exclusive occupation of E the premises and paid £1.20 per week periodic payment in respect of her occupation. That being the situation, and those being the circumstances of her occupancy, he submitted, when one applies the ratio of the decision in *Street* v *Mountford* the learned judge erred in concluding that there was no tenancy and consequently this appeal should be allowed.

In reply Mr Blackett-Ord, on behalf of the respondent, has advanced two alternative arguments. The first was based upon what he submitted was the clear intention of the parties or (to put it another way) the absence of any intention on the part of the parties to grant a tenancy on the facts of the instant case; what was contemplated was no form of tenancy, but an ultimate sale and purchase. In those circumstances he submitted that the decision in *Street* v *Mountford* F was irrelevant. At that stage, he contended, the fact that the intended contract for sale had never become sufficiently certain to be enforceable mattered not. The way he put it was that the whole flavour and colour of the arrangement was one of sale and purchase and it was therefore illogical and unjust to find that a tenancy had been created. He specifically relied upon what he described as the exceptions to the *Street* v *Mountford* principle, which I have already quoted from Lord Templeman's speech.

In the alternative, Mr Blackett-Ord submitted, the substance of the ultimate intentions of the parties had never been embodied in an enforceable contract and, although the appellant had entered into occupation and paid rent, it was not in law permissible to extract from an incomplete but potential contractual arrangement a term or terms and upon that term or terms to base the application of the principles adumbrated in *Street* v *Mountford* and hold that a tenancy was created.

In so far as Mr Blackett-Ord's submissions relating to intention are concerned, I think it only necessary to repeat the first sentence of the last quotation from Lord Templeman's speech to this effect:

> My Lords, the only intention which is relevant is the intention demonstrated by the agreement to grant exclusive possession for a term at a rent.

In so far as Mr Blackett-Ord's alternative submission is concerned, in other circumstances and as a general proposition of law his submission might have some substance. But I do not think it is applicable to the particular circumstances of the present case. There seems no doubt that, although the ultimate sale and purchase of the premises never reached enforceable form, there certainly was a contract between the appellant and the respondent under which the appellant went into possession and for that exclusive possession, as it was found to be and was not disputed, she undertook to pay not merely the rates but also £1.20 per week to the respondent.

That being the situation, the matter can be shortly disposed of. Those facts, as the learned judge below recognised, clearly fall within the principles reiterated by their Lordships' House in *Street* v *Mountford*. There was exclusive possession. There was the payment of rent. There was exclusive possession for a periodic fixed term. In those circumstances, on the law, there must have been a tenancy between the appellant and the respondent, and a tenancy, we are told, just within the protection of the Rent Act.

It is, I think, wrong to consider the circumstances referred to by Lord Templeman in the two passages that I have quoted as exceptions to the *Street* v *Mountford* principle. For my part I do not read that case as in any way making new law or requiring any hitherto unrecognised approach to this type of problem. Where a person is in occupation of land and the question arises between him and another who is alleging that that other is entitled to possession, the necessary inquiry is as to the nature if any of the occupier's right to occupy, ie as to the nature of the occupier's interest if any in the land. All that *Street* v *Mountford* decided was that, given the instance of the three indicia referred to, the answer to that inquiry is a tenancy. In other circumstances, such as a service occupancy, or an allowing into possession by a requisitioning authority, the situation will be different. The instant case is a clear case in which the ratio of *Street* v *Mountford* has to be applied.

In those circumstances, with respect to the learned judge, I differ from the conclusion to which he came and I would allow this appeal.

Agreeing, Sir George Waller said: I would only add a word or two out of respect for the argument which Mr Blackett-Ord, on behalf of the respondent, put before the court.

Taking the passage from *Street* v *Mountford*, quoted by my lord from p 826, where Lord Templeman had stated the essential requirements of a tenancy and then went on to deal with certain legal relationships which would not produce that result, Mr Blackett-Ord did not argue that there was no intention to enter into legal relations. But he did argue that, although arrangements made between the parties were far from agreement because each side intended a sale at some time, albeit at different prices, since occupation pursuant to a contract of sale would not give rise to a tenancy, as mentioned by Lord Templeman, a similar result would follow where there was merely an intention to enter into a contract. He submitted that, on the facts of this case, there was no general agreement and the amount of actual contractual agreement related only to the payment of £1.20 each week. He would therefore have brought this case within the second part of the paragraph of Lord Templeman's speech.

I do not find it possible to accept that argument. These parties entered into an arrangement with the ultimate intention of negotiating a sale. But, to quote the words of Lord Templeman, the intention which was demonstrated by the agreement was "to grant exclusive possession for a term at a rent", and therefore this case comes clearly within the decision of *Street* v *Mountford*.

I would also allow this appeal.

The appeal was allowed with costs and the order below set aside. Legal aid taxation of the appellant's costs was ordered.

Queen's Bench Division

March 19 1986

(Before Mr Justice McNEILL)

R v LONDON RENT ASSESSMENT PANEL, EX PARTE CHELMSFORD BUILDING CO LTD

Estates Gazette May 31 1986

278 EG 979-984

Rent Act 1977 — Decision of rent assessment committee of London Rent Assessment Panel challenged by application for judicial review — Landlords seek to have committee's decision quashed on the ground that the reasons given for it were inadequate — Committee in this case had upheld the determination of the rent officer that the rent of a flat should be £100 per month exclusive of rates — A main complaint made by the landlords was that the committee's reasons did not deal satisfactorily with the landlords' case in regard to the effects of inflation — The committee had not in fact commented in any detail on landlords' submissions as to percentage increases of costs and increases in average earnings, but had indicated that they did not accept their application — The court referred to the limitation on the committee's power and duty to give reasons in detail in determining an appeal on a fair rent, which was a matter of valuation, as explained by Lord Widgery CJ in *Metropolitan Holdings Ltd* v *Laufer* and *Guppys (Bridport) Ltd* v *Sandoe* — The committee had taken inflation into account and their reasons were sufficient to show how their decision was reached — The suggestion that the reference in their decision to their own knowledge and experience was merely a "common form of wording" which might cover weaknesses was rejected; the Court of Appeal in *Wareing* v *White* had recognised the phrase in which these words occurred as a permissible way for the committee to express themselves unless they were being untruthful — The judge agreed with comments made by Mann J in *Ellis & Sons Fourth Amalgamated Properties Ltd* v *Southern Rent Assessment Panel* as to the advantages of using the procedure of judicial review rather than statutory appeal to challenge the decision of a rent assessment committee — Application dismissed

This was an application by landlords, Chelmsford Building Co Ltd, for judicial review seeking an order of certiorari to quash the decision of a committee of the London Rent Assessment Panel. The decision related to a flat at 1 Longwood Court, Longwood Close, Corbets Tey Road, Upminster, of which the tenant at the time was a Mr Wakefield. The committee had confirmed the rent officer's registration of £100 per calendar month.

Derek Wood QC and Wayne Clark (instructed by Macdonald Stacey, agents for Elgar & Griffith, of Salisbury, Wilts) appeared on behalf of the applicants; John Laws (instructed by the Treasury Solicitor) represented the respondents.

Giving judgment, McNEILL J said: Chelmsford Building Co Ltd moves with leave for judicial review by way of certiorari to quash a decision of the London Rent Assessment Panel, reasons for which decision were given on October 23 1984.

The decision of the panel was upon an appeal by the applicant company from a registration by the rent officer of a fair rent of premises of which the applicant company was landlord, namely, flat no 1, Longwood Court, Longwood Close, Corbets Tey Road, Upminster. The tenant at that time was a Mr Wakefield. He has since left. By its decision the panel upheld the rent officer's determination that the fair rent for the premises (upon terms of the lease which are not material to this application) was £100 per calendar month.

No question arises as to the jurisdiction or as to the statutory functions of the rent officer and the panel under the Rent Act 1977. Moreover, although there is a statutory right of appeal from the panel under the Tribunals and Inquiries Act 1971 (section 13), Mr John Laws for the panel does not contend that these proceedings are improperly instituted. This is consistent with the opinion expressed by Mann J, which I accept and follow, in *Ellis & Sons Fourth Amalgamated Properties Ltd* v *Southern Rent Assessment Panel,* unreported, the judgment being given on February 8, 1984.* In the course of that judgment the learned judge said: G

> The appeal is under section 33 of the Tribunals and Inquiries Act and therefore attracts Order 55. The appellant is perfectly entitled to take that course, but as Mr Brown, who very helpfully appeared as *amicus* in this case, pointed out, a challenge to a decision of a Rent Assessment Committee by way of judicial review has three possible advantages. First, it enables the committee to be represented. The probability is that they would make use of that ability and thus avoid the delay such as occurred in the present case, which has occurred through the appointment of an *amicus.* Secondly, costs can follow the event in either direction. Thirdly, it permits the possibility of an appeal if a landlord's appeal against the committee's decision is successful. I of course postulate in looking at those advantages that the tenant does not appear. In the present case there is no tenant, but I understand that it is rare for tenants to appear on these occasions. So I would commend for consideration by those who are thinking of challenging a decision of a rent assessment committee that they should proceed by way of an application for judicial review. H

It is that course which has been followed in the present case. Those words of the learned judge appear to me to apply precisely to this application. Judicial review is particularly apt where what is under consideration is, as here, the procedure adopted by the panel and the reasons given by it for its decision. The present case is a good example of the task of this jurisdiction, to look not at the decision itself but at the decision-making process (see, for example, per Lord Brightman in *Chief Constable of the North Wales Police* v *Evans* [1982] 1 WLR 1155 at p 1174). J

Here the main thrust of the attack by Mr Derek Wood for the landlord applicant upon the decision of the tribunal is directed to the reasons given or the absence of any or any sufficient reasons given for the decision.

It is beyond question, as a matter of law, that a rent assessment panel must give reasons for its decision. This obligation arises under section 12(1) of the Tribunals and Inquiries Act 1971 and has been well recognised as applicable in this jurisdiction. The contrary is not argued. The matter is clearly expressed in the judgment of Lord Widgery CJ in *Metropolitan Property Holdings Ltd* v *Laufer* (1975) 30 P & CR 172 at p 176. Having stated that the obligation is derived from section 12(1) and having set out the relevant words of the subsection, the judgment continued: K

> So there is an obligation to state reasons, and it is necessary to say that this is a provision applicable to a wide range of tribunals, not merely the rent assessment committee, and it is, I think, obvious that the character of the reasons given, the nature of the reasons, the extent of the reasons, must in some measure be governed by the nature of the problem which the particular tribunal has been set to resolve. It is well established, so much so that I shall not refer to any authority, that rent assessment committees when fixing fair rents are entitled, indeed bound, to have regard to their own experience and knowledge. Initially they are not perhaps to be described as an expert tribunal, but they acquire a really remarkable knowledge of prevailing rents in their area from the simple experience which they gain from doing their job, and it has been, as I say, firmly established for many years that in considering what is a fair rent the committee can and should take into account its own experience and knowledge. That does not mean, as Lord Denning MR said, that they can throw away the expert evidence; they cannot throw away the expert evidence in the sense of declining to look at it. They have got to look at the evidence put before them, but if, as so often happens in the typical simple case, the landlords' experts say that the rent should be X and the tenant's experts say that the rent should be X minus Y, if the committee thinks that neither figure is the right one not only can it choose a figure in between the two extremes but it should do so. L
>
> Now, what reasons should the committee give in that situation? The answer is that in the simple typical case to which I have referred there are no reasons which the committee can give, save that it was not satisfied with either of the alternatives put forward in evidence and on its own expert knowledge preferred another figure. If that is all that the decision was, a simple matter of valuation opinion and nothing else, that is all that the decision involves, and if the decision is based simply on the committee's own views, having regard to the evidence put before them, then there are no reasons which the committee can give save to say that they think that their figure is right. Great complication and trouble would ensue if it were sought to make committees give reasons in such a situation beyond the simple and single reason to which I have referred. M

Here, says Mr Wood, the landlord advanced a substantial argument based upon what is by concession a relevant matter, namely, the effect of inflation itself and the effect of inflation upon the rentals of comparable properties, but he says the panel either

*Editor's note: The decision was in fact reported at (1984) 270 EG 39.

failed to give any reasons for rejecting the argument and the material supporting it or rejected it in such cryptic language as to make it impossible to ascertain whether it was rejecting it as inadmissible or irrelevant or upon some factual basis resulting from an assessment of conflicting evidence.

The factual background to the dispute can be set out shortly. Mr C B J Miller, a chartered surveyor and a Fellow of the Royal Institution of Chartered Surveyors, acts professionally for the landlord applicant, which was established by his late father. Since 1975 or thereabouts he has managed Longwood Court, which comprises six flats in one of the best parts of Havering. Four of the flats, including the subject flat, are conversions of the original building and the remaining two are in a more modern 1959 extension. A fair rent for the subject flat, no 1, was first registered in 1968. On April 16 1973 the panel determined the fair rent for the flat at £396 per annum, that is to say, £33 per month, effective from January 24 1972. This and all relevant rents hereafter referred to were exclusive of general and water rates. There had been no subsequent determination prior to the matters which I have to consider, but the tenant, Mr Wakefield, was in fact, prior to the dispute with which I am concerned, paying by agreement a monthly rental of £80 for the subject flat.

No issue of any significance arises as to the size, accommodation, facilities, state of repair and the like of flat no 1. The landlord's challenge to the rent officer's determination of a fair rent of £100 per month is, as appears from the proof of Mr Miller, which was before the panel, supplemented by his oral evidence or explanation. The landlord company was also represented by counsel. The tenant submitted written representations from Mr D J Parish, also a Fellow of the Royal Institution of Chartered Surveyors. Both Mr Miller's proof and Mr Parish's written representations are before me as exhibits to the landlord's solicitor's first and second affidavits.

According to Mr Miller, of the four flats in Longwood Court and the two then tenanted, only no 1 had a registered rent. However, at some stage, five of the six had been registered. No 3 was then tenanted at an unregistered rental of £108 per month, agreed in early 1983 on a three-year tenancy. In 1968 a fair rent for no 2 was registered at £23 per month. This flat was then occupied by Mr Miller's sister. Four other flats were subsequently registered and, on later references, he said that the fair rents registered were the same or, in most cases, more than the rental terms which had been offered to tenants.

Mr Miller also commented on four lettings referred to by the rent officer; one in particular, 7 Tadlows Close, Upminster, where the fair rent was registered at £123.50 per month. That was, said Mr Miller, the only one of the four referred to by the rent officer in close proximity to the subject flat.

Mr Miller advanced a number of contentions based on (i) the increased cost of complying with the landlord's obligations under the lease, put by him at an increase of 500/600% since the last registration in 1972; (ii) the average increase in income since that date of 480%, according to government statistics; (iii) The updating or phasing up of the rentals of the various flats referred to, to bring them to 1984 levels; (iv) A comparison of the accommodation and situation at 7 Tadlows Close and the fair rent registered for those premises in comparison with that registered for the subject flat. He concluded that on those bases the fair rent for flat no 1 should be registered at £160 per month.

Mr Parish, largely though not wholly, relied upon the condition and amenities of flat no 1 and upon seven rentals which he regarded as representing prevailing rentals in the district and concluded that the fair rent would be, as the rent officer had concluded, £100 per month. He regarded 7 Tadlows Close as "probably one of the best flats in the entire area with a number of advantages not shared by flat no 1".

That was the material before the tribunal, supplementary to the reports of and the particulars provided by the rent officer himself, and in addition to the inspection which the panel carried out prior to reaching their decision.

The method by which a fair rent is to be determined is laid down by section 70 of the Rent Act 1977. Subsection (1) reads as follows:

> In determining, for the purposes of this Part of this Act, what rent is or would be a fair rent under a regulated tenancy of a dwelling-house, regard shall be had to all circumstances (other than personal circumstances) and in particular to — (a) the age, character, locality and state of repair of the dwelling-house, and (b) if any furniture is provided for use under the tenancy, the quantity, quality and condition of the furniture.

Subsection (2), in effect, provides that scarcity value is to be disregarded and subsection (3) sets out a number of other matters which are specifically to be disregarded. These, then, are the provisions with which the panel had to comply when reaching its decision.

The panel set out its reasons for the decision as follows. The paragraphs in the decision are not in fact numbered, but can be so identified. The first three paragraphs are descriptive of the subject property. Nothing turns upon them or upon para 5, which records that the panel carried out its statutory duty under subsection (3). I set out the remaining paragraphs upon which the argument has turned:

> (4) We paid careful consideration to the written representations of Mr Parish on behalf of the tenant. We found his reference to recent registrations in the area to be of interest in showing the general level of rent officer registrations. We also noted his comments as to the state of repair of the property.
>
> (6) We also carefully considered the submissions made to us at the hearing by Mr Brock on behalf of the landlord, and by Mr Miller who presented his evidence in a convenient written proof.

The concluding sentence of that paragraph does not raise any controversial question.

> (7) We could not agree with Mr Miller's calculation of a rent of £160.00 per calendar month based on the updating of the unregistered rent of no 3 Longwood Court nor were we assisted to any extent by his argument based on phasing, having regard to the rental history in recent years of the subject property. We noted, but were not convinced by his submissions based on Government statistics of the index of average earnings from 1972-1984.

There again, I need not read the last sentence because no point is taken on that.

> (8) We inspected the exterior of 7 Tadlows Close, Upminster

and they give the reference of the decision

> which was drawn to our attention by both parties. We found this flat, which was situated close to the subject property, to be modern, of superior construction, and quite different from the subject flat.
>
> (9) In all those circumstances, we saw no reason to disturb the rent officer's determination.
>
> (10) Having regard to all the evidence put before us, to our inspection, to our own knowledge and experience, and to the provisions of section 70 of the Rent Act 1977, we confirmed the fair rent, to be registered exclusive of rates, at "100.00 per calendar month."

Mr Wood contended that although the theme running through the landlord's case before the panel was the effect of inflation and its impact upon other rents however established, whether by registration or agreement, the panel does not appear to deal with that in a manner which he said was legally satisfactory. Indeed, he said the way in which, in para (7), the panel dealt with Mr Miller's evidence, commencing with the words "We could not agree", appears to amount to dismissal of the validity of that evidence in respect of both inflation and updating (phasing), but gives no or no sufficient reason for dismissing it.

It is, in my view, clear that so far as no 7 Tadlows Close was concerned, the panel were not prepared to regard it as a proper comparable (see the last sentence of para (8)). Equally, the panel was (I consider) rejecting in clear terms Mr Miller's calculated rental of £160 per month, based on the updating of the unregistered rent at no 3 and his "phasing" arguments, having regard, as it was put, to the rental history of no 1. Finally, the panel expressed itself as not convinced by Mr Miller's argument based on the average earnings index. It could have been said, though it was not expressly spelt out, that the rental of £100 per month was broadly in line with registered rents of comparable properties as disclosed by the evidence. I am also satisfied, in particular from paras (4) and (6), that the panel did give consideration to all relevant material. Although para (4) of the grounds upon which relief is sought disputed this, Mr Wood did not rely on that ground at the hearing and it was not suggested on the landlord's behalf that the decision of the panel could be attacked on what can for convenience be called *Wednesbury* grounds.

Mr Laws submitted that the reasons given by the panel enabled an informed reader such as the landlord's professional adviser, Mr Miller, to know perfectly well what was and what was not being taken into account and how the decision was reached. I accept this. Moreover, he said the panel did take inflation into account. Although it expressly or by implication rejected or declined to apply Mr Miller's suggested percentage increases of rent to allow for inflation, and in particular for increased landlord's expenses and a proper return on capital, put variously at 480, 500 and 600%, the registered rent did in fact represent an increase of 200% over the last 1972 registered rent. Mr Wood had argued that a landlord such as the present applicant was entitled to guidance as to how this panel, adjudicating in a district where the landlord had a number of properties, would take into consideration inflation. I think it is sufficient for present purposes to say that in this case they did, though, of course, not to the extent claimed by the landlord. I am also mindful of the view of Lord Reid in *Mason* v *Skilling* [1974] 1 WLR 1437 at p 1439, commenting in relation to the valuation method of comparing with the subject premises the registered rent of comparable houses in the area, that ". . . as the number of comparable registered rents increases the more likely it will be that it will lead to a correct result".

I only mention, for example, that an unregistered rental for a superficially comparable house may reflect a scarcity value, which is of course an inadmissible factor in determining a fair rent. That inflation generally is a permissible consideration has been well recognised (see, for example, in the Court of Appeal *Wareing* v *White* (1985) 274 EG 488,* per Lawton LJ, with whom Fox LJ and Kerr LJ agreed, who, while accepting the relevance of inflation, affirmed the view expressed by Forbes J in *Kovats* v *Corporation of Trinity House* (1981) 262 EG 485 that it was a "fallacy to conclude that rents must rigorously be in line with inflation generally". See also Lord Widgery in *Metropolitan Holdings Ltd* v *Laufer*, to which I have already referred, and the decision of Woolf J (as he then was) in *Wareing* v *White* (1984) 270 EG 851, the decision which was upheld by the Court of Appeal in the case to which I have already referred.

Mr Laws reminded me that the nature of the panel's task in determining on appeal the fair rent of premises was one of valuation where there was more scope for the exercise of opinion and judgment than in other fields of adjudication. Panels are constituted with expert and professional members, who operate within a particular locality of which they would have or would gain experience. This panel, according to the papers, included a legal member and a Fellow of the Royal Institution of Chartered Surveyors.

*Editor's note: See also [1985] 1 EGLR 125.

In *Guppys (Bridport) Ltd* v *Sandoe* (1975) 30 P & CR 69, Lord Widgery CJ said at p 74: D

> There are plenty of judicial offices, not the least that of a High Court judge, where reasons undoubtedly have to be given for every decision, but that does not mean that if the judge has two conflicting opinions put before him in evidence he has to explain why he chooses one in preference to the other. Such explanations are not possible. They are matters of judgment, impression and sometimes even instinct, and it is quite impossible to give detailed reasons to explain how the system of decision has worked, and so with a rent assessment committee. If they have decided, having carefully weighed the evidence, that they must reject one approach and adopt another, then all they need to do is to say that in the exercise of their discretion and relying on their skill and judgment they prefer the method which in fact they do prefer. If they say that, it cannot be said against them that their decision is invalidated by the fact that E
> no further or more detailed explanation of why they prefer method A and reject method B is given.

In the present case, said Mr Laws, there is no real issue as to the primary facts, the location, condition, accommodation and so on of the subject premises. On the contested material, he said, the panel's reasons are unambiguous. This is, in my view, a correct analysis of the situation.

Mr Wood criticised what he called the "common form wording" of para (10) of the reasons and suggested that it might be there to cover the weaknesses of the earlier paragraphs of the reasons. However, the Court of Appeal in *Wareing* v *White*, to which I have already referred, recognised that phrase as a permissible way for the F
panel to express itself unless it could be said that the panel was being untruthful. That was not suggested there and is not here.

Having given careful consideration to all the arguments advanced by Mr Wood, I have come to the conclusion that upon the material before me the panel cannot be said not to have given sufficient reasons for the decision. In my view, adequate reasons were given. This application therefore fails and is dismissed.

The application was dismissed with costs

NEGLIGENCE

Chancery Division

October 29 1985

(Before His Honour Judge RUBIN, sitting as a judge of the High Court)

JGF PROPERTIES LTD v LAMBETH LONDON BOROUGH

Estates Gazette February 15 1986

277 EG 646-650

Action by property company alleging negligence against local authority — Claim that answers given to inquiries before purchase of properties failed to give warning of planning blight, with the result that, after unsuccessful attempts to sell properties privately, they had to be sold by auction at disappointing prices — The plaintiff property company had been created and was run by a developer who gave evidence in the action — The company's then solicitors had reported that, in reply to inquiries on form Con 29, the local authority had revealed that the property concerned was within a proposed future action area, which meant that there were controls on development — The answers, which the developer did not himself see, indicated more particularly that the area was programmed for light industry redevelopment — The answers set out what was in fact the true position — The developer decided to make a telephone inquiry himself and appeared to have received an answer that this was an action area for which certain grants were available — It was not clear from the evidence what questions were asked on the telephone or what precise reply was given or what was the identity or status of the person who replied — It seemed unlikely that anyone employed in the planning department would have given the reported answer if asked the right question — Australian case of *L Shaddock & Associates Property Ltd* v *Parramatta City Council* cited, where it was held that an unconfirmed answer given by an unidentified person to an inquiry made of a similar kind over the telephone could not give rise to a claim for negligence against the council — Held in the present case that the written replies given by the defendant council could not be faulted and that the telephone conversation could not be relied on as establishing negligence against the council — Action dismissed

This was an action for damages by JGF Properties Ltd, based on the alleged negligence of the defendant authority, the London Borough of Lambeth, in failing to disclose, in reply to inquiries, information as to planning blight affecting properties at 27 and 17 Rothschild Street, London SE27, purchased by the plaintiffs.

Michael Warren (instructed by Woolsey Morris & Kennedy) appeared on behalf of the plaintiffs; Brian Leech (instructed by Barlow Lyde & Gilbert) represented the defendants.

Giving judgment, JUDGE RUBIN said: This action arises out of the purchase of two properties by the plaintiff company in Rothschild Street, London SE27, the first one being no 27 and the second one being no 17. For the purposes of this judgment, I will deal solely with no 27, it not being suggested by either side that the result will be any different in respect of property no 17.

The plaintiff company at the time of the matters with which I am concerned was a newly created company, and the man behind it was a Mr Fishenden, who has given evidence in these proceedings. He quite clearly created this company and another companion company for entering into property development, no doubt in London, using this company as the vehicle to purchase properties and his other company as a vehicle to carry out repairs and improvements to those properties with a view at the end of the day to selling them at a profit.

Negotiations for the purchase of 27 Rothschild Street started in April 1982. It is a property situated in the London Borough of Lambeth. Mr Fishenden himself at the time was employed by another development company working in the neighbourhood and this property was brought to his notice. As a result of looking at it, he reached a conclusion that it was a property which could be developed with profit and accordingly he was minded to purchase it. On April 1 his then solicitors sent to the defendant, which is the borough council, the usual forms of inquiry, including *Form Common No 29 London*, and paid to the local authority the prescribed fee, which was then £10.20, for the supply of various answers by the defendant to questions raised on those forms. In a moment I will have to look at the form itself and the answers in a little more detail, but let me just deal first with the history of the matter. The local authority provided answers in the prescribed form. Whether they were accurate or not is a matter which I shall have to consider in a moment. As a result of those answers, the plaintiff's then solicitors wrote a letter, dated May 14 1982, to Mr Fishenden, which perhaps I should read at this stage:

> I am still awaiting to hear from the vendor's solicitors on some outstanding points regarding the missing deeds. In the meantime, I have the result of my local authority search and enquiries which reveals that the property is within a proposed future action area and direction awaited from the Department of the Environment that Norwood area be treated as an action area. This does, of course, mean that there are certain controls imposed regarding development but I presume that you are already aware of this.

As a result of receiving that letter, Mr Fishenden decided that he was going to embark upon some private inquiries himself and he telephoned Lambeth Council in order to make those inquiries. I have no reason to doubt that Mr Fishenden did indeed make a telephone call. The problem is: who answered that telephone, other than the telephonist? As to the time of this telephone call — his evidence is it was made in the afternoon, the third week in May of that year. At that time the planning department had an arrangement or rule that telephone calls would not be answered after midday, because the department was overworked dealing with written inquiries. Of course one knows that such regulations are made, but on occasions even civil servants who have kinder hearts will answer telephone calls during the prohibited period, so I do not think it is impossible that somebody in the department did indeed answer that telephone call. What the effect of that telephone call is, of course, it is now very difficult to determine. We are now three years after the event; Mr Fishenden is relying on his memory; and nobody can identify who the recipient of that telephone call was. Mr Fishenden told me that he thought there were two people in the room at the time, because the person answering the telephone call seemed to discuss his query with somebody else. Nor can one be certain with confidence as to precisely what was said and precisely what the answer was. What questions Mr Fishenden put to the person on the other end of the telephone I am not at all certain about. Did he simply ask what an "action area" was, and, if so, what was the reply? Was his question directed towards a particular street, Rothschild Street? Did the person who was receiving that telephone call really understand what the questions were? At the end of the day I am left in a state of doubt as to what happened during the course of that telephone conversation, and,

A where somebody is seeking to show negligence, the onus is at least on him to satisfy the court that something was precisely said. I would say my difficulty here is that on the evidence which I have heard I cannot be sure at all as to what was said during the course of that conversation.

I will come back to that in a moment. Mr Fishenden indeed decided to exchange contracts on this property, and that he did. He purchased it at a price of £15,000, and embarked upon the course of redeveloping 27 Rothschild Street, not for a moment aware of the problems which were going to face him. Having completed that redevelopment, he put it on the market and indeed a number of people were interested in purchasing. Two of them in fact made B offers subject to contract at something like the asking price. One of them offered £24,250. However, those purchases did not come to fruition, and I think the inference I can properly draw from the circumstances was that each of the purchasers, when they made further inquiries, realised that this property was in a development area and that in due course it was going to be of little value, in other words it was blighted, and very wisely took no further interest in the property and called off their respective purchases. At least that evidence seems to me to help in this regard, that it gives one a fairly shrewd idea of what the price of that property would probably have been at that time if it had not been the subject of planning blight.

After that, the property was put up for sale by auction and on the first auction it was not sold because the reserve was not reached, and C in the result, after further negotiations for private sales, it was put up for auction on the second occasion and this particular property was sold for the sum of £18,000.

Just shortly dealing with the second property, no 17, that again had the same history. Mr Fishenden purchased it soon afterwards. He redeveloped it. He attempted to resell it privately, with the same result, on that occasion having been offered something like £25,000. Again he put it up for auction and on the second auction it was sold on this occasion for £20,000.

His claim against the local authority is that they were negligent in failing to disclose to him, which he says they ought to have done, certain planning blight upon this property. Before I deal with that, it D is necessary to look shortly — and I am not going to deal with this at inordinate length, because I do not think much turns upon it — at the planning history of this matter. For those who are interested in it in more detail, it is set out in the Final Norwood Industrial Area Action Area Plan, which was adopted in September 1983. Basically, for present purposes, this property, under the London Plan, had in fact been zoned for light industry programmed for redevelopment 1960-1972. The Lambeth Council had proposals for an action area plan, which at the relevant time had not in fact been adopted, but they were in the proposal state. I need not go into the details of this plan, but when one looks at the area plan attached to the plan, it is perfectly clear that Rothschild Street fell way in the middle of one of these areas. We are concerned with a row of houses on the southern side of E Rothschild Street, well within this area.

Coming back, the solicitors, as I have already indicated, sent the form *Con 29* to the Lambeth Borough Council to be answered. I am only concerned with the question at 10 in its various forms. The draftsman of form *Con 29* I think was one of those draftsmen who like to make their questions as complicated as they can. I must say, to somebody uninitiated in this branch of the law, to see question 10(A) as it was drafted would cause a lot of mystification, but there it is. It asks:

> Do
> (1) the Initial Development Plan for Greater London
> (2) the Greater London Development Plan
> F (3) any current proposals for, or by way of, alterations to (1) and (2) which have been published for purposes of public participation
> (4) any current proposals for, or by way of:
> (i) a local plan
> (ii) any non-statutory plan
> (iii) alterations to (i) or (ii) which have been published for purposes of public participation
>
> either indicate the primary use for the area in which property is situated; or include any provisions for the property?
> If so, please give short particulars.

I am not going to read the details of those short particulars.

The answer to 10(A) was: "Light Industry programmed for redevelopment 1960-1972".

The answer to 10(A)(4) was "None", and I think it is accepted by the defendant that that technically may not have been entirely accurate.

Then to 10(B)(ii), which is important for present purposes: "Proposed future Action Area and Direction awaited from the Dept of the Environment that Norwood Area be treated as an Action Area".

Those answers of course were to a form sent by a solicitor seeking answers. During the course of the argument it seemed to me that when one is considering answers to a form of this nature one has to bear in mind the person to whom the answer is being made. A good analogy seems to me to be the case of a medical consultant who is seen by a patient and, after the patient has seen him, he is asked his advice. If that advice is being given to the patient's general practitioner, no doubt it can be properly given in a form which a normal general practitioner could be expected to understand but which might not be intelligible to a lay person. Whereas if he was advising his patient direct, the advice ought to be in language which an ordinary lay person can understand. Looking at these answers, they seem to me to be in a form which ought to be intelligible to the ordinary competent solicitor but may well be a total mystification to a layman seeing this form of answer for the first time.

They were sent to the solicitor in this case. I find it very difficult to see how one can possibly say that this local authority was negligent and failed in its duty of care in the answers which it gave. It quite clearly set out the light industry programme for redevelopment 1960-1972. I accept that it is long after 1972, but that was still a subsisting plan. Then one has that the "proposed Future Action Area and Direction awaited from the Department of the Environment that Norwood area be treated as an Action Area". Again it seems to me that that is accurately setting out what was in fact the true position. The other difficulty, it seems to me, that the plaintiff has in attempting to rely on these answers that were given by the local authority to his solicitor was that he never saw the answers. Indeed he said in the witness box that, if he had realised that this was indeed programmed for redevelopment for light industry, the warning bells would have sounded and when the warning bells sounded he would either have made proper inquiries through his solicitor or would not have in fact embarked upon this somewhat speculative development. So had the matter rested solely on the answers by the local authority on this form, I would without hesitation have come to the conclusion that negligence has not been established. It does not of course entirely rest on that, because the plaintiff is relying upon this telephone conversation.

I have been assisted by being referred in the course of argument to a case in the Supreme Court of Australia, *L Shaddock & Associates Property Ltd* v *Parramatta City Council* (1981) 36 Australian Law Reports 385. That was a case which was dealing with a slightly different point, because there had indeed been an oral inquiry first, followed by a written inquiry, and one could quite readily say that the oral inquiry ceased to be of importance because it was not relied upon and it was the written inquiry that was being relied upon. However, the court made certain observations about oral inquiries of this nature. Let me just deal very shortly with them. Chief Justice Gibbs said this at p 393:

> The nature of the inquiry — made by a solicitor, for conveyancing purposes, on a form commonly used and prepared by law stationers — made clear the gravity of the inquiry and the importance attached to the answer. The council therefore owed a duty of care to Mr Carroll's clients, the appellants, in answering the written inquiry. It would not, however, have been reasonable for the appellants to have relied on an unconfirmed answer given by an unidentified person in response to an inquiry made over the telephone. The council therefore owed no duty of care in making response to such inquiry.

Then Stephen J, at p 395:

> The appellants through their solicitor, Mr Carroll, made two inquiries of the council. The first, by telephone, should not, in my view, be regarded as giving rise to any consequences in law. It was marked by informality: the person in the council office to whom Mr Carroll spoke remained unidentified and the advice which that person then and there gave over the telephone remained unconfirmed by any writing. It must be but rarely that information conveyed by unidentified voices answering a telephone at the offices of municipal councils will render those councils liable in damages for negligence if the information should prove to be incorrect. In my view neither a council nor an inquirer would, in the absence of quite special circumstances, regard the response to such an inquiry as carrying with it liability in damages if incorrect: this must especially be the case when there exists a customary and more formal means of obtaining from a council the information which is sought.

I do not think I need to read any further citations from that judgment.

That accords with the view which I have been taking of this telephone conversation in the present case, before that case was indeed cited to me. One has here an informal telephone conversation to an officer of the council (if it was an officer), unidentified, who has answered a question whose precise terms are not clear, after consulting with apparently some unidentified other person in the same room, in circumstances where these matters are normally dealt with by way of formal inquiry made by solicitors to the council and indeed had been made in this very case and had in fact been answered. I cannot help thinking that the officer in the council, who was reported to have said that this was a housing action area for which certain grants become payable, either did not understand the question which was being put to her, or, if she did understand the question, was not a person who was competent to deal with these matters, because anybody in the planning department, I would have thought, would have known without a moment's hesitation what the action area meant in the context of the present case, bearing in mind what was going on in Lambeth at the time, which must have been extremely well-known to everybody in the planning department. One only has to see the various productions which have been produced before me, dealing with the draft action area plan, the action area plan (that was not in existence at that stage) and the adopted plan at the end of the day. So anybody employed in that department must have been very familiar with those matters and could not have given the answer which they are reported to have given if they had been asked the real question which ought to have been asked in the context.

For those reasons, it seems to me that that telephone conversation cannot be relied upon as proving negligence against the council. For these reasons, it seems to me that the plaintiff's claim is foredoomed to failure and must be dismissed. In those circumstances, I propose to say nothing about damages.

The claim was dismissed with costs.

Court of Appeal

February 18 1986

(Before Lord Justice PURCHAS, Lord Justice NICHOLLS and Mr Justice CAULFIELD)

KING v LIVERPOOL CITY COUNCIL

Estates Gazette May 3 1986

278 EG 516-523

Council tenant's action alleging damage caused by nuisance and negligence for which council were responsible fails — Damage was caused by escape of water, owing to actions of vandals, from an empty council flat above her own — Appeal by tenant against decision of county court judge who had rejected her claim — Question as to liability for wrongful actions of third parties — Appellant had warned the council as to the danger from trespassers and vandals presented by the unoccupied flat, and some ineffective steps had been taken by the council — Vandals, making more than one visit, had removed copper piping and other parts of the water-supply system, thereby causing the escape of water which flooded the appellant's premises and later forced her to move — There was evidence that vandals moved quickly into empty premises and that even sheeting, unless perhaps of steel, did not stop them — Appellant submitted that the council were under a duty of care to prevent the known hazard of the empty flat, with its vulnerable pipes and water system offering a familiar target for vandals, from materialising into damage in a predictable and reasonably foreseeable way — It was accepted on behalf of the appellant that, although her claim was framed in nuisance and negligence, it was necessary in any case to prove negligence — The county court judge, following guidance in *P Perl (Exporters) Ltd* v *Camden London Borough Council* and *Lamb* v *Camden London Borough Council*, had decided against the appellant's claim — The Court of Appeal considered these and a large number of other authorities, G
including *Dorset Yacht Co Ltd* v *Home Office*, *Anns* v *Merton London Borough Council* and *Paterson Zochonis Ltd* v *Merfarken Packaging Ltd* — The court concluded that it was too wide a proposition to suggest that a general duty of care was placed on a defendant to avoid providing third parties with the means of committing a tort against a plaintiff, although such a duty could arise in particular circumstances or in special relations — They agreed with the judge that in the present case the council owed no such duty to the appellant in respect of the actions of the vandals — The judge had found that it was not in fact possible for effective steps to be taken in H
the actual situation to defeat the activities of the vandals — This finding limited the area of the duty owed by the council in the circumstances of the case — Appeal dismissed

This was an appeal by Mrs Mary King, tenant of a council flat at 16 Great Mersey Street, Liverpool, against a decision of Judge Sachs, at Liverpool County Court, dismissing a claim by Mrs King, based on nuisance and negligence, against the council for damage caused by the escape of water from an empty flat, belonging to the council, immediately above.

David Marshall Evans QC and Graham Wood (instructed by Vauxhall Community Law Centre, Liverpool) appeared on behalf of the appellant; John William Kay QC and William Braithwaite J
(instructed by the City Solicitor, Liverpool City council) represented the respondent council.

Giving judgment, PURCHAS LJ said: This is an appeal by Mary King, the plaintiff, from a judgment of His Honour Judge Sachs given at the Liverpool County Court on January 8 1985. The learned judge dismissed the plaintiff's claim against the respondents, the Liverpool City Council ("the council") for damage caused to her property by water escaping from a flat owned by the council. The plaintiff was a tenant of a neighbouring flat owned by them. The premises were at 16 Great Mersey Street, Liverpool. The plaintiff's flat was no 16 and immediately above it was Flat 16a which, at the material time, was vacant but for which the council were responsible. K

The short facts leading up to this matter are as follows. The plaintiff, first under a joint tenancy and later, after her divorce, as sole tenant, had been in Flat 16 since 1978. At the end of August 1982 the tenants of Flat 16a left the premises and on September 2 the plaintiff, noticing that the flat above was unoccupied and not protected in any way against trespassers or vandals, telephoned the appropriate department of the council to this effect. That message was received and passed to the appropriate department, the depot at Shaw Street, Liverpool, where it was recorded and various steps were taken which, in the event, were not successful.

Between September 10 and 12 1982 vandals entered Flat 16a and removed copper piping and other parts of the water-supply equipment in that flat. In the result water escaped from the damaged L
system and flooded the plaintiff's premises. The council sent a plumber who did some repair work, but again the flat was not boarded up either by plywood or, as is sometimes done, with metal sheeting.

Two days later on September 14 1982 there was another flood and this was sufficiently serious to cause the plaintiff, understandably, to leave the flat. It appears very probable that that again was as a result of vandals invading Flat 16a. Thereafter the plaintiff lived with her sister until December 1982, when she was rehoused by the council.

In the meanwhile there had been a third flood, which appears to have occurred, again probably as a result of invasion by vandals, between September 16 and 17 1982.

Those are the facts, about which there is no dispute. Work records M
etc were produced by the council, indicating that boarding or sheeting work had been carried out to Flat 16a, although that was not seen by the plaintiff on her visits, and it was accepted by witnesses for the council that it was not certain that the sheeting work had been completed or effectively done. But their records certainly indicated that steps had been taken to that end.

The plaintiff brought an action in the Liverpool County Court in June 1983. She alleged negligence against the council in the first instance and the council counterclaimed for arrears of rent. Later, in February 1984, her particulars of claim were amended. Then the plaintiff alleged a breach of covenant for the quiet enjoyment of her flat (she was, of course, a tenant of the council); alternatively negligence, which was her original cause of action; or alternatively nuisance. The claim for breach of covenant was abandoned at the trial below, but the plaintiff continued her claim in nuisance and negligence.

The facts material to this appeal, as found by the judge, can be shortly summarised. It was accepted that the reason for the plaintiff's request, although not necessarily the same as the motive for the defendants' action, "was that unhappily, as in many areas of Liverpool, vandals readily discover that premises have been vacated, quickly descend on them and cause substantial damage". From documents supplied by the defendants' housing department there is a card relating to 16a Great Mersey Street which shows on September 2 a request for two sorts of work to be undertaken: "(1) to sheet up and (2) to remove services". The learned judge listened to a considerable body of evidence from, among others, the superintendent of the Shaw Street Depot and the assistant district housing manager.

The superintendent of the Shaw Street Depot, Mr Peter Gibney, said this in evidence:

> I know hundreds of tenants suffer from vandal damage. We sheet up and remove services. I know they have a rising main in a vulnerable position. We do not go looking for work. We used plywood albeit ineffective because we are told to. It looks like a burst on rising main

that is referring to the events in Flat 16a.

> Rising main would be made out of copper, basically the vandals go for the cylinder — sometimes for the rising main. There is a stop tap for each flat — can't remove pipe — would affect [occupants of the other flats].

The evidence of Stephen Alan Guy, the assistant district housing manager, was much to the same effect:

> Sheeting up regarded as urgent — accepted by both departments as such. Wouldn't expect 10/12 days to elapse. Many properties are not sheeted up — works department slow or not reported.
> If properties not boarded up — vandals would get to know in short time. Services would be interfered with in unprotected premises. No steps taken apart from boarding up and remove rubbish.

Then, in relation to flooding:

> Three floods uncommon — normally one. Local authority can remove services and secure premises. Particularly vulnerable if attacked once because not sheeted up.

Then in re-examination — a finding upon which the learned judge relied — he said:

> No guarantee sheeting up will stop vandals. Sheet steel is the answer. Quite often vandals get to know premises vacant within day or so.

The learned judge's findings on this evidence are recorded in the notes of his judgment in these terms:

> Further evidence was called by the defendants being the evidence of Mr Guy the assistant district housing manager. His is an administrative post and he explained the procedures to which I have referred. He accepted that at the time there would be no record sent to his Department that completion of the work had taken place. He agreed that unprotected premises would be likely to be the victims of vandals. He also told me that there was no guarantee that sheeting up would stop the vandals, although he would be more optimistic if steel sheeting was used and he accepted that vandals get to know of vacant premises within a day or so of them becoming vacant.
> I have to find as a matter of fact what occurred in respect of these vacated premises. I accept the evidence of the plaintiff bearing in mind that I find facts on a balance of probabilities. I accept she informed the department of the various events that there were the three floods in the sequence to which she refers and that certainly whenever she went to look at the flat 16a there was no sign of boarding up. If therefore boarding up was done (of which I am not persuaded) clearly it was ineffective as it was never observed by the plaintiff. I have no reason to doubt either her description of the floods that occurred or the damage done.
> Equally I accept the evidence called by the defendants that it is impossible to turn off the rising main supply of drinking water without affecting other tenants. I also accept that damage done to the flat 16a was damage done by vandals.

The learned judge's finding on nuisance was in the following terms. Having referred to the speech of Lord Wright in *Sedleigh-Denfield* v *O'Callaghan* [1940] AC 880, he said:

> The facts that I find are that regrettably boarding up is ineffective and further it is not possible to block off the rising main for the reasons I have given. I do not find therefore the defendants in this case have adopted or continued a nuisance. Regrettably in the circumstances I find that it is not possible for effective steps to be taken in a situation like this which could defeat the activities of vandals.
> I find on the evidence of the plaintiff that following the first flood a plumber did come to visit the premises.
> As the case is pleaded the plaintiff maintains that the nuisance was created by the failure to board up as opposed to the failure to rectify the plumbing. For these reasons I am satisfied that the defendants abated the situation by sending the plumber and did all that they could in that respect for the reasons I have given. Counsel for the plaintiff sought to say that the nuisance was not an escape of water but the condition of the premises. I was unpersuaded by the argument and found it of no significance.

As regards the claim in negligence, the learned judge based his finding on passages from the judgments in the case of *P Perl (Exporters) Ltd* v *Camden London Borough Council* [1984] 1 QB 342 and in particular an extract from the judgment of Waller LJ, dealing with liability for the acts of third parties. At p 349 he said:

> But no case has been cited to us where a party has been held liable for the acts of third party when there was no element of control over the third party. While I do not take the view there can never be such a case I do take the view that the absence of control must make the court approach the suggestion that there is liability for a third party who was not under the control of the defendant with caution.

The learned judge then referred to a passage from the judgment of Oliver LJ (as he then was) in *Lamb* v *Camden London Borough Council* [1981] QB 625 at p 644 to this effect:

> There may, for instance, be circumstances in which the court would require a degree of likelihood amounting almost to inevitability before it fixes a defendant with responsibility for the act of a third party over whom he has and can have no control.

The learned judge based himself on those judgments and found against the plaintiff on her claim in negligence.

Mr Evans, who has appeared for the plaintiff on this appeal, has based his submissions in negligence. He concedes — in my judgment rightly — that where nuisance was concerned, he could only establish the second of the three grounds of liability in nuisance, namely "if by a neglect of some duty he [the tortfeasor] allowed it to arise": see the judgment of Rowlatt J at p 338 in *Noble* v *Harrison* [1926] 2 KB 332. Mr Evans submitted that the judge was wrong in considering that he was bound to find for the defendants by the decision of this court in *Perl's* case. He distinguished that case on the basis that in *Perl's* case the failure to secure the premises had the result that mischievous third parties gained access to neighbouring premises and there caused the damage in a way that was not readily foreseeable. Here, Mr Evans submits, the failure to secure the premises did not allow access to neighbouring premises, but allowed vandals to come upon the very same premises as those occupied by the defendants, to behave in a foreseeable and wrongful manner, which led to water escaping from the water system in that council flat, thus doing damage to the plaintiff's flat and property.

Mr Evans relied upon the findings and evidence to which I have already referred, namely that, if premises were left unsecured, it was likely if not highly likely that vandals would invade those premises and would cause damage thereto. Further, Mr Evans submitted that it was foreseeable that they would be attracted to the valuable parts of the fittings of the flat, namely the copper piping and the cistern and metal attachments, and that in breaking these out from the premises they would release waters from the rising main; therefore, says Mr Evans, the damage to the plaintiff's property and the flat was entirely foreseeable.

Encapsulated (I hope without doing injustice to Mr Evans' able submissions) he puts his case in this way: these circumstances placed a duty upon the council, owed to the occupier of neighbouring premises — because there was a potential hazard in Flat 16a, constituted by the pipes and water system, in respect of which the occupier was under a duty of care to prevent that potential hazard materialising in a way reasonably foreseeable, ie as a result of the invasion of vandals if the premises were not secure or the water cut off.

With reference to *Perl's* case Mr Evans submitted that the effect of that case in no way bound the learned judge. He referred the court to a number of cases where it was submitted that liability was established for the act of a third party.

This is an area of the development of the law of tort and negligence which has received considerable attention in recent years. It is not necessary in this judgment to rehearse in detail the cases to which Mr Evans drew our attention and I hope that by dealing with them summarily that will not be considered a discourtesy to his submissions.

First of all, reference was made to the well-known case, *Dorset Yacht Co Ltd* v *Home Office* [1970] AC 1004. That case involved damage done by inmates of a Borstal institution, who escaped to a nearby yacht harbour, took a yacht and caused damage. The basis of liability in that case, however, was an established duty upon the warders employed in the establishment to control the acts of the wrongdoers (the third parties) and that, as a result of that breach of duty properly to contain the Borstal inmates, the damage caused by interference with the yachts was foreseeable and therefore recoverable. But, in considering that case, Oliver LJ in a later case drew the distinction between the proximity between the Borstal institution and the yacht basin and the more remote possibility of one of the inmates escaping to John o'Groats and there doing damage. It is in each case necessary for the court to look at the scope of the duty in relation to the context in which it arises, and in the *Dorset Yacht Club* case the circumstances were clearly established, showing a breach of duty on the part of the warders for whom the defendants were responsible.

In *Harris* v *Birkenhead Corporation* [1976] 1 WLR 279, as appears from the judgment of Megaw LJ on p 288, it was conceded by counsel for the corporation that there was a duty owed to the child trespasser who was injured as a result of his trespassing on property for which the defendant corporation were responsible.

Again, in the case of *Ward* v *Cannock Chase DC* [1985] 3 All ER 537, to which we were referred, the council admitted liability for negligence, as appears from the headnote in that case.

These cases were dealing with the remoteness of the damage sustained in the context of the facts and an established duty.

Davies v *Liverpool Corporation* [1949] 2 All ER 175, to which we were referred, does not, with respect to Mr Evans' submissions, assist. It is based upon the vicarious responsibility of the corporation for the negligent discharge of his duty by the conductor and, in my judgment, does not take the matter any further.

Newby v *General Lighterage Co Ltd* [1955] 1 Lloyds Rep 273 was also referred to us. There a barge had been moored, but not securely moored, on a fast-running waterway. There was interference, and predictable interference, by passers-by on the towpath, and liability was established.

As with *Haynes* v *Harwood* [1935] 1 KB 146, the leading case in this area, duties existed as between the users of either a highway or a waterway and the existence of that duty was accepted. In this case the court is concerned with the existence of the duty upon an occupier of land so to control access to that land by strangers as to prevent injury to neighbouring property. The authorities already mentioned in this judgment, in particular the judgment of Waller LJ, indicate that there is no direct authority on this point. I turn briefly to *Perl's* case to refer to the judgment of Waller LJ at p 352 C. The learned lord justice had been referring to the decision of this court in *Lamb* v *Camden London Borough Council* [1981] QB 625 and in particular to the judgment of Oliver LJ:

I agree with Oliver LJ that the foreseeability required to impose a liability for the acts of some independent third parties requires a very high degree of foreseeability. Adapting the words of Lord Atkin, "ought the defendants to have had the plaintiffs, as occupiers of no 142, in contemplation as being affected when directing their minds to the question of repairing the doors and locks of no 144?" It is not sought here to make the defendants liable for any act, it is sought to make the defendants liable for an omission to act. Can it be said that the defendants ought reasonably to have had in contemplation the fact that third parties would go into the empty basement of 144, make a hole in an 18 inch wall large enough for somebody to climb through and steal a large number of articles of clothing from within? I would unhesitatingly answer "No". Whether or not an occupier of a house can ever be liable to a neighbour for an omission to act is doubtful. I do not however have to consider whether such a case may possibly arise. It is sufficient to say that in this case I am satisfied that there was no breach of duty by the defendants to the plaintiffs and accordingly I would allow this appeal.

At p 355 Oliver LJ, referring to the *Dorset Yacht* case, said this:

The *Dorset Yacht* case does not, therefore, in my judgment support the conclusion at which the deputy judge arrived, unless it can be said that there was here some special relation taking the case out of the general rule which excludes liability for the acts of independent third parties. [Counsel for the plaintiffs] has submitted that that special relation is to be found from a number of factors combined that is to say (a) geographical propinquity; (b) the defendants' knowledge that the plaintiffs used their premises to store goods which might be attractive to thieves; (c) the defendants' knowledge or means of knowledge that there had been frequent incursions to trespassers (including burglaries in some of the flats in no 144) and (d) the relatively simple steps required to impede the entry of trespassers by fitting an effective lock on the front door. These factors, however, whilst they are, no doubt, relevant as regards remoteness of damage and may possibly be said to give rise to a relation between the defendants and the plaintiffs, go nowhere towards establishing the sort of relation referred to by the majority in the *Dorset Yacht* case and clearly envisaged by Dixon J in the passage from his judgment in *Smith* v *Leurs*, 70 CLR 256 . . . namely a relation between the defendant and G
the third party for whose act he is said to be responsible.

Oliver LJ then goes on to consider other cases, to which we were also referred by Mr Evans. Returning to the point at p 357 H, he said:

Speaking for myself, I am unable to see here any circumstances from which there could properly be inferred any duty upon the defendants so to protect their own premises as to prevent trespassers from entering the plaintiffs' premises beyond the fact that such entry was, as it plainly was, a foreseeable possibility. In my judgment that is not, by itself, sufficient to raise the duty for which the plaintiffs contend. Indeed the contrary proposition would, I think, lead to the most startling and far-reaching consequences.

Robert Goff LJ (as he then was) in his judgment refers to the question where the intervention of a third party wrongdoer is involved. At p 359 he said this: H

The vital feature in the type of case under consideration is, as I see it, that the plaintiffs are seeking to render the defendants liable in negligence for the wrongdoing of a third party. Now there may indeed be circumstances where a person may be liable for a third party's wrongdoings. He may of course be liable in contract (see *Stansbie* v *Troman* [1984] 2 KB 48); he may be liable under the Occupiers' Liability Act 1957 — for example, where he invites a crowd of persons onto his land and part of his premises, designed to control the crowd, are unfit for that purpose and collapse, with the result that the plaintiff is injured (see *Hosie* v *Arbroath Football Club Ltd* 1978 SLT 122); he may be liable in nuisance, if he causes or permits persons to gather on his land, and they impair his neighbour's enjoyment of his land (cf *Attorney-General* v *Corke* [1933] 1 Ch 89), though that case was expressed to be decided on the principle in *Rylands* v *Fletcher* . . .; and he may be vicariously liable for the J
third party's wrongdoing. He may even be liable in negligence, when the wrongdoer is a person who, by virtue of a special relationship, is under his control: see *Home Office* v *Dorset Yacht Co Ltd* . . . Speaking for myself, I do not rule out the possibility that there are other circumstances in which a person may be liable in negligence for the wrongdoing of a third party. This is a matter which this court considered recently in *Patterson Zochonis Ltd* v *Merfarken Packaging Ltd* [1983] 9 FSR 273, and which I need not therefore dwell upon in this judgment. In particular, I have in mind certain cases where the defendant presents the wrongdoer with the means to commit the wrong, in circumstances where it is obvious or very likely that he will do so — as, for example, where he hands over a car to be driven by a person who is drunk, or plainly incompetent, who then runs over the plaintiff . . . But such cases are very different from the present case, where the allegation is that the defendants failed to exercise reasonable care to prevent a third party from K
causing damage to the plaintiffs. In *Smith* v *Leurs*, . . . in a passage which was cited with approval in [the *Dorset Yacht Co* case] Dixon J said . . .:

"The general rule is that one man is under no duty of controlling another man to prevent his doing damage to a third. There are however, special relations which are the source of a duty of this nature."

It is of course true that in the present case the plaintiffs do not allege that the defendants should have controlled the thieves who broke into their storeroom. But they do allege that the defendants should have exercised reasonable care to prevent them from gaining access through their own premises; and in my judgment the statement of principle by Dixon J is equally apposite in such a case. I know of no case where it has been held, in the absence of a special relationship, that the defendant was liable in negligence for having failed to prevent a third party from wrongfully causing damage to the plaintiff.

L

Although we were not referred to the case of *Paterson Zochonis Ltd* v *Merfarken Packaging Ltd* it is helpful to look at that case in amplification of the passage from the judgment of Robert Goff LJ, to which both Mr Evans and Mr Kay (who appeared for the respondent council) referred. The facts of that case are entirely different from those in the present case. The main action was an action to strike out parts of the pleadings in a claim involving infringement of copyright. For the purpose of this appeal it is necessary only to consider the second claim, which based itself upon negligence. The summons was to strike out *inter alia* a claim for damages for negligence brought by the holders of a copyright against the printers whose products (ie the packaging and label) enabled a third party (the tortfeasor) to infringe the copyright. Whitford J M
struck out that part of the claim and on appeal the appeal was dismissed. The court consisted of Oliver, Fox and Robert Goff LJJ. All the judgments were to the same effect, but in the context of this appeal it is perhaps only necessary, with respect to the other members of the court, to refer to the judgment of Robert Goff LJ. I quote a few passages from that judgment. The first is at p 296. Having dealt with the general claim in negligence and referring to the classical statement of Lord Wilberforce in *Anns* v *Merton London Borough Council* [1978] AC 728 at pp 751-752 the learned lord justice at p 297 said this:

Even so, it is necessary to sound a note of caution. There is a not insignificant number of cases in which the courts have held that no duty of care exists; there are others where they have held that the duty must be limited. Moreover, in considering new cases as they arise the courts have, as must surely be right, recognised that a heavy duty of care rests upon them to ensure that they themselves do not overstep the bounds of reasonableness by imposing too onerous legal duties upon members of the public. Vivid examples of this circumspection are to be found in two of the leading cases, *Hedley Byrne & Co Ltd* v *Heller & Partners Ltd* [1964] AC 465, and the *Dorset Yacht* case itself. It is plain that, in considering whether the duty of care should be negatived or limited in any new situation, the courts are making what is usually called a decision of policy (see the *Dorset Yacht* case at p1059, per Lord Diplock). Presented with such a case, the courts have to do their limited best. They have no secretariat, or apparatus for enquiry. They have to derive all the guidance they can from the authorities, and from the writings of scholars; and they have the inestimable benefit of the assistance of counsel. But they are also fully entitled to, and do, draw on their own professional and practical experience. In the end, the choice must be one of judgment, in the balancing of conflicting interests; and in the exercise of that judgment the courts must strive neither unjustifiably to deprive potential plaintiffs of remedies for their injuries, nor unjustifiably to impose too heavy a legal responsibility on potential defendants.

At p 298 the learned lord justice said:

The first special feature is that there is here said to be a duty of care on the defendant to avoid providing a third party with the means of committing a tort against the plaintiff. Is there any general principle which excludes or limits liability in such a case? In my judgment, there is no principle which excludes liability. It is only necessary to give examples such as handing a loaded gun to a small boy or a simpleton, who then "has a go" and shoots the plaintiff, or entrusting a car to a man who is obviously drunk or even incompetent, to realise that there are cases in this category where there is a duty of care. It is implicit in the decision of the Privy Council in *Yachuk* v *Oliver Blais* [1949] AC 386 that there may be liability in the first of these examples: and the Canadian case of *Ontario Hospital Services Commission* v *Borsoski* [1974] 54 DLR 3d 339 provides strong persuasive authority that there may be liability in the second. But even so a note of caution has to be sounded. Take the case of a dealer in cars who sells a car to a man whom he knows to be an alcoholic, or to be the husband of an alcoholic who to his knowledge drives her husband's car; it is difficult to believe that he would be held liable when, a few weeks later, the husband or wife (as the case might be) was involved, while in a state of intoxication, in a collision with the plaintiff's car. Clearly there is a line to be drawn.

Furthermore, the examples I have so far given are of cases where the third party commits the tort of negligence, not where he is involved in deliberate wrongdoing. Once again, however, a duty of care cannot be negatived. A case such as *Stansbie* v *Troman* . . . suggests examples . . .

Robert Goff LJ then considers further such examples, and continues on the principle in these words:

It is tempting to conclude that, in the absence of some special relationship, there can be no liability for damage caused to the plaintiff by the deliberate wrongdoing of a third party. That this is so in cases where the defendant fails to prevent a third party causing loss or damage to the plaintiff is demonstrated by the *Dorset Yacht* case, and by the dictum of Dixon J (as he then was) in *Smith* v *Leurs* . . . cited with approval in that case, both primarily concerned with circumstances where the third party was under the defendant's control.

Robert Goff LJ then referred to Lord Sumner's judgment in *Weld-Blundell* v *Stephens*, and continued:

However, the analogy of cases where the defendant fails to prevent a third party from causing injury to the plaintiff is not exact; and in the changed climate of today, Lord Sumner's dictum must be regarded as too sweeping.

He then referred again to the *Dorset Yacht* case, to which I need not refer, and continued:

Of course, in the law of negligence the common criterion of foreseeability creates some overlap between the various constituent parts of the tort; and the point made by Lord Reid can perhaps be as legitimately made at the stage of breach of the duty of care as it can at the stage of causation. Even so, the effect of Lord Reid's observation is that there is no principle of law excluding liability in negligence for damage caused by the deliberate wrongdoing of a third party; but only in the limited circumstances indicated by Lord Reid will liability attach for such damage.

And I revert to the quotation from the *Dorset Yacht* case, from which the learned lord justice is referring:

These cases show that, where human action forms one of the links between the original wrongdoing of the defendant and the loss suffered by the plaintiff, that action must at least have been something very likely to happen if it is not to be regarded as *novus actus interveniens* breaking the chain of causation. I do not think that a mere foreseeable possibility is or should be sufficient, for then the intervening human action can more properly be regarded as a new cause than as a consequence of the original wrongdoing. But if the intervening action was likely to happen I do not think that it can matter whether that action was innocent or tortious or criminal. Unfortunately, tortious or criminal action by a third party is often the "very kind of thing" which is likely to happen as a result of the wrongful or careless act of the defendant.

Bearing those authorities in mind, I return to the findings of the learned judge, which I have already cited in this judgment. The learned judge considered the general system adopted by the council in the context of the extensive area of their responsibility and the regrettable, but established, social climate at present being experienced in that city, and indeed in other cities. The learned judge considered with great care the effectiveness, or lack of effectiveness, of "boarding up" as being one of the courses which admittedly the council from time to time took as they could, although not in this specific context of flooding; nevertheless flooding was a recognised result.

The learned judge's finding is, in my judgment, determinative of this appeal. Summarising his judgment, he said:

Regrettably . . . I find that it is not possible for effective steps to be taken in a situation like this which could defeat the activities of vandals.

Whether this finding, together with the established circumstances of the defendants, should operate to restrict the ambit of the duty to take any positive steps to secure the property, or duty arising in relation to an omission to take such steps; or whether it operates to break the chain of causation, may, as Robert Goff LJ suggested in the passage which I have just cited from the *Paterson Zochonis Ltd* case, not be essentially material. Personally I prefer the former approach and would limit the area of the duty itself in the circumstances prevailing in this case. In either event, in my judgment the learned judge was right to hold that the council owed no duty to the plaintiff in respect of the acts of the vandals in this case and accordingly I would dismiss this appeal.

Agreeing, NICHOLLS LJ said: I add observations only on two points:

The first is this. For the plaintiff Mr Marshall Evans sought to distinguish this case from the decision of this court in *P Perl (Exporters) Ltd* v *Camden London Borough Council* [1984] QB 342 on the ground that the *ratio decidendi* of that case was that an occupier of property owed no duty so to use his property as to prevent third parties from getting on to his property, and thence on to the plaintiff's property, because an occupier is not liable for the escape of burglars from his land. Contrast the present case, it was submitted, where the plaintiff does not have to set up the existence of a novel duty of care. An occupier is under a duty to take reasonable care to ensure that water on his property, which is a potential hazard to neighbouring properties, does not escape and damage those properties. Here it was reasonably foreseeable that a failure by the defendant to follow its normal practice of boarding up empty property would be very likely to result in vandals entering, and stripping away and removing the copper water pipes and cistern, with the escaping water then causing damage to the plaintiff's property.

I am unable to accept that any material ground of distinction exists between the two cases. In the *Perl* case, as in the instant case, the plaintiff sought to make the defendant occupier liable in negligence for the wrongdoing of a third party. In his judgment Robert Goff LJ set out at p 359, in a passage already cited by Purchas LJ, some examples of circumstances where there may be liability for a third party's wrongdoing, and concluded that those instances were very different from that case where, as in the present case, the allegation was that the defendant failed to exercise reasonable care to prevent a third party from causing damage to the plaintiff. In his preface to that passage Robert Goff LJ assumed that there might well be cases where the occupier could reasonably foresee that thieves might use the unprotected property as a means of access to neighbouring property. But he, in common with the other members of the court, rejected the existence of the broad duty of care contended for by the plaintiff's counsel, and his conclusion was to the effect that in the absence of a special relationship, there was no duty to prevent thieves from so using one's property.

I cannot see any distinction in principle between a case where the damage arises from the third party using the defendant's property as a means of obtaining unauthorised access to the plaintiff's property and there committing theft and one where the damage arises from the third party so conducting himself on the defendant's property as to damage the plaintiff's property by causing water to escape from the former property to the latter. Nor can I see that it is material that the defendants had a responsibility to take reasonable steps to prevent the escape from their property of water in an ordinary domestic water system. I do not consider that there is a greater responsibility on the defendants because the third party caused damage by creating an escape of water than if the damage had been caused by the third party lighting a fire on the defendants' property or, if the defendants' property had been on the top floor of the building, by the third party stripping lead from the roof and thereby permitting rain to enter and eventually to reach and damage the plaintiff's property.

Second, although in the *Perl* case the plaintiff seems to have been relying on a claim in negligence alone, and nuisance is relied on in the present case, before us it was accepted that, for the plaintiff to succeed in nuisance, she had to establish a breach of a duty of care. In this regard I add that this is not a case where the acts of a trespasser B have been continued or adopted by the defendants as occupiers of their land.

CAULFIELD J agreed with both judgments and did not add anything.

The appeal was dismissed. No order was made for costs. Leave to appeal to the House of Lords was refused.

For a further case on this subject see p 252

RATING

Chancery Division

October 31 1985

(Before Mr Justice MERVYN DAVIES)

TRENDWORTHY TWO LTD v ISLINGTON LONDON BOROUGH

Estates Gazette February 8 1986

277 EG 534-543

Rating — General Rate Act 1967, section 17 and Schedule 1 — Rating of unoccupied property — Action by plaintiff ratepayers in Chancery Division seeking declarations to the effect that they were not liable to pay unoccupied property rates in respect of hereditaments forming part of the Angel Centre, Pentonville Road — Defendant rating authority had served a completion notice under Schedule 1 on the plaintiffs — Plaintiffs' appeal against the notice had been dismissed by the county court, but an appeal to the Court of Appeal was pending — Meanwhile defendant authority were demanding unoccupied property rates on the basis of a proposal by the local valuation officer to enter the relevant hereditaments in the valuation list — Plaintiffs had objected and, in the absence of agreement or a determination by the local valuation court, no rateable values had been entered in respect of the centre — Plaintiffs submitted that in the circumstances no rate liability could arise and so no sum could be due in respect of rates — Defendant authority argued that as, according to para 1(1) of Schedule 1, the Act applied as if the hereditaments were occupied, section 6 was applicable during the relevant period of vacancy — It followed that as there had been a proposal, and as the hereditaments had come notionally into occupation when the relevant period began, the authority were entitled to add to or correct the rate — Held, rejecting this argument, that there was no notional coming into occupation within section 6(1), and that, in any case, there was authority that the provisions of Schedule 1 constituted a code which, within its terms, superseded the special rules relating to the liability of occupiers — The decision of the Court of Appeal in *Hastings Borough Council* v *Tarmac Properties Ltd* was binding on this point and must be preferred to views expressed by a divisional court in *Barr Hill Developments Ltd* v *South Cambridgeshire District Council* — It was strictly unnecessary to deal with an argument by the plaintiffs that their appeal against the completion notice had not been finally disposed of, but that argument was based on a misinterpretation of "appeal" in para 8(5) of Schedule 1 — Declaration accordingly that the plaintiffs were not liable to pay any unoccupied property rates until the relevant hereditaments and their rateable values were entered in the valuation list

This was an action commenced by writ in which the plaintiff, Trendworthy Two Ltd, sought against the defendant rating authority, the London Borough of Islington, declarations in respect of its liability to unoccupied property rates on hereditaments in Pentonville Road and St John's Street, Islington, London N1, forming part of the Angel Centre, a recently built office block. There was a counterclaim by the defendant authority for a declaration that the plaintiffs were liable to pay £976,407.06. F

W J Glover QC and G R G Roots (instructed by Michael Conn & Co) appeared on behalf of the plaintiff; M B Horton (instructed by Christopher Tapp, Borough Solicitor of Islington) represented the defendants.

Giving judgment, MERVYN DAVIES J said: This is a writ action in which the plaintiff, Trendworthy Two Ltd, seeks certain declarations respecting rates liability against the defendant rating authority, the London Borough of Islington. This is, to my mind, an unusual way in which to ascertain a rating liability under the General Rate Act 1967. However, counsel for the parties regarded a visit to the Chancery Division as appropriate and convenient. The hearing G proceeded without oral or affidavit evidence on the basis of admissions in the pleadings or of admissions made by counsel.

The facts are as follows. The plaintiff company own a recently built office block known as the Angel Centre in the Pentonville Road. There is a lease for part of the property but that fact is not of present concern. On June 1 1983 the defendant rating authority served a completion notice on the plaintiff pursuant to para 8 of Schedule 1 to the General Rate Act 1967. For present purposes the material parts of para 8 are as follows:

> (1) Where a rating authority are of opinion —
> (*a*) that the erection of a building within their area has been completed; or
> (*b*) that the work remaining to be done on a building within their area H is such that the erection of the building can reasonably be expected to be completed within three months,
>
> and that the building is, or when completed will be, comprised in a relevant hereditament, the authority may serve on the owner of the building a notice (hereafter in this paragraph referred to as "a completion notice") stating that the erection of the building is to be treated for the purposes of this Schedule as completed on the date of service of the notice or on such later date as may be specified by the notice.
>
> . . .
>
> (4) A person on whom a completion notice is served may, during the period of twenty-one days beginning with the date of service of the notice, appeal to the county court against the notice on the ground that the erection of the building to which the notice relates has not been or, as the case may be, cannot reasonably be expected to be completed by the date specified by the notice. J
>
> (5) If a completion notice served in respect of a building is not withdrawn and no appeal in pursuance of sub-paragraph (4) of this paragraph is brought against the notice or such an appeal is abandoned or dismissed, the erection of the building shall be treated for the purposes of this Schedule as completed on the date specified by the notice; and if the notice is not withdrawn and such an appeal is brought and is not abandoned or dismissed, the erection of the building shall be treated for those purposes as completed on such date as the court shall determine . . .

I have not seen the completion notice but it is accepted that it was to the effect that the Angel Centre be treated, for Schedule 1 purposes, as completed on September 1 1983. There was an appeal to the county court (see para 8(4)) against the completion notice. On December 21 1984 His Honour Judge Marder QC, sitting in the Clerkenwell County Court, dismissed the appeal. K

The plaintiff, desiring to question that dismissal, has given a notice of appeal dated January 11 1985 to the Court of Appeal. That appeal may be heard, so it is said, in February next*. Meanwhile the defendant authority has been seeking payment of rates in respect of the property. They do so despite the fact that the property is, or was at all material times, unoccupied.

In Islington rates may be claimed in respect of unoccupied property by virtue of section 17 of the 1967 Act and Schedule 1 to the Act. Demands totalling £976,407.06p for unoccupied property rates under Schedule 1 were served on December 21 1984. Those demands

*Editor's note: On March 24 1986 the appeal was allowed and the case remitted to the county court judge for reconsideration.

A were not based on any rateable values established in respect of the two parts of the Angel Centre as specified in (1) and (2) below. The demands were made by reason of a local valuation officer having, on March 28 1984, made proposals to enter in the valuation list hereditaments comprising parts of the Angel Centre:

(1) Offices and premises at 1 Pentonville Road and 403 St John's Street: gross value £695,000, rateable value £579,138; and

(2) Offices and premises at 401 St John's Street: gross value £52,500, rateable value £43,722.

That is to say, the rates demands were grounded on those proposals and not on any rateable values established as respects the Angel Centre.

Subsections (1) and (2) of section 6 of the 1967 Act are said to
B justify a rates demand being made on the mere proposals of a valuation officer. Subsections (1) and (2) are, so far as now material, as follows:

> (1) Subject to the provisions of this section, the rating authority may at any time make such amendments in a rate (being either the current or the last preceding rate) as appear to them necessary in order to make the rate conform with the enactments relating thereto, and in particular may —
> (*a*) correct any clerical or arithmetical error in the rate; or
> (*b*) correct any erroneous insertions or omissions or any misdescriptions; or
> (*c*) make such additions to or corrections in the rate as appear to the authority to be necessary by reason of —
> (i) the coming into occupation of any hereditament which has been newly erected or which was unoccupied at the time of the making of the rate; or
> C (ii) any change in the occupation of any hereditament

I need not read the rest of subsection (1). Subsection (2):

> Where the effect of the amendment would be either —
> (*a*) to alter, otherwise than by way of correction of a clerical or arithmetical error, the value on which a hereditament is rated; or
> (*b*) to charge to the rate a hereditament not shown, or not separately shown, in the valuation list,
> the rating authority shall not make any amendment of the rate unless either the amendment is necessary to bring the rate into conformity with the valuation list or a proposal for a corresponding alteration to the valuation list has been made by the valuation officer . . .
> (3) In the foregoing provisions of this section, other than subsection (1)(c)(i) references to a rate should be construed as references to that rate as it has been
> D applied in relation to particular hereditaments.

It is said that subsections (1) and (2) may be operated for the purpose of fixing a rate liability on the owner of unoccupied property. As appears from the Act, section 6 was enacted directly as respects occupied property. But the local authority says that it can also be taken into account in the operation of the Schedule 1 code that applies as respects unoccupied property.

I should say that on April 3 1984 the plaintiff company served notice of objection to the proposals of the valuation officer. Up till now the subject-matter of the proposals has not been settled by agreement or by any determination of the local valuation court. Accordingly, no rateable values are entered in the valuation list as respects Angel Centre.

E I have referred to steps taken by the local authority to obtain payment. Steps by way of petition to wind up were, as I understand, restrained. Then in June 1985 the authority took distress proceedings in the magistrates' court. Those proceedings, too, have been restrained, that restraint being imposed, so I was told, because by the time of the hearing before the magistrates the writ in this action dated February 15 1985 had been issued. Thus it was inappropriate to litigate the same liability in two courts.

Against this background of these disputes and proceedings there arise two submissions by the plaintiff company. First, it is said that the completion notice dated June 1 1983 is not yet of established validity, with the consequence that no rates liability as yet falls on the
F plaintiff. Second, that when operating Schedule 1 to the 1967 Act to determine what is an owner's liability for rates on unoccupied property, there can be no recourse to section 6 of the Act, in that Schedule 1 operates as respects liability independently of the rest of the Act.

These submissions seek respectively the following declarations set out in the statement of claim:

> 2 A declaration that the Plaintiff is not liable to pay any unoccupied property rates in respect of premises known as the Angel Centre, Islington, until the Plaintiff's appeal against the "completion notice" (within the meaning of paragraph 8 of Schedule 1 of the General Rate Act 1967) on appeal from a decision dated the 21st December 1984 of Clerkenwell County Court has been finally disposed of.
> 3 A declaration that the Plaintiff is not liable to pay any unoccupied property rates in respect of premises known as the Angel Centre, Islington, until relevant hereditaments and their respective rateable values have been entered in The Valuation List.

It is convenient first to consider 3 above. As I have mentioned, the Angel Centre is unoccupied, and in Islington rates may be claimed in respect of unoccupied property by virtue of section 17 of the 1967 Act and Schedule 1 to that Act. Mr Glover appeared for the plaintiff. He read para 1(1) and para 5(1) of Schedule 1, which are as follows:

> 1(1) Where, in the case of any rating area in which, by virtue of a resolution under section 17 of this Act, this Schedule is in operation, any relevant hereditament in that area is unoccupied for a continuous period exceeding three months, the owner shall, subject to the provisions of this Schedule, be rated in respect of that hereditament for any relevant period of vacancy; and the provisions of this Act shall apply accordingly as if the hereditament were occupied during that relevant period of vacancy by the owner.
> 5(1) Subject to the provisions of this Schedule, the rateable value of a hereditament for the purposes of paragraph 1 thereof shall be the rateable value ascribed to it in the valuation list in force for the area in which the hereditament is situated or, if the hereditament is not included in that list, the first rateable value subsequently ascribed to the hereditament in a valuation list in force for that area.

Mr Glover's simple submission then was that the plaintiff is not at present liable because there is no rateable value ascribed to the Angel Centre in the valuation list now in force (see para 5(1)). He accepted, as I understand, that if a rateable value was hereafter ascribed to the centre by an amendment or alteration in the current list, then a liability may arise and may relate back in time. But he says there is now no rateable value in the list and so no rate liability can be calculated and so no sum is due.

Mr Horton, for the authority, was not content to read Schedule 1 in isolation. He drew attention to the words at the end of para 1(1) of the schedule:

> and the provisions of this Act shall apply accordingly as if the hereditament were occupied during the relevant period of vacancy by the owner.

There, he said, was a thread leading the schedule back to the Act, and in particular to section 6 of the Act. I have already set out section 6. Fastening on to section 6(1)(c)(i), Mr Horton said that the authority was empowered to add to or correct the rate on "the coming into occupation" of a newly erected hereditament if and when there has been made by the valuation officer a proposal within section 6(2). That such a proposal has been made is admitted by the plaintiff. There was, said Mr Horton, a deemed coming into occupation of Angel Centre on the date when the relevant period of vacancy, as defined in para 15(1) of the schedule, began. In other words:

1 The Act applies as if the Angel Centre were occupied during the relevant period of vacancy;
2 So, during that period, section 6 applies;
3 There is in being a proposal pursuant to section 6(2);
4 So, the authority can add to or correct the rate (see section 6(1)(c)) when the Angel Centre comes into occupation; and
5 It notionally comes into occupation (presumably at the start of the relevant period of vacancy) because you apply the Act as if it were occupied (see para 1(1) of the schedule).

I find this circular argument difficult. First, the Act applies to the Angel Centre as if it were occupied (see para 1(1)). If the Act applies as if the property were occupied, I cannot see that there can be any coming into occupation notionally within section 6(1)(c)(i), since occupation, whether notional or actual, is not a relevant circumstance for rating liability as respects unoccupied land. The liability appears from the schedule, and in particular paras 1(1), 5, 7 and 8. Second, the centre is deemed to have been unoccupied on December 1 1983 (see para 7 of Schedule 1). According to Mr Horton, as I understand him, it is also deemed to have come into occupation on that date.

While those are my views it is not necessary that I should rely upon them as reasons for my decision on declaration 3. That is because there is authority that points the way. In *Hastings Borough Council* v *Tarmac Properties Ltd* [1985] RA 124 *Lawton LJ says at p 129:

> In my judgment the rating authority's construction is correct. The liability of the owners of unoccupied hereditaments to be rated has a statutory origin. It was a new concept in rating law and could not easily be clothed with the existing law based as it was on the concept of occupancy. Situations had to be dealt with which did not arise when there was an occupier of a hereditament. Examples are provided by para 2 of the schedule. Further, there has to be a means for determining the rateable value for the relevant period of vacancy

*Editor's note: Also reported at [1985] 1 EGLR 161; (1985) 274 EG 925.

when a newly occupied or altered building has been completed. Schedule 1 deals with all these matters. It is a code enacting when liability to be rated arises and how the quantum of liability is to be determined. On matters within its terms it supersedes any provisions of the Act which determine the liability of occupiers. The concluding words of para 1(1), which were expressed as a subordinate clause, are, in my judgment, intended to make applicable to the owners of unoccupied hereditaments the provisions of the Act which are not concerned with liability.

Kerr LJ and Fox LJ agreed with those sentiments. So it appears that the schedule supersedes any liability-imposing provisions of the Act. Section 6 appears to be such a provision. On the other hand, the *Hastings* case is, as I understand, at odds with a decision of a strong divisional court in *Barr Hill Developments Ltd* v *South Cambridgeshire District Council* [1979] RA 379†. In that case Eveleigh LJ said at p 389:

Counsel for the owners has conceded in this case that section 6 of the Act allows a rate to be recovered, although the premises are not in the valuation list, but he claims that it is only possible in respect of occupied property. I will not in this extempore judgment go into a detailed examination of the case which has caused counsel to make that concession, if indeed he is not constrained to make it on the very wording of section 6. It is the case of *Kettle & Co* v *Newcastle-under-Lyme Borough Council* decided in the Court of Appeal on March 21 1979. Suffice it to say that the wording of the section itself shows that not always is it envisaged that there must be actual entry in a valuation list before rates can be demanded. A proposal by the valuation officer is a sufficient foundation for the valuation. At a subsequent date it may be that valuation is challenged by one party or another and is then varied. The Act then provides that adjustments may be made, but it is clear that the rates based on that initial proposal must be paid in the first instance. Then, as I have said, the matter can be adjusted later in the event of a dispute as to the proposed rateable value. So that the actual recovery is provided for in the Act itself in the very detailed provisions of the Act particularly relating to what is to happen when a proposal is made. I do not think it is necessary in this case to go into that; I mention it simply to indicate that the Act itself contains the power to demand rates although the property is not in the list. I see nothing in the Act to indicate that that situation is confined to occupied property as opposed to unoccupied property. After all, section 17 is itself part of the Act and that of course relates to unoccupied property. The schedule should not be allowed to limit the effect of the sections of the Act itself. Nor do I think that the schedule in any way attempts to do so. What one finds in the schedule and in para 5 in particular is a number of different tests to be applied where appropriate to determine the rateable value.

So while Lawton LJ says that Schedule 1 in matters within its terms supersedes any provisions of the Act which determine the liability of occupiers, Eveleigh LJ says that the schedule should not be allowed to limit the effect of the sections of the Act itself. The situation is the more puzzling because I was informed that the *Barr Hill* case was read at the *Hastings* hearing.

However that may be, in this situation it is my duty to follow the Court of Appeal, and that I do. Accordingly I hold that the Islington Rating Authority cannot rely on section 6, with the consequence that the plaintiff is entitled to the declaration numbered 3 above.

I turn to declaration 2 sought by the plaintiff. Since I have found, as I have, as to declaration 3, it is not necessary for me to decide whether or not to make declaration 2. Accordingly I do not make declaration 2. But, since the matter was argued before me, I will say that if it were necessary to consider declaration 2, I should decline to make it. In brief, my reasons would be:

1 That the phrase "such an appeal" where it first appears in para 8(5) of Schedule 1 is a reference to an appeal to the county court (see para 8(4)). Since the plaintiff's appeal to the county court has been dismissed, the Angel Centre is now to be regarded as having been completed on the date specified by the completion notice.

2 Order 59, rule 19(5) provides, so far as now material, that an appeal from the county court is not to operate a stay of execution or of proceedings in the county court unless the judge of the county court or the Court of Appeal so orders. Judge Marder, on February 12 1985, dismissed an application by the plaintiff for a stay of proceedings on his dismissal on December 21 1984 of the appeal.

In these circumstances it seems to me that the rating authority can act on the completion notice unless and until it is set aside or amended by the Court of Appeal.

The plaintiff also sought declaration 1 in these terms:

A declaration that the Plaintiff is not liable to pay certain sums totalling £976,407.06 alleged by the Defendants to be due as unoccupied property rate in respect of two relevant hereditaments at premises known as Angel Centre, Islington, as particularised in two demand notes dated December 21 1984 served by the Defendants upon the Plaintiff. G

In the light of declaration 3 the plaintiff is not at present liable to pay the sum mentioned in declaration 1. However, I do not make declaration 1. It is sufficient for present purposes to make declaration 3. Furthermore, it may be that in future the plaintiff will become liable for the sum stated in declaration 1, as and when the machinery of Schedule 1 for fixing liability is eventually operated.

I add that there is a counterclaim by the rating authority for a declaration that the plaintiff is liable to pay the sum of £976,407.06 mentioned. It follows from my decision on the plaintiff's claim that the counterclaim must be dismissed.

Order made in terms of Declaration 3. Counterclaim dismissed. Defendants to pay plaintiff's costs. H

†Editor's note: Also reported at (1979) 252 EG 915.

Court of Appeal

March 25 1986

(Before Lord Justice FOX, Lord Justice NEILL and Lord Justice RALPH GIBSON)

DEBENHAMS PLC v WESTMINSTER CITY COUNCIL

Estates Gazette May 31 1986

278 EG 974-979 J

Rating of unoccupied property — Exemption for listed buildings under Town and Country Planning Act 1971 — Position where part of the unoccupied building is not mentioned in list but forms part of the hereditament for rating purposes — Definition of listed building in section 54(9) of the 1971 Act as covering a structure fixed to the building or comprised within the curtilage of the building — Application to the old Hamleys toy shop in Regent Street, part of which was on the further side of the adjoining Kingly Street, but was connected to the Regent Street shop by a tunnel at basement level and a footbridge at second-floor level — Appeal from decision of Hodgson J, who had held that relief from the unoccupied rate could be claimed in respect of the whole hereditament under the listed building exemption — Rating authority submitted that the exemption did not apply as it required the whole hereditament to be included in the list whereas in the present case part of it was not — Held, affirming the decision of Hodgson J, that at all material times the part of the hereditament on the further side of Kingly Street was a structure fixed to the remainder of the hereditament and also formed part of the curtilage of the larger building which constituted the remainder of the hereditament — The policy underlying the exemption of listed buildings from the unoccupied property rate in para 2(c) of Schedule 1 to the General Rate Act 1967 was the recognition of the fact that such buildings are subject to a statutory control which seriously affects the power of the owner to make full economic use of them — Such statutory control applied to the structures affixed to a building or comprised within its curtilage mentioned in section 54(9) of the 1971 Act — Hence the whole hereditament in the present case was within the exemption from the rate — Appeal dismissed

This was an appeal by Westminster City Council from the decision by Hodgson J, reported at [1985] 1 EGLR 159; (1985) 274 EG 826, allowing an appeal from the decision of a metropolitan stipendiary magistrate who had directed the issue of a distress warrant in respect of a claim by the council against Debenhams plc for unoccupied property rates on premises in Regent Street and Kingly Street, London W1, formerly occupied by Hamleys toy shop. M

Graham Eyre QC and Richard Hone (instructed by T F Neville, City Solicitor, Westminster City Council) appeared on behalf of the appellants; Mathew B Horton (instructed by Forsyte Kerman) represented the respondents.

Giving judgment, FOX LJ said: This appeal is concerned with the provisions of the General Rate Act 1967 regarding the rateability of an unoccupied hereditament if the hereditament is included in a list of buildings of special architectural or historic interest compiled under section 54 of the Town and Country Planning Act 1971.

The appeal is by the Lord Mayor and Citizens of the City of Westminster ("The council") from a decision of Hodgson J, on an appeal by way of case stated from an order of the metropolitan stipendiary magistrate and whereby the magistrate directed that a distress warrant should be issued in the sum of £68,696 relating to a period from February 1 1982 to March 31 1983 during which a hereditament owned by the respondents ("Debenhams") was unoccupied.

The council are the rating authority for the rating area of the City of Westminster. Debenhams were at all material times until October 31 1981 or thereabouts the rateable occupier of the hereditament in question, which consisted of Hamleys toy shop in Regent Street, W1. On or about October 31 1981 Debenhams ceased to be in rateable occupation. The premises thereafter were, at all material times, unoccupied but Debenhams remained the owner.

At the material times Hamleys consisted of three premises, namely 200-202 Regent Street, 50-52 Kingly Street and 27-28 Kingly Street. The hereditament was entered in the valuation list as "200-202 Regent Street (and 50-52 Kingly Street)". The magistrate found that the latter description was intended to include 27-28 Kingly Street.

Kingly Street is more or less parallel with Regent Street. Hamleys' shop front was in Regent Street (200-202); the shop ran through to the west side of Kingly Street (50-52). Nos 27-28 Kingly Street are on the east side of Kingly Street. The magistrate found as a fact that 50-52 Kingly Street and 27-28 Kingly Street were connected by a footbridge at second-floor level and by a tunnel at basement level; customers at the shop were thus able to walk through the shop from the Regent Street front to 27-28 Kingly Street. Works to fill in the tunnel were commenced in October 1982 and were completed in January 1983; the footbridge was removed in March 1983.

The magistrate also found that:

> During the relevant period within the part of the hereditament comprising 200-202 Regent Street and 50-52 Kingly Street there was no physical feature enabling one to identify where the Regent Street property ended and the Kingly Street property began. There was a central staircase serving both which had been provided at the time the property was constructed.

The absence of any distinguishing feature between 200-202 Regent Street and 50-52 Kingly Street was probably of long standing. The tunnel was constructed pursuant to a licence of January 27 1921 which authorises the construction of "a subway under the highway of Kingly Street and numbers 200-202 Regent Street".

Under section 17 of the General Rate Act 1967 a rating authority may, by resolution, adopt the provisions of Schedule 1 to the Act. By para 1 (1) of Schedule 1 it is provided that where in the case of a rating area where a resolution under section 17 is in operation:

> any relevant hereditament in that area is unoccupied for a continuous period exceeding three months, the owner shall, subject to the provisions of this Schedule, be rated in respect of that hereditament for any relevant period of vacancy; and the provisions of this Act shall apply accordingly as if the hereditament were occupied during that relevant period of vacancy by the owner.

Para 2 of Schedule 1 (as amended) is as follows:

> No rates shall be payable under paragraph 1 of this Schedule in respect of a hereditament for, or for any part of the three months beginning with the day following the end of, any period during which —
>
> (a) the owner is prohibited by law from occupying the hereditament or allowing it to be occupied;
>
> (b) the hereditament is kept vacant by reason of action taken by or on behalf of the Crown or any local or public authority with a view to prohibiting the occupation of the hereditament or to acquiring it;
>
> (c) the hereditament is the subject of a building preservation notice as defined by section 58 of the Town and Country Planning Act 1971 or is included in a list compiled or approved under section 54 of that Act;
>
> (d) the hereditament is the subject of a preservation order or an interim preservation notice under the Ancient Monument Acts 1913 to 1953, or is included in a list published by the Minister of Public Building and Works under those Acts;
>
> (e) an agreement is in force with respect to the hereditament under section 56 (1) of this Act; or
>
> (f) the hereditament is held for the purpose of being available for occupation by a minister of religion as a residence from which to perform the duties of his office.

The council duly made a resolution under section 17 (1) of the 1967 Act in consequence whereof the provisions of Schedule 1 to the Act were at all material times applicable in the council's rating area.

The "relevant hereditament" is defined by para 15 of Schedule 1 as meaning:

> any hereditament consisting of, or of part of, a house, shop, office, factory, mill or other building whatsoever, together with any garden, yard, court or other land ordinarily used or intended to be used or intended for use for the purposes of the building or part.

On November 22 1973 the Secretary of State for the Environment compiled a list of buildings of special architectural or historic interest in which were included a number of properties under the heading "Regent Street W1 (east side)". In the list was an entry "Numbers 172 to 206 (even)". Accordingly, nos 200-202 Regent Street were at all material times included in that list.

The position then is that the council claim rates in respect of the entire premises from February 1 1982 (being three months after the hereditament ceased to be occupied) to March 31 1983, during which period the hereditament was unoccupied and Debenhams were the owner. Debenhams resist the claim and rely on the provisions of para 2 (c) of Schedule 1 to the General Rate Act 1967 upon the ground that at all material times nos 200-202 Regent Street were included in a list compiled under section 54 of the Town and Country Planning Act 1971. That contention failed before the magistrate but succeeded before the judge.

Para 2(c) of Schedule 1 to the 1967 Act confers the exemption if "the hereditament is . . . included in a list compiled or approved under section 54 of [the Town and Country Planning Act 1971]".

The hereditament in respect of which Debenhams were liable to be assessed to rates, and which was, as the magistrate found, included in the valuation list, was 200-202 Regent Street, 50-52 Kingly Street, and 27-28 Kingly Street. That was the "relevant hereditament" referred to in para 1 of Schedule 1 and the "hereditament" referred to in para 2 of Schedule 1.

The first question is whether that hereditament was included in the list compiled under section 54 of the Town and Country Planning Act 1971. The council answer that the hereditament was not so included; only part of it was (namely 200-202 Regent Street and 50-52 Kingly Street). On that basis, say the council, the case does not come within para 2(c) of Schedule 1. For that the council rely on the decision of the Divisional Court in *Providence Properties Ltd* v *Liverpool City Council* [1980] RA 189. The hereditament in that case consisted of three warehouses some six storeys high, the two outer of which were linked to the central warehouse by a single-storey building and a detached gateway abutted by two gatehouses. Only the central warehouse together with the gates and gatehouses were included in the section 54 list. The Divisional Court (Lord Lane CJ and Boreham J) held that the hereditament was not exempted from rates under para 2(c) of Schedule 1. Boreham J (with whose judgment the Lord Chief Justice agreed) said, at p 193:

> In my judgment the meaning of paragraph 2(c) is clear. If Parliament had wished paragraph 2(c) to apply to part of a hereditament, it could have said so and said so very easily in the way counsel for the appellant demonstrated by the inclusion of those few words. They were not included. In my judgment the meaning is clear. One simply asks the question, what was the hereditament in this case? Was that hereditament in the list under section 54 of the Town and Country Planning Act 1971? The answer to that question seems to me to be clear. That hereditament was not in that list. A bit of it was. That is not enough.

In response to that approach Debenhams submit three answers. The first is that, on the facts of the present case, the whole of the rateable hereditament was included in the section 54 list. The listed premises are 200-202 Regent Street. On the facts as found by the magistrate it does not appear that, when the list was compiled, or indeed at any material time, there was any means of distinguishing between 200-202 Regent Street and 50-52 Kingly Street. I would accept, therefore, that 50-52 Kingly Street could properly be regarded as included in the description in the list. The same is not true of 27-28 Kingly Street. That property is on the east side of Kingly Street and, though connected by the bridge and tunnel, was at all times physically distinguishable from the rest of the premises between Regent Street and the west side of Kingly Street. Debenhams contend that what was listed was, in fact, Hamleys' toy shop. I do not feel able to accept that. Such a description cannot be extracted from the listed description "Numbers 172 to 206 Regent Street (east side)". The list does, in fact, contain some specific references to shops (for example, Peter Robinson's and Liberty's), but the listing in the present case contains no such reference.

Debenhams next (and principal) submission is based upon the provisions of section 54(9) of the Town and Country Planning Act 1971; those provisions were not referred to in the judgments in the *Providence Properties* case and Mr Horton doubts whether he referred the court to them.

Section 54 subsections (1) and (9) of the Town and Country Planning Act 1971 are in the following terms:

> (1) For the purposes of this Act and with a view to the guidance of local planning authorities in the performance of their functions under this Act in relation to buildings of special architectural or historic interest, the Secretary of State shall compile lists of such buildings, or approve, with or without modifications, such lists compiled by other persons or bodies of persons, and may amend any list so compiled or approved.
>
> (9) In this Act "listed building" means a building which is for the time being included in a list compiled or approved by the Secretary of State under this section; and, for the purposes of the provisions of this Act relating to listed buildings and building preservation notices, any object or structure fixed to a building, or forming part of the land and comprised within the curtilage of a building, shall be treated as part of the building.

It is submitted by Debenhams that the whole of the hereditament is to be treated as being comprised in the list because the part of it not specifically referred to in the listing (ie 27-28 Kingly Street) was:

(a) a structure "fixed" by the bridge to the building specifically included in the list.

and/or

(b) a structure forming part of the land and comprised within the curtilage of the building specifically referred to in the listing.

Debenhams rely on the decision of this court in *Attorney-General ex rel Sutcliffe and Others* v *Calderdale Borough Council* (1982) 46 P & CR 399, to which I will refer later. The judge accepted that approach, holding that 27-28 Kingly Street was fixed to the rest of the property.

I will consider first the general effect of section 54(9) of the 1971 Act. The subsection starts with a definition of "listed building" as meaning a building which is for the time being included in a list compiled or approved by the Secretary of State under section 54. Nos 200-202 Regent Street (including 50-52 Kingly Street) is such a building.

Section 54(9) goes on then to provide that for the purposes of the provisions of the Act related to listed buildings and building preservation notices "any object or structure fixed to a building, or forming part of the land and comprised within the curtilage of a building, shall be treated as part of the building".

The result, therefore, is that where a "structure" is fixed to a building and that building is for the time being included in a list compiled by the Secretary of State under section 54, then the structure is, for the purposes of the provisions of the Act relating to "listed buildings" (and building preservation notices), to be treated as part of the building to which it is fixed. The structure is therefore, to that extent, to be treated as part of a building which is included in the list.

The council's contention, however, is that section 54(9) has no relevance to the present case. Debenhams, it is said, must show that the rateable hereditament is included in a list compiled under section 54. The rateable hereditament (ie 200-202 Regent Street, 50-52 Kingly Street and 27-28 Kingly Street) is not so included; only part of it is. Section 54(9), it is asserted, does not alter that fact. The subsection does not provide that a structure which is fixed to (or within the curtilage of) a building which is included in the list is to be deemed for all purposes to be so included. The subsection merely says that for the purposes of the 1971 Act dealing with listed buildings and building preservation notices, the structure is to be treated as part of the building.

In resolving this matter it is, I think, necessary to begin with the policy of para 2(c) of Schedule 1 to the General Rate Act 1967. What is the reason for giving exemption to a building included in the list of buildings of special architectural or historic interest? It must, in my view, be the fact that such buildings are subjected to a statutory control which seriously affects the power of the owner to make full economic use of them. Thus, by section 55(1) of the 1971 Act:

> . . . if a person executes or causes to be executed any works for the demolition of a listed building or for its alteration or extension in any manner which would affect its character as a building of special architectural or historic interest, and the works are not authorised under this Part of this Act, he shall be guilty of an offence.

In general, works to a listed building require the consent of the local planning authority or the Secretary of State (section 55(2)).

So far as the present law is concerned, the list derives entirely from the provisions of section 54(1) of the 1971 Act. The purpose of the list is to impose on "listed buildings" the statutory control. If, therefore, a building is "a structure fixed to" a building which is itself included in the list compiled under section 54(1), then for the purposes of the statutory control it is part of the building included in the list. It is, in consequence, subjected to all the restrictions imposed by the 1971 Act upon the demolition, alteration or extension of buildings included in the list.

The council say that while that is the position in law, it is, in practice, much easier to obtain consent for works to a building which is brought into control by section 54(9) than it is for a building which is itself included in the list. That may be so in some, cases but it is not necessarily so in all. For example, the "fixed" building may be of some importance to the stability of the building included in the list; or its alteration or demolition might cause damage to the architectural features of the building so included. But, in any event, the statutory requirements for consent to works to the "fixed" building exists and is a restriction upon the ordinary powers of the landowner. In the present case, when the bridge was demolished in 1983, listed building consent had to be (and was) obtained.

Against that background, I return to the language of para 2(c) of Schedule 1 to the 1967 Act. Can a building which is treated as part of the listed building by the provisions of section 54(9) be properly regarded as "included in a list compiled or approved under section 54 of the Town and Country Planning Act 1971"? In my opinion it can. I accept that section 54(9) does not operate as a general deeming provision. But, for the purposes for which the Secretary of State compiles the list of buildings of special architectural or historic interest, the "fixed" building is treated as part of the listed building. And those purposes are, as it seems to me, the basis of the policy of Parliament in granting the exemption in para 2(c) of Schedule 1 to the General Rate Act. In those circumstances it is, in my opinion, unreal to suppose that Parliament did not intend the "fixed" building to have the benefit of the para 2(c) exemption. As a matter of language that conclusion, in my view, does not lead to any great difficulties in relation to para 2(c). That paragraph refers to a building included in a list compiled under section 54 of the 1971 Act. Section 54(1) directs the compilation of a list "for the purposes of this Act". Section 54(9) defines a listed building as a building for the time being included in the list and then says that for the purposes of the Act relating to listed buildings (ie buildings included in the list) "fixed" structures are to be treated as part of the building. They are therefore, to that extent, to be treated as part of a building included in the list and so included in the list themselves. The extent to which they are so treated is very substantial in that it relates to the principal purposes of the creation of the list and to the reasons for the grant of exemption in para 2(c).

It is said on behalf of the council that we must have regard to the fact that para 2(c) as originally enacted referred to the Town and Country Planning Act 1962, which contained no provision corresponding to section 54(9) of the 1971 Act. That is so, but what we have to construe is the amended statute, which in fact refers to the 1971 Act.

There remains the question whether 27-28 Kingly Street was a "structure fixed" to the remainder of the hereditament at all material times until the bridge was demolished in March 1983. Nos 27-28 Kingly Street, in my view, plainly constituted a "structure" which was "fixed" to the remainder of the rateable hereditament by the bridge. In *A-G ex rel Sutcliffe* v *Calderdale BC* (*supra*), there was a terrace of cottages, a mill and a bridge which linked the two. In 1971 the mill was listed under section 54 of the 1971 Act. The terrace was not included in the listing. "I have no doubt" said Stephenson LJ (who gave the leading judgment) at p 404 "that the linking bridge is fixed to the mill, although Mr Macleod withdrew a concession that it was. It was attached (on the other side) to No 1 and so to the terrace". Accordingly the Court of Appeal held that listed building consent was necessary to the demolition of the terrace, though it was not itself included in the list. The case was not concerned with the provisions of Schedule 1 to the General Rate Act 1967.

So far as the second limb of section 54(9) is concerned (ie "or forming part of the land and comprised within the curtilage of a building") it seems to me that, down to the time when the bridge was demolished in March 1983, 27-28 Kingly Street formed part of this curtilage of the larger building which was 200-202 Regent Street and 50-52 Kingly Street. The premises were in single ownership and had long been used as a single building and for the same purposes;

A further, they formed a single hereditament for rating purposes. Nos 27-28 Kingly Street was really an annexe fixed to the rest of the hereditament.

It appears from the case stated that the bridge was demolished so as to enable 27-28 Kingly Street to be sold separately. But during the short period in March 1983 between the demolition of the bridge and the end of March it was still owned by Debenhams; it still formed part of the rateable hereditament and was geographically close to the rest of the hereditament. I am not prepared to say that it ceased to be within the curtilage of the rest during that short period.

I should add this. It seems to me that, if the construction of para 2(c) and section 54(9) which I have adopted is wrong, the very
B unsatisfactory result follows that there will be no exemption for any part of the rateable hereditament even though a substantial part of it is listed. The unit of rating is "the hereditament or hereditaments comprising" the property liable to be assessed to rates (General Rate Act 1967, section 16). "Hereditament" means "property which is or may become liable to a rate, being a unit of such property which is, or would fall to be, shown as a separate item in the valuation list". Accordingly, the "rateable hereditament" and "the hereditament" referred to in Schedule 1, paras 1 and 2, is, in the present case, the entirety of the property (ie 200-202 Regent Street, 50-52 Kingly Street and 27-28 Kingly Street). That is the unit appearing in the valuation C
list. And it is that unit which must be shown to be included in the section 54 list in order to attract the exemption in para 2(c). That provision says "the hereditament" and not "the hereditament or any part thereof". On this aspect of the case I agree with the reasoning of the Divisional Court in the *Providence Properties* case.

The same considerations also, I think, dispose of the third argument advanced by Debenhams in the present appeal, namely that the exemption is available for all the buildings comprised in the rateable hereditament if *any* of the buildings is a listed building.

Only buildings can be listed. I say nothing about the position of a hereditament consisting of a house or other building with a garden or other unbuilt-on land where the house or building is listed.

For the reasons which I have given I conclude that the whole of Debenhams' rateable hereditament is within the exemption in para 2(c) of Schedule 1 to the 1967 Act. I would dismiss the appeal.

NEILL and RALPH GIBSON LJJ agreed that the appeal should be dismissed for the reasons given by Fox LJ and did not add anything.

The appeal was dismissed with costs. Leave to appeal to the House of Lords was refused, but on May 8 1986 leave to appeal was granted by the House of Lords Appeal Committee.

For further cases on this subject see p 213

TOWN AND COUNTRY PLANNING

Queen's Bench Division

October 30 1985

(Before Mr Justice WOOLF)

PENWITH DISTRICT COUNCIL v SECRETARY OF STATE FOR THE ENVIRONMENT AND ANOTHER

Estates Gazette January 18 1986

277 EG 194-196

Town and Country Planning Act 1971 — Application by planning authority under section 245 of Act to quash a decision of the Secretary of State, made on his behalf by an inspector, allowing an appeal by a company wishing to establish an amusement or entertainment centre — Proposed centre was to be located in St Ives, in an outstanding conservation area — Inspector in his decision letter said that the application site was within the shopping centre of St Ives in a predominantly seasonal trading zone with premises related to the tourist industry which were often closed outside the holiday season — He envisaged some problems, such as noise and the determination of opening hours, but he thought that these could be dealt with by conditions, and he considered that there was a shortage of entertainment facilities in the area which the proposed centre would help to supply — The inspector's letter was criticised by the appellant planning authority on the grounds that it gave no reasons for departing from Department of the Environment guidance on amusement centres and that it relied on the effect of planning conditions which were in fact valueless — The departmental guidance stated that "amusement centres are not acceptable near residential property; nor are they good neighbours for schools, churches, hospitals or hotels. They are out of place in conservation areas or other places of special architectural or historic character, except perhaps where these cover a really wide area" — Held that the decision letter did not indicate whether the inspector had taken this departmental guidance into account or, if he did, why he thought it right to depart from it — Held also that the bare reference to the need for conditions as to such matters as noise and opening hours was unsatisfactory — As regards opening hours, for example, no assistance was given of a practical kind for enforcement action — Decision quashed and case remitted to the Secretary of State for reconsideration.

This was an application by Penwith District Council under section 245 of the 1971 Act to quash the decision of the Secretary of State for the Environment allowing an appeal from the council's decision refusing permission to Mainstreet Marketing Ltd to establish an amusement or entertainment centre in St Ives, Cornwall.

J Steel (instructed by Sharpe Pritchard & Co, agents for L Rowland, of Penzance) appeared on behalf of the applicant planning authority; J Holdsworth (instructed by the Treasury Solicitor) represented the first respondent, the Secretary of State for the Environment; the second respondents, Mainstreet Marketing Ltd, were not represented and took no part in the proceedings.

Giving judgment, WOOLF J said: This is an application under section 245 of the Town and Country Planning Act. It is an application by the Penwith District Council in respect of a decision by F the Secretary of State for the Environment, made on his behalf by an inspector, to allow an appeal of Mainstreet Marketing Ltd, who wish to establish an amusement or entertainment centre in St Ives.

The area with which I am concerned is an outstanding conservation area. The fact that it is an outstanding conservation area means that more funds are available for its improvement than is ordinarily the case in a conservation area. The application was dealt with by written submissions. Obviously it was a matter of concern to the local planning authority that nothing should be allowed to take place in the conservation area which would be inconsistent with the policies which were applicable to it. Therefore, if the inspector was G minded to depart from any policies which applied to the conservation area, it was a situation in which the planning authority were entitled to have clear and intelligible reasons explaining why, in the circumstances of this planning decision, it was appropriate to depart from them.

The planning authority have got the task of ensuring that the treatment of planning applications in general is approached in a fair and reasonable manner. If the Secretary of State is going to make decisions without adequate reasons which depart from what has been laid down as the general policy for the area, the planning authority are not going to be in a position to perform their function properly. They need the appropriate assistance from the Secretary of State. H

The decision letter in this case, dealing with the written representations made by the parties, is admirably succinct. It sets out, as one would expect, a description of the area. It describes the application site as being within the shopping centre of St Ives. It specifies that it is in the St Ives Outstanding Conservation Area. It goes on to say that it is within Area 2. Under a non-statutory policy document, Area 2 is a predominantly seasonal trading zone, with premises related to the tourist industry, which often close outside the holiday season.

Apparently, there was a certain blurring of the area, to which the inspector has referred. There were submissions that the area has been modified. However, what the inspector says in his letter is that he is not greatly swayed by the council's attitude that this part of Area 2 J should now be regarded as part of the established shopping centre which provides mainly "resident" services. In other words, the inspector is indicating that he still regards it as an area which is providing premises for the tourist industry.

The inspector then goes on to say that having inspected the site, he has come to the conclusion that the main issues are:

1. Whether the proposed use would be damaging to the character and appearance of the conservation area? 2. Whether the proposal would be in keeping with the seasonal trading nature of Area 2 in the town centre, and would not harm its commercial vitality?

The inspector, in his decision letter, goes on to explain that the St Ives Conservation Area arises out of the narrow road width, the cobbled surface and the enclosed and intimate feel given by the two- K and three-storey buildings lining the street. He deals adequately with the fact that so far as external appearance is concerned, there is nothing to be feared by the proposal. He then goes on to deal with the effect of the proposal on the trading and shopping character of the street. Again, he finds that there is nothing adverse in that.

The inspector says that it appears to him that members of the public would be drawn to the entertainment centre in a similar way to those window shopping or visiting other outlets. He decided that the nature of the use is not one which would give rise to crowds congregating at particular times while leaving no activity at other times. He does recognise that there could be inappropriate noise, but he feels that that could be dealt with by conditions.

A Later, having dealt with other matters, he goes on to say that he thinks there is a shortage of entertainment facilities in the area and he is of the view that the proposal would be in keeping with the mixed function of shopping centre and holiday resort of this part of St Ives and that, subject to conditions, planning permission cannot reasonably be withheld. So far as opening hours are concerned, the inspector is of the view that that could give rise to problems but it could be dealt with by appropriate conditions.

In the course of dealing with the various submissions, the inspector refers specifically to the fact that he has borne in mind the various provisions of the development plan, including policy 10A of the structure plan, and the non-statutory policies of the St Ives Central
B Area Commercial Uses Study. The first ground of complaint arises out of the fact that he makes no mention of the Development Control Policy Note No 11. That is a development control policy note which deals specifically with amusement centres. It had been referred to and the relevant parts are set out in the written submissions of the planning authority. It is a Department of the Environment and Welsh Office circular and it does not set out rigid policies. It is intended to give general guidance to intending developers and, therefore, to planning authorities.

Among the matters which are dealt with in that note are the ordinary considerations which affect planning applications for amusement centres. So far as those are concerned, although there has been no reference to the development control policy note, I regard the
C decision letter as dealing quite satisfactorily with those general considerations. However, the note goes on to say in specific terms:

> As regards situation, amusement centres are not acceptable near residential property; nor are they good neighbours for schools, churches, hospitals or hotels. They are out of place in conservation areas or other places of special architectural or historic character, except perhaps where these cover a really wide area.

Although this must not be regarded as a rigid rule, here was the clearest guidance being given by the Secretary of State, indicating what should be the general and ordinary approach to amusement centres in areas, of which this is one, which are near residential property and hotels and which are conservation areas. Having regard to the fact that the inspector specifically referred to other policy
D documents, but made no reference to Policy Note No 11, which clearly should have had a considerable impact upon his decision, the planning authority say, first of all, that it looks as if the inspector failed to recall the importance of that note when coming to his decision. Alternatively, the planning authority submit that if he did have regard to the proposed requirements of the policy note, to which I have just made reference, then he gives no explanation or reasons as to why, in this case, he thought it right to depart from the general prohibition which is provided in the guidance. They say that without any explanation from the inspector as to why he thought it right to depart from that general prohibition, they cannot understand the basis of his decision and have been given no proper and intelligible
E reasons for it.

I am bound to say that I consider that there is great force in Mr Steel's submission. Notwithstanding Mr Holdsworth's assistance in the matter, it is my view that the inspector should, if indeed he did take the circular into account, have indicated why it was that he felt that this was a case in which he should depart from the general prohibition.

Looking at the decision, it seems to me that his general approach was one which would have been appropriate if this had not been a conservation area. On that ground alone, it seems to me that this is a matter where the applicants are entitled to say to this court that this decision is not an adequate decision and does not comply with the requirements of the Act. Therefore, they are entitled to have the
F decision quashed and remitted to the Secretary of State for reconsideration by him.

A second point is taken by Mr Steel on behalf of the planning authority. As I indicated when going through the reasoning of the inspector, he attached considerable importance to the fact that certain of the objections to the scheme could be overcome by the provision of conditions. The relevant condition that he proposes is in these terms:

> Before the development hereby permitted is commenced a scheme shall be agreed with the Local Planning Authority to cover the following points:
> A. Noise insulation. B. Shopfront type. C. Opening hours.

The condition, however, does not go on to say anything about the fact that the development should then be carried out in accordance with the scheme and that the amusement centre should be operated in accordance with the scheme. Mr Steel submits that a condition of this sort is valueless. It may be — and I do not propose to express a final conclusion on this — that with regard to noise insulation and shopfront type, it is implicit in the decision that the development should be carried out in accordance with the scheme and that the condition must be read in those terms. However, with regard to opening hours, that is not a practical approach. It would be of the greatest importance to the planning authority to be able to police the operation of this amusement arcade. That could require the service of an enforcement notice. An enforcement notice has to specify the breach of the condition which is relied upon. It is difficult to see how enforcement action can take place based on the condition proposed by the inspector. That is so, no matter what one seeks to read, by implication, into the condition, but in argument it is very undesirable that a condition, which may have to be enforced, should depend upon something being read into it. That will undoubtedly lead to arguments and problems when the matter comes before the magistrates. A planning authority are entitled to expect that a decision which complies with the requirements of the Act will provide conditions which are reasonably practical to enforce.

I have come to the conclusion that this condition does not comply with that standard. On that ground, too, I would regard this application as being one which is entitled to succeed. Accordingly, I quash this decision for the reasons given.

Costs were awarded against the Secretary of State.

Queen's Bench Division

November 6 1985

(Before Mr David WIDDICOMBE QC, sitting as a deputy judge of the division)

HALL AGGREGATES (THAMES VALLEY) LTD v SECRETARY OF STATE FOR THE ENVIRONMENT AND ANOTHER

Estates Gazette April 19 1986
278 EG 308-313

Town and Country Planning Act 1971, section 245 — Application by gravel company to quash a decision of the Secretary of State dismissing an appeal in respect of development of land for a hotel — Applicants wished to build hotel on surplus land left after working out of pit — Local planning authority had refused permission for the development, the site being within the metropolitan green belt — The inspector had, after the public inquiry, recommended to the Secretary of State that the planning authority should be overruled and approval given to the development — The Secretary of State did not accept this recommendation and dismissed the company's appeal from the authority's refusal — The company now applied to the High Court to quash the Secretary of State's decision — The inspector had taken a more flexible approach than the Secretary of State had considered justifiable in regard to the application of the green belt policy to the land in question — The applicants put forward a number of criticisms of the Secretary of State's decision, including an allegation that he had differed from the inspector on a finding or inference of fact without affording the applicants an opportunity of making representations; had failed to give reasons for his stricter approach to green belt policy; had failed to take account of the fact that permission had been given in 1964 for a motel near the present site; and had failed to give adequate reasons for his decision generally — The judge rejected all the criticisms as lacking in substance — He pointed out that the difference in the approach to the green belt question was a matter of planning policy or planning opinion, not an issue on a finding of fact — Application dismissed

This was an application by Hall Aggregates (Thames Valley) Ltd under section 245 of the Town and Country Planning Act 1971 to quash the decision of the Secretary of State dismissing an appeal by the applicants against the refusal of the Royal Berkshire County Council, the planning authority, to grant planning permission for a hotel on land belonging to the applicants at Wraysbury in Berkshire.

R Vandermeer QC and K J Lindblom (instructed by Freedman & Co) appeared on behalf of the applicants; N Pleming (instructed by the Treasury Solicitor) represented the first respondent, the Secretary of State for the Environment; the second respondent, Royal Berkshire County Council, the planning authority, were not represented and took no part in the proceedings.

Giving judgment, MR DAVID WIDDICOMBE QC said: This is an application under section 245 of the Town and Country Planning Act 1971 to quash a decision of the Secretary of State for the Environment dismissing an appeal in respect of the development of land for a hotel at Wraysbury in Berkshire. The applicants, to whom I shall refer as "the company", are a gravel company with a worked-out pit who wanted to build a hotel on the surplus land left after the workings. The Berkshire County Council, the local planning authority, refused planning permission. The company appealed. A local inquiry was held and the inspector reported to the Secretary of State recommending approval. However, this recommendation was not accepted by the Secretary of State, who dismissed the appeal.

The inspector, in his report, described the site as follows:

> 4 The appeal site is an area of partly worked gravel enclosed by the minor Wraysbury Road to the south and south-west, the Wraysbury slip road to the north and north-west, the elevated and newly constructed M25 roundabout and interchange which, with its approaches, dominates the skyline to the north-east and east. These roads curve around the appeal site whose sole access is from Wraysbury Road opposite a row of small country houses in the settlement known as Hythe End.
>
> 5 The appeal site is now a little over half lake following sand and gravel working, possibly some clay working, by the appellants. The remainder is partly filled land, land from which the topsoil has been removed, and extensive concrete standings for the plant which had been working the gravel. A few temporary buildings for contractors are now on the concrete.
>
> 6 The surroundings to the north, east and south-east are commanded by the M25 and A30 roads system which obliterates any Green Belt or rural aspect. The west and north-west Green Belt lands comprising old gravel workings which are now Keynes Lake, Heron Lake and Wraysbury Reservoir, and which with the appeal site and the roads system are within the area defined as the Colne Valley Regional Park, are beyond the slip road and not readily visible from the appeal site unless one goes to the western extremity where the slip road meets Wraysbury Road at a small roundabout at a level only a little above that of the appeal site. The hotel itself would be clearly visible from the north and west but largely out of view from the elevated motorway interchange from the south-east.
>
> 7 The south, south-west and west-south-west aspects comprise Wraysbury Road, Colne Way and Bell Weir Close and their modest country or village houses which are a substantial part of the Hythe End settlement.
>
> 8 Further south is the Staines Reservoirs Aqueduct with a gravel lake next to it at the A30 roadside. The latter is known as the greyhound stadium site for which there was an earlier permission for motel development.

He made findings of fact of which the only two I need read are nos 12 and 13:

> 12 The hotel building if sited as proposed near the elevated M25 interchange roundabout will be inconspicuous when the appeal site and extended Green Belt are viewed from the south-east.
>
> 13 The hotel building if sited as proposed will be conspicuous when viewed from the west, north-west and north, ie from the embanked Wraysbury slip road and the roundabout junction with the old Wraysbury Road.

His conclusions were as follows:

> 61 Strict adherence to Green Belt policy and Thames Water policy for the flood plain should mean dismissal of this appeal.
>
> 62 On the other hand the appeal site is dominated by the new roads system which almost surrounds it. Little open country can be seen from the site itself. The only aspects truly rural and open are to the west, north-west and north and these are unlikely to be affected by a 2-storey hotel tucked into an arm of the high interchange roundabout.
>
> 63 The brutish effect of this interchange and the converging system of roads with lorries thundering along them, is likely to be softened by an hotel almost immediately in front of it, if built as shown in the submitted drawings.
>
> 64 A flexible approach to Green Belt policy, as distinct from a purist view of the principles at stake, favours building the hotel. I commend this less than strict application of policy because the appeal site's contribution to Green Belt objectives is now muted; the hotel's visual impact on the open or rural scene will be modest or even negligible; and the employment, business and prosperity which is likely to result from this class of hotel development may be considerable. Similarly the policy or strategy for the Colne Valley Regional Park need not be unduly prejudiced by the proposed development and should not be a reason for refusing it.

Para 65 deals with the flood plain, and after considering certain G
technicalities the inspector concluded:

> ... I believe it is not a strong enough reason for dismissing the appeal although it is a technical matter of such complication that one cannot be certain about it.
>
> 66 Two other matters are important namely (1) the opposition from the District Councils and local residents; and (2) the possible weakening of Green Belt policy and development control which may result from allowing the appeal.
>
> 67 On the first point I believe it could be better for the local residents to have a modest hotel on a properly regulated site beneath the overhead motorway, hidden behind the trees, than have some 10 to 11 acres of reconstituted but vacant quarry land opposite their houses and for most practical purposes uncontrollable on a day to day basis. The presence of an hotel in the position
> proposed in the application would leave much of the open view across the H
> appeal site unchanged. The opposition by the District Councils is part of the local planning authority's case (the County Council) and has been well presented by their witnesses and advocate.
>
> 68 This leads to the second matter namely the possible weakening of development control throughout the county. I do not doubt that a decision to allow the appeal will be quoted by future appellants as a reason for allowing development on Green Belt land which they claim has been made redundant for Green Belt purposes by adjacent motorway interchanges and slip roads. The present case is, however, distinguished by the fact of a 1964 planning permission for a motel on similar Green Belt quarry land a few yards south of the old Wraysbury Road, loosely referred to as the greyhound stadium site. If the sand and gravel workings on this site had not taken place, or had been less extensive, a motel or hotel might have been built. The Green Belt policy
> circular 42/55 was then current as it is today. J
>
> 69 The 1964 permission might be a good reason for resisting a further weakening of Green Belt policy if this would result from allowing the present appeal. On the other hand it could favour the appeal by distinguishing it as an exception to the overall Green Belt policy which would not make it a precedent for weakening of Green Belt policy and development control at motorway intersections. This is a matter for decision. My recommendation is that allowing the appeal should not, and need not, weaken development control and Green Belt policy.
>
> 70 I finally conclude that bearing in mind the facts stated the appeal should be allowed subject to conditions.

The inspector then deals with the proposed conditions.

The Secretary of State gave his decision on the appeal by letter
dated February 28 1985. After citing the inspector's conclusions, he K
said:

> 3 The Secretary of State has considered the inspector's conclusions. He takes the view that the main issue in this case is that of the policies applying in the Green Belt. There is general agreement between the parties that the appeal site is included within the Metropolitan Green Belt in the approved East Berkshire Structure Plan and the deposited Green Belt Local Plan for Berkshire. It is noted from the latter plan, which the Secretary of State regards as a material consideration of some weight, that the Metropolitan Green Belt in Eastern Berkshire was formally approved as Green Belt in or before 1974. The policies of these plans provide that within the Green Belt permission will not be given, save in very special circumstances, for development other than that falling within certain specified categories; the appeal proposals do not fall within these categories.
>
> 4 The Secretary of State has said that the general presumption against L
> development in the Green Belt is not affected by the M25 and that there should continue to be the strongest restraint on development there. He notes the inspector's conclusions about the visual impact of the new roads system in the area of the appeal site, but in his opinion this does not invalidate the contribution of the appeal site in preventing the coalescence of the existing built up areas to the west and east and the smaller settlements of Hythe End and Horton. He also considers that the proposed use and its associated activities would be out of keeping with the open or rural scene. The Secretary of State recognises that there would be some commercial and employment advantages resulting from the appeal proposals but in his opinion these do not constitute sufficient justification for making an exception to Green Belt policy. It is noted that planning permission was granted in 1964 by the planning authority for motel development on a site to the south of the appeal
> site. However, every planning application must be considered on its merits M
> having regard to all the facts and the Secretary of State does not consider that the 1964 permission is any justification in this case for development now in conflict with Green Belt policy.
>
> 5 The Secretary of State notes the inspector's comments about the effect of the proposed development on the flood plain. He regards the evidence as inconclusive but as he considers the policy objections to your clients' proposals to be over-riding he is of the opinion that he does not need to take a view on this matter.

Mr Vandermeer, for the company, attacked his decision on five grounds. He referred, for general principles, to *Seddon Properties Ltd* v *Secretary of State for the Environment* (1978) 248 EG 950.

(1) He criticised the first sentence of para 4: "The Secretary of State has said that the general presumption against development in

the green belt is not affected by the M25, and that there should continue to be the strongest restraint on development there.'' He said that at the inquiry it was common ground that the general presumption against development in the green belt was not affected by the M25. The Secretary of State had so stated in a policy letter dated March 23 1984 and the company at the inquiry had not questioned that policy. That was why there was no reference to it in the inspector's report. However, the Secretary of State had introduced it without explanation of its significance, thereby inevitably giving rise to the suspicion that he had misunderstood the argument at the inquiry and/or that he had taken account of that policy in some unspecified way without giving the company a chance to make submissions on it.

(2) He contended that in the sentence in para 4, ''He also considers that the proposed use and its associated activities would be out of keeping with the open or rural scene'', the Secretary of State is differing from the inspector on a finding of fact without offering the company an opportunity of making representations in breach of rule 12(2) of the Town and Country Planning (Inquiries Procedure) Rules 1974. He referred to *Coleen Properties Ltd* v *Minister of Housing and Local Government* [1971] 1 All ER 1049 to show that findings of fact included inferences of fact.

(3) He contended that the Secretary of State had rejected the inspector's ''flexible approach to Green Belt policy'' (see para 64 of the report) in favour of a strict approach but without giving any reasons. He said that there was no, or no sufficient, evidence before the Secretary of State to lead him to a different conclusion from that of the inspector.

(4) He contended that the Secretary of State had failed to have regard to a material consideration, namely the grant of planning permission in 1964 for a motel on green belt quarry land a quarter of a mile to the south of the appeal site. He had failed to understand and deal with the inspector's view of this planning permission.

(5) This was a general ''mopping-up'' point, that ''The Secretary of State failed to give proper and adequate reasons for his decision and/or his decision is unintelligible.''

Mr Pleming, for the Secretary of State, joined issue on these contentions. He cited *Lord Luke of Pavenham* v *Minister of Housing and Local Government* [1968] 1 QB 172 in addition to the cases cited by Mr Vandermeer.

I have carefully read the whole of the inspector's report and the Secretary of State's letter in order to establish the context of the passages in issue. I will take Mr Vandermeer's points in order.

(1) I accept Mr Vandermeer's explanation that the M25 in a strategic sense was not part of the company's case at the inquiry, which is why the inspector did not refer to it in his conclusions. The Secretary of State has, however, referred to the policy statement on the M25. Mr Pleming invites me to read the first sentence of para 4 of the letter as ''Lest it be thought that the M25 makes any difference . . .''. However, I do not think that such rewording is necessary when paras 3, 4 and 5 of the letter are read as a whole. I think it is apparent that the Secretary of State is merely recording the undisputed fact that the presence of the M25 in its strategic, as opposed to its physical, sense is not a point in the company's favour. No doubt it was not necessary for him to say that, but the fact that he has done so is quite harmless to the company. I do not find any misunderstanding by the Secretary of State of the arguments at the inquiry or any possibility that he has taken the M25 policy into account in some hidden way. This point of Mr Vandermeer must therefore be rejected.

(2) As regards the sentence ''He also considers that the proposed use and its associated activities would be out of keeping with the open or rural scene'', Mr Pleming pointed out to me that this sentence comes from para 52 of the report, where evidence of the local planning authority is recorded to the effect that ''a hotel would be harmful to that open and rural character. Its use and the associated activities would be alien to this stretch of countryside.'' The inspector dealt, in the company's favour, with the visual aspect of the hotel (see para 64), but he did not refer in his conclusions to the use of the hotel. It appears that the Secretary of State has taken up this point against the company. He is not differing from the inspector's conclusion on the visual aspect, he is relying on a factor which was raised at the inquiry but not dealt with by the inspector. In my judgment he was fully entitled to do so and this point of Mr Vandermeer has no substance in it.

(3) As regards the issue of the ''flexible'' versus the ''strict'' approach to green belt policy, it is clear, as Mr Vandermeer points out, that the Secretary of State has not endorsed the inspector's flexible approach. In my judgment this is a matter of planning policy or planning opinion and not a finding of fact. There is no requirement that the Secretary of State should afford the parties an opportunity to make representations about it. I think it might have been better if the Secretary of State had said expressly that he disagreed with the inspector on this, but the fact he did so is clear enough, particularly from para 3 of his letter, where he stated the strict green belt policy which he was going to apply. Accordingly I reject this point of Mr Vandermeer.

The same reasoning applies to the other two sentences from the inspector's report referred to in ground 3 of the notice of motion; they are matters of planning opinion.

(4) I find no substance in Mr Vandermeer's fourth point, about the planning permission for a motel granted in 1964. The Secretary of State has had regard to this point and in my judgment he has not failed in any way to understand the inspector's view of it.

(5) Generally I find that the Secretary of State has given the necessary ''clear and intelligible'' reasons for his decision.

I can understand the company's disappointment at having won before the inspector but lost before the Secretary of State, especially now that almost all planning appeals are decided by inspectors alone, but that possibility is inherent in the planning appeals procedures and this court can interfere only if the Secretary of State has exceeded his powers or broken one of the procedural rules. That has not been established in this case and the application must therefore be dismissed.

The application was dismissed with costs

Queen's Bench Division

November 11 1985

(Before Mr David WIDDICOMBE QC, sitting as a deputy judge of the division)

DUDLEY BOWERS AMUSEMENTS ENTERPRISES LTD v SECRETARY OF STATE FOR THE ENVIRONMENT AND ANOTHER

Estates Gazette April 19 1986

278 EG 313-318

Town and Country Planning Act 1971, section 246 — Appeal against decision of inspector upholding an enforcement notice — Alleged breach of planning control was a material change in the use of the land in question without planning permission, namely, the holding of weekly Sunday markets with associated car parking — The steps required by the enforcement notice to be taken were to discontinue the use of the land for the holding of markets and associated car parking ''on such Sundays which fall within the period of summer time in any year'' — The main issue raised by the appellants, who were the leasehold owners of the land, was that the enforcement notice was a nullity because of the ambiguity of the expression ''the period of summer time'' — The inspector appeared to have recognised the ambiguity but did not go on to consider whether it rendered the notice a nullity — He proceeded to vary the notice, which he could not do if it was in fact a nullity, ie if there was no notice in existence to be varied — Held, applying the test laid down in *Miller-Mead* v *Minister of Housing and Local Government*, that the expression ''the period of summer time'' was ''hopelessly ambiguous and uncertain'' and consequently rendered the enforcement notice a nullity — That was sufficient to deal with the appeal, but the judge considered certain other criticisms of the inspector's decision letter in case he (the judge) was wrong on the nullity point — However, he concluded that he would not have wished to remit the case to the Secretary of State because of any of the other points — Appeal allowed and case remitted with the opinion that the enforcement notice was a nullity

This was an appeal under section 246 of the Town and Country Planning Act against the decision of an inspector acting on behalf of the Secretary of State for the Environment who had upheld an enforcement notice served by Cleethorpes Borough Council in respect of the use of land belonging to the appellants, Dudley Bowers Amusements Enterprises Ltd. The land in question was adjacent to "Wonderland", North Promenade, Cleethorpes.

A E Telling (instructed by D Jebb & Co, of Market Rasen, Lincolnshire) appeared on behalf of the appellants; P Havers (instructed by the Treasury Solicitor) represented the first respondent, the Secretary of State; the second respondents, Cleethorpes Borough Council, were not represented and took no part in the proceedings.

Giving judgment, MR DAVID WIDDICOMBE QC said: This is an appeal under section 246 of the Town and Country Planning Act 1971 against a decision of an inspector acting on behalf of the Secretary of State for the Environment, upholding an enforcement notice served by the Cleethorpes Borough Council in respect of the use of land adjacent to "Wonderland", North Promenade, Cleethorpes. An appeal lies to this court on a point of law.

The enforcement notice, dated August 5 1983, alleged a breach of planning control consisting of "the making of a material change in the use of the land to use for the purpose of holding weekly Sunday markets with associated car parking without the grant of planning permission". The steps required to be taken to remedy the breach were "to discontinue the use of the land for the holding of markets and associated car parking on such Sundays which fall within the period of summer time in any year".

The appellants, who are the leasehold owners of the land, appealed against the enforcement notice to the Secretary of State for the Environment. An inspector, who was appointed to determine the appeal, held a local inquiry and issued a decision by letter dated March 26 1984.

Validity of the notice.

Mr Telling for the appellants took four points, the first of which was that the enforcement notice was a nullity. It will be convenient if I deal with that ground of appeal first. Mr Telling referred to *Miller-Mead* v *Minister of Housing and Local Government* [1963] 2 QB 196 for the proposition, which was not disputed by Mr Havers for the Secretary of State for the Environment, that a notice which does not specify the steps to be taken to remedy the alleged breach of planning control is a nullity. Mr Telling contended that, in this case, the steps specified were hopelessly ambiguous, to use the phrase in the judgment of Upjohn LJ in *Miller-Mead*, and that the notice was therefore void. He also referred to *Metallic Protectives Ltd* v *Secretary of State for the Environment* [1976] JPL 166, in which an enforcement notice was held, with the consent of the Secretary of State, to be a nullity for just these reasons.

The inspector dealt with the ambiguity argument as follows:

4 The evidence given on behalf of your clients is that a market with associated car parking operated on the site on Sundays from some time in March 1983 to the end of September 1983, and on the Easter, Spring and August Bank Holiday Mondays in 1983. The evidence also shows that your clients seek to operate the market on those days from the beginning of March to the end of September each year. The period March to September is the interval not covered by a planning permission dated May 6 1982, details of which are given in para 16 below.

5 Whilst I do not consider that the scope of the enforcement notice could be extended to take in the Bank Holiday market use, it appears to me that the breach of planning control alleged in the notice should specify the period during which Sunday markets on the site have operated and during which it is the intention that they should operate, namely, in months other than January and February and October to December. The requirements of the notice, which as you have pointed out are ambiguous, in that the period of summer time is not defined, could then be varied to reflect the allegation.

6 With regard to the requirements, you have referred to the cases of *Metallic Protectives Ltd* v *Secretary of State for the Environment* [1976] JPL 166 and *Copeland Borough Council* v *Secretary of State for the Environment* [1976] JPL 304.

7 The Council, while pointing out that the authority to issue the notice required it to be issued "pursuant to all relevant statutory provisions", and saying that "summer time" should be interpreted as defined in the Summer Time Act 1972, agreed that the notice could be varied as proposed above.

8 I cannot see that there would be any injustice to your clients if the notice were varied as proposed. The scope of the alleged breach would be reduced rather than extended, and the requirements would relate to the breach as alleged. The Council's position would not be prejudiced by the variation in the allegation, since a Sunday market held on the site between October and February would be in breach of condition 5 attached to the planning G
permission dated May 6 1982 and appropriate action could be taken. The Council do not insist on, and you do not accept, their interpretation of "summer time", and I am not satisfied that, when issuing the notice, they intended the words to be interpreted in that way rather than as the period from March to September. I shall therefore vary the notice as proposed.

Mr Telling assured me, and I accepted from him, that he took the nullity point at the inquiry. The inspector appears to have held that the requirements of the notice were ambiguous, but he did not go on to consider whether such ambiguity rendered the notice a nullity. He proceeded to vary the notice, which of course he could not do if there was no notice in existence to be varied. He simply has not dealt with Mr Telling's initial point.

The test to be applied was stated by Upjohn LJ in *Miller-Mead* at H
p 232 as follows:

I repeat, therefore, that in my judgment the test must be: does the notice tell him fairly what he has done wrong and what he must do to remedy it?

Earlier at p 226, the learned lord justice had said:

Supposing then upon its true construction the notice was hopelessly ambiguous and uncertain, so that the owner or occupier could not tell in what respect it was alleged that he had developed the land without permission or in what respect it was alleged that he failed to comply with a condition or, again, that he could not tell with reasonable certainty what steps he had to take to remedy the alleged breaches. The notice would be bad on its face and a nullity.

The ambiguity to which Mr Telling draws attention lies in the phrase "the period of summer time". What does this mean? It is
common ground between the parties that the language of the J
enforcement notice must be given its ordinary or popular meaning and that extrinsic evidence, such as the earlier planning applications and decisions referred to by the inspector in paras 15 and 16 of his decision letter, are not admissible as an aid to construction. Mr Telling said the phrase "period of summer time" might mean (i) summer in the seasonal sense, from summer solstice to autumnal equinox, ie June to September or (ii) British Summer Time as defined in the Summer Time Act 1972 or (iii) summer in a larger sense as opposed to "winter". He contended that it was quite impossible to tell from the document itself which of these meanings was intended by the local planning authority.

Mr Havers for the Secretary of State said that the phrase meant K
British Summer Time as defined in the Act of 1972. He referred to the *Shorter Oxford English Dictionary*, where "summer time" is stated to be: "1. The season of summer; the time that summer lasts. 2. (as two words) The standard time (in advance of ordinary time) adopted in some countries during the summer months." He said that the absence of a hyphen between the word "summer" and "time" was significant.

The basic question I have to answer is, *per* Upjohn LJ: "Does the notice tell him", ie the landowner, "fairly what he has done wrong and what he must do to remedy it?" It must be borne in mind that the notice is not discharged by compliance but remains in force permanently (see section 93 of the Act) and that since 1981 it is
registrable under section 92A of the Act. The persons affected are, L
therefore, not just the actual recipients of the notice, but all those into whose hands the land may come in the future.

In my judgment, there is clearly ambiguity in the phrase "period of summer time". Can that ambiguity be resolved?

I cannot accept Mr Havers' submission that the ambiguity should be resolved by adopting the meaning in the Summer Time Act 1972. That Act provides for altering the clocks during specified dates in the year, the precise periods being subject to amendment in any year by Order in Council. In my view, this is a technical meaning of the phrase and not its meaning in ordinary language.

The inspector said, in para 8 of the decision letter, that he "was not satisfied that, when issuing the notice, [the council] intended the
words to be interpreted in that way" and I agree with him. It is true M
that there is no hyphen between the words "summer" and "time" and that the *Shorter Oxford Dictionary* suggests that that is significant. But dictionaries are only a guide to language and, for my part, I do not think persons affected by this notice can be assumed to know that a hyphen might make a difference to the meaning of the phrase "period of summer time". Nor do I think either of the other two possible meanings identified by Mr Telling is necessarily the correct one. Summer as a season does not seem a likely meaning to me; summer as opposed to winter is, in my view, much more likely to be taken as the meaning by a recipient of the notice but, if that is the meaning, when does it start and when does it end? That meaning itself is hopelessly vague.

A My conclusion is that the phrase is ambiguous and that the ambiguity cannot be resolved. In the words of Upjohn LJ, it is "hopelessly ambiguous" and the enforcement notice is, therefore, a nullity.

I add that the parties referred me to the recent decision of Nolan J in *Rhymney Valley District Council* v *Secretary of State for Wales and G Isaac* [1985] JPL 27, in which it was held that a decision by the Secretary of State that an enforcement notice was a nullity could not be challenged by appeal under section 246 of the Act and only by proceedings for judicial review. Mr Havers, for the Secretary of State, told me that he did not rely on that decision and was content that the question of validity should be dealt with on this appeal.

B Section 246 of the Act gives a right of appeal where the Secretary of State is alleged to have erred in law in a decision on an appeal against an enforcement notice. Here the Secretary of State, by his inspector, has given a decision on an appeal against an enforcement notice and, in my judgment, this court therefore has jurisdiction to consider whether that decision contains an error of law. In the *Rhymney Valley* case the inspector had held that the enforcement notice was a nullity. It seems doubtful to me whether that constitutes a ground for distinguishing between the cases. I may, therefore, be differing from Nolan J on this point. If so, I naturally do it with respect and diffidence.

Other points.

C Having concluded that the notice is a nullity, strictly speaking I do not need to consider Mr Telling's other points. But in case I am wrong, I think I should do so. I can take his other three points together. To explain them, I must refer again to the inspector's decision letter. After describing the appeal site and its surroundings, the inspector went on as follows:

Undisputed Facts.

11 For many years up to 1925 the land now containing the site and the Wonderland building was used for general amusement purposes.

12 The Wonderland building was erected in about 1925, whereafter it and the remaining land were used for various amusement purposes. For a long time attractions on the appeal site included a Big Dipper.

13 The land containing the Wonderland building and the appeal site is held D on a 42 year lease from January 1 1964 granted by the British Railways Board, the lease providing for a review after 21 years of the rent initially fixed at £12,500 per annum. Your client company acquired the lease and took control of the premises in December 1981.

14 On December 18 1981 planning permission was granted for an expansion of "existing activities, eg Sunday market sales and amusements", on land containing the Wonderland building and the appeal site. Conditions imposed included ones requiring the use to cease on February 28 1982, markets to be held on Sundays only from November 15 to February 28, and no trading to take place or market stalls to be erected other than within the building.

15 On February 19 1982 your clients submitted an application for planning permission for an "extension of permission to hold Sunday markets and amusements to apply all year. Occasional additional permission requested, ie bank holidays. Market to be outdoors during summer, indoors in winter". E The land was described as "Wonderland Amusement Park", and the submitted plan showed the appeal site as included in the Wonderland property and to be partly used as an outdoor market stall area and for car parking. A covering letter explained the proposed indoor winter market period as October to February annually, and the proposed outdoor summer market period as March to September annually.

16 In response to the application planning permission was granted on May 6 1982 for the operation of an "indoor Sunday market between October and February each year only". Conditions imposed required the permission to be for the period October 1 to February 28-29 in any one year, for the market to be held on Sundays only, and for no trading to take place or market stalls to be erected other than within the buildings.

17 On the same day planning permission was refused for the operation of "a Sunday market from March to September".

F 18 Your clients operated a Sunday market on the appeal site in the period March to September 1983 as stated in paragraph 4 above.

19 On July 8 1983 planning permission was refused for the operation on land within the appeal site of an "outdoor Sunday market during the period April 1 to September 30 annually plus occasional bank holidays, and use of land as car park".

The inspector held, in paras 20 to 22, that the market use of the land was a material change of use, which conclusion is not challenged. He then considered whether planning permission should be granted for the market use of the land. He said:

Planning Considerations.

23 It appears to me from the representations and my inspection of the site and surroundings that the main issue is the effects which the summer Sunday market use, if continued, would have on trade in the town's established shopping areas, on the character and attractiveness of this part of the town's sea front, and on the safety of pedestrians on North Promenade.

24 In the Town Map the site falls within an area allocated primarily for amusement, recreation and entertainment. A draft local plan states that the introduction of uses other than those supporting the resort function of the area will be resisted. It also states, with regard to shopping policy, that the further development of shopping uses outside defined areas centred on St Peter's Avenue and Sea View Street will only be permitted where there is an overriding local need. I have considered the relevant provisions of the local plan, and although the plan has not yet been approved, see no reason to doubt that in general terms the policies merit support.

25 Whilst I am mindful of the policy objection to the summer Sunday market use, and of the objections contained in letters received, I note that Sunday markets have been operating in winter in the Wonderland building for some time, and that the number of stalls at the markets is between 200 and 250. Sunday markets also operated on the appeal site throughout last summer. There is no firm indication that town centre trade has suffered as a result. Although I do not doubt that some persons make purchases at the Sunday markets which would otherwise be made in the town centre shops, I accept that the markets attract substantial numbers of visitors from outside the area, and that those of the town's businessmen and traders who provide services or facilities to the public on Sundays are likely to benefit from the spending of the visitors. I find myself not satisfied on the evidence before me that the holding of summer Sunday markets on the site would have a materially damaging effect on trade in the town's established shopping areas.

26 Your clients claim that the changed habits and demands of holidaymakers have resulted in use of the appeal site for the provision of amusements ceasing to be viable. They refer to the heavy cost of maintenance and repair work on their premises and equipment, of rates increases, and of the substantial increase in rent likely to result from the review at the end of this year. The Council say that, after a period of decline in the prosperity and appearance of North Promenade, investment is taking place in both existing and new entertainment and amusement type developments; that their policies have been drafted to take account of the upturn, and they now look forward to seeing further investment on North Promenade and uses of land there geared to fulfilling the needs of the holidaymaker and day tripper to the resort; that the Promenade may reasonably be expected to be reserved for walking; that the beach near the appeal site is very popular, and the quiet enjoyment of it should be protected as much as possible; and that a summer Sunday market on the site, while an attraction on that one day of the week, would result in the site being in a vacant, unused state for the other six days at the height of the holiday season.

27 Although the holding of Sunday markets in the Wonderland building appears to me to be a sensible and appropriate way to use the available space in winter, I am of the opinion that different considerations apply during the summer, when the uses of land on North Promenade should, in my view and in the words of the draft local plan, support the resort function of the area. Whilst a Sunday market on the site in summer would undoubtedly be an attraction to various persons using the sea front and the beach, I do not believe that the use is of a kind appropriate to the area. I consider that the activity, bustle, and vehicular movements generated by a market would be out of place and would harm the area's character at a time when the public may reasonably expect both to find relaxation on the beach and promenade, and to enjoy the customary attractions of a sea front. Furthermore, your clients have submitted no proposals for the use of the site on days when the market would not be operating, and in the absence of an appropriate use on those days it seems to me that the site would appeal as a lifeless, uninteresting and unattractive place. I am conscious of the financial problems confronting your clients, and of the advantages to them of holding Sunday markets on the site, but am convinced that the use does not represent a right long-term use for the site.

28 There is no dispute that in summer, especially at weekends, cars are attracted to North Promenade in considerable numbers, and that dangers to pedestrians arise from the combined effects of holiday attractions being on one side of the road, the promenade footway and the beach being on the other, and vehicles moving up and down between them and parking on the carriageway. The problem has resulted in the operation of a traffic control system at certain times in the summer. The system has varied over the years; in 1983 a barrier was operated at the south-eastern end of North Promenade, and the entry of vehicles was restricted from the Spring Bank Holiday weekend and on each weekend thereafter until the school summer holidays, when the restriction was imposed daily. The evidence is that the controls have been the subject of opposition from traders, and that no arrangements have been agreed for the 1984 season.

29 Although I accept that, with the inclusion of adjoining land owned by your clients, parking space at the site could be provided for traders' and patrons' vehicles, I also accept the likelihood that, if such accommodation were provided, it would not be feasible to maintain, at least on market days, a control system such as that which operated in 1983. In the absence of any control there would inevitably, in my opinion, be a large increase in vehicular flows along North Promenade as a result of the market, and, despite the existence of parking space at the site, in parking on the carriageway. I am in no doubt that this would lead to a substantial and unacceptable increase in dangers to pedestrians.

31 The matter of permitted development rights under Class IV, Schedule 1

to the General Development Order has been raised. I accept the matter as material consideration, and from the evidence and my inspection of the site and surroundings am of the view that the appeal site is curtilage land associated with the Wonderland building. If that is so, it does not enjoy Class IV rights for the holding of markets. But in any case the matter in my view is not sufficient to alter my conclusion that planning permission should not be granted.

32 I am conscious of the particular need at the present time to encourage business activity and promote employment, matters on which guidance is given in Circular 22/80, but I consider there to be sound and clear-cut reasons for refusal of planning permission in this case; and after taking into account all other matters raised in the representations, I find nothing to cause me to alter my decision. Accordingly the appeal fails on ground (a).

Then the inspector dealt with the variation of the notice.

Mr Telling contends that the inspector erred in law in respect of this part of his decision in three respects: (1) that he failed to have regard to a consideration material to the decision, namely the amount of traffic which would be generated by a resumption and/or intensification of the established use of the appeal site for amusement purposes; (2) that he took into account a consideration not material to that question, namely, the fact that the appellants had submitted no proposals for the use of the appeal site on days on which the Sunday market could not be operated; and (3) that on the evidence given at the inquiry into the appeal the inspector could not reasonably have concluded that the appeal site is a curtilage to a building.

Mr Havers for the Secretary of State joined issue on all three points.

As regards point (1), assuming in Mr Telling's favour that the traffic generated by a resumption of the existing use was a material consideration and assuming that the inspector left it out of account, both of which assumptions were disputed by Mr Havers, I nevertheless think that, in the circumstances of this case, the inspector was justified in disregarding it. In para 26 he stated: "Your clients claim that the changed habits and demands of holidaymakers have resulted in use of the appeal site for the provision of amusements ceasing to be viable." Apparently, it was part of the appellants' case for a market use of the site that it could no longer economically be used for amusements. It appears to me inconsistent that the appellants now complain that the inspector did not take into account traffic which would be generated by resumption of the use which they themselves say is unviable. They cannot blow hot and cold. I therefore reject this point.

Mr Telling's next point relates to the sentence in para 27 of the decision letter, where the inspector says: "Furthermore, your clients have submitted no proposals for the use of the site on days when the market would not be operating, and in the absence of an appropriate use on those days it seems to me that the site would appear as a lifeless, uninteresting and unattractive place." I think there is more substance in this point because there is no obligation under planning law for a landowner to use his land. He is entitled to leave it unused if he wishes. I think, therefore, that the inspector did take into account an irrelevant consideration. It may be that what the inspector intended to say is that the use of the land for a market on Sundays would necessarily preclude its use for anything else on other days of the week. That is a different point. In my judgment, that would be a material consideration from a planning point of view. That was the way the council put it at the inquiry. In para 26 it is recorded as part of their case that "a summer Sunday market on the site, while an attraction on that one day of the week, *would result* in the site being in a vacant, unused state for the other six days at the height of the holiday season" (my italics). But the inspector has not expressed it in that way and, in my view, he has therefore erred in law.

Would the decision have been any different if he had disregarded this factor? Mr Telling contends that it might be. I do not accept that. It is clear from the passage in the decision letter which I have quoted that the inspector had substantial other reasons for rejecting the appellants' proposals. He regarded the use as inappropriate to the area, contrary to the draft local plan and he held that the activity, bustle and vehicle movement generated by a summer market would be out of place and would harm the area's character, and that market use would cause a serious traffic problem.

Taken in the context of the decision as a whole, the sentence challenged by Mr Telling appears to me to be a subsidiary reason. I am in no doubt that the decision would have been the same without it. I would, therefore, in the exercise of my discretion not remit the case to the Secretary of State on this point.

Mr Telling's final point involves para 31 of the decision letter, as cited above. The Town and Country Planning General Development Order 1977, by Article III and Schedule I, Class 4, para 2, grants planning permission for the use of land other than the building or the curtilage of a building for, *inter alia*, market use on not more than 14 days in any calendar year. The appeal site has the benefit of this deemed planning permission. That, as the inspector rightly said, would be a material consideration. He held that the appeal site was part of the curtilage of the "Wonderland" building, so that there was no deemed planning permission for market use for 14 days in any year. Mr Telling elaborated his argument on this point at some length and cited several cases on the meaning of "curtilage". I think he made this issue unnecessarily complicated and I do not propose to follow him along that path. The question for me is really a short and simple one, as I think Mr Telling eventually agreed. Whether land is curtilage land is a matter of fact for the inspector and I cannot interfere with the decision, so long as there was evidence on which he could come to it and his conclusion is not obviously wrong. I have seen the plan; I have been taken through the evidence and, on that evidence, in my judgment, it was entirely open to the inspector to hold as he did, that this land was part of the curtilage of the building. This court does not have to decide whether this land is curtilage. It is enough that the inspector was entitled to conclude as he did.

I note that the inspector indicated that, in any case, the matter was not sufficient to alter his conclusion. That I find a little surprising because, if the land could be used for a market on 14 Sundays in the summer, it would seem to raise the question whether the addition of some more Sundays makes any difference. If I am wrong on the curtilage point, I would have been minded to remit the matter to the Secretary of State for reconsideration if it were open to me to do so; *a fortiori* because there would then be two errors of law in the decision, but that does not arise.

However, as stated, I do remit the case to the Secretary of State with the opinion of the court that the notice is a nullity.

The case was remitted to the Secretary of State, with costs for the appellant.

Court of Appeal

March 26 1986

(Before Lord Justice O'CONNOR, Lord Justice CROOM-JOHNSON and Lord Justice LLOYD)

BRADFORD CITY METROPOLITAN COUNCIL v SECRETARY OF STATE FOR THE ENVIRONMENT AND ANOTHER

Estates Gazette June 28 1986

278 EG 1473-1476

Town and Country Planning Act 1971, section 245 — Appeal by planning authority from decision of Farquharson J refusing to quash a decision of the Secretary of State for the Environment, who held that a condition imposed by the authority on giving planning consent to a residential development was *ultra vires* and unlawful — The condition in question was challenged as objectionable because in part relating to a piece of land neither owned nor controlled by the developers and in part requiring the cession of land by the developers — *Hall & Co Ltd* v *Shoreham-by-Sea Urban District Council* considered — Appellants submitted that the present case differed from *Hall* v *Shoreham UDC* because here the condition which was objected to had in fact been accepted voluntarily by the developers in the terms of their planning application — This suggested interpretation of their planning application was rejected by the developers, who asserted that they had never agreed to carry out road widening at their own expense or at all — The dispute as to this issue of fact was not resolved at any stage in the proceedings — The Court of Appeal held that the true principle to be applied was that conditions should not be manifestly unreasonable — If a proposed condition was manifestly unreasonable, then it was

A **beyond the power of the planning authority to impose it, even if the developer consented to it — The Secretary of State was entitled in the present case to take the view that the condition sought to be imposed was of this kind — Opinion expressed, although not necessary for the decision, that a condition which was manifestly unreasonable could not lawfully be incorporated in a section 52 agreement — Appeal by planning authority dismissed.**

This was an appeal by the planning authority, City of Bradford Metropolitan Council, from a decision of Farquharson J on May 9 1985 in which he dismissed an application by the authority to quash a
B decision of the Secretary of State for the Environment under section 245 of the Town and Country Planning Act 1971. The issue was as to a condition imposed by the authority in granting consent to a residential development, comprising 200 houses, by McLean Homes Northern Ltd outside Bradford. The first respondent to the present appeal was the Secretary of State and the second respondent was McLean Homes Northern Ltd. The decision of Farquharson J was reported at [1985] 1 EGLR 268.

D W Keene QC and Charles Cross (instructed by Warren Murton & Co, agents for A R Sykes, solicitor's department, City of Bradford Metropolitan Council) appeared on behalf of the appellant authority; John Laws (instructed by the Treasury Solicitor) represented the first respondent; Harry Wolton QC (instructed by
C T L Green) represented the second respondents.

Giving judgment, LLOYD LJ said: This is an appeal from a decision of Farquharson J refusing an application by the City of Bradford Metropolitan Council to quash a decision of the Secretary of State for the Environment under section 245 of the Town and Country Planning Act 1971. The point at issue in the appeal concerns the lawfulness of a condition imposed by the council, as local planning authority, in granting consent for a residential development by McLean Homes Northern Ltd. The appeal site lies about 3 miles from the centre of Bradford. It occupies an area of about 11½ acres. Along the northern and eastern boundary of the appeal site there runs a road known as Meadway. Meadway is a highway maintained at public expense by the highway authority, the West Yorkshire Metropolitan
D County Council. It is 4.8 m wide. According to the case submitted to the Secretary of State by the City of Bradford (to whom I shall refer as the planning authority) the normal minimum width of a local road of this type, as required by the highway authority and the Department of Transport, is 7.3 m. Not surprisingly, the Secretary of State found in para 4 of his decision letter dated May 16 1984 that Meadway is below the standard desirable for the amount of traffic which it was already carrying at the time of the application for planning consent.

On August 6 1982 McLean Homes Northern Ltd (to whom I shall refer as the developers) applied for permission to carry out a residential development comprising a total of 200 houses. There were in fact two separate applications, one for 160 houses and the other for
E 40. But nothing turns on that. Meanwhile on July 8 1982 the highway authority had written to the planning authority that Meadway must be widened before permission for development could be granted. Since the letter is important I will quote it verbatim:

> I confirm that Meadway in its present form is unacceptable for intensified use and must be widened before any redevelopment on the above land takes place. Meadway is narrow, a bus route and must be widened over the entire frontage of the land proposed for development up to the junction with Halifax Road. Since part of the land is in the ownership of the District Council I believe there is scope for negotiation with the developer and I advocate that this be pursued.

The junction with Halifax Road lies about a quarter of a mile beyond the extreme north-west tip of the proposed residential development.

F No doubt negotiations then took place between the parties, as is not unusual, in the light of the highway authority's requirements. On November 18 1982 the developers wrote:

> *Re proposed residential Development* . . . please find enclosed amended plans incorporating a revised drainage outfall to Royd's Hall Beck and the widening of Meadway by an additional 1 metre along the site frontage and through to the junction with Manorley Lane.

The junction with Manorley Lane is another way of referring to the junction with Halifax Road, which had been mentioned in the highway authority's letter of July 8 1982.

There were two amended plans included with the developer's letter on November 18 1982. One shows the addition of a strip one metre wide along the entire northern and eastern edge of the residential development, with the legend "One metre wide strip for future road widening". At the foot of the plan appears the comment "Drawings revised to incorporate Planning and Highways Department requirements". The other plan shows a narrow strip or tongue of land alongside Meadway from the point where it leaves the residential development to where it joins Halifax Road and Manorley Lane. Each of the strips is shown edged in red.

It would appear that shortly thereafter the developers approached the highway authority. They hoped to persuade the highway authority to include the widening of Meadway in their Development Access Road Programme. But they were not successful.

On January 12 1983 the planning authority granted consent subject to five conditions. The only condition to which I need refer is no 4, which was as follows:

> Before the dwellings hereby authorised are occupied, the highway improvements to Meadway ie the widening of the carriageway by 1 m together with the associated remedial works, shall be carried out in accordance with the submitted plans, and to the satisfaction of the Council.

The reason for that condition was stated to be "In the interests of highway safety".

On April 11 1983 the developers commenced development. On April 29 1983 the highway authority finally turned down the developer's proposal that the widening of the road should be carried out by the highway authority. I quote from the final paragraph of their letter:

> I appreciate that the additional highway works, as required by the condition attached to the detailed planning consent, are off-site works. However, in my view the highway requirements should not impede the development of the site. This issue clearly centres on the negotiations between your Company and the City Council regarding the strip of publicly owned land. In any event the estimated cost of the highway works quoted in your letter exceeds the total budget for developing access roads in the current financial year. For these reasons therefore the proposal is not considered to be eligible for inclusion in the programme.

On June 9 1983 the developers appealed to the Secretary of State, under section 36 of the 1971 Act, against the imposition of the condition. Each side submitted statements in support of their cases. I need not refer to the statements in detail, save to say that a serious dispute of fact emerged as to what had happened when the developers submitted their revised plans in November 1982. According to the planning authority the developers had, by submitting revised plans showing the widening of Meadway edged in red, "acceded to the requirement of the highway authority", by which they meant, presumably, that the developers had agreed to carry out the proposed widening of Meadway throughout its length at their own expense, and also to cede the land necessary for that purpose along the northern and eastern edge of the residential development. The developers, on the other hand, expressly denied any such suggestion. They never agreed to carry out the road widening at their own expense, or at all. They included the strips edged in red on their revised plans for no other reason than because that is what they were asked to do by the planning authority.

The Secretary of State did not attempt to resolve this dispute of fact. He discharged condition 4 on grounds which, in his view, made it unnecessary to decide which view of the facts was correct. He held that the condition was *ultra vires* so far as it related to what I have called the tongue of land extending as far as the junction with Halifax Road, since the land was not owned or controlled by the developers. The planning authority had no power to impose a condition under section 30(1)(a), since the power under that section extends only to land under the applicant's control. As for that part of the road widening which was to be carried out along the northern and eastern boundary of the residential development, the Secretary of State held, as indeed Mr Keene for the planning authority concedes, that the work would require the cession of land by the developers. Accordingly the condition fell within the vice described in Circular 5/68 to which the Secretary of State referred. That circular has now been cancelled. But it has been replaced in similar terms by Circular 1/85, which provides in para 63:

> No payment of money or other consideration can be required when granting a permission or any other kind of consent required by a statute, except where there is specific statutory authority. Conditions requiring, for instance, the cession of land for road improvements or for open space, or requiring the developer to contribute money towards the provision of public car parking facilities, should accordingly not be attached to planning permissions. Similarly, permission cannot be granted subject to a condition that the applicant enters into an agreement under section 52 of the Act or other

powers. However, conditions may in some cases reasonably be imposed to oblige developers to carry out works, eg provision of an access road, which are directly designed to facilitate the development.

Para 58 of the same circular provides under the heading "Highway conditions — cession of land":

Conditions may not require the cession of land to other parties, such as the highway authority.

In the Secretary of State's view the planning authority should either have refused planning permission altogether or attempted to negotiate an agreement with the developer under section 52 of the Act.

The learned judge upheld the Secretary of State's decision. He held that he was bound, in effect, by the decision of this court in *Hall & Co Ltd* v *Shoreham-by-Sea Urban District Council* [1964] 1WLR 240, on which the relevant provisions of Circular 1/85 are no doubt based. Before referring to *Hall & Co Ltd* v *Shoreham-by-Sea U D C*, it might be convenient to set out here the relevant provisions of the Act.

Section 29(1) provides:

Subject to the provisions of sections 26 to 28 of this Act, and to the following provisions of this Act, where an application is made to a local planning authority for planning permission, that authority, in dealing with the application, shall have regard to the provisions of the development plan, so far as material to the application, and to any other material considerations, and — (a) subject to sections 41, 42, 70 and 77 to 80 of this Act, may grant planning permission, either unconditionally or subject to such conditions as they think fit; or (b) may refuse planning permission.

Section 30(1)(a) provides:

Without prejudice to the generality of section 29(1) of this Act, conditions may be imposed on the grant of planning permission thereunder — (a) for regulating the development or use of any land under the control of the applicant (whether or not it is land in respect of which the application was made) or requiring the carrying out of works on any such land, so far as appears to the local planning authority to be expedient for the purposes of or in connection with the development authorised by the permission.

Section 36(1) and (3) provide:

(1) where an application is made to a local planning authority for planning permission to develop land, or for any approval of that authority required under a development order, and that permission or approval is refused by that authority or is granted by them subject to conditions, the applicant, if he is aggrieved by their decision, may by notice under this section appeal to the Secretary of State.
(3) Where an appeal is brought under this section from a decision of a local planning authority, the Secretary of State, subject to the following provisions of this section, may allow or dismiss the appeal, or may reverse or vary any part of the decision of the local planning authority, whether the appeal relates to that part thereof or not, and may deal with the application as if it had been made to him in the first instance.

Finally, section 52(1) provides:

A local planning authority may enter into an agreement with any person interested in land in their area for the purpose of restricting or regulating the development or use of the land, either permanently or during such period as may be prescribed by the agreement; and any such agreement may contain such incidental and consequential provisions (including provisions of a financial character) as appear to the local planning authority to be necessary or expedient for the purposes of the agreement.

I now return to *Hall & Co Ltd* v *Shoreham-by-Sea U D C*. The facts were that the plaintiffs applied for planning permission to develop part of their land for industrial purposes. The land adjoined a busy main road, which was already overloaded. The highway authority intended to widen the road at some future date and to acquire for that purpose a strip forming part of the plaintiffs' land. The planning authority granted planning permission subject to a condition that the plaintiffs should at their own expense construct an ancillary road parallel to the main road, such road to be 20 ft wide with a margin of 3 ft on either side, and should thereafter give right of passage over their ancillary road to and from such ancillary roads as might be constructed on either side. The Court of Appeal held that the condition was *ultra vires* and void, on the grounds that it was unreasonable in the *Wednesbury* sense (*Associated Provincial Picture Houses Ltd* v *Wednesbury Corporation* [1948] 1 KB 223). Willmer LJ said at p 248:

It remains to consider the contention that the conditions imposed are so unreasonable that they should be held to be *ultra vires* on that ground. In the first place, it is to be remarked that it is not sufficient merely to say that the conditions are unreasonable or unduly onerous, for that would properly be the subject for appeal to the Ministry under section 16 of the Act. In order to justify the court in granting a declaration that the conditions are *ultra vires* it must be shown that they are so unreasonable that no reasonable council could have imposed them . . . .

It appears to me that the object sought to be attained by the defendants in the present case is a perfectly reasonable one. I agree with the view of the judge that to require the construction of an ancillary road, and to limit the points at which traffic may enter or leave the main road, is "consistent with good traffic engineering." But it is said that the terms of the conditions actually imposed, particularly the requirement that the plaintiffs should give right of passage to and from the ancillary roads to be constructed on the adjoining land, are so unreasonable as to go beyond anything that Parliament can have intended, or that any reasonable authority could properly have imposed . . .

It has further been argued with force that if what the defendants desire to achieve is the construction of an ancillary road serving all the properties to be developed along the strip of land that is scheduled for development, for the use of all persons proceeding to or from such properties, they could and should have proceeded in a different way. What is suggested is that, in addition to the strip of land already earmarked for the proposed road widening, they could have designated a further strip 26 ft wide immediately to the southward, and could have imposed a condition that no building should be erected on this additional strip which would in any way interfere with its use hereafter for the building of the proposed ancillary road . . . .

Under the conditions now sought to be imposed, on the other hand, the plaintiffs must construct the ancillary road as and when they may be required to do so over the whole of their frontage entirely at their own expense . . . .The defendants would thus obtain the benefit of having the road constructed for them at the plaintiffs' expense, on the plaintiffs' land, and without the necessity for paying any compensation in respect thereof.

Bearing in mind that another and more regular course is open to the defendants, it seems to me that this result would be utterly unreasonable and such as Parliament cannot possibly have intended . . .

In these circumstances, although I have much sympathy with the object sought to be achieved by the defendants, I am satisfied that conditions 3 and 4 are so unreasonable that they must be held to be *ultra vires*.

Harman LJ said at p 255:

Next, the question for the court, in construing this kind of instrument, is not whether the court itself thinks the restriction a reasonable one, but whether the authority has exceeded its powers. If the authority has imposed a condition which no reasonable person should have imposed, the court can interfere, but not till that point is reached . . . .

On the question of the powers of the authority, I am of opinion that this condition 3 exceeds them. It may be that it is within the power of the authority to require an applicant to grant his neighbour a right of way over his land as a condition of its development. It is not in my judgment within the authority's powers to oblige the planner to dedicate part of his land as a highway open to the public at large without compensation, and this is the other possible interpretation of the condition. As was pointed out to us in argument, the Highways Acts provide the local authority with the means of acquiring lands for the purpose of highways, but that involves compensation of the person whose land is taken, and also the consent of the Minister.

Pearson LJ said at p 260:

There remains, however, the question of *vires*. By the general law a person whose land is taken for the purpose of making a highway is entitled to compensation. The effect of condition 3 is to convert a portion of Hall & Co Ltd's land into a quasi-highway, and to give them no compensation. Moreover, the resulting position as between the several frontages will be unsatisfactory . . . .

I agree with Willmer LJ that condition 3 is *ultra vires* because it is "unreasonable" in the sense which has been explained in *Kruse* v *Johnson* [1898] 2 QB 91 and other cases. I should, however, be inclined to say that the element of *ultra vires* is to be found in the conflict with the general law relating to highways. The general words of section 14(1) of the Town and Country Planning Act 1947 should not be interpreted as authorising a radical departure from the general law relating to highways.

It should be noticed that the question in *Hall's* case was not strictly the same as the question in the present case. In the *Hall* case the plaintiffs could proceed only by showing that the condition was unreasonable in the *Wednesbury* sense. As Willmer LJ pointed out, it was not enough for them to show that the condition was unreasonable in any other sense, for that argument should properly have been advanced by way of appeal to the Secretary of State under section 16 of the 1947 Act. In the present case, by contrast, there was an appeal to the Secretary of State. If the Secretary of State had discharged the condition on the simple ground that in his view it was unreasonable, there could have been no prospect of a successful appeal to this court under section 245 of the Act. But the Secretary of State did not discharge the condition on that ground, but on the ground that it was illegal. In para 8 of his decision letter he says there was "no legal basis" for the imposition of the condition. This has been treated by the parties as raising the same question as was raised in *Hall* v *Shoreham*. Thus the first ground of the planning authority's application, as stated in the amended notice of motion, is that the Secretary of State "erred in law in that he took the view that the condition the subject of the appeal to him was unlawful".

A Mr Keene, for the planning authority, seeks to distinguish *Hall* v *Shoreham* on two grounds. The first ground was that in *Hall* v *Shoreham* it was intended to widen the road in any event, irrespective of any development by the plaintiffs. The road was already overloaded. the condition was therefore imposed to obtain for the highway authority what Mr Keene called an extraneous advantage. It was not made necessary by the development itself. I need not decide whether Mr Keene's first ground provides him with a sound distinction in law, since it fails on the facts. I have already referred at the outset of this judgment to the Secretary of State's acceptance of the planning authority's submission that Meadway was below standard for the existing traffic, irrespective of any further development. No doubt the development would add to the
B congestion. But in that respect the position was no different from that in *Hall* v *Shoreham*.

Mr Keene's second ground of distinction was that in *Hall* v *Shoreham* the construction of the ancillary road formed no part of the planning application; the condition was imposed on the plaintiffs *in invitum*. Here, on the other hand, if the planning authority is right about what took place when the developers submitted their revised plan in November 1982, the widening of the road along its full length, including the cession of land along the northern and eastern perimeter of the residential development, was all part and parcel of the planning application. Admittedly the planning authority could not compel the developers to carry out works on land which was not
C within their control. Admittedly the planning authority could not force the developer to cede land for the purposes of the highway authority. But there is no reason why the developers should not volunteer to carry out the works. Having so volunteered, there is, according to Mr Keene, no good reason why the planning authority should not make it the subject of a condition. If the carrying out of the works could have been made the subject of an agreement under section 52 as Mr Laws conceded, it could be made the subject of a condition accepted voluntarily by the developers. There is no law against people giving away their land.

Mr Keene accepts that, even if he is correct on his second ground of distinction, he does not win outright. The case would have to go back to the Secretary of State in order that he may determine the issue of
D fact between the parties, whether the developers did indeed volunteer to carry out the works at their own expense. If they did not, then Mr Keene concedes that he will fail.

Mr Keene's second ground of distinction seems to raise a point of far-reaching importance in planning law. It is a point which has given rise to much controversy. It has usually been regarded as axiomatic that planning consent cannot be bought or sold. As a broad general proposition, that must be true. However, Mr Laws for the Secretary of State states the principle more narrowly when applied to the facts of the present case. He asks us to say, first, that a planning condition which requires the developer to carry out or fund a public function of a planning authority as the price of getting that planning permission
E is always unlawful, and, second, that the degree of acquiescence by the developer in the condition so imposed is wholly irrelevant.

I am bound to say that I have some doubt about the principle so stated. In the first place it does not explain the admittedly somewhat unlikely exception which Mr Laws himself would allow, namely, the "charitable volunteer" (in the judge's words) "who simply likes to build highways". Second, and more seriously, it does not explain why what is impermissible as a condition should be permissible if incorporated in a section 52 agreement. We were told that it is common practice for conditions such as condition 4 to be made the subject of a section 52 agreement; and Mr Laws conceded, as I have already mentioned, that if the condition in the present case had been incorporated in a section 52 agreement, it would have been perfectly
F lawful. I agree with the judge that that gives Mr Laws' principle a somewhat battered look.

What, then, is the true principle? In my judgment it is neither more nor less than the third of the three requirements identified by the House of Lords in *Newbury District Council* v *Secretary of State for the Environment* [1981] AC 578, namely, that the conditions imposed must not be manifestly unreasonable. If the proposed condition is manifestly unreasonable, then it is beyond the powers of the planning authority to impose it; and if it is beyond the powers of the planning authority to impose the condition, then it is beyond their powers to agree to impose it, *even if the developer consents*. As is stated in para 35 of Circular 1/85: "An unreasonable condition does not become reasonable because an applicant suggests it, or consents to its terms". If the condition is manifestly unreasonable, the willingness of the developer is irrelevant. *Vires* cannot be conferred by consent.

The fact that the applicant has suggested a condition or consented to its terms is, of course, likely to be powerful evidence that the condition is not unreasonable on the facts, since, as in the case of any commercial transaction, the parties are usually the best judges of what is reasonable. So I do not think there is likely to be any great rush of cases in which the developer obtains planning permission by consenting to a condition, and then appeals successfully against the imposition of the condition to the Secretary of State on the grounds that it is manifestly unreasonable. A successful appeal in such circumstances is likely to be rare. But I do not think it can be ruled out as a matter of law. For the analogy with an ordinary commercial transaction is not complete. The "parties" to a planning application are not in the same position as the parties to a commercial contract. For in addition to the interests of the "parties", there is the public interest in securing the fair imposition of planning control as between one developer and another. In *R* v *Bowman* [1898] 1 QB 663, a licensing case, the justices granted a liquor licence on condition, *inter alia*, that the applicant should pay £1,000 in reduction of the rates. Wills J, in giving the leading judgment in the Divisional Court, said:

> If the attachment of such a condition were allowed to pass without objection there would soon grow up a system of putting licences up to auction — a system which would be eminently mischievous and would open the door to the gravest abuses.

So there is a public interest as well as the interests of the "parties".

If the principle I have stated be correct and relevant, then two things follow. In the first place Mr Keene cannot distinguish *Hall* v *Shoreham* on the grounds that the carrying out of the works in the present case was included in the planning application, if indeed it was. His second ground for distinction therefore fails. For what it may be worth I would add that even in *Hall* v *Shoreham* the amended plan submitted by the plaintiffs showed a line 26 ft south of the proposed road widening, within the area coloured pink. Presumably that line represented the boundary of the proposed ancillary road. It does not appear to have been argued that that made any difference. So the fact that the proposed road widening is shown in the present case on the revised plans does not of itself distinguish the present case from *Hall* v *Shoreham*.

The second thing which follows is that the Secretary of State cannot be said to have erred in law in holding that condition 4 was unlawful, if he was entitled to take the view that the condition was manifestly unreasonable.

Was he, then, entitled to take that view? In my judgment he was. It is sufficient to say that I can see no relevant distinction between the present case and *Hall* v *Shoreham*. If the condition was manifestly unreasonable in *Hall* v *Shoreham*, so it was here. In addition, as Pearson LJ said in *Hall* v *Shoreham*, the element of *ultra vires* is to be found in the conflict with the general law relating to highways. "The general words of [section 29 of the Town and Country Planning Act 1971] should not be interpreted as authorising a radical departure from the general law". That comment is as true here as it was in *Hall* v *Shoreham*.

Next I should return to the role of the section 52 agreement in cases such as the present. Since the point does not arise directly for decision, and since it raises questions of considerable difficulty and importance on which we have heard only limited argument, I propose to confine myself to two observations, one general and one particular. The general observation is that the practice under section 52, convenient and beneficial though it undoubtedly is, may have gone beyond what the strict language of the section justifies. We were told that such agreements are now very common, much commoner than they used to be. It may be that in some future case it will be necessary for the court to consider the extent of the powers of planning authorities to enter into agreements under section 52. I am aware, of course, that such agreements are frequently entered into under combined powers, that is to say under powers contained in other statutory provisions as well as section 52

The particular observation is that I do not accept Mr Laws' submission that the present condition would have been lawful if incorporated in a section 52 agreement. If the condition was manifestly unreasonable, and so beyond the powers of the planning authority to impose it, whether or not the developers consented, it must follow that it was also beyond the powers of the planning authority to include the condition as "an incidental or consequential provision" of an agreement restricting or regulating the development or use of the land under section 52.

That is not to say that this might not have been a case for a more limited agreement under section 52. A contribution towards the cost of widening Meadway might well have been reasonable, due to the increased use of the road resulting from the development, and the benefit to the occupiers of the residential development: see Circular 22/83, "Planning Gain", where the considerations are well set out in paras 6-8. See also paras 3 and 6 of Appendix A. But I need not pursue that consideration further. For there is all the difference in the world between a provision of a section 52 agreement requiring a contribution from a developer towards the cost of widening the highway and a provision which requires the entire works to be carried out at his risk and expense.

There are three final points to be mentioned. Mr Keene submits that the Secretary of State was wrong to refer to section 30(1)(a) in connection with the so-called "tongue" and to hold, as he did, that the condition was *ultra vires* under that section, on the ground that the tongue was not land within the control of the developers. Mr Keene submitted that section 30(1)(a) is altogether irrelevant. Mr Laws conceded that the Secretary of State was wrong in that respect, and the judge has so held. This seems somewhat hard on the Secretary of State, since it was part of the planning authority's own case, as I follow it, that section 30(1)(a) was the relevant section, or at any rate a relevant section. Para 1.04 of the planning authority's response reads:

> So far as the *vires* of the appealed condition is concerned, the Town and Country Planning Act 1971 — section 30(1)(a) deals with the capability of a Local Planning Authority imposing conditions on land which is in the control of the applicant even though this may not be the subject area of that particular application.

If the tongue is not properly to be regarded as the "subject area of that particular application" then the Secretary of State was plainly right to regard section 30(1)(a) as the relevant section. But whether he be right or wrong, it cannot affect the result of this appeal. D

Second, if the condition had been construed as imposing a purely negative obligation, that the development should not proceed, or the houses not be occupied, until the road had been widened, then the condition could well have been valid: see *Grampian Regional Council* v *City of Aberdeen District Council* [1984] JPL 590. But that is not how this particular condition has been construed. It has been construed as imposing a positive obligation on the developers to carry out the widening at their own expense.

Finally I should mention a subsidiary argument, that the Secretary of State misconstrued his powers under section 36 of the 1971 Act. Under section 36(3) the Secretary of State may allow or dismiss an appeal, or may reverse or vary any part of the decision appealed E from. Mr Keene argues that, if one looks at para 9 of the decision letter, the Secretary of State appears to have given his mind to two options only, either to allow the appeal and discharge the condition or to allow the appeal and refuse consent, as happened in *Hall* v *Shoreham*. There was a third option open to the Secretary of State, to allow the appeal and limit the consent to the 50 or so houses which had been built and were already occupied at the time of the appeal.

I cannot accept this argument. It is true, as Mr Keene says, that there is no explicit reference to the third option. But it would not be right to infer that the Secretary of State had misconstrued his powers. He refers to section 36(3) in para 9 of the decision letter. We must assume that he had read it and was therefore aware of his powers under that subsection. It was not necessary for him to spell them out F in express terms, especially as it was never suggested that the consent should be limited in the way that is now argued. Nor, of course, was it suggested that consent should be refused altogether if the condition were discharged.

For the reasons I have given, I would dismiss this appeal.

O'Connor and Croom-Johnson LJJ agreed and did not add anything.

The appeal was dismissed with costs.

LANDS TRIBUNAL

COMPENSATION

TEXAS HOMECARE LTD v LEWES DISTRICT COUNCIL
(REF/147/1984)
October 17 1985
(V G WELLINGS QC)

Estates Gazette February 1 1986
277 EG 421-424

Stop notice compensation under s 177 of Town and Country Planning Act 1971 — Preliminary issue as to whether claim, in respect of warehouse units on Lewes estate, was made within 6 months' time-limit prescribed by regulation 14(2)(a) of Town and Country Planning General Regulations 1976 — Enforcement notice quashed by Secretary of State on January 27 1981 and claim for compensation dated August 12 and 13 1981 received by compensating authority on August 14 1981, ie more than 6 months after date of Secretary of State's decision — But claimant relies on passage in a letter from its solicitors to compensating authority dated January 28 1981 as constituting claim for compensation — After stating it seemed that their clients were entitled to compensation and costs and that they would be pleased to hear that the council "recognises its liability under these heads", the solicitors added: ". . . our clients will be formulating their various heads of claim with a view to a formal submission of the claim . . . in due course" — Held that the claim for compensation was not out of time — The relevant passage in the letter constituted a claim in itself and the "claims" made later, so far from being claims, were further and better particulars of a claim already delivered — In tribunal member's opinion, s 177 and reg 14 do not require claim to be in any particular form or that it should state sum claimed or be a detailed claim — Document alleged to constitute claim must, however, be unequivocal and make clear to compensating authority that a claim is being made, not that one is to be made in future — "It is right that a benevolent construction should be used" in construing relevant passage in letter in present case — *Thompson* v *Good & Co* cited by member in support

Guy Seward QC and George Bartlett (instructed by Titmuss Sainer & Webb) appeared for the claimants; Christopher Lockhart-Mummery (instructed by the solicitor to Lewes District Council) for the compensating authority.

Giving his decision, MR WELLINGS said: This is a preliminary issue as to whether the Lands Tribunal has jurisdiction to entertain a reference claiming compensation for loss alleged to be due to a prohibition contained in a stop notice. The answer to that question depends on whether or not the claim for compensation made by the claimants was made within the statutory time-limits.

Section 177 of the Town and Country Planning Act 1971 provides:

(1) A person who, when a stop notice under section 90 of this Act is first served, has an interest in or occupies the land to which the stop notice relates shall, in any of the circumstances mentioned in subsection (2) of this section, be entitled to be compensated by the local planning authority in respect of any loss or damage directly attributable to the prohibition contained in the notice...

(2) A person shall be entitled to compensation under subsection (1) of this section in respect of a prohibition contained in a stop notice in any of the following circumstances: F

(a) The enforcement notice is quashed on grounds other than those mentioned in paragraph (a) of section 88(2) of this Act;

(3) . . .

(4) A claim for compensation under this section shall be made to the local planning authority within the time and in the manner prescribed by regulations under this Act.

(5) . . .

(6) . . .

Under section 179 of the Act, any question of disputed compensation in respect of stop notices is to be referred to and G determined by the Lands Tribunal. It also provides that in relation to the determination of any such question, the provisions of sections 2 and 4 of the Land Compensation Act 1961 are to apply, subject to any necessary modifications and to the provisions of any regulations made under the Act (section 179(2)).

The procedure and time-limits for making claims for compensation of this kind are set out in regulation 14 of the Town and Country Planning General Regulations 1976. Regulation 14(1) says that a claim for compensation made to a local planning authority under *inter alia* section 177 of the Act of 1971 "shall be in writing and shall be served on that authority or council by delivering it at the offices of the authority or council, or by sending it by pre-paid post". Regulation 14(2) is as follows: H

The time within which any such claim. . .shall be served shall be —

(a) In the case of a claim for compensation, six months. . .

from the date of the decision in respect of which the claim or notice is made or given:

Provided that the period may be extended by the Secretary of State in any particular case.

There is an agreed statement of facts and issues. A copy of it is annexed to this decision. I can summarise the agreed statement as follows.

The claimant is the occupier of premises known as Units 2-5, Malling Brooks Estate, Lewes, East Sussex, and the Lewes District Council is the local planning authority for the area in which those premises are situated. J On July 21 1977 planning permission was granted by the council for the erection of the premises, which consist of warehouses with ancillary offices and other uses, but subject to a condition that the premises should be used only for warehousing and uses ancillary thereto.

On October 23 1979 the claimant applied for planning permission to use the premises without complying with the relevant condition. That permission was refused on November 23 1979 and on that date an enforcement notice was served alleging that there had been a breach of the relevant condition. On November 23 1979 a stop notice was also served by the council prohibiting the installation within the premises of various fittings in connection with the use of the said premises as a retail warehouse (see para 7 of the agreed statement). K The stop notice took effect on November 30 1979.

Enforcement notice quashed

Ultimately the secretary of state quashed the enforcement notice and allowed an appeal by the claimant under section 36 of the Act of 1971 and granted planning permission for the use of the premises for purposes stated in para 8 of the agreed statement without complying with the relevant condition. On February 18 1981 the compensating authority issued a notice of motion in the High Court for an order that *inter alia* the decision of the secretary of state might be remitted to him for rehearing and determination on the ground that he had erred in law.

Para 9 of the agreed statement says that on August 14 1981 the compensating authority received a claim from the claimant for compensation dated August 12 and 13 1981. Both parties treat the date (January 27 1981) of the secretary of state's decision letter as being the date from which the statutory six months' time-limit runs. Accordingly, if the claimant's claim for compensation was made as late as August 12, 13 or 14 1981, it was out of time.

Para 10 of the agreed statement shows that on September 22 1981 the claimant applied to the secretary of state for an extension of time under regulation 14(2) of the 1976 regulations and the point was taken in the application that the claim for compensation had not been made out of time because of the council's appeal to the High Court. That point has not been taken before me.

The secretary of state took the view that he had no power to extend the time (para 11). No application for judicial review of that decision was made. The council's notice of motion was withdrawn on January 24 1984. On May 31 1984 the claimant submitted to the council a revised claim for compensation (see para 13).

The claim for compensation referred to in para 9 of the agreed statement and the revised claim referred to in para 13 were detailed claims, quantified in terms of money.

Mr Guy Seward QC and Mr George Bartlett, appearing for the claimant at the hearing of the preliminary issue, did not rely on the claims referred to in paras 9 and 13 of the agreed statement, save as particulars of a claim already made. They relied on a passage in a letter not referred to in the agreed statement. The letter was dated January 28 1981, one day after the date of the secretary of state's decision, and was from the claimant's solicitors to the district solicitor of the council. The passage relied upon was as follows:

> Arising out of the quashing of the enforcement, and therefore of the stop notices, it seems to us that our clients are entitled to compensation and costs in relation to the appeal against the enforcement and stop notices and would be pleased to hear from you that your council recognises its liability under these heads. In the meantime our clients will be formulating their various heads of claim with a view to a formal submission of the claim to your council in due course.

It is implicit in the agreed statement that no claim for compensation was made by or on behalf of the claimant before August 12 1981. The correspondence which followed the letter dated Janaury 28 1981 gave no clue that the claimant was going to suggest that that letter constituted a claim for compensation. Thus in the claimant's solicitors' letter dated May 20 1981 to the council's agents, those solicitors said:

> Our clients have now given us details of their claim for compensation. We need some additional instructions but expect shortly to be able to submit it.

That letter, like others, contemplates a future claim. After the six months' period had expired on July 27 1981, the claimant's or its solicitors' attitude was consistent with its attitude before that date. Thus in those solicitors' letter dated November 13 1981 to the council's agent, the solicitors indicated that they took the view that the claim for compensation, having been made in August 1981, was three weeks out of time, explained the delay and stated "that the fact that we were only three weeks out of time is most material". Similarly, in a letter dated September 22 1981 to the secretary of state, the solicitors stated that in the letter dated January 28 1981 they intimated to the council "that a formal submission of a claim would be made in due course", and later in the same letter they made a further statement to the same effect. Even at the hearing, the claimant did not seek to make any amendment to the agreed statement.

Mr Seward accepted all these criticisms, but his case was quite simply that the question was one of the proper construction of the letter dated January 28 1981 and he submitted that on its true construction the passage in it which I have quoted fulfilled the requirements of the regulations as to the manner in which the claim was to be made. It fulfilled also the requirements as to time. He said that the claimant had wrongly thought that it had not made a claim prior to August 1981, but what the claimant had thought was not material, because the question was one of construction.

Mr Lockhart-Mummery appearing for the compensating authority said that the Act and regulations contemplated formality and required an unequivocal claim to be made. That was apparent from the use of the words in regulation 14(1): "claim in writing", "served" and "delivering". In his submission, a claim under this section must claim a liquidated sum or a sum which was, if not quantified, quantifiable. If that were done it might not be strictly necessary for the claim to include a breakdown of the sum claimed. Third, he said, any document which is alleged to have been a claim must have been intended to constitute a claim. The relevant passage in the letter dated January 28 1981 spoke of an intention to make a claim and did not in fact make one. The claim was, Mr Lockhart-Mummery said, made as stated in the agreed statement, at the earliest, on August 12 1981.

No particular form for claim

In reply, Mr Seward said that the relevant passage in the letter dated January 28 1981 indicated clearly to the council that a claim for compensation was being made and in that passage the council was asked to accept liability. There was no requirement in section 177 or the Regulations that a claim for compensation should state a sum and no requirement of formality. The question was one solely of construction.

In my opinion, section 177 and Regulation 14 do not require that a claim for compensation in respect of a stop notice should be in any particular form or that it should state the sum claimed or that it should be a detailed claim. Of course a document which is alleged to constitute a claim must be unequivocal: it must make clear to the authority in question that a claim is being made; not that a claim is to be made in the future. I agree with Mr Seward that the question in the present case is one of construction of the relevant passage in the letter dated January 28 1981. In my view, in construing that passage, it is right that a benevolent construction should be used.

In my judgment, the passage brought to the notice of the council that a claim was then being made against it and that the reference to a formal submission of the claim in due course was a reference to an intention to file particulars of it. If the intention in the passage were merely to give warning that in the future a claim was going to be made, then the request that the council should recognise its liability for compensation would hardly be apt until the claim had been made.

Accordingly, in my opinion, the relevant passage constituted a claim in itself and the claims referred to in paras 9 and 13 of the agreed statement, so far from being claims, were further and better particulars of a claim already delivered. The consequence is that the claim for compensation was not out of time and that the Lands Tribunal has jurisdiction to entertain the reference herein. I accordingly answer the preliminary issue in favour of the claimant.

In coming to the above decision, I have received some comfort from the decision of the House of Lords in *Thompson (Pauper)* v *Good & Co* [1910] AC 409. That decision was not cited to me, but there was, perhaps, no reason why it should have been cited, because it is concerned with a different Act, namely, the Workmen's Compensation Act 1897. In that case, Lord Loreburn LC said at p 410:

> This Act says that "the claim for compensation" must be made within six months of the accident. It has been held in the Court of Appeal that unless the amount claimed has been specified there has been no "claim for compensation". I cannot see why it should be so. There is no such provision in the Act itself; I mean in its actual terms. If a man says, as admittedly the appellant did say, "I claim compensation", why are we to conclude that he did not so claim? The reasoning for the respondent is that, unless a sum is named, the employer is deprived of an opportunity of settling the claim, and so avoiding proceedings under the Act. Surely, if he wants to know, he can ask the question, or he can make an offer himself... It is enough that the Act does not say the amount is to be specified, and, with all respect, we must construe it as it stands.

The compensating authority will pay the costs of the claimant such costs if not agreed to be taxed by the Registrar of the Lands Tribunal on the High Court Scale of Costs.

APPENDIX
Statement of facts and issues

1 The claimant is the occupier of premises known as Units 2-5, Malling Brooks Estate, Lewes, East Sussex.

2 The compensating authority is the local planning authority for the area in which the premises are situated.

3 On July 21 1977 planning permission was granted by the local planning authority for the erection of the premises consisting of warehouses with ancillary offices and a builders' merchant (with trade and retail counter) subject to, *inter alia*, condition 3: "The premises shall be used only for warehousing and uses ancillary thereto." That planning permission was implemented.

4 On October 23 1979 the claimants applied for planning permission to use the premises without complying with the said condition 3.

5 That permission was refused by the local planning authority by notice dated November 23 1979.

6 An enforcement notice dated November 23 1979 was served by the compensating authority on the claimant (then known as Home Charm Retail Ltd) alleging that there had been a breach of planning control by non-compliance with the said condition, in that following the carrying out of internal works for the conversion of the warehouse to a retail warehouse and the employment of staff on the premises required for the purpose of organising and operating a retail warehouse, all such activities amounting to a use of the premises not covered by the said condition.

7 A stop notice dated November 23 1979 was served by the compensating authority on the claimant (then known as Home Charm Retail Ltd) prohibiting the installation within the said land of retail display stands, units and shelving and customer pay-out desks in connection with the use of the said land as a retail warehouse. The stop notice took effect on November 30 1979.

8 Following an appeal, the Secretary of State for the Environment issued his decision letter on January 27 1981. The secretary of state quashed the enforcement notice on the ground contained in para (b) of section 88(2) of the Town and Country Planning Act 1971 that the matters alleged in the notice did not constitute a breach of planning control. Further, he allowed the appeal under section 36 of the Town and Country Planning Act 1971 and granted planning permission for the use of the premises as a retail builders' merchants/home-care centre without complying with the said condition, subject to certain further additional conditions. He also allowed one appeal and dismissed another appeal in relation to certain proposed advertisements. On February 18 1981 the compensating authority issued a notice of motion in the High Court for an order that, *inter alia*, the decision of the Secretary of State for the Environment might be remitted to him for rehearing and determination on the grounds that he had erred in law.

9 On August 14 1981 the compensating authority received a claim from the claimant for compensation, dated August 12 and 13 1981.

10 By letter dated September 22 1981 the claimant applied to the Secretary of State for the Environment under regulation 14 of the Town and Country Planning General Regulations 1976 for an extension of the time for making a claim for compensation under section 177 of the Town and Country Planning Act 1971. The application was expressed to be made without prejudice to the claimant's contention that the claim was not made out of time, by virtue of the facts that the compensating authority had appealed to the High Court and was stated to be made in order to avoid incurring considerable costs and an unnecessary controversy as to whether the claim was, or was not, out of time.

11 In a letter dated December 10 1981 the Department of the Environment took the view that the secretary of state had no power to extend the time under regulation 14 where the period for making the claim had expired.

No application for judicial review of this decision was made by the claimant.

12 On January 24 1984 the notice of motion referred to in para 8 above was withdrawn by the compensating authority with the consent of the Secretary of State for the Environment and the claimant.

13 On May 31 1984 the claimant submitted to the compensating authority a claim for compensation under section 177 of the Town and Country Planning Act 1971.

14 The claimant began trading from the premises on April 11 1981.

15 The claimant claims that by reason of its inability to trade from the premises between January 1 1980 and April 11 1981, it suffered loss as detailed in its summary of claim, which loss was, it alleges, directly attributable to the prohibition contained in the stop notice. The preliminary issue arises as to whether, on the basis of the above facts, the Lands Tribunal has any jurisdiction to entertain the claimant's claim.

16 The substantive issue which the claimant seeks to have determined by the Lands Tribunal is whether, by reason of the facts stated above, the claimant is entitled to compensation from the compensating authority under section 177 of the Town and Country Planning Act 1971 in the sum shown in its summary of claim.

MAHOOD v DEPARTMENT OF THE ENVIRONMENT FOR NORTHERN IRELAND

LANDS TRIBUNAL FOR NORTHERN IRELAND
(R/33/1983)

October 24 1985

(Judge R T ROWLAND QC (President) and A L JACOBSON FRICS)

Estates Gazette February 15 1986
277 EG 652-656

Surveyors' fees for negotiation of accommodation works — Whether fee should be based on amount of compensation for land taken (£30 in present case) plus cost of accommodation works (£234) or solely on value of land taken — Acquisition for road-widening purposes of part of frontagers' land at Newtownards, Northern Ireland — Accommodation works consisted of replacement of old wire mesh fence by new post and wire mesh bent arm fence — Firm of surveyors retained to act for some 60 residents, including present claimant — Contention for claimant that RICS scale 5 (1971 edition) fee based on compensation figure alone was quite inadequate for work involved — Acceptance by both parties that it was a question of fact and degree in every case whether there had been "negotiation" of accommodation works — DOE contended there were no negotiations sufficient to justify fee based on works — Some conflict of evidence on matters of detail — Surveyors' evidence that bundle of letters sent to DOE in 1974 informing them of each client's request for works, but in event no copies of such letters could be produced nor could they be found in DOE files — "Passage of time may have dimmed memories" — Tribunal finds works as originally offered in 1978 were not altered to any material extent when agreed in 1979 save that bull wire was substituted for barbed wire (proposal for the latter being "simply an error") and a 6-ft fence was to be provided instead of a 5-ft one, two matters which could not as a matter of fact and degree amount to negotiated works — Apart from those matters, no evidence before tribunal that any other negotiations or discussions between surveyors and DOE took place — DOE made simple proposals for replacement fence which were accepted without demur by surveyors subject to two minor amendments — On the facts there was no negotiation as required by scale 5 — Fee to be calculated on agreed figure of £30 compensation, ie minimum fee of £7.50

C J Millinson (instructed by J W Russell & Co, of Newtownards) appeared for the claimant; Mrs M Caher, solicitor, of the Department of the Environment's Solicitor's Branch, Belfast, for the respondent acquiring authority.

Giving their decision, THE LANDS TRIBUNAL FOR NORTHERN IRELAND said: This reference raises an important point relating to the scale of fees which a surveyor is entitled to charge when acting for a client whose lands have been compulsorily acquired for statutory purposes and for which compensation is payable.

The matter arises in this way. In 1973 the Newtownards Borough Council decided to widen and upgrade the Bangor Road just outside the town of Newtownards. For that purpose the council needed to acquire, either compulsorily or by agreement, part of the lands of a number of frontagers whose property bordered the existing main road. The council applied for and were granted, on August 21 1973, a vesting order vesting in the council the fee simple interest in portions of each frontager's land. The Department of the Environment for Northern Ireland ("the DOE") subsequently became responsible for all roads functions previously exercised by the council by virtue of SR&O No 504 of 1973 (NI Departments: Transfer of Functions & Adaptations of Enactments) Order 1973.

On January 7 1974, Russell Brothers FRICS, a well-known firm of Newtownards estate agents, surveyors and valuers, notified the local office of the DOE that they would be acting for a total of 57 named residents, in the matter of compensation and accommodation works, whose properties were to be affected by the scheme. The DOE, by letter dated February 7 1974, acknowledged Russell Brothers' interest and intimated that following a survey of the various entrances etc it was hoped to agree accommodation works in the near future. The work on the widening of the road was due to start some time in 1977, but the DOE wished to have the accommodation works agreed in writing in order that they could issue and complete tender documents.

One of the properties affected by the scheme was 68a Bangor Road, Newtownards, owned by the applicant Mr Mahood. Under the vesting order 40 sq yds of his frontage were acquired. This was part of a plot of vacant land which sloped down and away from the existing road so that the high cost of filling and levelling meant that the land was unsuitable for building. At the relevant time Mr Mahood was abroad for a period of months and he had given Mr Alfred Russell, principal of the firm of Russell Brothers, a power of attorney to enable his firm to carry out negotiations for compensation. Mr William Pitt FRICS, an associate in the firm, negotiated the compensation payable to Mr Mahood for land taken, which was agreed at £30.

A In addition to paying compensation of £30 the DOE carried out certain accommodation works which, broadly speaking, consisted of replacing the old fence with a new post and wire fence, the object being to reinstate the road boundary to an equivalent pre-vesting order condition.

Proper measure of fees

The present dispute arises as to the proper measure of agents' fees to be paid in respect of the compulsory acquisition. Mr Russell has contended that in road-widening cases such as this his fees ought to include a sum based on the value of the accommodation works and he relied on scale 5 of the Royal Institution of Chartered Surveyors (also B known as Ryde's Scale). He contended that, as a considerable amount of time had been spent in dealing with accommodation works on the scheme as a whole, the scale fee based on the compensation figure alone (£30 for land vested) was quite inadequate for the work involved. The relevant part of scale 5 reads:

> Where works are negotiated by the surveyor, the cost of such works, and any amount set off in respect of betterment, should be added to the amount of settlement for the purpose of calculating the fee chargeable under the scale set out above.

The DOE, on the other hand, contend that on the facts of this particular case there were no negotiations between Russell Brothers and Roads Service sufficient to justify a fee based on the cost of C accommodation works.

The facts in the instant case indicated a "straight-forward replacement works offered and agreed without negotiations", which would not come within scale 5.

Mr C J Millinson of counsel, for the claimant, posed the question before the tribunal in this way:

> Where a surveyor appointed by a claimant whose land is vested by a statutory authority and accommodation works are proposed by that authority in addition to payment of compensation for land vested, is the claimant entitled to recover estate agents' fees which he has paid in connection with work done by the agent for both the value of the compensation for land and for the negotiations conducted on his behalf by the agent in respect of the proposed accommodation works?

D The answer to that question, in the opinion of the tribunal, must depend upon the facts of the particular case, and in the face of the differing view of the facts taken by the respective parties the tribunal must look to see what was in fact done in connection with the accommodation works.

Mr William Thomas Pitt FRICS on behalf of the claimant testified that his firm had been retained to represent about 60 frontagers part of whose land was compulsorily acquired. At this rather remote point in time his recollection of detail was somewhat hazy. In or about 1977 or 1978 his firm had received from Mr Blaney (lands officer in the DOE from 1975) the DOE's land acquisition form on which the accommodation works proposed by the DOE were:

> E Provide and erect along new boundary standard 5 foot bent arm concrete post and wire fence with nylon covered wire mesh and 3 No strands of barbed wire on owners side of fence and 3 No strands of barbed wire running parallel on road side. Bent arms of fence to overhang owners property.

He spoke on site with one of the DOE's employees and agreed a 6-ft bent arm post and wire fence in lieu of the 5-ft fence as proposed. Apart from that he could recall no other discussions taking place about the proposed works. He had not previously negotiated any accommodation works for any claimant, but some time in 1974 his firm had sent a bundle of letters informing the DOE of the accommodation works requested for each of their clients in this road scheme.

F Mr Alfred Russell FRICS testified that in 1973, prior to the proposed road scheme, there was a dilapidated wooden fence bordering the property and children and vandals were able to enter the property and cause annoyance to an adjacent owner. Following representations from this owner and in order to prevent trespass a 1.8 m-high straight concrete post and mesh-wire security fence was subsequently installed. But he could not be certain precisely when this was done. He himself had not personally handled these road cases nor taken any part in the discussions. He had never before had occasion to submit a claim for fees calculated by aggregating the cost of accommodation works, but he had decided to do so in this instance as a matter of principle. He contended that once his firm had been retained to look after the owners' interests in the provision of accommodation works, his firm was thereby, like any professional firm, entitled to be remunerated for his retainer according to the value of the works.

Mr John Raymond Blaney and Mr Francis Malachy Skillen, both lands officers for the DOE, gave evidence, but neither officer knew of or could produce from departmental files any letters or memoranda from Russell Brothers indicating what accommodation works if any were requested by claimant. Mr Blaney confirmed that he sent out, in or about May 1978, the land acquisition agreement upon which was endorsed particulars of the accommodation works which were offered to Mr Mahood by the DOE. These were in the terms referred to earlier in this judgment. Although a number of inquiries were made by Mr Blaney and later by Mr Skillen, no replies were received from Russell Brothers.

Agreement signed

In August 1979, there was a meeting between Mr Pitt and DOE representatives to settle certain questions at sites 37 and 38 belonging to Mrs Doreen Russell and Miss Mary Russell. When these discussions were concluded the parties moved down the road to the claimant's site and took the opportunity of examining the claimant's land. The DOE suggested a fence to replace the existing fence. Agreement appears to have been reached without demur, and on October 17 1979 the agreement for accommodation works, which had been outstanding since May 1978, was signed and returned by Russell Bros with one amendment only, viz: the word "barbed" wire was struck out and the word "bull" wire was substituted, both parties realising barbed wire could not be used on the public side of a fence bordering the highway where it might cause serious injury to those lawfully using the road. Clearly, the proposal to use "barbed" wire was simply an error.

One other minor alteration to the proposed works became necessary. In November 1979 Mr Blaney discovered that the proposed fence was not available in a standard height of 5 ft. The nearest standard fence was a 6-ft concrete bent arm which Mr Pitt was only too happy to accept when he was told and this minor alteration was endorsed on the agreement.

Apart from these two items, there was no evidence of any other discussions about the DOE's proposals for accommodation works at the Mahood property.

Mr Millinson of counsel, for the claimant, submitted:

1. The accommodation works at the Mahood property remained a proposition until accepted. It was reasonable to instruct an estate agent to examine the proposal with a view to accepting or rejecting it.
2. Scale 5 of the Ryde Scale applies to this case because Mr Pitt took part in discussions, negotiations and settlement of the proposed works.
3. The negotiations relied on consisted of the following:
(a) The estate agent made written proposals for works some time in 1974.
(b) Some discussions between Mr Pitt and Mr Blaney took place prior to the DOE's form being sent out.
(c) As a result of subsequent discussions the DOE proposals were amended — a 6-ft fence replacing a 5-ft one and bull wire replacing barbed wire.
4. It was a question of fact and degree whether there were discussions and negotiations.
5. The fees of a surveyor who has prepared and negotiated a claim for compensation are payable as compensation within rule (6) of section 2 of the Acquisition of Land (Assessment of Compensation) Act 1919 being "any other matter not directly based on the value of land": *Minister of Transport* v *Lee* [1965] 2 All ER 986; *London County Council* v *Tobin* [1959] 1 All ER 649.

Mrs Caher, solicitor, for the respondent, submitted:

(a) Both parties have accepted that the accommodation works amounted to replacement works and did not involve either betterment or set-off from compensation for land taken.

(b) No negotiations took place between the DOE's land officers and Russell Brothers.

(c) The two items relied on by the claimant as constituting "negotiations", viz: "bull wire" for "barbed wire" and a 6-ft standard fence for a 5-ft standard fence, do not, as a matter of fact and degree, amount to "negotiation".

(d) It is accepted by both parties that the DOE does not normally pay fees based on the cost of accommodation works where, as here, such works are merely works of replacement.

(e) Scale 5 of the Royal Institution of Chartered Surveyors (applicable to compulsory acquisition of land) requires that works have to be "negotiated by the surveyor" before the cost thereof is included in the fee calculation. "Negotiated" in that context implies works of "betterment" before the cost of the works and any amount set off in respect of betterment can be added to the figure of compensation for the purpose of calculating the fee chargeable under the scale.

Decision

Where a claim for compensation, following a compulsory acquisition, is submitted to the Lands Tribunal for decision in default of agreement between the parties the tribunal has jurisdiction to award not only full compensation but also costs and witnesses' expenses to the successful party. See section 8 of the Lands Tribunal and Compensation Act (Northern Ireland) 1964 under which the tribunal has a full discretion as to the amount of such costs and expenses and is not necessarily bound by scale 5 of the RICS scales or any other scale. In such cases the tribunal may have to consider various items or headings of compensation such as are set out in the rules contained in section 2 of the Acquisition of Land (Assessment of Compensation) Act 1919. If a case is referred to the tribunal for decision the compensation awarded may include, in addition to the open market value of the land taken (under rule (2)), other matters stipulated in the other rules and full compensation will be the totality of the figures awarded (if any) in respect of these other matters.

In addition to the award of compensation, the tribunal may, in a proper case, award costs and witnesses' expenses which can include the valuer's fee for preparing and presenting the claim. Where, however, the case does not reach the tribunal and thus no question of an order for costs and expenses arises, it is clear from authority that the surveyor's fee in preparing, negotiating and settling the claim is a proper head of compensation under rule (6) of the 1919 Act rules, viz: "any other matter not directly based on the value of the land". See, for example, *London County Council* v *Tobin* [1959] 1 All ER 649 and *Minister of Transport* v *Lee* [1965] 2 All ER 986.

The measure of the surveyor's fee in those circumstances has generally been based on scale 5 of the RICS scale. What has happened in the present case is that the figure for the value of the land taken has been agreed, but the measure of the surveyor's fee has not and what has been referred to the tribunal for decision is one heading of compensation only — what is the proper measure of the surveyor's fee as part of the compensation to be paid? Both parties accept that scale 5 is applicable, but they are divided as to how it should be applied. The issue which the tribunal has to decide is — should the fee be based on the amount of compensation for land taken (£30) plus the cost of accommodation works (£234) or should it be based solely on the agreed value of £30 for the land taken?

Initially the case was submitted to the tribunal as a matter of principle which would be followed in many other cases, but during the course of the hearing it was accepted by both parties that it was a question of fact and degree in every case whether there had been "negotiation" of the accommodation works. The tribunal was informed that of the 60 or so cases in which Russell Brothers were retained there were some in which the cost of accommodation works had been allowed in calculating the fee.

The RICS scale 5 (July 1971 edition) was couched in the following terms: "(a) To the valuer who prepares the case (including negotiations for a settlement, where required): The scale set out below." Then followed the table setting out the amount and the fee from amounts of £100 to £20,000.

Then came the words giving rise to this dispute:

Where works are negotiated by the surveyor, the cost of such works, and any amount set off in respect of betterment, should be added to the amount of the settlement for the purpose of calculating the fee chargeable under the scale set out above.

In approaching this case it is pertinent to point out what has been the common practice in most road schemes, namely that accommodation works are usually those of replacement and reinstatement which are detailed and executed to the general approval of the landowner. While it is generally more satisfactory if accommodation works are agreed and carried out during the reconstruction of the road, it must be borne in mind that the acquiring authority is under no obligation to carry out such works and certainly ought not to do so without the owners' consent. In the absence of such consent the authority can leave the owners to make their own arrangements for reinstatement and claim the cost thereof (taking into account betterment, if any) as part of the compensation.

But in the majority of cases concerning strips of land compulsorily acquired for road-widening or realignment purposes the DOE proposes accommodation works which are designed to replace, as nearly as possible, the existing features on the ground. In most cases these are accepted either without alteration or with minor alterations of detail by the landowners or their advisers. However, if there are cases where, because of the inherent difficulties of the scheme or the topography or layout of the land, more elaborate negotiations are required, these may justify the payment of a higher inclusive fee. In cases, however, where merely the formal approval of the owner or agent is required to the DOE's proposals no aggregation of the cost of accommodation works and value of the land taken is made in order to calculate fees. The tribunal accepts that this is correct. Where accommodation works are negotiated and variations to the replacement works offered are made which result in betterment and thus a reduction in compensation, then the agent's fee would be calculated taking the betterment amount into account.

In the present case the replacement works carried out consisted of providing a new post and wire mesh bent arm fence for an old wire mesh fence. They were carried out on the basis of the agreement form returned on October 17 1979 following a letter written by Mr Blaney on October 12 1979. The works as offered and specified on the form were not altered to any material extent save that bull wire was specified in place of barbed wire and a 6-ft fence was provided in place of the 5-ft one specified. Mr Pitt had no difficulty in accepting that both these were standard specifications which did not require any specialist knowledge or techniques nor did they involve any complications. Neither of these two matters nor both of them taken together can, as a matter of fact and degree, amount to negotiated works.

Apart from those two matters, there is no evidence before the tribunal that any other negotiations or discussions between the agents and the DOE took place. The passage of time may have dimmed memories, as Mr Pitt conceded, but in the event no copies of letters from the claimant's sources setting forth particulars of the works requested in each individual case could be produced nor were they to be found in the DOE files. The case resolves itself into one in which the DOE made simple proposals for a replacement fence which were accepted without demur by the agents and subject to two minor amendments.

On the facts there was no negotiation as required by scale 5 and the result is that in this particular case the surveyor's fee is to be calculated on the agreed figure of £30 compensation — the minimum fee of £7.50

No order as to costs.

SMITH v WEST MIDLANDS COUNTY COUNCIL (REF/240/1983)

February 21 1986

(C R MALLETT FRICS)

Estates Gazette April 26 1986

278 EG 419-420

Residential backland (1,860 sq yd) in Small Heath, Birmingham, acquired for road improvement — Site, which was previously occupied by 42 garages and workshops, in predominantly residential area — Not disputed that outline planning permission would have been granted for three terrace houses with access by private road — Claim for £11,160 on assumption that land developable for residential purposes — Acquiring authority value at £1,850, contending that residential development not economically viable as the land could be developed to accommodate 37 single prefabricated garages at £50 a plot — Site described by tribunal as unique, with none of the transactions referred to by the parties being "even closely similar" although most were in same locality — Claimant's valuer's view, which was unchallenged, that £12 to £12.50 a sq yd would be fair figure for residential land in Birmingham inner areas — Authority's

A **valuer submitted three valuations based on residual method showing values varying between £1,088 and minus £13,312, but tribunal derives no assistance from them — "Results subject to very wide variations according to assumptions made" — Tribunal finds some indication of lively demand for infill development of nature proposed — Some 10 acres of backland in immediate vicinity acquired at figure in region of £6.50 a sq yd — Claimant's valuer's subjective view that, by comparison, the subject site was worth £6 a sq yd not shown to be in error — £11,160 awarded**

B Martin Kingston (instructed by Tyndallwoods, of Birmingham) appeared for the claimant; Roger Giles (instructed by the solicitor of West Midlands County Council) for the acquiring authority.

Giving his decision, MR MALLETT said: This reference is concerned with the value of residential backland in the centre of Birmingham. The land acquired was the subject of a compulsory purchase order, the West Midlands County Council (Small Heath Bypass Stage II) Compulsory Purchase Order 1981 made under the Highways Act 1980, the Acquisition of Land (Authorisation Procedure) Act 1946 and Acquisition of Land Act 1981. The order was confirmed by the Secretary of State for Transport on January 20 1982. Notice to treat and notice of entry dated May 27 1982 were duly served. The acquiring authority entered on December 20 1982, which becomes C the valuation date.

The site has an area of some 1,860 sq yd with access to Byron Road to the west of 13 Byron Road. The site is situated at the rear of 13, 15, 19, 23, 25, 27, 29, 31 and 35 Byron Road and abuts a railway line. There were 42 garages and workshops on the site but these have been demolished prior to the valuation date. The site is situated in a predominantly residential area of Small Heath close to the junction of Byron Road with Golden Hillock Road. Mainly the surrounding properties were constructed at the turn of the century.

It is not disputed that outline planning permission would have been granted for the erection of three terrace houses with access by private road to Byron Road.

D On the assumption that the land could be developed for residential purposes the claimants value the site at £11,160. The respondent acquiring authority contends that, despite the assumptions as to outline planning permission, the development of the site for residential purposes was not economically viable and they value the land at £1,850 on the assumption that the land could be developed to accommodate 37 single prefabricated garages.

Mr Martin Kingston of counsel appeared for the claimants and called Mr P J Frampton BSc MRTPI ARICS, a partner in Richard Wood & Partners, practising as town planners, surveyors and architectural consultants, and Mr Martin Webb ARICS ARVA, an associate in the firm of Jack Cotton & Partners. Mr Roger Giles of counsel appeared E for the acquiring authority and called Mr Jitendra V Patel MSc DipTP MRTPI, senior planning officer (Eastern District) in the City of Birmingham Planning Department, and Mr Godfrey Holder BSc ARICS, a senior valuer in the West Midlands County Council.

The dispute which arises as to the valuation of this land is largely due to the fact that it is unique and none of the transactions referred to by the parties are in respect of sites which are even closely similar, although most of them are in the same locality.

It is not disputed that if the site was developed additional expense would be involved in the construction and provision of adequate services and the laying of a private access drive with car-parking facilities and additional fencing and landscaping particularly on the railway frontage. The acquiring authority contend that the provision F of double-glazing and ventilation on some of the windows on the first floors would be essential in view of the noise created by rail traffic. The claimant contends that this would be unnecessary.

Mr Webb was of the view that £12 to £12.50 per sq yd would be a fair figure for residential land in the inner areas of Birmingham and this view was not challenged. He referred to a number of transactions on land in the area including a large site of over 10 acres at the rear of houses in Bolton Road and adjoining the railway line. He concluded that having regard to the nature of the subject land it should be valued at £6 per sq yd.

The acquiring authority sought to demonstrate that the subject land was incapable of being economically developed as a housing site by reference to three valuations based on a residual method. Mr Holder estimated the selling price of three houses that might be developed on the site and then deducted the estimated cost of construction including agent's and legal fees and developer's profit. He varied the size of the houses and the estimate of cost per sq ft and by this means his three valuations showed a variation for the value of the site between £1,088 and minus £13,312. He did not include in the estimate of cost any additional sums for the additional road work, services, double-glazing and screening which it was contended might be expected on this site.

Determination of plot value

From these calculations Mr Holder concluded that a residential development on this site would not be economically viable and he relied on a valuation on the basis that the site could be developed with 37 garages at £50 per plot, namely £1,850 or £1 per sq yd (£1,860).

The claimants contended that the site could contain a greater number of garages and at one time had contained 42 garages and workshops. In Mr Wells' experience £75 per plot was the standard rate adopted by the county council but some settlements had been negotiated at £100 per plot. There was no evidence to support a figure as low as £50 per plot.

I was referred to the following: *First Garden City Ltd* v *Letchworth Garden City Corporation* (1966) 200 EG 123; *Fairbairn Lawson Ltd* v *Leeds County Borough Council* (1972) 222 EG 561; *Bailey* v *Basildon Development Corporation* (1974) 230 EG 981; *Clinker & Ash Ltd* v *Southern Gas Board* (1967) 203 EG 735; *Viscount Camrose* v *Basingstoke Corporation* [1966] 3 All ER 161.

The subject site now forms part of the Small Heath by-pass development, but I visited the locality and was able to see those comparable sites that had not yet been developed.

It seems to me that any developer seeking to purchase the subject land and knowing that he would be likely to obtain outline planning permission for the development of three houses would wish to know whether such a development would be economically viable. He might well prepare his estimates on the lines suggested by Mr Holder, but the accuracy of the result would depend on his expertise in estimating the costs and the selling prices. Mr Holder has prepared his estimates from a variety of sources and not particularly from his own experience. The results are subject to very wide variations according to the assumptions made. He has not attempted to add anything for the extra cost that might be encountered at the site. If, therefore, his figures be right, it is difficult to see how any residential building plot in this locality could sell for as much as £1 per sq yd. The experience of the market, however, indicates that £12 to £12.50 per sq yd is the going rate for land with street frontage.

For these reasons I derive no assistance, in this instance, from the calculations based upon a residual method of valuation.

Nevertheless, it cannot be assumed that because outline planning permission for the development of three houses on this land would be forthcoming the land would have any value or would necessarily exceed the existing use value if developed for lock-up garages. The only evidence on actual transactions in the locality is in respect of lands which are not truly comparable and the variations cannot be adjusted by mathematical calculations.

There is some indication of a lively demand for infill development of this nature. Out of 101 suitable sites marketed by the county council towards the end of 1982 and the beginning of 1983 all but four had been placed under contract for sale, but none of these were said to be backland. Nevertheless, the acquisition of some 10 acres of land adjoining the railway in the immediate vicinity indicates that backland which was capable of development for housing purposes was acquired at a figure in the region of £6.50 per sq yd.

Mr Webb has taken a subjective view that by comparison the subject site is worth £6 per sq yd, and having heard the evidence I do not think that he has been shown to be in error. Therefore the claimant succeeds and I award the value of the freehold interest in the subject land at £11,160.

I received submissions as to costs. The acquiring authority is to pay the costs of the claimant if not agreed to be taxed by the Registrar of the Lands Tribunal on the High Court Scale.

For further cases on this subject see p 19

LANDS TRIBUNAL
LEASEHOLD REFORM

CUNNINGHAM AND ANOTHER v MORRISON-BELL AND ANOTHER
(LRA/2/1985)
January 21 1986
(W H REES FRICS)

Estates Gazette May 17 1986
278 EG 741-742

Freehold price of terraced two-bedroomed cottage in Shiremoor, Tyne and Wear, with some eight years of lease unexpired — Tenants appearing in person seek figure of £740 while landlords contend for £1,850 — Issue between parties the freehold value of the land — Tenants' approach is by reference to cleared site values, relying on prices of two sites, one (over six times the area of subject site) a sale price, the other (for two proposed terraced houses) an asking price — Landlords rely primarily on standing house approach supported by various sales of houses with vacant possession — Difference of opinion on costs of bringing roads and services to tenants' supporting sites — No deduction in a s 9(1)(a) case for improvements made by tenants — Tribunal derives little help from tenants' comparables and, on evidence, upholds entirety value found by LVT — Percentage deduction reduced having regard to disadvantages of situation — Tribunal values at £1,520, but price to be paid is £1,300 as determined by LVT, there being no appeal by landlords against that figure

Mr T Cunningham appeared on behalf of himself and his wife, the appellants; Mr C J Thompson, solicitor (of Wilkinson Marshall Clayton & Gibson, of Newcastle upon Tyne) for the respondents, Mrs Penelope Morrison-Bell and the Hon Mrs C S Beaumont.

Giving his decision, MR REES said: This is an appeal against the decision of a leasehold valuation tribunal determining the price to be paid by the lessees for the freehold of their terrace house at [3 Morley Place] Shiremoor, Tyne and Wear, at £1,300. The tribunal below gave a long and detailed written decision. Before me the appellants sought a figure of £740: the landlords contended for £1,850, but since they had lodged no appeal against the decision of the tribunal below I am limited to £1,300.

The difference between the parties comes down to one figure, the value of the freehold interest in the land on which the subject house stands. The appellants considered that that should be found by reference to cleared site values and sought to substantiate their figure on that basis by reference to the sale of land with planning permission for a bungalow in Shiremoor and by reference to the asking price for the site for two proposed terrace houses at Seghill, a few miles to the north. However, they also submitted a valuation on the "standing house" basis to support their figure. The landlords relied primarily on the latter method.

Mr Cunningham appeared for himself and his wife and gave evidence (I refer to Mr Cunningham hereinafter as "the appellant"). The solicitor for the landlords called Mr R P Robinson FRICS, a quantity surveyor, whose evidence related to the cost of bringing roads and services to the sites in Shiremoor and Seghill. He also called Mr I D Jefferson BA BSc (Estate Management), a probationer of the Royal Institution of Chartered Surveyors, awaiting election to professional associateship.

The subject house is a flat-fronted house on Earsdon Road, Shiremoor, which is a village a few miles north of Newcastle. Earsdon Road is a main road carrying fairly heavy traffic. Morley Place is a terrace of six cottages. One of these has been converted into a shop which at the relevant date was selling fish and chips (it is now an "Indian take away"). The subject house is opposite a garage. There is a secondary modern school to the north-east. The area to the south-east is mainly residential in character. The subject house is constructed of brick with a pitched slate roof. There is a small two-storey offshoot to the rear and a yard behind with access to a back lane. Accommodation on the ground floor comprises a living-room, kitchen/breakfast room and utility room. On the first floor there are two bedrooms and a bathroom. The site area is 106 sq yd. F

The appellant holds the premises on lease for 99 years from November 12 1893 at an annual apportioned ground rent of £1.54. The appellant served notice under the Leasehold Reform Act 1967, as amended, to purchase the freehold on July 13 1984. That is the relevant date for these proceedings and thus the lease had about eight years unexpired. G

The appellant's primary valuation was (in a shortened form) as set out below. The rates of interest used are not in dispute.

Value of ground rent of £1.54 per annum capitalised for 8 years		£ 8	(agreed)
Value of site of subject house	£1,270 (£3,150)		
PV £1 in 8 years at 7% (agreed)	0.582 (agreed)	739	(1,829)
		747	(1,837)
	Say	£740	(£1,850)

H

In the alternative the appellant used the standing house approach to support the figure of £1,270, the site value. This was as follows:

Value house as improved	£9,000	(£14,000)
Deduct tenant's improvements	3,000	(Nil)
	6,000	(14,000)
Take 20 ($22\frac{1}{2}$)%	£1,200	(£3,150)

Mr Jefferson's figures are those in parentheses in the above calculations.

To support his figure of £1,270 the appellant relied on the sale of a plot of land at Bridge Terrace in Shiremoor, a short distance south-east of the subject house but away from the main road. £3,000 was paid for this plot in July 1984. Its area was 675 sq yd. It is common ground that the cost of building a road within this site and the cost of bringing services to it have to be added to the figure of £3,000 for purposes of comparison. The appellant put these costs at £800. Mr Robinson's estimate was a minimum of £5,050 and a maximum of £8,400. Full details of the planning history of this site were put before me. There was detailed planning permission for one bungalow on this site, but there were indications that planning permission for two terrace houses might have been granted. J

The appellant also relied on the fact that a site for two proposed terrace houses at Seghill was on offer at £6,000 in September 1985. Again there was a difference of opinion between the appellant and Mr Robinson with regard to the costs of bringing services to that site, which has not been sold although it had been in the market for between two and three months. Reference was made by the appellant and by Mr Jefferson to a number of sales of houses with vacant possession, but in my judgment none of these support a figure as low as £9,000. K

In his evidence the appellant said that he had carried out a considerable number of repairs to the subject house since he bought the leasehold interest in 1979 for £2,000. In addition he had carried out further works which he regarded as improvements. These included the installation of a modern wc, wash hand-basin, bath with

A shower, kitchen units, a new sink, plumbing for a washing machine, central heating and an extension of the original electrical installation including a water heater. In his view, the value of these improvements, which he put at £3,000, ought to be deducted from his valuation of £9,000, as indeed had been done by the tribunal below. In his evidence the appellant criticised in considerable detail valuations put forward by Mr Jefferson both before me and at the tribunal below. The appellant also referred to a number of sites of former houses in Shiremoor owned by the landlords. These he said had been vacant for some time and that indicated that the site value of his house was low.

Mr Jefferson relied on a number of sales of houses with vacant possession in the area to support his figure of £14,000. Many of these
B were criticised by the appellant because of the differences between them and his house and between the relevant date here and the dates of the sales concerned. Mr Jefferson referred to three settlements under the Leasehold Reform Act 1967, as amended, to support his percentage of $22\frac{1}{2}$. He also referred to three modern ground rents of about £240 per annum each (on a 7% basis, that gives a capital value of £3,425) fixed in respect of houses in Walkergate in support of his site value of £3,150.

I viewed the subject house externally, the site at Bridge Terrace, that at Seghill and sufficient of the houses sold freehold with vacant possession to illustrate the evidence. Photographs of nearly all of these were put before me.

C I find the difference between the Bridge Terrace site and the site of the subject house to be so great as far as the characteristics of the two are concerned that I can place practically no reliance at all on it to support the appellant's figure of £1,270 being the site value of the subject house. Its area is over six times the area of the site of the subject house. There is a great deal of room for difference of opinion with regard to the cost of the services to be added as is shown from the above figures. The appellant's attempt to compare the sites at Bridge Terrace and Seghill with the site of his house on the basis of a unit price per square yard is in any event quite inappropriate here.

As regards the Seghill site, the figure of £6,000 (plus the cost of services) must be regarded as a maximum, but I find it to be in a totally different locality from the subject house; comparison is
D therefore very much a matter of opinion. The fact that the respondent landlords own the sites of former houses nearby and that these have stood vacant for some time is, in my judgment, of no significance. There could be many reasons for their not wishing to sell: there was no evidence that they have been offered in the market.

On the evidence I find for an entirety value of £13,000, as did the court below, for the subject house as improved. In coming to that figure I have had regard to the disadvantages of the layout of the accommodation of the first floor of the subject house, of the surrounding properties and of the situation on a main road. In my judgment no deduction should be made for the value of the improvements made by the tenant. There is no parallel provision in the Leasehold Reform Act 1967 as amended which relates to houses valued under section 9(1)(a) as there is in respect of houses valued under section 9(1A):

> (d) on the assumption that the price be diminished by the extent to which the value of the house and premises has been increased by any improvement carried out by the tenant or his predecessors in title at their own expense.

In assessing a modern ground rent an intending lessee would have regard to all the potentialities in the site ". . . for the uses to which the house and premises have been put since the commencement of the existing tenancy, . . . " (see section 15(2) (a)) and comparisons are made on that footing as was done in *H A Patten* v *Wenrose Investments Ltd*; *N M Lacklison* v *A A Davis (Liquidator of Finsbury Developments Ltd)*; *G E Thomas* v *Same* (1977) 241 EG 396 LT without deduction for the cost of improvements (see *Goldridge* v *Official Custodian for Charities and The Trustees of St Pancras Church Lands* (1972) 222 EG 1672 LT and *Barrable* v *City of Westminster* (a decision of this tribunal, my own) (1984) 271 EG 1273.

Having regard to the disadvantages of the situation of the subject house I prefer the appellant's 20% to Mr Jefferson's $22\frac{1}{2}$%. I therefore find for a site value of £2,600 which, deferred at the agreed rate for the agreed period, gives £1,513. To this I add £9 being the agreed value of the ground rent for eight years to give £1,522, say £1,520. The appeal fails. The price to be paid will be as determined by the tribunal below, £1,300, there being no appeal by the landlords against that decision (see *British Petroleum Pension Trust Ltd* v *Blake* (1982) 264 EG 721).

I was referred by the appellant to a considerable number of decisions of this tribunal and of leasehold valuation tribunals. I do not consider it necessary to refer to them in detail with the exception of those set out above.

In conclusion I record that I gave the appellant very considerable latitude in presenting his case, which I found difficult to follow at times, particularly since it was a mixture of evidence and submissions. He sought to criticise the composition of the tribunal below and to introduce a matter concerning an alleged non-enforcement of covenant by the landlords in respect of an adjoining property. As I pointed out to him during the hearing, the first could have no bearing at all on the question before me, which was the price to be paid by him for the freehold, and the latter was outside the jurisdiction of this tribunal.

I heard submissions as to costs. Having opened the sealed offer lodged by the respondents I find that the amount thereof is less than the amount of my award. The appellant will pay the respondent's costs of this appeal such costs unless agreed to be taxed by the Registrar of the Lands Tribunal on Scale 2 of the County Court Scales of Costs with certificate for two expert witnesses fit to qualify.

For further cases on this subject see pp 161 and 231

LANDS TRIBUNAL

RATING

WESTMINSTER CITY v BURTON GROUP PLC AND SAUL (VO)
(LVC/469/1983)
January 16 1986
(W H REES FRICS)

Estates Gazette February 22 1986
277 EG 757-758

Oxford Street shop — Part of shopping space on ground floor and whole of first floor in West One Centre complex over Bond Street Underground Station — Appeal by rating authority against LVC's decision of GV £98,250 — Main issue value of first floor — Factors examined included access to first floor (by 5 escalators) and effect of number of pillars — Ratio of ground-floor area to first-floor area at "in excess of 1 to 8" is highest in Oxford Street — Next highest is 1 to 3.5 — Evidence that first floors of Oxford Street retail properties assessed at between A/8 and A/10 of zone A values — Rating authority's valuer values at £60 (A/5) psm but agrees that only first floors of banking halls in Oxford Street assessed as high as A/5 — Fact that complex over underground station connected to two lines added, in his view, to value of premises — In tribunal's judgment, the ratio reduces the number of potential tenants — VO's figures of £30 psm for sales area and £20 psm for stock accepted — Tribunal derives no assistance from banking halls or from overall approach used by rating authority's valuer as a check valuation — VO's reliance on range of prices for first floors "much to be preferred" to rating authority's valuer's reference to assessments on shops on lower floors of complex — Additions for return frontage of 12½% (ground floor) and 5% (first floor) accepted — No deduction warranted for air shaft/fire shutter area — No support in analyses of comparables for addition for position, modernity and prominence of display — VO's valuation of GV £98,250 preferred and appeal dismissed

Richard Hone (instructed by the solicitor of Westminster City Council) appeared for the appellant rating authority; J A R Grove (instructed by Lee & Priestley, of Leeds) for the first respondents, the ratepayers; David Mole (instructed by the Solicitor of Inland Revenue) for the second respondent, the valuation officer.

Giving his decision, MR REES said: This is an appeal against the decision of a local valuation court determining the assessment on modern shop premises above Bond Street Underground Station, Oxford Street, W1 [379 Oxford Street (including first floor of 381 Oxford Street)], at gross value £98,250, rateable value £81,847. The appeal was made by the rating authority, who before me contended for a gross value of £154,000: since the originating proposal was for a gross value of £105,000, I am limited to that amount. The first respondent contended for a gross value of £87,000, but since it had not appealed against the decision of the court below I am limited to the gross value determined there. The second respondent sought to uphold the decision of the court below.

Counsel for the appellant called Mr W A Hampsher ARICS. Counsel for the first respondent called Mr R E Bedingfield FRICS. Counsel for the second respondent called Mr P Saul ARICS. Their valuations are summarised below.

On March 25 1982 the second respondent made a proposal to bring the subject premises into assessment for the first time at gross value £105,000. The grounds of that proposal were that the subject hereditament was not then included in the valuation list: a copy of the reasons for the decision of the court below are annexed to this decision as Appendix A.

The grounds of the appeal to this tribunal are that the assessment as determined by the court below is incorrect, insufficient, unfair and bad in law.

From an agreed statement of facts I note that the subject hereditament comprises retail selling space on ground and first floors, part of a modern office and shop complex redeveloped as part of Bond Street Underground Station. Trading commenced in 1981. There are 13 shops on the lower-ground-floor level close to the ticket sales and underground entrances and exits. There are 14 shops at ground-floor level with a main entrance in Oxford Street and a secondary access from Davies Street. The subject hereditament comprises shopping space on the ground floor and the whole of the first floor, which is reached by two escalators from the common parts of the shopping centre with a further three escalators giving access from within the ground-floor part of the subject hereditament. The floor areas of the subject hereditament are as shown below.

Above the first floor are four floors of offices with a separate entrance from Davies Street. Plans of the lower-ground, ground and first floors of the complex were put before me together with analyses of the assessments on the shops on the lower floors of the complex and on a number of other shops in Oxford Street which were referred to in detail by the experts.

Factors (including the presence or absence of loading facilities and escalators, including the latters' positions) which affect their various values were scrutinised closely.

The premises are held on a full repairing and insuring lease for 35 years from June 4 1981. The rent for the first five years is a peppercorn plus £200,000 per annum to be varied up or down quarterly depending on its being less or more than 6¼% of the gross turnover of the shop. Thereafter the amount of rent is to be calculated by reference to somewhat complicated provisions involving a base rent, the index of retail prices, the market value of the premises (as defined), floor areas and turnover. The amounts actually paid for 1982-85 (both inclusive) were given in evidence by Mr Bedingfield. (Although there was some reference to these terms by counsel and the experts — particularly by Mr Bedingfield — no great weight was put on them. Since the assessment has to be found by reference to section 20 of the General Rate Act 1967, I, too, place no great weight on them and do not refer again herein to them.)

The valuations spoken to may be summarised thus:

		Price per m² — £s					
Floor area in m²		*Mr Hampsher*	*Mr Bedingfield*	*Mr Saul*	*Total (Gross value - £s)*		
Ground	All three valuers put the zone A at £300 per m² and "halved back" to give				38,948	38,948	38,948
First							
Sales	1,524.71	60 (A/5)	25 (A/12)	30 (A/10)	91,483	38,118	45,741
Stock	321.0	30 (A/10)	15 (A/20)	20 (A/15)	9,630	4,815	6,420
					140,061	81,881	91,109
Add for return frontage:							
12½% of ground floor					-	4,868	4,868
5% of first floor					-	1,906	2,287
Deduct for air shaft/fire shutter area — 5% of first-floor sales					-	1,906	-
Add for position, modernity and prominence of display - 10%					14,006	-	-
					154,067	86,749	98,264
Gross value, say,					£154,000	£87,000	£98,250

A The value of the ground floor of the subject property having been agreed, it will be noted that there are four matters only in dispute and I consider them below.

Value of first floor

This dominates the other three items and is the crux of this appeal. I heard much about an agreement between a surveyor acting for the Multiple Shops Federation in respect of shops in Oxford Street, the agreement being between him and the valuation officer (the rating authority were not a party to it) with regard to zone A rates in Oxford Street and as to amounts to be attributed to basements, first and second floors. The parties were not *ad idem* as to the interpretation of this agreement, but I do not need to come to any conclusion on that,
B since the analyses of the assessments on retail properties in Oxford Street which were before me show that first floors are assessed between $\frac{A}{8}$ and $\frac{A}{10}$ of the zone A values.

It will be seen from the above summary of the valuations that Mr Saul takes $\frac{A}{10}$; Mr Bedingfield attributes a slightly lower figure. Mr Hampsher, however, took the view that the subject premises are not of the ordinary run of premises in Oxford Street: they have a great deal of display space, several entrances, ample frontages, a loading bay at the rear and there is easy access to the first floor by means of escalators. The fact that there are a number of pillars which interrupt the first-floor accommodation in his view affects the value of that part only to a minor degree. The other two experts considered that the
C pillars were a disadvantage.

In particular, Mr Hampsher was of the opinion that the ratio of the area of the ground floor to the area of the first floor, which is in excess of 1 to 8 ("the ratio"), was not a disability, mainly because of the presence of so many escalators, but in cross-examination he agreed that the highest ratio of the area of the ground floor to that of the first floor of any other shop in Oxford Street was 1 to 3½. He sought support for his £60 ($\frac{A}{5}$) per m² for the first floor in the analyses of the assessments on first-floor banking halls in Oxford Street and in the assessments of the shops on the ground floor and basement of the complex. In cross-examination he agreed that only the first floors of banking halls in Oxford Street were assessed as high as $\frac{A}{5}$. He considered the fact that the complex was over the underground
D station, which connected to two lines, added to the value of the subject premises. As a check he had considered comparison between the subject hereditament and a number of comparables on an overall basis.

Mr Bedingfield's view was that the subject hereditament was not without disadvantages, particularly with regard to access, and he attached great weight to the ratio. Starting with $\frac{A}{10}$ and having regard to the disadvantages and the ratio he came to £25 per m². In his view, Mr Hampshier's reference to the assessments on banks was not of any assistance: no bank would be in the market for the subject premises. Equally he thought that comparison on an overall basis was not appropriate here. He submitted a calculation designed to show what the assessment of the first floor would have been had there
E been no escalators, but I do not find that to be of great assistance. As one of the checks he had applied, he had considered the rent paid under the lease.

Mr Saul said that he had considered the comparable evidence in the light of his professional judgment. The subject premises were unique and difficult to value. There was no direct comparable because of the ratio. He had considered the range of prices which had been applied to first floors of shops in Oxford Street. In his opinion, the assessments on the smaller shops on the ground floor and the lower-ground floor of the complex were of no assistance nor were the assessments on banks, where a hybrid approach had been used. Weighing all the factors he came to £30 per m².

F Ratio the most important

I viewed the subject premises, the other shops on the ground floor and on the lower-ground floor. I saw sufficient of the comparables in other parts of Oxford Street to illustrate the evidence.

It seems to me that the ratio is the most important aspect which has to be kept in mind when the value of the first floor is being considered. In my judgment, it reduces the number of potential tenants. On the whole of the evidence I accept Mr Saul's figures of £30 per m² for the sales area and £20 per m² for the stock. I derive no assistance from either the banking halls or the overall approach: Mr Saul's reliance on the range of prices for first floors is much to be preferred to Mr Hampsher's reference to the assessments on the shops on the lower floors of the complex.

Addition for return frontage

Looking at the analyses of the comparables, it seems to me that the additions of 12½% and 5% made by Mr Bedingfield and Mr Saul are well supported and I accept those figures.

Deduction for shaft/fire shutter area

This is in my judgment such a small disadvantage that no deduction is warranted and I therefore prefer the approaches used by Mr Hampsher and Mr Saul.

Addition for position, modernity and prominence

There is no support in the analyses of the comparables for any addition for these items. In no other case has an end addition been made for any of them. I do not consider that there should be any addition here.

I was referred to the following: *City of Sheffield* v *T B and W Cockayne Ltd and Donmall (VO)* (1958) 4 RRC 258; 172 EG 711; *E Mayes & Son Ltd* v *Harding (VO)* (1958) 4 RRC 285; 173 EG 77; and *Reject Shop Ltd* v *Neal (VO)* (a decision of this tribunal — my own) [1985] 2 EGLR 219; 276 EG 89.

In summary I prefer Mr Saul's valuation to those of either of the other experts. The appeal is dismissed.

I heard submissions as to costs. The appellant will pay the costs of the first and second respondents such costs, unless agreed, to be taxed by the Registrar of the Lands Tribunal on the High Court Scale.

APPENDIX A

Decision of Local Valuation Court

The appeal arises out of a proposal by the valuation officer to assess the shop at £105,000 gross value. The shop was not previously assessed, and at the hearing, on an adjustment of measurement, he sought gross value £99,500. The appeal premises are a shop on ground and first floors of the West One Centre over Bond Street Station, with access from the station through the West One Centre and from Davies Street and from Oxford Street, and with a display frontage to Davies Street and to the Centre. The shop is occupied by the Burton Group Ltd trading as Top Shop and Top Man.

For the valuation officer Mr Perritt adopted £300 psm zone A, about which there was no real dispute between the parties. He adopted $\frac{A}{10}$ (£30 psm) for the first-floor sales area, which is 8¼ times the size of the ground floor, and explained that but for the disproportionate size of the first floor in relation to the ground, he would have adopted $\frac{A}{8}$. He considered the allowance of 20% which this represents in the value of the first floor sufficient to reflect the top-heavy nature of the shop. He described the access to the first floor as unparalleled and explained that it is served by six escalators, three from the Oxford Street entrance and three* from the ground floor of the West One Centre, which has escalator access from Bond Street Tube Station.

Mr Perritt added 12½% to the value of the ground floor to reflect the additional value attributable to the return frontage and to the second entrance. He valued the access area in front of the escalators at the Oxford Street entrance at zone B level and the ancillary offices and store on the first floor at $\frac{A}{15}$ (£20 psm). He introduced a schedule showing the valuations of other shops where the ratio of first-floor sales to ground floor was high, and where he had adopted $\frac{A}{8.6}$ and $\frac{A}{9.5}$ for the first floors according to size and whether there was escalator access. In 527-531, where there is no great difference in size between ground and first floors, with escalator access, he had adopted $\frac{A}{8}$.

Reference was made by the parties to a letter circulated by Mr Hilary Eve of Gerald Eve & Co indicating the general level of zone A prices agreed on shops in Oxford Street together with proportions agreed on upper and lower floors. This showed first floors at $\frac{A}{8}$ and $\frac{A}{10}$ according to quality and access.

For the Burton Group Mr Bedingfield described the appeal shop as unique. He outlined disadvantages of the ground floor which has to serve as an attraction and access to the very much larger first floor. To establish the value of the escalator access he submitted a check valuation transferring to the first floor the value of the area sacrificed on the ground floor in providing the escalators. Because there is no goods lift connection between the first-floor stockroom and the ground-floor sales he said goods have to be carried down by hand. He valued the ancillary space on the first floor at $\frac{A}{20}$ (£15 psm) and because of the disproportionate size of the first-floor sales area he adopted $\frac{A}{12}$ (£25 psm), making reference to the analyses of valuations of a number of other shops in Oxford Street. He sought an allowance of 5% in the first-floor value for obstruction caused by the air shaft and interruption of the sales space because of the provisions of fire shutters. Mr Bedingfield agreed with the valuation officer the addition of 12½% for the return frontage and the additional entrance, and he sought gross value £85,000.

For the rating authority, Mr Hampsher said the appeal shop was in a most desirable location, prominently situated on the corner of Davies Street and Oxford Street, with window space for advertising at first-floor level. He said that the air conditioning and the escalator approach brought the appeal shop outside the usual run of the mill and that the value of the whole is enhanced by its position, modernity and prominence of display.

*The Lands Tribunal in its decision correctly gives the number as two — Ed.

Referring to the decisions of the Lands Tribunal in the cases of *Trevail (VO)* v *C&A* and *Trevail (VO)* v *Marks & Spencer* he said that if an agreement offends common sense or rating principles then it should be ignored. He suggested that the $\frac{A}{8}$ to $\frac{A}{10}$ agreed for first-floor shops in Oxford Street was a low percentage to adopt and was not a true reflection of the value of the ancillary premises to zone A. He referred to improvements in the area since the completion of the Jubilee Line works suggesting that a zone A price of £300 might now be too low, although in his valuation he adopted £300 zone A, with $\frac{A}{5}$ (£60 psm) for the first-floor sales and $\frac{A}{10}$ (£30 psm) for ancillary space. He put no value on the area on the ground floor which forms the access to the escalators and he added 10% overall for position, modernity and prominence of display. He valued the premises at £154,000 and, this being outside the terms of the proposal, he sought the original value proposed by the valuation officer at £105,000.

Before arriving at its decision the court inspected the appeal shop.

The varying approaches to the valuation made for a complicated argument, but the court notes that despite some protest all three parties adopted a zone A price of £300 psm in their valuations. The court puts no value on the access area at ground-floor level in front of the escalators although accepting that it would enhance the value of the first-floor area.

Since the shop is unique it remains very much a matter of opinion as to the relative value of the first-floor sales, but looking for some guidance to the values adopted in other shops in Oxford Street the court considers the $\frac{A}{10}$ (£30) adopted by the valuation officer reasonable to allow for the disproportionate size and to reflect the splendid access, and certainly not excessive in relation to other shops.

Having seen the shop, the court agrees that it is prominently situated and adopts the 12½% addition to the ground floor sought by the valuation officer and by Mr Bedingfield for the value of the return frontage and the second entrance. With regard to the display potential at first-floor level, on inspection this seemed to the court a valuable asset and an addition of 5% is made to the value of the first-floor sales area to reflect this. No allowance is made for the clear area in proximity to the fire shutter since this did not appear to be a disability and makes reasonable and necessary circulation space.

Having made all these adjustments to the valuation, the court directs £98,250 gross value, £81,847 rateable value.

ALFORD DRAINAGE BOARD v MAPLETHORPE (LVC/1/1983)

November 7 1985

(V G WELLINGS QC)

Estates Gazette March 1 and 8 1986

277 EG 867-872 and 987-992

Drainage rates — Assessment to annual value under s67 of Land Drainage Act 1976 of agricultural land and buildings at Bilsby, Lincs — Drainage board's appeal from decision of LVC reducing AV from £600 to £364 — Last revaluation of AV alleged to have taken place in 1935 — Relevant year for purposes of present appeal 1980 — Agreement that 1980 annual rental value of hereditament was £7,700, to which board contend "factor" (more correctly, divisor) of 14 (derived from retail prices index) to be applied to give AV £550 expressed in 1935 terms — Before LVC, ratepayer adopted similar approach (applying "factor" of 16.82, which LVC accepted), but in tribunal different expert witness submitted valuation which adjusted Schedule A income tax assessment for 1962-63 of £179 to take account of such changes in circumstances which would have had meaning to a farmer in that year, namely erection of modern replacement buildings (additional 15%) and improvements in field drainage (5%) to give AV £215 — Tribunal finds advantages and disadvantages in both approaches — Submission for ratepayer that there was presumption that assessment for 1962-63 had regard to changes in circumstances occurring before that time not accepted — AV, in tribunal's judgment, was "fixed" at time 1935-36 assessment was made — Too much guesswork involved in adding piecemeal percentages to take account of particular features — Board's valuer's approach adopted — Tribunal's view that rental index not to be preferred to index based on retail prices — Most satisfactory course to apply factor based on latter — Factor of 14 would be reasonable to use if only purpose was to reflect change in value of money between 1935 and 1980 — But caution ought to be used in selecting factor in order to guard G against possibility that end result might not be equivalent of rent freely negotiated between landlord and tenant in 1935 and that too much of 1980 demand for agricultural land had been reflected — Highest factor figure in range, namely 16.82, applied to 1980 rent of £7,700 to give AV £460

John Grove (instructed by Cameron Markby) appeared for the appellant drainage board; Guy Roots (instructed by Epton & Co, of Lincoln) for the respondent ratepayer.

Giving his decision, MR WELLINGS said: This is an appeal by the Alford Drainage Board from a decision dated November 16 1982 of the Lincolnshire Local Valuation Court with respect to an assessment H to annual value by the board under section 67 of the Land Drainage Act 1976 of agricultural buildings and land at Manor Farm, Bilsby, in Lincolnshire. The appeal hereditament, which lies one-and-a-half miles to the east of the market town of Alford, consists of 175.12 acres of agricultural land, with agricultural buildings thereon, owned by Mr A M Maplethorpe and farmed in partnership by him and his son, Mr J L Maplethorpe.

Under section 64(1) of the Land Drainage Act 1976, a rate made by an internal drainage board in respect of agricultural land or agricultural buildings is to be assessed at a uniform amount per pound throughout the area on the annual value of the land or buildings. The annual value for drainage rate purposes in the case of any land in respect of which an assessment to income tax was made J under Schedule A for the year of assessment 1962-63 is to be its annual value for the purposes of Schedule A as shown in the assessment for that year signed and allowed under section 35 of the Income Tax Act 1952 or, in any case where an appeal was made against the assessment, its annual value for those purposes as finally determined in the proceedings on or in consequence of the appeal (subsection (2)). [The full terms of section 64 and other relevant sections of the Land Drainage Act 1976, namely sections 63, 65-67 and 79, were set out in Appendix A to this decision but are not reproduced here.]

The annual value of the land and buildings of the appeal hereditament for the purposes of Schedule A for the year 1962-63 was K £179. For the year 1963-64, the board varied the assessment of the appeal hereditament for drainage rate purposes to £182. For the year 1965-66 the assessment for drainage rate purposes became £181, at which figure the assessment stood until February 24 1982.

In addition to the appeal hereditament, the partnership also occupies two other agricultural hereditaments in the immediate vicinity: (a) 97.88 acres of land owned by Mrs R Maplethorpe; (b) 78.56 acres of land owned by Mr J L Maplethorpe.

The total area of land comprised in the three hereditaments is 351.56 acres. Further, the partnership also farms 20 acres as tenant of other landlords in other areas (the grand acreage being 371.56).

On November 18 1980, the board resolved to review the annual values of agricultural hereditaments in the board's district. On L February 24 1982, the board served on the respondents, as occupiers, three notices of determination of annual value under section 67 of the Act of 1976. The annual values so determined were:

(1) Of the appeal hereditament, £600;
(2) Of the 97.88 acres owned by Mrs R Maplethorpe, £230;
(3) Of the 78.56 acres owned by Mr J L Maplethorpe, £180.

The respondents duly objected to the determination in respect of the appeal hereditament and on November 16 1982 the local valuation court heard the resulting appeal, allowed it and determined that the annual value of the appeal hereditament should be reduced to £364. The board appeal from that decision to the Lands Tribunal. The respondents made no objections to either of the other two M determinations.

Section 67(1) of the Act provides that:

> If a drainage board are of the opinion that the amount of the annual value of any relevant land in their district should, for the purpose of securing that the burden of the drainage rates payable in respect of all land in their district is fairly distributed so far as reasonably practicable among the persons liable to pay those rates, be increased or reduced having regard to changes in the circumstances by reference to which the annual value of the land in question or of any other relevant land in the district was fixed, the board may make a determination of annual value under this section.

Subsection (2) enables the owner or occupier of any relevant land to request the drainage board in writing to make a determination under these sections in respect of his land. Subsection (3) provides that:

A determination of annual value under this section shall be a determination specifying as the annual value of the land in question such greater or smaller amount than the amount of the annual value as the board consider just for the purpose mentioned in subsection (1) above, having regard to the changes there referred to and to any other alterations of annual values under this section made or proposed by the board.

Agreement as to rental values

The agreed statement of facts contains the following agreement:

It is agreed, for the purpose of these proceedings only, and on the assumption that the landlord is responsible for all repairs and insurance, that:
(a) In 1980 the annual rental value of the hereditament was £7,700.
(b) The average annual rental value of agricultural land in the Alford Drainage District in the spring of 1935 was 19s per acre.
(c) The average annual rental value of agricultural land in the Alford Drainage District in the spring of 1980 was £35 per acre.

It was also, I understand, agreed between the parties at the hearing that the year 1980 was a relevant year for the purposes of the present appeal because it was the year in which the board resolved to revalue agricultural hereditaments in its district.

Summary of case presented to tribunal by board

The case put forward on behalf of the board was that:
(1) Under section 67 the board has power to determine a new annual value rather than merely adjust the previous Schedule A annual value;
(2) For the purpose of securing that the burden of the drainage rates payable in respect of all land in their district is fairly distributed so far as reasonably practicable among the persons liable to pay those rates, as mentioned in section 67(1), the best method of valuation is to assess the rental value of the appeal hereditament as at 1980 in the circumstances existing as at the date of assessment and, then, to adjust that rental value to what would have been payable by way of rent in 1935 (the last year, the board alleged, in which a revaluation of annual value took place) if the same circumstances had existed then.
(3) This method had the merit that it comprehends all the changes in circumstances which have occurred since the annual value was last fixed, without the need to identify them.
(4) The appropriate way of adjusting the 1980 rental value to 1935 values is to apply to the 1980 rental value a "factor" derived from the retail prices indices and the appropriate "factor" is 14 (strictly, the factor unless expressed as a fraction of 1 over 14 is a divisor, but, like the parties during the hearing, I will call it a factor).
(5) If the 1980 agreed rental value of £7,700 be divided by 14 the result is £550, which the board says is the annual value expressed in 1935 terms and the annual value for which they now contend.

Summary of respondents' case at hearing

It is the respondents' thesis that section 67 does not enable the board to determine a new annual value but only to adjust the last previous annual value. Accordingly, the respondents begin with what they call "the Schedule A tone", namely, the Schedule A assessment for 1962-63 and adjust that assessment to take account of certain changes in the circumstances existing in that year which occurred subsequently. Thus they begin with £179 and add to that sum 15% (£27) for new building substituted for previous buildings after 1962-63 and add 5% (£9) for improvements in field drainage of the hereditament. These additions provide a total of £215 adjusted annual value. The respondents' case involves the assumption that the Schedule A assessments for 1962-63 were up-to-date and took into account changes in circumstances which had occurred between 1935 and that year.

Local valuation court's reasons for its decision

It appears that the respondents in the local valuation court put forward a basis of valuation similar to that of the board in the Lands Tribunal (and in the court) but sought to apply a factor of 16.82. In the court the respondents called a different expert witness from the expert called by them in the tribunal. The court's reasons included the following:

(1) The years which are to be taken into account are 1935 and 1980. Whilst there was no real argument about these we think it is as well to state them.
(2) We were in great difficulty in deciding what figure the factor should be because there was no real hard evidence on this, but on the weight of evidence submitted by the appellant, we have decided the factor must be 16.82.

The court added that the land and buildings should be valued at £35 an acre and that "applying the above factor of 16.82 gives an assessment of £364".

Some findings of fact

The following facts were agreed or proved:

Generally

1 The board was constituted in 1937 under the provisions of the Land Drainage Act 1930. Under the Land Drainage Act 1976 the board continued in being as an internal drainage board.

2 The board is responsible for the internal drainage of approximately 55,000 acres of land between the foot of the Lincolnshire Wolds and the sea. The area of land in the board's district which is subject to drainage rates is 45,013 acres. The remaining area includes large areas of land developed for holiday, residential and light industrial purposes.

3 The board has 15 members elected by owners and occupiers of the hereditaments in the board's district. Mr A M Maplethorpe is a member of the board.

4 The respondents bought the freehold of the appeal hereditament, as to most of the land, in 1958 and, as to the remainder thereof, in 1959 and have farmed the land continuously since those respective years.

5 Most of the land is Grade II on the MAFF Land Classification Map, but some of it is Grade III. The soil is predominantly a medium to strong loam but with some strong clay.

6 Between the financial years 1935-36 and 1962-63 there were no valuations of annual value for the purposes of Schedule A income tax and the only changes which did occur in annual value were by reason of boundary changes. HM Inspector of Taxes himself made no amendments to annual value.

7 Prior to 1970 the board made no changes to annual value of agricultural land otherwise than on account of boundary changes.

8 After the enactment of section 30 of the Agriculture (Miscellaneous Provisions) Act 1968, the board made a review and uplifted the lowest annual values but did so on a mathematical basis, no professional valuer being involved.

9 Between 1970 and 1980 no changes to assessments to annual value were made by the board otherwise than on account of boundary changes.

10 The board's decision to review annual values under section 67 was pursuant to the advice given to it in a written report by the clerk of the board.

11 The revaluation was carried out by the district valuer, who received his original instructions in December 1980. The district valuer was given no instructions or advice as to the manner in which the revaluation should be carried out. The method of valuation was left entirely to his discretion.

12 The number of agricultural hereditaments existing at the time of revaluation was 1,186 whereas immediately prior to that time there had been 2,400 such hereditaments on the old Schedule A list. Some 25 units (such as broiler houses exempted from general rates by the Rating Act 1971) of the 1,186 units were valued under section 64 of the 1976 Act because they were being assessed for the first time. They were valued *rebus sic stantibus* in terms of 1935 values; they consisted mostly of buildings. The remainder of the 1,186 units were valued under section 67.

The number of determinations under section 67 of the 1976 Act served by the board on owners and occupiers was 922. Of those there were objections with respect to 311 hereditaments. 29 objections were withdrawn and 282 objections remain outstanding. Accordingly, 640 assessments were accepted.

Facts as to changes in circumstances between 1935 and 1982

13 At all material times there have been some farm buildings on the appeal hereditament. In the years hereinafter mentioned the following buildings were demolished:
(a) In 1966, a range of old loose boxes, chaff house and crewyard;
(b) In 1968, a shed with loose boxes;
(c) In 1976 a barn with loft and stabling.

14 As at the date (February 24 1982) of the board's determination in respect of the appeal hereditament, there existed on the appeal hereditament the following buildings and other things (the year

of construction in each case where known being stated in parenthesis):

(a) A block consisting of fertiliser store (1976), two corn stores (1966), potato/corn store (1966) and fan house;

(b) An implement store (about 1915);

(c) Corn store with three grain bins and with steel frame, asbestos cladding and roof, fully insulated, and concrete floor (1976);

(d) Fully insulated potato chitting house, with steel frame asbestos cladding and roof, concrete floor and electronically controlled environment (1976);

(e) Combine and implement store with steel frame asbestos cladding and roof and concrete floor (1968);

(f) Range of old buildings (various) with 10 open-front bays, timber frame, timber boarded and corrugated iron roof;

(g) A hardcore road (built between 1960 and 1978) with rubble from the demolished farm buildings runs through the main block of land.

The buildings which existed as at the date of the determination are situated in one farmyard adjoining the farmhouse of Manor Farm. Part of the yard and hardstanding is concreted.

15 An electricity supply suitable for lighting only existed at the farmhouse in 1955 and possibly earlier. A three-phase supply of electricity suitable for the purposes to which the modern buildings are put was provided in 1961 and those buildings have had such a supply since their construction.

16 Those buildings have had a main supply of water connected to them since their construction. A mains supply of water to the appeal hereditament was first provided after 1936.

17 Between 1958 and 1982 a system of field drainage of the appeal hereditament which had existed in rudimentary form for some 60 years prior to 1958 was reconditioned, improved and extended by the respondents.

18 Between 1958 and 1982 the respondents also filled in a number of ditches on the hereditament and grubbed up some of its hedges, thus improving the efficiency of the farming of the hereditament.

19 Between 1935 and 1982 there was continuing improvement in land drainage in the board's district. More particularly, a number of pumping stations were erected, in 1945, 1946, 1956 and 1966. Between 1937 and 1982 six large pumping stations were built. The effect of the operation of the pumping stations was to keep the outfall of the Wold Grift drain, which adjoins the hereditament land on its north-western side and to which the internal drainage system of the hereditament drains, clear, all the water in the drain being pumped out to sea. In 1970 a new concrete outfall to the Wold Grift drain was constructed. Between 1935 and 1982 the board's land drainage operations have become mechanised and more efficient.

20 Main sewerage came to Bilsby in the years 1962 and 1963. The watercourses of the district benefited from the construction of the sewers by the removal of pollution from the water.

21 Between 1935 and 1982, and more particularly after 1953, the defences against the sea on the east coast of England in the vicinity of the board's district were constantly being improved and at great capital expense. But for these sea defences the appeal hereditament would have been liable to flooding by the sea in the event that tides of a height corresponding to those of 1953 and 1978 should occur.

22 But for the improvement in field drainage of the appeal hereditament and drainage generally of the board's district, that part of the appeal hereditament which is adjacent to the Wold Grift drain would be flooded from time to time and a larger area of the appeal hereditament would from time to time be wet by percolation. But for these improvements the appeal hereditament could not have been used, as it is now used, as arable land, but it would have remained, as it was in 1935, mainly pasture land. By 1958 the hereditament had become mainly arable but with some cattle. The respondents have never kept any cattle, and in their occupation the hereditament has been arable apart from 1 acre of grass which is let for summer grazing.

23 Between 1935 and 1982 there were in the board's district improvements in farming equipment, methods and patterns of farming, farming machinery, increased yields, improvements in crop husbandry, pesticides, fertilisers and veterinary medicines, G
and improved road and transport facilities although railways virtually ceased to operate.

24 In the same period there were improved communications, improved security of tenure for tenant farmers, improved markets for farmers although some local markets closed down.

25 In the same period financial support for farmers was much improved.

26 In the same period there was a decrease in the number of holdings in favour of larger holdings.

27 During the same period farmers received assistance from the availability of guaranteed farm prices and minimum prices for crops produced.

28 Over the same period there was a great increase in demand for H
agricultural land. In 1935 such demand was virtually nil.

Evidence

Mr John Grove, for the board, called Mr James Richard Elkington ACIS, clerk of the board; Mr Kenneth Stacey ARVA, deputy district valuer and valuation officer in the Boston Area Valuation Office, who has had 37 years of valuation experience; Mr Harvey Richard Cole BA, an economic and development consultant; and Mr Thomas Bryan Balderston FRICS FSVA CAAV, senior partner of Thomas Balderston & Co, chartered surveyors, of Boston, Lincolnshire.

Mr Guy Roots, on behalf of the respondents, called Mr John J
Richard Trustram Eve MSc FRICS, senior partner of Messrs J R Eve, of London and elsewhere, who has had 25 years' experience of valuation including rating valuation and agricultural valuations.

In summarising the evidence of the witnesses I will not repeat such parts of their evidence from which I have derived my findings of facts. As to documentary evidence, I must mention that I was referred to extracts from the Jurors' Verdict of the Alford Court of Sewers given in 1926, which verdict shows that land in Bilsby, including the appeal hereditament or parts thereof, was below the level of the high-water mark and liable to damage from the sea if the sea defences were not forthwith rebuilt, heightened and strengthened and was liable to pay a rate to that end.

Mr Elkington has lived in east Lincolnshire since 1945 and since K
1948 has lived in Alford; he owns a small area of agricultural land about 45 miles from the appeal hereditament. He was appointed treasurer of the board in 1964 and in 1970 became clerk and treasurer. He said that after the board had carried out its review of 1970 and 1971 it appreciated that there were anomalies in the assessments of agricultural land. By 1980 the anomalies were being criticised by numbers of ratepayers. The main anomaly was that new farms which were highly improved with modern buildings were paying the same level of drainage rates as the older unimproved farms. The board was aware of this and of changes in circumstances which had occurred since 1935. Since then there had been a dramatic change in demand for agricultural land. Land which sold for £35 per acre in 1958 now let
at £100 per acre. Without the drainage works which had taken place L
in the area it would be marsh grazing land; it would have been let for a few pounds per annum for summer grazing. Mr Elkington added that the effect of the improvements in drainage of the area had been that the whole of it had increased in prosperity. The risk of damage from the sea had declined as a result of improvement in sea defences carried out over the years since 1926.

Mr Stacey's valuation is set out at Appendix B to this decision. Other appendices (see p 218) to it are:

(a) Appendix C: an analysis by Mr Stacey of original drainage rate assessments and also of the new list determinations;

(b) Appendix D: Mr Eve's valuation;

(c) Appendix E: Mr Stacey's alternative valuation using the same M
method as that used by Mr Eve.

Unlike Mr Eve, Mr Stacey said it was impossible to start with existing Schedule A values and add to them piecemeal on account of the various changes in circumstances which had occurred since the values were fixed. To do so would only perpetuate any anomalies already existing. He did not know precisely what the circumstances in 1935-36 were. He had taken March 31 1980 as the base date because it was the end of the last complete year before his review took place. All the hereditaments had been valued in the condition in which they had been found on inspection in 1981. They had been valued *rebus sic stantibus* applying 1980 values.

His method was to value as at March 31 1980 on the assumption

A that the rental value ascertained as at that date would reflect all the changes in circumstances which had occurred since 1935, that is to say, he said, from the era of the horse and cart to that of the combine harvester, but without the need to identify the specific changes which had taken place. He then sought a means of adjusting the 1980 rental value so as to convert it to a level appropriate to 1935. The method of adjustment adopted was to divide the 1980 rental by a factor of 14. He had selected that factor after considering the sources mentioned in notes 1 and 2 on his valuation (Appendix B to this decision).

He had heard the factor of 12.22 mentioned in note 2 on that valuation given in evidence by the valuer there mentioned in a local valuation court hearing. That evidence was that £1 in 1979 = £0.018 in 1935 or a factor of 12.22. Mr Stacey said that in 1980 retail prices B increased by 13.39% compared with 1979, which increased the factor from 12.22 to 13.85. He rounded that figure up to 14. As stated in note 1 on his valuation, Mr Stacey also had regard to the change in the purchasing power of the pound between March 31 1980 and March 31 1935 of 12.553 as supplied by the Bank of England and Barclays Bank. Since selecting the factor of 14 he had learned from the Central Statistical Office that if he had applied its "consumers' expenditure deflator" the appropriate factor would have been 15.86. While he was prepared to consider that figure, he thought that 14 was the right one; it was the middle of the range.

Mr Stacey said that his valuation resulted in the 1980 rental being expressed in terms of the 1935 purchasing power of the pound. That C was not necessarily the same as the rent which a tenant in 1935 would have been willing to pay having regard to the changes in

APPENDIX B
Mr Stacey's Valuation

Annual rental value at March 31 1980 of 175.12 acres of land and buildings occupied by the respondents

175 acres @ £40 per acre	=	£7,000
plus a 10% uplift to £44 per acre because the appeal property contains all the farm buildings serving the total holding of 351 acres.		700
Agreed March 31 1980 rental value		7,700
D Convert to 1935 rental value	÷	14
Annual value		£550

In arriving at the factor to convert 1980 rental values to 1935 values regard was taken of:

1 The change in the purchasing power of the £ between March 31 1980 and March 31 1935 of 12.553 as supplied by the Bank of England and Barclays Bank.
2 The factor of 12.22 adopted by the valuer acting for the Rivers Idle and Ryton Drainage Board to convert 1979 rental values to 1935 values.

APPENDIX C
1 Examples of analysis of original drainage rate assessments

Prices per acre range	*No of assessments in each range* Bilsby	Saleby	Alford	Ingoldmells
E 50p to £1.00	32	12	5	9
£1.01 to £1.50	44	6	7	20
£1.51 to £2.00	23	1	14	11
£2.01 to £2.50	5	1	10	8
£2.51 to £3.00	5	–	6	4
Over £3.00	5	–	12	1
No of assessments	114	20	54	53
Average price per acre	£1.01	£0.74	£1.45	£1.47

Average price per acre over all area is £1.04

Total number of assessments	–	2,413
Total acreage	–	45,686.832
F Total annual value	–	£47,471

2 New list determinations

Total number of assessments	–	1,186
Total acreage	–	45,013.15
Total annual value excluding intensive poultry units	–	£112,383
Average price per acre	=	£2.49

922 determinations served on occupiers and owners in respect of 1,186 assessments.
180 occupiers and owners objected in respect of 311 assessments.
29 objections were subsequently withdrawn, leaving 282 outstanding.
904 assessments accepted.

APPENDIX D
Mr Eve's valuation

1	Annual value of the land now comprising the hereditament for the purposes of the drainage rate assessment in 1965-66, being the last year in which it was reassessed.			
	Assessment no 8/9		£161	
	Assessment no 8/11		11	
	Assessment no 8/37		9	
			£181	
	Total of Schedule A assessment for 1962-63			£179
2	Additions for changes in circumstances by reference to which the 1965-66 assessment was fixed.			
	(1) Demolition of some previous buildings and erection of: (a) Concreting of yard and hardstanding (1965 onwards) (b) Corn and potato stores (1966) (c) Combine and implement stores (1968) (d) Lean-to extension to corn store for fertiliser storage (1976) (e) Potato chitting house and corn store (1976) (f) Hardcore farm road into main block of land (1965-78) Additional rental value @ 15%/16%	£27		£27
	(2) Renewal and extension of field drainage and filling in of ditches on ⅔rds of holding, say, 5%	9		9
			36	36
3	Net annual value for the purposes of section 67, Land Drainage Act 1976		£217	£215

APPENDIX E
Mr Stacey's alternative valuation

The old Schedule A values in the board's records comprise:

	Acres	*Annual value* £	*Per acre*
Reference no 43 01 035	163.36	161	
Reference no 43 01 038	5.94	11	
Reference no 43 01 041	5.29	9	
	174.59	181	£1.03
Add for:			
1. General changes in circumstances in agriculture which have taken place between 1935 and 1980 which have influenced rental values, eg improvements in public services such as water and electricity, road systems and transport facilities and major land drainage schemes carried out by drainage boards. The availability of modern farm machinery and fertilisers. Post-war legislation for agricultural holdings designed to produce stability and efficiency in the industry, providing security of tenure, guaranteed market and farm prices and the availability of various improvements grants. The post-war demand for farm tenancies in relation to the depressed 1930 era. Say 175 acres @ £1.00 per acre		175	
2. Field drainage on Manor Farm. Say 175 acres @ £0.50 per acre		87	
		443	£2.53*
3. The value of the farm buildings described at ASF para 4.4 plan ASF D. Say 175 acres @ £0.50 per acre		87	
		530	
Plus 10% uplift as agreed in 1980 rental valuation for the farm buildings serving 351 acres		53	
		583	say £580

*NB	The 97.88 acres of land without buildings owned by Mrs Maplethorpe assessed at £230, to which no objection was made, devalues at	£2.35 pa
	The 78.56 acres of land without buildings owned by the second respondent, J L Maplethorpe, assessed at £180, to which no objection was made, devalues at	£2.29 pa

A circumstances but that was what he was trying to achieve; he could not guarantee that he had achieved it. In his opinion, his valuation produced something near the appropriate rent for 1935. The average rent of £2.53 per acre (see Appendix E to this decision) would appear to be a reasonable rent in 1935 taking account of all the circumstances. There was, however, no way of judging the matter; it was a circular argument.

Mr Stacey agreed that the Schedule A assessments derived from rents which were agreed at a time of poor demand. He thought nevertheless that it was right that the increases in demand which came with the changes in circumstances should themselves be reflected in the adjusted rent for 1935; he was not departing from realism but was B adjusting for an increase in the value of the pound. As to the modern buildings, he thought that the 1935 tenant would be able to market the product for which the buildings were useful in 1980. His approach was not to be specific about the changes in circumstances since 1935 but to take the totality of them into account.

Mr Cole said that an attempt to assess a factor by reference to which the change in the general level of prices between March 1935 and March 1980 could be arrived at was not an exercise which could be carried out with a high degree of precision. There was no single retail price-index which covered the whole of the 45 years. The present retail prices index began in 1962, but a less sophisticated version of it, with different weighting, had in effect existed since 1947. Having regard to all the available evidence, it was his opinion C that the range for the appropriate factor lay between 12 and 14.5. Accordingly, Mr Stacey's choice of 14 was very reasonable. He was less happy with the consumers' expenditure deflator of 15.86. In his view, the CED was not appropriate for use over a long period and he had never known it used over a long period.

Mr Cole said that if one had regard to the average rents in 1935 and 1980 (19s and £35) there was an increase of 37 times. If 37 were divided by 14, the measurement of the extent to which rents had risen in the meantime by reason of changes in the value of money generally on the one hand and by reason of the particular circumstances affecting rents in the area on the other hand would be obtained = 2.64, which represented the ratio in which £35 minus 19s should be allocated to changes in the value of money and changes in D circumstances. The contribution of changes in the value of money simply could be ascertained by multiplying 19s by 14 = £13.30; but, Mr Cole said, the rent has become £35, an increase of £21.70, which increase was attributable to rental levels in the area which people were prepared to pay, that is to say, over 60% of the increase was attributable to matters other than changes in the value of money; accordingly over 60% was attributable to changes in farming methods, improvements in amenities and increases in demand etc. In Mr Cole's opinion, it was virtually impossible to identify the contribution to the increase of any particular item or even a group of items; the contribution had to be ascertained in broad terms.

If 19s were multiplied by 37, £35 would be obtained; if the agreed E rental of £7,700 were divided by 37, £208.10 would be obtained, the result in either case being very close to Mr Eve's conclusion of £217 or £215. However, Mr Cole said, it was impossible to accept that there had been virtually no changes other than changes in the value of money. There must have been some change on account of, for example, the potato store. A farmer in 1935 would not know what to do with it, but the 1980 farmer did; he was prepared to pay more for that facility than the mere change in the value of money justified. One had to apply a broad test which was a fair test. He thought that the factor of 16.82 applied by the local valuation court was off the edge of the picture.

When it was drawn to his attention that that factor originated in a letter from the Midland Bank Ltd dated November 8 1982 to the F expert witness who had been employed by the respondents at the local valuation court hearing (which letter said: "We confirm that the 1980 value of £1 in 1935 would be £16.82. Conversely the value of one pound in 1980 would be 6p in 1935"), Mr Cole said he did not know what weight to attach to it. He was persuaded that the lower part of the possible range of figures was more realistic but he would not entirely rule out the higher figures. His personal preference nevertheless was for the lower figures and for that reason he supported Mr Stacey's factor of 14.

Percentages not reasonable

Mr Balderston agreed with Mr Stacey that it was not a reasonable method to add percentages piecemeal to the 1962-63 Schedule A assessment for various improvements as Mr Eve had done; it was an G extremely difficult exercise and not an exact science in any way. He had been asked to carry out a valuation on such a basis but had not been able to do so. Having said that, there were additions which he would make to Mr Eve's valuation. He thought that an addition of 5% for improved drainage on the appeal hereditament was totally and unreasonably low. It was unreasonable to fix a firm and hard percentage but he thought that an addition in the range of 25-30% would be more accurate. For other matters he would add the following percentages:

(1) For the buildings on the hereditament: 20-25%;
(2) For the improvement in services, communications, roads and transportation: 15-20%; H
(3) For mechanisation of farming: 15-20%;
(4) For improvements in chemical weed control, fertilisers etc: 20-30%;
(5) For improvements in strains of plants and crop yields: 20-30%;
(6) For improvements in financial support (guaranteed prices, markets, increased security of tenure etc): 30-40%.

The range of percentage additions he would make totalled between 145% and 195%, the average being 170%. They were additions for the period from 1935 to 1980.

Mr Balderston said that to add such additions was an unreliable method because the 1935 tenant would not have the capacity to view the 1980 situation; he would not have believed you if you told him it was going to happen. In 1980 it was possible to look at matters in J retrospect; in 1935 it was not possible to look at matters in advance. The only accurate method was to look at things in retrospect and adjust back to 1935. He did not feel competent to say whether a factor based on changes in the value of the pound was the best method of adjustment retrospectively. He would accept the advice of the experts, although he did not think that changes in the purchasing power of the pound were necessarily the same as changes in rent; rents depend on bargaining strength, he said.

He had known the Alford district all his life. He had been a child in 1935 when, because of wetness, the prominent feature of local farming was grassland farming. Improved drainage began to come from 1940 onwards and it revolutionised the whole practice of local farming, enabling a change to arable farming to be made. K

Mr Eve's valuation (Appendix D to this decision) contains alternative starting points: £181, namely, the last assessment by the board in respect of the appeal hereditament (for the year 1965-1966) and £179, namely the last Schedule A assessment (for the year 1962-63). I understand that only the latter starting point is now relied upon and that I may therefore now disregard that part of Mr Eve's valuation which refers to the former.

Mr Eve's thesis, backed up by the formidable weight of Mr Root's advice, was that under section 67 of the 1976 Act it is necessary to begin with annual value as defined for purposes of Schedule A income tax and adjust it in accordance with such changes of circumstances which can be identified as having occurred since the annual value was "fixed". L He began, as he said, with the Schedule A tone and adjusted it to take account of only those changes in circumstances which he believed would have had meaning to a farmer in 1962-63, namely, the erection of the modern buildings and the improvement of drainage on the hereditament. He understood the Schedule A annual value to have been fixed in that year because it was to be assumed that HM Inspector of Taxes would have required the assessment of annual values to be kept up to date and that they would therefore comprehend changes of circumstances which had occurred before that year.

Mixed metaphors

As to Mr Stacey's approach, Mr Eve saw the advantages of M beginning with the 1980 rent, but the world had changed since 1935. If you applied to the 1980 rent a retail prices index, you did not end up with anything which could be described as the Schedule A tone; you ended up with mixed metaphors. Mr Eve had tried to look at matters in 1935 economic terms rather than in 1980 technological terms: 1935 was in an era of depression and there was no demand then for agricultural land. It was difficult to unscramble the 1980 rent and transport the element of demand in it back to 1935. What was needed was a rental index, because the best guide to rents was rents. A rental index almost existed because the average agricultural rents in the area payable in 1935 and 1980 were known: 19 shillings per acre and £35 per acre respectively. The fraction 19 over 35 x 20 resulted in a factor

A of 1 over 36.8. If that factor were applied to £7,700, the result was £209, which was very near his valuation of £215. Such a rental index had the advantage that it left the effect of inflation out of account, but he conceded that, while he had nearly all the information necessary to enable that index to be used, more information was nevertheless still needed.

Mr Eve found Mr Cole's evidence extremely interesting, but it was, he said, totally unrelated to rents and rental values. Agricultural rents were not taken into account in retail prices indices.

Mr Eve said that if it was right to start with 1935, then he would add certain percentages to his valuation:

(1) For three-phase electricity: 1 or 2 or even $2\frac{1}{2}$%;

B (2) For improvements in drainage on the hereditament: $2\frac{1}{2}$% (that is to say he would increase the allowance of 5% already made in his allowance to $7\frac{1}{2}$%, that is to say, by 50%).

It was to be borne in mind that farming in 1935 would have been with horses which could get on the ground when wet, so that drainage was not so critical as with modern machinery. He thought that the percentages which Mr Balderston sought to add to his valuation were grossly excessive.

It was Mr Eve's understanding that, while the board set out to create fairness between one assessment and another, not all hereditaments in the district had been revalued. Thus gas terminal hereditaments were assessed by statutory order; gas pipes by reference to cumulo value; telephone installations were revalued C every five years.

Mr Roots submitted in answer to a question from me that in law it was not possible to take the simple course and treat the agreed 1980 rent of £7,700 for the appeal hereditament as a new annual value and ignore the earlier situation. He said that under the Land Drainage Act 1976 annual value must be derived from Schedule A assessments for the year 1962-63. There were two reasons for this:

(1) Because of the proper construction of section 67;

(2) In order to ensure fair distribution of the burden of payment of drainage rates having regard to the way in which sections 65 and 66 deal with properties which have a rateable value.

Taking the second point first, Mr Roots said that Parliament had D laid down the relativities. The purpose of the relevant fraction defined in section 66 was that the value of each hereditament which had a rateable value should be adjusted in proportion to the ratio which the general level of original Schedule A assessments bore to the general level of rateable values. For this purpose the annual value of any hereditament was expressly to be taken to be its annual value for the purposes of section 64(2). Annual value for this purpose did not include cases where annual value was derived from comparable land under section 64(4). The absence of reference in section 66 to section 64(4) was of paramount importance. The board was not permitted to reassess annual value at modern rates for the purposes of the relevant fraction; this was the effect of section 64(2). If properties were assessed under section 67 at 1980 values they would stand in the list at E those values unadjusted and the agricultural occupiers would in effect pay substantially more than was their due.

As to the first point, Mr Roots said that the language of section 67 presupposes that an annual value already exists which may fall to be increased or reduced. That cast one back to section 64, from which the term was derived. Section 67 must be construed with section 64. Accordingly, where a hereditament has a 1962-63 Schedule A assessment, one is obliged to start with that figure: see section 64(2).

As to the meaning of the word "circumstances" in section 67, Mr Roots said that the word was limited by the context in which it was used. If all the 1980 circumstances were taken into account, the role of the landlord and tenant negotiating in 1935 would be eliminated. If one had to refer to the 1935 tone, that must place a limit on F "circumstances". The relevant circumstances were the circumstances of the hereditament itself, its intrinsic qualities, its advantages and disadvantages. The word did not extend to general market considerations; demand for land at different dates etc which create the market which establishes value. In Mr Roots' submission the relevant circumstances were "fixed" at the time of the Schedule A assessment for 1962-63 and that was the correct starting point. Having regard to the proper construction of the income tax legislation and the scheme of the land drainage legislation, there was a presumption that the Schedule A assessment for 1962-63 had regard to the circumstances prevailing in that year. If that presumption were to be rebutted the onus of proof lay on the board to do so.

Finally, Mr Roots submitted that Mr Eve's valuation was well founded on the statute. Any higher valuation would, as he said, involve mixing one's metaphors. There was no separate and independent way in which the board's approach could be checked. Retail prices indices were extremely imprecise tools. It was impossible to conclude that the index of 16.82 applied by the local valuation court was actually wrong.

Mr Grove submitted that there was no presumption that the Schedule A assessment for 1962-63 took into account the circumstances then existing. It was wrong to say that the annual value of the appeal hereditament was fixed in that year; it may have been apportioned or adjusted mathematically, but it was not fixed then; it was fixed in respect of the year 1935-36. There was, further, no requirement in section 67, Mr Grove said, that the valuation had to begin with the pre-existing annual value. The section permitted the determination of an entirely new annual value. If Mr Eve's approach were applied generally, it would perpetuate the anomalies which section 67 sought to avoid. Mr Eve's valuation was wrong also because he had regard only to the physical changes with respect to the hereditament itself. There was no restriction on the meaning of the word "circumstances" in section 67. The great advantage of Mr Stacey's approach was that, by starting from 1980, all changes in circumstances which had occurred since the annual value was fixed were comprehended without the need specifically to identify them. Identification of them might be difficult or impossible.

As to the appropriate factor to apply in order to get back to 1935 values, on the evidence, the best choice was 14. Mr Eve disapproved the method; he did not put forward an alternative factor.

Decision

This is a difficult case. I accept the submission of Mr Roots that it is not possible to adopt the commonsense course, namely, to treat the agreed 1980 rent of £7,700 as itself annual value. It would seem that legislation to expand the definition of "annual value" in section 62(2)(b) of the 1976 Act would be necessary before that course could be taken. I therefore have no discretion but to make a choice between the valuation approaches adopted by Mr Stacey and Mr Eve.

Both approaches have their advantages and disadvantages. The main advantages of Mr Stacey's method are that it makes unnecessary the identification of the particular changes in circumstances which have occurred since 1935 and that, when applied generally to agricultural hereditaments, it will provide for consistency between one assessment and another. The disadvantages of this method are that:

(1) There is no guarantee, nor indeed any means of checking after applying the appropriate factor, that a rent which the 1935 landlord and tenant would have agreed on in free negotiations has been ascertained;

(2) It transports back to 1935, a time of depression and low demand for agricultural land, some portion of the demand for it in a time of high profitability;

(3) Some of the circumstances existing in 1980 may have had little meaning to a farmer in the horse and cart era as Mr Stacey described it; *per contra*, if improved land drainage, which after 1940, according to Mr Balderston, revolutionised farming in East Lincolnshire, had been available in 1935, it is not difficult to draw the inference that the demand for agricultural land at that time would have increased notwithstanding the depression.

Wrong starting point

Mr Eve's valuation, too, is readily applicable to a large number of hereditaments but, if applied generally, would perpetuate the inequalities which section 67 is designed to remove. Moreover, in my opinion, Mr Eve's valuation has the wrong starting point. I do not accept that there is any requirement in section 67 that the determination under that section shall be a mere adjustment of a pre-existing annual value; on the contrary, the determination under that section is intended to be a new assessment of annual value: see subsections (3) and (5). Moreover, I do not accept that there is any presumption that the assessment for 1962-63 had regard to changes in circumstances which had occurred before that time. There is no evidence that it did so. If the question is one of onus of proof, the onus is on the respondents, because it is they who in effect make the affirmative allegation; they have not discharged the onus. In my judgment, the annual value was "fixed" at the time when the assessment for the year 1935-36 was made.

For these reasons, adjustments to Mr Eve's valuation must be made in order to take account of changes in circumstances which occurred before 1962-63. More particularly, a rather greater allowance for improved drainage of the hereditament must, in my opinion, be made than Mr Eve admits. I doubt whether he has allowed a sufficient percentage for the modern buildings and I do not think it right that only those changes in circumstances which took place on the hereditament itself should be taken into account; improvements in external drainage and sea defences must have benefited the hereditament and increased its value as must some at least of the other changes in external circumstances mentioned in my findings of facts.

Mr Stacey and Mr Balderston did not look kindly on the exercise of adding piecemeal percentages to Mr Eve's valuation to take account of particular features, although they did attempt the exercise. I prefer not to make the attempt. There is, in my opinion, too much guesswork involved in it. It is true that the ascertainment of the appropriate factor to be applied as part of Mr Staccy's method also involves a degree of guesswork, but it is guesswork within a reasonably defined range and with respect to one feature only of the valuation. I prefer, therefore, to adopt Mr Stacey's approach and so decide. It does not appear to me to be a wrong approach, novel though it may be.

I do not think that a rental index is to be preferred to one based on retail prices. Mr Eve did not recommend a rental index. He said that further information was needed before it could be used. Moreover, a rental index would seem to have no more validity than a retail prices index because it would be based on average rental values whereas the appeal hereditament is not an average farm; the agreed rental value of it in 1980 amounted to £9 per acre more than the average.

On the evidence, the most satisfactory course, in my opinion, is to apply a factor based on retail prices. I think that a factor of 14 would be a reasonable factor to use if the only purpose were to reflect the change in the value of money between 1935 and 1980. Some caution, nevertheless, ought, in my judgment, to be used in selecting the factor in order to guard against the possibility that the end result might not be the equivalent of a rent freely negotiated between landlord and tenant in 1935 and that too much of the 1980 demand for the agricultural land had been reflected in it. For these reasons I shall apply as a factor the highest figure in the range, namely, 16.82, which itself has a respectable pedigree, having been put forward by the Midland Bank and, I believe, by the respondents in the local valuation court and having been applied by that court. The court applied it to £35 per acre. I shall apply it, of course, to the 1980 rent agreed before me, namely £7,700 (£44 per acre) and the result is £457.79, which sum I shall round up to £460.

I therefore determine the annual value of the appeal hereditament pursuant to section 67 of the Land Drainage Act 1976 in the sum of £460 and the appeal is allowed to that extent.

The respondents will pay the costs incurred by the appellant in the Lands Tribunal such costs if not agreed to be taxed by the Registrar of the Lands Tribunal on the High Court Scale.

LONRHO TEXTILES LTD T/A BRENTFORDS v TRIPPIER (VO)

(LVC/1042/1984)

February 6 1986

(J H EMLYN JONES CBE FRICS)

Estates Gazette April 12 1986

278 EG 165-168

Shop premises on ground floor and basement in Piccadilly Plaza shopping centre, Manchester — Largest retail unit in this centre, but basement largely unused and suitable only for storage — Rear of ground floor of more than double width, with part partitioned off to form storeroom — 1973 valuation list assessment of GV £16,750 reduced by agreement in 1977 to GV £15,075, reflecting effect of then recently completed phase one of Arndale Centre — Following ratepayers' proposal in 1982 LVC further reduced GV to £13,650, the VO's figure — Ratepayers' valuer contends for GV £9,000 based on zone A of £47.50 psm, deriving support from relativities between gross values and rents negotiated since 1977 in shops in the area, adopting lower values than zone C for large rear parts and allowing added value for return frontages — VO applies zone A of £57 psm as agreed for two immediately adjoining premises without end allowance — Member's comment that comparing gross values with later rents does not commend itself to him as basis for valuation although useful as a check — Ratepayers' valuer's approach of lower values for rear parts preferred by tribunal because total floor area far too great to support sales space throughout whole of ground floor — Tribunal also prefers to see added value existing in return frontages expressed in valuation — Nevertheless, end allowance large enough to take account of inherent disabilities in layout — Finding that zone A of £57 psm not too high — Opening of Arndale Centre had not affected relative values of appeal and adjoining premises within terms of s20 of 1967 Act, but it had affected relative values of different parts of appeal premises — Basement together with extensive rear space of ground floor has lower value in relation to whole than before advent of Arndale Centre — Tribunal's opinion that basement now quite unsuitable for retail sales and small downward adjustments in value made — Ratepayers' valuer's end allowance of 33.33% preferred to VO's 31.5% — GV reduced to £13,000, representing in round terms two-thirds of rent unadjusted for repairs and insurance, which seemed to tribunal not inconsistent with general pattern in Piccadilly Plaza

Nicholas Burton (instructed by Sharpe Pritchard & Co) appeared for the appellant ratepayers; the respondent valuation officer in person.

Giving his decision, MR EMLYN JONES said: This is an appeal by Lonrho Textiles Ltd, formerly Brentford Nylons Ltd and now trading as Brentfords, who are the occupiers of a shop at 45 Piccadilly Plaza, Manchester.

Piccadilly Plaza is a major commercial development built in the 1960s on an island site extending to some 2½ acres. It is situated to the south of Piccadilly Gardens, which is an important focal point in the city centre, lying to the south of Piccadilly. The development is bounded on the north by Parker Street, which adjoins Piccadilly Gardens, on the south by York Street, on the east by Portland Street and on the west by Mosley Street. The development comprises approximately 30 shop units on the ground-floor level with frontages to Mosley Street, Parker Street and Portland Street or, in some cases, frontages to arcades within the development. There is a further 27 shop units on first-floor piazza level, served by separate escalator access from the arcades. Some of the piazza units are occupied together with a ground floor and are intercommunicating. Some of the ground-floor units also have basement accommodation. In addition to the shop units, the development contains two large office blocks, a five-star hotel, a petrol filling station and both public and private car parking. The appeal property is situated at the extreme north-eastern corner of the development, with an entrance from Parker Street and a long return frontage to Portland Street.

The history of the appeal is as follows. The subject property was included in the 1973 valuation list as a shop and premises with a rating assessment of gross value £16,750, rateable value £13,930. At that time Manchester's principal shopping street was Market Street, which formed a westward continuation of Piccadilly. The northern side of Market Street, however, was in course of redevelopment and in due course there emerged on this site the Arndale Shopping Centre, said to be the largest covered shopping centre in Europe, containing over 200 retail units and a number of major walk-round stores. The development took place in three phases, phase one being completed in 1976, phase two in 1978 and phase three in September 1979.

Following a proposal made on behalf of Brentford Nylons (1976) Ltd, dated September 13 1977, the rating assessment was reduced by agreement to gross value £15,075, rateable value £12,534. It is common ground that this reduction reflected the effect at that time of phase one of the Arndale development on the value of the appeal hereditament. Other shops in the Piccadilly Plaza development were similarly reduced in value at or about the same time.

On March 24 1982, a further proposal was made on behalf of the appellant. Following an objection by the valuation officer for the Manchester area the matter was referred to the Greater Manchester (South) Local Valuation Court. Following a hearing on July 3 1984, the court, by a decision on October 8 1984, further reduced the assessment to gross value £13,650, rateable value £11,347. These figures were put forward by the valuation officer at the hearing. The appellant, being dissatisfied with this decision, has appealed to the Lands Tribunal, and expert evidence called on behalf of the company seeks to support an assessment of gross value £9,000, rateable value £7,472. It is common ground that the value of the appeal property has been reduced by the coming into existence of the Arndale Centre, but in the opinion of the respondent valuation officer the figures which he put forward and which were accepted by the local valuation court adequately reflect that reduction.

The premises

The appeal hereditament is the largest retail unit in the Piccadilly Plaza. The main construction is of reinforced concrete and the display windows are of aluminium-framed plate glass, standing on three to five courses of brick-work. There is a concrete canopy extending approximately 4.25 m over the pavements along both the Parker and Portland Street frontages. The frontage to Parker Street is 9.14 m (30 ft) including a splay 2.9 m wide by 2.95 m deep on the corner. The frontage to Portland Street, which runs the full depth of the premises, is slightly bowed in plan and is 51.82 m (170 ft) long including a splay of 2.95 m. Both these frontages are fitted with shop windows along their full length. There is also a display frontage of 16.33 m (53 ft 6 in) to an arcade within the development, known as the Portland Arcade, which runs parallel to the Portland Street frontage. There are four double entrance doors along Parker Street and the splayed corner and two double entrance doors in the Portland Arcade, which were, however, not in use at the time of my inspection.

There is one staircase at the rear of the shop giving access to the basement and another towards the front of the shop which has been floored-over. At a depth of 17.5 m in the shop there is a drop in floor levels of about 1 ft which is overcome by two parallel sets of three shallow steps with a ramp in between. At that point the width of the shop is more than doubled where the rear area runs behind the adjoining properties, 43 and 44 Piccadilly Plaza. At a point about 16 ft further back a light partition separates the part of the shop which is used as sales space from the rear area which is used as a storeroom. The rear part of the shop itself is obstructed by the basement staircase to which I have already referred and by an area partitioned off containing a substantial duct which forms part of the structure of the building.

The basement is largely unused but is described on the agreed plan as used for stock. The area at the rear is of rather inferior quality and is separated by a solid wall; there is also a small store and staff-room with toilet accommodation. The front part of the basement could well have been used as sales space at one time but, as I have already stated, the staircase at the front of the shop has now been blanked-off and in its present condition the basement is suitable only for storage purposes.

The premises are heated, ventilated and sprinklered and the agreed areas are as follows:

		Area in m²
Ground floor		
Sales zone A		64.75
Sales zone B		72.00
Sales remainder — above ramp	19.4	
beyond ramp	277.9	
Stock	358.9	
Display to Portland Street	33.8	
Display to Portland Arcade	5.0	695.00
		831.75
Basement		
Front	498.0	
Rear	121.5	
Staff	28.0	
Stores	16.0	663.5

The property is held on lease for 21 years from June 14 1966 at an initial rent of £12,500 per annum, exclusive of rates, rising after seven years to £15,000 per annum and subject to further review after 14 years. Under the lease the lessee is responsible for repairing the shop fronts and interior of the premises and pays a further 5% of the rent for insurance of the premises and the maintenance of the service areas and common parts. The landlord is responsible for repairing the exterior, the main fabric and the structural columns. The lease was assigned by the former lessees to Brentford Nylons Ltd on May 17 1971 for the sum of £2,000. The second rent review, which took effect on June 14 1980, was determined by arbitration at £19,378.94.

This rent was determined in accordance with clause 2A of the lease, which provided that to the then yearly rent of £15,000 should be added the lower of the following two amounts:

(i) the yearly proportion in addition to the £15,000 of the yearly average percentage increase in the yearly rents of those ground floor shops in the building which were let prior to 1st January 1966 including only those rents which shall have been reviewed in the 3 years immediately preceding the expiration of the 14th year of the term; and
(ii) such a yearly sum (if any) as shall after adding thereto the said yearly rent of £15,000, equal the rent at which having regard to the terms of this Underlease (other than those relating to the rent) the premises might reasonably be expected to let in the open market by a willing Lessor for the term of 7 years with vacant possession there being disregarded any effect on rent of the fact that the Lessee has been in occupation of the premises, any goodwill attached to the premises by reason of the carrying on thereat of the business of the Lessee, and any effect on rent of any improvement carried out by the Lessee otherwise than in pursuance of an obligation imposed by this Underlease and also disregarding the rental value of the shop front in this Underlease mentioned.

It is agreed that the rent was determined by the arbitrator on the first alternative method, but the appellant does not accept that the rent determined in the second alternative would have been materially higher.

The issue between the parties is one of valuation within the terms of sections 19 and 20 of the General Rate Act 1967. In view of the actual rent being paid at the date of the proposal, it is not suggested that the gross value as determined by the local valuation court was excessive when calculated in accordance with the provisions in section 19 at that date. The valuation arrived at in accordance with section 20 is therefore to be taken as imposing a ceiling and in round terms justifying a valuation on the level of values current in 1972/1973 having regard to the state of the hereditament and the locality at the date of the proposal in March 1982. The parties are agreed that the development of the Arndale Centre is a factor to be taken into account in arriving at the value and the dispute concerns the amount by which the 1973 assessment is to be reduced to give effect to that reduction in value.

I have been presented with a mass of agreed information relating to other retail shop units in the Piccadilly Plaza, in the Arndale Centre and in other adjoining streets. Details of the rents and the rent reviews have been given together with the gross values as featured in the new valuation list in 1973, and the subsequent reductions made in 1977, following the completion of phase one of the Arndale development, and further reductions made after the completion of the Arndale Centre. All the shop units referred to have been valued on the zoning method traditionally adopted by rating surveyors and the pattern of values adopted for the front parts of the shops, zone A, has been set out and marked on street plans of the area.

Valuation evidence

For the appellants, Mr Burton called Mr B N Furniss FRICS, a partner in the firm of Storey Sons & Parker, who has had 34 years' experience as a valuer and has acted for the ratepayer on rating matters for more than 20 years. Mr Furniss considers that the gross value of £13,650 determined by the local valuation court is excessive on two counts. In the first place he considers that the zone A price applied at the eastern end of Piccadilly Plaza is too high and that the value drops materially the further one moves from the Arndale Centre. In the second place, he considers that there are a number of unusual features about the appeal hereditament which justify a further reduction. He values the appeal hereditament at £47.50 per m² for zone A and is of the opinion that the large remainder area of the ground floor should be taken at a lower price than the normal zone C. The rear parts of the shop and the basement no longer have any value as retail sales space and he values them as storage. I set out below the details of his valuation together with that of the respondent valuation officer, Mr J M Trippier ARICS:

	Mr B N Furniss Area m^2	@£ per m^2	Gross value	*Mr J M Trippier* Area m^2	@£ per m^2	Gross value
Ground floor						
Zone A	64.75	47.50	3,075	64.75	57.00	3,691
Zone B	72.00	23.75	1,710	72.00	28.50	2,052
Zone C	19.40	11.87	230			
Remainder (beyond ramp)	277.90	9.50	2,640			
Stock areas	358.90	7.92	2,842	695.00	14.25	9,904
Extended display to Portland St	33.80	9.50	321			
Display to Portland Arcade	5.00	9.50	47			
Basement						
Front storage	498.00	4.00	1,992			
Rear storage	121.50	3.50	425	663.50	6.50	4,313
Staff	44.00	4.00	176			
			£13,458			£19,960
Less for size and disabilities	@ 33.33%		4,486	@ 31.5%		6,287
			£ 8,972			£13,673
Say gross value	£9,000			Say gross value		£13,650
Rateable value	£7,472			Rateable value		£11,347

Mr Furniss supports his valuation by drawing attention to the relativities between gross values and rents negotiated since 1977 on shops within the Piccadilly Plaza, where the factors vary between 1.33 and 2.86. He carries out a similar exercise in respect of the Arndale Centre, where the highest factor is 4.05 in respect of a 1982 rent, and a further 16 shops in adjoining streets of which all, save four, show factors between 1.6 and 2.1. The assessment of £13,650 shows a factor of 1.42 on the 1980 rent and his valuation of £9,000 shows a factor of 2.15.

With regard to the particular factors which affect only the appeal hereditament, Mr Furniss considers that the area to the rear of the ramp has a lower value because much is to the side and obscured from the front in what he describes as a "wrap-round" situation. He also points to the fact that the appeal hereditament is substantially larger than any other shops in the Piccadilly Plaza and is the last shop lying on the corner of Portland Street, which is not a shopping street.

The respondent valuation officer, Mr Trippier, appeared in person and gave evidence. He has had 17 years' experience as a valuer, mainly in north-west England, and for the last five years has been responsible for rating matters in the Manchester city centre. It will be seen from his valuation that Mr Trippier applies an appreciably higher zone A at £57 per m^2 and takes the whole of the rear of the ground floor as zone C at a quarter of the zone A price. He adds nothing specifically for the display frontages on Portland Street and in the Portland Arcade, but suggests that in so far as the "wrap-round" factor has any relevance it is offset by the additional value of the display frontages. He supports his basic zone A value by referring to the general pattern of agreed assessments in Piccadilly Plaza, which show figures of £81 per m^2 at the western end falling by stages to £57 at the eastern end. These figures compare with zone A values in the 1973 valuation list of £90 at the western end, £80 in the centre portion and £70 at the eastern end. He also relied on the pattern of agreement in Mosley Street, where it appears to have been accepted that shops on the western side were not adversely affected by the Arndale Centre whereas those on the eastern side forming part of the Piccadilly Plaza development had been reduced in value, in one case by the local valuation court and in other cases by agreement.

In general, Mr Trippier submits that the revised zone A price of £57 per m^2, having been accepted in respect of the adjoining properties, is the proper figure to apply to the appeal hereditament. They were all taken at the same zone A price of £70 per m^2 in the 1973 valuation list and the relativities of the different shops in the Piccadilly Plaza were well established in the valuation list at that time. The reductions agreed in 1977 were on the basis of a 10% reduction applied generally, but for the purposes of assessing the proper reductions in the light of later proposals Mr Trippier had looked afresh at the valuations and had come to the conclusion that no further reduction was justified at the western end.

At the eastern end, where the appeal hereditament lay, Mr Trippier saw no reason to differentiate, in terms of zone A, between no 45 and the adjoining properties, nos 43 and 44, both of which had been agreed on the basis of £57 per m^2. The factors, such as the coming into existence of the Arndale Centre, which were to be taken into account in arriving at values under section 20, provided no reason for drawing a distinction between the three properties. Similarly the relative values applied to different parts of the hereditament on the 1973 valuation list were still valid in his opinion and he saw no reason why there should be any variation now. As to Portland Street, Mr Trippier conceded that this was not a shopping street, but he pointed out that whereas many of the buildings on the opposite side of Portland Street had been vacant at the time when the valuation list was being prepared, they had now been refurbished, there were two first-class hotels and the building directly opposite the appeal hereditament was now occupied as the Greater Manchester County Hall.

Decision

I have inspected the appeal hereditament, looked at the other shops in the Piccadilly Plaza and in the surrounding streets and also spent some time within the Arndale Centre. In considering the evidence of two experienced valuers I have to say first of all that the exercise of comparing gross values with later rents does not commend itself to me as a basis for valuation. It seems to me that there are too many variable factors, as for example the difference between rent reviews, lease renewals and new lettings on the one hand, and different dates and uncertainty about the basis adopted for the rent reviews on the other. Nevertheless, this evidence can be useful in providing a check and I propose therefore to look again at the evidence of Mr Furniss on this point when I have considered the rest of the evidence.

I start with the proposition that the relativities of the different shops in the Piccadilly Plaza were properly reflected in the 1973 valuation list as finally settled. The list was well established at that time and I have heard nothing which persuades me to the contrary. I also accept that the zone A price of £57 per m^2 is now accepted for the adjoining properties, 43 and 44 Piccadilly Plaza. It is therefore for the appellant to satisfy me that the relative values of the appeal hereditament and those of the immediately adjacent shops have been altered by factors properly to be taken into account in a valuation under section 20, such as the opening of the Arndale Centre.

I have to say that I prefer the approach adopted by Mr Furniss in adopting lower values for the rear parts of the shop and the area behind the partition used for storage. I say this not because of the existence of the light partition but because the total floor area of the shop in this situation is far too great to support sales space throughout the whole of the ground floor. I would also prefer to see the added value, which undoubtedly exists in the return frontages, to be expressed in the valuation as Mr Furniss has done.

Having said that, however, I must say that the end allowance of 33.33% as suggested by Mr Furniss is large enough to take account of any inherent disabilities in the layout of the appeal premises, including the large areas at the rear of the shop, even after taking into account any added value attributable to the return frontages. In effect, it represents a zone C value of £9.50 per m^2 overall when related to a zone A price of £57, as taken by Mr Trippier. On that basis, I do not consider £57 per m^2 can be said to be too high, having been applied to the adjoining properties 43 and 44 Piccadilly Plaza, and agreed, without any end allowance.

There is one matter, however, where I think the advent of the Arndale Centre has affected the relative values of the different parts of the appeal hereditament. In my opinion, the basement extending to some 663 m^2 or thereabouts, taken together with the extensive floorspace at the rear of the ground floor, has a lower value in relation to the whole than it did in 1973 before the Arndale Centre had arrived. It seems to me that Mr Trippier accepts that the effect of the Arndale Centre today is not constant over the whole of the Piccadilly Plaza development by his expressed opinion that the shops at the western end fronting Parker Street are not entitled to any further reduction than the 10% allowed in 1977. I was not told the date when the front staircase leading from the ground floor to the basement was blanked-off. The evidence of Mr Furniss was that little use of the basement had been made for the last 10 years. That takes us

A back to 1975, but whenever the staircase disappeared, it was an indication of a diminished demand for sales space in the basement and in my opinion the basement as it exists today is quite unsuitable for retail sales.

I therefore accept the approach of Mr Trippier with regard to the valuation of the ground floor, but I consider that his values in the basement are too high. I also agree with Mr Furniss that the rear part of the basement is of inferior quality. In my opinion the correct value for the basement is £6 per m^2 on the main part and £5.50 on the rear portion. With this adjustment the valuation before consideration of an end allowance comes to £19,567. Mr Trippier adopts an end allowance of 31.5%. It was pointed out by Mr Furniss, however, that B at the local valuation court, where the same price per m^2 was applied to rather larger areas, an allowance of 33.33% was adopted. I propose to adopt a similar proportion, which produces a gross value of £13,045, say £13,000.

Finally, I have looked at this figure in relation to the 1980 rent and I find in round terms this gross value represents two-thirds of the rent unadjusted for repairs and insurance. That seems to me to be not inconsistent with the general pattern in Piccadilly Plaza. It certainly represents a higher proportion of the rent than is to be found in a number of units in the Arndale Centre, but I accept the opinion of Mr Trippier that lease rents in a situation of rapidly rising values need to be discounted rather more than those in other areas. In any case the correctness of the Arndale assessments is not a question which is C raised before me in this case.

The appeal is accordingly allowed to the extent which I have indicated, and an order will be issued directing that the assessment in the valuation list be amended to gross value £13,000, rateable value £10,805. I have heard submissions as to costs. The respondent will pay one-half of the costs of the appellant, such costs failing agreement to be taxed by the Registrar of the Lands Tribunal on the High Court Scale.

WILLACRE LTD T/A DAVID LLOYD SLAZENGER RACQUET CLUB v BOND (VO)

D (LVC/838/1984)

March 13 1986

(J H EMLYN JONES CBE FRICS)

Estates Gazette May 10 1986

278 EG 629-632

Purpose-built tennis centre at Hounslow — Valuation under terms of s20 of 1967 Act — Ratepayers' valuer argues for GV £11,500, applying unit prices to the different courts and price psm for ancillary parts and relying in support on assessments E of other similar hereditaments in various parts of country — VO adopts contractor's method, starting with actual cost of construction (agreed at £1,692,404), to produce GV £29,400 — Evidence of disposals of three comparables within 2 to 4 years of erection at figures substantially below construction costs — Valuation on profits basis not feasible because ratepayers had been trading for only few months at relevant date in March 1983 — Tribunal by no means satisfied that at that date the hypothetical landlord and tenant would have been able to rely on centre's financial success — Sports Council grant to be deducted from construction costs, tribunal decides, contrary to VO's view, as hypothetical F tenant could reasonably expect such grant — Commenting on three forms of index used by VO to adjust costs some 10 years back to 1972 and to ratepayers' valuer's criticisms of them, tribunal says that fine distinctions sought to be drawn between them "introduce a refinement which leads to an illusory sense of accuracy" — Same to be said of arguments about appropriate dates to be adopted in the valuation — VO's factor of 0.25 adopted — "Pioneering" allowance of 10% made to reflect uncertainties facing hypothetical tenant, there being rather more than normal expectations of risk and uncertainty in setting up the undertaking — Values applied by ratepayers' valuer to constituent parts do not, tribunal considers, adequately reflect superior quality and location of appeal hereditament — Values properly to be applied, derived from comparables, set out in decision — Point made that VO's attribution of greater value to racquetball courts (which had not proved popular) merely because they are larger in area than squash courts when on evidence they are less valuable illustrated mistakes which are likely to occur by equating cost of construction with value — Tribunal prepares figures adopting each party's approach to arrive at same result, namely GV £25,250, indicating that "two methods of valuation if properly applied must arrive at same answer" — Determination accordingly

Gerald Moriarty QC and Guy Roots (instructed by Ashbys) appeared for the appellant ratepayers; Solicitor of Inland Revenue for the respondent valuation officer.

Giving his decision, MR EMLYN JONES said: This appeal arises out of a proposal made by the respondent valuation officer dated March 18 1983 for the alteration of the valuation list for the London Borough of Hounslow by the inclusion therein of a hereditament described as Tennis and Sports Club and Premises at Southall Lane in the borough. The assessment proposed was gross value £37,000, rateable value £30,805. Following objection by the occupiers, a company known as Willacre Ltd, trading as the David Lloyd Slazenger Racquet Club, the matter was referred to the Greater London (South Western) Local Valuation Court on June 4 1984, when the assessment proposed by the valuation officer was confirmed. The occupiers now appeal to the Lands Tribunal.

The appeal hereditament is located on the east side of Southall Lane at its junction with Spitfire Way. It is bounded on the south by an industrial estate, and the "Airlinks" golf course lies to the north and east. The site is close to junction 3 of the M4 motorway.

At the date of the proposal the appeal hereditament comprised a purpose-built tennis centre completed in October 1982. It consisted of 12 indoor tennis courts laid out in three halls, four squash courts with glass backs, two racquetball courts, a small swimming pool, changing rooms with jacuzzis and saunas, bar, restaurant, gymnasium, office, shop, and crèche. Outside are nine all-weather tennis courts and three small-size junior courts. There are parking spaces for 102 vehicles.

The three tennis halls consist of laminated arched timber framing mounted on reinforced concrete columns approximately 3 m high. The timber arches are covered in continuous timber boarding insulated and covered with roofing felt and forming barrel vaults. The dwarf walls below the roofing area are of cavity stock-brick construction. The floors are of tarmac laid on a graded limestone base and treated with two proprietary types of carpet surfacing. At the date of the proposal two of the halls had "Actionweave" carpet surfacing, which is agreed to form part of the hereditament and is accordingly rateable. The third hall was surfaced with "Supreme" matting, which it is agreed is non-rateable.

The ancillary accommodation is contained in a single-storey building connected to and open to the tennis halls. It is of cavity stock-brick and concrete framed construction under a felt roof with a concrete raft floor. The small swimming pool is contained in a cedar-wood cabin constructed at the end of the ancillary block. Since the date of the proposal a fourth hall has been added with three tennis courts and badminton facilities and an extension to the bar. The courts set aside for racquetball, which I am told is a ball game which has arrived from America and is to be distinguished from the ancient game of racquets, have not proved popular and these courts are currently being adapted for other purposes.

The whole centre is run as a proprietary club mainly concerned with the provision of high-quality indoor tennis with a number of ancillary features and at the present time is open 24 hours a day. It is purpose-built and it is common ground that the hereditament is quite unsuitable for any alternative industrial or warehouse use. The net internal area of all the buildings amounts to rather less than 8,000 m^2, and the site has an area of 3 ha (7.36 acres).

For the appellant company Mr Moriarty called Mr Andrew Eric Gibson ARICS FRVA, a partner in the firm of Collier & Madge. In Mr Gibson's opinion, the assessment determined by the local valuation court was excessive and he put forward a valuation of gross value £11,500, of which figure the biggest single item relates to the 12 tennis courts, each taken at £600 or £1.13 per m^2, producing a value of

£7,200. He values the squash courts and the racquet ball courts at £300 each; outside courts at £40; and the junior courts at £20.

The main part of the ancillary accommodation, including the restaurant and the bar, he values at £2 per m² and makes no specific addition for the car-parking facilities. In support of his valuation and of the method which he has adopted, he relies on the assessments of a number of other similar hereditaments in different parts of the country.

The respondent valuation officer, Mr Patrick Howard Bond BSc ARICS, no longer supports the assessment as confirmed by the local valuation court. He puts to the tribunal an assessment of gross value £29,400, which he arrives at by adopting the contractor's method of valuation. Mr Bond also relies on a number of comparables, but these are referred to primarily in order to support the method of valuation which he has adopted rather than the actual figures. He starts with the actual cost of construction of the hereditament agreed at £1,692,404 and discounts this sum by 75% to reflect the downward indexation of building costs from the fourth quarter of 1981 to the first quarter of 1972. To the resultant figure he adds an agreed amount of £67,000, being the estimated capital value of the land in 1973, and decapitalises the total at 6% to produce a gross value of £29,400.

Matters agreed

It will be seen that there is a substantial difference of opinion as to the correct value to be attributed to the appeal hereditament. Certain matters, however, are agreed. First, it is agreed that the hereditament should properly be valued to gross value in accordance with section 19(2) and as defined in section 19 (6) of the General Rate Act 1967. It is also agreed that an assessment arrived at in accordance with those provisions would be not less than a valuation having regard to the provisions of section 20 of the Act.

What has to be determined, therefore, is the rent at which the hereditament might reasonably be expected to be let from year to year, the tenant paying the rates and the landlord the costs of repairs and insurance, as that value would have been properly entered in the valuation list if the hereditament had been subsisting throughout 1972; and on the further assumption that at that time the hereditament was in the same state and the mode or category of its occupation was the same as at March 18 1983, being the date of the valuation officer's proposal. It is also to be assumed, in the words of section 20(1)(b) of the Act, that:

> The locality in which the hereditament is situated was in the same state, so far as concerns the other premises situated in that locality and the occupation and use of those premises, the transport services and other facilities available in the locality, and other matters affecting the amenities of the locality, as at the time of valuation [that is at March 23 1983].

The provision of leisure centres with an emphasis on indoor tennis is a comparatively recent development, but it is not suggested in this case that in valuing the hereditament in accordance with the terms of section 20 of the Act it is necessary to consider the question of demand for hereditaments of this sort in 1972. From this I deduce that both valuers in putting forward their valuations assume that the demand existing at the date of the proposal is a factor properly to be taken into account in a valuation under section 20. This assumption is not disputed.

Also agreed are the actual costs of construction as related to the items forming the constituent parts of the hereditament. Of the total sum of £1,692,404 it is agreed that an amount towards construction costs of £90,000 was made by the Sports Council. The figure of £67,000 for the site is agreed as being its value cleared of all buildings with public services available for connection as they existed at the date of the proposal and with planning permission for development as a tennis centre as at April 1 1973.

I was supplied with details of the ground lease under which the appellant company holds the property from the London Borough of Hounslow, but since neither valuer relied in any way on the terms of the lease as an indication of the value of the hereditament arrived at in accordance with the statutory terms I need make no further reference to it.

Different valuation methods

It will be seen that the two experienced valuers employ two different methods of valuation. The valuation officer relies on the contractor's basis, starting with the firm evidence of the actual costs of construction. For the appellant, Mr Gibson criticises this approach principally on the grounds that no attempt is made to convert the figures of actual cost into value. He points out that of the comparable indoor tennis centres which are referred to by the two valuers three of them have been developed by companies who have subsequently gone into liquidation, in one case by two operators in succession. The evidence in one case showed that where building costs in 1980 to 1981 amounted to £1,697,838 the property was subsequently sold, or the lease assigned, in May 1985 for £435,000. In another case where the 1982 costs amounted to £340,000 the property was sold in July 1985 for £241,000 including fixtures and fittings, and another centre built in 1982 at a cost of £380,000 was sold in September 1984 for £230,000 including fixtures and fittings.

Mr Gibson arrives at his valuation, as I have indicated, by applying unit prices to the different courts and price per m² for the ancillary parts of the premises. In support of his figures he relies on the assessments of a number of comparables where he has broken down the valuations in the same way. The valuations of the centres at Ipswich, Paddington and Darlington, Bournemouth and the Vanderbilt Indoor Tennis Club in West London show values per court varying between £450 and £650. In the case of the Coventry Tennis Centre, which is a larger centre, the values attributable to each court vary between £1,300 and £1,500, depending on the way in which the assessment is analysed.

It is difficult to adopt Mr Gibson's approach as a primary method of valuation, if only because when a new valuation list is being prepared there is no evidence available upon which such a valuation can be founded. It is nevertheless of assistance to the valuer in a case like this as a check on a valuation produced on the contractor's basis. I have already indicated that no reliance is placed on the actual rent payable under the ground lease. It is also common ground that a valuation on the profits basis is not feasible because at the relevant date the appellant had been trading for a matter of a few months and there was accordingly no accurate information available leading to an estimate of annual value.

It is therefore necessary to look at the approach adopted by the valuation officer and I have to state that I accept to some extent the criticisms advanced by Mr Gibson. All the evidence available today seems to indicate that the operations at the appeal hereditament are well established and that the undertaking will in due course fully justify the capital expenditure which has been incurred. I am by no means satisfied, however, that at the date of the proposal the hypothetical landlord and the hypothetical tenant in negotiating terms for a tenancy of the centre on the statutory basis would have been able to rely on such an outcome. This was a new venture, and there must have been some element of risk as to its financial success, particularly in its initial stages.

I have inspected the appeal hereditament and also the comparable properties to which I was referred at Coventry, Ipswich, Paddington and Darlington and the Vanderbilt. I have also been shown round Queen's Club in Hammersmith and the West Cross Squash Club in Hounslow, where both valuers suggested that the agreed assessment there might be of some assistance in determining the values appropriate for the squash courts which form part of the appeal hereditament. In the light of this evidence, the general impression which I have formed is that, although perhaps not surprisingly there is some inconsistency in the valuation of the comparable properties spread over such a wide area, the appeal premises at Hounslow are the most valuable having regard to their size, their location and the general quality of the facilities provided. More specifically I find that the actual cost of construction as agreed compares very closely with that at the Coventry centre, and this was a cost incurred at approximately the same time.

In considering a valuation on the contractor's basis the first question which arises, following the adoption of the agreed cost of construction as a starting point, is the effect if any to be given to the grant of £90,000 made by the Sports Council. The valuation officer considers that this sum is irrelevant and would form no part of the hypothetical tenancy negotiations. For the appellant, Mr Gibson considered that this was an actual amount received by the occupiers and must properly form part of the value judgment to be made.

I have come to the conclusion that Mr Gibson is right and that the figure of £90,000 should be deducted from the agreed construction costs. The valuation for rating purposes by way of the contractor's method is a theoretical exercise which is deemed to have taken place prior to the negotiations between a hypothetical landlord and a

A hypothetical tenant. In this case the actual occupier has received a grant and I consider it most probable that the hypothetical tenant in his place could reasonably expect such a grant and would have it in mind when considering what rent he could afford to pay. Thus the amount of the grant is, in my opinion, to be taken into account in estimating the deemed actual capital cost.

The next stage is the adjustment of this figure of cost, in order to apply the provisions of section 20 of the Act relating to the date of valuation. The valuation officer looked at three forms of index, which gave factors of 0.23, 0.28 and 0.32 respectively, taken from the fourth quarter of 1981 back to the first quarter of 1972, and had adopted 0.25 as his own factor for the purposes of his valuation. Mr Gibson, although not accepting that the contractor's basis as applied
B by the valuation officer could produce the right answer, nevertheless looked at the question of indexation and pointed to the fact that in an agreed analysis of the Coventry assessment an index factor of 0.235 had been adopted. He produced what he described as an "alternative contractor's test valuation", which arrived at a gross value of £23,700. I am satisfied, however, from his evidence taken as a whole that this was more a mathematical exercise than a valuation, and, as I have already stated, his valuation was in the figure of gross value £11,500.

I have to say that these fine distinctions which are sought to be drawn between a firm price index on the one hand and a tender price index from *Spon's Building and Architect's Price Book* and an index
C produced from the Business Statistics Office related to all new construction output, introduce a refinement which leads to an illusory sense of accuracy. The same is to be said of arguments about the appropriate dates to be adopted at either end of the period of adjustment, for example the date when the costs were to be agreed in the deemed mental exercise leading to agreement on a rental value on April 1 1973; and the actual date to be adopted for the agreed cost of construction, for example the date when the contract was signed or the date when payment was actually made.

As I have said, I do not find it necessary to arrive at any conclusion on the basis of this sort of fine tuning. For present purposes in the instant case I propose to adopt a factor of 0.25. In this way the adjusted cost of construction comes to £400,601. Adding on the
D agreed value of the land I arrive at an adjusted figure of cost of £467,601.

The next question is to consider whether this figure of cost is to be taken as a proper reflection of value. At this stage it is necessary to draw a comparison between the annual value of the hereditament on a year-to-year basis in March 1983 and the annual equivalent of a capital sum notionally adjusted to agree with the provisions of section 20. Lord Denning MR said in *Cardiff City Council* v *Williams (VO)* (1973) 18 RRC1; 226 EG 613 that this difference is one of the factors to be taken into account in deciding the appropriate rate per cent to apply to the adjusted capital cost.

It seems to me in the instant case, however, that there is a further
E matter which has to be reflected. At the date of the valuation officer's proposal there was rather more than the normal expectation of risk and uncertainty in the setting up of this undertaking. The devaluation of the agreed assessment at Coventry showed a deduction of 12½% for what was described as a pioneering allowance. Similar allowances appear to have been made in at least two other cases. With hindsight we now know that the ratepayers' business has prospered, but we also know that in a number of other cases operators have gone into liquidation. In my opinion, some allowance is justified even though 6% is to be applied in order to convert to the annual equivalent on the basis of an annual tenancy. An allowance of 10% to reflect the uncertainties facing the hypothetical tenant would produce a gross value, on the figures
F which I have taken, of £25,250.

It is at this stage that I turn to a more detailed analysis of the assessments of the so-called comparable properties. Queen's Club and Bournemouth are not directly comparable because the indoor courts form a comparatively small part of the whole premises. At Queen's Club some of the buildings are very much older and less convenient and have been added to and extended from time to time. Vanderbilt is adapted from railway premises and the hall is at first-floor level, in a building on stilts; and the Darlington centre is in a converted aircraft hangar. Finally, the Ipswich centre is no longer used for tennis, but has been converted — without structural alteration — to indoor cricket. In the light of this evidence, I consider that the values applied by Mr Gibson do not adequately reflect the superior quality and location of the appeal hereditament.

I set out below the values which in my opinion are properly to be applied to the constituent parts of the hereditament in the light of the pattern of values, such as it is, which I derive from the comparables.

3 tennis halls (6,354 m²)		
8 courts with "actionweave" matting @ £1,350	£10,800	
4 courts with non-rateable carpet surface @ £1,300	5,200	
		£16,000
9 outside courts @ £50 each		450
3 outside junior courts @ £30 each		90
4 squash courts @ £400 each		1,600
2 racquetball courts @ £300 each		600
Indoor swimming pool, say		500
Reception, restaurant, bar, shop, crèche, changing rooms, kitchen 782.4m² @ £5		3,912
Display and games area 100m² @ £4		400
Stores and plant rooms 145.5m² @ £1		145
Swimming pool changing rooms 16.8m² @ £2		34
Store 3.4m² @ £1		3
Car parking		1,520
		£ 25,254

The principal item is the value attributable to the courts, and the courts at Hounslow looked at by themselves are certainly as good as those at the Coventry centre. I do not follow the valuation officer in attributing a greater value to the racquetball courts merely because they are larger in area than the squash courts. On the evidence which I have heard, they were in fact less valuable and this small item clearly illustrates the mistakes which are likely to occur if one equates cost of construction with value.

I have considered it right to add a specific figure in respect of the car parking, which in my opinion adds appreciably to the value of the whole hereditament. The evidence was that the premises were inadequately served by public transport, and access in the main for those visiting the premises was by private car.

At one stage it appeared that I was being invited to determine the correct method of valuation to be adopted in the case. For my part I regard the issue as one of valuation, and the figures which I have now prepared indicate that two methods of valuation if properly applied must arrive at the same answer. It seems to me that adopting Mr Gibson's approach I am rejecting his values on the grounds that they are not supported by the evidence of comparables, having regard particularly to questions of quality, size, and location.

The valuation which I have set out appears to confirm that a "pioneering" allowance of 10% or thereabouts is appropriate and I therefore determine the assessment of the appeal hereditament at gross value £25,250, rateable value £21,013. The appeal is accordingly allowed to that extent.

The respondent valuation officer will pay the costs of the appellant, such costs failing agreement to be taxed by the Registrar of the Lands Tribunal on the High Court Scale.

HAYS BUSINESS SERVICES LTD v RALEY (VO) (LVC/881/1984)

March 26 1986

(J H EMLYN JONES CBE FRICS)

Estates Gazette June 7 1986

278 EG 1101-1104

Thamesside six-storey former tea warehouse at Limehouse used for storage of documents, archives, records, tapes, films and similar small items — Rateability of goods lifts and fire-protection equipment under s21 of and Schedule 3 to the General Rate Act 1967 and the Plant and Machinery (Rating) Order 1960 as amended — Parts of ground and first floors protected by special form of fire-fighting equipment regarded as essential for some archive material — Tribunal rejects VO's submission that although this equipment protects stored material its main function is to protect the building and that the storage cannot properly be described as a trade process within proviso to para 1 of Schedule 3 — The "trade", so far

as material on the point, was storage of sensitive materials and the "process" was provision and maintenance of suitable temperature and environment involving application in emergency of non-toxic and harmless gas (Halon) which had effect of extinguishing fire — Two one-ton electric goods lifts in brick-walled shafts serve each floor — Not mainly or usually used for passengers and not disputed that deemed not to be part of hereditament — Ratepayers' valuer arrived at GV £12,300 on assumption that value of three lower floors substantially reduced without goods lifts (although could be serviced by fork-lift trucks) and there was no value on higher floors — VO's valuation of GV £37,000 assumed lift shafts were part of structure and of hereditament and were available for hypothetical tenant for installation of goods lifts as tenant's equipment — Tribunal upholds VO's view — In hypothetical world no hereditaments fitted with goods lifts, and tenants must make what use of premises they can by installing own equipment — Held the goods lifts clearly plant and deemed to be part of tenant's plant — Indistinguishable in principle from any other items of tenant's plant such as furniture, typewriters and, particularly in present context, fork-lift trucks — Tribunal sees no reason to distinguish between putting value on second floor because tenant can introduce fork-lift truck and putting value on higher floors because tenant can introduce goods lift — Heavy onus on ratepayers to show GV agreed in 1977 by then occupiers no longer appropriate was not discharged, and tribunal not persuaded that values attributed to different parts of hereditament were excessive — Ratepayers' appeal dismissed

Christopher Cochrane (instructed by Clifford-Turner) appeared for the appellants; Alan Sainer (for Solicitor of Inland Revenue) for the respondent valuation officer.

Giving his decision, MR EMLYN JONES said: This is an appeal by Hays Business Services Ltd against a decision of the Greater London (North Eastern) Local Valuation Court given on July 11 1984 confirming the assessment of a hereditament at Buchanans Wharf, Emmett Street, London E14, in their occupation, described as warehouse, wharf, stores, offices and premises at gross value £37,000, rateable value £30,805. The respondent is the valuation officer duly appointed in respect of the Tower Hamlets rating area in which the hereditament is located.

The premises were originally assessed in the new valuation list in 1973 as two separate hereditaments the assessments of which, following proposals made by the then occupiers, were agreed at gross value £9,000 in respect of the ground floor and gross value £28,000 in respect of the upper floors. Following a proposal made by the valuation officer on July 12 1977 these two assessments were combined in a single assessment of gross value £37,000, rateable value £30,805, without objection.

On March 23 1983 a proposal was made on behalf of the present occupiers seeking a reduction in the assessment on the grounds that the assessment was excessive, unfair and incorrect. As I have said, the figures now under appeal were confirmed by the local valuation court.

The hereditament comprises a six-storey former tea warehouse at Buchanans Wharf on the banks of the Thames in Limehouse. The warehouse lies flush with Emmett Street, which is restricted to one-way traffic running from north to south and lies to the south of Commercial Road near its junction with West India Dock Road. It is approximately 3 miles from the centre of the City of London.

The warehouse was built at the beginning of this century of brick walls with a slated and under-boarded roof running on light steel trusses. The original part of the building has substantial timber floors supported on rolled steel stanchions, but following war damage the greater part was rebuilt in 1952 with concrete floors supported on concrete stanchions. There are two one-ton electric goods lifts contained within brick-walled shafts serving each of the six floors, which are divided up internally by structural brick walls to form eight different-sized chambers each fitted with double fire-resistant steel doors. The clear height under the joists of each floor is 8ft 6in with the exception of the top floor which is 7ft 6in clear. The loading and unloading doors on each floor level above ground level are located on the south side of the building and are approached from a 20ft-wide G
access way from Emmett Street. They were used at one time when goods were raised by external hydraulic hoists. The hydraulic power is no longer available, and loading and unloading is now effected by the use of a fork-lift truck at yard level into or out of two doors at first-floor level some 2m above the level of the yard. All main services are connected.

The existing goods lifts were installed in or about 1952, each comprising a steel cage with sliding steel-slatted gates suspended on steel cables from a winding drum driven by a 12-horsepower, 3-phase, 415-volt electric motor with reduction gearing and electronic controls. Each cage slides between two vertical steel rods attached to the brick walls of the shafts. The shafts measure internally approximately 8ft square and extend from the ground H
floor to the lift motor-room at roof level.

The warehouse is used for the storage of important documents, archives, records, data banks, video and audio-magnetic tapes, films, maps, dry core samples of geological drillings and other small items that can be stored in small containers capable of being readily retrieved at any time. Because of the sensitive nature of some of the materials stored on the premises, and at the request of the owners of the materials, parts of the ground, first and second floors are required to be maintained at a temperature of 70° ± 5° fahrenheit with a humidity of 55% ± 5%. Parts of the ground and first floors are also protected by a special form of fire-fighting equipment employing the use of Halon gas. The advantage of this method is J
apparently that it cuts out the flame without causing any damage to the goods stored and is regarded as essential for some of the archive material, particularly film, which is irreplaceable.

The dispute in this case turns principally on the question of how far, if at all, the assessment of the hereditament for rating purposes is to take account of the goods lifts and the fire-fighting equipment. As originally put forward, the case for the valuation officer also submitted that the heating plant and the humidifiers were also to be valued as part of the hereditament, but in the light of evidence called on behalf of the appellants this claim was not proceeded with, with the exception of the central heating in one of the offices, which is accepted to be rateable and is reflected in the value attributable to that part of the premises. K

Statutory provisions

At this stage I think it appropriate that I should set out the statutory provisions which govern the issues raised. Section 21 of the General Rate Act 1967 so far as material provides as follows:

21 — (1) For the purpose of the valuation of any hereditament under section 19 of this Act otherwise than on the profits basis —
(a) subject to any order under subsection (5) of this section, all such plant or machinery in or on the hereditament as belongs to any of the classes set out in the statement for the time being having effect under subsection (4) of this section shall be deemed to be a part of the hereditament;
(b) except as provided in the foregoing paragraph, no account shall be taken L
of the value of any plant or machinery in or on the hereditament.
(2) . . .
(3) From time to time, at such intervals as the Minister may direct, a committee consisting of five persons appointed by the Minister shall transmit to the Minister a statement setting out in detail all the machinery and plant which at the date of the preparation of the statement appears to the committee to fall within any of the classes specified in Schedule 3 to this Act.
(4) The Minister shall cause any statement transmitted to him under subsection (3) of this section to be published in such manner as he thinks fit and, after considering the statement and any representations which may be made to him with respect thereto, may if he thinks fit make an order, to come into operation on such date as may be specified therein, confirming that statement with or without modifications; . . . .

Schedule 3 to the Act, so far as it is material, contains the M
following:

Classes of Machinery and Plant Deemed to be Part of Hereditament

CLASS 1

1. Machinery and plant (together with the shafting, pipes, cables, wires and other appliances and structures accessory thereto) which is used or intended to be used mainly or exclusively in connection with any of the following purposes, that is to say —
(a) the generation, storage, primary transformation or main transmission of power in or on the hereditament; or
(b) the heating, cooling, ventilating, lighting, draining, or supplying of water to the land or buildings of which the hereditament consists, or the protecting of the hereditament from fire:

A Provided that, in the case of machinery or plant which is in or on the hereditament for the purpose of manufacturing operations or trade processes, the fact that it is used in connection with those operations or processes for the purpose of heating, cooling, ventilating, lighting, draining, supplying water, or protecting from fire shall not cause it to be treated as falling within the classes of machinery or plant specified in this Schedule.

CLASS 2

2. Lifts and elevators mainly or usually used for passengers.

By virtue of section 117(3) of the Act, the Plant and Machinery (Rating) Order 1960 as amended has effect as if made under section 21(4). The order contains in the schedule headed "Classes of Machinery and Plant to be Deemed to be Part of the Hereditament" B the following items under Class 1B. Class 2 of the schedule repeats without variation the words used in Class 2 of Schedule 3 to the Act.

CLASS 1B

Machinery and plant specified in Table 1B (together with the appliances and structures accessory thereto specified in paragraph (2) of the List of Accessories) which is used or intended to be used mainly or exclusively in connection with the heating, cooling, ventilating, lighting, draining or supplying of water to the land or buildings of which the hereditaments consists, or the protecting of the hereditament from fire.:

Provided that, in the case of machinery or plant which is in or on the hereditament for the purpose of manufacturing operations or trade processes, the fact that it is used in connection with those operations or processes for the purposes of heating, cooling, ventilating, lighting, draining, C supplying water or protecting from fire shall not cause it to be treated as falling within the classes of machinery and plant specified in this Schedule.

Included in Table 1B under "(h) Protection from Fire" are the following items: "Tanks; pumps; hydrants; sprinkler systems; fire alarm systems; lightning conductors."

Fire-protection equipment

For the valuation officer, Mr Sainer accepts that he has three hurdles to surmount in order to bring the fire-protection equipment into the valuation. First of all, he asks, is it in the table? The answer is yes, and that is not in dispute. Second, is it used mainly or exclusively in connection with the protection of the hereditament from fire? D Third, is it taken out by the proviso contained in paragraph 1 of Schedule 3 on the grounds that it is in or on the hereditament for the purpose of trade processes? In Mr Sainer's submission, although the fire-protection equipment protects the materials stored on the premises its main function is to protect the building which is housing the material and the storage of that material cannot properly be described as a trade process. In his submission these words are intended to apply in cases where some manufacturing operations involve some particular degree of risk such as the handling of molten metal.

I cannot accept these submissions. In my judgment, the fire-fighting equipment, like the heating plant and the dehumidifiers, is on the hereditament primarily to protect the material that is stored E there. Even if it were to be found that this could only be done by the protection of the building and therefore that that was the main use of the equipment, it would nevertheless not be included within the schedule because it was there expressly for the purpose of the trade process being carried on. In my opinion, the trade so far as material on this point was the storage of sensitive materials and the process was the provision and maintenance of a suitable temperature and environment involving the application in emergency of a non-toxic and harmless gas which had the effect of extinguishing a fire.

The valuation officer in his evidence suggested that the assessment determined by the local valuation court included nothing in respect of this equipment and the correct valuation reflecting the added value F attributable to it was gross value £38,400. I will need to look further at the assessment of the £37,000, but for the time being I am prepared to accept that nothing is included in that figure for these items which I have now held to be non-rateable.

Goods lifts

I turn now to the second main area of disagreement to consider how the goods lifts on the hereditament are to be regarded for purposes of valuation. As will be seen from the order, lifts mainly or usually used for passengers are to be deemed to be part of the hereditament. It is common ground that the lifts in this case are not mainly or usually used for passengers although no doubt from time to time passengers travel in them. They are, however, goods lifts and it is not in dispute that they are deemed not to be part of the hereditament.

For the appellants, Mr Cochrane called Mr A T Stow BSc FRICS, an experienced rating surveyor and a consultant to Fuller Peiser. Mr Stow's valuation was gross value £12,300, rateable value £10,222, and was arrived at on the assumption that without goods lifts the value of the three lower floors was substantially reduced and that there was no value on the third, fourth and fifth floors because no use could be made of them. The respondent valuation officer, Mr D J Raley ARICS, accepted that the goods lifts were not to be included as part of the hereditament, but his valuation of gross value £37,000 was arrived at on the assumption that the lift shafts were part of the structure and part of the hereditament and were to be considered as available for the hypothetical tenant for the installation of goods lifts as part of the tenant's equipment.

Mr Stow was of the opinion that it was quite wrong to value the premises on the statutory basis on the assumption that the tenant had installed the goods lifts himself. No tenant, he suggested, would be prepared to pay for the cost of installation in the context of a tenancy from year to year, as required in the rating hypothesis, and then pay a rent to the landlord on the basis that the goods lifts were available.

In support of his valuation he referred to a number of other warehouses which were not fitted with lifts which showed that both rents and rating assessments attached no value to upper floors. He also referred to some sale particulars of two warehouse blocks, one with lifts and the other without, where the asking rents quoted for the building without lifts was 23p per sq ft and for the building with lifts 43p per sq ft.

In support of Mr Stow's approach, Mr Cochrane referred me to *Edmondson (VO)* v *Teesside Textiles Ltd* [1984] RA 247; (1983) 267 EG 263, where it was held that process plant and machinery was to be ignored and treated as if it were not there, whether the actual effect of its presence in the hereditament was that a tenant would pay more or less. Mr Cochrane submitted that the agreed assessments on which the valuation officer relied in his comparables were all arrived at before the decision of the Court of Appeal in the *Edmondson* case and were therefore unreliable. He did not suggest that that case laid down any new law but merely that it produced a clarification of the law in an area where previously there had been some misunderstanding.

I do not draw from the *Edmondson* case the conclusion which Mr Cochrane invites me to adopt. The hereditament which falls to be valued in the instant case is a warehouse building, provided with two lift shafts, deemed to be vacant and to let within the terms of the rating hypothesis. It is true that in this hypothesis the tenant pays a rent on a letting from year to year, but as Willmer LJ said in *Humber Ltd* v *Jones (VO)* (1960) 6 RRC 161 at p171:

> The fact is that it is impossible to get away from the situation that the statute postulates not only a hypothetical tenant but also a hypothetical landlord, and, as the Lands Tribunal said in the passage read, in the context of a hypothetical world in which the hypothetical tenant cannot become the owner of the premises and cannot get a lease for a term of years. Moreover, one has to postulate a world in which not only this hypothetical tenant is in the position, but everybody else is in the same position. In the end, therefore, we are in a world of make-believe.

The passage in the decision of the Lands Tribunal to which Willmer LJ referred and which was also approved by Hodson LJ is to be found in (1959) 5 RRC 23 at p 33 in the decision of Mr Hobbs:

> Rating is not the everyday world, it is the domain of a hypothetical tenant who is faced with the fact that, try as he may, he cannot become the owner of the premises he occupies, neither can he get a lease for a term of years, short or long.

In this hypothetical world, therefore, no hereditaments are fitted with goods lifts and it must be left to all tenants to make what use of the premises they can by the installation of their own equipment. Plant was defined in *Yarmouth* v *France* (1887) 19 QBD 647 by Lindley LJ at p 658 as including:

> . . . whatever apparatus is used by a business man for carrying on his business, — not his stock-in-trade which he buys or makes for sale; but all goods and chattels, fixed or moveable, live or dead, which he keeps for permanent employment in his business.

The goods lifts are clearly plant and not being part of the hereditament are not deemed to be provided by the landlord. In so far as they may be brought on to the hereditament it follows that they are deemed to be part of the tenant's plant. In that way they are indistinguishable in principle from any other items of tenant's plant such as furniture, typewriters, office materials and, more particularly in the context of this case, fork-lift trucks which are used by the occupiers in moving the stored materials about the premises and, indeed, for lifting them into the premises. Mr Stow in his valuation accepts that the first and second floors have a value because they can be serviced by the fork-lift trucks. He therefore accepts that the value which is directly attributable to the use of the tenant's plant is properly to be taken into account. I can see no reason, therefore, to distinguish between putting a value on the second floor because the tenant can introduce a fork-lift truck and putting a value on the third, fourth and fifth floors because the tenant can introduce a goods lift. Indeed, to exclude from the mind entirely the possibility of the use of goods lifts so that the hypothetical tenant would pay no rent for the upper floors must logically lead to the conclusion that the introduction by the tenant of any plant, including furniture and office equipment, is also to be disregarded. I cannot accept a conclusion which leads to this *reductio ad absurdum*.

The comparables put forward therefore by Mr Stow of hereditaments without lifts do not help because they are also hereditaments without lift shafts and clearly it is not possible to assume the installation of lifts in new lift shafts without offending the principle of *rebus sic stantibus*. I am quite unable, therefore, to accept Mr Stow's valuation, which seems to me to have been prepared on a false basis.

I am required, finally, to review the assessment determined by the local valuation court even on the assumption that the existence of the lift shafts does lead to a rejection of Mr Stow's valuation. In other C words, whether an assessment of gross value £37,000 is, in itself, correct. Mr Raley produces his own valuation in support of this figure and observes that it was an assessment on behalf of the occupiers in 1977. In support of this valuation, he refers me to a number of other warehouses one of which is on the riverside at Wapping and three others in the Whitechapel and Spitalfields area. Mr Stow did not consider that these three assessments supported the figures adopted by Mr Raley for the appeal premises because they were all situated in what he described as "the rag trade area", where there was a continuing demand for warehouse space.

I have inspected all the comparables externally and I have to say that in the light of the assessment previously agreed on the appeal hereditament there is a heavy onus on the appellants to show that that D figure is no longer appropriate. I do not consider that they have discharged that onus and, after making due allowance for the difference in locality of the various comparables, I am not persuaded that any of the values attributed to the different parts of the appeal hereditament are excessive.

The appeal accordingly is dismissed.

I have, however, found against the valuation officer on the rateability of fire-protection equipment. In these circumstances the appellants will pay two-thirds of the costs of the respondent valuation officer, such costs failing agreement to be taxed by the Registrar of the Lands Tribunal on the High Court scale.

For further cases on this subject see p 187

LEASEHOLD VALUATION TRIBUNALS

BURFORD ESTATE & PROPERTY CO LTD v CREASEY

LONDON RENT ASSESSMENT PANEL (LON/LVT/203)

January 14 and March 31 1985

(LADY FOX MA (chairman), M CLARK CBE LLB FRICS and G W PAYNE-BUTLER CIPFA)

Estates Gazette January 11 1986

277 EG 73-77

Rent payable under extended lease of house at Bowes Park, N London — Extended lease for 50 years granted to tenant's predecessor in title from December 1978 and assigned to tenant in December 1982 — Present landlords acquired freehold in June 1983 — Evidence that rent at old rate (£6.75 pa) continued to be paid for some four years after commencement of extended lease — Preliminary issue whether tribunal's power to determine modern ground rent is exercisable after commencement of extended term and whether such rent had been agreed at £6.75 by parties — S15(2)(b), contrary to tenant's submission, has, in tribunal's view, no relevance as a time-limit within which parties or tribunal must determine rent — Alternatively, s15(2)(c) provides a commencement limit for period during which parties or tribunal may determine rent but gives no termination limit — Held that power to determine rent payable on extended lease is exercisable after commencement of extended term — Tribunal rejects tenant's submission, relying on Sched 1, para 10(4), that on expiry of original lease no obligation accrued to tenant to pay modern ground rent — If on correct construction of extended lease a modern ground rent is determinable, such determination binds tenant — Further submission by tenant that by demanding and accepting rent at old rate for four years landlords had agreed rent for first 25 years of extended lease not upheld — Clear and unambiguous evidence of an agreement or conduct required to establish landlords prepared to continue rent at level originally fixed in 1879 — Finding that there was no agreement between parties as to modern ground rent and that tribunal had jurisdiction, in default of such agreement, to determine rent payable.

Determination of rent payable — Rents charged for modern properties, in tribunal's view, depressed by reason of developer's wish to obtain high price for leasehold and no reliance placed on them — Level of fair rents, on which tenant's valuer based his valuation, not relevant *vide* Lands Tribunal's decision in *Carthew's* case — Standing house approach adopted by landlords' valuer the correct approach, but his 38% for site value not accepted — Percentages of this order applicable only to sites nearer centre of London with immediate access to all amenities of metropolis — Tenant's proposed 30% more appropriate for London suburban sites, but 27½% adopted by tribunal as site unattractive, overlooked and awkward to develop — Determination of £435 pa against tenant's figure of £150 to £225 pa and landlords' £585 pa

Wayne Clark (instructed by Green Daniel Conway & Co) appeared for the applicants, the landlords; D C Gerrey (instructed by Frank E C Forney & Partners) for the respondent, the tenant, Mr C J Creasey.

This decision is made on an application by the landlords for a determination of rent payable in accordance with section 15(2) of the Leasehold Reform Act 1967 for the house and premises at 13 Whittington Road, London N22. The tenant's predecessor in title, Mrs D E Ashby, held the said property under a lease dated July 27 1880 for a term of 99 years with effect from December 25 1879 at a ground rent of £6.75 per annum, and by notice dated June 9 1972 given to the landlord's predecessor in title, the Hearts of Oak Benefit Society, she claimed the right to an extended lease of the house and premises under the Leasehold Reform Act 1967. Pursuant to such notice, on December 5 1972 an extended lease for the statutory period of 50 years was granted to Mrs D E Ashby to commence on December 25 1978, being the date on which the original term expired. Mrs D E Ashby assigned her leasehold interest under the said lease in 13 Whittington Road to the tenant, Mr C J Creasey, which was registered at HM Land Registry on December 6 1982. On June 1 1983 the landlords acquired the freehold interest in the said house and premises from the Hearts of Oak Benefit Society.

At the hearing on December 10 1984 the tribunal was requested by counsel for the landlords to determine the rent payable in accordance with section 15(2) of the Act and clauses 1 and 6 of the extended lease. The landlords proposed the sum of £585 pa as the modern ground rent. Counsel for the tenant submitted that the parties had agreed the modern ground rent at £6.75 pa, that the landlords had accepted rent in that sum from the former tenant, Mrs D E Ashby, and the present tenant and that section 15(2)(b) of the 1967 Act excluded revision of that rent within the period of 25 years. Accordingly, on behalf of the tenant he submitted that the landlords' application to the tribunal was not valid. The tribunal adjourned and gave its reasons in writing as set out below on January 14 1985.

Decision as to jurisdiction

By section 21(1) of the 1967 Act as amended we have power to determine the amount of the modern ground rent payable on an extended lease but only in default of agreement. It is, therefore, necessary to decide the preliminary issues raised by counsel for the tenant whether the power of determination is exercisable after the commencement of the extended term and whether the modern ground rent has been agreed at £6.75 by the parties.

By the extended lease dated December 5 1972 under which the tenant now holds, the original ground rent of £6.75 is payable until the expiry date of the original lease at December 25 1978. Thereafter, by clause 6 of the extended lease, a modern ground rent is to be payable (subject to 25-year review) calculated in accordance with section 15(2). Counsel for the tenant contended that by operation of section 15(2)(c) of the Leasehold Reform Act 1967 and clause 6(2) of the extended lease, the period for determination of the modern ground rent was restricted to a 12-month period immediately preceding (and terminating at) the date of expiry of the original lease. He also relied on section 14(2) of the 1967 Act in support of this construction.

In our view, this contention is ill-founded for the following reasons:

(a) Section 15(2)(b) of the 1967 Act provides the date by reference to which the valuation of the modern ground rent has to be calculated, and subsection 2(c) does no more than permit such valuation to be made, or more accurately "forecast", 12 months, but no more, in advance of that valuation date. The subsection accordingly has no relevance as a time-limit within which the parties or the tribunal must determine the rent.

(b) Alternatively, subsection 2(c) provides a *commencement* limit for the period during which either the parties or the tribunal may determine the rent but gives no *termination* limit to that period. It is open-ended in the sense that either party may apply for its determination subsequent to the expiry date of the original lease. This construction is supported by section 15(6), which contemplates a situation where not only the determination of a modern ground rent is subsequent to the expiry of the original term but also the grant of an extended term. It is also borne out by the requirement in section 15(2)(c) relating to written notice of the landlord's right to revise the rent after the first 25 years: for the giving of such written notice a period is defined both at commencement (12 months prior) and at termination (the expiry of the 25th year of the lease); but, as with the original determination of the modern ground rent, no such defined period is made for the determination of rent payable for the second 25-year period.

(c) Section 14(2)(c) relied on by the tenant is a provision relating to recovery of costs and implies no *termination* point for the period in which the parties or tribunal may determine the rent.

(d) A period limited to 12 months as contended by the tenant's counsel for determination of the modern ground rent appears unreasonably short for all valuation purposes. The tenant is not prejudiced by the absence of any termination date for the determination, since he himself may always apply to the tribunal at any time for its determination if he cannot obtain the landlord's agreement; further, the valuation date for the modern ground rent is fixed at the commencement of the extended term, subsequent rises in price being irrelevant; and until the modern ground rent is determined, the tenant by paying at the lower rate enjoys the use of his money.

(e) The situation is analogous, despite the 1967 Act being designed to give the tenant the right to enfranchisement, to the rent review cases where, on the authority of the House of Lords' decision in *United Scientific Holdings Ltd* v *Burnley Borough Council* [1977] 2 All ER 62, time is not of the essence in implementing review clauses.

We accordingly hold that the power to determine the rent payable on the extended lease is exercisable after the commencement of the extended term.

Counsel for the tenant also submitted that on the expiry of the original lease, no obligation accrued to the tenant when he acquired the leasehold interest in December 1982 to pay a modern ground rent and he relied on Schedule 1, para 10(4) in support. That schedule, in our view, relates to extension by subtenants and has no application here. The tenant, as assignee of the original tenant, holds on the terms of the extended lease, and his rights and obligations are governed by that lease. If on the correct construction of that lease a modern ground rent is determinable, such a determination binds the tenant.

The final submission of counsel for the tenant was that the original and present landlords, by demanding and accepting rent at the old rate of £6.75 from the original tenant and the tenant for the period from the commencement of the extended lease to March 23 1983, had agreed that the rent for the first 25 years of the extended term should be at that rate. As supporting evidence, Mr Gerrey produced a demand for rent for £3.37 for a six-month period dated March 23 1983 from the original landlords, the Hearts of Oak Benefit Society. In answer to questions from the tribunal, he said that the tenant's cheque in payment of that demand and a further payment of £6.75 made in March 1984 had not been accepted by the landlords.

On behalf of the landlords, Mr Clark denied that there had been any agreement between landlords and tenants as to the modern ground rent. He relied on a statement of Mr R F Cheshire FRICS, partner in Walker Son & Packman, chartered surveyors to the former landlords. Mr Cheshire said that there had been no agreement as to rent reached with the former tenant, Mrs D E Ashby. Mr Clark also relied on the evidence of Mr Rosen, managing director of the landlords, who produced two letters, dated June 28 1983 and March 23 1984, in which he demanded payment of rent at a rate higher than £6.75 pa from the tenant. The tenant had not agreed to pay rent at the higher rate.

While the tribunal places no great reliance on the written statement of Mr Cheshire, the evidence produced on behalf of the tenant establishes, at most, that rent continued to be paid at the old rate for some four years after the commencement of the extended lease. In our opinion this falls far short of establishing an agreement between the parties to modify the term of the lease for determination and payment of an increased rent, a term which is clearly set out in a written agreement and which constituted, as provided by the 1967 Act, the consideration for the grant of the extended term. Nor, in our view, does the evidence adduced by the tenant support any estoppel debarring the landlords from relying on the written lease and its terms. Clear and unambiguous evidence of an agreement or conduct is required to establish that the landlords were prepared to continue a rent beyond the original 99 years for another 25 years at a level which was originally fixed in 1879. Accordingly, we find that there is no agreement between the parties as to the modern ground rent and that we have jurisdiction, in default of such agreement, to determine the rent payable for the house and premises in accordance with section 15(2). Our jurisdiction is, of course, confined solely to the assessment of the amount of the modern ground rent; we are not concerned with questions relating to its recovery or for what amount or period.

It follows that we find the application submitted to the tribunal on behalf of the landlords to be a valid application. The matter can now proceed to a further hearing when we will consider evidence on valuation.

Second hearing on March 4 1985

The tribunal reconvened to determine the amount of the modern ground rent in accordance with section 15(2) of the 1967 Act as amended. Counsel for the tenant requested the tribunal to assess the rent at a figure of £150 to £225 per annum. He produced a valuation from Mr D H Barnfield FSVA MRSH, of Stephen Gracey, estate agents, Palmers Green, London N13, dated March 4 1985 in which he stated that the building as it stood commanded a registered rent of £750 pa exclusive and that the ground rent to the site would be in the region of £225 pa exclusive, calculated at 30%. He applied this ratio, 30%, as being that used for the purpose of calculating the value of the freehold interest under the Leasehold Reform Act 1967, where the site value would represent approximately 30% of the total value of the vacant possession value of the property as a whole. In his opinion, the vacant possession value of the property as a freehold was £45,000. He asserted that any increase in the proposed figure for the rent would have a serious effect upon the saleability of the property in the future, bearing in mind that most modern ground rents were at a level of between £50 and £150 pa for properties in urban areas.

On behalf of the landlords Mr F D Jones FRICS, senior partner in Sturt & Tivendale, chartered surveyors, submitted a valuation as follows:

Capital value of the property on a freehold basis with vacant possession	£23,500
Site value — London suburb —	
Say 38% of capital value or	
6 rooms at £1,500 per room,	
Land value for a serviced site in an established road	9,000
Modern ground rent as at December 25 1978	
6½% of standing house site value	£585pa

In support of his valuation, he submitted a plan of Whittington Road showing the location and shape of the site of the subject property, a list of four comparable properties, a plan showing their location by reference to the subject property and its location in a well-established neighbourhood in Bowes Park, predominantly residential but with some commercial use, very convenient for British Rail and London Underground stations, bus routes and local amenities. The date of valuation was December 25 1978 and was crucial to the assessment, as in 1977-78 prices of secondhand old houses rose more than in the whole of the four previous years. There was a plateau at December 1978, which lasted for only a short time as prices rose even more during 1979. The Nationwide Building Society Index of prices for older secondhand houses showed the following rise:

1973	4th quarter	100
1977	4th quarter	134
1978	4th quarter	171
1979	4th quarter	229
1984	4th quarter	341

Since December 1978, values had doubled as at the present day, giving on his valuation of £23,500 at December 1978 a current value of £47,000, which was reasonable. He supported his entirety value by reference to four valuations for mortgage purposes carried out by his firm in February 1979, at a time when transactions would reflect conditions at Christmas 1978, making due allowance for preliminaries between the property being put on the market and the acceptance of an offer. The particulars of these transactions, concerning Victorian or Edwardian houses with approximately similar accommodation, were as follows:

	Internal measurement	*Value*	*Remarks*
Subject Property	1,223 sq ft	£23,500	Estimated with repair covenants satisfied
66 Dunbar Road, N22	827 sq ft	£21,500	Modern extension, small plot, small floor area, satisfactory condition
40 Sirdar Road, N22	1,280 sq ft	£23,500	Satisfactory condition
22 Selborne Road, N22	1,167 sq ft	£27,000 After repairs	Modernised internally but exterior in poor condition
20 Dorset Road, N22	700 sq ft	£24,000 After repairs	Central heating, modernised, satisfactory condition; small floor area

Mr Jones also referred to the sale of a site of a rectangular plot of land adjacent to 52 Lewes Road, where contracts had been exchanged in January 1978 for a sale at £9,750. The site was for development of one six-roomed house. He estimated that by December 1978 this value would have increased to £12,500, which gave a value of £2,000 per room. Taking into account that Lewes Road was located in Finchley in a better locality, this site transaction supported his site value of £9,000 or £1,500 per room. In commenting on the valuation submitted on behalf of the tenant, Mr Jones said that fair rents bore no relationship to modern ground rents under section 15(2) of the 1967 Act. The level of ground rents charged by developers related to modern properties and had regard to the high prices which were sought for the purchase of the leasehold.

In answer to questions from the tribunal, Mr Jones said that his figure of 38% for the proportion of the site value to the entirety was based on leasehold valuation tribunal decisions. Bowes Park was a Victorian suburb with good transport some 8 miles distance from central London and in his view justified such a percentage.

Inspection

Following the hearing, the tribunal inspected the subject property the same day. It comprises a two-storey, mid-terrace house and premises, built of brick with slate roof in 1879-80, of typical Victorian style and appearance. Much of the immediate neighbourhood is of the same period and is mainly residential but with some commercial development also. Situated some 7 or 8 miles north of central London, it is conveniently placed for London Transport and British Rail services and for the usual urban facilities.

The site of the property, which has a frontage of 18 ft approximately to Whittington Road, is of irregular shape and may be loosely described as "dog legged". The front portion occupied by the house and front garden is more or less rectangular but the back garden runs off at a slight angle and narrows to approximately 9 ft at the rear boundary. The total depth is about 119 ft. Beyond the rear boundary is a strip of land a few yards wide and then a railway cutting. There is no rear access to the subject premises. The front garden is fenced on the road boundary and has a single hand-gate. There is no off-street parking facility.

The house itself contains, on the ground floor, a recessed porch and entrance passage, front sitting-room with splayed bay window, double doors leading to a middle living-room, a back living-room, and a kitchen with door to an outbuilding. On the first floor there are a front bedroom, middle bedroom, separate wc, a rear bedroom and a bathroom formed by partitioning. All main services are connected. On visual inspection the house appears to be in average condition for a house built a century ago.

The tribunal also inspected externally all the properties referred to on behalf of the landlord. No 20 Dorset Road is a small, early Victorian cottage in a back lane and is no way comparable. No 66 Dunbar Road is a small mid-terrace house situated in a much wider road near playing fields. No 40 Sirdar Road, of slightly different G
style, and 22 Selborne Road are properties of broadly similar size and location comparable with that of the subject property.

The tribunal also inspected the site adjacent to 52 Lewes Road. Lewes Road is a quiet, pleasant, residential road located in a good area of Finchley with semi-detached houses built in the 1930s or later. The site, rectangular in shape, remains undeveloped and appears to have contained a hard tennis court at one time.

Decision as to amount of rent

The tenant's valuer based his valuation on the level of fair rents for houses such as the subject property and also had regard to the level of £50 to £150 per annum which developers charged for modern H
properties. In our view, these modern ground rents are depressed by reason of the developer's wish to obtain a high price for the leasehold and we place no reliance on them. As to an assessment based on the level of fair rents, we do not regard these as relevant and rely on a decision of the Lands Tribunal, *Carthew* v *Estates Governors of Alleyn's College* (1974) 231 EG 809, which rejected a similar argument. The Lands Tribunal in that case set out their determined figures for rent and said:

> We are conscious of the fact that the above determinations of rental value for sites are in each case higher than the figure agreed between the parties as representing a fair rent of the site including the house on it. This is in one sense anomalous, but the disparity stems essentially from the gulf . . . between house prices on the one hand and house rents on the other, and it illustrates J
> that under current legislation the fair rent that a landlord receives for a dwelling-house may well represent less than a fair return on the capital value of the site alone.

Discarding the fair rent comparison and in the absence of evidence of site sales in the immediate vicinity, we agree with the landlords' valuer that the standing house approach is the correct approach. We noted that both valuers were not greatly at variance on the current figure for the freehold value with vacant possession, Mr Barnfield giving £45,000 and Mr Jones £47,000. Mr Barnfield, however, gave no help as to the level of prices at December 1978, which is the date of valuation.

From our own knowledge and experience, supported by the K
evidence of both parties, we consider that for houses of the type and character of the subject property, in good repair and with no site disadvantage, £45,000 represents a realistic price likely to be achieved at the present time. We think Mr Jones' £47,000 is more of an asking price or a top value for a fully modernised property. The indices of price increases support Mr Jones' submission that prices of houses similar to the subject property have doubled since December 1978. Half the figure which we have determined gives £22,500 and this seems to us an acceptable figure for the entirety value, and supported by Mr Jones' evidence relating to transactions in February 1979, particularly in respect of 40 Sirdar Road and 22 Selborne Road.

As to the appropriate proportion for the value of the site, we cannot accept the 38% proposed by the landlords. Percentages of L
this order, in our view, are applicable only for sites situated nearer the centre of London with immediate access to all the amenities of the metropolis. We regard the figure of 30% advanced on behalf of the tenant as the more appropriate percentage for sites in London suburbs and one supported by leasehold valuation tribunal decisions. However, the particular site is an unattractive one; the "dog-leg" shape of the rear garden abuts on to the separately owned small piece of land, which itself adjoins the railway cutting, and the site is overlooked by properties on three sides. These factors in our view make it an awkward one to develop and are reflected in the $27\frac{1}{2}$% for the site value which we have adopted. Finally, in making our valuation we have had some regard to the price achieved for the M
site at 52 Lewes Road; the higher price there achieved reflects that it is a better site, of rectangular shape, located in a much superior street and neighbourhood.

The tribunal's valuation is therefore as follows:

Section 15(2) ground rent
Valuation date: December 25 1978.

Standing house value	£22,500	
Site value 27.5%		£6,187.50
Ground rental value @ 7%		0.07
		433.12
		Say £435 pa

We accordingly determine that the annual rent payable by the tenant in accordance with section 15(2) of the Leasehold Reform Act 1967 for 13 Whittington Road, N22, is £435 per annum, payable quarterly in accordance with the terms of the lease.

PILGRIM AND ANOTHER v CENTRAL ESTATES (BELGRAVIA) LTD AND ANOTHER

LONDON RENT ASSESSMENT PANEL
(LON/LVT/253)
February 28 1986

(Lady FOX MA (chairman), M CLARK CBE LLB FRICS and G W PAYNE-BUTLER CIPFA)

Estates Gazette June 21 1986
278 EG 1373-1375

Price of end-terrace Pimlico, London, house on four floors — Lease with $4\frac{1}{2}$ years unexpired — Two superior interests: freeholder's interest and head lessees' reversionary interest with 118 years to run — Total absence of any evidence of sales of sites — Standing house approach adopted — Figures derived from agreed settlements approached with caution by tribunal but provided some guidance to level of values — Site value at 35% adopted, relying on general practice in previous decisions of Lands Tribunal and London LVT and bearing in mind site's awkward shape (a trapezium) and its proximity to adjacent higher buildings on one side — Tribunal consider it unnecessary to include reversion to rack rent stage, as in head lessees' valuation — In their view, where there is a lengthy period before expiry of head lease the realistic approach is to value in perpetuity — Evidence of settlements of freehold interests on basis of 15 to 24 YP of ground rent, but tribunal adopt 10 YP — Some "Delaforce" effect, particularly where, as in present case, sums involved are relatively small — Head leasehold interest price determined at £23,055 against tenant's £15,163 and head lessees' £30,650 — Freehold interest price determined at £156

Mr J W Pilgrim appeared for himself and Mrs B M Pilgrim as applicants; Mr T E Job FRICS, of King & King, chartered surveyors, appeared for the head lessee respondents. The freeholder did not appear and was not represented.

Giving their decision, THE TRIBUNAL said: This is a decision on an application by the tenants for a determination under section 9 of the Leasehold Reform Act 1967, as amended, of the price payable for the freehold interest in a dwelling-house and premises at 36 Charlwood Street, London SW1.

The tenants hold the subject property under a lease dated November 25 1925 for a term of 63 years from June 24 1925 at a ground rent of £16. The tenants' notice of claim to purchase the freehold, which was admitted, was dated December 16 1983, which is accordingly the date of valuation. At that date the unexpired term of the lease was four years and six months.

The parties further agreed the following facts. By a head lease entered into subsequent to the above tenancy on March 11 1952 and made reversionary to that tenancy, The Royal Liver Friendly Society, as the then freeholder, granted to St George's Square Property Co Ltd a term of 150 years of the Pimlico estate which included the subject property. The said term commenced on December 25 1951 and at the date of valuation the unexpired term was 118 years. At the date of valuation the freeholder was Routestone Ltd, and Central Estates (Belgravia) Ltd (hereinafter referred to as the head lessees) were the successor in title of St George's Square Property Co Ltd.

By letter dated January 13 1986 Albany Property Management Services Ltd, agents on behalf of Routestone Ltd, stated that the freeholder would not be attending the hearing but was happy to rely on the fairness of the tribunal after considering the evidence supplied in their letter dated December 16 1985. By letter dated January 13 1986, King & King, chartered surveyors acting on behalf of the head lessees, stated that they had no instructions to act for the freeholder and would act at the hearing solely on behalf of the head lessees. Mr T E Job, of King & King, repeated this statement at the hearing.

The valuations presented on behalf of the parties were as below:

Valuation by tenant, Mr J W Pilgrim			*Amended at hearing*	
4.5 years at rent £0.40	(YP 5.6)	£2.24	(YP 3.74)	£1.49
Open market value	£80,000		£68,500	
Site value 30%	£24,000		£20,550	
Section 15 ground rent 7%	£ 1,680		£ 1,438.50	
Capitalised 7% perp def 4.5 years	10.54	17,707.20	10.54	15,161.79
		£17,709.44		£15,163.28

Valuation for head lessees by Mr T E Job FRICS			
Stage I			
Profit rent £16-£15.60 per annum for $4\frac{1}{2}$ years		£0.40	
YP $4\frac{1}{2}$ years at 7% per annum		3.74	£1.50
Stage II			
Reversion to a modern ground rent (section 15 rent) for term of 50 years from June 24 1988 deferred $4\frac{1}{2}$ years			
Take estimated value of freehold in possession			
"Entirety value" at	£100,000		
Take site value at 40%	40,000		
Take section 15 rent at 7%	2,800 pa		
Less head rent	16		
	£ 2,784 pa		
YP for 50 years def $4\frac{1}{2}$ years at 7% 13.80 × 0.738	10.13		28,202.00
Stage III			
Reversion to rack rent for the remainder of the head lease from June 2038 to December 2101 for $63\frac{1}{2}$ years deferred $54\frac{1}{2}$ years			
Take estimated rent at 7% of £100,000	£7,000		
Less	16		
	£6,984 pa		
YP 14.09 × 0.0252	0.35		2,444.40
		Total	£30,647.90
		Say	£30,650.00

In a letter dated December 16 1985 from Albany Property Management Services Ltd, the tribunal's attention was drawn to nine sales by Routestone Ltd, the freeholder, of freehold interests in cases where the right to enfranchise had been admitted. All these related to houses in the same neighbourhood as the subject premises, the prices ranging from 15.7 years' purchase to 25 years' purchase of the freehold ground rent.

The tenant, Mr J W Pilgrim, gave evidence in support of his valuation. The subject property on four floors was typical of smaller properties in Pimlico except that it was built on a small, irregular-shaped site. The property was surrounded on two sides by the private garage owned by the adjoining block of flats, Kinsey House, which somewhat overshadowed the property. There was no leasehold valuation tribunal decision relating to a property in SW1, so it was impossible to make a comparison of previous leasehold valuation tribunal decisions. He had selected for consideration a settlement made with the head lessees relating to 36 Denbigh Street, occupied by his son and held on almost identical terms under a long lease, where the tenant's notice to enfranchise was dated January 18 1982. That property was double-fronted, being on five floors and substantially larger than the subject property with a site area of 1,250 sq ft. The valuation of the site of 36 Denbigh Street had been put forward by the head lessees and the enfranchisement price agreed on that basis. The figures relating to the settlement at 36 Denbigh Street were as shown opposite.

Mr Pilgrim submitted that the evidence for the head lessees, despite King & King's acting for them over a period of years, related only to a selective list of seven properties covering a period of six years. With the exception of 60 Denbigh Street, it did not include any of the properties listed by the freeholder in Albany Property Management Services Ltd's letter of December 16 1985 nor 36

36 Denbigh Street			
Term expiring	29.9.90		
Date of notice	18.1.82		
Unexpired term 8.75 years.			
Ground rent reserved £38.00 per annum			
Ground rent passed to freeholder £27.20			
1. Profit ground rent	£10.80		
YP 8.75 years at 7%	6.38		£68.90
2. Open market value		£61,875.00	
Site value at 40%		£24,750.00	
Modern ground rent at 7%		£ 1,732.50	
Less head ground rent reserved		27.20	
Net ground rent		£ 1,705.30	
Capitalisation at 7% deferred 8.75 years 8.17			13,932.30
			14,001.20
		Total, say	£14,000.00

Denbigh Street relied upon by the tenant. The head lessees' list was inaccurate in that the site area of 25 Charlwood Street was 1,375 sq ft, not 915 sq ft, and 38b Charlwood Street was the Richmond House Hotel. However, on the basis of the head lessees' figures given for the entirety values, in particular £80,000 for 160 Tachbrook Street, with a site area of 1,032 sq ft, the tenant calculated a value of £77 a sq ft which, applied to the site area of 890 sq ft of the subject property, produced an entirety value of £68,500 for 36 Charlwood Street. As the dates of valuation were only four months apart, he put forward this entirety value of £68,500 as the basis for an amended valuation. He maintained that the resulting figure of £15,163.28 was the appropriate sum to be paid for the head lessees' interest. He had at the time instructed Messrs Dauntons to settle for £17,000 or so but had withdrawn the offer when the head lessees had asked for more. The £15,163.28 was exclusive of the sum to be paid for the freeholder's interest, the assessment of which he left to the tribunal.

In answer to a question from Mr Job, the tenant said he had chosen 36 Denbigh Street as a comparable settlement because he had personal knowledge of the facts. In answer to questions from the tribunal, he supported the 30% adopted in his valuation for the site value on the basis of the percentage adopted in the majority of the decisions of London leasehold valuation tribunals. He accepted that building costs might be higher the more central the site, with consequent decrease in value of the site. The subject site was difficult to rebuild upon and unusual in that it narrowed to nothing at the rear. He considered 7% to be a fair rate for use throughout the valuation, having regard to the property market generally. The third stage of Mr Job's valuation for the head lessee would not take effect until 54 years hence and was, in his view, too remote to be included in the calculation.

Mr T E Job, of King & King, chartered surveyors, gave evidence in accordance with a written proof in support of his valuation. Since 1970 he had acted for the head lessees in a number of settlements relating to the enfranchisement price under the Leasehold Reform Act 1967. He knew of no market evidence relating to sale and purchase of vacant sites relevant to the determination of the price to be paid for the subject site. Accordingly, his valuation was based on the standing house approach, whereby he sought to arrive at a market value for the premises on the basis of their being in good repair with accommodation as existing and with the kind of amenities consistent with the neighbourhood. In doing so, values based on seriously dilapidated or elaborately modernised property were to be disregarded. Prices varied but the site value should remain steady. The subject property was located in a convenient and attractive residential area within easy walking distance of Victoria Station and Pimlico Station on the Victoria Tube line.

He produced a list of comparable properties where amounts had been agreed for the head leasehold interest with the tenant, who in every case had been professionally advised by a surveyor. He had restricted the list to premises of comparable size and frontages near to the subject property which he had personally dealt with since 1981. The *Delaforce* effect had played no part in any of these settlements. The agreed amount was a negotiated figure between the two values and did not represent the exact mathematical calculation of the section 15 rent multiplied by the years' purchase. These settlements supported a vacant freehold possession value for the subject property of £100,000, which he had accordingly adopted as the entirety value G
in his valuation.

Mr Job relied on the London tribunal's own decision in *Middleton-Smith* v *Harrow School Road Trust* (1981) 262 EG 73 relating to 19 Hamilton Close, NW8, to support a percentage of 40% for the site: 40% had been used and accepted by surveyors in the neighbourhood of the subject property. The use of the rate of 7% had for some years been a generally accepted formula. His valuation included a stage III relating to the reversion to rack rent for the remainder of the head lease and he relied on *Collins* v *Jones* (1981) 261 EG 1001, a decision of the West Midland Leasehold Valuation Tribunal relating to 25 Gower Road, Halesowen, for such a method. He finally referred to a letter from Messrs Dauntons, acting on behalf H
of the tenants, dated July 1 1985, in which a figure of £27,000 subject to contract was mentioned as the price.

In answer to questions from the tenant, Mr Pilgrim, Mr Job said that in the settlements, variations in overall site dimensions were not important; the significant factor was the width of the frontage of the property. He agreed that there had been no stage III in the landlords' valuation in respect of 36 Denbigh Street relied upon by the tenant. In answer to further questions about stage III by the surveyor member of the tribunal, Mr Clark, Mr Job said that an alternative rough and ready method was to disregard stage III and adopt a slightly higher years' purchase in respect of the deferment in perpetuity of the section 15 ground rent. He accepted that, in respect of the settlements J
which he relied upon, no stage III of valuation had been advanced. With regard to the 40% adopted for the site value, he maintained that the site was relatively easy to build on and compared favourably as to transport and accessibility with sites in the City. He had little experience whether sites were more expensive to develop in the centre of London. He considered 7% the appropriate rate to apply but agreed that, if a stage III valuation were adopted, the risk after 50 years was likely to be greater and ought properly to be reflected in a higher rate of interest such as 8%. In answer to a question from the chairman, Mr Job said that he relied on no evidence of actual market sales of properties although his valuation was made against a background of sales. He knew that 38-40 Charlwood Street sold together in late 1983 achieved £200,000. K

As regards the value of the freehold interest, in answer to questions from the tribunal Mr Job confirmed that settlements had been made on the basis of 15 to 24 years' purchase of the ground rent. As the sums involved were small — £150 to £200 — and tenants anxious to settle, he considered that there might be some *Delaforce* effect by which tenants were settling at sums higher than the true value. He suggested that a mean of 20 years' purchase might be appropriate. Although he accepted Mr Clark's suggestion that an investment in gilts with a return of 10% represented a better return, he nevertheless considered tenants would settle at 20 years' purchase (on 5% basis). Some tenants wished to escape onerous obligations under the lease.

L

Inspection

Following the hearing, the tribunal inspected the subject property. It comprises a mid-Victorian end-terrace house, typical of its age in appearance and style, with three storeys plus basement, the latter being separately occupied as a flat. The site is a trapezium shape with a frontage of about 18 ft 3 in and a maximum depth of 50 ft 6 in, plus the basement front area. The rear of an adjacent modern block of flats overlooks the site on one side. The location is about a half-mile distant from Victoria Station, convenient for all urban amenities, and the immediately adjoining area is predominantly residential but with some commercial user also.

The construction is of brick, partly rendered, and all main services M
are installed. The main accommodation is modernised and comprises on the ground floor an entrance passage, dining-room open to a modern kitchen at the rear, a small study and a wc. On the first floor, at the front, is a sitting-room with windows to a small balcony (the latter in some disrepair) with a communicating room behind. On the top floor there are two bedrooms. There is a rather cramped bathroom with modern bath, washbasin and wc. Off a landing is a small patio roof with a low retaining wall.

We were unable to gain access to the basement but noted that the railings and steps leading to it required maintenance. We also inspected externally 25 and 38b Charlwood Street; 36 and 60 Denbigh Street; 60, 106 and 160 Tachbrook Street, and 38 Lupus Street.

A **Decision**

In the present case there are two interests superior to the tenancy in possession vested in Mr and Mrs Pilgrim under the lease of November 25 1925, namely the interest of the freeholder, Routestone Ltd, and the head lessees, Central Estates (Belgravia) Ltd, who hold a reversionary lease dated March 7 1952, which at the date of valuation had 118 years to run. By paragraph 7(1)(b) of the First Schedule to the 1967 Act, a separate price is payable in accordance with section 9 for each of the interests superior to the tenancy in possession and it is accordingly the duty of the tribunal to determine a separate price for each of these interests. Provision is made in para 1 of the First Schedule for the proceedings before the tribunal to be conducted on
B behalf of all superior interests by one of them, which in the present case would be the head lessees. Mr Job, on behalf of the head lessees, having expressed himself unwilling to act on behalf of the freeholder, the freeholder was invited to appear. By letter of January 13 1986 the freeholder's agents notified the tribunal that they were willing for the assessment of the freeholder's interest to proceed in their absence on the basis of information contained in the letter of December 16 1985.

We accordingly proceeded to our determination of the sums to be paid for the enfranchisement of the subject property by the tenants to the head lessees and to the freeholder. With regard to the price to be paid to the head lessees for their interest, our task is made difficult by the total absence of any evidence of sales of sites. Further, no evidence was produced to us of sales of freehold reversions of
C comparable properties. We accordingly have no choice but to adopt the standing house approach. We have done our best to derive an entirety value from the evidence of settlements produced by the parties but have supplemented this with our own knowledge and experience of property values in the area at the valuation date.

While we approach with some caution figures derived from agreed settlements, we considered the entirety values for the properties set out in the schedule produced by Mr Job, in particular 60 Denbigh Street and 106 and 160 Tachbrook Street, provided some guidance to the level of values at the relevant time. No 38 Lupus Street was located in a busy main street and was less directly comparable. No 25 Charlwood Street was a corner site and 38b a hotel. No 60 Tachbrook
D Street was a larger property, with no fourth floor.

We consider £90,000 to be a more realistic figure than the £100,000 put forward by the head lessees and find some support for this lower figure in the entirety value of £80,000 proposed by the tenant in his original valuation. As to the correct percentage for the site value, we consider that the site of the subject property presents features similar to other sites in districts convenient to central London but not commanding the highest rentals. We also bear in mind the shape of the site and its proximity to the adjacent higher buildings on one side. Relying on the general practice as set out in previous decisions of the Lands Tribunal and the London Leasehold Valuation Tribunal we adopt 35%. Seven per cent was agreed by both parties for deriving the section 15 rent and its capitalisation and we adopt that rate,
E particularly as we do not consider it necessary to include a stage III reversion to rack rent in our valuation where considerations of higher risk in respect of a remote future might have justified a variation from 7%. The head lessees included in their valuation stage III a valuation of the reversion to a rack rent for the remainder of the head lease, which produced a rent of £7,000 and an additional amount of £2,444.40 to be added to the amount payable for the head lessees' interest. In our view, where there is a lengthy period — over 100 years in the present case — before the expiry of the head lease, the realistic approach is to value in perpetuity. We note that this approach was followed by the parties in all the settlements relied on by the head lessees.

We now turn to the sum to be paid to the freeholder, which we are required to value separately. The parties agreed that of the total rent received under the head lease by the freeholder, £15.60 per annum was attributable to the subject property. We take 10 years' purchase of this rent. We note that, according both to the letter of December 16 1985 from the freeholder's agents and to Mr Job, considerably greater years' purchase on ground rents have been paid by tenants by agreement with the freeholder. Mr Job agreed that the settlements at these higher rates might be affected by factors noted in the *Delaforce* decision, whereby tenants, being anxious to complete the purchase without delay and the trouble of resort to a tribunal, are willing, particularly where as in the present case the sums involved are relatively small, to settle for a price higher than the value if the tenant's bid were excluded.

The valuation of the tribunal is accordingly as follows:

Head leasehold interest			
Term			
Rent receivable	£16 pa		
Less head rent	£15.60		
Profit rent	0.40		
YP $4\frac{1}{2}$ years 7%	3.756		£1.50
Reversion			
Entirety value	£90,000		
Site value 35%	£31,500		
Section 15 rent 7%	£ 2,205 pa		
Less head rent	15.60		
	£ 2,189.40		
YP perp 7% def $4\frac{1}{2}$ years	10.530		23,054.38
			£23,055.88
		Say	£23,055
Freehold interest			
Rent receivable for 118 years	£15.60 pa		
YP perp 10%	10		
	£156		

The tribunal accordingly determines:

the sum to be paid by the tenants of 36 Charlwood Street, SW1, for the acquisition of the head leasehold interest in accordance with section 9 of the Leasehold Reform Act 1967, as amended, to be £23,055;

the sum to be paid by the tenants of 36 Charlwood Street, SW1, for the acquisition of the freehold interest in accordance with section 9 of the Leasehold Reform Act 1967, as amended, to be £156.

For further cases on this subject see pp 161 and 211

Note: The next case appears on p 238 of the green section

SUPPLEMENT

CASES NOT REPORTED IN "ESTATES GAZETTE"

ESTATE AGENTS

Queen's Bench Division

November 13 1985

(Before Sir Douglas FRANK QC, sitting as a deputy judge of the division)

JOHN D WOOD & CO v BARDIGER

Estate agency — Actions by agents for damages for anticipatory breach of contract and for commission — Claim based on sole agency — House sold to purchaser introduced by another firm of estate agents — Original sole agency instructions resulted in an agreement subject to contract for the sale to a purchaser introduced by the plaintiffs, but the defendant vendor then withdrew for reasons connected with his wife's health — Contact between the parties was, however, maintained — There was a question as to whether the sole agency had been terminated by the withdrawal or merely suspended — The judge decided that it had been terminated, although it appeared to be anticipated by both parties that, if the defendant should decide to remarket the property, the plaintiffs would be reappointed — A few months later the defendant made contact with the plaintiffs and there was a conflict of evidence (one of many in this case) as to whether a telephone conversation resulted in a new sole agency appointment — There were further telephone conversations and after one of them the plaintiffs wrote to the defendant to record an arrangement that if the other agents introduced an applicant before a certain date the defendant would pay their commission "and you have also agreed to pay our fee as sole agents as previously agreed" — The letter also mentioned that the plaintiffs had sent photographs of the property to *Country Life* and referred to further contact in regard to advertising — No reply was received to this letter — Contracts were exchanged before the date mentioned with a purchaser introduced by the other firm — Plaintiffs submitted their account for commission on the sole agency basis and this litigation resulted — Held that the evidence pointed to a new sole agency agreement having been made when the contact between the parties was resumed — Among the indications were conversations on photographs, a request to "feel the market", preparation by plaintiffs of brochures, expenses incurred in advertising to the knowledge of the defendant, a contemporaneous note by plaintiffs of a material telephone conversation, and the letter above referred to "which screamed out for a reply" but received none — Plaintiffs were entitled to commission on the sole agency basis of $2\frac{1}{2}$ per cent of the sale price plus VAT

In these actions John D Wood & Co sued the defendant J D Bardiger for damages for anticipatory breach of contract, and alternatively for commission, in respect of the sale of a house at 145 Gloucester Road, London SW7, by the defendant.

Richard Slowe (instructed by Jaques & Lewis) appeared on behalf of the plaintiffs; Edward Cohen (instructed by Salter Cartier & Co) represented the defendant.

Giving judgment, SIR DOUGLAS FRANK QC said: In these actions the plaintiffs claim respectively for damages for anticipatory breach of a contract for estate agency and estate agents' commission in respect of the sale of a house at 145 Gloucester Road, London SW7. The house was, in fact, sold to a purchaser introduced by another firm of estate agents, Robert Bruce & Partners, but the plaintiffs allege that at all material times the plaintiffs were sole agents.

In July 1982 the defendant appointed the plaintiffs agents to sell the property for £350,000. It was agreed that the plaintiffs would act as sole agents in consideration for which their commission would be two and a half per cent instead of three per cent. That arrangement, which was made over the telephone, was confirmed in writing by the plaintiffs and is not disputed. Thereupon the plaintiffs carried out the usual services such as the printing of brochures and advertising required for the marketing of a property.

On October 4 1982 one Teltscher offered the asking price which was accepted, and solicitors were instructed. The agreement was subject to contract. On December 2 the defendant withdrew because his wife was pregnant and, as she had a history of miscarriages, it was important for her to rest and not concern herself with moving home.

In a letter dated December 2 Mr A A Buchanan, the partner in the plaintiffs dealing with this matter, said "I am very much looking forward to helping you with the house in the future and I shall keep all my sales particulars and photographs on the file".

I pause there in the narrative to deal with the plaintiffs' first contention that the effect of the withdrawal of the house was to suspend and not terminate the agency agreement. In my opinion, it is necessary to look at and only at the circumstances at that time and not to posterior events unless they throw light on what was intended at that time. The material facts were: first, that the property had been withdrawn from the market; and, secondly, that no decision had been made whether it would be reoffered in the future. In my judgment, *prima facie* those factors amount to the termination of the agency, albeit it was thought on both sides that should the defendant decide in the future to remarket the property the plaintiffs would be appointed as selling agents. I do not think that any subsequent event, such as the present of champagne at Christmas, shifts the *prima facie* assumption.

The next material date was March 17 1983. It is common ground that the defendant telephoned Mr Buchanan from London Airport. It is the plaintiffs' case that if the previous agency did not continue in force, then a new agreement on identical terms was entered into as a result of that conversation. It is common ground that the defendant stated that he was considering moving from the house in the autumn of that year. The defendant says that he asked Mr Buchanan to let him know if Teltscher was still interested in buying and that Mr Buchanan replied that he would speak to Teltscher and then inform the defendant of the position. The defendant denies that any agreement was entered into. Mr Buchanan's version of the conversation is that the defendant asked him to get ready to reoffer the house and that he inquired whether Teltscher was still in the market. He understood that he had been reinstructed to sell the house by preparing sale particulars and advertisements. A few days later Mr Buchanan showed the Teltschers and another potential purchaser over the property, but neither took the matter further.

The next conversation was on March 30 when Mr Buchanan told the defendant that the Teltschers were no longer interested. It seems to be common ground that Mr Buchanan thought the asking price should be just below £400,000 whereas the defendant had in mind about £425,000. It was left, to use the defendant's expression, that the plaintiffs would "feel the market out". That was to be done by stating that the price would be given on application. There was a discussion about taking photographs of the house. The plaintiffs prepared amended sale particulars and an advertisement which appeared in *The Observer* newspaper on April 10.

Mr Buchanan said that on April 18 the defendant telephoned and they had a further discussion. The defendant denies that this conversation took place. Mr Buchanan's version of it is that the defendant said that he had seen a magazine called *Portrait* showing properties for sale in advertisements by Chesterfield & Co and Robert Bruce & Partners of whom he had inquired whether they had comparable properties for sale and were asking higher prices. He told Mr Buchanan that he had been unethical in speaking to the other agents, but Mr Buchanan had replied that that was not so but that any discussions about the sale should take place through him as sole agent with a reduced fee. He was not told that the defendant had appointed other agents. The defendant denies there was any conversation on that day, but he agrees that he had heard of Robert Bruce from advertisements seen in *Portrait* magazine.

On April 19 Mr Buchanan telephoned the defendant about taking photographs of the property — that had been postponed pending the relaying of the lawn. During the conversation there was a discussion about the other two agents. Mr Buchanan's version is that the defendant said he had allowed Chesterfields and Robert Bruce to see the property, and that Robert Bruce had shown two people over it but he (the defendant) had then withdrawn his instructions. He confirmed that he wished Mr Buchanan to prepare the particulars and to advertise in *Country Life*. Mr Buchanan said that he asked the position of the plaintiffs if one of Bruce's people bought and was told that they would get their fees irrespective of arrangements with other agents. He (Mr Buchanan) explained that there was a fee-sharing agreement between agents, but the defendant replied that he thought Bruce's applicant was not interested. On the same day Mr Buchanan, at the suggestion of his partner, wrote a memorandum of the conversation as follows:

> AB had telephone conversation with JDB on 19 4 82. He agreed to pay our fee if other agents were charging separate fee for introducing a purchaser. JDB also agreed to pay for Country Life Adv. June 9 — (?) — we would take payments out of eventual fee.

The defendant's version of that telephone conversation is that Mr Buchanan was clarifying his position as he was perturbed about his sole agency. He said he concluded an arrangement with Mr Buchanan that after 10.15 that day the plaintiffs would become sole agents and that Robert Bruce would become subagents. He denied that he had agreed to pay two commissions although he accepts that he agreed to reimburse the cost of advertising if the property was sold to a purchaser introduced by Robert Bruce before 10.15 that day. There was a letter of that day from Robert Bruce to the defendant in which they recited the arrangement I have just described.

On April 20 Mr Buchanan wrote a letter to the defendant in which he said:

> Following our conversation on the telephone on Tuesday, I understand that if Robert Bruce & Partners introduce an applicant to 145 Gloucester Road, prior to 10.15 a.m. on 19th April, you have agreed to pay them a fee of 3%, plus VAT at 15%, and you have also agreed to pay our fee as sole agents as previously agreed.
>
> With regard to *Country Life*, I have now sent our copy together with two good photographs for their issue dated 9th June. In the meantime, I will keep in contact with you regarding response to our advertising.

There was no reply to that letter. On the same day the plaintiffs paid *Country Life* £600 for an advertisement which appeared in a later issue.

On June 9 Mr Buchanan learned that a contract for the sale of the house had been exchanged with a purchaser introduced by Robert Bruce prior to 10.15 on April 19. Accordingly, they (the plaintiffs) submitted their invoice for £10,375, being $2\frac{1}{2}$% of the sale price plus VAT. On July 28 the plaintiffs issued the writ in the first action, claiming damages for anticipatory breach of the contract of agency, and on January 30 1984 they issued their writ in the second action for the fees alleged to be due on completion of the sale.

I have already held that the admitted sole agency granted in 1982 was terminated. Therefore, the next question is whether a new sole agency was granted on March 17 1983. I start with the premise that it is unlikely that a person would telephone a house agent and discuss arrangements for selling his house unless that agent had already been retained or was then being retained. The defendant's explanation for the telephone call was that he merely wanted to know if the Teltschers were still interested, but that inquiry could not have been made out of idle curiosity. Indeed, the defendant conceded in cross-examination that he instructed Mr Buchanan to "feel the market" and to find out what a purchaser would pay, and said that if he accepted an offer he would feel a moral obligation to pay the fee. He also thought there was a discussion about photographs for the purpose of preparing particulars and advertisements. He said that that was one of the outgoings of an agent, from which I deduce that what he meant was that the plaintiffs paid those expenses at their own risk in anticipation of being appointed agents. That seems to me to be inherently unlikely. The evidence given by the defendant seems necessarily to demonstrate that the defendant was asking Mr Buchanan to carry out the functions normally carried out by an estate agent.

The defendant had a reason for wanting to sell his house, namely, that he had been offered another house in the area in which he wished to live and had made an offer for it which was subsequently accepted. Mr Buchanan's subsequent conduct, such as advertising at considerable expense, was consistent with the assumption that his firm had been appointed agents. The defendant, knowing what the plaintiffs were doing, did nothing to disabuse them. I, having heard both Mr Buchanan and then the defendant give evidence and having considered that evidence, have come to the conclusion that whatever were the precise terms of the conversation on March 17, there was no doubt in the minds of the parties to it that the plaintiffs were being instructed to market the house. By the verb "market" I mean to try to find somebody willing to pay a price acceptable to the defendant.

The subsequent events tend to confirm that the plaintiffs had been appointed sole agents. I have already mentioned that Mr Buchanan took two potential purchasers round the property, that he prepared particulars for a brochure, that he advertised in *The Observer* and *Country Life* at considerable cost, and that the defendant stood by and allowed him to do these things without demur. Then there was the conversation of April 19. Any doubt as to whose recollection is most reliable seems to me to be removed by the memorandum made by Mr Buchanan on the same day and the fact that there was no reply to his letter of April 20, a letter which screamed out for a reply if what it related was inaccurate.

Mr Buchanan concedes that no terms were discussed on March 17, but that is not surprising having regard to the admitted agreement in the previous year that the plaintiffs would be appointed sole agents at a reduced commission of $2\frac{1}{2}$%. As Mr Slowe said, had the plaintiffs introduced a purchaser and then charged commission at 3%, there is little doubt that the defendant would have protested that he was relying on the sole agency agreement for commission at $2\frac{1}{2}$%. I think that by the previous agreement between the parties it necessarily implied that the reappointment was to be on the same terms as the previous agency. If an officious bystander had asked them what were the conditions of the agency, that is what they would have replied.

That, I think, is sufficient to dispose of the matter, since it is not pleaded or argued that if the plaintiffs were appointed agents the agency was to be terminated.

For the other cases on this subject see p 29

LANDLORD AND TENANT

GENERAL

Court of Appeal

December 20 1985

(Before Lord Justice KERR, Lord Justice DILLON and Sir George WALLER)

MANCETTER DEVELOPMENTS LTD v GARMANSON LTD AND ANOTHER

Liability for waste — Removal of fixtures without making good damage — Appeal from decision of deputy official referee — Unusual circumstances — Respondents (plaintiffs in the action) let industrial premises to company P which used them for a chemical business, installing pipes and extractor fans which involved piercing holes in the outside walls — Company P got into financial difficulties and its assets were transferred to company G (the first defendants in the action) which was let into possession but was not an assignee of company P's lease at the time when the fixtures were removed — In due course company G also got into difficulties and eventually into liquidation, but before leaving the premises they removed the pipes and extractor fans without filling up or making good the holes in the walls left by the removal — The present appellant (the second defendant in the action), who was the last active director of company G, gave instructions for the removal but without ordering that the holes should be filled up — Company G, being insolvent and in liquidation, took no part in this appeal — The deputy official referee found company G liable for waste and the appellant personally liable for directing or procuring the tortious acts — The issue in the appeal, upon which the appellant's personal liability depended, was whether company G had been properly found liable for waste — Held, Sir George Waller dissenting, that, on the authorities, the removal of the fixtures by company G without filling up and making good the holes in the walls, constituted waste and the appellant was personally liable for procuring the tort of the company — When fixtures are removed the premises must be made good to the extent of being left in a reasonable condition — Where, as here, there was no obligation under contract, the liability was that of the person who removed the fixtures, not that of the person who installed them — The deputy official referee had come to the right conclusion — In dissenting, Sir George Waller said that when company P made the holes they had been guilty of waste and if they had remained in occupation and left without making good they would have been guilty of continuous waste — But all that company G had done was to remove their own property without actually damaging the walls and there was no authority that voluntary waste could be committed by the omission to do something — Kerr LJ, however, agreed with Dillon LJ and the appeal was dismissed — Discussion of authorities on waste in relation to fixtures

This was an appeal by Philip Givertz, the second defendant in an action brought by the respondents, Mancetter Developments Ltd, against Garmanson Ltd, the first defendants, and the appellant, from a decision of Sir Ian Percival QC, sitting as a deputy official referee. The action and appeal concerned allegations of waste made by the respondents as plaintiffs in respect of industrial premises known as 33 Willow Road on the Brackley Industrial Estate in Northamptonshire, of which premises the respondents were owners.

Nicholas Nardecchia (instructed by Robbins Olivey & Blake Lapthorn, agents for Leeds Smith, of Sandy, Beds) appeared on behalf of the appellant; Oliver Smith (instructed by F B Hancock & Co, of Banbury) represented the respondents.

Giving judgment, DILLON LJ said: This is an appeal by the second defendant in the action, Mr Givertz, against an order made by Sir Ian Percival QC, sitting as a deputy official referee, on May 25 1984. By that order the plaintiffs were given judgment against Mr Givertz in the sum of £524.87 with the costs of the action on the county court scale; the plaintiffs were also given judgment against the first defendant in the action, Garmanson Ltd ("Garmanson") in the sum of £771.58 with like costs, but Garmanson, being insolvent and in liquidation, does not appeal.

The plaintiffs are the owners of industrial premises known as Unit 33, Willow Road, on the Brackley Industrial Estate at Brackley in Northamptonshire. On January 29 1973 the plaintiffs let these premises to a company called Pilot Chemical Co Ltd ("Pilot") by a lease for a term of 42 years from June 24 1972 at a rent of £3,160 pa for the first five years and then as reviewed, and with full repairing obligations on the tenant. Pilot used the premises for a chemical business and to that end put in various pipes which pierced the outside walls of the building and also put extractor fans in the outside walls. The outside walls were of brick to about halfway up; above that there was a steel frame covered with sheets of cladding on the outside and lined with sheets of plasterboard on the inside. The putting in of the extractor fans involved cutting holes in the cladding and lining; the putting in of the pipes involved making holes either in the cladding and lining or, if the pipes were lower, in the brickwork. It is to be inferred that the putting in of the extractor fans and pipes was done with the plaintiffs' consent or that any ground of objection the plaintiffs may have had was waived by the acceptance of rent from Pilot. It is common ground that the extractor fans and pipes were tenant's fixtures which Pilot would have been entitled to remove during the term of the lease.

Unfortunately Pilot failed to keep the premises in repair and got into financial difficulties. Ultimately, on January 6 1978, a receiver of the undertaking and assets of Pilot was appointed. By an agreement for sale of January 30 1978, which seems to have been a usual form of hiving-down operation, the receiver agreed to transfer the assets of Pilot, including the lease, the tenant's fixtures and Pilot's stocks of chemicals, to a company formed or acquired for the purpose, which was Garmanson, and a few days later, on February 6 1978, the receiver sold and transferred the entire issued share capital of Garmanson — two shares of a nominal value of £1 each — to a company called Food Drugs and Cosmetic Colours Ltd, of which Mr Givertz was managing director. Mr Givertz thereupon became a director of Garmanson and he was the only active director of Garmanson until it, too, went into liquidation as hereinafter appears.

Garmanson went into occupation of the premises on or immediately after February 6 1978. It seems that Mr Givertz originally hoped that Garmanson might be able to trade profitably from the premises, and from March to July 1978 there were negotiations between the plaintiffs and Garmanson with a view to the grant to Garmanson of a new lease of the premises. During this period Garmanson paid, and the plaintiffs accepted, the instalments of rent which fell due under the existing 42-year lease on the March and June quarter days in 1978. However, Garmanson decided that it was not viable to continue trading in the premises, and accordingly, having given some notice to the plaintiffs, Garmanson left the premises and delivered up possession to the plaintiffs on October 5 or 6 1978.

Before delivering up possession of the premises, however, Garmanson removed all the tenant's fixtures installed by Pilot, and in particular the extractor fans and pipes which I have mentioned. This was done by employees of Garmanson and removal contractors, instructed in each case by Mr Givertz. It involved reopening the holes in the brickwork and in the cladding and lining above which had been made when the fixtures were installed by Pilot. No attempt was made to fill up or make good the holes. I would infer that Mr Givertz never gave any instructions, and never had any intention, that the holes should be filled up or made good.

In September 1978 the plaintiffs gave Pilot formal licence to assign the lease to Garmanson, and it was in fact so assigned on March 26 1979. In 1981 Garmanson went into liquidation; it is insolvent and any judgment against it is of little more than academic interest.

Immediately after Garmanson gave up possession in October 1978 the plaintiffs sent in surveyors to prepare a comprehensive schedule of dilapidations. There is no doubt that the plaintiffs have suffered damage from the disrepair of the premises, since when, after several months, they relet to new tenants, they had to allow the new tenants a seven-month rent-free period because of the disrepair.

It does not of course follow, because the plaintiffs have suffered damage, that Mr Givertz is personally liable to the plaintiffs for any of that damage. In this court we only have to consider the personal liability of Mr Givertz to the plaintiffs; the claims that the plaintiffs have against Garmanson are relevant only in so far as they provide the basis for a claim by the plaintiffs against Mr Givertz personally as the sole active director of Garmanson who gave the instructions for what was done by Garmanson. It is important to keep in mind that Mr Givertz is under no liability in contract to the plaintiffs; in particular he was never a guarantor of the original lease of the premises or of any arrangements under which Garmanson were let into possession of the premises.

There are, however, cases which establish that if a director of a company gives instructions for the commission of a tort by the company, the director may be personally liable in damages to the injured party for the tort although the tort was the act of the company. It is sufficient to refer to the speech of Lord Buckmaster in *Rainham Chemical Works Ltd* v *Belvedere Fish Guano Co Ltd* [1921] 2 AC 465 at p 476 and to the judgment of Atkin LJ in *Performing Rights Society Ltd* v *Ciryl Theatrical Syndicate Ltd* [1924] 1 KB 1 at pp 14-15 where he refers to directors being liable if they have themselves directed or procured the commission of the tortious act.

Although a full schedule of dilapidations had been prepared for the plaintiffs, the case at the trial of this action turned only on the damage caused by the removal of the tenant's fixtures without making good the holes. The learned deputy official referee held that, in removing the fixtures without making good the holes in the brickwork, cladding and lining, Garmanson committed acts of waste which were tortious. He further held that as Mr Givertz had himself directed or procured the tortious acts, he was personally liable in damages to the plaintiffs. The learned deputy official referee fixed those damages at the sum of £524.87, which represents the cost of making good the holes; in fact, the cost of making good the holes in the brickwork is very small, and by far the greater part of the sum represents the cost of providing and fixing new sheets of cladding and lining in the place of those left with holes in them. The learned deputy official referee held Garmanson liable in the higher sum of £771.58, which represents the £524.87 plus a further sum for damage, which went beyond Mr Givertz's instructions, caused to the premises by those who actually did the removal, eg damage to other parts of the premises caused by carelessness in taking out parts of the chemical plant which were tenant's fixtures. This appeal is, however, an appeal on principle and is not concerned with precise figures.

If the judge were right that the acts of Garmanson in removing the tenant's fixtures were acts of waste, it seems to me that the personal liability of Mr Givertz must follow, on the authorities mentioned above, because he directed and procured those acts. So far as the personal liability of a director is concerned, I can see no distinction between giving instructions for those acts, if they are acts of waste, and giving instructions for any other acts of waste, such as breaking down walls or removing landlord's fixtures. The key question on this appeal is therefore whether the acts of Garmanson were acts of waste, ie were tortious.

The judge delivered his judgment on the basis that by the end of the argument it had been conceded by the plaintiffs that Garmanson had the right to remove the tenant's fixtures. He says at one point that the question of "removability" was not an issue, and at another point he describes as central to the case a proposition of Mr Oliver Smith for the plaintiffs that, though Garmanson had a right to remove the equipment in question as trade fixtures, it had no right to do injury or damage to the property. When this appeal was called on, Mr Oliver Smith asserted that the judge was mistaken in supposing that he ever conceded that Garmanson had any right to remove tenant's fixtures; his main point had been, from first to last, that, though Pilot had had the right to remove the fixtures, Garmanson, being in occupation as licensee (on Mr Smith's submissions) and not as assignee of the lease, had not. I can well appreciate that there may have been a misunderstanding, but no objection was taken by Mr Smith in the lower court after the judgment was delivered. The point only surfaced in his skeleton argument delivered just before the hearing in this court. In view of this, and as it seemed that, if this court had to decide, in the absence of concession, the exact basis of Garmanson's occupation of the premises, we might need transcripts of the oral evidence at the trial which had not been bespoken, we ruled that Mr Smith was bound by the concession which the judge — understandably — supposed him to have made. I would add, as my own preliminary reaction, that, in the circumstances of this case, as rent was paid by Garmanson under the lease, a licence to assign was granted and the lease was in fact later assigned to Garmanson. I would be surprised if Garmanson did not have the same right to remove the tenant's fixtures as Pilot had had.

"Waste" is defined in *Woodfall's Law of Landlord and Tenant*, vol 1, at p 644, as being "a spoil or destruction to houses, gardens, trees, or other corporeal hereditaments, to the injury of the reversion of inheritance." It is divided into two main categories, voluntary waste and permissive waste. Permissive waste, which is not in question here, is a matter of suffering buildings to fall into disrepair by neglect. Voluntary waste, which is in question, is said in *Woodfall* to be actual or commissive, as by pulling down houses or altering their structure. Waste is a somewhat archaic subject, now seldom mentioned; actions in respect of disrepair are now usually brought on the covenant. But I see no reason to doubt the continuing validity, as a statement of the current law, of the following passage in *Woodfall* at p 679, viz:

> When fixtures which have become part of the realty and irremovable according to law have been removed, the landlord may sue as for waste: for such removal amounts to an injury to the reversion, which the law considers waste. Where such waste amounts to a breach of covenant, the landlord may sue either in tort, or on the covenant, at his election.

If the removal of fixtures which have become part of the realty and irremovable according to law — ie of landlord's fixtures — is actionable as for waste, what is the position where tenant's fixtures are lawfully removed but the places where they were installed are not made good?

The law of fixtures is of very ancient origin. Originally the position was simply that anything affixed to the land became part of the land and passed with the land — *quicquid plantatur solo, solo cedit*. The development of a tenant's right to remove tenant's or trade fixtures was a mitigation of that rule; it came about not by any change in the concept of what is a fixture but by a qualification of the tenant's obligation not to remove fixtures once they had been affixed to the land. See per Kindersley V-C in *Gibson* v *Hammersmith and City Railway Co* (1862) 2 Dr & Sm 603 at p 608. It is possible that, when a tenant's right to remove trade fixtures was first recognised by the courts, that right was upheld irrespective of any damage caused to the premises by the removal; thus in *Poole's* case as reported in (1703) 1 Salk 367 at p 368 it seems to have been held that the sheriff levying execution to enforce a judgment against a subtenant was entitled to remove trade fixtures installed by the subtenant in premises notwithstanding that the premises were thereby rendered ruinous and the liability to make good the damage was left to fall on the sublessor under his own covenants in his head lease. Whatever the initial view of the courts, however, the position that developed was, as put in *Amos and Ferard on the Law of Fixtures*, 3rd ed at p 124, that:

> It appears to have been generally understood in practice that, as well where

A trading as where ornamental fixtures are taken down, the tenant is liable to repair the injury the premises may sustain by the act of removal.

The extent of the liability is expressed in *Foley* v *Addenbrooke* (1844) 13 M & W 174 at pp 196 and 199 as being that the tenant must leave the premises in such a state as would be most useful and beneficial to the lessors or those who might next take the premises and must not leave the premises in such a state as not to be conveniently applicable to the same purpose. I would interpret this as a requirement of the law that if tenant's fixtures are removed the premises must be made good to the extent of being left in a reasonable condition.

The right of a tenant to remove tenant's or trade fixtures arose by B the common law independently of contract, though the right might be confirmed or excluded by contract. So equally, in my judgment, the obligation on the tenant to make good the damage, if tenant's fixtures were removed, arose at common law irrespective of contract, although there might also in a particular case be a relevant contract. This is demonstrated by the fact that, as between tenant for life of land and the remaindermen, the tenant for life had a similar right to remove tenant's fixtures and a similar obligation to make good damage, even though there would not have been any contract between the tenant for life and the remaindermen: see *Re De Falbe, Ward* v *Taylor* [1901] 1 Ch 523 at p 542.

The liability to make good the damage, or to repair the injury the premises may sustain by the act of removal of tenant's fixtures, must, C in so far as it is a liability at common law and not under a contract, be the liability of the person who removes the fixtures, and not of the person, if different, who originally installed the fixtures and left them there. The analysis of the liability at common law is, in my judgment, that the liability to make good the damage is a condition of the tenant's right to remove tenant's fixtures; therefore, removal of the fixtures without making good the damage, being in excess of the tenant's right of removal, is waste, actionable in tort, just as much as removal by the tenant of a landlord's fixture which the tenant has no right to remove is waste. As appears from the passage in *Woodfall* quoted earlier in this judgment, the landlord has an election whether to sue in tort or on the covenant if the damage is also covered by a relevant covenant. The act of the tenant, or in the present case D Garmanson, in removing the tenant's fixtures without making good was commissive rather than permissive, as I understand those terms used in the definitions of "waste", "permissive waste" and "voluntary waste" in *Woodfall*.

The main argument of the appellant on this part of the case was to the effect that Garmanson's liability for injury sustained by the premises by the removal of the fixtures was limited to extra damage caused by the removal and did not require Garmanson to fill up or make good the holes in the brickwork and cladding which had been lawfully made by Pilot. It seems to me, however, contrary to common sense and contrary to my understanding of the concept of voluntary waste, as something that applies even in the absence of any E contract, or repairing covenant, that a tenant who for his own convenience installs tenant's fixtures, and who has, in order to do so, to make holes in the walls of his landlord's building which are filled by those fixtures while they remain installed, should be allowed to remove the fixtures without filling the holes. So long as the fixtures remain installed, the building is wind and weather proof and there is no damage to the reversion. Once affixed, the fixtures become part of the freehold. If they are removed, albeit legitimately, and the holes are not filled, the reversion then suffers damage. To make good that damage by filling in the holes (which involves, in part, replacing sheets of cladding and lining with holes in them) is part of the condition attached by law to Garmanson's right to remove the fixtures.

F The filling of screw holes or nail holes where a fixture is removed which has been screwed or nailed to a wall may be a matter *de minimis*, and the comments of Rigby LJ in *Re De Falbe* at p 542 show that the tenant's obligation does not extend to redecorating a wall where a fixture has been removed. But the leaving of holes, such as those in question here, affects the structure and is not a matter of mere decoration.

I agree with the reasoning of the learned deputy official referee in his very careful judgment, and for my part I would dismiss this appeal.

Dissenting, SIR GEORGE WALLER said: This appeal is concerned with the liability of a director of a company for the actions of the company's servants. If a director authorises an employee of the company to do something which is a tortious act he, the director, will be personally liable. If, however, the instructions given do not authorise a tortious act, the director will not be liable for the consequences (see *Wah Tat Bank Ltd* v *Chan Cheng Kum* [1975] AC 507).

The facts have been fully set out in the judgment of Dillon LJ and there is no need to repeat them. When the occupation by Garmanson Ltd ("Garmanson") ended, the landlord had a schedule of dilapidations prepared. There was a dispute as to the status of Garmanson, who occupied the premises for six months having purchased the trade fixtures from Pilot Ltd ("Pilot"), the previous tenants. Whether Garmanson was a tenant or licensee, it was certainly not an assignee of Pilot's tenancy at the time when Garmanson left. When Pilot became the tenant, it installed a ventilation system for its chemical manufacturing. This included exhaust fans in the walls and it made holes in the walls to hold these exhaust fans and also holes in the walls for piping. When Garmanson left, it removed these trade fixtures which it had purchased. This was done, according to the finding of the learned judge, without causing any damage to the fabric, but nothing was done to fill the holes left behind. I agree with Dillon LJ's view that Garmanson probably had the same right to remove the tenant's fixtures as Pilot had. In my view, whether it was as tenant or licensee, the right to remove would include an obligation to restore the realty to its condition before the original lease was entered into. See *Re De Falbe* [1901] 1 Ch 523.

The judge found that Garmanson was not liable for many of the items but found that it was liable for leaving the holes unfilled. The judge further said:

> It is my view and conclusion that Garmanson, in removing each of the items in question from its hole, and leaving the hole, did do injury and damage to the buildings and therefore to the plaintiff and that each was a wrongful act. It therefore had all the ingredients of a tort. I do not think it necessary to endeavour to put a label on it.

The judge has refrained from giving any name to the tortious act, but in my judgment, unless he were creating a new tort, the only possible tort is the tort of waste. Dillon LJ has already quoted the definition of "waste" from *Woodfall* and I do not need to repeat it here. In essence there has to be a "spoil or destruction" to the building "to the injury of the reversion". No question arises of permissive waste which is caused by long-term neglect. The question here is whether Garmanson committed voluntary waste. The earlier authorities dealing with voluntary waste are all concerned with substantial damage to the reversion. In *Yool on Waste* (1869), for example, there is no mention anywhere of damage as slight as a hole in the wall. In more modern cases, however, it is different. In *Re De Falbe* (*supra*) the question arose as to whether tapestries fixed to the walls passed to the freeholder on the death of the tenant for life or whether they could be removed by the executors. It was held that they could be removed but that the executors must pay the cost of any damage done in the course of removing them, but not for the consequential cost of redecorating. So there is no doubt that if damage is caused to the fabric when fixtures are removed, the person who causes that damage will be liable to put that damage right or to pay damages being the cost of doing so. A similar case was *Spyer* v *Phillipson* [1931] 2 Ch 183, where valuable panelling was held to be a tenant's fixture but damage to the realty caused by its removal had to be paid for. It follows from the other passage quoted from *Woodfall* by Dillon LJ that if the panelling had become part of the realty, its removal would have been waste quite apart from any damage caused in the course of removal. In *Marsden* v *Edward Heyes Ltd* [1927] 2 KB 1 the tenant removed an internal partition wall. Bankes LJ, relying on the words of Lord Loreburn in *Hyman* v *Rose* [1912] AC 623 at p 632, "It is a question of fact whether such an act changes the nature of the thing demised . . . ", held that it was an act of waste, that which had been a shop with living rooms behind had been made into a large shop. Although the partition wall had been removed 10 years before so that the original act was statute barred, it was held to be a continuing act to maintain the building in the same condition.

In the present case there can be little doubt that Pilot when they made the holes were guilty of waste. What they had done was a "spoil or destruction" to the building. If Pilot had remained in occupation and then removed the fittings and left the premises without repairing the holes, they would have been guilty of continuing the waste up to the time of leaving. Indeed, even if they had left the fittings in place contrary to the wishes of the landlord, they might well have been said to be continuing the act of waste. The basic tortious act was making the holes.

The question which this court has to decide is whether or not Garmanson committed a tortious act when it removed the fittings from the holes. All that Garmanson did was to remove its own property without doing any damage to the walls. The realty was a wall with a hole in it and was not a wall with a hole filled with an exhaust fan. In the course of argument counsel were unable to draw our attention to any case where it was held that voluntary waste could be committed by an omission to do something and I have not been able to find any such case. Since it is accepted that no damage was caused to the walls when the fittings were removed, I have come to the conclusion that it is not possible to say that Garmanson was guilty of waste in removing the fittings. There was no tortious act at that time. Accordingly it follows that the appellant cannot be guilty of authorising a tort and I would allow this appeal.

Agreeing that the appeal should be dismissed, KERR LJ said: My mind has wavered during this appeal, and having had the opportunity of reading the judgments of Dillon LJ and Sir George Waller I confess that I remained in doubt for some time. In the end, however — and contrary to my earlier impression — I have come to the conclusion that the appeal should be dismissed for the reasons stated in the judgment of Dillon LJ. To these I would respectfully add the following comments, particularly in the context of the contrary views expressed by Sir George Waller.

First, it is important to bear in mind the highly unusual facts of this case, in relation to which Mr Givertz's liability for the tort of waste falls to be considered. The issue whether or not this tort results from a failure to reinstate or repair premises arises in the present case independently from any contractual stipulations which normally cover the same ground, usually in the form of covenants in leases. Such covenants will bind not only the original tenant but also any assignee of the lease and any licensee occupying the premises with the freeholder's consent subject to an express or implied condition that he is to be subject to the same obligations. In most of these situations I doubt — without expressing any concluded opinion — whether there would be much scope for the application of the tort of waste, because the act or omission complained of will usually be governed by contractual stipulations. I doubt whether the law goes so far as to permit an alternative claim in contract or tort in the majority of such cases. Prior to the passage in *Woodfall* concerning the wrongful removal of landlord's fixtures which Dillon LJ has quoted, stating that the landlord may sue either in tort or on the covenant, there is a reference in para 1-1514 at p 644 to *Jones* v *Hill* (1817) 7 Taunt 392. In that case a tenant was sued for permissive waste although his lease contained a full repairing covenant. It was held that the facts did not amount to permissive waste, but in delivering the judgment of the court Gibbs CJ also said at p 396: "Waste can only lie for that which would be waste if there were no stipulation respecting it . . . ", and this is the situation in relation to the tort of waste as summarised in para 1-1514 of *Woodfall*. In this connection one should also bear in mind the restricted meaning which was given in the authorities mentioned below to acts constituting waste by tenants who were subject to the usual covenants.

If that is the correct approach, two consequences follow. First, in the great majority of cases the present problem would not arise, because the owner of the premises would be bound to sue on the reinstatement of repairing covenants, or contractually binding conditions *vis-à-vis* licensees, and not on the basis of the tort of waste. What I have in mind is that a failure by a tenant or licensee company to reinstate or repair could not normally be visited upon a director of the tenant or licensee, by seeking to make him personally liable in tort on the ground that he directed or procured what would in effect have been a breach of contract by his company. This issue would arise only in exceptional cases like the present where the company — in this case Garmanson Ltd — was not subject to any contractual requirements, because its occupation of the premises occurred without the attachment of any relevant contractual stipulations, and where the act complained of is properly to be regarded as constituting the tort of waste.

Second, and more directly material to the problems raised by the present case, it would follow that when Pilot made the holes or apertures in the fabric of the building in order to accommodate the pipes for their machinery and the extractor fans, Pilot were not there and then committing any act of waste. With respect to the contrary view expressed by Sir George Waller, I cannot agree with his conclusion on this point. The decision of this court in *Marsden* v *Edward Heyes Ltd* [1927] 2 KB 1 proceeded on the basis that the structural alterations made by the defendant were such as to show that he had not used the premises "in a husbandlike manner" (per Bankes LJ at p 6, quoting Gibbs CJ in *Horsefall* v *Mather* (1815) Holt NP 7) or "in a tenantlike manner" (per Scrutton LJ at p 7) and Atkin LJ at p 8 used both these expressions. Moreover, Bankes LJ relied upon the test laid down in *West Ham Central Charity Board* v *East London Waterworks Co* [1900] 1 Ch 624, which is correctly summarised in *Woodfall* (para 1-1522) by saying that "an act which alters the nature of the thing demised is waste". The wrongful removal of landlord's fixtures, as referred to by *Woodfall* in the passage quoted by Dillon LJ, may fall within this category and may therefore give alternative rights of suit in contract and tort. But I cannot accept that the same could be said of Pilot's acts in making the holes. We have not seen the lease between the plaintiffs and Pilot or any other material showing the purposes for which Pilot were entitled to use the premises. But we know that they were used for a chemical business, as Pilot's name indicates, and it is an overwhelmingly probable inference that in making these holes Pilot were not acting so inconsistently with the terms of their lease that they could be said thereby to have acted in an untenantlike manner. One can test this by asking oneself whether the plaintiffs could possibly have sued Pilot for the tort of waste there and then, as soon as the holes had been made. In my view, the answer would clearly be in the negative. It would have been held that although the fabric of the premises had been damaged, there was no act of waste, since the interference with the fabric of the premises did not go so far as to have this effect in the circumstances. The holes clearly required to be made good under the covenants in the lease by reinstatement or repair if the machinery and extractor fans were removed at the termination of the tenancy. But making the holes was not waste, since these covenants — and any other terms in the lease dealing with Pilot's right to occupy and use the premises for a chemical business — should be treated as "stipulations respecting" acts of this kind, to use the words of Gibbs CJ in *Jones* v *Hill*.

If the making of the holes by Pilot was not an act of waste, then it necessarily follows that Pilot were not guilty of any act of waste at all. With the machinery pipes and extractor fans in position, the deputy official referee rightly said: "I do not think that anyone, landlord, tenant or passer-by, would have said at that point that any injury or damage had been caused to the building."

When one then turns to the acts of Garmanson Ltd or their removal contractors, as directed by Mr Givertz, there is virtually nothing which I can usefully add to the analysis in the judgment of Dillon LJ. I respectfully agree with Sir George Waller than an omission to do something can hardly — and perhaps never — constitute an act of voluntary waste. But in the present case there was a voluntary act, the removal of the machinery and the extractor fans. Admittedly, there was a right to remove them, although they had been affixed to the realty. But if their removal caused the fabric of the premises to cease to be wind and water tight due to the exposure of the holes, as was the case, then the act of removal also had the effect of causing the building to become damaged when it could not be considered to have been relevantly damaged before. I therefore see no escape from the conclusion that a direction to carry out the removal without reinstating the holes was an act for which a claim in the tort of waste must lie. Accordingly I would dismiss this appeal.

The appeal was dismissed with costs.

For the other cases on this subject see p 39

LANDLORD AND TENANT

BUSINESS TENANCIES

A
Chancery Division

December 20 1985

(Before Mr Francis FERRIS QC, sitting as a deputy judge of the division)

HAMISH CATHIE TRAVEL ENGLAND LTD v INSIGHT INTERNATIONAL TOURS LTD AND ANOTHER

Landlord and tenant — Rent review clause in underlease — Whether certain alterations, assumed for the purpose of the B decision to be improvements, were to be taken into account or disregarded in determining the review rent — The material part of the review clause provided that the rental value should be determined on the supposition (if not a fact) that the tenant had complied with all the obligations imposed by the tenancy; and that no account should be taken of, *inter alia*, the effect of improvements (to which the landlord had given written consent) carried out by the tenant otherwise than in pursuance of an obligation to the landlord — In fact in the course of a long history of complex negotiations written consent had not been given to the alterations in question, which included the installation of a new shop front and ground-floor entrance, C the installation of a toilet and shower and the infilling of an internal staircase — Plaintiff landlords eventually sought a declaration that, as no improvements had been carried out for which the necessary consent had been obtained, account should be taken of all improvements in determining the rental value — Defendant tenants claimed that consent had been unreasonably withheld and made four submissions — These were (1) that the review clause should be read as if it provided that improvements to which the landlords had unreasonably refused consent should be disregarded; (2) that in view of the unreasonable refusal the plaintiffs would be taking advantage D of their own wrong if the improvements were taken into account; (3) that, as the review clause required it to be supposed that the tenants had complied with all their obligations, it must be assumed that they had not carried out the improvements without consent; and (4) that the landlords were estopped from denying that the tenants had obtained such consent, on the ground of proprietary estoppel or estoppel by convention — Held, rejecting the claim that the landlords had unreasonably refused consent and the four submissions mentioned above, that account should be taken of all the improvements in determining the rental value — E Declarations made as sought by plaintiff landlords

This was an originating summons issued by the plaintiffs, Hamish Cathie Travel England Ltd, seeking declarations the effect of which was that, for the purpose of the relevant rent review clause, account should be taken of the effect of all improvements to the demised premises carried out by the first defendants, Insight International Tours Ltd or their predecessors in title — the second defendants were Black & Edgington Ltd, of which the first defendants were a subsidiary. The rent review clause was in an underlease of premises at 25 and 25A Cockspur Street, London W1. The freehold of the building was vested in the Crown and the plaintiffs were head lessees for a term ending in 1995. There was a counterclaim by the defendants that consent to the improvements had been unreasonably withheld.

Peter Crampin (instructed by Malkin, Cullis & Sumption) appeared on behalf of the plaintiffs; Michael Driscoll (instructed by Kagan Morris & Co) represented the defendants.

Giving judgment, MR FRANCIS FERRIS QC said: The question raised by these proceedings is whether certain alterations made by a tenant to demised premises, some or all of which alterations may be classified as improvements, are to be disregarded in assessing the amount of the rent payable in respect of the demised premises under a rent review.

The premises in question are the ground floor and basement of a building known as 25 and 25A Cockspur Street, London W1. The building as a whole consists of a basement, ground floor and three upper floors. The freehold has at all material times been vested in the Crown on whose behalf it is administered by the Crown Estate Commissioners ("the Commissioners"). The plaintiff company, which as its name suggests is engaged in the travel business (and to which I shall refer as "Travel England"), has been in occupation of part of the building since 1975. Between 1975 and the summer of 1979 it occupied the ground floor and basement under tenancy arrangements of which no details are before the court. During that period the entrance to the ground floor was set back some way from the street frontage and was reached through an area open to the street at the front and lined with display windows on each side. The entrance to the upper part was by means of a door and staircase situated to one side of this area and reached by passing through the area. This arrangement was not very satisfactory, because it did not provide a good means of access to the upper parts and the open entrance area attracted rubbish and sometimes vagrants. Moreover the ground floor and basement of the building were in a poor state of repair, the basement being particularly bad and almost unusable.

In 1979 Travel England agreed with the Commissioners to take a new lease of the entire building on terms which I shall mention in a moment. Presumably in anticipation of the grant of such lease, Travel England moved its offices in 1979: first (but only for a short time) to the fourth floor of the building, and then in September 1979 to the third floor, which it still occupies. As part of the negotiations for the new lease, Travel England proposed to alter the entrance to the building so that the upstairs parts would be reached by a new door facing directly on to the street. This was agreed to by the Commissioners and, as will be seen, the making of this alteration became an obligation of Travel England under the new lease.

The new lease ("the headlease") is dated December 10 1979. By it, the Commissioners demised the whole of the building to Travel England for a term beginning on July 19 1979 and ending on October 10 1995 at a rent of a peppercorn until September 22 1979 and thereafter a rent of £33,500 per annum plus additional rents payable from each of two review dates, namely October 10 in the years 1985 and 1990. These additional rents were to be ascertained by reference to the market rent of the demised premises at the review dates. I need not describe the machinery for ascertaining such market rent except to say that there was no provision for leaving out of account the value of improvements made by the tenant.

The headlease contained two covenants on the part of the tenant which call for mention. First, by clause 3(3)(a) the tenant covenanted that it would, before July 19 1980 or such later date as the landlord might agree in writing, complete the closure of the existing entrance to the upstairs part of the building at ground-floor level and the installation of a new entrance in such a position and of such design as should be approved by the landlord and all such works as were required to be done to the demised premises to put the same into good and substantial repair, order and condition. Second, by clause 3(13) the tenant covenanted (*inter alia*) that none of the principal timbers, iron or steelwork or walls of the demised premises should be altered, cut or damaged and that no alteration whatsoever should be made in the plan elevation, architectural decorations or external colour of the demised premises. It will be observed that this last mentioned covenant was an absolute covenant against the making of alterations, not a covenant against the making of alterations without the landlord's consent.

By an underlease dated January 18 1980 and made between Travel England as landlord and a company named Avco Financial Services Ltd ("Avco") as tenant, Travel England sublet to Avco the ground floor and basement of the building for a term of 16 years and 9 days from September 29 1979 at the initial rent of £15,000. The question with which I am concerned arises under this underlease and I must refer to a number of its provisions.

First, the term of the underlease will expire on October 8 1995: so that Travel England reserved only a nominal reversion of two days. Second, the underlease provided for the rent to be reviewed on September 29 1982, September 29 1985 and September 29 1990. The last two review dates are practically the same as those on which the rent payable under the headlease is to be reviewed, but the review date of September 29 1982 has no counterpart in the headlease. The underlease provided that on each review date the rent should be increased to an amount equal to the best rack rental value of the demised premises, the meaning of which is set out in clause 2(C) of the underlease which I must read in full:

The rack rental value of the demised premises at each date of review shall be such an amount as may be agreed between the Landlord and the Tenant or determined in accordance with sub-clause (D) of this Clause as representing the rental value of the demised premises for a term of years equal to the terms of years granted by these presents as between a willing lessor and willing lessee with vacant possession but upon the supposition (if not a fact) that the Tenant has complied with all the obligations on the part of the Tenant imposed by these presents (but without prejudice to any rights of the Landlord in regard thereto) and taking no account of:

(i) any goodwill attributable to the demised premises by reason of any trade or business carried on therein by the Tenant or any undertenant; and
(ii) any effect of any improvements to the demised premises (to which the Landlord shall have given written consent) carried out by the Tenant or any undertenant otherwise than in pursuance of an obligation to the Landlord; and
(iii) any effect on rent of the fact that the Tenant or any undertenant may have been in occupation of the demised premises; and in all other respects on the terms and conditions of these presents.

The question in this case is whether certain assumed improvements which I refer to later are to be left out of account by virtue of para (ii) of clause 2(C).

The next provision of the underlease which I must mention is clause 3(J), relating to alterations or additions. This contains two separate subclauses. Subclause (i) provides that if the tenant desires to make any structural alterations or additions it is to supply full particulars to the landlord which, if it agrees in principle to the making of such alterations or additions, is then to have the option to carry out the works itself and to charge an increased rent calculated by reference to the cost of such alterations or additions. Subclause (ii) is of more immediate relevance to the issues with which I am concerned. It is a covenant on behalf of the tenant and reads as follows:

Subject as aforesaid not to cut remove divide alter maim or injure the demised premises nor merge the demised premises with any adjoining premises nor make any alterations or additions to the demised premises except with the previous consent in writing of the Landlord and in accordance with drawings and specifications previously submitted to and approved in writing by the Landlord's Surveyors (such consent not to be unreasonably withheld)...

There follows a proviso enabling the landlord to attach certain conditions to any such consent, but none of the permitted conditions is material to the issue in this case.

By clause 3(P) the tenant covenanted (*inter alia*) at all times during the term to comply in all respects with the Town and Country Planning Acts and not to make any application for planning permission or give any notice to any authority of an intention to commence or carry out any development or any step related thereto without the previous consent of the landlord, such consent not to be unreasonably withheld. By clause 3(S) the tenant covenanted to observe and perform the covenants, other than the covenants for payment of rent, contained in the headlease so far as the same related to the demised premises. One effect of this last mentioned covenant **G** was that the tenant became liable to perform the obligations assumed by Travel England under clause 3(3)(a) of the headlease in respect of the alteration of the entrance to the upstairs part and the putting of the ground floor and basement into good and substantial repair and condition.

It appears that Avco entered into the underlease with a view to obtaining the ground floor and basement for the use of one of its subsidiaries, a company named Cartan Travel Ltd ("Cartan"). Cartan went into occupation and carried out some of the works of repair and alteration which the headlease required to be done. In particular, the entrance to the upstairs part of the building was repositioned as required. But the necessary work was not finished, a newly constructed wall not being properly rendered and the work not **H** being done to a very high standard. Some repair work was also carried out but evidently not to the satisfaction of the Commissioners.

At some time during or before September 1980 the Commissioners served on Travel England a schedule of dilapidations specifying various work which the Commissioners required to be done in order to comply with the covenants in the headlease. On September 15 1980 Travel England served on Avco a schedule of dilapidations incorporating such of the works mentioned in the Commissioners' schedule as related to the ground floor and basement.

At about this time, or shortly afterwards, Avco sought an assignee for the underlease. By October 20 1980 the first defendant, Insight **J** International Tours Ltd ("Insight"), appeared on the scene as a prospective assignee. As its name suggests, Insight is also engaged in the travel business. It was and remains a subsidiary of the second defendant, Black & Edgington Ltd. Before the end of November 1980 the solicitor to Travel England, Mr K D Bichard, a partner in the firm of Cripps Harries Willis & Carter, was in contact with Insight's solicitors, Kagan Morris, and had submitted to them a draft of the proposed licence to assign. In its original form this draft contained covenants on the part of Insight to complete the installation of a new shop front at ground-floor level within six months and within one month to do all such works as were necessary to put the demised premises into good and substantial repair, order and condition to the satisfaction of Travel England and the Commissioners. **K**

Between the submission of the draft licence and the end of 1980 there were several communications about the transaction between the solicitors to the parties. I need not go into these in any detail. It will suffice to say that on December 1 1980 Kagan Morris sent to Mr Bichard a plan which was said to show Insight's proposals for the premises. Travel England was asked to confirm that it was agreeable to these proposals, that it did not wish to carry them out itself (it appears to have been thought that the proposals involved structural work to which the option contained in clause 3(J)(i) of the underlease applied) and that the works would be excluded on a rent review. Shortly afterwards Mr Bichard informed Kagan Morris over the telephone that Travel England did not wish to carry out the works **L** itself and was not agreeable to the alterations being excluded on a rent review. On December 23 1980 Mr Bichard indicated to Kagan Morris that Travel England was not content with the plan which had been submitted on December 1. He asked Insight to submit three drawings showing more clearly the structural alterations which were proposed, and he suggested that a meeting be held, to be attended by the surveyors to the parties, to discuss the proposals. He informed Kagan Morris that Travel England was not able to give immediate consent to the proposals, that the Commissioners had not been approached for their consent to the alterations and that Travel England would only approach the Commissioners for such consent when a more satisfactory plan had been supplied.

There matters remained until a meeting on January 13 1981. This **M** meeting was attended by Mr Grant (the managing director of Travel England) and Mr Bichard on behalf of Travel England; by Mr Tarsh (a director of Insight) and Miss Shirley Morris (a partner in Kagan Morris) on behalf of Insight; and by representatives of Cartan. There was a wide-ranging discussion of all the issues. The evidence indicates that there was an air of urgency about the matter. Avco, through Cartan's representatives, indicated its desire to divest itself of the underlease; Insight was eager to complete the assignment and to begin work on the premises; and Travel England was concerned that the ground floor and basement should be put into good repair as soon as possible. There was some discussion as to the alterations which Insight was going to carry out, but it is not clear how far these

A alterations were explained in detail. Evidently there was a great measure of agreement about the matters which were discussed. Mr Grant said in evidence that he came away from the meeting with the feeling that matters were sufficiently resolved and that Insight would be taking an assignment of the underlease. Mr Tarsh evidently had much the same feeling, because the transaction with Avco was rapidly finalised.

One of the matters which were discussed at the meeting on January 13 was whether the alterations would or would not be left out of account on a rent review. On the affidavit evidence there was some conflict as to the course of this discussion. However, in cross-examination Mr Grant, and I think Mr Bichard as well, agreed that Mr Tarsh's version of the discussion was probably the most accurate.
B What Mr Tarsh said in his affidavit was that the question of exclusion of the alterations on a rent review was not resolved at the meeting, that there was no agreement to reconsider the matter in due course, but that Mr Grant and Mr Bichard simply did not commit themselves at the meeting to the alterations being excluded. I accept this evidence as substantially correct.

After the meeting on January 13 1981 matters proceeded with some speed. The form of licence to assign was agreed between the solicitors to the parties. The covenant which had been contained in the draft as to completion of a new shop front within six months was omitted, and the period for carrying out the work necessary to put the demised premises into good and substantial repair was extended to
C two months from the date of the licence. It was agreed that the second defendant should join in the licence and become surety for Insight. The assignment of the underlease from Avco to Insight was actually completed on January 21. At that date the licence to assign had not yet been executed, but it was completed on February 5 and was then backdated to January 21. Nothing seems to turn on this irregularity. I ought, however, to say that, despite the absence of any reference to the new shop front in the licence, I am satisfied that both Travel England and Insight had it in mind that work on the new shop front would at least be begun while the repair work was being carried out.

In a letter of January 21 1981, which was delivered by hand to Kagan Morris and was received by them before the assignment had been completed, Mr Bichard said among other things the following:

D We certainly confirm that in principle our clients would like your clients' proposed works carried out, but we must make it clear that as yet we have certainly not yet granted a Licence for these works. As mentioned in previous correspondence, if you wish to apply to us for such a Licence in due course, then perhaps you would be kind enough to let us have an undertaking to be responsible for our charges in this matter, when the application can be considered further.

Kagan Morris replied immediately on January 22 giving their undertaking in respect of costs.

In connection with the schedule of dilapidations, the Commissioners had for some time past been represented by Cluttons (on whose behalf the person concerned was a Mr Sankey) and Travel England had been represented by Hillier Parker May & Rowden
E ("Hillier Parker"), where the person concerned was a Mr Crowther. In about the middle of February 1981, Insight instructed a firm of architects, who were not the architects who had prepared the plan submitted in December 1980, namely Trehearne & Norman, Preston & Partners ("Trehearnes"), to prepare new plans for their proposed alterations (the person dealing with the matter being a Mr McKenzie). On February 25 1981, Trehearnes wrote to Hillier Parker saying that Insight had instructed Trehearnes to obtain agreement from Travel England to alterations and improvements to the premises. Trehearnes enclosed new drawings, one of which was stated to indicate the proposals which Insight wished to implement without delay. A copy of the same letter with the drawings was sent to Cluttons with a covering letter stating that Trehearnes would like, on
F behalf of Insight, the Commissioners' approval to the proposed alterations.

On February 26 1981, Cluttons replied to Trehearnes' letter, stating that they saw no objection in principle to the proposal and commenting that they would advise the Commissioners to consent to the proposal provided certain works, being the outstanding items from the schedule of dilapidations, were carried out. These works were then set out under five heads. In the concluding paragraph of the letter Cluttons said:

If you could confirm that these conditions will be met and subject to any other comments which the mesne landlord might wish to make, we would be prepared to recommend the Commissioners to grant consent to your proposals.

At the time when the letter of February 25 was received by Hillier Parker, Mr Grant was abroad, where he remained until March 9. Hillier Parker were therefore unable to obtain direct instructions from him, but Mr Crowther discussed the matter on the telephone with Mr Bichard. In a letter of February 27 1981 to Hillier Parker, Mr Bichard recorded his view as being that no acknowledgement of consent to the alterations should be given because Mr Bichard had not yet himself approached the Commissioners and because Mr Grant ought to see the plans as he had a keen interest in the new entrance. Mr Bichard remarked that the consent must in any event be dealt with by deed and not on an informal basis. On February 27 Mr Crowther spoke to Mr McKenzie of Trehearnes on the telephone. It seems likely that he had not at that time received Mr Bichard's letter. According to a letter which Mr McKenzie wrote to Mrs Hoffman (another director of Insight) on March 2, which is consistent with a note which Mr McKenzie appears to have written immediately after the telephone conversation, Mr Crowther informed Mr McKenzie that approval of Travel England would be given subject to the approval of the Commissioners and of the Westminster City Council as planning authority. In that same letter to Mrs Hoffman, Mr McKenzie referred to Cluttons' letter of February 26 as approving the proposals shown in the drawings. This of course was not entirely correct. On March 4 Mrs Hoffman wrote to Mr McKenzie in connection with the five outstanding items mentioned in Cluttons' letter of February 26. Two of these items were regarded by Mrs Hoffman as being of no concern to Insight. As regards the other three items, Mrs Hoffman indicated that remedial work was well under way.

On March 17 Mr Sankey of Cluttons wrote to Mr Crowther, saying that he had passed the premises and noticed that Insight had commenced the proposed work without the Commissioners' formal consent and without planning consent. He concluded his letter by saying:

We can see no reason why consent should not be granted in principle to allow these works to proceed, subject to the conditions set out in our letter of the 26th February. We should appreciate hearing whether or not your clients agree these proposals so that consideration can be given to the granting of consent to the works proposed.

Mr Sankey sent a copy of this letter to Trehearnes and it appears that they sent a copy of it to Mrs Hoffman on March 19.

Also on March 19 a meeting took place between Mr Tarsh and Mr Grant at which the new shop front was discussed. It appears that Mr Grant had not liked some aspects of the designs shown on the drawing which Trehearnes had put forward on February 25. After the meeting on March 19 Mr Tarsh at once went back to Trehearnes and got them to produce a revised drawing, a copy of which he sent to Mr Grant on March 20, asking Mr Grant for his comments on a number of matters. It does not appear what comments, if any, Mr Grant made. He said that he thought that he had already made his comments to Mr Tarsh on March 19. Mr Tarsh did not point to any specific subsequent discussions, although he said he thought that Mr Grant had approved the detailed points mentioned in his letter of March 20 as the work was carried out.

It appears from an account submitted by Insight's builders on March 31 1981 that the bulk of the work, both of repair and of alteration, had been carried out by that date, although the work to the new shop front might still have been uncompleted.

On April 2 1981, Hillier Parker wrote to Insight saying that they had visited the premises and noted that alterations to the ground floor and basement were going on apace. The letter continued:

We must formally record the fact that our clients, Travel England Ltd, have not yet given you their formal approval as you have not yet furnished them with evidence that you have obtained all the necessary statutory and landlords approvals.

Therefore you must recognise that the works currently in progress are being executed entirely at your own risk.

Mr Bichard seems to have been unaware of the plans which had been submitted to Hillier Parker on February 25 and of the revised plan of the frontage supplied to Mr Grant on March 20. On March 24 he wrote to Kagan Morris, referring to earlier correspondence in which the question of submission of new plans had been discussed, and asked whether or not fresh plans had been prepared. Kagan Morris seem to have been equally unaware of what had been going on, because on March 26 they replied that they had asked Insight about this matter. On April 8, having received from Mr Crowther a copy of Hillier Parker's letter of April 2 to Insight, Mr Bichard wrote to Kagan Morris saying:

> Whilst our clients are naturally pleased that progress has been made so quickly may we respectfully remind you that our clients cannot permit the proposed alterations to the Ground Floor entrance until firstly our clients consent to this is given and secondly the consent of the Freeholders is obtained through ourselves. We must emphasise that we ourselves need to contact the Freeholders concerning these proposals and informal contacts between your own clients and Messrs Cluttons are not sufficient as far as we are concerned, quite apart from the fact that as yet of course our clients have not agreed to your own clients proposals. Before we approach the Freeholder's solicitors we naturally require to see full plans showing the proposal alterations and if your clients wish to proceed with this no doubt they will forward such plans to us as soon as possible.

On April 9 Kagan Morris replied that they had reminded Insight of the position and sent Insight a copy of Mr Bichard's letter of April 8.

Thereafter correspondence between the solicitors on this aspect of the matter fell into abeyance. On April 30 Westminster City Council gave planning consent for the new shop front. Consent for an illuminated sign was refused and Insight decided not to install such a sign. The evidence shows that all the work was completed at about the time the planning consent was granted or quite soon afterwards. There was, however, an exchange of correspondence between Mr Grant and Mr Tarsh in August 1981 in which Mr Grant complained about certain minor matters concerning the new entrance to the upstairs part of the building, but this correspondence appears to be of no great significance and Insight duly attended to all these matters.

In June 1981 Cluttons drew attention to the fact that the pavement light, which had been removed by Avco or Cartan and replaced by a concrete slab, must be replaced by a new pavement light. Unsuccessful attempts were made to persuade Cluttons to withdraw this requirement and eventually Insight did the necessary work in November 1981.

It thus appears that by November 1981 the five matters mentioned in Cluttons' letter of February 26 as being conditions of the Commissioners' consent under the headlease had all been dealt with, although neither the Commissioners nor Cluttons had ever been formally notified that these conditions were accepted and no formal consent had been obtained from the Commissioners.

On January 20 1982, Mr Bichard had to write to Kagan Morris about another matter concerning the property, and he reminded Kagan Morris that Insight had carried out various works to the ground-floor entrance without any formal licence or consent from Travel England. He referred back to the letters of April 8 and 9 1981 which I have already mentioned and continued:

> May we take this opportunity of reminding you again that the headlease of this property, the covenants of which our clients agree to observe, contains restrictions upon alterations and therefore any works which your clients have carried out are entirely at their own risk. Having said this, we do not wish informal contacts between your clients and our clients' landlord's agents to take place and any future dealings should be direct with our clients or with ourselves and not with Messrs Cluttons. It does appear that Messrs Cluttons are aware of the alterations to the ground floor entrance and have not raised any objections to these but we must naturally reserve our client's position with regards these generally for reasons which we are sure you will appreciate.

On February 11 1982, Kagan Morris replied:

> In connection with the alterations we have sent a copy of your letter to our clients who tell us that their proposals were approved some time ago by Cluttons on behalf of the Superior Landlords and that the works have been inspected and approved by Philip Grant on behalf of the Landlords. We do, however, agree that a formal Licence should be completed for which purpose we have asked our clients to let us have sets of the final drawings and we trust that as soon as we can produce these to you we will be able to settle the outstanding Licence.

On February 19 1982, Kagan Morris sent Mr Bichard two drawings and asked for a formal licence for the alterations shown in these drawings. One of the drawings then submitted included ground-floor and basement plans which were for all practical purposes the same as those shown in the corresponding drawing previously submitted by Trehearnes to Cluttons and Hillier Parker under cover of the letters of February 25 1981. The other drawing showed the front and side elevations of the new shop front, the front elevation being similar in all important respects to the drawing which Mr Tarsh had sent to Mr Grant on March 20 1981. For reasons which are not explained, no formal licence was granted in response to the request contained in the letter of February 19 1981 and the request was not followed up.

In October 1982 the first rent review under the underlease became due. Apparently the valuers to Travel England and Insight produced figures for the revised rent which were very wide apart, a major cause of the divergence being the exclusion of the value attributable to the alterations carried out in 1981 by one valuer and the inclusion of this G value by the other. This led to a further exchange of correspondence between the solicitors to the parties in March 1983 in which Kagan Morris claimed that Travel England had already given written consent to the alterations and Mr Bichard denied this. On December 9 1983 Travel England issued the originating summons in these proceedings seeking a declaration that as at September 29 1982 neither Avco nor Insight had carried out any improvements to the demised premises for which they had the written consent of Travel England other than improvements carried out in pursuance of an obligation to Travel England; and a further declaration that for the purpose of the determination of the rack rental value of the demised premises at September 29 account is to be taken of the effect of all H improvements, if any, to the demised premises by whomsoever carried out. The defendants responded by raising a number of contentions which I mention below and by making a counterclaim, by summons dated March 21 1984, for a declaration that as at September 29 1982 Travel England were unreasonably withholding the grant of a licence in respect of the improvements referred to in the letters dated February 11 and 19 1982 from Kagan Morris and shown in the two drawings enclosed with the letter of February 19 1982; and further or alternatively a declaration that Travel England is still unreasonably withholding the grant of such a licence. In the course of argument it became apparent that the counterclaim, as so formulated, might not correspond with the contentions put forward by Mr Driscoll for Insight, but no issue was raised as to this and it is J common ground that I can consider the submissions as made and, if appropriate, give any necessary directions as to the amendment of the counterclaim after I have ruled on the submissions.

It is appropriate to note at this stage that the improvements which, according to Insight, are to be left out of account on a rent review fall under four heads, namely (i) the installation of a toilet and shower in the basement of the premises; (ii) the installation of a new shop front and ground-floor entrance; (iii) the infilling of an internal staircase to provide additional floorspace on both ground-floor and basement levels; and (iv) the installation of air conditioning in the basement. Travel England does not accept that all these items are necessarily improvements, but I am not asked to decide this question: I am only K concerned to decide whether, on the assumption that all these items are improvements carried out by Insight, the effect of them is or is not to be left out of account under clause 2(C)(ii) of the underlease.

The basic submission made on behalf of Travel England is that under the terms of clause 2(C) of the underlease what has to be considered is the rack rental value of the demised premises at the relevant date of review; that the reference to the demised premises means the demised premises as they exist at that date, including all alterations and improvements not expressly directed to be left out of account; that under clause 2(C)(ii) improvements carried out by the tenant are only to be left out of account if (*inter alia*) the landlord has given its written consent to them; and that no written consent has been given here. L

Although the correspondence and the affidavits had indicated that Insight would submit that there was a written consent to the alterations which are said to constitute the improvements in this case, Mr Driscoll, on behalf of Insight, did not in the end pursue this point. The submissions which he made can be summarised as follows. First, he said that on the true construction of the underlease the words in parenthesis in clause 2(C)(ii) should be read as if they referred not only to improvements to which the landlord had given written consent but also to improvements in respect of which the landlord's consent had been unreasonably withheld. He contended also that Travel England has unreasonably withheld its consent to the improvements in this case. Secondly, he said that, in view of this M unreasonable withholding of consent, improvements must in any event be left out of account on a rent review because not to do so would enable Travel England to take advantage of its own wrong. Thirdly, he said that even if he were wrong about the first two points, this would not matter because clause 2(C) requires it to be supposed that the tenant has complied with all its obligations under the underlease and this means that it has to be supposed that Insight has obtained the written consent of Travel England to the improvements in question. Fourthly, he argued that in the circumstances of this case Travel England is estopped from denying that it has given written consent to the improvements.

It appears to me that the submission made on behalf of Travel England is correct unless it is displaced by one or more of the arguments advanced by Mr Driscoll. By this I do not in the least mean that the burden of satisfying me as to the correct answer lies upon Insight rather than upon Travel England. I merely mean to indicate that, as I see it, the convenient course will be to consider the submissions made on behalf of Insight before returning to those made on behalf of Travel England. As will be seen, there is a certain degree of overlap between the component parts of Mr Driscoll's first three submissions.

The first submission has two limbs, the first being an argument as to the construction of clause 2(C) (ii) of the underlease and the second being a contention that consent to the improvements has in fact been unreasonably withheld. As to the first limb, Mr Driscoll submitted that the words in parenthesis in clause 2(C)(ii) should be read as if they referred to improvements carried out by the tenant "to which the landlord shall have given written consent or in respect of which the landlord's consent has been unreasonably withheld". It appears to me that this cannot be done by any form of interpretation of the words which are actually present in the clause and that if it is to be done at all it involves the implication into clause 2(C)(ii) of a new condition which constitutes an alternative to the condition actually expressed in the parenthesis. In other words the form of construction relied upon is that form which, where applicable, implies into an instrument terms which are not actually there.

The circumstances in which the court will make such an implication have, of course, been frequently considered in reported cases, the most authoritative recent instance being *Liverpool City Council* v *Irwin* [1977] AC 239. In that case at pp 253-4 Lord Wilberforce said this:

> But there are varieties of implications which the courts think fit to make and they do not necessarily involve the same process. Where there is, on the face of it, a complete, bilateral contract, the courts are sometimes willing to add terms to it, as implied terms: this is very common in mercantile contracts where there is an established usage: in that case the courts are spelling out what both parties know and would, if asked, unhesitatingly agree to be part of the bargain. In other cases, where there is an apparently complete bargain, the courts are willing to add a term on the ground that without it the contract will not work — this is the case, if not of *The Moorcock* (1889) 14 PD 64 itself on its facts, at least of the doctrine of *The Moorcock* as usually applied. This is, as was pointed out by the majority in the Court of Appeal, a strict test — though the degree of strictness seems to vary with the current legal trend — and I think that they were right not to accept it as applicable here. There is a third variety of implication, that which I think Lord Denning MR favours, or at least did favour in this case, and that is the implication of reasonable terms. But though I agree with many of his instances, which in fact fall under one or other of the preceding heads, I cannot go so far as to endorse his principle; indeed, it seems to me, with respect, to extend a long, and undesirable, way beyond sound authority.

Lord Wilberforce went on to mention a fourth case, of which the *Liverpool City Council* case was itself an example, where the court is concerned to establish what the contract is, the parties not having themselves fully stated the terms. Clearly, we are not concerned with a case of this type today. Lord Cross of Chelsea, addressing himself to a case where the court is being asked to insert into a contract a term which the parties have not expressed, said at p 258:

> Here it is not enough for the court to say that the suggested term is a reasonable one the presence of which would make the contract a better or fairer one; it must be able to say that the insertion of the term is necessary to give — as it is put — "business efficacy" to the contract and that if its absence had been pointed out at the time both parties — assuming them to have been reasonable men — would have agreed without hesitation to its insertion.

Lord Edmund Davies said trenchantly at p 266 "the touchstone is always *necessity* and not mere *reasonableness*".

Mr Driscoll contended that the written consent which is referred to in the words in parenthesis in clause 2(C)(ii) is the written consent without which, by virtue of clause 3(J)(ii), the tenant is precluded from carrying out any alterations to the demised premises. This last mentioned consent is a consent which the landlord may not unreasonably withhold. Mr Driscoll contended that it would be illogical if the landlord were precluded from withholding consent unreasonably for the purposes of clause 3(J)(ii) but were allowed to prevent an improvement being left out of account under clause 2(C)(ii) by the simple expedient of withholding consent, albeit unreasonably.

While I see the force of this argument, it is not enough, in my judgment, to give rise to the suggested implication here. The provisions of clause 2(C) of the underlease will work perfectly well without the suggested implication. Mr Driscoll's complaint is in reality a complaint that it will operate unfairly or unreasonably and *Liverpool City Council* v *Irwin* has shown that this is not enough. Moreover, if the well-known officious bystander test were applied I think that, in deciding what answer the parties would have given to the question whether the matter was to be governed by the proposed implication, the parties should be treated as having had their attention drawn not only to the absence of a reference to the unreasonable withholding of consent in clause 2(C)(ii) but also to the absence of a corresponding disregard in the rent review provisions of the headlease and to the presence in the headlease of an absolute covenant against the making of alterations. If the parties had all these matters in mind, it is, in my judgment, by no means apparent that they would answer the officious bystander by saying that of course they intended the suggested implication to be made. I consider it just as likely that one at least of them would have said either that clause 3(J)(ii) must be reformulated so as to correspond with the equivalent covenant in the headlease or that clause 2(C) must contain no provision at all for disregarding improvements.

My decision on this first limb of the argument would, by itself, be enough to dispose of Mr Driscoll's first submission, but I think I must consider also the contention as to unreasonable withholding of consent, in case I am wrong about the first limb and also because this contention forms part of Mr Driscoll's later submissions.

The counterclaim as formulated in the summons dated March 21 1984 seemed to be based upon an unreasonable withholding of consent pursuant to the request for a licence which was made in the correspondence in February 1982 to which I have already referred. But Mr Driscoll accepted in argument that he could not rely upon this request because it was made after all the relevant work had been completed and the consent required by clause 3(J)(ii) of the underlease is "the previous consent in writing of the landlord"; that is to say consent given before the work is done. Instead Mr Driscoll relied upon the failure of Travel England to grant consent pursuant to the request made by Trehearnes in their letter of February 25 1981 to Hillier Parker.

I am prepared to treat that letter as a letter asking for formal written consent to the alterations, other than the installation of air conditioning in the basement, although I see the force of Mr Crampin's submission that all that was being asked for was an informal indication of approval. I make an exception as regards the air conditioning because the alterations for which agreement was sought are said to be set out in certain plans enclosed with the letter of February 25 1981 and I cannot find in those plans any indication that there was to be air conditioning in the basement. Treating the letter in this way, the answer which Trehearnes received from Mr Crowther in the telephone conversation of February 27 1981 as recorded by Mr McKenzie was not the agreement which had been asked for. It was merely an indication that approval would be given in the future if the Commissioners gave their own approval and if planning permission was obtained. It is not suggested that it was unreasonable for this attitude to be taken on behalf of Travel England at that stage. Likewise the Commissioners did not give their own immediate approval. Cluttons indicated that they would advise the Commissioners to give approval if certain conditions were met. They asked for confirmation of the acceptance of these conditions but never received it. The indications are that Insight was prepared to accept these conditions, but Cluttons were never informed of this and the last of the work required to satisfy the conditions, namely the replacement of the pavement light, was not in fact completed until November, by which time all the alteration work had been done for some time. The Commissioners have indeed never actually given their formal consent to the alterations. As to planning consent, this was not obtained until April 30, but neither Hillier Parker nor Travel England appear to have been informed of this. In the meantime both Cluttons and Hillier Parker had drawn the attention of Insight to the fact that they were proceeding with the work without consent and, as between the solicitors, there had been correspondence in April 1981 on the footing that an application for consent had yet to be made on behalf of Insight. An attempt was made to criticise Travel England because it made no effort itself to obtain the consent of the Commissioners. In my judgment such criticism is unjustified. Hillier Parker thought that Insight was dealing direct with Cluttons on behalf of the Commissioners and Mr Bichard made it clear to Kagan Morris in his letter of April 8 that he was not going to approach the Commissioners until full plans were submitted. He was not met with the response that such plans had already been submitted. Moreover, when the question of a licence was revived in January and February

1982 the initiative was taken by Mr Bichard, not by Insight or its advisers, and Insight made no attempt to suggest that the fault lay with Travel England for not attending fully to Trehearnes' request made in the letter of February 25 1981.

Mr Driscoll drew my attention to *Lambert* v *F W Woolworth & Co Ltd (No 2)* [1938] 1 Ch 883 and submitted that, having regard to that case and to section 19(2) of the Landlord and Tenant Act 1927, Travel England could not properly have refused consent on the ground of diminution in the value of its reversionary interest, but would have been limited to demanding reasonable compensation for such diminution. That may well be so. But in my judgment the matter never proceeded to that stage. I think that Trehearnes accepted the telephone conversation with Mr Crowther on February 27 1981 as sufficient for their purposes, even though it fell a long way short of what I am prepared to treat them as having asked for. If, notwithstanding this, the request remained outstanding to any extent after February 27 the parties nevertheless effectively treated it as being in abeyance and did not regard Travel England as being under an obligation to do anything further. No attempt was made to raise the matter again until Mr Bichard drew attention to the absence of consent early in 1982 at which date it was too late to obtain "previous consent in writing". If, contrary to my view, Travel England was under an obligation to do something further after February 27 1981 it was, in my judgment, reasonable for it to await the formal consent of the Commissioners, either in response to the approach made direct by Trehearnes or in response to an approach which Mr Bichard indicated that Travel England was prepared to make if and when Kagan Morris supplied him with particular material.

I therefore reject the submission that Travel England has unreasonably withheld from Insight such consent as was made necessary by clause 3(J)(ii) of the underlease. Moreover, in respect of the air conditioning in the basement I consider that such consent was never sought by Insight.

I turn to Mr Driscoll's second main submission. This has two elements, namely (i) that Travel England has unreasonably withheld its consent to the improvements; and (ii) that this constitutes a "wrong" on the part of Travel England for the purposes of the principle that no man may take advantage of his own wrong. Having regard to what I have already said about unreasonable withholding of consent, this submission must fail because the first element is not present. Nevertheless I think I ought to state my views on the second element as well, particularly because the principle that no man may take advantage of his own wrong has a bearing on the third main submission.

The principle in question was considered by the House of Lords in *New Zealand Shipping Co Ltd* v *Société des Ateliers et Chantiers de France* [1919] AC 1. In the recent case of *Cheall* v *APEX* [1983] 2 AC 180, at pp 188-9, Lord Diplock described what the House of Lords had said in the earlier case in the following way:

> In the course of the speeches [in the *New Zealand Shipping* case], which are not entirely consistent with one another, reference was made by all their Lordships to the well known rule of construction that, except in the unlikely case that the contract contains clear express provisions to the contrary, it is to be presumed that it was not the intention of the parties that either party should be entitled to rely upon his own breaches of his primary obligations as bringing the contract to an end, ie as terminating any further primary obligations on his part then remaining unperformed. This rule of construction, which is paralleled by the rule of law that a contracting party cannot rely upon an event brought about by his own breach of contract as having terminated a contract by frustration, is often expressed in broad language as: "A man cannot be permitted to take advantage of his own wrong". But this may be misleading if it is adopted without defining the breach of duty to which the pejorative word "wrong" is intended to refer and the person to whom the duty is owed.

Here also it is necessary to consider a similar point, bearing in mind that the "wrong" relied upon is the unreasonable withholding of consent to alterations pursuant to clause 3(J)(ii). For this purpose I assume, contrary to what I have already decided, that consent has been unreasonably withheld. It is clear, however, that this would not render Travel England liable to an action for damages. In the well-known case of *Treloar* v *Bigge* (1874) LR 9 Ex 151 a covenant by a tenant not to assign leasehold premises without the consent of the lessor, "such consent not being arbitrarily withheld" was under consideration. In that case Kelly CB said at p 154:

> The words, taken grammatically, do not seem to me to amount to an undertaking by the lessor, but are a part of the same sentence as that containing the lessee's covenant, and qualify its generality. They prevent that covenant operating in any case of arbitrary refusal on the part of the lessor, that is, in any case where, without fair, solid, and substantial cause, and without reason given, the lessor refuses his assent.

Amphlett B said at p 156:

> . . . I think they [the words relied upon] ought to be construed as a qualification on the covenant of the lessee. That covenant is in derogation of his common law rights, and it is more convenient and reasonable to hold that the words were introduced to limit the generality of the covenant than to hold them to impose an obligation on the lessor. The true interpretation of the words, I think, is to release the plaintiff from his covenant not to assign without the plaintiff's assent, if that assent is arbitrarily withheld.

It appears to me that this reasoning leads inevitably to the conclusion that where a lessee covenants not to do something without the consent of the lessor, such consent not to be unreasonably refused, an unreasonable refusal of consent by the lessor does not amount to any kind of wrong on his part. It merely produces a situation in which the lessee's covenant no longer applies and leaves the lessee free to do the act in question as if he had never entered into a covenant not to do it.

In these circumstances I consider that Mr Driscoll's attempt to invoke the principle that no man may take advantage of his own wrong fails, whether this principle is used in support of the implication which I have already refused to make or whether it is relied upon in support of a free-standing argument that, even without such implication, the improvements in this case are to be left out of account on a rent review.

The next submission which I must deal with is that which is based upon the supposition required to be made by clause 2(C) of the underlease. What clause 2(C) requires is that the rental value of the demised premises shall be agreed or determined as if they were let for a particular term of years as between a willing lessor and a willing lessee with vacant possession, subject to two qualifications. The first qualification is expressed in the words "but upon the supposition (if not a fact) that the Tenant has complied with all the obligations on the part of the Tenant imposed by these presents". I will refer to it as "the supposition". The second qualification is made by providing that no account shall be taken of certain matters, including the effect of improvements to which the landlord has given written consent.

It is obvious that the main purpose of the supposition was to prevent the rental values being depressed as a result of a breach by the tenant of its obligations under the underlease, for example by the tenant's failure to repair the demised premises. However, there seems to be no reason why the supposition should not have a wider effect if, on the language of clause 2(C) as a whole, it is capable of doing so.

The first point to be noticed is that the language of clause 2(C) requires that the supposition shall only be made where it is not the fact that the tenant has complied with all its obligations. Accordingly it is not necessary to make the supposition in relation to alterations where the landlord has in fact given his written consent to them. In my judgment this is also the case where the landlord has been asked for his consent to the alterations and has unreasonably withheld it. The reason why this is so is that, as I have already stated, where a tenant's covenant provides that something shall not be done without the landlord's consent, such consent not to be unreasonably withheld, the unreasonable withholding of consent by the landlord frees the tenant from the burden of the covenant in respect of the matter for which consent was sought. If in such a case the tenant proceeds to do what the covenant originally prohibited him from doing without consent, he commits no breach of any obligation.

There remains, however, the case where there is no unreasonable withholding of consent. For example, the tenant may carry out alterations without asking for consent or without giving the landlord a reasonable opportunity to consider whether to give or withhold consent or the alterations may be such that the landlord can reasonably refuse his consent to them. On the findings I have already made, there was no unreasonable withholding of consent in the present case, so that this case falls within the category now under consideration. Is the supposition applicable in this category of case and, if so, what precisely does it require to be supposed?

It is convenient to deal with the second part of this question first, making the assumption at this stage that the supposition applies. There appear to be three possibilities. The first is to assume that the tenant has obtained the landlord's written consent, although it is known that he has not. The second is to recognise that the tenant has not obtained written consent to the alterations and to assume that he has complied with his covenant so that he has not in fact made any alterations, with the result that those alterations which constitute improvements are to be left out of account on a rent review because they are to be treated as not having been made at all. The third possibility is to recognise both the absence of written consent and the presence of the alterations, with the result that the alterations must be assumed to have been made by somebody other than the tenant. This would result in alterations which are improvements being left out of account because they would be deemed not to have been carried out by the tenant.

Mr Driscoll contended primarily for the first of these possibilities. However, I cannot accept that this is correct. The relevant obligation of the tenant under the underlease was not a positive obligation to obtain the written consent of the landlord to the proposed alterations but a negative obligation not to carry out alterations if it did not have such consent. Mr Driscoll's second preference was for the second possibility and in this he was joined by Mr Crampin. The third possibility might have suited Mr Crampin's case better, but he felt unable to contend for it because of its artificiality and because it would lead to absurd results if the alteration had been such as to reduce the rental value of the demised premises. I am not altogether convinced that the second possibility is any less artificial than the third but, faced with a measure of consensus between opposing counsel, I think I must proceed on the footing that, if it applies at all, the supposition requires it to be assumed that Insight has made no alterations at all.

Nevertheless, returning to the first part of the question which I posed above, I feel bound to hold that the supposition does not apply at all in the circumstances of this case. I reach this conclusion for two reasons. First, if it were to apply in the manner contended for it would virtually destroy the effect of the requirement, imposed by clause 2(C)(ii), that improvements carried out by the tenant are to be left out of account only if they are improvements to which the landlord has given written consent. In a case where consent has actually been given clause 2(C)(ii) would apply. In a case where consent had not been asked for or had been reasonably withheld the supposition would, if allowed to operate, result in the improvements being left out of account without resort to clause 2(C)(ii). The only circumstance in which improvements would not be left out of account would be when the tenant has quite properly carried them out after an unreasonable refusal of consent by the landlord. I cannot accept an interpretation of clause 2(C) which means that a tenant who has not acted in breach of covenant is in a worse position than one who has acted in breach. In my judgment the express requirement for written consent imposed by clause 2(C)(ii) overrides the supposition in determining whether improvements are or are not to be left out of account.

Second, my conclusion is based on the principle, already mentioned, that no man may take advantage of his own wrong. In the passage from *Cheall* v *APEX* which I have already quoted, Lord Diplock referred "to the well-known rule of construction that . . . it is to be presumed that it was not the intention of the parties that either party should be entitled to rely upon his own breaches of his primary obligations as bringing the contract to an end". Cases where this rule of construction have been applied were cited in the *New Zealand Shipping* case. I do not see why this rule should be limited in its application to cases where a party seeks to rely on his own default as bringing a contract to an end. I would hold that it applies also where a party seeks to assert that his own breach of contract enables him to take advantage of a contractual presumption. In my judgment, on the true construction of clause 2(C) of the underlease, the supposition that, contrary to the true fact, the tenant has complied with its obligations is only to be made in favour of the innocent party, namely the landlord.

Finally, I must deal with the arguments founded upon estoppel. Mr Driscoll relied primarily on proprietary estoppel, saying that Travel England represented to Insight that if Insight incurred expense in carrying out the proposed alterations, Travel England would grant a formal written licence for these alterations. He said that this raised an equity in favour of Insight which should be given effect to by requiring Travel England to act as if such a licence had been given and therefore to accept an assessment of rent on the rent review as if such a licence was in existence. Alternatively I think he contended that Travel England should be required to grant a licence now with retrospective effect. In support of this proposition he referred to *Taylors Fashions Ltd* v *Liverpool Victoria Trustees Co Ltd* (Note) [1982] 1 QB 133 and to *Snell's Equity*, 28th ed, p 558. In the alternative Mr Driscoll relied on estoppel by convention as discussed in *Amalgamated Investment & Property Co Ltd* v *Texas Commerce International Bank Ltd* [1982] QB 84. In the further alternative, Mr Driscoll invoked promissory estoppel.

In my judgment estoppel by convention cannot assist Mr Driscoll here. The principle of this variety of estoppel is, I think, accurately summarised in the headnote to the *Amalgamated Property* case in the following words:

> When parties in their course of dealing in a transaction have acted upon an agreed assumption that a particular state of facts between them is to be accepted as true, each is to be regarded as estopped as against the other from questioning as regards that transaction the truth of the statement of facts so assumed.

On no view of the facts of this case can it be said that the parties have acted on the footing that a particular state of facts exists when this is not in fact the case. The highest that the matter can be put, although I shall have to examine the correctness of this in a moment, is that they acted on the footing that a particular instrument would be brought into existence in the future.

Mr Crampin contended that the claim based on proprietary estoppel must also fail because proprietary estoppel could not apply in circumstances such as those of the present case. He cited *Western Fish Products Ltd* v *Penwith District Council* [1981] 2 All ER 204, where Megaw LJ, giving the judgment of the Court of Appeal, discussed the law relating to proprietary estoppel and said, at p 218:

> We know of no case, and none has been cited to us, in which the principle set out in *Ramsden* v *Dyson* and *Crabb* v *Arun District Council* has been applied otherwise than to rights and interests created in and over land. It may extend to other forms of property: see per Lord Denning MR in *Moorgate Mercantile Co Ltd* v *Twitchings* [1975] 3 All ER 314 at 323-324, [1976] QB 225 at 242. In our judgment there is no good reason for extending the principle further. As Harman LJ pointed out in *Campbell Discount Co Ltd* v *Bridge* [1961] 2 All ER 97 at 103, [1961] 1 QB 445 at 459, the system of equity has become a very precise one. The creation of new rights and remedies is a matter for Parliament, not the judges.
>
> In his reply counsel for the plaintiffs seemed to recognise that the reported cases did put limits to the application of the so-called concept of proprietary estoppel. He submitted that the plaintiffs' case was within that concept because what the defendant council, by their officers, had represented had, to their knowledge, caused the plaintiffs to spend money on or in connection with their own land which they would not otherwise have spent. On their own case they have spent money in order to take advantage of existing rights over their own land which the defendant council by their officers had confirmed they possessed. There was no question of their acquiring any rights in relation to any other person's land, which is what proprietary estoppel is concerned with.

In my judgment the possible application of proprietary estoppel cannot be excluded on these grounds here. The dispute in this case does concern rights or interests in land. The question, in effect, is whether the value of the improvements, which have become part of the land, is to be treated on a rent review as attributable to the landlord's interest in the demised premises or to the tenant's interest. Insight, by asserting the latter, is in effect claiming that, by virtue of an estoppel, what would otherwise be part of the landlord's property is to be treated, for valuation purposes, as part of the tenant's property. If I am satisfied on the facts that a representation to the necessary effect was made, I see no reason why I should not give effect to this by recognising an equity and implementing it in whatever way appears to be most just.

Accordingly I consider that the outcome of the arguments based on proprietary estoppel and indeed on promissory estoppel depends on the evaluation of the facts of this case. I have already stated most of the primary facts, but there is one matter to which I must return. Mr Tarsh said in his oral evidence that, although the question whether or not the alterations would be excluded on a rent review was not resolved at the meeting on January 13, he thought it was resolved a few days later. The basis for this was the exchange of letters in which Mr Bichard had indicated that he had in mind a grant of a written licence and had asked for Kagan Morris' undertaking to be responsible for his charges in relation to such a licence and Kagan Morris had given their undertaking in this respect by their letter of January 22 1981. Mr Tarsh said he appreciated at the time that such a licence would be a written consent for the purposes of clause 2(C)(ii) and that after Insight, through its solicitors, gave its undertaking to pay the costs of such a licence he thought there was no need to get express agreement to exclude the effect of improvements on a rent review because this would be automatic under the terms of the underlease itself.

I am afraid that I cannot accept Mr Tarsh's evidence on this point.

Although Mr Tarsh is clearly an able businessman who is quite capable of understanding legal documents such as the underlease, I cannot believe that this was how he appraised the matter in January or February 1981. If he had looked at the matter in this way it would have been a most important piece of evidence, yet I see no trace of it in either of his two affidavits and, although he supplemented these affidavits by evidence in chief, the statement of his belief did not emerge until he was cross-examined. I am sure Mr Tarsh was trying to give his recollection of what happened and what he thought at the time to the best of his ability, but I fear that what he said about his reasoning in January or February 1981 does not represent the state of his mind at the time. I think that what has happened is that unconsciously Mr Tarsh has come to believe that some of the points which have been discussed in the course of preparation of this case represent what actually happened at the time.

What representation, if any, did Travel England make to Insight concerning the alterations? I have no doubt that Travel England represented that it had no objection to the alterations being carried out. Had these proceedings been based upon an alleged breach of the prohibition contained in clause 3(J)(ii) of the underlease I would have had no hesitation in holding that Travel England is estopped from complaining of any such breach. But it appears to me that in order for Mr Driscoll's arguments on clause 2(C)(ii) to succeed on the basis that Travel England is estopped from denying that it has consented in writing to the improvements, something more than mere non-objection to the alterations must be proved. In my judgment it would be necessary for Insight to show either that Travel England promised to grant a written licence (in which case one would have expected Insight's claim to have been formulated as a contractual claim) or that it represented that, notwithstanding the absence of written consent, it was content for the improvements to be left out of account on a rent review. In my judgment nothing of this kind can be spelt out of the facts. I regard it as of the utmost significance that the last occasion before this dispute arose on which the leaving of the improvements out of account was discussed between the parties was on January 13, when everyone agrees that the matter was left undecided. Thereafter all the discussions concerning a licence were, in my judgment, discussions in relation to the consent to alterations required by clause 3(J)(ii) and the parties did not address their minds to whether improvements were or were not to be left out of account on a rent review. No form of licence was ever discussed or agreed. Although Travel England clearly indicated that it had no objection to the alterations it did not, in my judgment, make any representation as to leaving out of account on a rent review any of these alterations. The fact that it was contemplated that the matter would be dealt with by a written licence does not, in my view, amount to a representation that there would be a written consent for the purposes of clause 2(C)(ii). It may be that if a formal licence had been prepared Travel England would have forgotten about the point and would unwittingly have given a written consent for the purposes of clause 2(C)(ii). On the other hand, if a formal licence had been prepared and considered the matter might have been reconsidered and a provision altering the effect of clause 2(C)(ii) might have been agreed. At the highest it appears to me that there is no more than a suspicion that, if the matter of formal consent for the purposes of clause 3(J) had been pursued, Travel England might have forgotten about the clause 2(C)(ii) point. In my judgment it is impossible to elevate this into a representation that Travel England was prepared to grant a licence which would resolve the clause 2(C)(ii) point in favour of Insight.

I would add that it appears to me that the conclusion which I have just expressed would not be any different even if I had felt able to accept Mr Tarsh's evidence as to his belief that the matter had been resolved by the exchange of correspondence in January 1981. If Mr Tarsh had held that belief it did not, in my judgment, arise from anything said or done on behalf of Travel England. In particular Mr Bichard cannot, in my view, be taken to have represented that if a licence was granted as envisaged it would resolve the clause 2(C)(ii) point by its silence.

In the result, therefore, I am unable to accept any of the arguments advanced on behalf of Insight. In my judgment the arguments on behalf of Travel England accord with the natural meaning of the words of clause 2(C)(ii) and are not in any way displaced. Subject to any submissions as to the precise form of order, it appears to me that I ought to grant Travel England the declarations which it has claimed and to dismiss the counterclaim.

Declarations were made as sought in plaintiffs' amended summons, the counterclaim as amended was dismissed. The defendants were ordered to pay the costs including costs of counterclaim.

For the other cases on this subject see p 97

NEGLIGENCE

Court of Appeal

December 20 1985

(Before Lord Justice SLADE, Lord Justice ROBERT GOFF and Sir John MEGAW)

INVESTORS IN INDUSTRY COMMERCIAL PROPERTIES LTD v SOUTH BEDFORDSHIRE DISTRICT COUNCIL

No duty of care owed by local authority, in the exercise of their supervisory powers over building operations, to building owners not in occupation where the authority had approved plans and carried out inspections of a site on which warehouses were built on totally inadequate foundations — Review of cases from *Dutton* v *Bognor Regis UDC* and *Anns* v *Merton London Borough Council* to *Governors of Peabody Donation Fund* v *Sir Lindsay Parkinson & Co Ltd* — In the present case property developers, who had been advised by architects and structural engineers, claimed damages for negligence and breach of statutory duty against a district council when warehouses built on a filled-in site disintegrated and had finally to be demolished and rebuilt on proper foundations — The council appealed against the decision of the judge, sitting as an official referee, who held the council liable — The judge's decision was given before the *Peabody* case reached the House of Lords, at a time when "the tide of judicial authority was blowing strongly against local authorities" — The judge had also found the structural engineers at fault, but proceedings against them had been discontinued because they were uninsured — He found in favour of the architects, who had also been joined as third parties — The Court of Appeal, in allowing the council's appeal, held that the owners had themselves been in breach of the building regulations in regard to the foundations of the warehouses — The court commented on earlier cases, including *Acrecrest Ltd* v *W S Hattrell & Partners*, which must now be taken to have been overruled, and laid down five propositions of law, derived from the *Anns* and *Peabody* decisions considered together — These propositions, summarised somewhat, are as follows: (1) the purpose of the authorities' supervisory powers is to protect occupiers and members of the public against dangers to health and personal safety, not to safeguard the building developer against economic loss; (2) it may well be that an authority will be held to owe a duty to a subsequent occupier other than the original building owner; (3) an occupier to whom the duty is owed may be entitled to recover from the authority the cost of restoring his property; (4) an authority will normally owe no duty to the original building owner because it is his duty to see that the building is erected in accordance with the relevant regulations; (5) *a fortiori* an authority will owe no duty to the owner who has relied on the advice of architects, engineers and contractors — As the developers in the present case had failed to establish their claim against the council, the latter's third party claim against the architects did not arise — However, the court expressed the view that no breach of professional duty had been proved against the architects — Council's appeal allowed

This was an appeal by the South Bedfordshire District Council against a decision of Judge Sir William Stabb QC, sitting as an official referee, in favour of the plaintiffs in the action, Investors in Industry Commercial Properties Ltd, who had claimed damages from the council for alleged negligence and breach of statutory duty, and who were the first respondents to the appeal. The second respondents, Hamilton Associates, the architects, had been the second third party; Ellison & Partners, the structural engineers, had been the first third party, but proceedings against them had been discontinued and they took no part in the appeal.

Piers Ashworth QC and Adrian J N Brunner (instructed by Barlow Lyde & Gilbert) appeared on behalf of the appellants; Stanley E Brodie QC and Andrew C Pugh (instructed by Hunt & Hunt, of Romford) represented the respondents, Investors in Industry Commercial Properties Ltd; Stephen C Desch QC and H de Lotbiniere (instructed by Hextall, Erskine & Co) represented the respondents, Hamilton Associates.

Giving judgment, SLADE LJ said: This is the judgment of the court on an appeal by the District Council of South Bedfordshire from a judgment of His Honour Judge Sir William Stabb QC, sitting as an official referee, delivered on April 5 1984. In reaching his decision he followed the decision in *Acrecrest Ltd* v *W S Hattrell & Partners* [1983] QB 260 ("*Acrecrest*") in which this court had purported to apply the decision of the House of Lords in *Anns* v *Merton London Borough Council* [1978] AC 728 ("*Anns*"). Later in 1984, however, *Acrecrest* was overruled by the House of Lords in *Governors of the Peabody Donation Fund* v *Sir Lindsay Parkinson Ltd* [1985] AC 210 ("*Peabody*").

This appeal raises important and difficult questions as to the frontiers of the principle of *Anns*, particularly having regard to the later decision in *Peabody*.

The appellants are the successors to the Leighton Linsdale Urban District Council. No differentiation falls to be drawn between either of these bodies for the purposes of this appeal and we shall use the expression "the council" to refer to either or both of them. The plaintiffs in the proceedings ("Anglia") are property developers, who were formerly called Anglia Commercial Properties Ltd and are now called Investors in Industry Commercial Properties Ltd.

In the action Anglia claimed damages for negligence or breach of statutory duty under the Public Health Acts 1936 and 1961 against the council, as the responsible local authority, in connection with the approval of the plans for and the inspection of the foundations of four warehouses, which had been erected for Anglia as developers, on a site known as the Spinney Pool Estate, Billington Road, Leighton Buzzard, Bedfordshire. In addition to Anglia, there were also concerned in this building project, for at least part of the material time, Hamilton Associates as architects and Ellison & Partners ("Ellisons") as structural engineers.

Anglia's claim arose out of the rapid disintegration and subsequent demolition of two of the four warehouses after they had been built, due to what the learned judge fairly described as "the totally inadequate design" of the foundations, in the light of the nature of the ground on which the warehouses had been erected. Though the council formally denied any negligence or breach of their statutory duties, the learned judge concluded that "they woefully failed to discharge the responsibilities imposed upon them by statute".

Having served their defence in the action, the council joined Ellisons as first third party and Hamilton Associates as second third party, claiming against them, in the event of the council being held liable to Anglia, an indemnity against Anglia's claim on the grounds that such loss and damage (if any) as Anglia might prove had been caused by the negligence of the two third parties or their respective servants or agents.

Though the learned judge at the trial found that Ellisons were "clearly at fault", the proceedings against them had been discontinued before the trial, because they were uninsured. He was told that they had made a contribution towards Anglia's costs. By his judgment he ordered that the defendants should pay to Anglia damages to be assessed. He also gave judgment for Hamilton Associates against the council and ordered that the council should pay the costs of Anglia and of Hamilton Associates.

The council now appeal from the whole of this judgment. By their notice of appeal they seek an order that judgment may be entered for the council or, alternatively, that Anglia may be found contributorily negligent, the proportion contended for being 80%. Further or alternatively, they seek judgment against Hamilton Associates in the third party proceedings.

The learned judge did not have the advantage of hearing any evidence from witnesses as to the facts, the oral evidence adduced being confined to evidence from expert witnesses. He was therefore obliged to reach his conclusions as to fact from contemporaneous correspondence and documents. These conclusions he set out in a full and careful judgment.

Events up to September 1973

The site in question, the Spinney Pool Estate, which is about 6 acres, had formerly contained an exceptionally large pool used for swimming. In 1972 the site was acquired by a haulage contractor, Mr Pentus Brown. He applied for permission to tip soil, hardcore and other innocuous material at the site. In April 1972 he applied for planning permission to erect a so-called "transit building" on it. Hamilton Associates, who had by that time been engaged as architects on his behalf, lodged this application. In a letter of April 25 1972 to the building contractors then employed in connection with this project, they expressed their belief that "it may well be necessary for you to incorporate piled foundations". Since this was a filled-in site, their anxiety at that time in relation to the proposed foundations was well justified. As is stated in clause 2.2.1.1 of the *British Standard Code of Practice for Foundations CP2004: 1972* ("the Civil Engineering Code of Practice"):

> Before foundations are designed and a method of construction determined, it is desirable to carry out a site exploration to ascertain the character and variability of the soil strata underlying the site of the proposed structure . . .

Clause 2.2.3.5 states:

> All made ground should be treated as suspect, because of the likelihood of extreme variability. Any proposal to found a structure on made ground should be investigated with extreme care . . . Loading tests may be completely misleading . . .

As appears from clause 2.2.1.3, trial pits have certain advantages for the purposes of examination to fairly shallow depths, but where the ground is unsuitable for trial pits, or deeper exploration is needed, bore holes are required. The evidence shows that in the building trade there is a well-recognised difference between trial pits and bore holes, which can go much deeper than trial pits and can provide a greater range of undisturbed samples.

On April 27 1972, Hamilton Associates wrote to the contractors suggesting that they should write direct to their consulting engineers, Conner & Heath ("Conners") instructing them to prepare details and calculations of the foundations for the transit building. It appears from this letter that a filling and levelling operation on the site was in progress or in contemplation.

By June 1972 Anglia had become interested in purchasing the site, which by then had planning permission both for tipping and for the transit building and at least an indication from the council that permission would be granted for the erection of four warehouses. On August 19 1972 contracts providing for the purchase of the site by Anglia were exchanged. It was contemplated that Anglia would complete the construction of the transit building and then lease it back to Mr Brown. At the same time Anglia inherited Hamilton Associates as their architects.

The learned judge found as facts that by this time

(1) both Anglia and Hamilton Associates knew that levelling and refilling of the site was going on;
(2) the council must also have known of this because they had given the permission for tipping and, as the local authority, G
must have known something of the history of the site.

Meantime, on July 7 1972 Conners had sent to Hamilton Associates a foundation plan for the erection of the transit building. On August 10 1972 the council gave building regulation approval for this project. In October 1972 the site was conveyed to Anglia. Thereafter plans began to be prepared for the erection of the four warehouses.

On December 13 1972 a revised planning application was submitted seeking permission to erect the transit building on the site. On December 14 1972 a letter was written to Ellisons on behalf of Francis Parker Contracts Ltd ("Parkers"), who were the main contractors for this project, giving them instructions as consulting engineers in this connection. H This was Ellisons' first involvement with the site.

On January 2 1973 a site meeting was held at which were present representatives of the council, Hamilton Associates, Ellisons, Parkers and Peter Lind Ltd, the subcontractors of Parkers. The object of the meeting was stated to be "to inspect trial holes" in the vicinity of the proposed transit building and to discuss foundation proposals. Notes of the meeting record that satisfactory bottoms were reached at depths of 6 ft, 12 ft and 16 ft in three pairs of trial pits. We infer that these trial pits may well have been made in the region of what had been the shallow end of the former pool. Later events show they gave no true picture of the site as a whole.

On January 15 1973 Hamilton Associates dispensed with the J services of Conners so as to allow Ellisons to design both the retaining walls for the transit building and its foundations. On January 16 Hamilton Associates submitted to the council a further revised detailed planning application in respect of the proposed resited transit building.

Also on January 16 1973 Hamilton Associates wrote to the surveyor and engineer of the council, referring to the proposed transit building and to the making of four trial pits in the position of the proposed new foundations, which they suggested had been unexpectedly reassuring and had been inspected both by the council's building inspector and by Ellisons. On February 13 1973 planning permission was granted in respect of the transit building.

K Meantime, on February 10 1973, an application had been submitted to the council by Hamilton Associates for permission for the erection of four warehouses (commonly referred to respectively as Units A-B, C-D, E-F and G-H). This project was to be known as Phase II. The main contractors were to be GTE Construction Ltd, the foundation subcontractors Peter Lind Ltd, the architects Hamilton Associates and the consulting engineers Ellisons. Anglia's claim relates to two of these warehouses (Units E-F and G-H).

On July 4 1973 Hamilton Associates submitted to Anglia the tenders received by them relating to the Phase II development. By July 19 1973 the transit building had been more or less completed. On July 25 1973 a site meeting took place. The notes of this meeting record that it was agreed (*inter alia*) that arrangements would be L made for two trial pits to be dug in selected positions on each of the proposed remaining units. A representative of Hamilton Associates was present, but the council and Ellisons were not represented on this occasion.

On September 11 1973 Anglia wrote to GTE Construction Ltd accepting their fixed price tender and emphasising that the start of Phase II had become extremely urgent.

On September 12 1973 a site meeting took place at which the representative of Hamilton Associates confirmed that he would be approaching Anglia to ask them to appoint Ellisons as consulting engineers and it was agreed that Ellisons would contact the council's building inspector to ascertain whether he would accept a design certificate in lieu of full structural calculations to support the M application for approval of the plans. On September 14 1973 Hamilton Associates wrote to Anglia suggesting that they should write to Ellisons confirming their appointment as consulting engineers. They further stated: "The four buildings are being pegged out this week and two trial holes per area are being dug in the presence of the local building inspector." The confirmation of Ellisons' appointment was duly given on or about September 17 1973. Following a telephone conversation with a Mr Chapel of the Engineers and Surveyors Department of the council, Mr Loveday informed Hamilton Associates by a letter of September 17 1973 that "although he [Mr Chapel] will accept a design certificate he also requires full working drawings and calculations".

The relevant statutory provisions

At this point it is convenient to summarise the most relevant statutory provisions relating to building operations. By section 4(1) of the Public Health Act 1961 the minister is empowered to make building regulations for all or any of the matters set out in sections 61 and 62 of the Public Health Act 1936, which include (*inter alia*) "the construction of buildings and the materials to be used in the construction of buildings". Section 4(6) of the 1961 Act provides: "If a person contravenes or fails to comply with any provision contained in building regulations he shall be liable to a fine", as therein provided. Section 64(1) of the 1936 Act (as amended) provides:

> Where plans of any proposed work are, in accordance with building [regulations], deposited with a local authority, the local authority shall, subject to the provisions of any other section of this Act which expressly requires or authorises them in certain cases to reject plans, pass the plans unless they either are defective, or show that the proposed work would contravene any of those [regulations], and, if the plans are defective or show that the proposed work would contravene any of those [regulations], they shall reject the plans.

Section 65(1) of the 1936 Act (as amended) provides:

> If any work to which building [regulations] are applicable contravenes any of those [regulations], the authority, without prejudice to their right to take proceedings for a fine in respect of the contravention, may by notice require the owner either to pull down or remove the work or, if he so elects, to effect such alterations therein as may be necessary to make it comply with the [regulations].

Section 65(5), so far as material, provides:

> Nothing in this section shall affect the right of a local authority, or of the Attorney-General, or any other person, to apply for an injunction for the removal or alteration of any work on the ground that it contravenes any [regulation] or any enactment in this Act, but if the work is one in respect of which plans were deposited and the plans were passed by the local authority, or notice of their rejection was not given within the prescribed period after the deposit thereof, and if the work has been executed in accordance with the plans, the court on granting an injunction shall have power to order the local authority to pay to the owner of the work such compensation as the court thinks just, . . .

The relevant building regulations are the Building Regulations 1972 (SI 1972 no 317). Regulation A10(1), so far as material, provides: ". . . any person who intends to — (a) erect any building; . . . shall if any provisions of these regulations apply to such operation . . . give notices and deposit plans, sections, specifications and written particulars in accordance with the relevant rules of Schedule 2". Regulation D3, so far as material, provides that "the foundations of a building shall (a) safely sustain and transmit to the ground the combined dead load, imposed load and wind load in such a manner as not to cause any settlement or other movement which would impair the stability of, or cause damage to, the whole or any part of the building or of any adjoining building of works; . . . ". By virtue of regulation D4 the requirements of regulation D3 are deemed to be satisfied if the foundations of a building are constructed in accordance with the relevant recommendations of the Civil Engineering Code of Practice. It is common ground that, in the event, neither these recommendations nor the provisions of regulation D3 were complied with in relation to the foundations of the warehouses.

There is no regulation specifying whose duty it shall be to comply with these various regulations. Regulation A6 simply provides that, subject to immaterial exceptions, "Parts A to L of these regulations shall apply to the erection of a building".

We think it clear that the obligations described in regulation A10(1) were ones to which Anglia, as the building owner, was subject, whether or not other persons were subject to them as well. Though it has been submitted on behalf of the council that Anglia did not in fact comply even with this regulation, we find it unnecessary to deal with this submission.

Much more material for present purposes is regulation D3, which specifies the requirements which the foundations of a building must satisfy. There can, in our judgment, be no doubt that an owner who personally or by his agents erects a building is under a statutory duty to comply with them and is liable to a fine under section 4(6) of the 1961 Act if he is in breach of this duty. On behalf of Anglia, however, Mr Brodie submitted that an owner who employs an independent contractor to erect a building on his behalf is himself under no obligation to comply with regulation D3 and is accordingly under no liability if the contractor negligently fails to comply with that regulation. Though for the purposes of its decision in *Acrecrest* this court assumed that owners who were not the builders were under a statutory duty to comply with the relevant building regulations, it left this point open. As Donaldson LJ pointed out in *Acrecrest* (at p 279G), regulation D3 in terms appears to operate in rem: ". . . the foundations of a building shall . . . ". Section 4(6) of the 1961 Act, quoted above, makes it clear that someone has to comply, but not who.

In our judgment, on the true construction of these two provisions, any owner-developer of land whose actions lead to a breach of regulation D3, in the course of building done on his behalf, is liable to a fine and correspondingly under a duty to comply with the regulation. In such circumstances, in our opinion, it is unlikely to have been the intention of the legislature that having, so to speak, put into circulation a building which is potentially unsafe because of a breach of the building regulations, he should be able to shelter behind the builder whom he was employed to do the work on his behalf merely because the builder is an independent contractor; he will himself still have "failed to comply" with the building regulations within the meaning of section 4(6) of the 1961 Act, whether or not the authority will choose to proceed against him rather than the contractor. Our conclusion in this context is fortified by two points. First, the provisions of sections 65(1) and 65(5) of the 1936 Act, quoted above, appear to us to contemplate that in the event of a work contravening any of the regulations, the owner himself will be in breach of statutory duty and will correspondingly be vulnerable to a notice requiring him to pull down the work and to civil proceedings requiring him to do so, "without prejudice to" the authority's right to take proceedings against him for a fine. Second, Lord Keith in *Peabody's* case [1985] AC 210 at p 244H, clearly regarded the contravention of the relevant byelaws relating to foundations as having rendered the plaintiff building owners in the case of *Dennis* v *Charnwood Borough Council* [1983] QB 409 *themselves* in breach of those byelaws, even though those owners had employed a contractor to do the work for them.

To anticipate a little, we think it plain that the plaintiffs, Anglia, were themselves in breach of the building regulations in regard to the foundations of the warehouses which were in the event erected.

Events from September 1973 onwards

We now revert to the history of the matter. On receipt of the letter of September 17 1973, Hamilton Associates, as architects on behalf of Anglia, submitted to the council a notice of intended building operations dated September 18 1973 pursuant to the Building Regulations 1972. The notice was on a standard printed form with manuscript insertions. It gave notice of the intention "to carry out the following building operations namely the erection of 4 factory units with toilets at [the site] in accordance with the plans and sections and the particulars hereinafter contained". The respective names and addresses of Anglia as "Owner" and of GTE Construction Ltd as "Builder" were given. There then followed a number of particulars of the site and the proposed building. Next to a printed heading "Description and Size of Foundation" appeared in manuscript the words "to structural engineer's design details". The notice was accompanied by drawings but these did not show the precise design of the foundations, since Ellisons had not yet designed them.

On November 15 1973 the council gave planning permission in respect of the four warehouses. Also, on November 15 1973, before any design certificate or full working drawings in respect of the foundations had been supplied, the council gave Hamilton Associates written notice "in pursuance of section 64 of the Public Health Act 1936" that the plans submitted on September 18 1983 had been "passed".

On November 20 1973 Anglia and GTE Developments Ltd signed articles of agreement relating to the construction of the four warehouses.

Ellisons prepared a drawing no 7323/4 showing the foundation details in respect of the four proposed warehouses. They sent a copy of it to Hamilton Associates who, under cover of a letter of December 3 1973, returned it saying: "This drawing has been examined by us and would appear to be in order." Ellisons' design provided for foundations for each portal frame consisting merely of a 6 ft × 3 ft × 3 ft deep concrete block based upon a lean mix of concrete at a depth at which Ellisons decided that the bottom was capable of providing a safe bearing pressure of 1 ton per sq ft. This depth, they thought, was approximately 6 ft in each case.

A When a proper soil investigation was subsequently carried out in June 1975 the depth of the fill was found to be some 30 ft to 40 ft, with a high ratio of voids. Nevertheless, in December 1973, all interested parties appeared content with Ellisons' design for the foundations.

Under cover of a letter of December 20 1973, Ellisons submitted to the council copies of their drawing showing details of the foundations and retaining walls, their calculations sheet, together with a copy of their design certificate covering the relevant structural work. By the certificate they certified that (*inter alia*) all structural work to Phase II would conform as to the foundations in every respect to the current editions of the *British Standard Code of Practice 2004*.

So far as the documentation before us shows, this letter of B December 20 1973 elicited no explicit response from the council. Nevertheless, on the basis of their inspections of the site and of various trial pits, the plans and drawings and calculations which had been presented to them and of Ellisons' design certificate, the council raised no objections to the proposed foundations or to the assumptions in regard to bearing pressures upon which they were based. They knew that there were to be no piled foundations, but that, instead, there were to be mass based foundations taken down in concrete to bottoms thought to be capable of providing the required bearing pressures. During the weeks which followed, as work on the foundations proceeded, their inspector approved all the foundation bases on the site, before the concrete was poured in.

C We pause to say that at the trial the learned judge heard evidence from four independent experts who were all agreed that where building work was to be carried out on made ground or fill, such as the site was known to be, prudence demanded that a proper soil investigation should be carried out in order to determine how the foundations should be designed so as to comply with the building regulations. While a number of trial pits had been dug, we have not thought it necessary to do more than refer in passing to a few of these digging operations. For on the basis of the expert evidence, the learned judge found (at pp 28-29), and it is not now disputed, that "the digging of trial holes was quite useless". He said: "The only possible way of investigating the nature of the ground was by means of bore holes and subsequent laboratory analysis of the undisturbed D samples." No investigation by means of bore holes was ever carried out on this site. If it had been, the subsequent trouble might well have been avoided. He found as a fact (p 18D) that the units should have been constructed with pile foundations. He did not think it surprising that Mr Akroyd, the independent consulting engineer called on behalf of the council, branded everyone concerned with this project as "crazy".

We now revert to the history. In December 1973, at the time when the foundations were in the course of being laid, a firm of architects, Lockhart, Smith & Kersey Galc ("Lockharts"), acting for a potential purchaser, intervened to ask Anglia for information about the warehouses in general and the foundations in particular, having E regard to the fact that at one time the whole area had been under water. Correspondence ensued between Lockharts, Hamilton Associates and Ellisons, to whom the architects directed queries concerning the foundations. Lockharts eventually withdrew from the scene. However, still no trial borings were made.

During the next few months work was proceeding on the site, with visits from time to time by council inspectors.

On November 21 1974 there took place a "practical completion meeting" relating to Phase II. On November 26 1974 Hamilton Associates issued a certificate of practical completion certifying that Units A-B, C-D and E-F (representing three of the four warehouses) were "taken into possession" under the building contract on November 15 1974.

F By December 4 1974 ominous cracks had appeared in the external brickwork of two of the warehouses (Units E-F and G-H).

Ellisons inspected the site on December 6 and on December 13 1974 sent a letter to Hamilton Associates in which they reported that diagonal cracking had occurred in the rear flank wall radiating from the centre of the building Unit E-F and that cracking had occurred to the gable and the front flank walls to Unit H. They gave it as their opinion that the cracking had resulted through "uneven foundation settlement", and that the foundations were structurally sound. As the learned judge pointed out, this opinion was plainly wrong and was subsequently shown to be wrong.

On December 20 1974 Hamilton Associates issued a certificate of practical completion certifying that the works to the four warehouses were practically completed on that date. G

By January 27 1975 even the service road was showing signs of settlement and cracks. Anglia themselves expressed their concern in a letter to Hamilton Associates of March 10 1975, saying that they thought that the problems arising from the partial collapse of the access road and the settlement that had occurred in Units E, F and H required urgent attention.

On April 16 1975, following a site visit, Ellisons sent a report to Hamilton Associates concerning "noticeable settlement resulting in severe cracking to the floor slabs" which was taking place in Unit H. They said they had no reason to anticipate any problems with the foundations, but suggested that bore holes should be put down to enable them to recommend specific remedial proposals for the H architects' consideration.

On April 23 1975 Hamilton Associates wrote to Ellisons pointing out that the settlement problem was not confined to Units G and H, but extended to Units E and F. They referred to evidence in both blocks that "seems to stem from settlement of the foundations under certain frames and edge beams for example brickwork, cracking and jamming of doors in frames". They said that a very full and comprehensive report was required, also giving recommendations for cure of the settlement.

As a result of this letter a ground exploration by means of bore holes was arranged and carried out by Contract Drilling Ltd. Their report, which was sent to Hamilton Associates by Ellisons under cover of a letter of June 20 1975, was not reassuring. It indicated that J the made ground went down to a depth of 10 to 11 metres with a high voids ratio. They gave their opinion that the basic remedies would appear to be either "the improvement of the state of compaction of the soil by the injection of chemical grout or the transfer of the structural loads down to the natural ground by means of piles". However, they mentioned that one possible course would be to carry out remedial works and adjust the levels as natural consolidation progressed.

In their covering letter of June 20 1975 Ellisons recommended the third course as being far less costly than piling or grouting. Their tone remained optimistic. They suggested it was reasonable to anticipate little trouble from natural consolidation in the future.

However, Rocket Precision Ltd ("Rocket") which had by then K entered into occupation of the warehouse E-F took a less sanguine view. On June 27 1975 they wrote to Anglia saying:

> There is ample evidence in the structure of progressive subsidence which is manifested by the fissures in the walls and the sinking of the floor levels. From our observations the condition is progressively worsening and it appears possible that some collapse is imminent . . . . The impact of these conditions is having its effect on personnel, and the operation of our plant which requires frequent relevelling and whilst these conditions obtain you will appreciate that we are prevented from enjoying the full benefits we expected under the terms of our lease.

By the end of July 1975 Hamilton Associates had made it clear that they considered that the blame for what had occurred lay with Ellisons. L

In September 1975 Anglia engaged independent consulting engineers, Oscar Faber & Partners, to advise them. Mr Burke, on that firm investigated the matter and later gave evidence before the learned judge. In response to questions asked in a letter from Mr Burke, of October 23 1975 Hamilton Associates said (*inter alia*) that "no site investigation was carried out to our knowledge other than rudimentary digging of trial holes and agreeing the load bearing capacity of the ground with the local authority"; that so far as they were aware no letter advising Anglia of any alternative foundation existed; that "the contractor on site was asked by Mr Loveday of Ellisons & Partners to excavate each base until satisfactory strata had been reached, and thereafter obtain approval of the building inspector". M

In November 1975 Rocket issued proceedings against Anglia claiming damages in respect of the serious cracking which had appeared in Unit E. On March 18 1976 Anglia's solicitors wrote to Ellisons informing them they would look to them to indemnify them against any judgment recovered by Rocket. Rocket's action was subsequently settled.

Eventually, in March 1979, Anglia decided to demolish the warehouses E-F and G-H and rebuild them on proper foundations. The work of demolition began on March 13 1979. Having instituted separate proceedings against Ellisons, Anglia's solicitors on September 4 1979 wrote to the council a letter before action in which they said:

> A We are instructed that in accordance with your statutory duties you approved the design for the foundations and that your building inspector was present when they were laid. The foundations were in fact so designed and laid as to rest in what your officials knew or ought to have known as "made" ground. Accordingly the foundations were wholly inadequate to support the buildings and were not in accordance with the Building Regulations.
>
> Our clients have been advised that you were negligent and accordingly are liable to our clients for the substantial damage which they have sustained.

On September 19 1980 Anglia issued a writ against the council and this was followed by the service of a statement of claim on February 19 1981. Nothing turns on the precise form of the pleading. The basis of Anglia's claim for damages was more or less the same as in its letter before action. Particulars of damage were given which included B items totalling about £660,000, of which the greater part represented expense incurred in or about demolishing and rebuilding units E, F, G and H, but about £11,500 was said to represent a sum paid to Rocket in settlement of their claim and about £145,000 was said to represent loss of rent.

A defence was served by the council on March 31 1981 and later amended. Some 18 months later, on September 8 1982, the council obtained leave to serve the third party notice to which we have already referred. The defence had implicitly sought to cast part of the blame on to Hamilton Associates, who also received warning by a solicitor's letter shortly before the third party notice was served. With these exceptions, however, it appears that neither Anglia nor the council itself sought to place any blame whatever on the architects until the C third party notice was served. The correspondence shows that after the cracks in the building appeared all the parties were directing their criticisms to Ellisons and looking to them to find a solution to the problems, until they realised that no solution would be forthcoming from this source.

*The pre-*Peabody *authorities*

Before turning to the learned judge's decision, it may be convenient to consider the state of the relevant authorities as they stood in 1983, relating to the duty owed by a local authority to owners and occupiers of buildings in exercising its supervisory powers over building operations.

In *Dutton* v *Bognor Regis Urban District Council* [1972] 1 QB 373 D this court held that a local authority, in exercising its statutory power under the Public Health Acts to inspect the foundations of a building to see if they complied with the building byelaws made under those Acts, owed a duty to the plaintiff who had bought the house from the owner-builder and was occupying it at the time when serious defects in its internal structure developed. The inspection of the house and the subsequent approval of the plans was held to have been negligent and the plaintiff succeeded.

In *Anns'** case a number of occupiers of flats under long leases from the owner-builder sued the owner-builder and the local authority in respect of damage from defective foundations. The plaintiffs relied not only upon negligent inspection but also upon a failure to make any inspections. Among the questions which the E House of Lords had to consider (see p 751D) was whether the defendant council was under a duty to the plaintiff occupiers to take reasonable care to see that the byelaws made pursuant to the powers conferred by the Public Health Act 1936 were complied with, as held in *Dutton*'s case. Their Lordships held that it was under such a duty. The ratio of their decision on this point perhaps sufficiently appears from the following passages in the speech of Lord Wilberforce, with which Lord Diplock, Lord Simon of Glaisdale and Lord Russell of Killowen agreed. Lord Wilberforce (at pp 751-752) said:

> Through the trilogy of cases in this House — *Donoghue* v *Stevenson* [1932] AC 562, *Hedley Byrne & Co Ltd* v *Heller & Partners Ltd* [1964] AC 465, and *Dorset Yacht Co Ltd* v *Home Office* [1970] AC 1004, the position has now F been reached that in order to establish that a duty of care arises in a particular situation, it is not necessary to bring the facts of that situation within those of previous situations in which a duty of care has been held to exist. Rather the question has to be approached in two stages. First one has to ask whether, as between the alleged wrongdoer and the person who has suffered damage there is a sufficient relationship of proximity or neighbourhood such that in the reasonable contemplation of the former, carelessness on his part may be likely to cause damage to the latter — in which case a *prima facie* duty of care arises. Secondly, if the first question is answered affirmatively, it is necessary to consider whether there are any considerations which ought to negative, or to reduce or limit the scope of the duty or the class of person to whom it is owed or the damages to which a breach of it may give rise: see *Dorset Yacht* case [1970] AC 1004, per Lord Reid at p1027.

*Editor's note: *Anns* v *Merton London Borough Council* [1978] AC 728.

G At p 753G-H he said:

> It must be in the reasonable contemplation not only of the builder but also of the local authority that failure to comply with the byelaws' requirement as to foundations may give rise to a hidden defect which in the future may cause damage to the building affecting the safety and health of owners and occupiers. And as the building is intended to last, the class of owners and occupiers likely to be affected cannot be limited to those who go in immediately after construction.

At p 758D-E Lord Wilberforce defined the persons to whom the duty is owed:

> There is, in my opinion, no difficulty about this. A reasonable man in the position of the inspector must realise that if the foundations are covered in without adequate depth or strength as required by the byelaws, injury to H safety or health may be suffered by owners or occupiers of the house. The duty is owed to them — not of course to a negligent building owner, the source of his own loss. I would leave open the case of users, who might themselves have a remedy against the occupier under the Occupiers' Liability Act 1957. A right of action can only be conferred upon an owner or occupier, who is such when the damage occurs (see below). This disposes of the possible objection that an endless, indeterminate class of potential plaintiffs may be called into existence.

At p 758F he defined the nature of the duty:

> This must be related closely to the purpose for which powers of inspection are granted, namely to secure compliance with the byelaws. The duty is to take reasonable care, no more, no less, to secure that the builder does not cover in foundations which do not comply with byelaw requirements.

J At p 759F-H Lord Wilberforce dealt with the question of recoverable damages:

> The damages recoverable include all those which foreseeably arise from the breach of the duty of care which, as regards the council, I have held to be a duty to take reasonable care to secure compliance with the byelaws. Subject always to adequate proof of causation, these damages may include damages for personal injury and damage to property. In my opinion they may also include damage to the dwelling-house itself; for the whole purpose of the byelaws in requiring foundations to be of a certain standard is to prevent damage arising from weakness of the foundations which is certain to endanger the health or safety of occupants.
>
> To allow recovery for such damage to the house follows, in my opinion, from normal principle. If classification is required, the relevant damage is in my opinion material, physical damage, and what is recoverable is the amount K of expenditure necessary to restore the dwelling to a condition in which it is no longer a danger to the health or safety of persons occupying and possibly (depending on the circumstances) expenses arising from necessary displacement.

He concluded (at p 760) that *Dutton's* case had in the result been correctly decided.

Dutton's case and *Anns'* case considerably extended the established frontiers of the law of negligence as regards the duty owed by a local authority to subsequent occupiers, in exercising its supervisory powers over building projects under the Public Health Acts. However, apart from Lord Wilberforce's passing reference (at p 758E) to "a negligent building owner, the source of his own loss", L both cases left open the question whether the local authority would owe any duty to the original builder-owner himself. This question fell to be considered by different divisions of this court in *Acrecrest,* in which judgment was delivered on July 22 1982, and *Dennis* v *Charnwood Borough Council* [1983] QB 409 in which judgment was delivered on July 29 1982. It would appear that neither division was aware of the case before the other. We will take the summaries of these two cases more or less verbatim from the summaries which are to be found in the speech of Lord Keith in *Peabody* [1985] AC at pp 243-244.

In *Dennis'* case the plaintiffs commissioned a builder to construct a house for their own occupation upon a site which consisted partly of an unfilled sandpit. They apparently did not have the assistance of an M architect or a structural engineer. Plans were submitted to the local authority under byelaws which corresponded in material respects with para 13 of Part III of Schedule 9 to the London Government Act 1963 ("the 1963 Act"). The plans showed the foundation to consist of a concrete raft and were duly approved by the local authority. This was in fact an unsuitable foundation for the site and many years later subsidence led to serious cracking of the structure. The plaintiffs sued the statutory successors of the local authority for damages on the ground of negligence in passing plans indicating an inadequate foundation. A judgment in their favour was affirmed by this court. Templeman LJ, delivering the leading judgment, regarded the matter as concluded in the favour of the plaintiffs by the decision in the *Anns* case.

In *Dennis'* case the plaintiffs were both the original owner-builders of the house and the occupiers when the trouble arose. In *Acrecrest's* case the plaintiffs were the original owner-builders but, at least so far as the report shows, not the occupiers of any part of the buildings. They had employed independent architects and contractors to build a block of flats and garages which they subsequently leased out. The architects had prepared plans which provided for foundations of inadequate depths. The local authority's building inspector required some increases of these depths, but the requirements were still inadequate. As a result, defects developed in the buildings in respect of which the tenants claimed against the owners. The owners sued the architects for damages for negligence and breach of contract and the architects joined the local authority as third parties claiming contribution. The action against the architects was compromised, but their claim for contribution proceeded to trial. The trial judge, purporting to apply *Dutton's* and *Anns'* cases, held the local authority were liable for breach of a duty of care owed to the owners, and assessed their contribution at 25%. His decision was affirmed by this court, which held that a building owner who has delegated the building work to an independent contractor and has not himself been negligent, is included in the class of persons to whom a local authority owes a duty to take reasonable care in the inspection and approval of building works.

Apart from the decision of this court in *Peabody,* this was the state of the authorities at the time when the learned judge gave his decision.

The judge's decision as to the liability of the council

The learned judge, unquestionably correctly, held that there had been a breach of the relevant building regulations in regard to the erection of the four warehouses. He accepted that it was the duty of Ellisons to decide what site investigation should be carried out in this case, where it was known that the site was on made ground. He regarded Ellisons as clearly at fault in failing to see that a proper investigation was carried out by means of bore holes and subsequent laboratory analysis of the undisturbed samples. Instead of this, he said, they seemed to have accepted the foundation subcontractors' statement that virgin ground had been reached, which clearly was not the case, and accordingly designed the simplest foundation which would have been suitable only for good intact clay.

As to the council, he pointed out that it was their duty under section 64 of the 1936 Act to pass the plans if they complied with the building regulations or to reject them if they did not. He considered it plain beyond peradventure that the council were at fault in allowing the plans to pass and further in expressing their positive satisfaction with the bases excavated without having satisfied themselves at least that a soil investigation report had in some way confirmed or corroborated the validity of the design which the engineers were submitting. His conclusion as to the council's liability was expressed as follows:

> The local authority owed a duty of care to Anglia even though Anglia were not the occupiers of the buildings and even though they as the building owners engaged independent contractors to build and architects and engineers to design. This was established in *Acrecrest.*
>
> . . . I hold that the local authority were in breach of their statutory duty and in breach of their duty of care and that these breaches were causative of the damage which Anglia have suffered as a result of the necessity to demolish and rebuild the two warehouses.

The learned judge did not refer in his judgment to the decisions of this court in *Peabody* (which had not yet been heard on appeal in the House of Lords) or in *Dennis.* The latter decision, if cited to him, would no doubt have reinforced his conclusions.

As to the position of Anglia themselves, he found that they were not personally negligent. He thought that the building regulations gave some ground for thinking that unless the owner or the architect or anyone else deliberately intervenes to cause a non-compliance, the only person under a statutory duty to comply with the regulations is the builder himself. However, he found it unnecessary to decide whether Anglia were or were not under any such duty because he was satisfied that any such duty, if it existed, was not of the "absolute non-delegable" type of duty so as to leave them liable in the event of a breach of the building regulations having been caused by the fault of an independent contractor. On this basis he found that Anglia were not a "negligent building owner the source of his own loss", within Lord Wilberforce's description in *Anns,* and also acquitted Anglia of contributory negligence.

In the result, he gave judgment for Anglia on their claim against the council. He also gave judgment for Hamilton Associates against the council in the third party proceedings. We will consider the third party claim in a separate section at the end of this judgment.

Meantime, we find it convenient at this stage to refer to two matters arising out of the learned judge's decision. First, while Mr Ashworth, on behalf of the council, accepted that in broad terms the council had been careless, he did not formally concede that they had been in breach of their statutory obligations under section 64 of the Public Health Act 1936. We think it clear that the council were in such breach on account of one or more of the following matters, namely (a) their passing of the relevant plans on November 15 1973; (b) their failure to object to the design details submitted to them by Ellisons under cover of their letter of December 20 1973; (c) their approval of the foundation bases on the site. For the purposes of our decision, we do not find it necessary to deal with this point with greater precision or in greater detail. Second, however, as we have already indicated, we think it clear that Anglia themselves were in breach of the relevant building regulations. The all-important question of law is whether the council, in exercising their functions under the 1936 Act, owed a duty of care to Anglia as the building owners. The council's principal submission on this appeal is that they did not.

The Peabody *decision*

At the time when the learned judge delivered his judgment in this case the tide of judicial authority, save for the decision of this court in *Peabody* delivered on July 29 1983, was flowing strongly against local authorities, as exemplified in the *Acrecrest* and *Dennis* cases. In October 1984 the House of Lords gave its decision in *Peabody* which appeared to call a halt to this process. The essential facts of that case appear from the headnote to the report:

> Plans, submitted by architects acting for the plaintiffs and approved by the local authority, an inner London borough, provided for the construction of a flexible system of drainage for a housing development on a site owned by the plaintiffs. The drainage system subsequently installed by the plaintiffs' building contractors was, on instructions from the architects, of a different, rigid design. The departure from the approved design came to the knowledge of the local authority drainage inspector while the installation was proceeding but he took no action. When the drains were tested two years later they were found to be unsatisfactory and had to be reconstructed, causing the development to be delayed with consequent substantial loss to the plaintiffs. In proceedings brought by the plaintiffs against, *inter alia,* the local authority, alleging negligence in the discharge of their functions under paragraph 15 of Part III of Schedule 9 to the London Government Act 1963, the judge held that the local authority were liable to the plaintiffs in respect of the local authority's failure to ensure, on discovering that rigid drains were being constructed, that the drainage system being installed complied with the design already approved, and he made a declaration accordingly.

The Court of Appeal, in allowing the appeal, distinguished *Acrecrest's* case on grounds which perhaps sufficiently appear from the following passage in Slade LJ's judgment, which was later cited with approval by Lord Keith in *Peabody:*

> Can it have been the intention of the legislature, in conferring on a borough council power to enforce against a defaulting site owner requirements made by it in accordance with paragraph 13 of Part III of Schedule 9, to protect such owner against damage which he himself might suffer through his own failure to comply with such requirements? In my opinion, this question can only be answered in the negative. This particular power exists for the protection of other persons — not for that of the person in default. I say nothing about the case where a local authority have failed to make known their requirements or where they have made requirements of an inadequate or defective nature. However, I can see no justification for extending the law of negligence by imposing on a local authority, over and above their public law powers and duties under paragraphs 13 and 15, a duty to exercise their powers of enforcement under paragraph 15(2), owed in private law towards a site owner, who, whether with or without personal negligence, disregards the proper requirements of the local authority, duly made under paragraph 13 and duly communicated to him or persons authorised to receive them on his behalf. The practical implications of giving the defaulting owner a right to sue the local authority for damages in such circumstances need consideration, but no elaboration.

The House of Lords unanimously dismissed the appeal. Lord Keith, who delivered the leading speech, summarised the nature of Peabody's argument as follows:

> Peabody's case, in substance, is that Lambeth were under a duty of care, owed to Peabody, when they became aware through the letter of 4 May 1973 that the contractors were installing drains with rigid joints, to invoke their powers under paragraph 15 of Part III of Schedule 9 to the Act of 1963 so as to require a reversion to the flexible joint design which they had previously approved. The duty is not said to be a statutory one, but one that arose on common law principles because Lambeth ought to have foreseen that, if they did not take steps to stop the installation of drains with rigid joints, severe economic loss would ensue to Peabody through the necessity at some future date of taking up these drains and replacing them with others of the approved design, as in

fact happened . . . So the issue really comes to be whether Lambeth owed a duty to Peabody to warn them that they were heading for financial disaster.

Lord Keith, having drawn attention to (*inter alia*) the first passage from Lord Wilberforce's speech in *Anns* which we have cited, said (at p 240F):

There has been a tendency in some recent cases to treat these passages as being themselves of a definitive character. This is a temptation which should be resisted. The true question in each case is whether the particular defendant owed to the particular plaintiff a duty of care having the scope which is contended for, and whether he was in breach of that duty with consequent loss to the plaintiff. A relationship of proximity in Lord Atkin's sense must exist before any duty of care can arise, but the scope of the duty must depend on all the circumstances of the case.

A little later, having cited a passage from the speech of Lord Morris of Borth-y-Gest in *Dorset Yacht Co* v *Home Office* [1970] AC 1004 at p 1038, Lord Keith said this (at p 241C):

So in determining whether or not a duty of care of particular scope was incumbent upon a defendant it is material to take into consideration whether it is just and reasonable that it should be so.

Lord Keith then proceeded to consider whether on the particular facts of the case it would be just and reasonable to impose on Lambeth a duty of care of the particular scope alleged in that case. His conclusion was that it would not. As he put it (at p 241C-F):

In the instant case Peabody, the owners of the building site and the undertakers of the development thereon, bore responsibility, under paragraph 13 of Part III of Schedule 9 to the Act of 1963, for securing that the drains conformed to the design approved by Lambeth. Mr Dyson, for Peabody, brought to the attention of the House certain drainage byelaws made by the Greater London Council in 1962 under the statutory predecessor of the Act of 1963. These undoubtedly place certain very specific obligations upon contractors carrying out building operations, but they do nothing to detract from what is clearly the proper construction of paragraph 13(1), namely, that observance of its provisions is incumbent upon any person who puts in train a house building project. Peabody no doubt had no personal knowledge or understanding of what was going on. They relied on the advice of their architects, engineers and contractors, and in the event they were sadly let down, particularly by the architects. But it would be neither reasonable nor just, in these circumstances, to impose upon Lambeth a liability to indemnify Peabody against loss resulting from such disastrous reliance.

Lord Keith then went on (at p 241F-G) to consider the powers for which the relevant statutory powers had been conferred on Lambeth. He considered that this was "not to safeguard building developers against economic loss resulting from their failure to comply with approved plans". He said:

The purpose for which the powers contained in paragraph 15 of Part III of Schedule 9 have been conferred on Lambeth is not to safeguard building developers against economic loss resulting from their failure to comply with approved plans. It is in my opinion to safeguard the occupiers of houses built in the local authority's area, and also members of the public generally, against dangers to their health which may arise from defective drainage installations. The provisions are public health measures.

Lord Keith (at pp 241G-242E) then cited the paragraph from Lord Wilberforce's speech in *Anns'* case (at p 758) beginning with the heading "To whom the duty is owed". He pointed out that the damages which the plaintiffs were held entitled to recover consisted of such sums as each required to expend in order to put his house in such a state that it was no longer a danger to safety or health. He said (at p 242A):

It is important to notice that these sums were not recoverable as economic loss pure and simple, but as representing expenditure necessary to avert injury to safety or health.

Lord Keith then referred to certain difficulties to which this passage from Lord Wilberforce's speech gives rise. In particular he found it difficult to see how a "non-resident owner" could fall within the scope of it, since he would not be subject to any "possible injury to safety or health". He pointed out that the plaintiffs in *Anns* as long lessees were in substance both owners and occupiers and considered (see p 242C) that the decision "should be treated as restricted to that situation".

However, as Lord Keith observed, these difficulties were not necessary to the determination of the instant appeal. He continued (at p 242E):

It is sufficient to hold that Lambeth owed no duty to Peabody to activate their paragraph 15 powers, notwithstanding that they might reasonably have foreseen that failure to do so would result in economic loss to Peabody, because the purpose of avoiding such loss was not one of the purposes for which these powers were vested in them.

Lord Keith then cited with approval the passage from the judgment of Slade LJ quoted above. Finally, he dealt with the *Acrecrest* and *Dennis* cases. Having summarised the facts of *Acrecrest*, he observed (at p 243E-F):

Stephenson LJ, giving the leading judgment, construed the passages which I have quoted from the speech of Lord Wilberforce in the *Anns* case [1978] AC 728 as intended to lay down that a local authority, in the exercise of its supervisory functions over building projects, owed a duty of care to a building developer to see that his property did not suffer damage, even though there was no question of apprehended injury to the health or safety of the developer. Donaldson LJ expressed a similar view, as did Sir David Cairns. In my opinion they failed to appreciate correctly the course of Lord Wilberforce's reasoning and consequently misapplied the decision in *Anns*. In the result, the *Acrecrest* case . . . was wrongly decided and should in my opinion be overruled.

Having summarised the facts of *Dennis* and cited a passage from Templeman LJ's judgment in that case [1983] QB 409 at pp 414-415), he concluded (at pp 244G-245B):

The decision is in my opinion to be justified on the basis that the plaintiffs, as owners who were the intended occupiers of the house, were within the ambit of the duty of care laid down in *Anns*. They were persons injury to whose safety or health might necessarily be expected to occur if the foundations of the house were inadequate. There can be no doubt that, under the *ratio decidendi* of *Anns*, a remedy against the local authority would have been available to any subsequent occupier who had purchased the house. The plaintiffs were in breach of certain material provisions of the relevant byelaws dealing with the adequacy of foundations, but the fact remains that plans showing the intended foundations had been submitted with their authority and had been approved. This approval might reasonably be taken as an indication that the foundations were satisfactory, and considering that the plaintiffs themselves had no technical knowledge nor understanding of the position and that their own safety and health were in issue, it would be unreasonable and unjust to hold that the local authority owed them no duty. The decision does not, however, assist Peabody in the present case, because not only was there no question of injury to health or safety at issue so far as they were concerned, but they were proceeding to install drains with fixed joints in flat disregard of Lambeth's requirements.
My Lords, for these reasons, I would dismiss the appeal.

Anglia's case against the council

The arguments addressed to this court have shown us that, on minute analysis by able advocates, the speeches of Lord Wilberforce in *Anns* and of Lord Keith in *Peabody* are open to a number of possible interpretations and are not entirely easy to reconcile in every respect. Of particular significance for present purposes is Lord Keith's reference to *Acrecrest*, since this was the case on which the learned judge founded his decision. Mr Brodie, for Anglia, submitted that the overruling of *Acrecrest* by their Lordships was obiter. In the course of other criticisms of the *Peabody* decision, he submitted that the distinction drawn between the duty owed by a local authority to owners and that owed to occupiers is an artificial one; so too, he said, is the distinction drawn between "economic loss" and other kinds of loss. In truth, he suggested, expenditure incurred to remedy defective buildings must always represent economic loss. Where their defective state has arisen through the default of a local authority, common sense, in his submission, demands that the person who has to pay the bill should be entitled to sue, without the drawing of nice distinctions between owners and occupiers.

For all this, we think there is no doubt that this court must regard *Acrecrest* as having been overruled by the House of Lords in *Peabody* and must proceed on this footing. On this assumption, Mr Brodie submitted that *Acrecrest* had been overruled solely on the question whether a duty of care is owed to an owner who is not an occupier. The *Acrecrest* decision, he suggested, remains intact in so far as it decides that a local authority, in exercising its statutory function to approve or reject plans, owes a duty of care to the building owner who is also an occupier when the injury occurs and, though in breach of the building regulations, has not been personally negligent. In the present case, he submitted, subject to one exception, Anglia was in possession or occupation of the site and buildings at all material times and continued to be so until they were demolished in July 1979; the sole exception was the period from April 1975 until July 1976 when Rocket occupied Unit E as lessees. From December 1974 onwards the unsafe foundations of the buildings, he contended, involved ever-present danger to the health and safety of their occupants. Thus Anglia, their servants or agents, were included in the class of person whom the council could reasonably have foreseen were likely to suffer injury if unsafe foundations were constructed upon the basis

A of defective plans approved by it. This, Mr Brodie submitted, brought Anglia within the class of persons to whom, in accordance with the principles of *Anns*, the council owed a duty of care in discharging their statutory duty under section 64 of the Public Health Act 1936 to approve or reject plans. Anglia were not, in his submission, "negligent building owners" within Lord Wilberforce's description in *Anns* (at p 758F): the House of Lords in *Peabody* did not overrule *Acrecrest* in so far as it decided that the negligence of an independent architect or consulting engineer is not to be attributed to the building owner (see, for example, *Acrecrest* [1983] QB 260 at p 273D per Stephenson LJ).

Of all the problems presented by the various authorities which
B have been cited to us, perhaps the most formidable is raised by Lord Wilberforce's brief statement in *Anns* that the relevant duty is owed "not of course to a negligent building owner, the source of his own loss". We think there can be no doubt that (whether or not he is the intended occupier) the building owner who, with personal carelessness on his part, permits the foundations of his building to be constructed in breach of the relevant building regulations is owed no duty of care by the local authority under *Anns* principles. But what of the building owner who does so with no personal carelessness, because he has employed apparently competent independent experts?

We recognise that, on one reading of *Peabody*, a building owner guilty of no personal carelessness could be regarded as falling within
C the relevant duty of care, even if he is in breach of the relevant building regulations. Some support for this view is in particular to be derived from Lord Keith's reference to *Dennis'* case without disapproval.

Nevertheless, in deciding whether or not the alleged duty of care exists in the present case, we are, on the authority of *Peabody*, entitled and bound to take into consideration whether it is just and reasonable that it should exist. We feel no doubt that it is not just and reasonable. *Mutatis mutandis*, the passage which we have cited from Lord Keith's speech at pp 241C-F seems to us fully to apply to the facts of the present case. Observance of the provisions of the Building Regulations 1972 was incumbent on (among others) Anglia who put in train the building project. We are prepared to assume in favour of
D Anglia that, despite their previous experience as building developers, they had no proper understanding of the risks which they were running in proceeding with the building operation on this unfilled site without a proper soil investigation and without piled foundations. They no doubt relied on the advice of their architects, consulting engineers and contractors and in the event were badly let down, at least by their consulting engineers. In our opinion, however, it would be neither reasonable nor just to impose upon the council a liability to indemnify Anglia against loss resulting from such disastrous reliance.

We can see no difference in principle, and it has not been submitted that there is any material difference, between the powers to secure or
E enforce compliance with building regulations which have been conferred on a local authority by the Public Health Act 1936 and those conferred on it by para 15 of Part III of Schedule 9 to the Act of 1963. We will refer to all these powers collectively as "supervisory powers" and venture to think that the following five propositions of law can be derived from the *Anns* and *Peabody* decisions when considered together, in the light of the overruling of *Acrecrest*:

(1) The purpose for which the legislature has conferred the supervisory powers over building operations on local authorities is to protect the occupiers of buildings built in the local authority's area and also members of the public generally against dangers to health or personal safety. It is not to safeguard the building developer himself against economic loss incurred in the course of
F a building project, or indeed anyone else against purely economic loss.

(2) In view of the statutory purpose of these supervisory powers (though this point does not arise for decision in the present case) it may well be that on the basis of *Anns* a local authority, in exercising their statutory powers, will be held to owe a duty to a subsequent occupier other than the original building owner to take reasonable care to ensure that the building is erected in accordance with the building regulations, so as not to cause danger to health or personal safety of the occupier.

(3) Where this duty of care has been broken and, as a result, the condition of the property gives rise to danger to the health or safety of persons present on the premises, an occupier to whom the duty is owed may be at liberty to restore the property to a G
condition in which such danger is eliminated, and having done so, to recover the amount of any such necessary expenditure from the local authority. Whether or not he will have such right of recovery must depend on the particular facts of the case.

(4) On the basis of *Peabody*, however, a local authority, in exercising these supervisory powers, will normally owe no duty to an original building owner, because it is normally incumbent on the building owner himself to ensure that the building is erected in accordance with the relevant building regulations, and it cannot have been the intention of the legislature that, save perhaps in exceptional circumstances, a local authority could owe a duty to a person who is in such breach.

(5) *A fortiori* (again on the basis of *Peabody*) a local authority, in H
exercising these supervisory powers, will normally owe no duty to an original building owner who has had the benefit of the advice of architects, engineers and contractors and has relied on it. In such circumstances it will normally be neither reasonable nor just to impose upon the local authority a liability to indemnify the building owner against liability resulting from such reliance.

Dennis is, in our judgment, to be regarded as a decision on its own special facts. The original building owners to whom the duty was held to be owed had no technical knowledge or understanding of the position, they had no professional advisers other than the building contractor, they were known to the local authority to be the intended occupiers, and the local authority, in approving the plans, could (as
Lord Keith said) reasonably be understood as having indicated that J
they involved no risk to the health or safety of the occupier.

In the light of the *Anns* and *Peabody* decisions there has been much debate in this court as to (a) whether, on the evidence, Anglia could properly have been regarded as "occupiers" of the site at the material time; (b) whether there can ever be said to exist imminent danger to the health or safety of a *corporate* occupier; (c) whether there was ever any imminent danger to the health or safety of Anglia's agents and employees; (d) whether the loss suffererd by Anglia is to be regarded as purely economic loss. On the basis of our conclusions as to the relevant law, we find it unnecessary to decide any of these points.

However, as to (a) above, we should mention that Lord Keith's
speech in *Peabody* may be regarded as giving support to the argument K
addressed to us on behalf of the council, to the effect that the use of the word "occupiers", in the wide senses which it has in other branches of the law such as the Rating Acts and the Occupiers' Liability Act, is inappropriate in the present context. It may be thought unrealistic to suggest that any substantial danger to the safety or health of a non-resident owner, even though for other purposes of the law he may be called an "occupier", would be reasonably foreseeable by the local authority at the relevant time. Hence comes Lord Keith's exclusion, from the scope of the duty, of a "non-resident owner". It cannot be suggested that Lord Keith overlooked the possibility that a "non-resident owner" might well,
for certain purposes, be an "occupier". L

As to (d) above, so far as this may be relevant, we must observe that in our view it would be unrealistic to regard Anglia's loss in this case as anything other than purely economic loss.

We are firmly of the view that the legislature, in imposing on local authorities for the general protection of the public the relevant statutory obligations under section 64 of the Public Health Act, cannot have been intending to protect a building developer such as Anglia against damage which he himself may suffer through his failure to comply with the relevant building regulations — or to entitle him to an indemnity from his fellow ratepayers against the conseqences of any such failure.

We can see no sufficient reason for regarding the facts of the
present case as exceptional in any relevant respect. For the reasons M
already stated, we consider that Anglia at all material times has itself been in breach of the building regulations. Though we think the learned judge was right to conclude that the council were under a duty and consequent liability to Anglia on the basis of the law as it appeared from the *Acrecrest* decision to be, we think that numbers (4) and (5) of the five propositions listed above, which we derive from *Peabody*, fairly and squarely apply to the present case. In our judgment, therefore, the council, in considering whether or not to approve the plans and in their subsequent inspections of the site, owed no duty of care to Anglia, whatever duty they may have owed to other persons. We therefore propose to allow the council's appeal, so far as it relates to Anglia.

The council's third party claim

Since in our judgment Anglia has failed to establish its claim against the council, the council's third party claim against Hamilton Associates for an indemnity or contribution does not arise. Nevertheless, this claim has been fully argued and we have formed clear views about it. We therefore think we should express these views, albeit relatively briefly.

The third party notice pleaded the appointment of Hamilton Associates by Anglia "on a date unknown but prior to February 15 1973 . . . to act as their architects in connection with the design and construction" of the four warehouses. It asserted that they had acted in breach of the implied terms of their retainer "and/or were negligent".

The existence or otherwise of any valid claim in Anglia against Hamilton Associates, whether in contract or in tort, would have largely depended on the precise terms of the contract under which Hamilton Associates were engaged by Anglia to act as architects in connection with the Phase II project. (It has not been suggested that on the facts of the present case the architects could be liable in tort for negligence even if there were no contractual liability.) The council may well have had great difficulty in ascertaining the terms of this contract. We therefore intend no criticism of their legal advisers in saying that the evidence adduced at the trial as to such terms was of the flimsiest nature. While it was common ground that Hamilton Associates had indeed been appointed architects in connection with the Phase II project, no written contract or letter of appointment was adduced. No oral evidence was given by any representative either of Anglia or of Hamilton Associates (or by any other person) as to the nature of the specific duties which the architects agreed to perform for Anglia or as to the other terms of the contract. The learned judge was left to draw inferences as to these matters as best he could from the correspondence or other documentary evidence before him. It was, however, common ground before him (as it has been before us) that the contract was governed by the Conditions of Engagement of the Royal Institute of British Architects ("the RIBA") then in force, even though the pleadings contain no reference to these conditions.

We have had the benefit of a careful and detailed analysis of these conditions in the course of argument. For present purposes we think it necessary to refer to only very few of them.

Clause 1.22 provides: "The architect will advise on the need for independent consultants and will be responsible for the direction and integration of their work but not for the detailed design, inspection and performance of the work entrusted to them." We read this reference to "integration" as referring to co-ordination (compare clause 3.101). Clause 1.23 provides: " Independent consultants . . . should be nominated or approved by the architect in agreement with the client. They should be appointed and paid by the client."

On the basis of the evidence before us, we are satisfied that Ellisons were nominated as independent consultants in connection with Phase II by Hamilton Associates and were appointed by Anglia, in accordance with clause 1.23, and that the work entrusted to them included the "detailed design" of the foundations for the four warehouses, within the meaning of clause 1.22.

The learned judge summarised the argument submitted to him in relation to the role of Hamilton Associates as follows:

> It is suggested that because they must have known the nature of the site from the outset and because, when initially acting for Mr Pentus Brown, they envisaged poor ground conditions and suspected that piling might be necessary, they ought to have realised that the foundations designed by the engineers would not be suitable and they should presumably have queried or objected to the design.

The learned judge, after referring to clause 1.22, said that he would have thought that Hamilton Associates "who had certainly advised on the necessity for engaging consulting engineers to design the foundations, were absolved from responsibility for the design of the foundations . . .". In the light of certain oral expert evidence, to which he referred, given by Mr Robert Foster FRIBA and a consulting engineer, Mr Winders, he concluded:

> I am satisfied that the architects, once the engineers had been appointed to design and supervise the foundations of these buildings, were entitled to rely upon the engineers' judgment as to what was suitable. They invited them to send the necessary structural information in support of the application for building regulation approval direct to the local authority, and in the result building regulation approval was given.
>
> Whatever the risk presented by the nature of the ground, I think that the architects properly met that risk by seeing that consulting engineers were appointed to deal with it, and in those circumstances I acquit the architects of any responsibility for the failure of the foundations.

Subject to what is said below, we think that the learned judge was right to regard clause 1.22, read together with the other relevant conditions, as defining and limiting the scope of Hamilton Associates' contractual duty to Anglia in respect of the foundations.

The only authority to which we have been referred concerning the liability of an architect who has delegated specialised tasks to qualified contractors is *Moresk Cleaners Ltd* v *Hicks* [1966] 2 Lloyd's Rep 388. That, however, was a case very different from the present, since it concerned the design of a building and it was held that the defendant had no authority to delegate the function of design. The effect of the RIBA Conditions does not appear there to have been in issue.

Whether or not in any given instance these conditions apply, it must generally be the duty of an architect to exercise reasonable care in the work which he is engaged to perform. However, clauses 1.20, 1.22 and 1.23 of the RIBA Conditions, in our judgment, clearly contemplate that where a particular part of the work involved in a building contract involves specialist knowledge or skill beyond that which an architect of ordinary competence may reasonably be expected to possess, the architect is at liberty to recommend to his client that a reputable independent consultant, who appears to have the relevant specialist knowledge or skill, shall be appointed by the client to perform this task. If following such a recommendation a consultant with these qualifications is appointed, the architect will normally carry no legal responsibility for the work to be done by the expert which is beyond the capability of an architect of ordinary competence; in relation to the work allotted to the expert, the architect's legal responsibility will normally be confined to directing and co-ordinating the expert's work in the whole. However, this is subject to one important qualification. If any danger or problem arises in connection with the work allotted to the expert, of which an architect of ordinary competence reasonably ought to be aware and reasonably could be expected to warn the client, despite the employment of the expert, and despite what the expert says or does about it, it is in our judgment the duty of the architect to warn the client. In such a contingency he is not entitled to rely blindly on the expert, with no mind of his own, on matters which must or should have been apparent to him. In this context the observations of May LJ in *Davy-Chiesman* v *Davy-Chiesman* [1984] Fam 48 at p 66A-B, though relating to the duty of a solicitor who has acted on counsel's advice, are in our opinion apposite.

We do not think that Mr Ashworth for the council seriously challenged the correctness of the propositions of law set out in the immediately preceding paragraph, which are substantially based on Mr Desch's submissions on behalf of the architects. Reduced to its essentials, Mr Ashworth's forceful attack on the architects can be summarised in a few sentences.

The whole trouble in the present case arose, he suggested, because the foundations for the warehouses had been designed without an adequate prior soil investigation of the site by bore holes, which would have shown that piled foundations were essential. Hamilton Associates, he submitted, in approving this design on December 3 1973, knew, or ought to have known, five crucial matters, namely that (1) the site consisted of made ground which had once been a sandpit; (2) a proper investigation of the nature of the ground was esssential; (3) investigation by means of trial pits was useless; (4) bore holes were the only reliable method of investigation; and (5) no bore holes had been bored and no site investigation had been carried out other than the rudimentary digging of trial pits. In these circumstances, he submitted, Hamilton Associates were in breach of their professional duty to Anglia in approving the plans for the foundations; they should, instead, have withheld their approval and warned Anglia that foundations designed without a proper prior soil investigation might well be totally inadequate.

It was conceded on behalf of Hamilton Associates that at the material time, December 3 1973, they had knowledge of points (1) and (2). While this is rather less clear on the evidence before us, we are also prepared to assume in favour of the council that at that date the architects also had knowledge of point (5) — they certainly had it by the time they wrote their letter to Oscar Faber & Partners of October 23 1975.

As to points (3) and (4), we accept that, as the learned judge found, the digging of trial pits was *in fact* useless and that *in fact* the only reliable way of investigating the nature of the ground was by means of bore holes and subsequent laboratory analysis of the undisturbed samples. The all-important question, however, is whether at the material time, December 1973, Hamilton Associates knew, or ought to have known, of these two points. Any suggestion that they had actual knowledge is quite unsustainable on the evidence. The question, therefore, resolves itself as to whether despite the advice given by the experts, Ellisons, they ought to have known of these two points and given their clients the appropriate consequential warning.

Expert evidence from suitably qualified professional persons is, in our judgment, admissible to show what competent architects in the position of Hamilton Associates could reasonably have been expected to know and do in their position at the relevant time. Indeed, in our judgment, there could be no question of the court condemning them for professional negligence on account of their failure to appreciate points (3) and (4) and to take the suitable consequential action, unless there were appropriate expert evidence to support the allegation that their conduct fell below the standard which might reasonably be expected of an ordinarily competent architect (compare *Worboys* v *Acme Investments Ltd* [1969] 4 BLR 133 at p 139 per Sachs LJ).

The expert evidence before the learned judge consisted of evidence from three engineers and one architect, Mr Foster. The questions put to the engineers and answered by them included questions relating to the nature and extent of the professional duties owed by Hamilton Associates to Anglia. However, we think that little reliance can be placed on their answers to these particular questions which related to a profession other than their own. The only directly relevant evidence in this context was that of Mr Foster, who was called on behalf of Hamilton Associates. In the course of his cross-examination, he specifically accepted that they would bear some professional responsibility if they had known, or ought to have known, that the building to be erected was not going to be satisfactory (see transcript March 6 1984 at p 5B). He accepted that the architects did run a risk (*ibid* at p 5C). Nevertheless, he considered that, having employed the structural engineers for the specific task of designing the foundations, the architects, as professional men, were entitled to assume that sufficient investigations of the soil had been made and that the foundations would be suitable for the proposed structures. Since this was not obvious to the structural engineers, he clearly did not think it should have been obvious to the architects that the proposed buildings, or some of them, were doomed to failure (see *ibid* at pp 5-6). At an earlier stage in his cross-examination (transcript March 5 1984 at pp 37H-38A) he had given evidence that nowadays few architects would advise on foundation questions. On the other hand, it was not suggested to Mr Foster, or any other witness, that there was reason to doubt the apparent competence of Ellisons to deal with such matters as qualified experts.

In the face of this, the only directly relevant expert evidence, we do not think that the council have come near to establishing that Hamilton Associates were in breach of their professional duty in failing to reject the plans produced or to question the advice given by the structural engineers whom they had instructed to produce plans for the foundations. We might add that, even if the relevant expert evidence had been to the contrary effect, we would have felt great difficulty in finding the architects guilty of professional negligence in the absence of any oral evidence as to the many other oral conversations between the interested parties, which must have punctuated the correspondence and the conversations referred to in that correspondence; it seems inconceivable that there were not a number of conversations between one or more of Hamilton Associates, Ellisons and Anglia relating to what action should be taken in regard to soil investigation. It has to be borne in mind that the burden of proving professional negligence on the part of the architects falls throughout on the council. In our judgment this onus has not been discharged.

Conclusions

However, since we have concluded that Anglia has no valid claim against the council, our decision in relation to the liability of Hamilton Associates has no practical relevance, save possibly in relation to costs. For the reasons already stated, the appeal of the council as against Anglia will be allowed.

The appeal of the district council against the judge's decision was allowed with costs in the Court of Appeal and below. The council's appeal against the order in favour of the second third party (Hamilton Associates) was dismissed, the council to pay the latter the costs of the appeal; the judge's order as to the costs of the second third party to stand. Leave to appeal to the House of Lords was refused.

For the other cases on this subject see p 179

REAL PROPERTY AND CONVEYANCING

A Queen's Bench Division

May 21 1985

(Before Mr Justice HIRST)

DEEN v ANDREWS AND ANOTHER

Conversion — Whether a large greenhouse was a building or a chattel — Dispute between vendor and purchasers of a house — The conveyance conveyed the land in question "together with the farmhouse and other buildings erected thereon" — The greenhouse which was the subject of the dispute was very
B large — It was of prefabricated construction and was placed in position on a previously installed concrete base — It was of cedar frame with a series of ridged roofs supported in the centre by steel supports which went down into concrete posts — The wood at the foot was bolted to large concrete plinths which rested unfixed on the concrete base — The purchasers of the house refused to allow the vendor to remove the greenhouse — Held, on the question of construction, after reviewing the authorities, that the greenhouse in question was a chattel — Although a large structure, it was not affixed to the land but lay on it by its own weight and fell within the
C definition of a chattel given by Blackburn J in *Holland* v *Hodgson* — The onus of showing that in these circumstances it was intended to be part of the land was not discharged by the purchasers — The judge also rejected a submission on behalf of the purchasers based on alleged estoppel by representation — Principles governing assessment of damages for conversion discussed, the action being based on conversion of the greenhouse by the defendants' refusal to allow the plaintiff to remove it — Judgment for plaintiff vendor

This action was brought by Mrs Leonie Geraldine Deen, vendor of a large cottage called Quays Farm, Toppesfield, Essex, against the
D purchasers, the defendants, Mr and Mrs Andrews.

H Lloyd (instructed by Ian Guyster & Co) appeared on behalf of the plaintiff; K Garnett (instructed by Irving Spellman & Co) represented the defendants.

Giving judgment, HIRST J said: This case is the saga of a greenhouse. In 1979 the plaintiff, Mrs Leonie Geraldine Deen, was the owner of a large cottage called Quays Farm in Toppesfield, Essex, and during the course of that year she and her husband caused to be erected a very large greenhouse which cost some £2,700 plus VAT.

The photographs which are in evidence show very clearly its construction. It was of cedar frame with a series of ridged roofs across the building, supported in the centre by steel supports which went down into concrete posts. The wood itself at the base was bolted
E to a series of large concrete plinths which ran all the way round the building consecutively, and they themselves were resting unfixed on a concrete base, which was put in before the building was erected.

The erector himself, Mr David Windley (whom I shall call hereafter "David" for short), gave evidence and he explained how the building arrived on a lorry in flat sections like a huge piece of Lego or Meccano and how, after it arrived, the concrete base was installed, the plinths were then placed in position on top of it, and the woodwork was erected — it came in smaller sections and they had to be bolted together. The roof, of course, was higher up, but also, as I understand it, of the same sort of prefabricated construction, and then, finally, the building was glazed and the glass itself not puttied in but fixed with plastic clips.

There was also some oil heating on the premises and there was at one stage a dispute about some oil, but that has now disappeared and so I shall say no more about the oil.

Early in the 1980s the plaintiff decided to sell the property. Mr Deen was, I think, already working in Florida in the United States and he has come from there for this case to give evidence. The property was put in the hands of two agents, Bairstow Eves, of Braintree, and also Savills, but Savills do not figure further in the case. It is the plaintiff's case and Mr Deen's evidence that he instructed Mr Adie, who was the Bairstow Eves' representative, to make it clear from the outset that the greenhouse was supposed to be separate.

Now, I say at once I have no doubt that was, in fact, said or intended to be conveyed. But in fact much of the discussion seems to have taken place on the telephone from America, and it is by no means clear whether and to what extent Mr Adie fully understood that, and Mr Adie himself has not given evidence. The particulars of sale are at p A1, and in the description of the outside there is a fairly neutral inventory, headed "Numerous out-buildings". I say "neutral" because it says nothing either way about the greenhouse.

There was, incidentally, another much smaller, older greenhouse on the premises also, of conventional, inter-war construction, which can be seen through the big greenhouse on one of the photographs.

The defendants, Mr Christopher Frederick Andrews and his wife, Mrs Gloria May Andrews, got the particulars of this property for sale from the agents. Having gone to the estate agent's offices, they were interested in the appearance and particulars of this building. Undoubtedly, even in a photostat of a photograph, it looks quite an attractive house, and so on that same day they went to the house to have a look at it, accompanied by Mr Adie. There they also met David, whom I have already referred to, who was the gardener.

There is a dispute, which is not of any importance save on credibility, as to whether or not David accompanied Mr and Mrs Andrews and Mr Adie round the house. What is, however, clear, and common ground, is that there was a conversation between David and Mr and Mrs Andrews out in the garden, at which the greenhouse was discussed. There is a dispute — an absolutely fundamental dispute, in fact — as to what was said in that conversation. Mr and Mrs Andrews' version is that they asked Mr Adie whether this big greenhouse was included or excluded; that Mr Adie referred the matter to David, and that David then said it was included, in other words part of the property on sale. David's evidence, on the other hand, is that when the matter was referred to him he made it quite clear that the greenhouse was separate. This conversation, of course, was right at the outset of the whole affair, but I shall of course come back to it later, as it is relied on as founding an estoppel.

Solicitors were instructed, and in the usual enquiries before contract there was the following question:

Fixtures and fittings. Does the sale include all of the following items now on the property: Trees, shrubs, plants, flowers, garden produce

then the important line

Greenhouses, garden sheds and garden ornaments, aerials, fitted furniture and shelves, electric switches, points and wall and ceiling fittings.
Answer: Yes, but the purchaser should check with the vendor as to the precise details.

Certainly this was a qualified answer, though the word "yes" at the beginning is a fairly strong affirmative, subject to the recommendation in the qualification.

On April 10 1980 the estate contract was entered into on the

National Conditions of Sale, and the property and interest therein sold is described as follows:

> ALL THAT piece and parcel of land situated in the parish of Toppesfield including hedges et cetera delineated on the plan attached hereto . . .

and then the vital words

> together with the farm houses and other fieldings

which everybody agrees was a misprint for "buildings."

The conveyance, which is dated May 12, is worded in almost exactly the same terms and I read only the important phrase, "together with the farmhouse and other buildings erected thereon." Between the conveyance and the contract there was some correspondence. On May 6 the plaintiff's solicitors wrote to the defendants' solicitors:

> Our client has informed us, as we believe your clients are aware, the greenhouses on the property are not included in the purchase price. Before our client dismantles same we should be obliged if you would kindly contact your clients to see if they would be interested in purchasing them accordingly.

There was clearly a telephone conversation following that letter, because three days later, on May 9, the plaintiff's solicitors write again to the defendants' solicitors, saying:

> With regard to the telephone message to the effect that your clients considered the greenhouses were included in the purchase price on the basis of the required reply to enquiry number 11, we would inform you that firstly we stated that your clients should check with our client as to the precise details, and secondly in any event our client instructed the agents, whom we understand informed your clients that the greenhouses were not included in the purchase price.

On the day of completion itself, three days later on May 12, the completion meeting was held at the plaintiff's solicitor's office, Ian Guyster & Co, with a representative of the estate agents present and also, of course, attended by the defendants' solicitor, Mr Spellman, of Irving Spellman & Co. The plaintiff's solicitor's attendance note of that meeting reads as follows, after the usual "attending upon" phrase:

> Our client did not wish to complete until it was agreed that the one greenhouse which was removable could be removed, and also he had arranged to drain off the oil, neither of which the purchaser wished to buy. There was considerable argument as to whether or not the greenhouse was included in the purchase price, but finally the purchaser agreed to allow our client to remove both the oil and the greenhouse, so that the matter was at an end.

There is no dispute about the accuracy of that note as an account of that meeting other than about the comment, "so that the matter was at an end", which indeed was a somewhat optimistic comment, since here we are today, five years and more after the event.

That was followed by a letter on May 12 from the plaintiff's to the defendants' solicitors, confirming the conversation in terms I do not need to read.

The situation on that day was as follows: The defendants, Mr and Mrs Andrews, who had been living in rented accommodation, were actually at the premises with their removal men and their furniture in tow when they got the message about the ultimatum from the plaintiff. Mr Deen in evidence said, quite frankly, that he was digging his toes in about this and therefore I do not think the word "ultimatum" is unreasonable. Mr and Mrs Andrews were therefore in a complete dilemma. To call the whole thing off and send their removal men away, perhaps with the idea that later on they would bring an action, was one possible course of action; indeed, that was what Mr Andrews first resolved upon. But having given it a little more thought he acceded, though he said very unwillingly indeed, to the ultimatum and agreed about the greenhouse.

The disproportion between the two aspects of the agreement — if there are two aspects — was, of course, noteworthy because the contract price was over £82,000 for the property as a whole, and if indeed the greenhouse is correctly to be treated separately (which I shall be deciding in a moment) then its value was certainly under £2,000 by May 1980, allowing for depreciation, costs of removal, etc.

There was an ensuing correspondence which I can go through very quickly. Mr Spellman replied on May 20 to the letter of the 12th, saying:

> You must be aware there was a genuine dispute as to the question of the greenhouses. When completion took place at your offices both Messrs W H Matthews and myself were aware of your clients tactics in trying to take advantage of my clients' situation with their removal vans and my clients must reserve their rights.

The plaintiff's solicitors responded with a further account of the meeting, which does not really add anything, on May 26 disputing the "trying to take advantage" point. G

Then some time just after that, before the 27th, the Deens attempted to go on to the premises to get the greenhouse but were denied access, as is shown by a letter of May 27 from Mr Deen to Bairstow Eves, giving them that information and saying that he would be relying on Mr Adie to give testimony about the instructions originally given from America. His solicitors also recorded the refusal of access, in letters, both May 29, written respectively to the agents and to the defendants' solicitors, and then there was further correspondence about it which I do not need to read.

The first issue is a pure question of construction. I wish to make it quite clear to all the parties who are sitting in court, as I am sure their own lawyers have told them, that when it comes to construing a legal document like a contract or a conveyance — and here the words are H the same in both — it is a pure matter of law for the judge to interpret what the words mean, having regard to any legal authorities which bind him in our system of legal precedent, and has nothing whatsoever to do either with what evidence the parties give about it, which is inadmissible anyway on this topic, or what they may have wanted or intended. It is purely a matter of construction. The vital words are, as I have said, "together with the farmhouse and other buildings."

If I may put the defendants' argument first, which was short, crisp, clear and to the point, Mr Garnett submits that it is a plain matter of English language. Why, he asked, as a matter of plain English, should not a big greenhouse like this, of such solid construction, with J the heavy concrete plinths at the bottom, and built clearly for a commercial purpose, be regarded in ordinary parlance as a building within the ordinary meaning of that word, just as the other structures on the site, such as the other greenhouse, the barns and the timber shed and I think some others, are agreed to be buildings. He submits that the matter is to be tested in particular by the fact that (unnecessarily if there was no separate intention) the additional words "the farmhouse and other buildings" are added. He submits that in this context, with them explicitly added to the general description of the land, those words must appropriately be treated as covering this greenhouse.

Mr Lloyd, on the other hand, submits that that is not the proper approach. He relies by analogy on the concept of fixtures in landlord K and tenant law, and submits that the same principle should apply here as applies there. He cited paras 1545 and 1547 from *Woodfall's Landlord and Tenant* and also a number of passages from the 5th ed (1984) of the *Law of Real Property* by the present Vice-Chancellor, Sir Robert Megarry and Professor H W R Wade. There is no conflict between the two textbooks and I am going to draw, for the purpose of illustrating and explaining Mr Lloyd's submissions, on *Megarry and Wade.* I start with the relevant passage under the heading "Distinction between Fixtures and Chattels" on p 731, and that starts off:

> A physical object will usually be either land or a chattel but its nature may change according to the use made of it. L

They then give some examples and say at the bottom of the page:

> A tenant for example who attaches fixtures to the demised premises may thereby make them the property of his landlord and a purchaser of land may claim as part of his purchase all objects which were fixtures at the date of the contract for they form part of the land sold to him.

I pause there. The learned authors there are, as Mr Lloyd submits is the correct approach, assimilating the concept of fixtures in landlord and tenant law with what does or does not pass on a purchase contract between purchaser and vendor such as we have here. They then say:

> The first need therefore is to be able to decide what is a fixture and what is not. In borderline cases this is often difficult but in principle it depends upon two tests namely — (1) the degree of annexation, and (2) the purpose of annexation. M
>
> (1) Degree of annexation. This is the primary test. An article is prima facie a fixture if it has some substantial connection with the land or a building on it. An article which merely rests on the ground by its own right such as a cistern or a "Dutch barn" standing in sockets let into the ground is prima facie not a fixture. On the other hand, a chattel attached to the land or a building on it in some substantial manner, eg by nails and screws, will prima facie be a fixture even if it would not be difficult to remove it.

Then under "Purpose of Annexation" there is quoted a well-known passage of great authority from the judgment of Blackburn J in the case of *Holland* v *Hodgson* (1872) LR 7CP 328 at p 335, which reads as follows:

> Perhaps the true rule is that articles not otherwise attached to the land than by

their own weight are not to be considered as part of the land, unless the circumstances are such as to show that they were intended to be part of the land, the onus of showing that they were so intended lying on those who assert that they have ceased to be chattels, and that, on the contrary, an article which is affixed to the land even slightly is to be considered as part of the land, unless the circumstances are such to show that it was intended all along to continue a chattel

I think it is clear that "purpose" in this context means objective purpose, to be judged from the evidence, and not the particular purpose in the mind of the person who brought the object there, which might vary, and in any event might well not be known to the other party. Examples of such objective purpose are given in the textbook; for instance, for a drystone wall which, though not actually cemented, would none the less of its very nature indicate an objective purpose of permanency; and statues, which form part of an elaborate landscaping project on the property.

I say at once that there is no evidence here at all relied on by either side of any such purpose which would remove this building from whatever general category it is properly placed in, following the principles laid down by Blackburn J, which of course I regard as authoritatively binding on me and which indeed have been applied for years.

Finally, Mr Lloyd cited a fairly recent Court of Appeal case of very considerable significance, in my judgment; the case of *H E Dibble Ltd* v *Moore, West Third Party* [1970] 2 QB 181. That was a rather complicated case involving a plaintiff, a defendant and a third party, in relation to a company which had gone into receivership, and its facts are not by any means dissimilar to the present case. There were two greenhouses on some land which was sold — and I am quoting from the headnote — "standing on their own weight on dollies not affixed or attached to the ground." The conveyance of the land is set out on p 188 in the judgment of Harman LJ, and the wording was as follows:

All those pieces or parcel of land now used as a market garden . . . .

there being no reference to buildings in addition to the general reference to the market garden.

Section 62 of the Law of Property Act 1925, provides:

A conveyance of land shall be deemed to include and shall by virtue of this Act operate to convey with the land all buildings, erections, fixtures, ...

So by operation of law, that conveyance of the market garden was deemed to include all buildings and erections, and the learned county court judge, who decided the case at first instance, held that these two greenhouses were erections and therefore passed with the land. This was unanimously reversed by the Court of Appeal. Megaw LJ said, without elaboration, that they were not within the section. Harman LJ said:

But it is said,

and I am reading from p 189

and the judge so held, that they were erections. So in a sense they were. But are they "erections" within the meaning of the general words of section 62? In my opinion, clearly not. Nothing I think, is an erection within the meaning of that section which is not something which is part and parcel of the land conveyed. It does not operate to convey chattels at all.

Russell LJ, as he then was, said at p 189:

In my judgment that

ie what the county court judge held

was wrong in law. In section 62(1) the word "erection" does not embrace something that is not a part of the land, either as a fixture or otherwise. I know of no authority to the contrary. No text book even discusses the possibility. In *Hood and Challis's Property Acts*, 8th ed (1938), at p 141, I find this comment:

"general words as used in conveyances of lands and manors denote things divisible into three classes;
(1) Things which, in construction of law, are parcel of the thing conveyed . . . ."

Then

I note the phrase "things which, in construction of law, are parcel of the thing conveyed", and with that I wholly agree.

It is manifest that in *Dibble's* case the greenhouses were not erections within the meaning of that word in section 62 of the Law of Property Act 1925. *A fortiori* it is quite plain that they were not buildings within the meaning of section 62 of the Law of Property Act 1925.

This greenhouse here, bolted to part of its own structure, which was delivered with it — though admittedly fairly hefty bits of structure, namely the plinths — is not in any way affixed to the land. It lies on its own weight and therefore *prima facie* falls within the definition of a non-fixture or a chattel as given by Blackburn J in the case of *Holland* v *Hodgson*.

In my judgment, the word "building" in the conveyance here cannot, in all commonsense, be given a different construction from the word "building" in section 62 of the Law of Property Act, although it is perfectly true, as Mr Garnett submits, that the inclusion of buildings is expressed in the words of the contract rather than implied by operation of law under section 62. I can see no difference in principle between construing a word which is expressed as it is here and which is implied by virtue of section 62. In my judgment, this structure, this greenhouse, was in truth and in fact a chattel rather than a building and was, therefore, I hold, not covered by the terms of either the conveyance or the contract.

I next consider the issue as to the conversation between David and Mr and Mrs Andrews. All three were absolutely sincere, honest and conscientious witnesses. I saw them all give evidence and it is always invidious to have to choose between such honest witnesses, particularly when one of them, Mrs Andrews, for reasons which are perfectly understandable, feels rather fraught about the whole matter, as was plain at the end of her evidence. I do not criticise her for that.

Mr and Mrs Andrews are quite firm that David said the greenhouse was included in the conveyance, and, as I said, they were fair witnesses, and Mr Andrews particularly fair in his concession in cross-examination that as this was at the very beginning, his very first visit to the house, he was not very worried at that stage whether he got the greenhouse at once and the matter only surfaced in his mind when the conversation took place in the garden.

David said quite firmly the opposite and he said also two other very significant things which were not challenged. The first was that he had arranged for a local farmer to transport it away on his tractors and trailers, and he gave the name of Mr Chinnery; secondly, he said that he had placed a number of advertisements on instructions in the local paper, though he could not remember which one, for the sale of this greenhouse.

Those were two matters which, in my judgment, were fully consistent with his evidence, and were impressive in helping me to decide what really happened. He had no axe to grind and although, of course, it has become a matter of importance now to the defendants, it was not, as Mr Andrews admitted, of much importance at the time till it was mentioned. On the other hand, it was a matter of importance to David, since we know that apparently he did some market gardening himself, and indeed he now runs a produce business.

I am absolutely satisfied that Mr and Mrs Andrews genuinely believe their version of what David said, and in the finding I am about to give there is no implied criticism of them whatsoever. None the less, the burden being on them, I am not satisfied on the balance of probabilities that David did make the representation which is attributed to him. Consequently, the whole factual basis of this particular conversation as relied on by the defendants to support an estoppel fails. I would add, however, that even if I had been in the defendants' favour on the course of the conversation, I think it would have been extraordinarily difficult for Mr Garnett to establish an estoppel, despite his very ingenious submissions to that end.

The pleading simply asserts that David, as the plaintiff's agent, made the representation. Mr Garnett frankly accepts that David himself was not in a position of authority to be capable of binding his employer, either the plaintiff or her husband, to make it a binding representation. However, he submits that the true position is that Mr Adie, knowing what the situation was, should have spoken up and corrected what David said (if I had held that David had said it) and that by failing to do so and lying-by he, in effect, confirmed it by his silence, so that the defendants cannot go back on it. I find that very difficult as a concept of fact, even if I allowed an amendment to cover it, without Mr Adie's evidence as to what he knew; because if he did not know, even though the plaintiff thought he had explained it to him, then it would be quite impossible to attribute any culpable lying-by to him under the principles laid down in the well-known passages on estoppel cited by Mr Garnett from *Spencer, Bower and Turner on Estoppel by Representation*.

There would be further difficulties also, in my judgment, in the defendants' way. Firstly, it seems to me very difficult to establish an estoppel of this kind when the whole position is subsumed into a written contract, and I do not accept that the rather extraordinary facts of *Amalgamated Investment and Property Co Ltd* v *Texas Commerce International Bank Ltd* [1982] QB 84 (where the Court of Appeal held there was an estoppel in the form of an underlying assumption to be derived from the course of business) can be equated to the informal conversation which is relied on here.

Finally (though I do not found any conclusion on it, since it was not pleaded), it seems to me very difficult to apply the estoppel concept in view of section 40 of the Law of Property Act 1925, on which Mr Garnett himself relies, as I shall show in a moment.

I now deal very briefly with the effects of the conversation on May 12, that is the date of completion, which of course is also strictly unnecessary because I have already found in favour of the plaintiff on the issue of construction. If, however, I had been against the plaintiff on that, I should have found it impossible to hold that the events of May 12 changed the situation back again in the plaintiff's favour.

Firstly, in my judgment, there was no consideration for the promise on which the plaintiff relies, since what the plaintiff was agreeing to do was no more than what she was bound to do already, namely to sell the land, and that is very clearly borne out by old authorities and also the recent decision by Mocatta J in *North Ocean Shipping Co Ltd* v *Hyundai Construction Co Ltd* [1979] QB 705. Secondly, it seems to me that there would have been a very strong argument, quite apart from lack of consideration, that this was a case where the agreement of the defendants on May 12, if it had mattered, was obtained by duress. Of course we have got to be very careful before finding duress, as is shown by the Privy Council case of *Pao On* v *Lau Yiu Long* [1980] AC 614.

But, whereas in the Hong Kong case there under consideration the party who was disadvantaged had plenty of time to consider the matter, take legal advice and reach a mature resolution as to his line of conduct, here Mr and Mrs Andrews had absolutely no time at all. They were completely up against it with their removal vans at the door and I think, in those circumstances, the threatened breach of contract would probably amount to duress, but none of that matters, since on the key point, namely the construction of the contract, I am in the plaintiff's favour and there is nothing in the estoppel point either in fact or in law.

I now come to consider the damages here. I first want to specify what I understand to be the legal principles involved. Here, as a result of the decision I made earlier, it follows that when on May 25 the Deens sought access to take away what I have held was still their greenhouse, they were refused permission to come and remove it. In my judgment, that was clearly a denial of the plaintiff's title to it, though of course a perfectly sincere one so far as the defendants were concerned in view of what they believed the position to be; that was a denial of the plaintiff's title which was an act of conversion, and in fact the writ claiming damages for conversion was issued on September 22 1980.

Therefore, the questions are (i) what damages has the plaintiff established as a result of that act of conversion, and (ii) has the plaintiff failed to mitigate those damages in a way she should reasonably have done?

Mr Garnett first relies on a letter written in December 1980 where there was an unconditional offer from the defendants' solicitors to the plaintiff's solicitors to let the plaintiff come and remove the greenhouse. The words are:

> I would confirm that my instructions are that the plaintiff be at liberty to remove the greenhouse from the land.

Because of that statement, he submits that the only question to be decided is what loss did the plaintiff suffer during a seven months' withholding? I do not think that is the correct approach having regard to my findings on mitigation.

In my judgment, the first question to be decided in considering the damages is Mr Deen's evidence as to the sale which he said he managed to secure of the greenhouse for a sum of £1,900 plus VAT. His evidence on that was rather vague as he was not able to remember the purchaser's name and address or very many of the details, and not unnaturally that evidence was challenged — and very fairly challenged — by Mr Garnett on behalf of the defendants. It is, however, true that Mr Deen has been in the United States all these last five years — perhaps beforehand, too — and so, not entirely surprising, he does not remember with crystal clarity the details of the subsale, which he avers took place. There is also the fact that as early as January 3 1981, in answer to the letter to which I have just referred, the plaintiff's solicitors answered that letter by saying: G

> However, as we informed you
>
> there had obviously been a telephone conversation
>
> owing to the delay in respect of this our client has lost the purchaser she
>
> that is Mrs Deen
>
> had for the greenhouse and although to mitigate her loss she is trying to find another purchaser, nevertheless at this moment no purchaser has been obtained and therefore the matter must proceed.

In other words, there was immediate reliance on that alleged subsale the moment the matter was raised at the New Year 1980/1981.

I accept the plaintiff's evidence that there was such a sale. The whole thing was obviously oral and there was no documentary evidence to confirm it, H and whether it was just on the point of becoming a binding contract or actually had done so may not be very important. What I am quite satisfied of, and I find as a fact, was that the plaintiff either got a price or had an imminently prospective price of £1,900 plus VAT, which seems to me in fact to be a sensible sort of price for a greenhouse which was about a year or more old and, therefore, still in pretty good condition; though of course any piece of perishable, structural equipment like this tends to depreciate fairly quickly the moment it is installed on its first site, and therefore it is not surprising that it would have gone down by getting on for a third or 30 per cent or so even after a year.

Did the plaintiff fail to mitigate the damages? There is evidence from both Mr Deen and from David, which I accept, that Mr Deen placed advertisements, J and indeed those are referred to by the plaintiff's solicitors in the letter which I have just read. It is perfectly plain that those advertisements were unfruitful. Now, in those circumstances, is it reasonable to say that Mr Deen ought to have come in and got the greenhouse when it was offered back to him in December 1980, and then either held on to it indefinitely or dealt with it in some other way? In my judgment, such a suggestion, with all respect to Mr Garnett, is unreasonable. There he is in America. It is a very difficult matter for him to deal with. He has no buyer in prospect, and therefore would be obliged to store this very easily breakable equipment somewhere, no doubt at very considerable expense, insure it and so on and so forth. Therefore, I do not think that it was reasonable to expect him to accept the offer. K Consequently I hold that the plaintiff has not failed to mitigate the damages.

However, there is a final question of importance as to the proper valuation. David's evidence was that it would have taken him and a mate seven working days to dismantle that greenhouse, and I should think, looking at the size of it and bearing in mind that screwed-in parts are sometimes difficult to dislodge after a period of time, that is the very minimum it could possibly have taken. Over and above that, it is inevitable when dismantling a structure of this kind that there will be some breakages, certainly of the glass, and there was also the question of transportation, which was put in by one witness at £100, to carry the very big pile of material which would accumulate, once this greenhouse was totally dismantled, to the purchaser. L

I can only do the best I can, but even allowing for the fairly low wages which David appeared to be willing to accept — though whether his mate would have been so charitable we do not know — I think that the correct figure for the expenses, allowing for dismantling, breakages and transport, is the sum of £400, and consequently there will be judgment for the plaintiff for £1,500, subject to any additional matter such as interest.

The plaintiff was awarded in addition £360 in respect of interest, and costs on the highest county court scale, the latter not to be enforced for the period after the defendants became legally aided without leave.

M

Court of Appeal

January 16 1986

(Before Sir Nicolas BROWNE-WILKINSON V-C and Lord Justice MUSTILL)

WATSON v GOLDSBROUGH AND OTHERS

Fishing rights — Dispute between members of angling club and landowners as to rights to fish in certain ponds — No lease, licence or other legal document had been completed conferring fishing rights on the anglers, although at one time a

form of lease had been executed by the plaintiff, who brought the present proceedings on behalf of himself and the other members of the club — This lease was never executed by the owners, but the plaintiff claimed that there was an informal agreement that the club could have a lease or option to purchase the ponds in question — In these circumstances the club proceeded to spend substantial sums of money in dredging and improving the ponds and stocking them with fish — Matters came to a head when the first and second defendants, who had originally been licensees, became the freehold owners of the land which included the ponds and entered into negotiations to sell the ponds to the third defendant — The sale of the ponds was never completed — In the proceedings brought by the plaintiff in the county court, based on a claim that there was a binding agreement or alternatively that the defendants were estopped on the principle of proprietary estoppel from denying the club's interest in the ponds, the judge decided in favour of the plaintiff on the estoppel ground — The first defendant appealed (his wife, the second defendant, did not appeal and the third defendant had gone out of the picture) — On appeal the court rejected submissions that there had been insufficient evidence of the club's exclusive right to fish or of its work and expenditure in creating valuable fishing grounds — They also rejected an argument that the principles of proprietary estoppel could not be established against persons who were mere licensees at the material time — The estoppel was "fed" when the licensees, who had stood by watching the work being done, acquired the legal estate — Principles of proprietary estoppel outlined — The judge was right in deciding in favour of the club, but the relief given in his order was too wide — The order should be varied by shortening the period of security and safeguarding the preservation of a sufficient stock of fish

This was an appeal by William Maurice Goldsbrough, the first defendant in an action brought by the plaintiff, Bernard Watson, the respondent to the appeal, from the decision of Judge Gill at Middlesbrough County Court. The action had been brought by the plaintiff on behalf of himself and all other members of the Middlesbrough Angling Club. The second defendant, Constance Heather Goldsbrough (the first defendant's wife), and the third defendant, Kenneth Millward, took no part in the appeal.

Miss Susan Cooper (instructed by Person & Watson, of Middlesbrough) appeared on behalf of the appellant; Ian Lamb (instructed by Goodswens, of Middlesbrough) represented the respondent.

Giving judgment, SIR NICOLAS BROWNE-WILKINSON V-C said: This is an appeal from the order of His Honour Judge Gill sitting at Middlesbrough County Court. The action concerns fishing rights over certain ponds. At the time of trial they were owned by the first defendant, Mr Goldsbrough, and his wife, Mrs Goldsbrough. The plaintiff is a Mr Watson, who is suing on behalf of himself and all other members of the Middlesbrough Angling Club. The trial took place on June 25 1985. At the trial Mr Goldsbrough acted in person, he having previously instructed legal advisers whose instructions had been terminated. Before the conclusion of the hearing Mr Goldsbrough apparently left the court. In those circumstances we have not had the benefit of a full judgment from the learned judge, the matter proceeding thereafter in the presence of only the plaintiff's legal advisers and the learned judge. It has therefore been quite difficult for us to reconstitute the relevant facts.

They seem to be as follows. In 1965 the land in question was owned by the parents of the second defendant, Mrs Goldsbrough. On the land were two ponds. The farmland, the farmhouse and the ponds were in the occupation of Mr and Mrs Goldsbrough, apparently as licensees of Mrs Goldsbrough's parents. At some time between 1970 and 1972 Mr Watson approached Mr Goldsbrough on behalf of the angling club to ask if they could fish the ponds and do works of improvement to them. There was a conflict of evidence between Mr Watson on the one hand and Mr Goldsbrough on the other; it appears the judge clearly accepted Mr Watson's evidence and rejected that of Mr Goldsbrough. At some stage around that date Mr Watson on behalf of the angling club sent to Mr Goldsbrough a draft fishing lease, the exact terms of which I shall have to return to shortly. Briefly, it provided that Mr Watson was to have the exclusive rights of fishing over the ponds, together with a right to kill, take and dispose of the fish, for a term of 10 years at a rent of 5 pence per year. There was also to be an option to renew for a further term of 10 years. But clause 5 contained a provision that the lease might be determined at any time by either party giving to the other not less than 12 months' notice in writing expiring on the first day of February in any year.

The lease as sent to Mr Goldsbrough had been executed by Mr Watson. The evidence before the judge was that Mr Goldsbrough declined to execute that or any other legal agreement, since he was not the owner of the land and in addition there was apparently a mortgage on the land. However, Mr Watson's evidence was that Mr Goldsbrough did agree that the angling club could have a lease or an option to purchase the ponds in question. In those circumstances, the club went ahead and expended substantial sums of money in dredging the ponds and improving them and in stocking the ponds, which were apparently then either totally or virtually free of fish, with fish caught elsewhere and put into the ponds. The angling club incurred considerable expenditure and put in considerable effort in taking those steps.

The major expenditure and effort was incurred during the years 1972 to 1974. Thereafter the club continued to do certain work, the later work being more by way of maintenance of the ponds and looking after them as angling grounds rather than by further major capital expenditure. The result of the angling club's efforts was that the ponds became well stocked with coarse fish and were sound fishing grounds for the club.

On July 17 1980, Mr and Mrs Goldsbrough became the freehold owners of the land. The club continued to fish the waters. They apparently paid a small amount of money and made small gifts to the Goldsbroughs. Right from 1972 onwards there was the expectation that they had some security in their use of the ponds for fishing. However, after the Goldsbroughs became the owners of the freehold they set about selling certain parts of the land and also negotiating for the sale of the ponds. According to Mr Watson's evidence and the documentation before the learned judge, on a number of occasions prices were agreed at which the club could purchase the ponds, but Mr Goldsbrough failed to give his solicitors instructions to go ahead with such sales. Over the years the price crept up from £9,000 to, at the end of the story, £20,000.

Shortly before the proceedings began, it emerged that Mr Goldsbrough was in negotiation with the third defendant (Mr Millward) for the sale of the ponds to Mr Millward at a price in the region of £20,000. In those circumstances, these proceedings were started in order to stop the sale to Mr Millward. Ex parte relief was obtained; the sale to him was never completed and he has gone out of the picture. It was in those circumstances that these proceedings were brought by the club through Mr Watson as plaintiff.

By the pleadings, the plaintiff claimed either that there was a binding agreement to grant a lease or, alternatively, to grant an option to purchase or, alternatively, that in the circumstances I have related the defendants (Mr and Mrs Goldsbrough) were estopped from denying that the plaintiffs had some form of interest in the ponds and the fishing in the ponds, the claim in estoppel being a claim to a proprietary estoppel.

The judge heard the evidence of Mr Watson and the evidence of Mr Goldsbrough. He had before him a written statement from Mrs Goldsbrough. He also had before him a most voluminous body of papers. As I have said, at the end of the hearing Mr Goldsbrough left without awaiting to hear the final outcome and the notes of the learned judge's judgment simply record this:

> Mr Lamb, I need not trouble you further upon the law. . . This is an increasingly familiar type of case — proprietary estoppel: equity stepping in to provide a remedy where the legal formalities and requirements have not been complied with but where the plaintiff has been induced to expend money or effort on the defendant's land in the expectation of receiving some appropriate interest to satisfy the justice of the case. The court has to consider what form of order would achieve this. I need hardly say I entirely accept the evidence of Mr Watson. It was modest and careful and supported up to the hilt by the documentary evidence: the invoices for hire of heavy equipment for clearing the ponds (I entirely reject the notion that this could all have been done without being noticed by Mr Goldsbrough, or that all that happened was a few people came and did damage to the ponds) and the letters to and from the various landowners and gentry who were canvassed for help in allowing the members to fish out coarse fish from their rivers so they could stock the ponds. All this indicated prolonged and dedicated work by the club, and it was outrageous to deny it and to try to sell the resulting fishing grounds to

A someone else. The average price per acre for this 50 acre holding was £1,500, and yet the valuation of these ponds is now £10,000 per acre. Mr Goldsbrough would have it that this was purely due to an increasing interest in land covered by water. It is obvious to me that the plaintiffs have by their efforts created this valuable asset, and are entitled to some security in it. (I had in mind at this point to grant a lease for a further 10 years at a rent and asked counsel how I should value it: but at this point he stated that the club had lost their enthusiasm for a prolonged relationship or contact with Mr Goldsbrough, and would prefer a sufficient licence to enable them to fish out most of their stock of fish and to transfer it to other waters).

The order actually made reads as follows:

> It is declared that the Middlesbrough Angling Club do have a right to security commensurate with their work and expenditure in having established
> B valuable fishing grounds. And in order to effectuate the said right it is ordered that:— 1. The Plaintiffs do have an exclusive right to fish the ponds for a term of 5 years from 1985 to 1990 and to terminate on March 26 1990. And to remove their stocks of fish therein. And subject thereto to leave the ponds in a proper order for fishing. Paying therefor a token rent of 5 pence per annum, in the terms of and with the landlords and tenants covenants contained in a proposed lease dated September 1 1972 but omitting clauses 4 and 5 thereof, and the option to renew, so far as the same are consistent with this order. 2. The counterclaim be dismissed. 3. There be liberty to apply.

It is against that order that Mr Goldsbrough (but not Mrs Goldsbrough) appeals.

Miss Cooper has presented the appeal on behalf of Mr Goldsbrough under substantial difficulties. She, of course, was not present at the hearing and the note of the judgment (as I have
C indicated) is scanty. Her first attack is on the findings of fact made by the learned judge. She sought, but we refused, an adjournment to enable her to apply to put in further evidence which she said, I have no doubt with some force, would have had an impact on the judge if it had been before him at the trial. We refused that adjournment, because there were no grounds for putting in such further evidence on the appeal, since it was all available at the time of the hearing and therefore could not properly be admitted on the appeal. In the absence of such evidence, she was facing an uphill task in seeking to persuade us that the judge's findings of fact as contained in his judgment were not justified.

The first two findings she attacks were the judge's finding implicit
D in the order that the club had had an exclusive right to fish the ponds during the period up to the date of the hearing; she submitted that the judge failed to take into account that several other groups of people had fished the ponds. She says there was no evidence that the rights were exclusive and on that basis there was nothing to justify the judge's findings that the club's rights to fishing were exclusive. On that I am unable to accept her submission. The draft lease sent to the first defendant, Mr Goldsbrough, plainly showed that the intention (at that stage anyway) was that the club should have exclusive rights of fishing. That was certainly sufficient evidence for the judge to go on, backed as it was by Mr Watson, and it would be quite impossible for us to upset that finding merely on the basis of the evidence of Mr Goldsbrough, which the judge obviously found not acceptable.

The next attack on findings of fact was that there was no sufficient evidence for the judge's finding that the club had carried out prolonged and dedicated work on the ponds or that the club had established valuable fishing grounds in the ponds. Again, I cannot accept that. There was Mr Watson's evidence that they had done such work. As the judge himself said, there was the documentary evidence showing expenditure. True, it was not in every way complete in that they were not receipts for expenditure incurred so much as tenders to do work. But there was plainly evidence which the learned judge could properly accept from what Mr Watson said, namely that plant had been hired and fish had been brought in. In those circumstances, the attack on the findings of fact, in my judgment, must fail and we must proceed on the basis that the learned judge found the facts correctly.

There is a further ground of appeal, which suggests that no estoppel of the kind the judge dealt with could exist, since at the time of the alleged agreement (namely, 1972) Mr and Mrs Goldsbrough were mere licensees of the land against which the proprietary estoppel is alleged. In my judgment that is not well founded. True it was that at that date they did not have the legal estate. But if they entered into the agreement and stood by watching the work being done and subsequently became the owners of the land, then the estoppel established against them was "fed" when they acquired the legal estate and thereby became binding on the land itself.

I move to what in my judgment is much the most difficult point on this appeal. Assuming, as I do, that the judge was right in finding that the acts done by the club were done in reliance on an assurance by Mr G
Goldsbrough that they should have some form of permanent or semi-permanent right to fish, what is the appropriate remedy to afford to them in the circumstances that occurred? Without in any sense wishing to state definitively what are the principles of proprietary estoppel, in outline they are these. An owner of land cannot stand by and watch somebody else expend money on his land in the knowledge that the person expending the money either mistakenly thinks that he already has an interest in that land or has an expectation encouraged or known to the owner that he will obtain such an interest from the owner. If an owner does so behave, in equity an estoppel is raised against him and the court makes an order securing to the person who establishes the estoppel the rights in expectation of which he incurred the expenditure or did the work. The doctrine is one of comparatively H
recent development (if not recent origin), but so far as I am aware it has never been applied so as to give the person establishing the estoppel any right greater than that which he either mistakenly thought he had or thought he was going to be afforded.

The attack launched by Miss Cooper in the present case is to the effect that the learned judge has given to the club a right greater than that which they could properly have expected to get when they incurred the expenditure. In my judgment, that attack is well-founded. For this purpose one must look again at the proposed lease, a copy of which, dated September 1 1972, was sent by the plaintiffs to Mr Goldsbrough and of which he was aware. I have already indicated that it was a lease initially for a period of 10 years at 5 pence per year J
of fishing rights which were exclusive and enabled the club to kill, take and dispose of any or all fish in the above ponds. It also contained an option for a further renewal of 10 years on the same terms to be exercised by not less than 12 months' notice in writing given by the club.

In clause 2(c) the tenants (that is to say, the club) under the draft covenanted with the landlord "to use their best endeavours to preserve a proper and sufficient stock of fish in the said ponds and to leave such ponds as well stocked at the determination of as at the commencement of his Lease." Finally, in clause 5 it was provided that the lease can be determined at any time by either party giving to the other not less than 12 months' notice in writing expiring on the first day of February in any year. K

There are certain somewhat unusual features of that document. The first is that, although the initial term was for 10 years and there was an option to renew for a further 10 years, the final clause (clause 5) enabled the landlord to put an end to the rights by giving one year's notice expiring on February 1. Second, it is not clear (and I will return to this point) exactly how many fish the club were entitled to take out in exercise of the rights granted under that lease if it had been granted. In my judgment, in seeing what is equitable to award to the plaintiff in this case it cannot be right to give the plaintiffs a right greater than that which they themselves expected to receive or thought they were going to receive when they incurred the expenditure which gave rise to the estoppel. The best evidence of that must be the lease which the L
club itself put forward under which their maximum security of tenure was for a period determinable by one year's notice expiring on February 1 in every year. To give the club, as the learned judge did, five years' security of tenure is to give the club more than they could ever have expected to get even if Mr Goldsbrough had behaved in the way which the learned judge plainly thought he ought to have behaved.

Second, the order as made by the learned judge appears to give the right to the club to fish out their stock of fish in the ponds within the five-year period that he has awarded. First, it seems to me obscure what is the meaning of the order made. Is it to give a right to take all the fish out that were in the ponds (it being the evidence that there were few, if any, fish in 1972) or is it to remove a lesser amount of fish M
and if so how many? Second, when one looks at the lease, it is to my mind clear that there is an obligation at all times to preserve a proper and sufficient stock of fish in the ponds; ie there should be no right there to fish out the ponds completely, although there is a right to take out fish up to the extent of leaving a proper and sufficient stock.

In those circumstances, in my judgment the learned judge erred in giving the club such wide relief. I believe the right order is that the club should have one year's security of tenure terminating on February 1 after the material date from which notice could have been given and during that period have the right to fish the ponds by line, killing, taking and disposing of any or all of the fish in the ponds, provided that in so doing they preserve a proper and sufficient stock

A of fish in the said ponds. In my judgment, the right period for a commencement of that equitable right under declaration of the court should be the date of the learned judge's judgment. The first date by which notice given on that date could expire on February 1 would be February 1 1987. I would myself, therefore, vary the learned judge's order to declare that the plaintiff on behalf of himself and the other members of the Middlesbrough Angling Club have a right to fish the ponds down to February 1 1987, together with the right to kill, take and dispose of any or all fish in the ponds during that period provided that the fishing of the ponds is such as to preserve a proper and sufficient stock of fish in the said ponds. I would vary the learned judge's judgment to that extent.

MUSTILL LJ agreed with the order proposed and the reasons and did not add anything further.

The judge's order was varied as mentioned in the Vice-Chancellor's judgment. There was no order as to costs of appeal and the order for costs below was not disturbed.